Handbook of Downstream Processing

Handbook of Downstream Processing

Edited by
ELLIOTT GOLDBERG
Director of Process Engineering
Lockwood Greene Engineers, Inc.,
New York

BLACKIE ACADEMIC & PROFESSIONAL
An Imprint of Chapman & Hall
London · Weinheim · New York · Tokyo · Melbourne · Madras

Published by Blackie Academic & Professional, an imprint of Chapman & Hall, 2–6 Boundary Row, London SE1 8HN, UK

Chapman & Hall, 2–6 Boundary Row, London SE1 8HN, UK

Chapman & Hall GmbH, Pappelallee 3, 69469 Weinheim, Germany

Chapman & Hall USA, Fourth Floor, 115 Fifth Avenue, New York NY 10003, USA

Chapman & Hall Japan, ITP-Japan, Kyowa Building, 3F, 2-2-1 Hirakawacho, Chiyoda-ku, Tokyo 102, Japan

DA Book (Aust.) Pty Ltd, 648 Whitehorse Road, Mitcham 3132, Victoria, Australia

Chapman & Hall India, R. Seshadri, 32 Second Main Road, CIT East, Madras 600 035, India

First edition 1997

Typeset in 10/12pt Times by AFS Image Setters Ltd, Glasgow

Printed in Great Britain at the University Press, Cambridge

ISBN 0 7514 0364 4

A catalogue record for this book is available from the British Library
Library of Congress Catalog Card Number: 95–81419

∞ Printed on permanent acid-free text paper, manufactured in accordance with ANSI/NISO Z39.48-1992 (Permanence of Paper)

To my wife
Shirley

Elliott Goldberg

Contents

Contributors

D.S. Alterman Chief Process Engineer, E.I. Associates, East Orange, NJ 07018, USA

P.M. Armenante Professor, Department of Chemical Engineering, Chemistry and Environmental Science, New Jersey Institute of Technology, University Heights, Newark NJ 07102, USA

R. Chen Engineering Associate, Merck & Co. Inc., PO Box 2000, RY 7A-215, Rahway NJ 07065, USA

J.A. Covey Principal Process Engineering, John Brown E&C, 440 Rt22E, Bridgewater, NJ 08807, USA

K.E. Crowell Senior Environmental Engineer, Schering-Plough Corporate Engineering, 60 Orange Street, Building B-214, Bloomfield NJ 07003, USA

W.C. Czander Manager of Pharmaceutical Engineering, Lockwood Greene Engineers, 440 Ninth Avenue, New York NY 10001, USA

T.A. Davis Davis Consulting, 5 Davis Farm Road, Annandale NJ 08801, USA

V.C.F. DiChiaria Vice President and General Manager, Sani-Tech Silicones Group, 2 Wilson Drive, Suite 3, Sparta NJ 07871, USA

B. Eckman Manager of Validation Services, Torcon Inc., 214 East Grove, Westfield NJ 07091, USA

J. Edwards Project Director, Genetics Institute, One Burtt Road, Andover MA 01810, USA

W. Eykamp Adjunct Professor of Chemical Engineering: Tufts University, Medford, MA 02155, USA

S. Fishkin Consultant, 519 Sergeantsville Road, Flemington NJ 08822, USA

D.A. Glassner Bio-Processing Branch Manager, National Renewable Energy Laboratory, PO Box 27609, Lansing MI 48909, USA

A.C. Kirpekar Merck & Co. Inc., US Highway 340 South, Elkton VA 22827, USA

W.S. Kletch Senior Consultant, Vectech, 24543 Indoplex Circle, Farmington Hills MI 48335, USA

R. Lesnik Director of Pharmaceutical Technology, John Brown E&C, 440 Route 22 East, Bridgewater, NJ 08807, USA

R.V. Leung Project Manager, Lockwood Greene Engineers, 440 Ninth Avenue, New York NY 10001, USA

S. Litman Regional Manager, John Brown E&C, Grande Commons Building, Suite 210, 440 Route 22 East, PO Box 6884, Bridgewater NJ 08807-6884, USA

R.J. McDonough Mixing Specialist, Fluor Daniel, 301 Lippincott Center, P.O. Box 950, Marlton, NJ 08053, USA

J.R. Millis President and Director of Process Development, Bio-Technical Resources, 1035 South 7th Street, Manitowoc WI 54220, USA

A.S. Mujumdar Professor, Department of Chemical Engineering, McGill University, 3480 University Street, Montreal, Quebec H3A 2A7, Canada

S. Norwood General Manager, Certco Inc., 2 Audobon Road, Suite 5, Wakefield MA 01880, USA

D.J. Odde Assistant Professor, Department of Chemical Engineering, Michigan Technological University, Houghton, MI 49931, USA

D.P. Petrides Associate Professor, Department of Chemical Engineering, Chemistry and Environmental Science, New Jersey Institute of Technology, University Heights, Newark NJ 07102, USA

E.S. Sapidou Applications Engineer, INTELLIGEN, Inc., 2326 Morse Avenue, Scotch Plains NJ 07076, USA

C.A. Schall Assistant Professor, Chemical Engineering Department, University of Tulsa, 600 South College, Tulsa, OK 74104, USA

J.W. Snowman BOC Edwards Calumatic, Manor Royal, Crawley, West Sussex RH10 2LW, UK

A. Tergevorkian Senior Project Engineer, Lockwood Greene Engineers, 440 Ninth Avenue, New York NY 10001, USA

J.M. Wiencek Associate Professor, Department of Chemical and Biochemical Engineering, University of Iowa, Iowa City, IA 52242, USA

R. Wilcox Process Engineering Section Manager, Jacob's Engineering Group Inc., Houston TX 77025, USA

R. Wolthuis Manager, 3V COGEIM SpA, PO Box 219, Via Friuli 19, 24044 Dalmine (BG), Italy

P.W. Yang Senior Engineering Consultant, John Brown E&C, Grande Commons Building, Suite 210, 440 Route 22 East, PO Box 6884, Bridgewater NJ 08807 6884, USA

S.K. Yu Senior Process Engineer, Ortho–McNeil Pharmaceutical Corporation, PO Box 300, Route 202 South, Raritan NJ 08869 0602, USA

Preface

The last two decades have seen a phenomenal growth of the field of genetic or biochemical engineering and have witnessed the development and ultimately marketing of a variety of products–typically through the manipulation and growth of different types of microorganisms, followed by the recovery and purification of the associated products.

The engineers and biotechnologists who are involved in the full-scale process design of such facilities must be familiar with the variety of unit operations and equipment and the applicable regulatory requirements.

This book describes current commercial practice and will be useful to those engineers working in this field in the design, construction and operation of pharmaceutical and biotechnology plants. It will be of help to the chemical or pharmaceutical engineer who is developing a plant design and who faces issues such as:

Should the process be batch or continuous or a combination of batch and continuous?

How should the optimum process design be developed?

Should one employ a new revolutionary separation which could be potentially difficult to validate or use accepted technology which involves less risk?

Should the process be run with ingredients formulated from water for injection, deionized water, or even filtered tap water?

Should any of the separations be run in cold rooms or in glycol jacketed lines to minimize microbial growth where sterilization is not possible?

Should the process equipment and lines be designed to be sterilized-in-place, cleaned-in-place, or should every piece be broken down, cleaned and autoclaved after every turn?

Handbook of Downstream Processing focuses on the operations which promise to produce genetically engineered materials for use in producing end products such as enzymes, antibiotics, hormones, peptides, polypeptides, proteins, amino acids, veterinary drugs, steroids, herbicides, growth promoters, liposomes, diagnostic kits for infectious diseases, and many other materials that focus on pharmaceutical needs.

Contributors to this volume come largely from industry—from both engineering and pharmaceutical/biotech firms and from equipment manufacturers. The remainder come from academia.

The choice of chapters for this volume was dictated by the editor's experience and modified by recommendations and suggestions by knowledgeable associates.

The chapters have been grouped into three categories: Unit operations; Specialized techniques; Engineering and design.

Many chapters belong in two categories, some in all three, and some do not fit comfortably in the categories in which they have been included. All the chapters are excellent reviews of the areas of technology currently used in the field of interest.

Unit operations

Descriptions of unit operations currently in use for the separation and purification or downstream processing of products developed by biochemical processing, including lysing, filtration, drying, extraction, affinity adsorption, membrane technology, electrodialysis, chromatography, precipitation, mixing and distillation.

Specialized techniques

Some of the specialized techniques used for processing materials being purified are included in the unit operations above and in the engineering and design sections below, and include chapters on CIP/SIP systems, clean room design, clean room testing and certification, sterilization and pure water facilities.

Engineering and design

Biotech facility project execution, controls and automation, bulk pharmaceutical plant design, optimization of downstream processing of protein, pharmaceutical finishing operations, wastewater treatment, offsite construction, regulatory considerations and validation.

Most of the chapters include references to current papers, which are not intended to be exhaustive.

Many contributed in the development of this book. Firstly, thanks to the chapter authors who are listed on p. xix. Then there are the unsung reviewers who made their critiques and offered suggestions for each chapter. We owe a depth of gratitude to the following reviewers: Richard Lang, Sheldon Finkler, Joel Kirman, and Robert Torregrossa, all of Lockwood Greene; Robert Clement and Peter Notwick of Jacobs Engineering; Jack Kearns and Anthony Dearning of Liposome Inc.; Francessa McBride of Triad Technologies; Mark Milano of Merck and Co.; John Bergen of E&V Services; Tom Davis, Consultant; Jim Banafato of Plastic Engineered Products; Dave Alterman of E.I. Associates Inc.; Richard Lesnik of John Brown E&C and Louis Bernero.

My debts are numerous to the associates and others who patiently put up with many requests for information, comments, questions, etc. Special gratitude is due to Kevin Sowter, Senior Vice President of Lockwood Greene, who had much to do with the origin of this book.

Thanks to Dr. Bruce Eckman for his useful contributions and direction during the formative stages of this volume, and appreciation to Ed Bomba for his patience and useful recommendations. Many thanks are due to the administrative personnel who performed the typing and corrections, including Peggy Burkhard, Courtènay Adams and Sarah Davis (who was helpful on the research end). Special appreciation is due to the Lockwood Greene and Jacobs Engineering Management, who were encouraging and generous with their time during the development of this volume. Thanks to David Bursik Esq. for legal advice.

Lastly, thanks to my wife, Shirley, for putting up with the many hours I spent during weekends working on this handbook instead of enjoying the company of our family.

E.G.

Conversion of units

Length

1 in = 2.54 cm	1 m = 39.37 in
1 ft = 0.304 m	1 m = 3.280 ft

Area

1 ft^2 = 0.092 m^2	1 m^2 = 10.763 ft^2

Volume

1 ft^3 = 0.028 m^3	1 m^3 = 35.314 ft^3

Capacity

1 gal = 4.546 l	1 l = 0.220 gal

Pressure

1 psi = 6.894 KNm^{-2}	1 KNm^{-2} = 0.145 psi
1 Pa = 10.2×10^{-6} $Kgcm^{-2}$	

Energy

1 BTU = 1.055×10^3 J

Power

1 hp = 745.7 W

Temperature

x°C = (1.8x + 32)°F

1 Mechanical disruption of cells

J.R. MILLIS

1.1 Introduction

The disruption of microorganisms is often required in the large-scale production of microbial products, such as enzymes, toxins, and diagnostic or therapeutic proteins. Many alternatives for release of product from cells exist for the process engineer: for example, mechanical disruption, autolysis, and chemical or enzymatic lysis. Genetic advances have also led to microbial host-vector systems, which secrete products from the cell, thus lowering contamination and lessening the risk of proteolysis. However, for many products of interest, these systems are either not available or not applicable.

Methods for selection and operation of mechanical disruption techniques are presented in this chapter because they represent a unit operation applicable to all types of cells and processes. Also, these techniques have been scaled to commercial production with great success, in contrast with other lytic techniques, which have unique problems in large-scale production.[1,2]

The ideal technology for cell disruption may be characterized by:

- Maximum release of the product of interest
- No mechanical or thermal denaturation of the product during disruption
- Minimal release of proteases which may degrade the product
- Minimal release of particulates or soluble contaminants that may influence downstream processing.

Also, the desirable attributes of mechanical cell disruption equipment may be described as:[3]

- Low capital and operating costs
- Sterilizable
- May be cleaned in place
- May be validated for cGMP requirements
- Scalable
- May be automated.

Two technologies dominate the market for mechanical disruption; (i) homogenization; and (ii) milling (using bead mills). Both have their origin in other industries and have been adapted successfully for biotechnology-related manufacturing. Each technique is based on the principle of high mechanical shear, and each has features that make them unique in the treatment of bacterial, fungal, yeast, algal and eukaryotic cells.

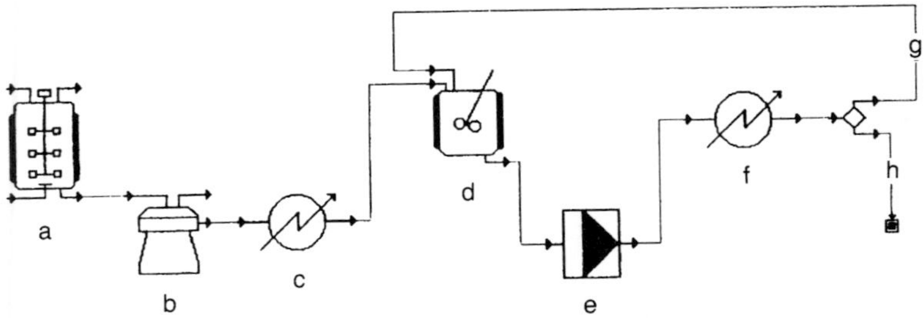

Figure 1.1 Typical process flow diagram: (a) fermenter; (b) centrifuge or ultrafilter; (c) heat exchanger; (d) surge vessel; (e) homogenizer or mill; (f) cooling process; (g) recirculation; (h) further processing.

A typical process flow diagram, shown in Figure 1.1, illustrates the basic engineering and placement of homogenizers or mills. The fermenter (a) is harvested by centrifugation or ultrafiltration (b). The cell concentrate is cooled by a heat exchanger (c) and stored in a surge vessel (d). The cells are pumped continuously to a homogenizer or mill (e). Homogenate is cooled (f), then recirculated (g) or sent for further processing (h). Note the need for cooling of the homogenate to remove heat that is generated by the dissipation of mechanical energy.

Heat removal is a common feature for any lytic system and is necessary to prevent denaturation of biological materials.

1.2 Homogenizers

1.2.1 Principle of operation

A homogenizer consists of a positive-displacement pump, which supplies a cell slurry at high upsteam pressure through to a restricted nozzle or orifice valve (Figure 1.2). Upsteam pressure is controlled by altering the spring tension on the valve piston either manually or automatically, generating extremely high hydrodynamic shear as the cell slurry passes through a restricted orifice. Disruption results from the combination of shear force and impingement on the valve.[4,5] Dominant factors which influence operation are the pressure of operation and valve design. Temperature and cell concentration of the feed stream also influence performance. Each is described in detail below.

1.2.2 Influence of pressure

Cell disruption follows first-order kinetics as first described by Hetherington *et al.*[16] by the equation:

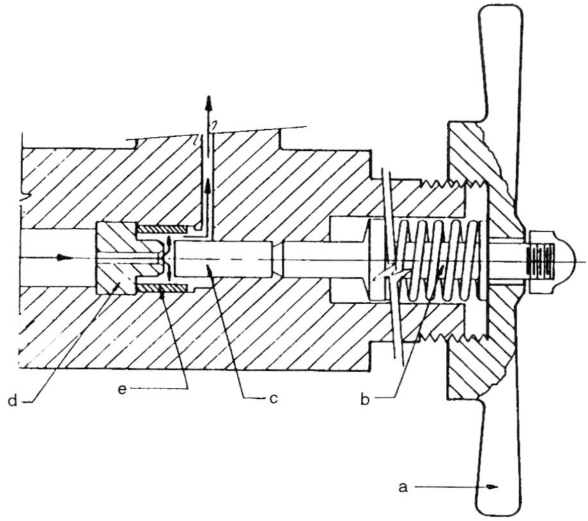

Figure 1.2 Homogenizer valve: (a) hard wheel for pressure control; (b) spring-loaded valve rod; (c) valve; (d) valve seat; (e) impact ring.

$$\log \frac{R_m}{R_m - R} = kN \tag{1.1}$$

where R_m = maximum protein release or enzyme activity; R = measured protein release or enzyme activity after N passes; k = rate constant (1/s); and N = number of passes.

From experience, this relationship is valid below approximately 90–95% cell lysis. Beyond this value, the rate of cell lysis and the value of the rate constant may decrease as a protein is released from organelles.[7] Also, the maximum obtainable protein release has been reported to be influenced by pressure.[5,8]

Further characterization of the relationship between the rate constant and pressure has been attempted by various authors according to the equation:

$$k = k_1 P^n \tag{1.2}$$

The results, presented in Table 1.1, indicate that the value of n is highly dependent on the type of cell and the conditions under which the cells were grown. Fermentation conditions that are known to influence the cell physiology and, hence, the ease of disruption include: the media; growth rate; age of the cells; and reactor engineering (continuous versus batch).

Most applications operate within the range of 500–1000 bar (7250–14 500 psi). The maximum allowable operating pressure is often dictated by the mechanical stability of the valve design, which may differ due to design or material of construction.

Table 1.1 Value of the exponent used in equation (1.2) for various organisms as described in the literature

Organism	Exponent	Reference
Candida utilis (batch)	1.17	5
(continuous)	1.77	5
Candida lipolytica	2.2	5
Escherichia coli	2.2	1
Saccharomyces cerevisiae	2.9	9
	2.3	10
	2.7	7

1.2.3 Influence of valve design

Valve design and selection optimize the influence of shear and impingement on cell disruption. Operational factors such as temperature and pressure stability may also influence selection. A variety of valves are available from each manufacturer for evaluation, including valves specifically designed for cell disruption. Selection of the optimal valve for each new application is somewhat empirical and should be specified only after testing. Designs that have been shown to be effective and versatile include a flat valve, a knife-edge valve, and a ball valve. The flat and knife-edge valves have been compared by several authors,[6,10,11] with the knife-edge valve being favored for most cell disruption applications. Both designs include an integral impact ring and are shown in Figure 1.3. Materials of construction include stainless steel, tungsten carbide, and ceramic.

Historically, homogenizers that have been used in the dairy industry for milk processing have used a two-stage valve. For most biotechnology applications, however, a single stage at an elevated pressure will yield equivalent performance at lower capital investment.

1.2.4 Influence of temperature

The rate of cell disruption increases with temperature.[12] As a rule, the rate at 30°C is approximately double that at 5°C. However, in selecting the inlet

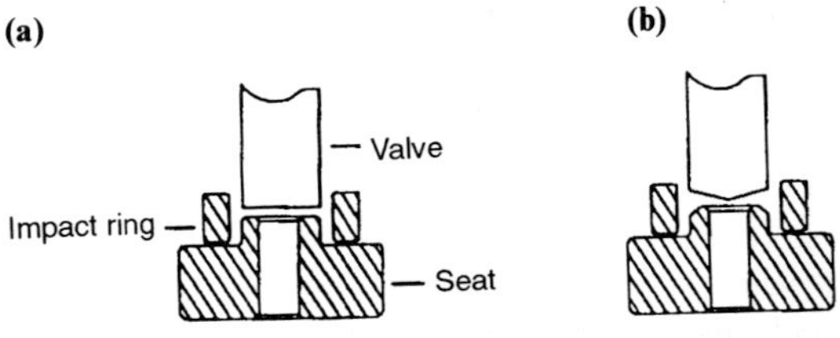

Figure 1.3 Valve designs: (a) standard valve; (b) knife-edge valve.

temperature for a homogenizer, one must consider both the temperature rise that occurs during processing and the maximum allowable temperature of the product. Homogenization leads to a rapid rise in temperature from adiabatic compression and viscous dissipation. Typical values are 2.2–2.4 °C/100 bar.[4,10] Using a typical operating pressure of 800 bar, a temperature rise of 9.6°C may be expected

The maximum allowable temperature for many proteins is considered to be 30–35°C.[10,11] Using this value, it becomes necessary to feed cell slurry to the homogenizer at 5–15°C. For recovery of more fragile proteins on a large scale, many manufacturers fit the nozzle with a cooling element to minimize the temperature rise. In all applications, it is essential that external cooling be provided for multiple-pass treatments.

1.2.5 Influence of cell concentration

The cell concentration, expressed as percent dry solids, has been found to exert a weak influence on performance of a homogenizer. Several authors have reported little or no influence of yeast concentration over the range of 4–24%.[9,10,12] In practice, the cell concentration is determined by the harvesting step. For example, most yeast lysis is conducted directly following centrifugation or ultrafiltration at a solids concentration of 12–18% (dry weight).

1.3 Bead Mills

1.3.1 Principle of operation

Bead mills use a horizontal, jacketed grinding chamber filled with grinding media, such as glass beads (Figure 1.4). A cell slurry is introduced into the supply side of the chamber on a continuous basis. Kinetic energy is imparted to the beads by a variable speed shaft equipped with multiple disks, which leads to cell disruption through the combined forces of cavitation, generation of high shear forces, grinding between the beads and by direct collision with the beads.[13] The cell homogenate is then separated from the grinding media by mechanical means, using either an annular disk or a screen. Heat dissipation is removed by passing cooling water or refrigerant through the jacket.

Performance of the bead mill is influenced by agitator speed, bead size and volume, flow rate, cell concentration, temperature and equipment design.

1.3.2 Kinetics

The kinetics of cell disruption in a bead mill follow the first-order reaction presented earlier. In contrast to homogenizers, however, mean residence time

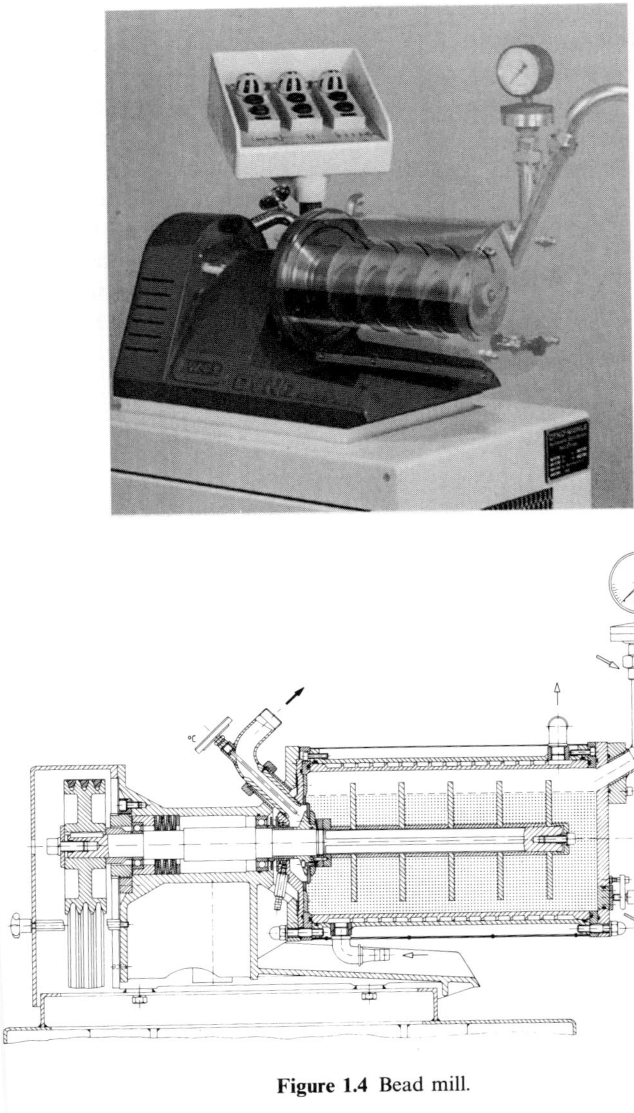

Figure 1.4 Bead mill.

of the cell slurry in the chamber is introduced to yield the modified equation:

$$\log \frac{R_m}{R_m - R} = kNt \tag{1.3}$$

where R_m = maximum protein release or enzyme activity; R = measured protein release or enzyme activity after N passes; k = rate constant (1/s); N = number of passes; and t = mean residence time (s) per pass.

1.3.3 Influence of agitator speed

Tip speed of the disk is used to optimize and scale-up agitator speed in a bead mill. Tip speeds lie within the broad range of 5–15 m/s, and more commonly 8–10 m/s. Below these values, lysis is slow and inefficient; at higher values heat generation is excessive. The relationship between the rate constant and tip speed is presented in Figure 1.5 (adapted from Schutte *et al.*[10]).

1.3.4 Influence of bead size and volume

Lead-free glass or porcelain beads are the medium of choice for most biotechnology applications because of their inertness, abrasion resistance and low cost. Bead diameters are available in the range of 0.2–1.5 mm. Smaller sized beads, for example 0.2–0.5 mm, are used efficiently in smaller mills and are often selected for bacterial cells. Smaller beads generate more impacts but at reduced energy. They also require very narrow slots or screens for separating the grinding media from the homogenate at the exit of the chamber. Slot diameters are restricted to approximately one-third bead diameter.[14]

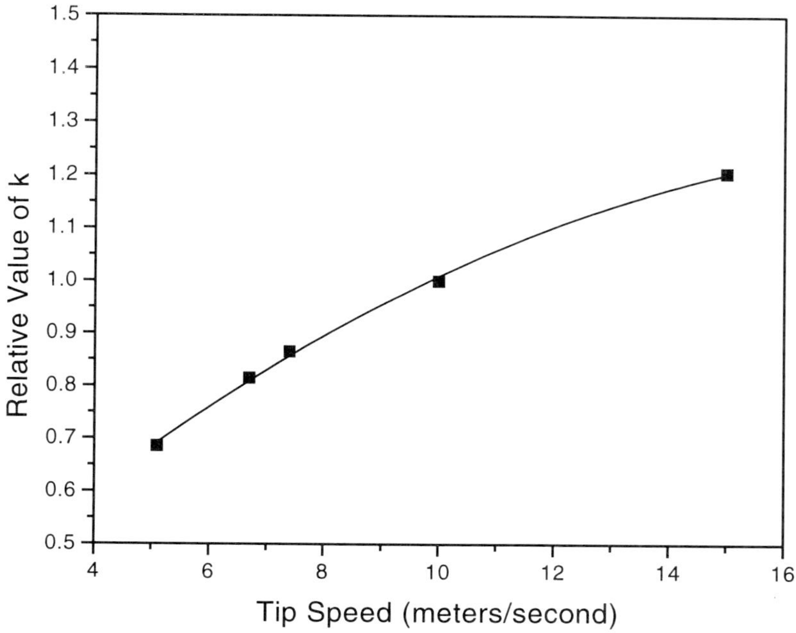

Figure 1.5 Graph of rate constant versus tip speed. From Schutte *et al.*[10]

Also, smaller beads (less than 0.8 mm) may float, thereby reducing effectiveness.[15]

For use on larger cells and in larger mills, bead sizes of 0.5–1.5 mm are preferred. Above 1.5 mm, effectiveness drops rapidly.[15] The larger beads allow larger separator dimensions, hence higher flows. Schutte *et al.*[10] noted that larger beads (> 1 mm) are preferred for periplasmic enzymes or proteins, which do not require complete disintegration of the cell, whereas smaller beads are preferred for cytoplasmic enzymes or proteins. Larger beads also suffer less reduction in diameter from abrasion. Useful lifetimes for beads are 100–200 h for beads less than 0.5 mm but extend to 500–700 h for beads in the range 0.5–1.0 mm. Abrasion increases in low-viscosity fluids.[10]

Bead volume is expressed as the fraction of chamber volume occupied by beads. Typical values are 0.75–0.85 mm^3. Below 0.70 mm^3, disruption is inefficient; above 0.85 mm^3, power requirement and heat dissipation increase dramatically.[16,17]

1.3.5 *Influence of flow rate*

Cell disruption is not a simple function of residence time, particularly at very high degrees of lysis. The first-order rate constantly decreases with flow rate.[14,15,17] Nevertheless, it is often preferable to operate at higher flows if losses may be accommodated. This relationship was demonstrated by Schutte *et al.*,[16] who were able to increase flow rate by a factor of 9 with only a 10% decrease in disruption. Energy requirement per unit of cell mass treated also decreased. Rehacek and Schaefer[17] noted a 3.5-fold decrease in energy utilization with 4-fold increase in flow.

The flow rate influences the residence time distribution within the chamber.[18] Typical residence time distributions indicate considerable backmixing. Modeling the residence time as a series of continuous stirred tank reactors (CSTRs) suggests between 1.5 and 5 theoretical stages, depending on design.[10] Residence time distribution may be narrowed by using multiple passes.

1.3.6 *Influence of cell concentration*

The efficiency of disruption is not strongly influenced by cell concentration over the range of 4–20%. For yeast, 17–18% is considered optimal,[16] with slight reduction in effectiveness at higher or lower concentrations.

1.3.7 *Influence of temperature*

Operating inlet temperature for bead mills is typically in the range of 5–10°C. The amount of heat generated in the chamber of a bead mill is influenced by the tip speed and the bead volume and may be estimated directly from the power input, assuming 80–90% of the power is dissipated as heat.

Temperature rise is controlled by circulating cooling water or a refrigerant through the jacket. The need for adequate heat-exchange surface dictates that the length to diameter ratio of larger volume units be increased.

Temperature exerts a weak influence on the rate constant, with a 20% decrease as the temperature increases from 5–40°C.[15]

1.3.8 Influence of equipment design

The major variables in the design of bead mills lie in the geometry and spacing of the disks mounted on the shafts, and the means of separating the homogenate from the media.

Disk designs are primarily of the two types shown in Figure 1.6. Notched or slotted disks are designed to act like a centrifugal pump, imparting movement on the beads. Disks may be mounted either concentrically or eccentrically on the drive shaft. The latter is believed to prevent uneven compaction of the grinding media. Schutte and Kula[19] reported best results with both *Saccharomyces* spp. and *Brevibacterium* spp. using notched, cone-shaped disks mounted concentrically on the drive shaft.

Separators retain the media in the chamber while allowing the homogenate to exit. Design of separators may also influence operation. Microslit, dynamic

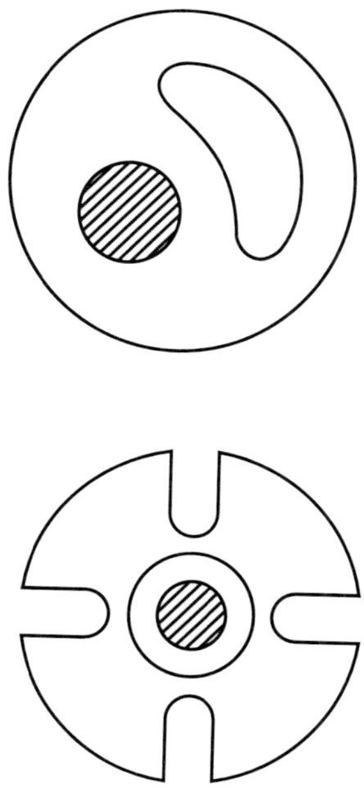

Figure 1.6 Bead mill disks.

separators are self-cleaning designs but impose restrictions on flow and are mechanically complex, subjecting the user to higher maintenance costs. Sieve plates are more common on large mills, offering a 3–10 fold increase in flow. However, these units are subject to blinding.

1.4 Microorganisms

1.4.1 Literature

A summary of literature references for disruption of specific microorganisms and products is presented in Table 1.2. These data may be used as a benchmark for performance. A summary of the information for major cell types is presented below in sections 1.4.2 and 1.4.3

Table 1.2 References (in square brackets) for disruption of specific microorganisms and products using homogenization and milling techniques

Microorganism	Homogenization	Milling
Yeast		
Candida boidinii	Total protein[10] Formate dehydrogenase[12]	Formate dehydrogenase[12,16]
Candida lipolytica	Total protein[20]	G6PDH*[8]
Candida tropicalis	Total protein[21]	
Candida utilis	Invertase[5]	Total protein[17]
Saccharomyces carlsbergensis	G6PDH[12]	G6PDH[12,16]
Saccharomyces cerevisiae	G6PDH[7,10,12] Multiple enzymes[8,11] Total protein[4,6,8,20,22]	G6PDH[12,16] Total protein[15,17,23,24]
Bacteria		
Brevibacterium ammoniagenes	Fumarase[12]	Fumarase[16]
Bacillus sphaericus		Leucine dehydrogenase[16]
Escherichia coli	Total protein[10,23] Penicillin acylase[12]	Isoleucine-tRNA synthetase[16] Penicillin acylase[12] Alcohol dehydrogenase[26] Eco Ri restriction endonuclease[16]
Lactobacillus confusus		Lactate dehydrogenase[16]
Providencia stuartii		PstI restriction endonuclease[4]
Pseudomonas aeruginosa	Amidase[1]	
Filamentous organisms		
Aspergillus niger		Alcohol dehydrogenase[24]
Basidiomyces spp.		Alcohol dehydrogenase[24]

*G6PDH = glucose-6-phosphate dehydrogenase.

1.4.2 *Yeasts*

Yeasts are relatively large microorganisms, with dimensions of 4–7 microns (μ). The size makes them well-suited to lysis in a bead mill. Cell disruption of greater than 90% is achievable in a single pass of a bead mill. Capacities of mills for *Saccharomyces* spp. are 1.9–3.1 kg dry weight/h/l chamber volume.[12,15,16] *Candida* spp. have more durable cell walls, and capacities of 20–30% less may be expected. Optimal bead size for treatment of yeast is 0.55–0.85 mm, with smaller beads more effective at lower cell concentrations.[15]

Homogenizers are also effective for yeast but require multiple (three to four) passes to achieve greater than 90% disruption.[12] Typical throughputs (in kg/h) are 3–6% of the rated capacity of the homogenizer, tending towards the lower figures for lysis efficiencies approaching 95%. A knife-edge valve is reported to be superior for yeast.[6]

1.4.3 *Bacteria*

Bacteria are generally much smaller than yeast and less efficiently lysed in a bead mill. Two or more passes are common to achieve 90% disruption. Expected capacities are 0.5–0.9 kg dry weight/h/l chamber volume. Optimal bead size for bacteria tends to be smaller than that for yeast, and the minimum practical size for the scale of operation is recommended.

Homogenizers by contrast, offer throughputs for bacteria comparable to those achievable for yeast (3–6% of rated capacity) at 85–90% disruption.[10,12,23] No preference has been published for valve design, though a knife-edged valve is most commonly used in industry. Operating pressures are typically 700–1000 bar. However, some bacterial species, such as *Brevibacterium ammoniagenes*, *Micrococcus luteus* or *Nocardia*[12,19] are difficult to lyse effectively even at high pressures.

1.5 Economics of cell disruption

1.5.1 *Capital investment*

Purchase costs for homogenizers and bead mills may be estimated for a known flow rate and cell concentration as shown in Figures 1.7 and 1.8 (May 1991 data). Note that homogenizer capacities are traditionally stated for milk homogenization. A rough estimate of capacities for different cell lines may be derived from sections 1.4.2 and 1.4.3. Alternatively, the residence time and/or number of passes may be estimated from rate constants.

The logarithmic relationship may be used to estimate equipment of different capacities. The use of factors is particularly useful when developing spreadsheet-based cost estimates. Exponents for equipment cost versus

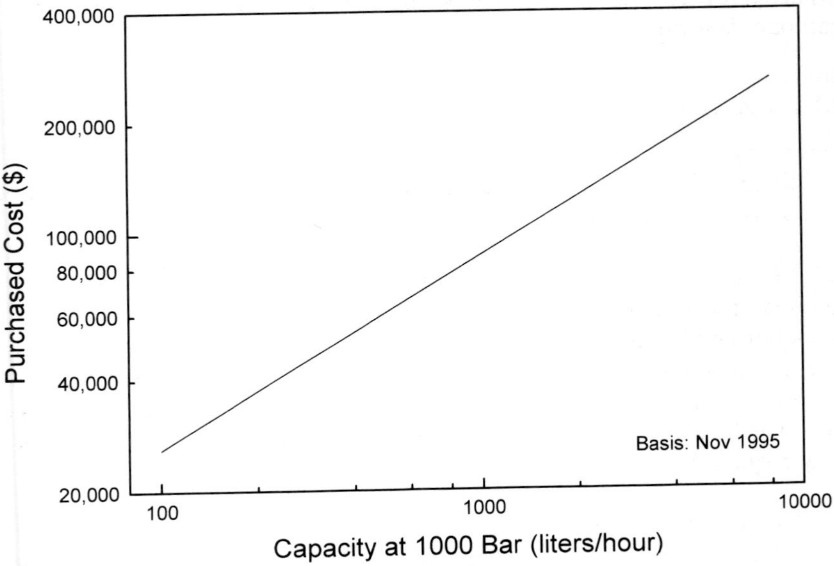

Figure 1.7 Purchase cost versus capacity for homogenizers and bead mills.

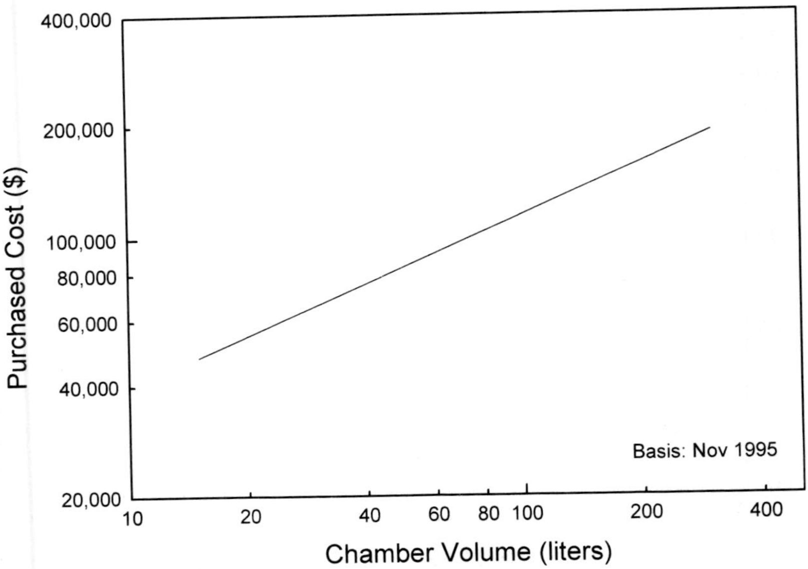

Figure 1.8 Purchase cost versus chamber volume.

capacity are 0.5 for homogenizers and 0.47 for bead mills. These values compare favorably with previously reported values.[25]

1.5.2 Operating costs

Operating costs for bead mills and homogenizers may be itemized as: (i) utilities; (ii) labor, and (iii) consumables or replacement parts. A direct comparison is presented in section 1.5.2.4.

1.5.2.1 Labor. Operating labor for both bead mills and homogenizers is minimal for large-scale batch and continuous operation. Assignment of one-quarter of an operator may be assumed for cost estimation of small scale operations, one-half for larger scale ones. Labor requirements for cleaning the equipment after use or for sterilization are much greater for the bead mill, which requires dismantling and rebuilding. Production scale equipment may require 4–8 h of preparation before reuse.

1.5.2.2 Utilities. Power requirement for a bead mill may be estimated from the chamber volume according to Figure 1.9. Alternatively, power requirements for various organisms as a function of the degree of disruption may be estimated from Figure 1.10 (adapted from Schutte and Kula[18]).

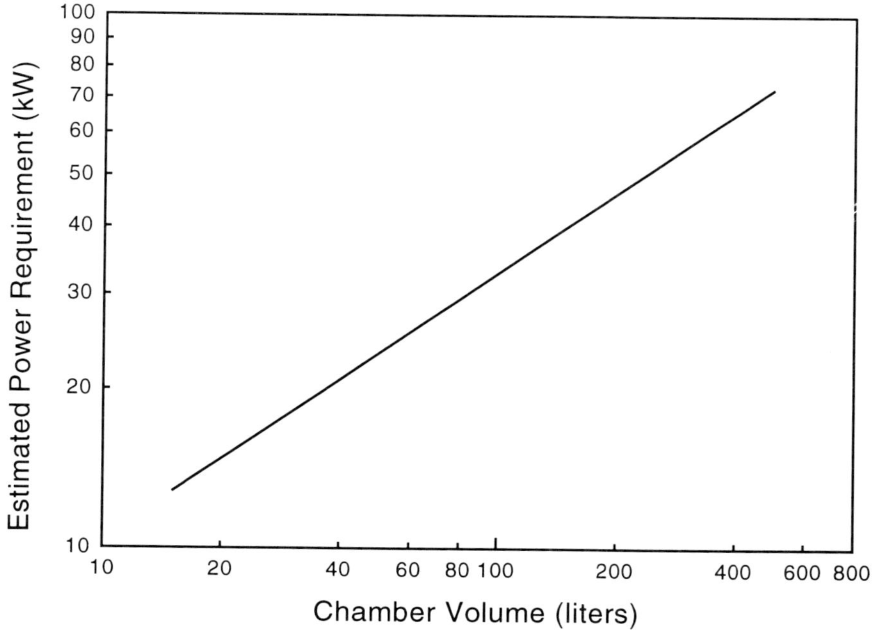

Figure 1.9 Power requirement for a bead mill estimated from chamber volume.

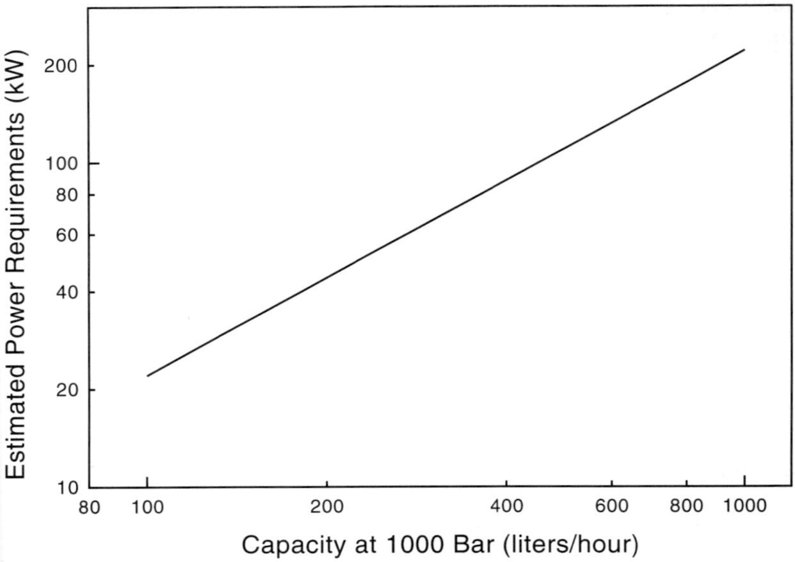

Figure 1.10 Power requirement for various organisms estimated from disruption.

Power requirements for homogenizers may be likewise estimated from rated capacities according to Figure 1.10.

In addition, cooling water is required to maintain temperature within acceptable limits. Assuming that temperature is controlled by a recirculating chiller, the heat load may be calculated from the flow, and calculation of the temperature difference and heat capacity of the homogenate using traditional methods. Alternatively, cooling demand may be calculated for homogenizers based on the expected 2.2–2.4°C/100 bar temperature increase. For bead mills, it is necessary to assume 80–90% power is dissipated as heat.

1.5.2.3 Consumables/replacement parts. Bead mills require periodic replacement of the grinding media. Typical life-expectancies of the media are 500–700 h on a large scale. Media cost for lead-free glass beads is $2–3.50 per pound. Rotating seals are typically replaced on a yearly basis.

In addition to replacing the beads, mills require periodic replacement of agitator and chamber components. Typical lifetimes of disks are 1.5–2 years, 6 months for sieve plates, and 2–3 years for the chamber liner.

Homogenizers have fewer moving parts and require less routine maintenance.

Valves suffer from wear and must be replaced on a yearly basis. Routine maintenance is also required for the piston pump.

1.5.2.4 Economic comparison. Table 1.3 presents a comparison of purchase and variable costs for large-scale production of an intracellular protein by *Saccharomyces* spp. and the small-scale production of a protein by a recombinant *E. coli* (May 1991 basis).

1.6 Selection and optimization of cell-disruption equipment

1.6.1 Manufacturers

Selected manufacturers of equipment for cell disruption are shown in Table 1.4. Each supplier provides technical assistance and guidance, and offers laboratory services and pilot equipment for data generation.

Table 1.3 Comparison of costs of large-scale production of protein using *Saccharomyces* spp. and large-scale production compared with small-scale production using recombinant *E. coli*

Basis	Protein production by *Saccharomyces* spp.		Protein production by *E. coli*	
Production	5 000 000 kg cells/year		50 000 kg cells/year	
Cell concentration	18%		15%	
Disruption	90%		90%	
Operation	2000 h/year		6000 h/year	
	Homogenizer	Bead mill	Homogenizer	Bead mill
Capacity	5000 l/h	100 l	170 l/h	12 l
Capital investment[1]	$210 000	$125 000	$39 000	$46 000
Annual costs				
Depreciation (10 year, straight-line)	$18 300	$10 900	$3 400	$4 000
Labor	$10 000	$12 000	$10 000	$15 000
Power	$17 600	$5 600	$2 400	$5 800
Refrigeration	$2 200	$7 000	$100	$200
Maintenance	$10 000	$13 600	$3 400	$8 500
Total annual cost	$58 100	$42 800	$19 300	$33 500
Unit cost ($/kg)	$0.116	$0.086	$0.39	$0.77

[1] Includes 15% for installation.
NB $ = US dollars.

Table 1.4 Selected US manufacturers of cell-disruption equipment

Homogenizers	APV Gaulin, 500 Research Drive, Wilmington, Massachusetts 01877, USA 508-988-9300
	Centrico, Inc., 100 Fairway Court, Northvale, NJ 07647, USA 201-767-3906
Bead mills	B. Braun Biotech, 999 Postal Road, Allentown, Pennsylvania 18105, USA 800-258-9000
	Glenn Mills, 395 Allwood Road, Clifton, New Jersey 07012, USA 201-777-0777
	Netzsch, 199 Pickering Way, Exton, Pennsylvania 19454, USA 215-363-8010

1.6.2 Criteria for monitoring degree of lysis

Many analytical techniques are available for evaluating the performance and optimization of cell disruption equipment and protocols. The engineer may optimize the value of the parameter directly or, preferably, as a percentage of total lysis. The latter value for total lysis is best derived from a gentle, thorough treatment such as may be achieved in a small laboratory mill.

1.6.2.1 Protein activity. The most applicable means is by direct assay of the biological activity of the protein released. For example, measurement of enzyme activity in the homogenate is the preferred method for enzyme manufacture. Note that the results may differ substantially from measurements of total protein for reasons such as the intracellular location of the desired protein or the degree of denaturation.[14]

1.6.2.2 Total protein. Total protein content in the cell-free homogenate may be estimated by such techniques of the Lowry assay,[26] or direct Kjeldahl measurement. This test should be performed on samples as quickly after lysis as possible to minimize denaturation. Denaturation of samples may be inhibited by cooling or the addition of a reducing agent such as 2-mercaptoethanol.[20] Variable or inconclusive results may occur with extensive disruption from protein denaturation.[1]

1.6.2.3 Total dissolved solids. Total protoplasmic dissolved solids may be used if protein determination is difficult but this technique requires a correction for the increase in aqueous-phase water fraction.[6]

1.6.2.4 Count of intact cells. The determination of the fraction of undisrupted cells may be accomplished by using a Coulter counter,[1] or by counting the cells microscopically using a hemacytometer or equivalent device. The microscopic technique may be aided by a permeability-based cell stain. The technique tends to be tedious and requires a large amount of time for a large number of samples.

Plate counts have occasionally been cited, but their use is not recommended because viability is often lost before lysis occurs.

1.6.2.5 Conductivity. The release of electrolytes from the cytolasm of the cells increases the conductivity of the homogenate. Luther[27] has shown a linear relationship between conductivity and the degree of cell lysis.

1.6.3 General considerations

The principle of *more is better* does not always apply to cell disruption. Excessive disruption may lead to protein denaturation. Marffy and Kula[6] found a 10–20% decrease in activity when residence time exceeded 30 min in a bead mill. Furthermore, excessive disruption of the cell walls and internal structures decreases the mean particle size, which, in turn, makes separation of solids by centrifugation or ultrafiltration more difficult. Also, viscosity may increase with excessive treatment. Homogenates tend to behave as non-newtonian and pseudoplastic fluids. The result is that excessive processing may lead to problems in downstream processing, such that optimization of

Table 1.5 Disadvantages and advantages of bead mills and homogenizers

Bead mills	
Advantages	Disadvantages
1. May achieve disruption in a single pass	1. Difficult to clean and sterilize
2. Better temperature distribution and control	2. Broad residence time distribution
3. Single pass often sufficient	3. Introduces colloidal silica into homogenate from abrasion
4. Aerosol generation is minimized	4. Performance and capacity varies greatly with cell type
Homogenizers	
Advantages	Disadvantages
1. More flexible with regard to cell types	1. Aerosol generation
2. Requires less maintenance	2. Heat generation may lead to denaturation
3. More amenable to cleaning and sterilization	3. Multiple-pass operation is standard

cell lysis should be accomplished in conjunction with subsequent processing steps whenever possible.

Aerosol generation is a major issue in containment of recombinant organisms and pathogens, particularly with homogenizers. Bead mills tend to generate less aerosols than homogenizers. Biosafety of mills has, however, been reported to be improved by enclosure in a soft film cabinet.[28]

Other factors that influence disruption, independent of technique, are the age of the cell and the conditions of production. Mild use of enzymatic or chemical means may also be advantageous. These treatments weaken the cell and render it amenable to disruption under milder conditions.[29]

A summary of the advantages and disadvantages of bead mills and homogenizers are presented in Table 1.5.

References

1. Lilly, M.D. and Dunnill, P. (1969). Isolation of intracellular enzymes from microorganisms – the development of a continuous process, in *Fermentation Advances* (Ed. D. Perlman), Academic Press, London, pp. 225–248.
2. Edebo, L. (1969). Disintegration of cells, in *Fermentation Advances* (Ed. D Perlman), Academic Press, London, pp. 249–272.
3. Foster, D. (1992). Cell disruption: breaking up is hard to do. *Biotechnol.*, **10,** 1539.
4. Brookman, J.S.G. (1974). Mechanism of cell disintegration in a high pressure homogenizer. *Biotechnol. Bioeng.*, **16,** 371.
5. Engler, C.R. and Robinson, C.W. (1981). Disruption of *Candida utilis* in high pressure flow devices. *Biotechnol. Bioeng.*, **23,** 765.
6. Hetherington, P.J., Follows, M. and Dunnill, P. (1971). Release of protein from Baker's yeast (*Saccharomyces cerevisiae*) by disruption in an industrial homogenizer. *Trans. Inst. Chem. Engr.*, **49,** 142.
7. Madsen, F.S. and Ibsen, C.I. (1987). *Cell Disruption by Means of High Pressure*. APV-Rannie.
8. Whitworth, D.A. (1974). Assessment of an industrial homogenizer for protein and enzyme solubilization from spent brewery yeast. *Compt. Rend. Trav. Lab. Carlsberg.*, **40**(2): 19.
9. Dunnill, P. and Lilly, M.D. (1975). Protein extraction and recovery from microbial cells, in *Single Cell Protein II*. (Eds Tannenbaum and Wang), MIT Press, p. 179.
10. Schutte, H., Kroner, K.H. and Kula, M.-R. (1984). Recent studies of mechanical disintegrators for the large scale disruption of microorganism, in *Third European Conference on Biotechnology*, Verlag Chemie, p. I-621.
11. Follows, M., Hetherington, P.J., Dunnill, P. and Lilly, M.D. (1971). Release of enzymes from Baker's yeast by disruption in an industrial homogenizer. *Biotechnol. Bioeng.* **13,** 549.
12. Schutte, H and Kula, M.-R. (1986). Application of stirred ball mills and high pressure homogenizers for microorganism disintegration in technical scale. BTF Biotech-Forum Nr. 23. argang.
13. Patton, T.C. (1970). *J. Paint Technol.*, **42,** 626.
14. Marffy, F. and Kula, M.-R. (1974). Enzyme yields from cells of brewer's yeast disrupted by treatment in a horizontal disintegrator. *Biotechnol. Bioeng.*, **16,** 623.
15. Currie, J.A., Dunnill, P. and Lilly, M.D. (1972). Release of protein from baker's yeast (*Saccharomyces cerevisiae*) by disruption in an industrial agitator mill. *Biotechnol. Bioeng.*, **16,** 725.
16. Schutte, H., Kroner, K.H., Hustedt, H. and Kula, M.-R. (1983). Experiences with a 20 liter industrial bead mill for the disruption of microorganisms. *Enzyme Microb. Technol.*, **5,** 143.

17. Rehacek, J. and Schaefer, J. (1977). Disintegration of microorganisms in an industrial horizontal mill of novel design. *Biotechnol. Bioeng.*, **19,** 1523.
18. Schutte, H. and Kula, M.-R. (1990). In *Separation Processes in Biotechnology*. (Ed. J.A. Asenjo), Marcel Dekker, New York, p. 107.
19. Schutte, H. and Kula, M.-R. (1990). Pilot- and process-scale techniques for cell disruption. *Biotechnol. Appl. Biochem.*, **12,** 599.
20. Engler, C.R. and Robinson, C.W. (1979). New method of measuring cell-wall rupture. *Biotechnol. Bioeng.*, **21,** 1861.
21. Cunningham, S.D. *et al.* (1975). Rupture and protein extraction of petroleum-grown yeast. *J. Food Sci.*, **40,** 732.
22. Doulah, M.S. and Hammond T.H. (1975). A hydrodynamic mechanism for the disintegration of *Saccharomyces cerevisiae* in an industrial homogenizer. *Biotechnol. Bioeng.*, **17,** 845.
23. Anon. (1987). *Cell Disruption by a Rannie Lab 12.51 Hyper High Pressure Homogenizer.* Obtained from APV-Rannie.
24. Rahacek, J., Beran, K. and Bicik, V. (1969). Disintegration of microorganisms and preparation of yeast cell walls in a new type of disintegrator. *Appl. Microbiol.*, **17,** 462.
25. Langlykke, A.F. and Kalk, J.P. (1986). In *Manual of Industrial Microbiology and Biotechnology.* (Eds A.L. Demain and N.A. Solomon), American Society of Microbiology, p. 363.
26. Lowry, O.H., Rosebrough, N.J., Farr, A. and Randall, R.J. (1951). Protein measurement with the folin phenol reagent. *J. Biol. Chem.*, **193,** 265.
27. Luther, H. (1980). Charakterisierung des mechcanischen aufschlusses von mikroorgansimen – zellen mit der leitfahigkeitmessung. *Nahrung* **24,** 165.
28. Trexlor, P.C. (1971). *Vet. Rec.*, **88,** 15.
29. Knorr, D., Shetty, K.J. and Kinsella, J.E. (1979). Enzymatic lysis of yeast cell walls. *Biotechnol. Bioeng.*, **21,** 2011.

2 Conventional Filtration

R. WOLTHUIS and V.C. DICHIARIA

2.1 Introduction

In many pharmaceutical processes, separation of solid and liquid is required. Simple decantation is inadequate when the process is slow, when large volumes are involved and when a low liquid content in the solid is required.

The general problem of separating solids from liquids may be executed using a wide variety of operations, depending upon the character of the solids and the proportion of solid to liquid in the mixture to be separated. Much research on solid–liquid separation technology has been performed by Ladislav Svarovsky. When the amount of solid is relatively small compared with the liquid, the process is usually called filtration. As the percentage of solid in suspension increases, the operation either pressing or centrifugation are required. (Centrifugation is also used, however, when the amount of solid to be removed is almost infinitesimal.)

The operation properly known as filtration is far more important than any of the other operations for separating solids from liquids. The aim of the separation is:

- to recover the solid
- to recover the liquid, or
- to recover both the solid and the liquid.

Pressure filters are efficient in the separation and recovery of fine particles, as well as large crystalline solids from slurries. Batch-pressure filters are used widely instead of vacuum or gravity filters because their high driving force achieves the most economical filtering rates and highest product quality from a totally enclosed process equipment. Continuous-pressure filters are still more preferable but equipment is generally mechanically more complex.

Filtration process equipment has been developed almost entirely from practical considerations and experience and with only minimal relation to theory. Consequently, the apparatus used for filtration consists of many types, each dictated by the specific mechanical considerations needed in the industries where filters have been used.

More recently, the mathematical development of the theory of filtration has greatly improved; however, the practical application of this theory is still in its early stages. This is partly because it is difficult to define the parameters such as size, shape and properties of the particles to be filtered, and because of the wide batch variation even in this carefully controlled process.

The primary factor in the design of filters is the cake resistance or cake permeability. The value of the cake resistance can be determined only on the basis of experimental data. Laboratory or pilot-plant tests are almost always necessary to yield the required information. After the basic parameters for the filter cake have been determined experimentally, the theoretical concepts of filtration can then be used to establish the effects of changes in operating variables such as filtering area, slurry concentration or pressure-difference driving force.

Furthermore, recent market demand for high-quality products has forced manufacturers of filter equipment to improve not only the filter design and filtration performance, but also the cake-washing, reslurrying, cake-squeezing and partial or complete drying operations. For this purpose, a better understanding of the basic mechanisms of solid–liquid separation has been required.

A further important aspect of pressure filter application in the solid separation process concerns its application for aseptic and lethal service, as well as for environmental and operator protection. Both design engineers and manufacturers working in concert are required to achieve high standards of safety and good process facilities.

2.2 Theory of filtration

Although filtration processes have been developed as a practical art rather than a theoretical science, filtration theory has been developed by many authors, and each solid–liquid separation technique has its own application. Filtration theory, although usually not used as the sole basis for filter design, aids the engineer when: (i) interpreting laboratory and pilot-test results; (ii) evaluating the optimum conditions for industrial filtration; and (iii) understanding the effects caused by changes in operating conditions. The use of filtration theory is limited by the fact that filtering characteristics must always be determined on the actual slurry in question and for the specific operating conditions; moreover, complex interactions between the different parameters are difficult to predict.

In general, filtration results in the formation of a layer (or cake) of solid particles on the surface of a porous support, frequently a fabric or wire, on which the actual filtering medium is formed. This cake is delicate and should be handled with care to avoid compromising subsequent processing. Once the initial layer has formed, its structure will operate as the actual filter medium; solids being deposited and adding to the thickness of the cake, while the clear liquid passes through. The cake is therefore composed of a bulky mass of particles, usually of irregular shape, supporting each other and held together by the mother liquor.

Depending upon the chemical and physical characteristics of both solid and liquid components, the mother liquor, as well as subsequent washing liquids, will run freely through small channels within the cake structure.

A filtration process is generally defined by its filtration rate, which is a ratio between the driving force against the cake resistance, and can be expressed simply as:[1]

$$\text{Filtration rate} = \frac{\text{Driving force}}{\text{Cake resistance}} \tag{2.1}$$

The specific filtration rate (q) is the volumetric liquid flow (V) that has passed through the filtering cake area (A) in time (t):

$$q = \frac{\mathrm{d}V}{A\,\mathrm{d}t} \tag{2.2}$$

The driving force is total available pressure drop across the cake (P). The applied pressure drop may be constant or may vary with the time, in such case the function $P = f(t)$ has to be applied.

The total cake resistance is a complex function of different variables and it also varies in time depending upon following parameters.

- Liquid viscosity (μ), is constant provided that the temperature remains constant and that the liquid is newtonian.
- Specific cake resistance (r), which is the resistance of the amount of cake deposited on the filter support preventing flow (W/A). Ideally constant, in practice it depends on the approach velocity of the slurry, the degree of flow consolidation of the cake with time, the feed solid concentration and, most important, the applied pressure drop, which changes according to the compressibility of the cakes. The specific cake resistance often decreases with velocity and feed concentration, and the strongest effect on it is caused by the applied pressure drop. Since each layer in a cake is subject to a different pressure drop, the only simple way to deal with the problem is to define an average value. Practically the resistance of the product layer can be expressed as $r(W/A)$.
- Filter system resistance (R) includes the resistance of the filter medium as well as other losses in the system across which (P) is the pressure drop.

The filtration rate can therefore be expressed as follows:

$$\frac{\mathrm{d}V}{\mathrm{d}t} = \frac{AP}{\mu[r(W/A) + R]} \tag{2.3}$$

From pilot studies, the specific cake resistance (r) can be calculated as a variable for different cake thicknesses or as an overall figure to be related to the applied pressure drop. The filter size calulated using this conventional approach is almost always larger than necessary and thus on the safe side.

2.2.1 *Practical considerations*

The specific cake resistance (r) is an average value and is related mainly to the pressure by the expression:

$$r = r' \cdot P^s \tag{2.4}$$

where r' is a constant determined chiefly by the size of the particles forming the cake, and s is the cake compressibility, varying from 0 for rigid, incompressible cakes, such as fine sand, to 1 for highly compressible cakes.

For most industrial slurries, s has a value between 0.1 and 0.8. When the cake is composed of hard granular particles (where $s = 0$), an increase of pressure will result in no deformation of the particles and, neglecting the filter system resistance (R), equation (2.3) becomes:

$$\frac{dV}{dt} = \frac{AP}{\mu r'(W/A)} \tag{2.5}$$

Therefore, for incompressible cakes, the flow rate is directly proportional to the filter area and applied pressure but inversely proportional to the viscosity, the total amount of the cake and the specific cake resistance.

When the cake consists of extremely soft, easily deformed particles, such as ferric or other metal hydroxides, $s = 1$, and the equation becomes:

$$\frac{dV}{dt} = \frac{A}{\mu r'(W/A)} \tag{2.6}$$

where the rate is independent of the pressure.

2.3 Classification of separators

Solid–liquid separation processes are usually grouped into two broad categories: (i) sedimentation; and (ii) filtration. Such a classification is not strictly correct if it is to include, for example, flotation or solids screening. A more general mode of classification is to define two kinds of separations:

1. The liquid is constrained in a vessel and particles can move freely within the liquid:
   (a) flotation (dispersed air, dissolved air, electrolytic);
   (b) gravity sedimentation (thickeners, clarifiers);
   (c) centrifugal sedimentation (fixed-wall hydrocyclone, rotating-wall sedimenting centrifuges).
2. The particles are constrained by a permeable medium through which the liquid can flow:
   (a) cake filtration (vacuum, pressure, centrifugal);
   (b) deep-bed filtration (sand and coke);
   (c) screening (dewatering, vibrating screens).

The pharmaceutical industry in particular uses the following classifications of separators:

- Filters for intermediate product or bulk material versus final product
- Filters for clean operation versus aseptic and/or cleanroom operation
- Filters for dedicated product versus multipurpose operation
- Filters for batchwise versus continuous operation
- Filters versus filter/dryers.

The choice of a filter depends on many factors, the most important of which are:

- Fixed and operating costs
- Quantities and value of materials to be handled
- Specific process requirements and conditions (e.g. concentration, temperature and pressure of the slurry, use of toxic or aseptic products)
- Ambient, environmental and operator protection condition
- Properties of the fluid (viscosity, density, corrosive nature)
- Whether the valuable product is the solid, liquid or both
- Particle size and shape, surface characteristics of the particles and compressibility of the solid material
- Extent of washing for the filter cake
- Extent of drying for the final product
- Operating conditions (batchwise or continuous, manual or automatic).

2.4 Filtration performance

Solid–liquid separation is never a complete process. Some of the solids invariably leave with the liquid stream, and some of the liquid will always leave with the solid flow. To compare the performance of different separators, there must be measurement of this imperfection of separation. A common way of describing the efficiency of a separation is to consider the solid and liquid fractions separately to derive the following two independent factors:

1. The mass fraction of the solids recovered — called the separation efficiency. This factor does not take the liquid recovery into account.
2. The moisture content in the recovered solids, which is a measure of the imperfection of the liquid separation.

Since the relative importance of both factors varies from application to application, it is probably best to consider each separately.

The filtration operation is always followed by a cake-washing and cake-squeezing process, required to improve the filtration performance itself and to allow to obtain the highest quality of the final product.

The purpose of this cake-washing operation is to replace the mother liquor from the solid cake with a wash liquid. This process is becoming increasingly important because of heightened demands for product purity in the face of declining raw material quality. In the economics of a process involving multistage washing, this step may represent a dominant fraction of the total installation cast.

Several models exist to predict washing performance using a minimum of experimental measurements. Frequently, the operation requires several concurrent or countercurrent stages. Compression can be used to enhance performance.

If the cake is compacted or has a resistance, reslurrying the cake during the washing may be advantageous. The dewatering operation is often used to reduce the moisture content of cake discharged from a solid–liquid separation. It is achieved either by mechanical compression or by air displacement in a vacuum, pressure or centrifugal system.

The cake-washing process identifies mainly two main procedures: (i) reslurry washing; and (ii) displacement washing.

2.4.1 Reslurry washing

This procedure requires an agitated filter where the wet cake is resuspended into a washing liquid, followed by a new filtration operation. This washing step eliminates the impurities contained in the mother liquor and allows eventual solid impurities to redissolve by using a specific washing solvent. The reslurry washing operation allows the cake to be washed with a liquid different from the original mother liquor, which has the advantage of dissolving the impurities, without the cake structure collapsing.

2.4.2 Displacement washing

Displacement washing is the most frequent and simplest procedure. It can also provide a very efficient washing except when the cake presents cracks, when the cake structure collapses or when chemical or physical interactions between the solid product and the mother liquor or its impurities obstruct the washing process. Displacement washing requires the washing liquid to be pressed through the cake, thus substituting the mother liquor in the cake. Usually cracks occur when the mother liquor within the cake is substituted by a different liquid or gas, or when solid impurities are washed out by dissolving. Cracks cannot be avoided but can be limited by:

- Operating with a reduced cake thickness (using continuous filters or centrifuges).
- Operating with a properly designed stirrer, which smoothes the cake surface during the washing process and fills up the cracks with product (e.g. agitated Nutsche filters).

A very important washing procedure is however, blowing the cake with nitrogen or air, which blows out both mother liquor and/or washing liquid. The result is a cake with a lower moisture content. The formation of cracks in the cake prevents a correct blowing if not assisted by smoothing the cake surface.

Usually a sequence of reslurry washing and displacement washing together with cake blowing is operated to perform the required washing process to obtain the highest quality of product.

2.5 The effect of pressure

In general, the filter fabric itself is rarely, if ever, the true filtering medium. The real filtering medium is the initial thin layer of the precipitate itself, which is entangled in the surface of the fabric. Therefore, the formation of this first layer is of prime importance in securing a satisfactory filtration. Usually the average particle is considerably finer than the average opening between fibres (Figure 2.1).

A pressure filter may be operated in any one of several ways. The simplest method is to apply the full pressure at the start of the filtration and to maintain the pressure constant throughout the run (Figure 2.2a).

A serious disadvantage of this method is that, if the initial pressure is high, the first particles caught will be compacted into a tight mass that largely fills the pores of the cloth and results in a low rate of filtration for the rest of the cycle. If the sludge is non-homogeneous, that is, if it contains both crystalline and colloidal particles, high initial pressure tends to force the colloidal portions of the sludge into the interstices of the granular portions and greatly decreases the rate of filtration. Conversely, if the initial pressure is low, the initial layer of precipitate will be more open so that the rate of filtration will be higher, and this layer will not be crowded into the fibres of

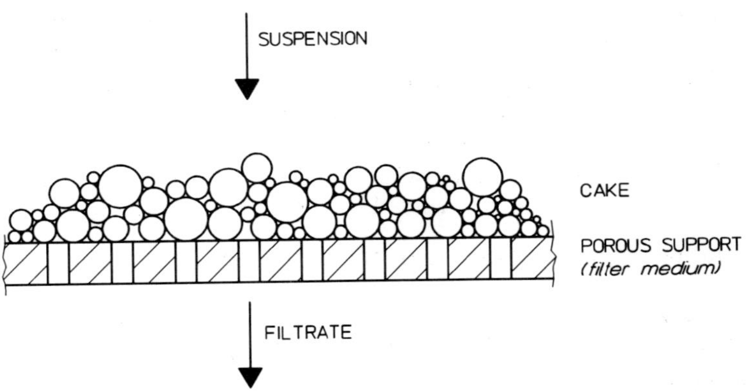

Figure 2.1 Cake filtration.

the fabric, with the result that the cake will separate more cleanly from the cloth.

A low initial pressure may mean that the first runnings from the press will not be clear, but this is more than compensated for by the more rapid filtration rate and consequently larger capacity. Another method is to maintain a nearly constant rate by starting at low pressure and continuously increasing the pressure to overcome the increasing resistance of the cake until the maximum pressure is reached at the end of the run (Figure 2.2b). The difficulty with this method is that the maximum pressure is reached only at the end of the run, and hence the whole cycle is at less than maximum capacity.

A common method of meeting the difficulties of constant-pressure filtration is to operate at constant rate during the first part of the filtration. As soon as the cloths are well coated with cake and the filtrate is clear, the pressure is then increased to the maximum, and filtration is completed at constant pressure (Figure 2.2c).

The relative advantages of operation at constant pressure or constant rate depend to a large extent on whether the precipitate is granular, colloidal, or a mixture.

In Figure 2.3 the related flow rates against filtration time for the different

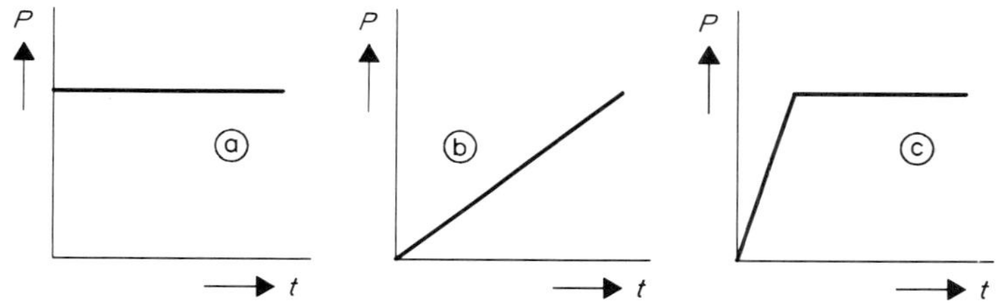

Figure 2.2 Time–pressure graphs during the filtration process where: (a) full pressure is applied at the start of filtration; (b) there is a constant increase in pressure during the filtration; and (c) there is an initial constant pressure followed by full pressure after the initial cake layer is formed.

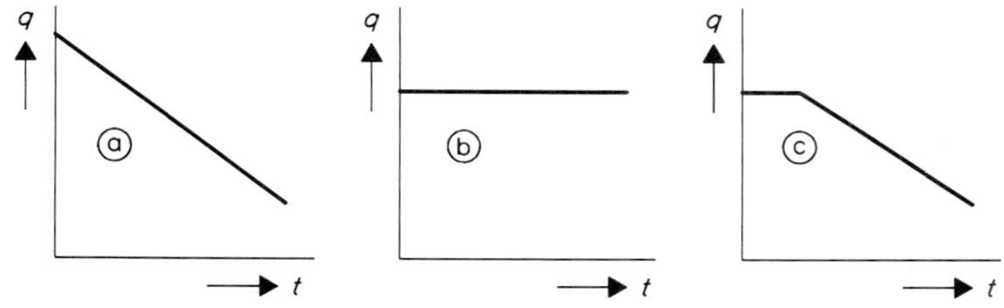

Figure 2.3 Graph of flow rate with time during filtration where: (a) shows constant full pressure operation (decreased flow-rate); (b) shows constant flow-rate operation (constant increased pressure); (c) shows initial constant rate followed by full pressure operation.

pressure operation procedures are shown. The graphs show ideal curves, where time is equivalent to the obtained cake thickness.

For reasonably compressible sludges, however, rate of flow may not increase with pressure. At low pressure, the increase in rate of filtration accompanying a slight increase in pressure may be greater than the decrease in rate because of the increase in resistance caused by the compression of the sludge. As the pressure increases, however, these two factors tend to become equal and, at a certain optimum pressure, the rate of flow is a maximum. For pressures higher than this, the added resistance due to the compression of the sludge is greater than the added driving force represented by the increase in pressure, and the net effect is a decrease in the rate of flow with increase in pressure. Filter operation should be maintained at a pressure below the optimum. This phenomenon is absent when filtering non-compressible sludges.

2.6 Limitations of filtration theory

A word of caution should be added about the theory of filtration. The mathematical derivations and manipulations seem straightforward, and there has been a tendency to elaborate filter theory beyond its practical usefulness.

All the equations reported in the literature contain constants whose numerical value is difficult or impossible to determine directly and parameters that are variable during the filtration process itself.

This is not an obstacle in itself for many useful applications of theory, for it might be supposed that, if an experiment on a given slurry gave workable numerical values to be used in the equations, such numerical values may then be used for predicting the effect of changes in operating conditions or predicting scaling-up calculations. Within acceptable limits, this is true; however, tests on precipitates prepared from small-scale laboratory runs may not represent the properties of the precipitate produced in a large-scale operation. Shipping a sample of slurry to a laboratory equipped to perform filtration tests is out of the question since agitation during shipping may change the degree of flocculation, while temperature variations may change the size and properties of the precipitate.

Nevertheless, the conclusion that filter theory is of no practical value is quite incorrect. Carefully run laboratory filter experiments, made with an understanding of the highly temperamental behaviour of the precipitate, can be of real and practical value. Laboratory experiments should be followed by carefully scheduled pilot tests in order to select the more qualified type of filter with the required filter area.

Pilot tests also have to be performed to confirm the effect of operating variables in the process by which the precipitate is prepared or of variations in filtration procedure.

2.7 Filter aids

For sludges that are difficult to filter, various filter aids are used. A filter aid is a solid material, finely divided but consisting of hard, strong particles that are, *en masse*, incompressible. The most important filter aid used commercially is diatomaceous earth. It consists of siliceous skeletons of minute marine organisms known as diatoms. These skeletons are characterized, first, by the fact that they consist practically of pure silica and, second, by the fact that the individual skeletons, though of very small size, are exceedingly complex in their structures and therefore offer an enormous surface for the adsorption of colloids.

Although this material, because of its enormous surface and rigid structure, is highly effective as a filter aid, any granular, incompressible material, such as precipitated calcium carbonate or any other granular crystalline precipitate, may be added to a compressible sludge to advantage.

2.7.1 Precoating with filter aid

Filter aids can be used in either of three ways. The first method is the use of filter and as precoat where a thin layer of the material is hand drawn on the filter before the sludge is pumped to the apparatus.

A precoat prevents the colloidal particles of the sludge from becoming so entangled in the filter cloth that the resistance of the cloth itself becomes high. It also facilitates the removal of the cake at the end of filtration. With a precoat, the filtering medium is really the precoat material rather than the filter cloth itself.

This method of operation is common where the filtration is merely a finishing operation, removing small amounts of turbidity that would never build up into a real filter cake. This method is usually adopted for filtration processes on pressure leaf filters or rotary vacuum-drum filters.

2.7.2 Incorporating filter aid with the sludge

The second method of using filter aids is the incorporation of a certain percentage of the material with the sludge before sending it to the press. The presence of the filter aid increases the porosity of the sludge, decreases its compressibility, and reduces the resistance of the cake during filtration.

2.7.3 Using a special precoat filter

A third method of using filter aid is the use of a special precoat filter. This is essentially a rotary-drum vacuum filter. A slurry of filter aid only is fed to the filter until a layer of precoat has been laid down, then followed by the sludge to be filtered.

A scalpel is positioned so that it peels off the sludge and an extremely thin layer of the precoat. The scalpel has an automatic micrometer feed so that it advances continuously until the layer of precoat is exhausted. Filtration is then interrupted, a new thick layer of precoat deposited, and filtration continued.

This method is often used for slimy or gelatinous precipitates, such as those characteristic of fermentation broths, that can never be built-up into a cake but must be removed when the filter is covered with a very thin layer of the precipitate.

The nature of the filter aid usually limits the use of this method to those cases where the cake is discharged and it cannot usually be applied when the cake is an intermediate or final product. Separation of the precipitate from the filter aid would be possible only where chemical methods were feasible.

2.8 Filtration drying technology

Based upon a specific demand from the market, a few filter manufacturers have developed very sophisticated agitated Nutsche pressure filters, which can be used also as very efficient vacuum dryers. The main advantages of these filter/dryers over the traditional filter with separate dryer are that there is/are:

- No transfer of wet product from filter to dryer equipment
- No product contamination or operator contact during wet product transfer
- Easy cleaning procedure of a Nutsche filter/dryer alone instead of two pieces of equipment (with fewer product loss)
- Lower installation, operator maintenance costs.

The application of a filter/dryer is even more economical when this unit can be used in a production cycle where the filtration plus drying time is similar to the longest operation time within the total production cycle.

The decision to proceed with a filter/dryer usually provides the answer whether to use a centrifuge or a filter, taking the maximum advantages of the high efficiency and maximum flexibility of the filter/dryer.

Best performances from a filter/dryer are obtained when a specific design with a full heated agitator is provided. Such a specific agitator has to be installed in the Nutsche pressure filter but does not limit or reduce the original function of filtration and washing.

Numerous applications have demonstrated that only a specific customized design of the agitator allows maximum performance of the filter/dryer unit, usually resulting in drying times similar to those obtained from traditional agitated vacuum dryers. Furthermore, the specific design of the mixing characteristics of the heated agitator allows a homogeneous dried product to be obtained.

2.9 Comparison and selection of filter types

A considerable variety of filter constructions is available, and the question of how one chooses the best type of filter for a given situation naturally arises. The choice is often influenced by tradition or accepted practice in a particular industry. When selecting filters, factors concerned with job demand must be weighed against equipment (including filter media) characteristics. The important job-related factors are: (i) slurry character; (ii) production quantity; (iii) process conditions; (iv) performance requirements; and (v) permissible materials of construction. The important equipment-related factors are: (i) type of cycle (batch or continuous); (ii) driving force; (iii) production rates of the largest and smallest units; (iv) separation sharpness; (v) washing capability; (vi) dependability; (vii) construction materials; and (viii) cost. The cost estimate must include depreciation (installed cost plus life), maintenance operating cost (labour, services and filter media), and penalty of product loss (if any).

Factors of slurry preconditioning and the use of filter aids must also be considered. Slurry characteristics determine whether a clarifying or a cake filter is appropriate and, if the latter, they determine the rate of formation and nature of the cake. They also affect the choice of driving force and cycle as well as the specific design of machine.

Cake-washing requirements also may significantly affect equipment choice, as the washing capability varies greatly with filter type. Continuous filters are most attractive, of course, when the process to which they contribute is a steady-level continuous one, although the rate at which the cake forms and the magnitude of production rate are likely to be overriding factors. Often the desirability of flexibility and relatively high filtering pressures will rise above considerations that would otherwise dictate the use of continuous filtration equipment. Also, the necessity of predrying or complete drying of the product within the same filter unit must be evaluated. For estimating filtration rate (therefore operating pressure and size of filter), washing characteristics, and other important features, pilot-scale tests are usually essential.

2.10 Small-scale filtration

For most applications, the industrial filter cannot be selected and sized correctly by applying only the theory. This is mainly because of the nature of the filter cake structure, which cannot be defined mathematically. Therefore, unless exact data have already been established for the same cake and same filtration conditions, small-scale tests must be performed.

The purpose of small-scale tests is first to select the proper filter type, then to determine the optimum filtration procedure and operation conditions and, finally, to determine the optimum filter sizing and predict its performance.

Selection of the best type of filter is usually determined using laboratory tests and a bomb filter. Bomb filters with a filter area of about 0.01 m^2 yield reliable data on the 'filtrability' of the product, and allow the filter requirements to be determined. Subsequent small-scale tests are performed on pilot equipment to confirm the correct type of filter and to provide all filtration variables and constants to allow a correct upscaling calculation.

Pilot experiments are performed using pressure filters of similar design to the industrial filter. For example, for agitated Nutsche pressure filters, pilot units are available in the range of 0.2–0.4 m^2. In particular, when tests are performed on Nutsche filter/dryers, the agitator of the pilot unit should be heated in the same way as on the industrial unit to obtain reliable results for upscaling. For continuous-pressure filters, pilot units of 0.1 m^2 are available for test work. A pilot test will yield all required information for correct upscaling in a short time, provided that representative samples and correct pilot units are used.

In any event, before undertaking filtration tests, it is advisable to consult an experienced and qualified filter manfacturer in the industry who can provide as much information as possible about the slurry and the objectives of filtration in particular. The assistance of such a vendor will be invaluable to the inexperienced person carrying out a small-scale filtration test programme.

2.11 Filtration equipment

Many commercially available filters exist; a description of several of the more commonly applied batch and continuous-pressure filters is supplied here. These filters all have in common the possibility of keeping separate the slurry, cake and filtrate from the external environment. This is often an important consideration in pharmaceutical processing.

2.11.1 Discontinuous Nutsche-type pressure filter

The Agitated Nutsche-type pressure filter (Figure 2.4) is manufactured in two standard models, on which some variations can be made to adapt it to treat different products.

In the fixed-bottom version, the filter basically consists of: (i) a vertical cylindrical casing with dished head and flat filtering bottom; (ii) a filtrate draining structure, on which the filtering cloth is installed; (iii) a mixer-scraper, which can rotate in both directions and can be raised and lowered (to the central shaft are fixed two arms to which blades are attached having a purposely designed slant); (iv) a rotation and axial movement unit of the stirrer, formed by a variable-speed motor and gear box. A discharge hatch fixed to the cylindrical ferrule and usually a heating system on the cylindrical

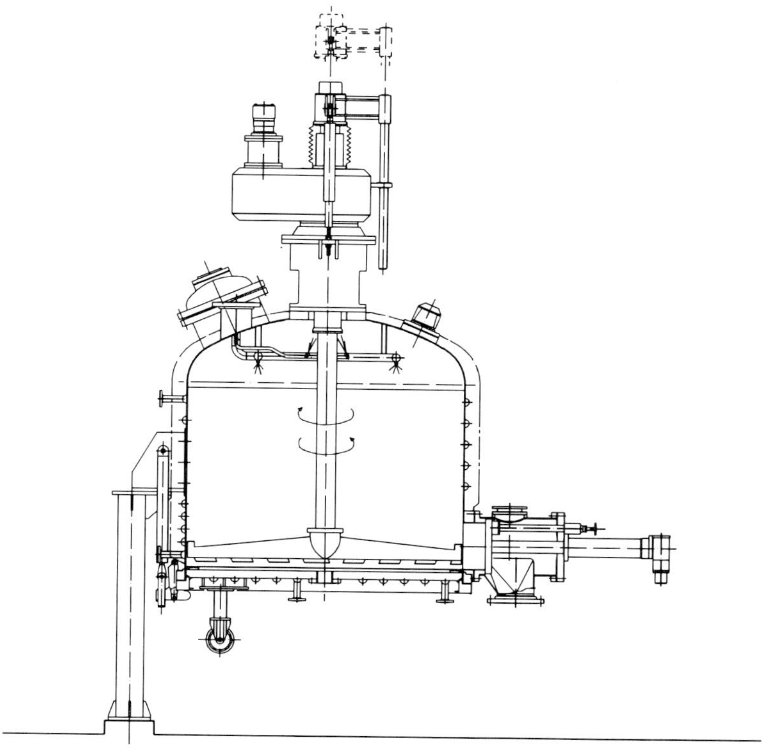

Figure 2.4 Agitated Nutsche-type pressure filter (by courtesy of 3V-COGEIM SpA).

casing as well as an optional heating system on the bottom in case heating (or cooling) is also necessary.

In the alternative model, the bottom can be removable and is fastened to the vessel by clamps or bayonet system; opening of the bottom is operated hydraulically.

The discharge hatch on both models consists of a plug, operated by means of one hydraulic system, which moves within a fully-sealed hatch. The rotation of the mixer-scraping unit is carried out either by a gear motor equipped with a hydraulic motor or with an electrical motor provided with frequency variator, allowing a wide variation of revolutions, while its raising–lowering movements are obtained through two hydraulic cylinders. All electric and hydraulic systems are operated using a control panel, provided with full interlock system for safe operation. The filter can also be controlled using any computerized system that allows for sequence control as well as fully automatic operation of the unit.

2.11.1.1 Operation. The slurry is introduced into the filter in batches, and the liquid filtrate passing through the filtering cloth covering the bottom is discharged through a nozzle located in the centre of the bottom itself. By creating pressure using compressed air or an appropriate gas, the filtration is completed and the mother liquor eliminated.

Washing is accomplished by introducing another liquid, always working under pressure. When the required washing performance is obtained, the cake is squeezed by compressed air or gas. During these last phases, the agitator keeps the cake surface smooth and free from cracks, so that washing and squeezing is homogeneous.

Finally, the cake is discharged by opening the side hatch, and slowly lowering the agitator, now rotating in the opposite way, into the cake; the scraping teeth make breaking the cake easy, allowing its conveyance to the discharge port.

2.11.1.2 Fields of application. The Agitated Nutsche-type pressure filter (see Figure 2.4) is suitable for filtering slurries with middle–high solid concentration (up to 40–50%). It allows a perfect separation of mother liquor from the solid, with the possibility of performing more washes with different liquids. Washing can be taken to the extreme and can be carried out by percolation and/or re-suspending the cake in the liquid using the mixer-scraper.

When washing by percolation (displacement), preferential paths formed by flaws is avoided by smoothing and pressing the cake surface continuously.

The main characteristics of this filter are:

- Automatic discharge of the cake
- Reduced quantity of washing liquid
- Drainage made by a fluid under pressure, which squeezes the cake perfectly
- Completely sealed construction, allowing the treatment of toxic or easily oxidable products without environmental and product pollution
- Possibility of sterilizing the filter for full aseptic operation
- Possibility of drying the cake
- Possibility of heating and/or cooling the filter
- Low-energy consumption
- Great ease in cleaning and replacing the cloth in the case of the filter with the removable bottom.

Owing to the abovementioned characteristics, the Agitated Nutsche-type pressure filter is used increasingly in place of other methods of filtration, above all in the chemical intermediates and pharmaceutical industries for increased productivity and environmental hygiene.

2.11.2 *Nutsche-type filter/dryer*

When the Agitated Nutsche-type pressure filter is used also for the subsequent drying process it is provided with an adequate heating system; in this case it is called a filter/dryer (Figure 2.5). The pressure filter becomes a highly sophisticated piece of process equipment by providing it with a full heating jacket on bottom, cylindrical wall and dished head. The mixer-scraper is provided with a full internal heating system and a second special designed stirrer is added. This second agitator is provided with vertical blades, which are also heated and move through the product shares. The combined action between the two stirrers guarantees optimum mixing and heat transfer.

The filter/dryer is also provided with a dust filter on the vapour outlet line; an automatic reverse-jet cleaning system allows for a continuous operation. The filter/dryer is usually operated through an automatic control system.

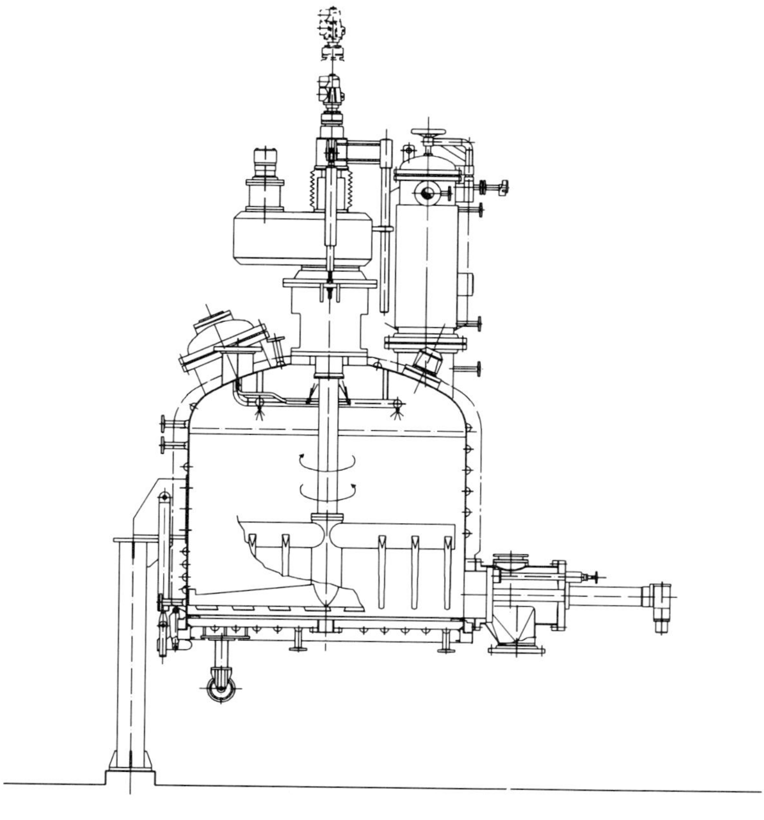

Figure 2.5 Agitated Nutsche-type filter/dryer (by courtesy of 3V-COGEIM SpA).

2.11.2.1 Operation. The wet cake, after its filtration and washing process, is mixed by the special mixer-stirrer, which is lowered into the cake while rotating in the scraping direction. Vacuum is applied and all heating systems are providing for the solvent evaporation. The special agitator design mixes efficiently and allows high heat-transfer rates. The final homogenized dry product is discharged through the side hatch.

2.11.2.2 Fields of application. The filter/dryer is particularly suitable for the pharmaceutical industry where high-value products are handled, frequent product changes have to be operated and where contact of the product with the environment and operator must be avoided.

The main advantages of the filter/dryer are:

- Operation of all filtration and drying processes in one single closed vessel (no product transfer)
- No exposure of the product to the environment
- No handling of products by the operator since final product can be discharged directly into drums
- Easy cleaning procedure for frequent product changes.

2.11.3 *Continuous rotary pressure filter*

Filters that work in a continuous way were used originally in the production of calcinated soda to separate sodium carbonate from its solution. Today they are employed in different branches of the chemical and pharmaceutical industry, whereas, historically, only continuous vacuum filters were known.

The continuous rotary pressure filter (Figure 2.6) is a continuous filter, consisting basically of an external casing, inside which a drum rotates at adjustable speed. Between the casing and the drum, some annular space is left (which is gas- and liquid-tight by means of special side sealing). It is divided into chambers or sectors by longitudinal seals. Each sector is a single consecutive working cycle of operations such as filtration, cake-washing, cake-squeezing, cake discharge and cloth-washing. Therefore it is considered to operate like a continuous Nutsche-type pressure filter.

The drum surface is subdivided in cells, separated by shaped ribs, and the filtering frames, lined with the real filtering cloth, are fixed on the cell bottom. By raising or lowering the filtering frames (adjusting the distance from the filtering frame top and cell bottom) it is possible to obtain cakes with a thickness of approximately 20 mm. The rotational speed of the drum is adjusted in order to obtain a cake as thick as the free space of the cell.

Each filtering sector is connected to a draining pipe and all draining pipes combine in a rotary valve, which handles the fluids coming from the different filtration phase separately. The longitudinal seals are kept pressed onto the drum using compressed air, thus guaranteeing perfect separation between

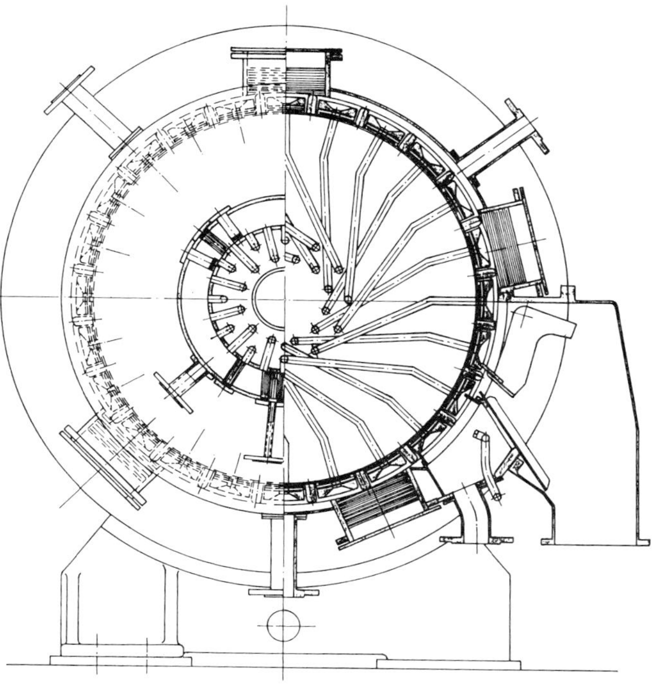

Figure 2.6 Continuous rotary pressure filter (by courtesy of 3V-COGEIM SpA).

the different sectors. The ordinary pressure of filtration reaches 2–3 bar; the maximum is 6 bar.

2.11.3.1 Operation. The slurry is pumped into the filtering chamber, the filtered liquid passes through the draining pipes and is discharged by the valve, while the solid deposits on the filtering cloth until the cells are completely filled. Then it is pressed and passes from the filtering chamber to the washing chamber, where it is washed. The number and procedure of washings can be varied according to requirements; the washing fluids can be fed under different pressures and temperatures. A countercurrent washing system further reduces the quantity of used washing liquid.

As the cells are completely filled, washing is obtained by forced percolation upon a perfectly even cake, which is devoid of cracks and therefore without any preferential path.

After this, the cake passes into the squeezing chamber where it is drained by means of air, gas or steam, resulting in perfectly washed cake and with a highly reduced final moisture content. The cake is discharged subsequently by means of a sprung blade, which penetrates into each cell until it almost scrapes the cloth; it is then discharged continuously into a hopper, which can be sealed completely.

A counter jet of compressed air is used in order to facilitate detaching the cake from the cloth. Before entering again into the filtering chamber, each cell is washed with fluid or steam, keeping a light blast of air in counterflow, to avoid the phenomenon of progressive clogging of the cloth.

2.11.3.2 Fields of application. The rotary pressure filter is particularly suitable for filtering low–medium solid content slurries, when a very good washing of the cake with a minimum quantity and a low residual of washing liquid moisture are required. Output is usually from 4- to 10-fold greater than the traditional continuous-vacuum filter.

The principal advantages of the continuous rotary pressure filter are:

- Continuous working
- High hourly output
- Reduced filtering surfaces and then reduced overall dimensions
- Highly efficient cake-washing
- Use of a low quantity of washing liquid
- Good squeezing of the cake, obtained through pressure, with a substantial reduction of the residual moisture content
- Accurate recovery both of the liquid and of the solid
- No need for pumping to recover both the filtered material and the washing liquid as it operates under pressure
- Automatic discharge of the cake
- Fully sealed and tight construction, which allows the filtering of harmful solvent or substances without environmental pollution (it is sufficient to close the hopper discharging the cake)
- Particularly low energy consumption.

2.12 Ultra-fine filtration

Filtration equipment and processes in the pharmaceutical industry have commonly been classified on the basis of size of the smallest particle reliably retained by the filtration media (refer to Figure 2.7). Conventional filtration is capable of removing particles down to approximately 70 μ in diameter, fine filtration refers to removal of particles in the 70–2 μ range but has only limited applicability to pharmaceutical processing. In contrast, ultra-fine filtration, practised extensively in the pharmaceutical industry, is normally accomplished by the use of cartridge-type filters, which are capable of removing bacteria and debris. In particular, 0.2 μ filters are widely applied in the industry to maintain sterility of process liquids and gases.

Ultrafiltration and reverse osmosis are membrane processes that are widely applied in the pharmaceutical industry.

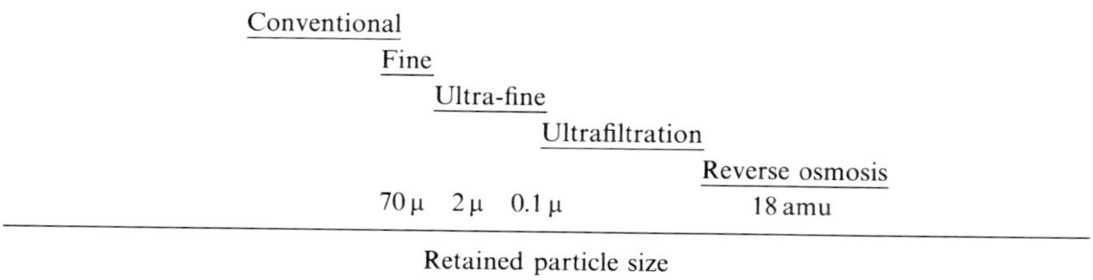

Figure 2.7 Types of filtration and retained particle size.

Ultrafilters are commonly used for the isolation and concentration (but not purification) of protein products and are classified according to molecular weight cut-off point, based on the approximate size of the protein that will pass through the membrane. Cut-off is actually somewhat of a misnomer, as the shape and chemical nature of the protein, as well as its molecular weight, will affect its ability to pass through the membrane. Separation, therefore, is not exact since some proteins with molecular weights less than the rated cut-off point will be retained and some heavier ones will pass. The process is useful for isolating a fraction containing a desired product protein prior to true purification by chromatography or other means. The objective is to start with a clarified fermentation broth containing many different proteins and metabolites and process it such that only a limited number of proteins remain. This usually greatly simplifies the ultimate purification process.

Reverse osmosis membranes pass only water molecules retaining metabolites, salts and other contaminants. Although reverse osmosis has sometimes been used in product purification processes, its typical use in the pharmaceutical industry is for the preparation of USP pure water. It is very widely used in the industry for this purpose.

2.12.1 Mechanisms of filtration

Cartridge filtration of fluids occurs by three principal mechanisms: (i) direct interception; (ii) inertial impaction; and (iii) diffusional interception (Figure 2.8). In general, direct interception (Figure 2.9) is the major mechanism occurring in liquid filtration. Inertial impaction (Figure 2.10) and diffusional interception (Figure 2.11) play a greater role in gas-phase filtration. When considering filtration mechanisms, it should be remembered that practical filtration occurs with irregularly-shaped particles rather than ideal and perfectly rigid spheres.

Direct interception occurs when the particle is larger than the largest pore opening of the filter and hence is retained. With irregularly-shaped particles, the orientation of the particles to the surface may affect the passage or retention of any given particle. If the irregularly-shaped particles are close

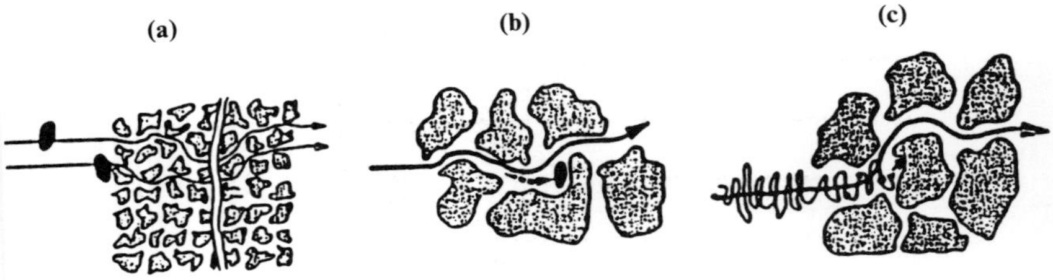

Figure 2.8 Mechanisms of filtration.

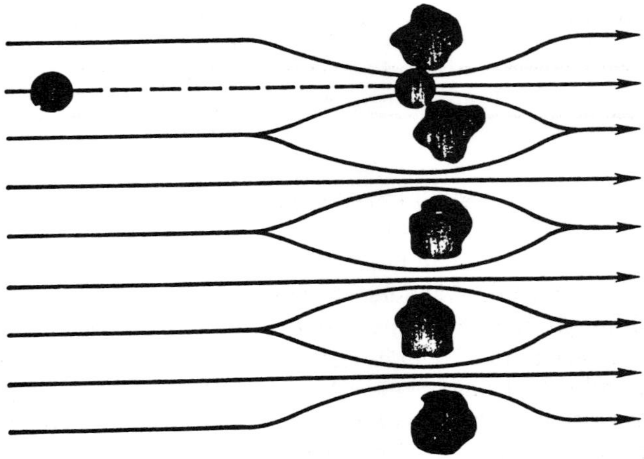

Figure 2.9 Direct interception.

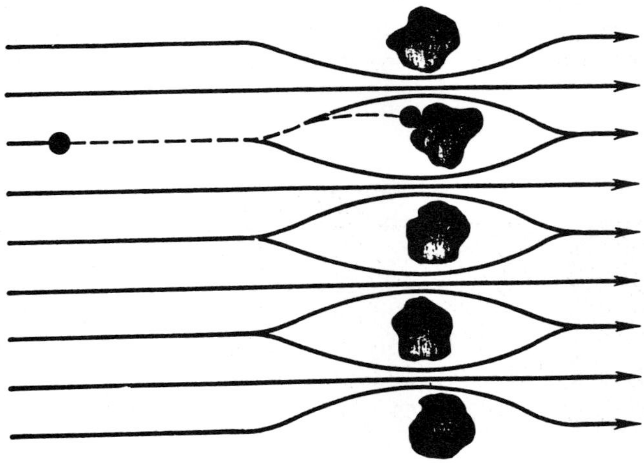

Figure 2.10 Inertial impaction.

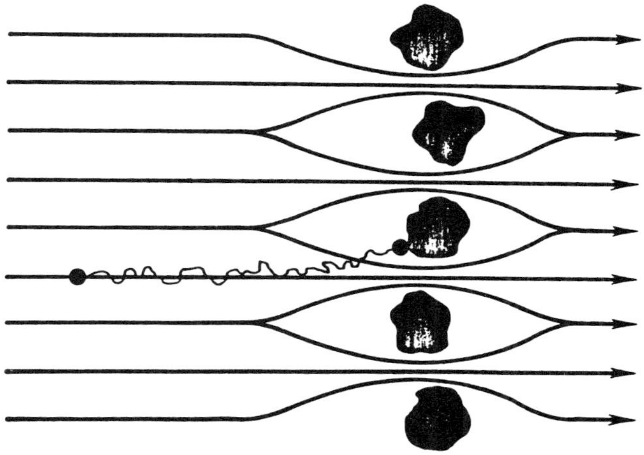

Figure 2.11 Diffusional interception.

enough in size to the pore diameter they may be filtered out in depth through the filter although most may pass the filter surface. If smaller particles are agglutinated, the clump of bound particles may be retained by the filter even though the individual particles are far too small to be retained. A similar effect, termed bridging, occurs when two or more particles are attached end to end producing an elongated particle that cannot pass through the filter (Figure 2.12).

Inertial impaction is significant only in the filtration of gaseous fluids, where a very large density difference always exists between the fluid and the particles. The gas will readily flow through the tortuous pathway of pores in the filter media, while the solid particles will not so readily make the

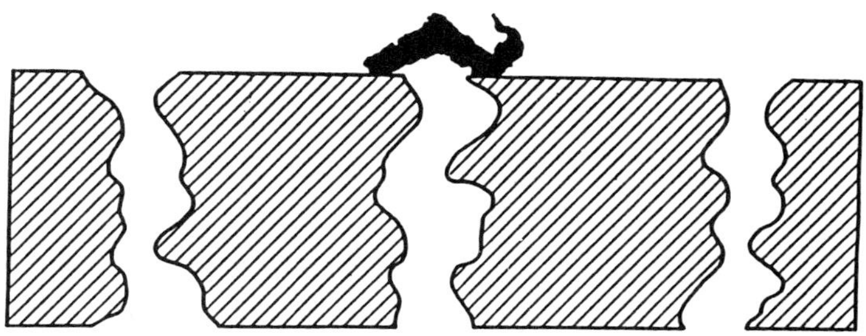

Figure 2.12 Bridging caused by an irregular-shaped or elongated particle. The particle cannot pass through the filter.

Figure 2.13 Inertial impaction/diffusion interception.

numerous directional changes. Eventually, the particles may be impacted into relatively stagnant cavities in the media from which they are unlikely to be discharged (Figure 2.13).

Diffusional interception is significant only for the case of filtration of very fine particles from a gaseous stream. Diffusion of these fine particles, by random Brownian motion, into stagnant areas of the filter will occur. This mechanism can result in good entrapment of fine particles in a filter even though flow of the gaseous fluid through the media is fairly linear.

2.12.2 Depth versus surface removal.

In general a membrane filter separates particulates at the membrane surface while a depth filter removes particulates through the complete depth of the filter medium. In practice this characterization is really not absolute as any filter with a depth dimension will perform some filtration below the surface while the bulk of the particulate load is typically removed relatively near the surface. Surface filtration is accomplished if 95% or greater of the removal efficiency of a given filter occurs within the first 10% to 15% of the filter's depth.

Cartridge-type depth filters may, in turn, be classified non-fixed or fixed random pore media filters.

String wound and compressed fibre filters are typical examples of non-fixed random pore filters (Figure 2.14). These filters are most effective within a relatively narrow pressure drop range for which they were designed. At higher pressure drops or under pressure pulsation the pores may become distorted or enlarged. This may result in unloading (Figure 2.15) and migration of previously retained particles causing fluid contamination.

Non-fixed random pore media characteristics are as follows:

- Filtration depends largely upon inertial impaction and/or diffusional interception

- Many tortuous flow passages are present
- Pore size increases with pressure drop
- Subject to media migration.

Fixed pore media depth filters, in contrast, have non-deformable pores and are effective with much higher pressure drops. Filtrate flow rate will decrease drastically as the filter becomes plugged and its maximum useful life is approached; previously retained particles, however, will not be displaced. These filters are available in pleated, spun fibre or cast media form.

Fixed random pore media characteristics may be summarized as:

- Filtration depends largely upon direct interception
- Consists of layers of media or single layer having depth
- Contain many tortuous flow passages
- Pore size does not increase with pressure drop
- Pore size is controlled during manufacture so that quantitative removal of particles larger than a given size may be ensured.

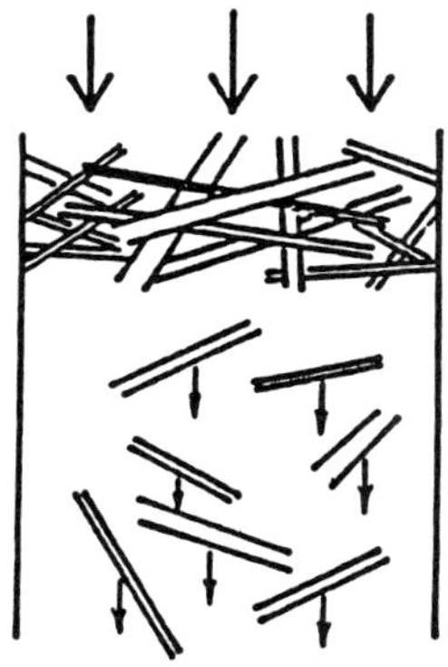

Figure 2.14 Media migration through a non-fixed random pore filter.

Figure 2.15 Unloading and migration of previously retained particles caused by higher pressure or pressure pulsation.

Membrane type filters also have non-deformable pores of a specific size, which is controlled carefully during manufacture of the filter. Common materials of construction for membrane filters include cellulose acetate, cellulose nitrate, polysulfone, nylon 66, polytetrafluoroethylene, polyvinyl-denefluoride and ceramics. For a given filtration application it is desirable to use as few materials for construction as possible. This minimizes the difficulty of selecting materials compatible with the process fluid.

2.12.3 *Filter ratings.*

Filters are rated on the basis of their efficiency in removing particles of a given size. Ratings may be absolute or nominal. An absolute rating for a filter is based on testing, which indicates the diameter of the largest hard spherical particle that the filter would allow to pass or equivalently the largest pore openings within the filter matrix. A nominal filter rating, on the other hand, expresses the filtration capabilities of the filter as the percentage of particles of a given size that are retained by the filter. While, in principle, either type of rating is equally effective in expressing the capabilities of a given filter, different manufacturers have different standards and procedures for establishing nominal filter ratings. There is often therefore some degree of uncertainty as to precisely what a given nominal rating means, and this is particularly true when using ratings to compare the offerings of different manufacturers. When using a nominally rated filter, an engineer should always therefore request detailed efficiency and procedure information from the manufacturer. In general, nominal ratings are applied to less expensive filters intended for non-critical or pre-filter applications. Such filters are often used upsteam of more expensive filters to protect them and to reduce the particulate loading.

In the pharmaceutical industry, absolute rated cartridge filters are most often used to remove bacteria and maintain sterility in process lines. To receive a bacterial retentive rating a filter must be challenged with a specific organism at a specific organism suspension density. To rate a filter for inert particulates spherical particles of a given size are used. Both of these testing procedures are termed destructive. In general, the filter manufacturer performs sufficient destructive testing of a given filter model to assure the FDA of its quality and then takes and tests sufficient samples from production runs to

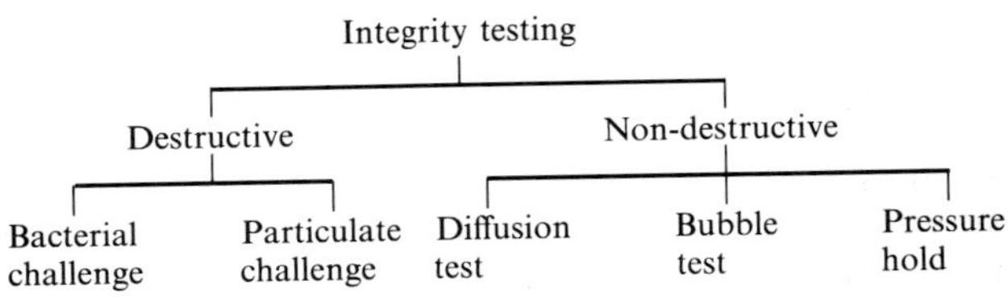

Figure 2.16 Flow-chart of integrity testing of filters.

ensure filter consistency (Figure 2.16). This information is supplied to the user when he in turn validates his process using a specific filter.

One of the smallest known bacterial species is *Pseudomonas diminuta* and a culture of this species at a given suspension density is recognized by the industry and FDA as the test organism for 0.2 μ filters. It is generally accepted that bacterial sterility is maintained by such a 0.2 μ filter. In the case of 0.45 μ filters the test organism is *Serratia marcescens*, again at a specific suspension density. It must also be demonstrated that the suspensions used for integrity challenges are composed of viable organisms.

The user also typically performs non-destructive testing of his filters prior to and after running a production batch, using them as a part of his CGMP procedures. The most commonly applied non-destructive tests are bubble, diffusion and pressure-hold tests. These tests are correlated to the results of bacteriological testing but allow the user to check the integrity of the filters used for each batch of a pharmaceutical product simply and reliably.

2.12.4 Hydrophilic and hydrophobic filters.

The two principal subcategories of sterilizing filters are hydrophilic and hydophobic filters: a hydrophobic filter is one that may readily be wetted by water or aqueous solutions; while a hydrophobic filter resists wetting.

Hydrophilic filters allows a free flow of the liquid into the pores of the filter with a minimum amount of pressure. This is desirable in liquid filtration applications where water is the fluid medium or solvent. Running a filtration at low pressure usually results in longer and more reliable filter operation.

Gas-stream filtration is the primary application for hydrophobic filters. Sterile filters are used for pressurized air and nitrogen lines to bioreactors and for vent lines to allow vessels to maintain pressure by breathing without losing sterility. Should these filters become wetted, pressure would rapidly build up on the upsteam side of the filter as a result of the greatly reduced area available for gas flow. This in turn would cause the vessel to become over or under pressurized the next time fluid was pumped into or out of the vessel. By using hydrophobic filters this situation is avoided.

For the purpose of performing certain non-destructive filter tests it is sometimes desirable to wet a hydrophobic filter. This may be achieved first by exposing the filter to a low-surface-tension fluid, such as alcohol, and then displacing the alcohol with water.

2.12.5 Filter validation.

As for most equipment used in pharmaceutical applications, process validation considerations are a major influence on the selection of cartridge filters. One aspect that is somewhat unique to filter validation is the degree to which both manufacturer and user share responsibility.

Key validation considerations include filter materials, extractables, batch records, integrity tests, sterilization procedures, operating conditions, bacterial removal and particulates. A typical division of responsibility for these parameters between vendor and user is shown in Table 2.1

Table 2.1

	Filter manufacturer's responsibility	Filter user's responsibility
Filter materials	Established master file with FDA Provide data on biological safety per USP tests for plastics	Reference vendors in drug application Perform biological safety testing, as appropriate, on final product
Filter extract-ables	Provide information required on flush volume to yield zero oxidizables for USP test Provide data on level of extractables with various solvents	Set standards for flushing filter and downstream piping Include filter assembly in product preparations for stability tests and product assays for 'safety, identity, strength, quality and purity'
Filter batch records	Provide lot and serial number	Record lot and serial number used for each batch Record non-destructive integrity test values prior to and at end of each filtration
Filter integrity tests	Provide correlation of integrity test values with bacteriological retention test results Provide written instructions and specifications for integrity testing Recommend appropriate testing equipment	Perform integrity testing according to manufacturer's specification prior to and at end of each filtration Record integrity test values
Filter sterilization procedures	Recommend general procedures Provide time/temperature limits of filter	Validate procedures with thermo-couples and/or pore strips for filter assembly and downstream equipment Establish cumulative sterilization time limits
Filter operating conditions	Provide operating limits, particularly with regard to maximum allowable pressure drop	Operate within filter limits Pre-test for process compatibility Use appropriate filter housing Establish effective maintenance procedure Incorporate change control procedure
Filter bacterial removal	Provide lab test data, including titer reduction for appropriate test organism, e.g. *Pseudomonas diminuta* (ATCC 19146)	Monitor bioburden of material to be sterile filtered; maintain low bioburden Select appropriate filtration system to provide assurance of sterile effluent Use conforming filter elements and housing
Filter particu-lates	Provide information on flush volume required to yield zero fibres in effluent	Set standards for flushing of filter assembly and downstream piping

References

1. Perry, R.H. and Chilton, C.H. (Eds) 1973. *Chemical Engineers' Handbook*, 5th ed. McGraw-Hill, New York, pp. 19-65–19-73.
2. Svarovsky, L. (1977). *Solid–Liquid Separation.* Butterworths, London.
3. Svarovsky, L. (1979). Advances in solid–liquid separation-I, *Chem. Eng.*, **July 2,** 63–66.
4. From Pall Corporation, Glencove, New York, USA.
5. From Sartorius Corporation, Shirley, New York, USA.
6. From Millipore Corporation, Bedford, Masacchusetts, USA.

3 Pharmaceutical applications of liquid–liquid extraction

K.E. CROWELL

3.1 Introduction

Liquid–liquid extraction is a unit operation based on differential solubility of a consolute in two immiscible solvents. This separation technique, also known as solvent extraction, has many successful applications in the pharmaceutical industry because of its inherent flexibility and its suitability for processing heat-sensitive products.

It is the dependence of extraction on the chemical solubility properties of the components that leads directly to the strengths and limitations of the operation. The many possible choices of solvents results in the unmatched versatility of extraction as a mass-transfer operation. The need to separate the product from the extracting solvent, and the consequent recovery and purification of the solvent, make extraction a complex and expensive process, compared to, for example, simple distillation. Characteristics of the operation of particular relevance to bioseparations are the effectiveness of extraction at low temperatures, and the tendency of organic solvents to denature most proteins irreversibly.

Extraction was developed in the 1920s and 1930s in the petrochemical industry, with new developments in basic understanding, design procedures, applications and equipment evolving continuously from that time. The range of different extractor configurations and types today is probably greater than for any other class of mass-transfer equipment.[1] Antibiotics, because of their thermal lability, were among the earlier successful applications of extraction outside of the petrochemical field.

Extraction in the pharmaceutical industry has been used primarily in the isolation of antibiotics from fermentor broth, in the preparation of natural and synthetic vitamins, and in the preparation of drugs from naturally occurring materials.[2,3,4] Contrary to extraction applications in most other industries, very dilute aqueous solutions are usually fed into the process.[5]

Extraction may be carried out in a batch or staged-batch sequence (i.e. mixer–settler units); however, most pharmaceutical extractions have been carried out in differential units. Centrifugal extractors, in particular, of which the Podbielniak extractor is perhaps the best known example, have been related intimately to the development of commercial antibiotic production processes.[4,6–8]

Fractional extraction processes involve two or more consolute components, and are designed to enrich the extract in the desired product component and to isolate the product from the aqueous feed solution. General pharmaceutical fractional extractions have been commercially practised, usually in cases where antibiotics together with intermediates, complexes or isomers are present in the fermentation broth.[2]

Conventional liquid–liquid extraction is not normally useful for the isolation of protein products because of the irreversible denaturation of these macromolecules that typically occurs in organic solvents. However, a considerable body of research is accumulating on the use of two immiscible aqueous phases for protein extraction.[9–11] Immiscible aqueous-phase solutions can be prepared by the addition of polymers or polymers plus salts to water. The use of reverse micellar-extraction systems has also been considered for this application by several investigators.[12–15] It is possible that these types of extraction systems will become commercially significant in the biotechnology industry of the future.

3.2 Fundamentals

When a given solute is exposed to two different solvents it often will be more soluble in one of them. The unit operation of liquid–liquid extraction exploits this differential solubility of a consolute component in two immiscible, or partially immiscible, solvents.

Successful application of extraction in a separation process hinges upon a careful consideration of the underlying thermodynamic equilibrium contraints, rates of mass transfer from phase to phase, hydrodynamics and relative flow rates of the contacting fluids, choice of solvent and of processing equipment and of the economics of the proposed process.

While many excellent texts and review articles[2,3,5,16–19] deal with all of these topics in much detail, a brief overview is provided here for the convenience of the reader.

3.2.1 Equilibrium

Thermodynamics sets the fundamental limitations of proposed extraction process. The concept of a partition coefficient is basic. This is a quantitative measure of the relative solubilities of the consolute in the two liquid phases and is defined as:

$$K = C_{I}/C_{II} \tag{3.1}$$

In chemical or petrochemical extractions the solubility data are typically represented in triangular coordinates at a given temperature.[16] The compositions in the two phases are represented by the points on the phase

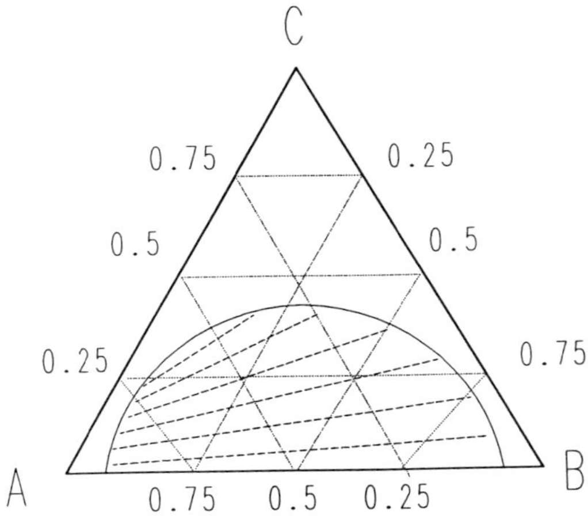

Figure 3.1 Phase envelope at a given temperature showing solubility data for a typical ternary, two phase mixture.

envelope joined by the tie line passing through the feed composition (Figure 3.1). This form of representation of data is not typically useful for pharmaceutical extractions because of the extreme dilutions of the solutions. Data are instead typically represented as plots of concentration, mole, or mass fraction in one phase versus that in the other phase (Figure 3.2).

The partition coefficient is a function of composition, temperature, pH, and ionic species and strength. Manipulation of these variables can be used

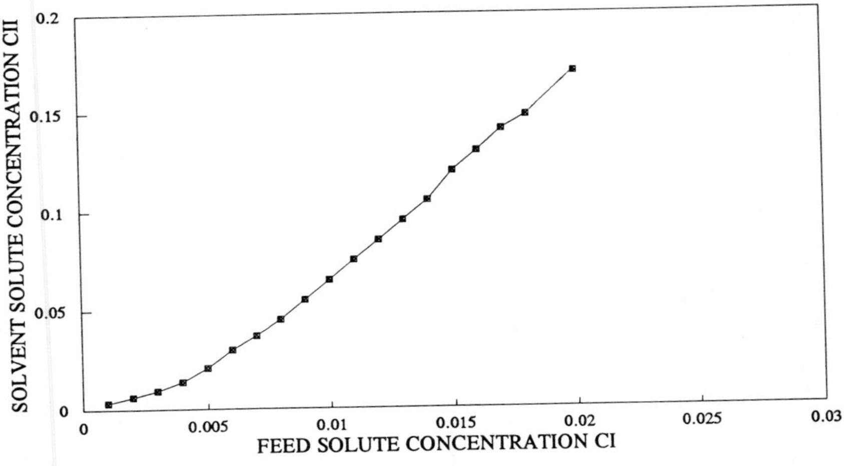

Figure 3.2 Equilibrium curve showing solute concentration in the solvent compared with solute concentration in the feed.

to separate the purified product from the extracting solvent, as well as to optimize the extraction process itself. In the case of biological products, however, careful consideration must also be given to product stability under the conditions of the extraction.

While partition coefficients must always be determined experimentally, thermodynamic theory can provide guidance to possible solvents and conditions for a desired separation. Fundamentally, it must be that the chemical potential of the solute in both phases is equal at equilibrium:

$$\mu_{\mathrm{I}} = \mu_{\mathrm{II}} \tag{3.2}$$

or:

$$\mu_{\mathrm{I}}^{0} = RT\ \ln(C_{\mathrm{I}}) = \mu_{\mathrm{II}}^{0} + RT\ \ln(C_{\mathrm{II}}) \tag{3.3}$$

which can be rearranged to the form:

$$\mathrm{K} = C_{\mathrm{I}}/C_{\mathrm{II}} = \exp[(\mu_{\mathrm{I}}^{0} - \mu_{\mathrm{II}}^{0})/RT] \tag{3.4}$$

In words, the partition coefficient is an exponential function of the difference in standard-state chemical potentials of the consulute in the two liquid phases, and of the temperature. Any factor tending to lead to a standard-state chemical potential difference will produce a large change in the partition coefficient.

The most direct approach to optimizing K is selection of a good solvent. While predictions of equilibrium concentrations, using for example, Scatchard-Hildebrand, UNIFAC, or ASOG models, can be useful guiding experimental work, final selection of a solvent must be based on good laboratory data.

A good solvent should be inexpensive, readily available, have a high affinity for the product, be highly selective for the product, be immiscible with the aqueous feed phase, of substantially different density from water, of low viscosity and vapour pressure, non-degrading to the product, non-toxic, non-corrosive, non-flammable, non-explosive, environmentally benign and easily recoverable from the product.[1,3] Clearly, the choice of solvent is one of the more challenging aspects of designing an extraction process. In practice, past experience of either the owner's engineers or of potential equipment vendors is usually invaluable in providing initial direction to the search for the best solvent for a given application.

Solvents that have commonly been used for antibiotic extractions in the past include alcohols (*n*-butanol), ketones (methyl isobutyl ketone), and acetates (butyl and pentyl acetates).

The equilibrium distribution of a solute between feed and solvent can be altered dramatically by a change in pH of the aqueous phase.[9] This fact has been greatly utilized in pharmaceutical separations, for example in the manufacture of penicillin. In order for this technique to be applicable, the solute to be extracted must be either a weak base or a weak acid. Such a solute will partially ionize (dissociate) in water but not in an organic solvent. Only the unionized solute can migrate from the aqueous phase to the solvent. The degree of ionization of the solute will be controlled by the pH of the aqueous phase. In the case of a weak base for example, the partition coefficient will be:

$$K = \frac{[RNH_2]_I}{[RNH_2]_{II} + [RNH_3]_{II}} \tag{3.5}$$

The intrinsic partition coefficient for such a solute is defined as:

$$K_I = \frac{[RNH_2]_I}{[RNH_2]_{II}} \tag{3.6}$$

(i.e. considering only the unionized solute) pKa, which is a measure of an acid's tendency to dissociate, is defined as:

$$\log_{10} \frac{[RCOO^-][H^+]}{[RCOOH]} \tag{3.7}$$

and pKb, which is a measure of a base's tendency to ionize, is defined as:

$$\log_{10} \frac{[RNH_3^+][OH^-]}{[RNH_2]} \tag{3.8}$$

Finally the equations relating pH to the partition coefficient are:

$$\log_{10}[K_I/K - 1] = pH - pKa \tag{3.9}$$

$$\log_{10}[K_I/K - 1] = pKb - pH \tag{3.10}$$

These relationships may also have implications for the purification of a solute in the case of fractional extraction processes. Given two solutes, I and II of equal intrinsic partition coefficient (at a given degree of ionization), the one that is less ionized under extracting conditions will have a higher effective partition coefficient and therefore will be selectively extracted. The selectivity of such an extraction process, where both solutes are, for example weak acids, may be expressed as:

$$S = [K_I(I)/K_I(II)][(1 + Ka(II))/(1 + Ka(I))] \tag{3.11}$$

When processing fermentation products it is frequently the case that several closely related solutes will be present in the feed. Fractional extraction exploiting different pH solubility behaviour of the related species may be especially beneficial in such cases.

A related method of improving the partition coefficient is the use of organic counter ions.[5] As already mentioned, ionized species are not normally soluble in organic solvents and, therefore, will not be extracted into them. However, it has also been found that, if both members of an ion pair are organic species, the ions tend to be soluble in organic solvents. Such species are referred to as greasy salts, with examples of those commonly used as counter ions being butyrate, tetrabutylammonium, dodecanoate, acetate and citrate. Replacing an inorganic ion such as a chloride or hydroxide, with such a counter ion offers another means of enlisting the assistance of thermodynamics in the improvement of the extraction process.

3.2.2 Mass transfer

While thermodynamics sets the feasibility of a proposed separation, the rate of separation and therefore the size and cost of the required equipment is determined by mass-transfer rates. This is particularly true of processes carried out in differential extraction equipment.

In an extraction process, one phase is chosen to be the dispersed phase, which exists as relatively fine droplets within the other phase, which is then called the continuous phase. Choice of which phase is to be continuous is a significant design variable and will be discussed at greater length later.

Mass transfer in extractors is usually analysed within the framework of the two-film theory.[2,16,17] This theory postulates the existence of two thin films of stagnant fluid, one on either side of the interface between the dispersed droplets and the continuous phase. The two phases are taken to be at equilibrium at the interface so that there is no resistance to mass transfer at the interface itself. This is a reasonable assumption, except for cases such as the presence of a surface active contaminant or reaction product in which a barrier forms at the interface. Diffusional mass-transfer resistance occurs in both boundary layers in accordance with Fick's law, which states that the mass flux is proportional to the negative of the concentration gradient driving force across each film:

$$J = DC \tag{3.12}$$

The bulk of both phases, outside of the boundary layers, is assumed to be perfectly mixed.

The thickness of the boundary layer films is dependent on the properties of the fluids and on hydrodynamic conditions in the extractor but a thickness of about 10 μ is probably typical.[2] Owing to the great experimental difficulties in measuring the thickness of such a thin fluid layer, the mass flux is usually expressed in terms of mass transfer coefficients that incorporate the effects of both film thickness and of molecular diffusivity into one measurable parameter:

$$J = k_{\mathrm{I}}(C_{\mathrm{I}} - C_{\mathrm{Ii}}) = k_{\mathrm{Ii}}(C_{\mathrm{IIi}} - C_{\mathrm{II}}) \tag{3.13}$$

With C_{Ii} and C_{IIi} the concentrations at the interface which will normally be at equilibrium with one another.

To avoid the inconvenience of measuring concentrations at the interface, mass transfer coefficients based on overall bulk concentration difference driving forces are most often used in practice:

$$J = \mathrm{K}'_{\mathrm{I}}(C_{\mathrm{I}} - C_{\mathrm{I}*}) = \mathrm{K}'_{\mathrm{II}}(C_{\mathrm{II}*} - C_{\mathrm{II}}) \tag{3.14}$$

with $C_{\mathrm{II}*} = \mathrm{K}C_{\mathrm{I}}$ (the phase II concentration that would be in equilibrium with the bulk concentration in phase I).

The bulk mass transfer coefficients can be related to the film coefficients by the equations:

$$1/\mathrm{K}'_{\mathrm{I}} = 1/k_{\mathrm{I}} + 1/(\mathrm{K}k_{\mathrm{II}}) \tag{3.15}$$

and

$$1/\mathrm{K}'_{1} = 1/k_{\mathrm{I}} + 1/\mathrm{K}k_{\mathrm{II}} \tag{3.16}$$

Mass transfer coefficients are influenced by diffusivity, interfacial turbulence, film thickness, the densities, viscosities and interfacial surface tensions of the fluids, and by whether the direction of transfer is from dispersed to continuous phases or vice versa. These variables, in turn, are determined by the composition of both phases, by the temperature of the system and by the degree and type of agitation.

The rate of mass transfer per unit volume of extractor is the product of the mass flux and the interfacial surface area per unit volume:

$$Ja = \mathrm{K}_{\mathrm{I}}a(C_{\mathrm{I}} - C_{\mathrm{I}*}) \tag{3.17}$$

with

$$a = 6f/d_{\mathrm{D}} \tag{3.18}$$

where f is the volume fraction of the dispersed phase, termed the holdup, and the droplets of the dispersed phase have diameter d_{D}.

The interfacial area also depends on the interfacial surface tension, the densities of the phases, the phase compositions, the temperature, and the degree and type of agitation. The interfacial area can be maximized (within limitations set by hydrodynamic phenomena) by increasing the holdup and decreasing the mean droplet diameter.

Mass transfer, in practical applications, is complicated by the fact that perfect mixing in the bulk phases is often not even approached; it is also complicated by droplet coalescence and redispersion phenomena. In particular, if the interfacial surface tension is sensitive to, and lowered by, an increase in consolute concentration at the interface, the dispersed phase may rapidly coalesce (the droplets grow in size) as a result of mass transfer from the droplets to the continuous phase. The effect is interestingly much less pronounced if the direction of mass transfer is reversed. This is called the Miragoni effect.

Chemical reaction simultaneous with extraction is frequently found in pharmaceutical processes (e.g., in the extraction of penicillin) and has a significant impact on mass-transfer rates. The actual consequence of reaction, however, depends on the specifics of reaction order and rate and is best covered in more specific articles.

3.2.3 *Hydrodynamic factors*

The third set of variables influencing extractor performance are general inter-related hydrodynamic factors. These relationships include: agitation; dispersion, drop size and coalescence; holdup, fluid velocities and flooding; axial mixing; and emulsification effects.

As already described, turbulence and interfacial area are two of the most significant variables determining mass transfer rates. Both of these variables, in turn, are determined by the type and degree of agitation occurring in the extractor. As turbulence increases the boundary layers at the phase interface, the dispersed phase droplets become thinner, the mass-transfer resistances are reduced, and the transfer rate is increased. Simultaneously, the mean droplet diameter is reduced and therefore the interfacial area is increased. However, as the droplet diameter continues to decrease with further additional energy input, several other factors, including decreased bulk mixing within the droplets and reduced interfacial turbulence, begin to adversely affect mass transfer. Coalescence time also increases as the droplets become very fine. Therefore, there will be an optimum degree of agitation for a given extraction in a particular apparatus that maximizes volumetric throughput and mass-transfer rates.[2]

Coalescence rates are higher when the interfacial surface tension is high, and are reduced when the continuous-phase viscosity is high.

Solids or surface active chemical species may concentrate at the interface causing the droplets to coalesce. This emulsion or 'rag' layer may build up in the extractor and cause unstable performance.[19] In some cases, this problem may be solved by switching dispersed and continuous phases. A second potential solution is continuous draw-off of the emulsion as it forms; however, this creates a waste disposal problem.

In most extractors coalescence occurs under the influence of gravity and, therefore, the coalescence rate depends on the density differences between the phases. Centrifugal extractors have been developed, however, to accelerate the dispersion and coalescence of the fluids by applying centrifugal forces to the phases in the extractor.[21] These extractors were developed for pharmaceutical applications and are still much used.

Hold-up is the fraction of the column volume taken up by the dispersed phase. Flooding occurs when the two liquid phases are unable to pass through one another, and represents an upper limit on column capacity for a given extraction process. This happens when the hold-up exceeds the capacity of the column and the dispersed phase coalesces at some point within the column. Extraction efficiency drops off precipitously at flooding, and entrained droplets or slugs of dispersed phase often leave the column with the continuous-phase stream.

Prediction of hold-up and fluid velocities at flooding requires consideration of the relative volumetric flow rates, mean droplet diameter, and characteristics of the type of extraction equipment in service.[16]

The slip velocity is a function of flow rates only and is defined as:[18]

$$V_s = \frac{V_c}{(f - 1)} + \frac{V_d}{f} \tag{3.19}$$

where V_c is the superficial velocity of the continuous phase, V_d is the superficial velocity of the dispersed phase, and f is the holdup.[10] The characteristic velocity is a function of the extractor configuration and is defined as:

$$V_0 = \frac{V_s}{(1 - f)^K} \tag{3.20}$$

where the exponent K depends upon the type of extractor. The characteristic velocity is related to the maximum droplet velocity.

Flooding velocities of the phases can then be calculated from:

$$V_{cf} = \frac{V_0 f_f (1 - f_f)^{K+1}}{[f_f + (1 - f_f)R']} \tag{3.21}$$

and from:

$$V_{df} = R' V_{cf} \tag{3.22}$$

where:

$$R = \frac{V_d}{V_c} \tag{3.23}$$

and the hold-up at flooding from:

$$f_f = \frac{[R'(K + 2) - (R'^2(K + 2) - 4R'(R' - 1)(K + 1))^{0.5}]}{2(R' - 1)(K + 1)}$$

Most commercial columns are operated at 50–75% of flooding velocity for operating safety and flexibility, and to minimize axial mixing effects.[1]

In summary, mass transfer is improved by increased agitation but coalescence time is increased and volumetric throughput decreased. Extraction processes must be designed for the optimum trade-off between mass transfer and hydrodynamic requirements.

Axial mixing reflects non-ideal mixing in the continuous phase, axially, along the height of the column.[2] This produces a reduced concentration gradient driving force and hence reduced interphase mass transfer. Axial mixing may result from vertical circulation currents or eddies, entrainment of continuous-phase fluid in the dispersed-phase droplets, forced backmixing from pulsation, or from channelling flow.[19] Axial mixing can reduce extractor efficiency significantly with as much as 60–75% effective height of some commercial extractors being required to compensate for it.[1]

Finally, it is necessary for column internals to be wetted preferentially by the continuous-phase fluid; otherwise the dispersed phase droplets will tend to coalesce on the column internals, with a consequent reduction of interfacial area and mass transfer efficiency. With an aqueous continuous phase the internals may be metal or ceramic. If the organic solvent phase is continuous the internals are normally non-metallic (e.g. PTEFE).[18]

3.3 Commercial extractors

An extremely diverse range of extractors has been developed over the past 70 or more years. Many types have been developed for specific applications but, even when considering only those configurations that have found wide acceptance in several industries, the range of choice of equipment is quite broad. Many excellent articles have been written on this topic.[1,3,18–20,22]

Parameters useful in the selection of extraction equipment for a given process include: (i) the required feed flow rate and composition; (ii) product stability and residence time constraints; (iii) physical properties of the fluids; (iv) settling time characteristics; (v) required number of equilibrium stages (for ease of separation); (vi) capital, maintenance and operating costs; (vii) desired mode of operation (batch or continuous); (viii) available space in the plant; and (ix) prior experience of operating, maintenance and engineering personnel.[1,18]

All extractors must create sufficient interfacial mass transfer area by forming dispersed phase droplets, allow sufficient residence time for extraction of the consolute into the solvent and provide additional residence time for coalescence and separation of the phases into raffinate (spent feed) and extract (consolute-bearing solvent) streams.

Extractors can be classified as stagewise or differential units. Stagewise extractors consist of distinct units or stages in which the phases are contacted and then separated. An approach to equilibrium is achieved in each stage. Differential units keep the phases in contact along the height of the extractor, separating them only at entrance and exit. Equilibrium is never achieved at any point in the unit; hence mass-transfer considerations are particularly important in the design of such an extractor

3.3.1 Stagewise extractors

Stagewise extractors, although well-suited to batch operation, are currently of only limited importance to the pharmaceutical industry and hence will be only discussed briefly here.

Small-scale extractions can be carried out by crossflow contacting in a single mixer-settler tank or tank plus decanter. In this type of operation, feed and solvent are added to the tank and mixed. Agitation is stopped and sufficient time is allowed to permit the phases to separate. The raffinate and extract are then separated by decanting one from the other. Depending on the desired degree of product recovery and the solute's partition coefficient in the system, the raffinate may be returned to the tank and extracted with fresh solvent. Such units are obviously inefficient in the use of solvent; however, for small-scale extractions the simplicity of operation and low capital and maintenance costs can compensate for the added operating costs of additional solvent.

Solvent extraction in a single stage unit can be analysed using equations representing the equilibrium:

$$C_{\mathrm{I}} = \mathrm{K}C_{\mathrm{II}} \tag{3.24}$$

and material balance:

$$FC_{\mathrm{IIin}} + SC_{\mathrm{Iin}} = FC_{\mathrm{IIout}} + SC_{\mathrm{Iout}} \tag{3.25}$$

constraints. A parameter called the extraction factor, E, which is defined as:

$$E = \mathrm{K}S/F \tag{3.26}$$

plays an important role in extraction calculations. For a single-stage extraction, the solute concentrations leaving the extractor will be:

$$C_{\mathrm{I}} = (\mathrm{K}C_{\mathrm{IIin}})/(1 + E) \tag{3.27}$$

$$C_{\mathrm{II}} = C_{\mathrm{IIin}}/(1 + E) \tag{3.28}$$

and the fraction in the aqueous feed extracted into the solvent phase, p, can be calculated from:

$$p = (SC_{\mathrm{I}})/(FC_{\mathrm{IIin}}) = E/(1 + E) \tag{3.29}$$

If an analytical expression for phase-distribution equilibrium is unavailable, the calculations may be performed graphically by plotting experimental equilibrium data points and the proposed operating line or curve on the same xy plot. The operating point of the extractor will be the intersection of the two curves (Figure 3.3). The operating line can be plotted from the material balance:

$$C_{\mathrm{I}} = (F/S)(C_{\mathrm{IIin}} - C_{\mathrm{II}}) \tag{3.30}$$

A more efficient configuration is to use multi-stage extraction. The required number of mixer-settler units are placed in series and feed and solvent are run countercurrent to each other in a continuous or semi-continuous operation. Properly designed mixer-settlers provide 0.8–0.9 theoretical stage units per actual unit.[18] Units like this were employed in the pharmaceutical industry in the past but are used very little today.

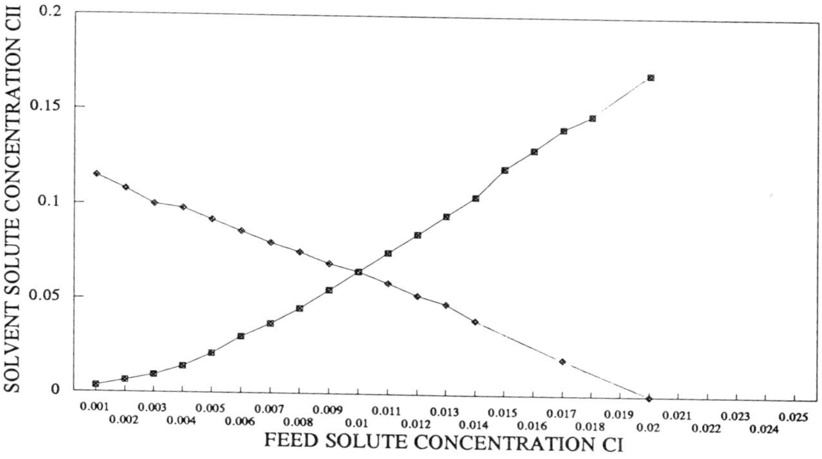

Figure 3.3 Graphical means of obtaining the phase-distribution equilibrium point.

The number of stages required for a given extraction can be estimated from:

$$C_{\text{II}}n + 1 = \frac{(E^{n+1} - 1)C_{\text{II}}^{1}}{(E - 1)} \tag{3.31}$$

and the fraction extracted from:

$$p = \frac{E(E^{n} - 1)}{(E^{n+1} - 1)} \tag{3.32}$$

Calculations can also be performed graphically.[2,5,16]

A second small-scale stagewise extractor found only in biological applications where fractional extraction is required is the Craig extractor.[23,24]. This is basically a laboratory device that has sometimes been scaled up and applied to production scale fractional extraction processes. Automated versions specifically designed for production scale are also commercially available.[1] The apparatus, in its simplest form, consists of a series of tubes connected by glassware. Fluid in the tubes is mixed by shaking, much as with a separatory funnel. The system consists of $n + 1$ tubes, all but the first of which initially contain pure solvent of greater density than the feed. The first tube contains feed plus solvent. After mixing, the lighter aqueous phase is decanted from the first tube to the second. Consolutes having greater affinity for an organic phase will tend to be left behind, while those with partition coefficients favouring the water phase will move forward. Pure light phase (water) is added to the first tube, the apparatus is shaken, and the first tube light phase is transferred to the second tube, while the second tube light phase is transferred to the third tube. The process is repeated many times (up to hundreds of transfers) and a concentration profile of the consolutes is developed across the tubes. Again this technique is specific to small-scale fractional extraction applications where the objective is product purification as well as isolation.

The balance of the extraction equipment used in pharmaceutical or biotechnological applications are differential units.

3.3.2 *Differential extractors*

Differential extractors may be classified as unagitated columns, mechanically agitated rotary columns, mechanically agitated reciprocating columns, pulsed-flow columns, or centrifugal units. Of these, mechanically agitated (both rotary and reciprocal) and centrifugal extractors have been by far the most frequently applied in pharmaceutical processes.

Various types of rotary agitator extractors have been used to isolate naturally derived pharmaceuticals and vitamins and in bulk pharmaceutical plant-extraction processes.

Reciprocating extractors have found application in the fractional extraction of antibiotics from isomers and closely related compounds. They are also a good choice for extracting products directly from fermentation media because of their ability to handle foaming and frothing solutions, emulsifying solutions, and solutions containing solids.[1,18]

Centrifugal extractors were developed for penicillin purification and remain the most widely used extractors in antibiotic manufacturing processes.

In a differential extractor phase, partition equilibrium is never reached and therefore mass-transfer rates are important and are incorporated into the design equations.

In most pharmaceutical applications, the quantity of solute transferred is small relative to the volume of both feed and solvent, and hence the flow rates of the two phases do not appreciably change over the height of the extractor. Given that this assumption holds, a relatively straightforward analytical solution, similar to the calculation of stages for stagewise units, is available to determine the required height of the column.[2,16]

A relationship for phase equilibrium is also required and it is usually assumed that this is a linear relationship of the form:

$$C_{\mathrm{I}} = KC_{\mathrm{II}*} \tag{3.33}$$

where $C_{\mathrm{II}*}$ is the concentration of the solute that would exist in the heavy phase if the two phases were in equilibrium.

The height of the column is represented by the variable z, with $z = 0$ at the bottom of the extractor and dz the differential height, or equivalently, differential volume per unit cross sectional area. Fresh solvent is fed and aqueous raffinate exits at the column bottom (assuming the solvent is lighter than the aqueous feed).

The steady-state mass balance is then:

$$S\,\mathrm{d}C_{\mathrm{I}} = \mathrm{F}\,\mathrm{d}C_{\mathrm{II}} = \mathrm{K}_{\mathrm{I}}a(C_{\mathrm{I*}} - C_{\mathrm{I}})\,\mathrm{d}z \tag{3.34}$$

or

$$\mathrm{d}z = [S/\mathrm{K}_{\mathrm{I}}a]\frac{\mathrm{d}C_{\mathrm{I}}}{(C_{\mathrm{I*}} - C_{\mathrm{I}})} \tag{3.35}$$

$$Z = [S/\mathrm{K}_{\mathrm{I}}a]\int_{C_{\mathrm{Iin}}}^{C_{\mathrm{Iout}}}\frac{\mathrm{d}C_{\mathrm{I}}}{(C_{\mathrm{I*}} - C_{\mathrm{I}})} \tag{3.36}$$

where $[S/\mathrm{K}_{\mathrm{I}}a]$ is called the height of a transfer unit or HTU. The HTU is a measure of the efficiency of the extractor. For clarity, one must be careful to specify the phase and driving force to which the HTU applies. Here, for example, the HTU is specified according to the overall concentration gradient in the solvent phase. The HTU does not change along the height of the extractor with concentration changes but may vary in response to changing interfacial area.

Referring to the earlier discussion of mass-transfer coefficients, it can be seen that the overall mass-transfer resistance is determined by the film coefficients of both phases. An analogous expression can be written for the overall HTU:

$$\mathrm{HTU}_{\mathrm{OA}} = \frac{S}{k_{\mathrm{I}}a} + \frac{\mathrm{K}S}{k_{\mathrm{II}}a} = \frac{S}{k_{\mathrm{I}}a} + \frac{EF}{k_{\mathrm{II}}a} = \mathrm{HTU}_{\mathrm{S}} + E(\mathrm{HTU}_{\mathrm{F}}) \tag{3.37}$$

where E is the extraction factor previously defined as:

$$E = \mathrm{K}S/F \tag{3.38}$$

This dimensionless group is normally chosen to be in the range 0.5–2.0 for economical operation. The integral:

$$\int_{C_{\mathrm{Iin}}}^{C_{\mathrm{Iout}}}\frac{\mathrm{d}C_{\mathrm{I}}}{(C_{\mathrm{I*}} - C_{\mathrm{I}})} \tag{3.39}$$

can be evaluated analytically if the equilibrium and mass-balance relationships are linear, or graphically or numerically if non-linear. This integral is termed the number of transfer units, or NTU. NTU is a measure of the difficulty of the extraction. For the analytical case:

$$\mathrm{NTU} = \frac{1}{1-E}\ln\frac{(C_{\mathrm{IO}} - C_{\mathrm{IZ*}})(1-E)}{(C_{\mathrm{IZ}} - C_{\mathrm{IZ*}})} + E \tag{3.40}$$

The required column height is then:

$$Z = (\mathrm{HTU})(\mathrm{NTU}) = \frac{S}{\mathrm{K}_{\mathrm{I}}a}\,\frac{1}{1-E}\ln\frac{(C_{\mathrm{IO}} - C_{\mathrm{IZ*}})(1-E)}{(C_{\mathrm{IZ}} - C_{\mathrm{IZ}})} + E \tag{3.41}$$

The number of theoretical stages, while not physically meaningful for a differential contactor, can be related to NTU by the expression:

$$\frac{N}{\text{NTU}} = \frac{E - 1}{\ln(E)} \tag{3.42}$$

For real extractors, axial mixing effects, as previously mentioned, may require two or more times the height predicted from purely theoretical considerations. Axial mixing effects are more likely to be significant in larger diameter columns, in short, squat columns, and in columns with low superficial velocities. Correction factors for different extractor configurations have been developed.[1] When scaling-up a differential column, it is important to consider that axial mixing effects may be much more significant in plant scale equipment than in the pilot unit.

3.3.3 Centrifugal extractors

Centrifugal extractors, as the name implies, uses centrifugal force to reduce the time required for dispersion and coalescence of the two phases. These extractors are thus able to achieve high throughput and short residence times. They are a good choice for extractions involving chemically unstable compounds or where very small density differences exist between the two phases.[1] Centrifugal extractors readily handle emulsifying fluids, are compact, and are easily adaptable to CIP requirements.

Disadvantages of this design include high capital and maintenance costs and plugging difficulties when processing feeds containing solids. Rotational speeds of 10 000 rpm or greater can be achieved, and frequent bearing and seal maintenance is required.

The short residence time requirements and emulsion inhibiting tendencies of centrifugal extractors have been the key attributes leading to the widespread use of this type of equipment in the antibiotics manufacturing industry. Antibiotics often undergo degradation under process conditions required for isolation, and hence minimum residence time results in minimum loss of valuable product. The fermentation broth from which these drugs are extracted also tends to form emulsions, which also favours the selection of a centrifugal extractor for these applications.

While little information on the specifics of most pharmaceutical production processes is available in the open literature, penicillin manufacture is a notable exception and a good illustration of a centrifugal extraction process.[1,4,25–27]

Penicillin is a strong organic acid. As mentioned in the discussion of equilibrium fundamentals, the partition coefficient for penicillin between aqueous and organic phases is altered dramatically by changes in the pH of the solution. This behaviour is exploited in the separation. Pentyl acetate, butyl acetate and methyl isobutyl ketone (MIBK) have all been commercially used as solvents.

At low pH, penicillin is easily extracted into the organic phase and the fermentation broth is typically adjusted to a pH of 2.0. Under these acidic conditions, however, the product degrades at an appreciable rate. The broth–solvent system also readily forms stable emulsions, and substantial quantities of product may be lost with the emulsion. It is actually for this particular extraction that centrifugal extractors were originally developed.[21,27]. Product recoveries of 90% using a single extractor, or of up to 99% using two extractors in series can be achieved.

The extract is then pH adjusted to 7.5 where the partition coefficient strongly favours the aqueous phase. The penicillin is then extracted into water using another centrifugal extractor. The penicillin can be purified at this stage by precipitation and washing; however, yet a third centrifugal extraction is commonly carried out, once again at low pH, into an organic solvent. The penicillin is then crystallized from the solvent, washed, tested and packaged.

Solvent extraction is an important part of the economics of such a process.

Most commercial antibiotic processes are very similar to the penicillin process, although operating conditions and solvents are obviously chosen to produce favourable partition coefficients for the particular product. Erythromycin, for example, is an organic base and is therefore extracted into organic solvents at high pH (9.0–10.0).[4]

Several different configurations of centrifugal extractor are commercially available. The original version was the Podbielniak extractor. This extractor consists of several concentric, cylindrical, perforated plates, which rotate rapidly about a horizontal axis. The feed is introduced along the centre of the unit, while the solvent is fed near the outer radius. The denser phase is pushed toward the outer shell of the unit by the developed centrifugal forces, displacing the solvent and forcing it toward the centreline. The fluids thus countercurrently contact one another but rapidly coalesce under the influence of the applied centrifugal force.

Proper positioning of the interface within the unit, and choice of which phase is to be dispersed can improve greatly the efficiency of the extraction. The interface position is controlled by adjusting the backpressure of the effluent solvent. Increasing the backpressure causes the interface to move radially outwards, and the solvent to become the continuous phase. Decreasing the backpressure has the opposite effect.

Rotor speed, flow ratio of solvent to feed and throughput are also significant process variables in centrifugal extraction. Increased rotor speed inhibits emulsion formation but simultaneously reduces residence time and the available number of theoretical stages. Flow-ratio selection involves consideration of partition coefficients as well as emulsion formation. Throughput involves a trade-off between turbulence and improved mass transfer and reduction of theoretical stages at higher flows.

While the Podbielniak and Quadronic extractors have been most commonly used in North America, the Alfa Laval extractor has dominated centrifugal extraction in Europe. This extractor rotates about a vertical axis. Operational factors are similar to those discussed in connection with the Podbielniak design, although more theoretical stages are achievable within one extractor — up to 20 stages versus three to five for Podbielniak units.

Discrete stage centrifugal extractors are also manufactured but have a much more limited range of applicability to pharmaceutical processes.

3.3.4 Agitated columns

The Scheibel column, invented in 1948,[28,29] was originally developed for pharmaceutical applications requiring many theoretical stages, and was the first agitated column extractor to gain widespread commercial acceptance.

The column is vertically divided, by baffles, into several mixing and settling zones. Each mixing region has a turbine impeller between an inner set of baffles that both promote mixing and also direct the fluids to an outer set of baffles, next to the shell, that allow phase separation prior to entering the next mixing zone. The outer baffles also prevent channelling of the fluids along the wall.

Scheibel columns are a good choice for difficult extractions where many theoretical stages are required. For small columns, up to three stages per foot (30 cm approx.) of height may be achieved. HTU is proportional to the square root of the diameter for these columns. Columns up to 2.5 m in diameter have been built, however. Emulsifying fluids can pose serious operating problems in a Scheibel column and should probably be processed in another extractor configuration unless a chemical control solution is feasible and cost-effective.

Bench and pilot-scale units are readily available and scale-up is reliable as so many columns of this type have been designed. Column throughput is typically in the range of 350–600 gal/h/ft^2.

Many other configurations of rotating agitated columns have been developed, most of which have found limited application in pharmaceutical or biological separations.

Kuhni columns are very similar to Scheibel columns but the mixing compartments are separated by perforated plates, which are attached to the rotating shaft. The area and pattern of the holes in the perforated plates are important variables in the design and application of this type of unit.

Oldshue-Rushton columns are very similar to Scheibel columns but the size of each mixing section is much larger. In a Scheibel column, each mixing section normally extends along the height of the column up to 30 cm. An Oldshue-Rushton mixing stage is usually several feet in height. The behaviour of such a column is actually closer to stagewise performance and can be thought of as a vertically stacked series of stagewise mixer-settler units.

Enhanced coalescence columns are conceptually similar to Kuhni columns, except that the plates separating the mixing stages are stationary rather than rotating. An interesting aspect of these columns is that the coalescence plates are fabricated of a material that is preferentially wetted by the dispersed phase.[18] Normally, of course, columns and their internals are designed to be preferentially wetted by the continuous phase to avoid premature coalescence. Here, however, wetting of the plate by the dispersed phase leads to desired coalescence, with redispersion accomplished in the next mixing stage. Owing to this continual coalescence and redispersion, these columns also approach stagewise, rather than differential operation.

Rotating-disc contactors, or RDC columns, use discs rotating inside an array of concentric stator rings to interdisperse the phases by shearing action.

Asymmetrical rotating-disc contactors, or ARD columns, unlike RDCs, have been used in pharmaceutical applications, particularly in Europe. These columns augment the original RDC concept by providing for separate mixing and settling compartments. To achieve this, the rotating shaft, with its discs, has been moved off the axial centre of the extractor. This leaves one side of the column relatively unagitated, as a series of mixing stages divided by horizontal baffles or stators. Throughput is typically 250–750 gal/h/ft^2.

In reciprocating agitated columns, dispersion of the phases is accomplished by reciprocating plates. The two major variations are the Karr column and the vibrating-plate extractor. The principal differences between these units are the diameter of the holes of, and the open area of, the reciprocating plates.

Karr columns have plates with relatively large holes (7.5 mm to 14 mm) and free areas (50%–60%). Vibrating plate extractors have smaller hole diameters and lower free areas (20%–30%). They also have 'downcomers' to allow passage of the continuous phase despite the low free area. These columns have sometimes been manufactured with two separate sets of plates, on two separate shafts, reciprocating 180° out of phase with one another.[1] Capacity and efficiency is improved in such countermotion plate columns but the columns are mechanically more complex and require greater maintenance.

Karr columns have frequently been applied in pharmaceutical separations and have several attractive advantages in such processes. Of particular advantage is the inherent tendency of the reciprocating plates to minimize emulsification. For this reason, Karr columns are well-suited to products from fermentation broths, when residence time requirements are not so critical as to require a centrifugal extractor. This is particularly the case if many theoretical stages are also required for the separation. These columns offer high throughput, turndown, and mass-transfer rates relative to other agitated extractor configurations. Bench and pilot-scale columns are widely available and scale-up procedures are well-established.[19]

3.4 Process design and scale-up

Although extraction processes have been practised for many years and many good design correlations are available in the literature, one should not place undue reliance on theoretical calculations during the design of a new extraction process. Conversely, a purely empirical approach is almost guaranteed to be wasteful of time and money and is likely to result in a suboptimal process.

Calculations and past experience are useful for preliminary design work. An estimate of likely stage requirements, solvent/feed ratios, temperature and pH conditions, and potential solvents can help guide and focus laboratory work. The stage requirements and process conditions will also suggest possible extractor configurations for the process.

Bench-scale experiments will establish mixing and settling characteristics and equilibrium data. It is important to conduct this work, as well as the later pilot-scale testing, with a feed that is as close as possible to that being used in the commercial application. Subtle differences between laboratory and full-scale feeds are a frequent cause of failed processes and embarrassed engineers.

Pilot-scale work is required to confirm preliminary calculations and experimental data, and to obtain the further data required for dependable scale-up. Process optimization work is best completed at this stage of the design process.

From the pilot-scale results, the production scale column can be designed reliably using proven scale-up correlations. The active assistance of a reliable equipment vendor can be extremely helpful during scale-up. Many experimental and operating results, data and correlations are proprietary to the equipment vendors. The extra cost and effort required for a joint development effort with a vendor is usually justified if potential suppliers are qualified and selected carefully.

3.5 Potential for protein purifications

Historically, extraction has been commercially applied only to small- and medium-weight biomolecules. Most proteins denature irreversibly in organic solvents, hence conventional solvent extraction is not normally useful for isolating or purifying protein products.

Proteins are usually isolated and purified using chromatography. Partially this reflects the sensitivity of proteins to operating conditions, and the great specificity and flexibility of chromatographic separation processes. It also reflects a very natural bias on the part of the scientists who have developed most such processes to date, to use technology with which they are familiar.

There are several factors developing in the biotechnology business that would favour the use of extraction processes, at least in some applications. These factors include increasing scale of production, increased competition leading to increasing emphasis on process economics, increasing importance of process reproducibility for validation and to meet CGMP, and the increasing number of separations that are being attempted. Two technologies show particular promise for these applications.

Aqueous two-phase extraction systems, so called A2PS systems, were developed in the late 1970s and early 1980s, principally by Dr M.R. Kula and her co-workers, in Germany.[10,11,23,30–34] Laboratory work has continued on these systems to the present time, although few, if any, significant commercial applications have yet been publicized.

The basic concept is to create two immiscible aqueous phases with different affinities for protein consolutes. Such systems can be created by the addition of polyethylene gycols of varying molecular weight (PEGs) and either another polymer (e.g. dextran), or a salt (e.g. potassium phosphate) to the aqueous feed solution. Protein partition coefficients betwen the two phases as high as 100 have been observed. The PEG–salt variant, in particular, would appear to be economically viable. Polymer–polymer A2PS tend to have very low interfacial tension, which allows for easy and rapid dispersion; however, the viscosity of such systems is often quite high (up to several thousand cP) and the density difference is often very small, all of which leads to very slow coalescence. Polymer–salt systems are also easy to disperse but have lower viscosities and greater density differences and hence coalesce readily. In general polymer–polymer A2PS extraction has been investigated using centrifugal extractors while polymer–salt systems are gravity separable.

A second approach to protein extraction is the use of reversed micellar systems. This technology uses surfactants to create stable droplets of aqueous protein solution surrounded by a single layer, the surfactant molecular shell. These reversed micelles are one to several nanometers in diameter. The micelles are readily soluble in organic solvents, but because the protein is in the protected interior, it never comes in contact with the solvent. This technology has been largely developed in Dr T. Hatton's laboratories at MIT since the mid 1980s. The partitioning behaviour of such systems is largely the result of electrostatic interactions between the protein and surfactant and can be influenced by the control of system pH. It is also possible to improve the efficiency of this type of process by incorporating a biospecific ligand into the micellar-phase formulation

Nomenclature

a	specific interfacial area per unit volume	ft^2/ft^3
C	solute molar concentration	lb mole/ft^3
C^*	solute molar concentration at equilibrium	lb mole/ft^3

D	diffusivity of solute in phase	ft^3/h
d_p	dispersed-phase droplet diameter	ft
E	extraction factor	KS/F
F	feed	ft^3 or ft^3/h
f	hold-up = dispersed-phase volume fraction	
HETS	height equivalent to a theoretical stage	ft
HTU	height of a transfer unit	ft
J	mass flux	lb mole/h
K	partition coefficient (phase equilibrium constant)	ft/h
K_I	intrinsic partition coefficient	
k	mass-transfer coefficient	ft/h
K'	overall mass-transfer coefficient	ft/h
NTU	number of transfer units	
R	gas constant	1.987 BTU/lb mole °R
S	solvent	ft^3 or ft^3/h
T	temperature	°R
V_c	continuous-phase superficial velocity	ft/s
V_d	dispersed-phase superficial velocity	ft/s
V_0	characteristic velocity	ft/s
V_s	slip velocity	ft/s
z, Z	height	ft
μ	chemical potential	BTU
μ°	standard rate chemical potential	BTU
z	interfacial property	

Subscripts

I	solvent phase
II	feed (aqueous) phase
c	continuous phase
d	dispersed phase
f	flooding

References

1. Lo, T.C. (1988). Commercial liquid–liquid extraction equipment, in *Handbook of Separation Techniques for Chemical Engineers*, 2nd edn (Ed. P.A. Schweitzer), McGraw-Hill, New York.
2. Lo, T.C. and Baird, M.H.I. (1980). Liquid–liquid extraction, in *KirkOhmer: Encyclopedia of Chemical Technology*, 3rd edn, John Wiley, New York.
3. Cusack, R.W. and Fremeaux, P. (1991). *Chem. Eng.*, **98,** 66.
4. Hatton, T.A. (1987). Liquid–liquid extraction of antibiotics. In *Comprehensive Biotechnology, Vol. 2, Ch. 27*, p. 439.
5. Belter, P.A., Cussler, E.L. and Hu, W.S. (1988). *Bioseparations*. John Wiley, New York.
6. Hafez, M. (1983). Centrifugal extractors, in *Handbook of Solvent Extraction* (Eds T. Lo, M. Baird and C. Hanson), John Wiley, New York.
7. Todd, D. (1972). *Chem. Eng*. **79,** 152.

8. Todd, D. and Davies, R. (1973). *Filt. Sep.*, **1**, 663.
9. Robinson, R.G. and Cha, D.Y. (1985). *Biotech. Progress*, **1**, 18.
10. Kula, M. (1979). Extraction and purification of enzymes using aqueous two-phase systems, in *Applied Biochemisty and Bioengineering, Vol. 2* (Eds L. Wingard, E. Katchalski-Katzir and L. Goldstein), Academic Press, New York.
11. Kroner, K. and Kula, M. (1978). *Process Biochem.*, **13**, 7.
12. About, N. and Hatton, T. (1988). *Chem. Eng. Prog.*, **84**, 31.
13. Rahaman, R., Chee, J., Cabral, J. and Hatton, T. (1988). *Biotech. Prog.*, **4**, 218.
14. Goklen, K. and Hatton, T. (1985). *Biotech. Prog.*, **1**, 69.
15. Woll, J., Hatton, T. and Yarmush, M. (1989). *Biotech. Prog.*, **5**, 57.
16. Treybal, R.E. (1979). *Liquid Extraction*, 3rd edn, McGraw-Hill, New York.
17. Treybal, R.E. (1980). *Mass-transfer Operations*, 3rd edn, McGraw-Hill, New York.
18. Cusack, R.W. and Fremeaux, P. (1991). *Chem. Eng.*, **98**, 132.
19. Cusack, R.W. and Karr, A. (1991). *Chem. Eng.*, **98**, 112.
20. Lo, T., Baird, M. and Hanson, C. Eds. (1983). *Handbook of Solvent Extraction*. Wiley Interscience, New York.
21. Podbielniak, W. (1953). *Chem. Eng. Prog.*, **49**, 252.
22. Hanson, C. (1971). *Recent Advances in Liquid–Liquid Extraction*. Pergamon Press, New York.
23. Kula, M., Kroner, K.H., Hustedt, H. and Schutee, H. (1981). *Ann. N.Y. Acad. Sci.*, **341**.
24. Craig, L. and Sogn, J. (1975). Isolation of antibiotics by countercurrent distribution, in *Methods in Enzymology, Vol XLIII*. (Ed. J. Hash). Academic Press, New York.
25. Porter, J. (1976). *Antibiotics in Industrial Microbiology*. (Eds J. Miller and W. Litsky). McGraw-Hill, New York.
26. Queener, S. and Swartz, R. (1979). Penicillins: biosynthetic and semisynthetic. In *Economic Microbiology, Vol 3* (Ed. A Rose). Academic Press, New York.
27. Todd, D. and Podbielniak, W. (1965). *Chem. Eng. Prog.*, **61**, 69.
28. Scheibel, E. (1948). *Chem. Eng. Prog.*, **44**, 681.
29. Scheibel, E. and Karr, A. (1950). *Ind. Eng. Chem.*, **42**, 1048.
30. Kroner, K., Hustedt, H., Granda, S. and Kula, M. (1978). *Biotechnol. Bioeng.*, **20**, 23.
31. Hustedt, H., Kroner, K., Stach, W. and Kula, M. (1978). *Biotechnol. Bioeng.*, **20**, 37
32. Grossman, P. and Gainer, J. (1988). *Biotech. Prog.*, **4**, 6.
33. Chen, C., Zhu, Y. and Evans, L. (1989). *Biotech. Prog.*, **5**, 111.
34. Evans, L., Zhu, Y. and Chen, C. (1990). *Biotech. Prog.*, **6**, 266.

Further reading

Albright, L. (1966). *Chem. Eng.*, **73**, 169.
Bailes, P., Hanson, C. and Hughes, M. (1976). *Chem. Eng.*, **83**, 86.
Bibaud, R. and Treybal, R. (1966). *AIChE J.*, **12**, 472.
Hoover, J. and Nash, C. (1978). Antibiotics — B lactams, in *Encyclopedia of Chemical Technology*. 3rd edn. (Eds M. Grayson and D. Eckroth). John Wiley, New York.
Humphry, J., Rocha, J. and Fair, J. (1984). *Chem. Eng.*, **91**, 9.
Karr, A. and Lo, T. (1976). *Chem. Eng. Proc.*, **72**, 68.
Karr, A. and Scheibel, E. (1954). *Chem. Eng. Prog. Symp. Ser.*, **50**, 73.
Kim, S. and Baird, M. (1976). *Can. J. Chem. Eng.*, **54**, 81.
McCullough, J. (1976). *Chem. Eng.*, **83**, 101.
Oldshue, J. (1966). *Biotech Bioeng.*, **8**, 3.
Perry, R. and Green, D. (1984). *Perry's Chemical Engineers' Handbook*. 6th edn, McGraw-Hill, New York.
Reissinger, K. and Schroter, J. (1978). *Chem. Eng.*, **85**, 109.
Scheibel, E. (1966). *Chem. Eng. Prog.*, **62**, 66.
Skelland, A.H.P. and Tedder, D.W. (1987). Extraction — organic chemical processing, in *Handbook of Separation Process Technology*. (Ed. R.W. Rousseau). John Wiley, New York.
Treybal, R. (1966). *Chem. Eng. Prog.*, **62**, 9.
Walas, S. (1985). *Phase Equilibria In Chemical Engineering*. Butterworth Publishers, Stoneham, MA.
Walas, S. (1988). *Chemical Process Equipment*. Butterworth Publishers, Stoneham, MA.
Wankat, P. (1973). *Ind. Eng. Chem. Fund.*, **12**, 372.

4 Affinity adsorption

D.J. ODDE

4.1 Introduction

Affinity adsorption is a powerful technique for the purification of biological molecules. While other separation techniques rely primarily on molecular size, charge, or solubility, affinity adsorption relies on highly specific binding interactions. Specificity is the hallmark of affinity-based purifications. It allows the capture of a single molecular component from a complex mixture. The high binding affinity allows high recovery of products in the μM to nM concentration range. An affinity adsorbent is prepared by immobilizing a ligand onto a solid support. The immobilized ligand is then contacted with a solution containing a product that binds specifically to the ligand. Traditionally, beaded supports are packed into a column and the solution is passed through it. While the product binds to the ligand, the remainder flows through the column. The column is then washed and the adsorbed material eluted, usually by altering the buffer conditions such that binding is less favorable. The properties of affinity adsorbents make them especially suited for recovery of biomolecules from fermentation broths where the product is often dilute and the broth heterogeneous. Although commonly used at the bench-top scale, large-scale processing has only become common in the last 10 years. Published examples of large-scale affinity-based purifications include interferon purification using immobilized antibodies[1] and anti-human tissue plasminogen activator antibodies purification using immobilized Protein A.[2]

4.2 Affinity adsorption

4.2.1 Adsorbent preparation

4.2.1.1 Support materials. Choice of support material is an important consideration in the design of an affinity adsorbent. A support should be derivatizable, rigid, have low non-specific adsorption, low mass-transfer resistance, high capacity per void volume, and high resistance to chemical and biological degradation (Table 4.1). Traditionally, supports have been made in the form of beads which are then packed into a column or agitated as a slurry in a stirred tank. Some of the more common support materials used are listed in Figure 4.1 along with their chemical structure.

Table 4.1 Criteria for support selection

Derivatizable
Rigid
Low non-specific adsorption
Low mass transfer resistance
High capacity per void volume
Resistance to chemical degradation
Resistance to biological degradation

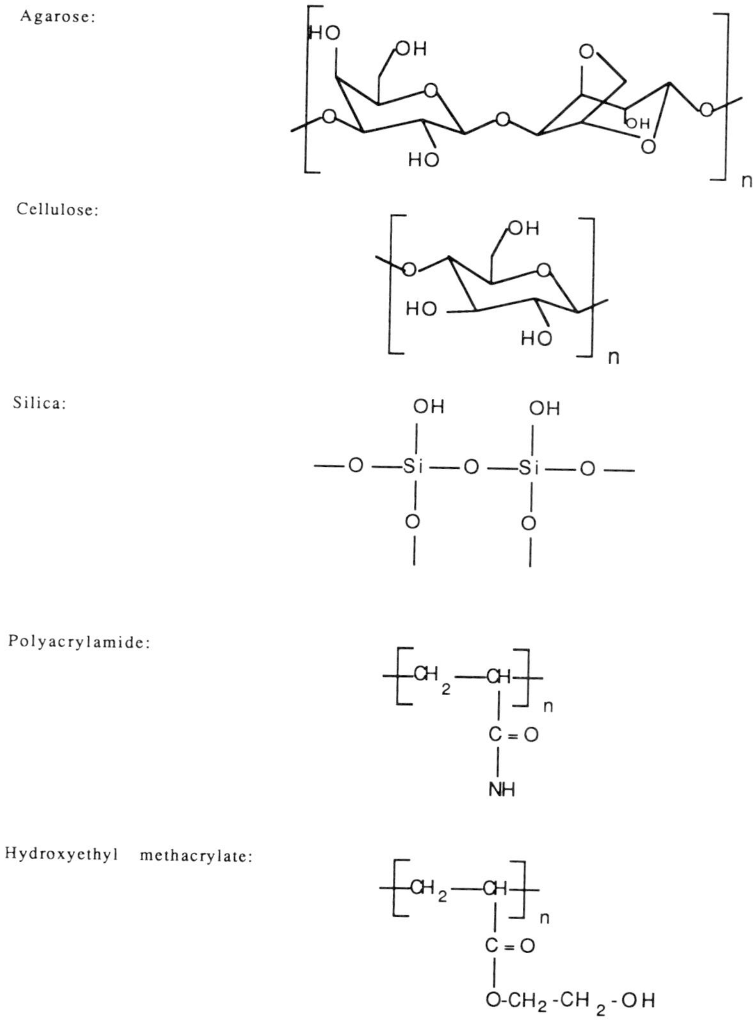

Figure 4.1 Support materials.

Agarose is one of the most widely used supports for affinity adsorption. It is a polysaccharide derived from agar and has free hydroxyl groups, which can be modified using several activation chemistries. It has very low non-specific binding and is resistant to microbial degradation making it very attractive as a support. Although it possesses some sulfate and carboxy groups, they do not contribute to non-specific adsorption at ionic strengths above 0.02.[3] The gel is formed spontaneously by cooling a hot agarose solution. It is made with various concentrations of agarose, usually 2–6% (w/v). Agarose is commercially available from several suppliers including Pharmacia, Schleicher and Schuell, and Bio-Rad. The gel is very porous and can become compressed when high-pressure gradients exist in a packed bed. The rigidity of the beads can be increased by crosslinking of the gel;[4] however, this results in a reduction in the number of hydroxyl groups available for activation. The crosslinking also increases the resistance to chemical degradation. Agarose gels were originally used for gel filtration and, as such, possess greater resistance to mass transfer than supports designed specifically for affinity adsorption. The greater the mass transfer resistance, the longer the loading and elution steps. Longer loading and elution results in lower productivity and, if the elution conditions are harsh, product and/or ligand degradation.

Controlled pore glass and silica have also been used as support materials.[5] They are available commercially through chomatographic media suppliers. Again, the support can be derivatized via hydroxyl groups. Although porous inorganic supports have excellent rigidity and are inexpensive, they can have high non-specific adsorption[6] and are susceptible to degradation under alkaline conditions.[7] Surface modification by silanes with uncharged hydrophilic character can reduce non-specific adsorption.[8]

Polyacrylamide, polymethacrylamide, and hydroxyethyl methacrylate beads are some of the more prominent organic polymers that have been used as supports.[9] One popular polymethacrylamide support is Eupergit®, manufactured by Röhm Pharma. These matrices are hydrophilic and stable under a wide range of conditions but are often mechanically unstable at the porosities required for effective macromolecule immobilization and adsorption.[10]

All of the above materials are very porous, allowing for larger surface areas per volume of bed. However, since the beads have diameters on the order of 100 μm, considerable diffusional resistance during loading and elution can occur. For this reason, large porous beads may not be the optimal support structure for affinity adsorption of proteins. It has been suggested that non-porous supports offer better transport properties[11] but the trade-off is a lower capacity per volume of bed. This is offset, however, by the fact that the void volume (liquid phase volume) is lower with non-porous beads. In this case, the volume of eluent required will be lower, while the total bed volume remains constant. Although capacity per cycle may also be lower, the cycle time is much shorter and therefore more cycles can be run in a given time period.

Membranes have also been used to avoid the mass-transfer limitations found in porous beads. Silica-impregnated polyvinylchloride is one example and is available commercially from FMC Corporation. Short cycle times on the order of 20 mins have permitted purification of gram quantities of protein using a membrane with total bed volume of 16 ml.[12] These membranes have channels that are about 1 μm in diameter, which reduces diffusional resistance. Regenerated cellulose has been used as a membrane support in addition to organic polymer membranes.[13] So-called perfursion beads attempt to combine the high porosity of beads with the transport properties of membranes by creating macro channels in the beads through which convective transport can occur.[14] Such beads are commercially available from PerSeptive Biosystems. It appears the trend is toward supports that provide the best transport properties, since there are several materials that meet the other criteria.

4.2.1.2 Affinity ligands. Numerous molecules have been used as affinity ligands and an extensive listing of examples in the literature is given elsewhere.[15] However, they can be broken into categories of ligands that encompass known biological binding interactions. They include dye/protein, enzyme/substrate/substrate analog, antibody/antigen, protein/co-factor and receptor/ligand. In principle, one member of a pair can be used as the affinity ligand to purify the other. However, affinity adsorption is rarely used to purify dyes, substrates, or co-factors.

Since the product is frequently a protein, the antibody/antigen interaction is the most versatile as antibodies can be made against virtually any protein antigen. This type of adsorption is called immunoadsorption. It should be noted that the antigen can be used as the affinity ligand to purify the antibody although the reverse situation is more common. When antibodies are purified by affinity adsorption, they are more frequently purified by immobilized Protein A or Protein G, depending on the type of antibody. In fact, many support manufacturers provide beads or membranes with Protein A or Protein G already immobilized. The primary cost of immunoadsorption (where Abs are used to purify antigens) is the cost of the antibody itself.[6] The cost of monoclonal antibodies is roughly $1000/g. Therefore a 100 ml column with an immobilized antibody loading of 10 mg/ml will cost at least $1000 for the antibody alone. Thus the stability of the immunoadsorbent becomes an important consideration while attempting to distribute the capital cost over as many cycles as possible. However, with the continuing advances in monoclonal antibody production,[16] this cost should decrease over time. Also, recombinant antibodies and antibody fragments may open other avenues for improving the properties of immunoadsorbents.[17,18] These factors should make the cost of antibodies lower and make affinity adsorption more efficient and attractive as a purification process.

4.2.1.3 Immobilization chemistries. Cyanogen bromide (CNBr) activation, first introduced by Axen *et al.*,[19] is the most widely used activation method. This method requires treating the support with CNBr dissolved in an organic solvent under alkaline conditions (pH 11–12) and then reacting the protein with the activated support under mildly alkaline conditions (pH 8–9). The principal reactions are shown in Figure 4.2. It has been demonstrated that the cyanate esters are the primary reactive moiety, while imidocarbonates are less reactive.[20] Carbamates are essentially inert with respect to primary amines. The resultant linkage is an isourea bond, which, at neutral pH, has a net positive charge and therefore imparts ion exchange character on the support. This bond is also susceptible to attack by nucleophiles such as hydroxide ions and primary amines[21–23] making the bond less stable under alkaline conditions. However, the bond is stable under neutral and slightly acidic conditions — those frequently employed in affinity adsorption.

The high toxicity of CNBr has motivated the development of reagents that are less noxious. Cyano-transfer agents have been employed as a replacement for CNBr. Whereas the CNBr method requires the use of an inorganic base such as NaOH as the catalyst, cyano-transfer agents employ a tertiary amine, such as triethylamine. Some common cyanotransfer agents are *N*-cyanotriethyl ammonium tetrafluoroborate (CTEA) and 1-cyano-4-(dimethyl)-pyridinium tetrafluoroborate (CDAP).[15]

Organic sulfonyl chloride activation was introduced by Nilsson and Mosbach[24] in 1981 and has gained popularity because of its stability of the resulting linkage. Of these, 2,2,2-trifluoroethanesulfonyl (tresyl) chloride has

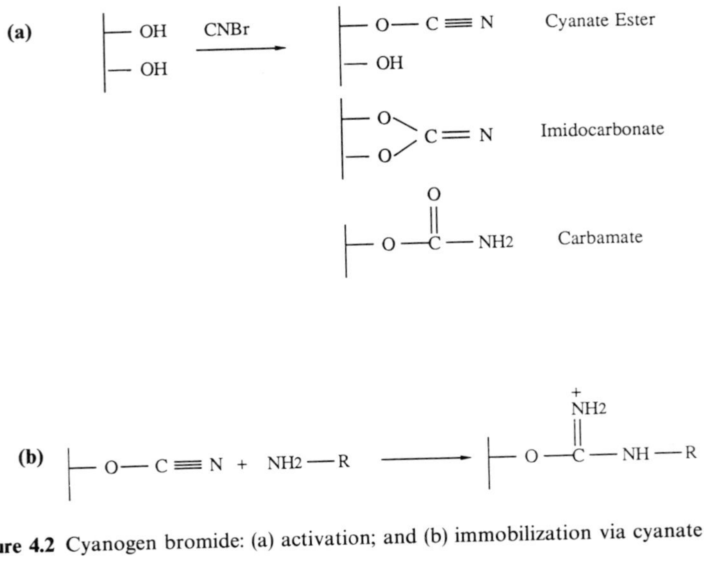

Figure 4.2 Cyanogen bromide: (a) activation; and (b) immobilization via cyanate ester.

become the most popular since it allows the use of milder coupling conditions relative to other sulfonyl chlorides. The tresyl group is added to a hydroxyl on the support, and is subsequently replaced by an amine- or sulfhydryl-containing ligand, thus releasing the tresyl group. The enhanced stability achieved with this method has been attributed to the ligand coupling directly to the support (Figure 4.3).

Bisoxiranes, such as 1,3-butanediol diglycidyl ether, have been employed

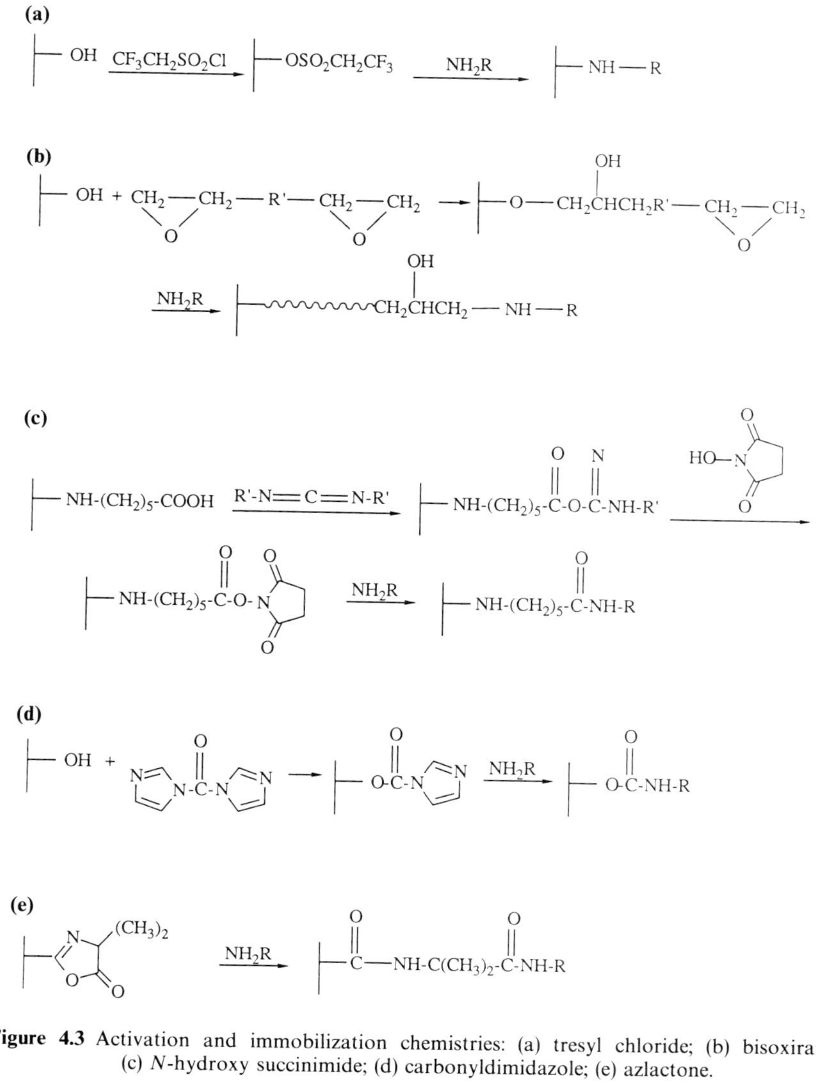

Figure 4.3 Activation and immobilization chemistries: (a) tresyl chloride; (b) bisoxiranes; (c) *N*-hydroxy succinimide; (d) carbonyldimidazole; (e) azlactone.

as activating agents.[25] This method converts the hydroxyl to an epoxide, which can subsequently be reacted with hydroxyls, amines, or sulfhydryls on the protein. By using a long-chain bisoxirane, a spacer arm is incorporated between the ligand and the support. This technique is used with other coupling chemistries as well and is intended to reduce the steric effects of support/ligand interaction. Reaction times are generally longer with bisoxiranes and are performed at higher temperatures than with CNBr-activated supports. The chemistry of coupling is shown in Figure 4.3. These supports have been found to have excellent stability but require longer activation and immobilization times.[26] Pre-activated supports using the above techniques are commercially available from many suppliers including Pharmacia, Schleicher & Schuell, and Bio-Rad.

N-hydroxy succinimide has also been used for activation.[27] The chemistry of activation and immobilization is shown in Figure 4.3. The primary bond formed is an amide bond, which is very stable. However, ligand leakage has been found under alkaline conditions because of side reactions occurring during activation.[28] Methods have been proposed to eliminate this effect but the trade-off is a lowered coupling efficiency.

Activation with 1,1-carbonyldiimidazole yields an activated carbamate, which reacts with a primary amine to give a substituted carbamate. The carbamate bond is more stable than the isourea and has no net charge (see Figure 4.3). The reagents used are less noxious than CNBr.[26]

Recently, a novel activation based on an azlactone functionality has been demonstrated.[29] Beads were formed by co-polymerization of vinyldimethyl azlactone and methylene-*bis*-acrylamide to form beads with diameters of 50 – 70 μm and surface areas of 170–220 m^2/g. The structure of the azlactone ring and its subsequent opening upon attack by nucleophiles is shown in Figure 4.3. The authors claimed high coupling efficiencies and recoveries from affinity columns, up to 56 mg of IgG per ml of column bed volume from a Protein A column. This appears to be an interesting new type of pre-activated support and is being developed by 3M Corporation.

All of the above methods are non-specific in the orientation of the immobilized protein. For instance, CNBr-activated supports react specifically with primary amines, either ε-amino groups on lysine residues or amino termini. Proteins usually contain many lysine residues (IgG has on the order of 100), which are distributed throughout the molecule. This random or non-oriented immobilization allows the possibility of immobilizing ligands such that their binding sites are inaccessible.[30] To overcome this inefficiency, various schemes have been employed to achieve a correctly oriented ligand.

One example is the immobilization of Protein A, binding of IgG, and subsequent cross-linking by glutaraldehyde.[31] Protein A binds IgG in the Fc region of the antibody, away from the antigen building site. Presumably this type of immobilization reduces the chance that a binding site will be blocked. Higher capacities on a per-ligand basis have been achieved but at the cost of introducing more reagents and additional steps in the preparation of the affinity adsorbent.

Another technique employed is the oxidation of carbohydrate moieties on IgG by periodate to yield aldehyde groups.[25] These groups are subsequently reduced in the presence of a hydrazine-activated support. The motivation for this method is based on the fact that the carbohydrate moieties are primarily located around the hinge region of the antibody, away from the antigen binding sites. However, it is known that carbohydrate moieties can be distributed throughout the antibody molecule.[32] Therefore this approach may not have an advantage over conventional immobilization techniques.

4.2.2 Adsorbent stability

Many phenomena have been found to diminish affinity adsorbent performance. Those cited include the following: (i) matrix degradation,[33] (ii) proteolysis,[34] (iii) denaturation of the affinity ligand,[35] and (iv) immobilizing bond lysis and subsequent leakage.[21–23] These are discussed individually below.

When designing an affinity-based process, matrix degradation must be considered. For example, Sepharose™ 4B (an agarose support available from Pharmacia, Uppsala, Sweden) is stable in the range from pH 4–9. Sepharose™ CL, a cross-linked agarose, has a stability range from pH 3–14.[3] In general, the matrices frequently used tend to be more stable than the affinity ligands or the bonds immobilizing the ligands to the matrix. Proteolysis is a potential problem when the feed solution is serum or cell lysate. Supports with minimal mass-transfer resistance can offer protection against proteolysis since the loading times are much shorter.

Ligand denaturation can be a problem when repeated cycles of various buffer conditions are used for loading, washing, and especially, elution, which is often the harshest step for immobilized proteins. In the case of immunoadsorption, it is difficult to predict how sensitive antibody-antigen binding will be to various elution conditions and therefore screening is required. In studying the effect of repeated cycles of operation on immunosorbent stability, Antonsen *et al.*[35] found that antibody capacity, defined as moles product bound per mole antibody immobilized, decreased to 50–75% of initial capacity after 100 cycles of operation. While capacity decreased, the affinity constant remained unchanged and the presence of serum had no effect on degradation, implying that proteolysis was not a significant mechanism for this system. The conclusion was that degradation was due to the action of the eluent on the antibody and was an all-or-none phenomenon whereby a given binding site was either fully active or completely inactive. The authors proposed that reversible denaturation was primarily responsible for effecting elution. The loss of capacity was also dependent on the concentration of antibody immobilized where higher antibody loadings resulted in greater degradation. The interpretation was that antibodies at higher concentrations formed fewer immobilizing bonds at a constant level.

Immobilizing bond lysis has been the best documented mode of deactivation. Although denaturation may account for the bulk of degradation, leakage is still of concern when producing therapeutics and may affect results when the adsorbent is used in binding assays. While the FDA recommends that no detectable leaked ligand be present in the effluent,[36] concentrations as low as 10 ng/ml can be detected by ELISA assays. In reality, there will always be at least one step required downstream to remove leaked ligand and non-specifically adsorbed materials. This is often accomplished by ion exchange, gel filtration or hydrophobic interaction chromatography.

Many leakage studies have been concerned with the stability of the isourea bond formed during immobilization on CNBr-activated supports. CNBr is still the most widely used and best characterized immobilization chemistry. One of the first reports of leakage was by Tesser *et al.*[21] They found that the isourea bond was stable under slightly acidic and neutral conditions, but labile under slightly alkaline conditions. Specifically, the rates of Sepharose™-bound cyclic 3′,5′-adenosine monophosphate leakage increased as the pH increased from 5 to 8. The maximum rate observed was at pH 8 where approximately 200 ppm of the total immobilized ligands were released in 2 h at room temperature. Since these molecules were immobilized by a single immobilizing bond, it was suggested that polyvalent handles be used to create multiple attachments to the support, thereby alleviating the leakage problem. Wilchek[37] proceeded to demonstrate the feasibility of this approach by synthesizing polylysine, polyornithine, and polyvinylamine derivatives. He found that the polyvalent handles enhanced stability and that polylysyl-Sepharose™ showed no 'detectable' leakage, even after storage in a pH 8 buffer at room temperature for 3 months. However, ε-DNP-lysine coupled directly to Inactivated Sepharose™ showed leakage of 15% of the ligand under the same conditions. The polylysine tails ranged from 100 to several thousand residues and all were effective in reducing leakage.

Tesser *et al.*[22] characterized the chemical nature of the leaked species and proposed mechanisms for immobilizing bond lysis under various conditions of pH and amine concentrations. Their work demonstrated that nucleophiles, such as hydroxide ions and primary amines, were attacking the isourea bond and causing the leakage. Wilchek's analysis of insulin-Sepharose™ leakage products[23] confirmed these results. From these observations, Gribnau and Tesser[38] and Lasch[39] formulated mathematical models to describe the leakage of ligands with multiple attachments to the support. Gribnau and Tesser assumed that the immobilizing bonds break in a specific order while Lasch assumed that they break randomly. Both models assume that all immobilizing bonds break according to a single rate constant and that all the ligands in the matrix have the same number of immobilizing bonds initially. As a consequence, both predict no leakage at early times for ligands with multiple immobilizing bonds. These models are inconsistent with observed behavior, indicating there probably is a distribution of the number of immobilizing bonds per ligand.

Lasch and Koelsch[40] investigated the leakage behavior for leucine aminopeptidase (LAP) immobilized onto CNBr-activated Sepharose™. They found the percent of LAP lost versus time showed linear behavior for 25 days, which represented 10% of the total ligand. This was interpreted as being caused by heterogeneity of the initial number of immobilizing bonds. The number of immobilizing bonds as measured by amino acid analysis was 22 (out of 190 lysine residues). Koch-Schmidt and Mosbach[41] measured 1 to 8 immobilizing bonds for Ribonuclease A-Sepharose™ by varying the amount of CNBr used for activation and therefore presumably changing the degree of activation.

Another approach to stabilizing the ligand was proposed by Schnapp and Shalitin[42] who activated amine-carrying supports with cyanogen bromide and subsequently immobilized the protein. They found markedly reduced leakage rates and proposed that hydrazine and hydrazide supports could be used for this purpose as well. Wilchek and Miron[43] also tested polyglutamic hydrazide and linear polyacrylic hydrazide polyvalent handles, which increase the number of immobilizing bonds to the support without incorporating the ionic character associated with polylysine tails.

Attempts have been made to measure the relative stability of various immobilization chemistries under various buffer conditions.[33,44] However, the systems used have made interpretation of results difficult because of variable activations, multiple immobilizing bonds, and significant contributions of non-covalently desorbed protein. It appears that most chemistries are susceptible to attack by nucleophiles to varying degrees and that, at least in therapeutic purification, this small amount of leaked ligand poses a problem that must be resolved.

4.2.3 *Equilibrium models*

Ideally, the immobilized ligand will possess the same binding activity as the soluble molecule. However, this has not always been the case in practice. For example, immobilized enzymes have been found to have reduced enzymatic activity relative to soluble enzymes.[41] Similarly, immobilized antibodies have been found to have reduced affinity and capacity (on a molar basis) for binding compared with soluble antibodies.[11,45,46] Reduced enzymatic activity has been attributed to phenomena such as immobilizing bond-induced conformational changes,[41,47] increased immobilized enzyme density,[48] and support/substrate ionic repulsion.[49]

Immobilized antibodies have also been shown to have reduced capacity (on a per mole basis) at higher antibody concentrations and higher activation levels.[11] The implications are that antibody crowding and increased number of immobilizing bonds per antibody result in decreased capacity. Possible mechanisms for this have been suggested by Chase.[30] Since most immobilization procedures presumably result in random orientation of the ligand and therefore its binding site, it can be argued that there will be a variety of possible fates for each ligand in the system. Some may be immobilized directly through their binding sites, while others may have their binding sites immobilized sufficiently close to the support to reduce binding affinity without destroying the site completely. In addition to these ligand/support effects, there may be ligand/ligand effects on capacity and affinity where binding at one site affects binding at another site. These phenomena would be expected to be dependent on the concentration of antibody immobilized and the number of immobilizing bonds per antibody. This model is consistent with the results obtained by Eveleigh and Levy.[11]

The affinity is important in designing the adsorbent as it will determine the concentration at which the antigen will bind the antibody, the time required to load the adsorbent, and the time required to elute the product. A model commonly used for affinity adsorbents is the Langmuir isotherm.[6,50] The equilibrium reaction can be written as:

$$A + B \rightleftarrows AB \tag{4.1}$$

where A is the product to be purified, B is the immobilized ligand, and AB their complex. An equilibrium constant, K, can be defined such that

$$\mathrm{K} = \frac{[AB]}{[A][B]} \tag{4.2}$$

Combining with a mass balance on A, the Langmuir isotherm can be obtained:

$$r = \frac{n\mathrm{K}[A]}{1 + \mathrm{K}[A]} \tag{4.3}$$

where r = moles A bound per mole B, and n = maximum moles A bound per mole B (capacity). Two assumptions in this model are that all the sites in the system bind the product with equal energy and that there are no cooperative effects where the binding at one site affects the binding at another site. Given the potential heterogeneous nature of immobilization, however, it is reasonable to expect that there may be a heterogeneity of binding energies and possibly cooperativity. If there is heterogeneity of binding in the system, then the time required to load and elute a given percentage of the product will be longer because of the presence of lower affinity binding sites. In addition, a significant fraction of the binding sites may be lost on the first cycle because of incomplete elution from higher affinity binding sites. This phenomenon has been observed and lends further credence to the hypothesis of binding heterogeneity.[11]

Another approach that has been taken is to use the empirical Freundlich isotherm.[51] This isotherm has been found to work well in heterogeneous systems, especially catalysts. The isotherm has the following form:

$$r = K'[A]^{a'} \tag{4.4}$$

where K′ and a′ are constants of empirical origin.

By making the Freundlich a saturable isotherm, the so-called Sips isotherm[52] is obtained, which has the form:

$$r = \frac{n(K[A])^{a}}{1 + (K[A])^{a}} \tag{4.5}$$

This isotherm has been used to describe polyclonal antibody-antigen binding, which is inherently heterogeneous.[53] Sips showed that this isotherm has a corresponding distribution of energies closely approximated by a Gaussian distribution. The value of a varies between 0 and 1 for heterogeneous systems and is related to the 'broadness' of the distribution, with the distribution becoming broader as a decreases. In the limit where a equals 1, the Langmuir isotherm is obtained. This equation was proposed by Hill[54] to describe the binding of oxygen to hemoglobin, which is a positively cooperative system where $a = 2.8$.[55] Isotherms for various values of a are shown in Figure 4.4.

Olson *et al.*[56] showed that for a heterogeneous system, the time required to eletrophoretically elute a given fraction of antigen increased as a decreased from 1. Their model showed that they would take twice as long to recover 80% of the antigen when $a = 0.625$ as it would when $a = 1.00$. While these results were for an electrophoretically eluted product, they are still valid for

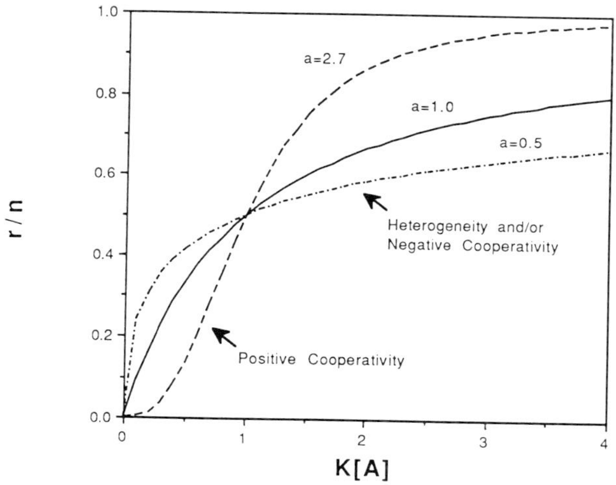

Figure 4.4 Sips isotherm.

other elution techniques so long as there is a distribution of affinities under the elution conditions used.

4.2.4 *Operation*

Affinity adsorption operations are commonly run either as fixed beds or stirred tanks with fixed beds being the more common of the two. The primary advantages of the stirred tank configuration are the following: (i) more rapid mass transport for conventional porous beads; (ii) ability to process feed streams containing insolubles, which might clog a fixed bed; (iii) ability to process a larger volume in a shorter time. The disadvantages of the stirred tank are that the beads are less concentrated than they are in a fixed bed and the processing is more complex because of the required agitation and solid–liquid separations.

Fixed beds, on the other hand, are easier to clean and maintain.[6] The potential disadvantage of slower transport can be offset by using fixed beds with better transport properties, such as membranes,[12] perfusion beads,[14] or non-porous beads.[11] A disadvantage of fixed beds is that some packed bead columns can be compressible under pressure gradients depending on the mechanical strength of the beads.[6] Pre-activated beaded supports are commercially available and cost roughly \$1000/100 ml of bed volume including the column. Automated control systems are also available and cost from \$10 000–\$20 000.[57] Operating costs depend on the buffers used and required maintenance which can vary considerably depending on the system and performance requirements.

4.2.4.1 Fixed beds. As mentioned previously, the two forms of support used are beads and membranes. Beads are usually packed into columns and supported by porous frits. Packed bead columns allow greater flexibility in geometric configuration and are therefore more easily scaled up. Membranes, on the other hand, generally have better mass transport properties, fewer channeling problems, and are more resistant to compressing under pressure gradients.

Packed beads. Loading
The product loading step is controlled thermodynamically by the equilibrium constant, such that the concentration of the product in the feed stream, c_0, needs to be greater than the inverse of the equilibrium constant

$$c_0 > 1/K$$

to use a significant fraction of the binding capacity. In practice, however, it is not efficient to wait for equilibrium to be reached for the reasons cited above regarding mass-transport limitations. Generally the diffusion in the pores will be the limiting transport step for purifying proteins using porous beads of 50 – 100 μm in diameter, assuming that the ligand is homogeneously distributed throughout the bead.

As a consequence, columns are often run under non-equilibrium conditions. One way to assess column performance is by analyzing breakthrough curves, which are simply plots of column outlet concentration versus time (or column volumes). A column loaded at equilibrium will have a breakthrough curve that is a step function as shown in Figure 4.5. When non-idealities are present, such as mass-transport limitations, the normalized outlet concentration, X, will approach 1 at a finite rate. From these curves column utilization and feed wasted can be calculated.

Arnold *et al.*[58] analyzed loading, washing, and elution steps in column and stirred tank affinity adsorption. They defined a dimensionless number, called the number of transfer units N, for loading in a fixed bed. A different N was defined for each potential transport limiting case. N can be thought of as a ratio of the column residence time to the characteristic action time of the transport phenomenon. The transport step with the smallest N is the controlling step. Transfer units were defined for pore diffusion (N_{pore}), diffusion through a homogeneous solid medium (N_p), fluid mass transfer (N_f) and axial dispersion (N_d). The values of N can be estimated from known mass-transfer correlations, column geometry, operating conditions and diffusivities.

An important case to consider when loading a column is the pore diffusion limited regime. In this case, N_{pore} is the smallest transfer unit and the breakthrough curve is described by:

$$X = 1 - (\tfrac{2}{3} - 0.273N_{\text{pore}}(T - 1))^2 \tag{4.6}$$

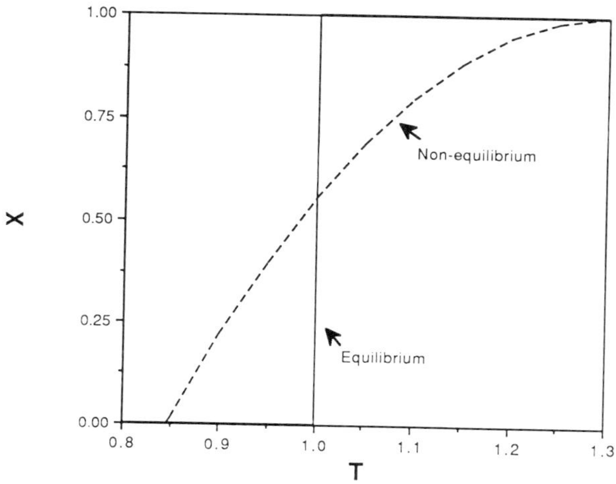

Figure 4.5 Fixed-bed breakthrough curve.

where

$$N_{\text{pore}} = \frac{15\, D_i (1 - \varepsilon)\, L}{R^2 u_0} \tag{4.7}$$

$$T = \frac{V - \varepsilon v}{\Gamma v} \tag{4.8}$$

$$\Gamma = \frac{\rho_b q_0}{c_0} \tag{4.9}$$

$$X = \frac{c}{c_0} \tag{4.10}$$

where c = outlet concentration; c_0 = feed concentration; V = throughput volume; v = total bed volume; ε = column void fraction; ρ_b = particle bulk density; q_0 = product concentration in equilibrium with c_0; D_i = effective intraparticle diffusivity; L = bed length; R = particle radius; u_0 = liquid superficial velocity.

Additional assumptions used in the derivation of this expression are that: (i) axial dispersion is negligible; (ii) the liquid phase concentration in the pores is negligible relative to the solid phase; (iii) both the liquid and solid concentration profiles have constant shape; and (iv) fast, irreversible adsorption. Although not stated explicitly, the authors also assume uniform bead structure and size and uniform ligand concentration within the support. Nevertheless, heterogeneous bead structure and ligand distributions have been found to occur in some porous media.[59,60] For example, it has been reported that there can be much higher concentrations of ligand near the outer edges of porous beads, which can be attributed to the immobilization chemistry and immobilization process.[60] In this case, the entire bead volume is not used and the effective diffusion length is shorter. The net result is that the mass transport resistance is diminished while the bed volume is underused.

Washing

The function of washing is two-fold. First, soluble material present in the column is removed. Second, non-specifically bound impurities are removed. In both cases, there may be loss of product. Tarnowski and Liptak[1] reported a loss of 24% of the product during the cumulative washings of an immunoaffinity column used to purify interferon. Therefore it is apparent that the washing protocol can greatly affect yield. Since downstream purification steps are inevitable even with the high purity obtained from affinity adsorption, it seems that non-specifically adsorbed materials may be more effectively dealt with using other techniques such as gel filtration, ion exchange, or hydrophobic interaction chromatography. In this approach, the most important function of washing is to eliminate the soluble material in the column.

Elution

The advantages of affinity adsorption as a purification process are primarily specificity for the product and ability to purify dilute feed streams. These advantages are because of the high binding affinity inherent in affinity adsorption. During elution, however, the high binding affinity becomes an obstacle and therefore elution often poses the greatest challenge in process design. The affinity needs to be lowered so that the product will be released into the liquid phase. While ligands have been used that bind with equilibrium constants, K, of $10^8\ M^{-1}$ and higher, K needs to be reduced to 10^4–$10^6\ M^{-1}$ to attain high recovery with a reasonable number of column volumes.[35] It is difficult to predict *a priori* what conditions will be effective and mild. Many different conditions need to be screened for suitability in a given application.

There are many techniques by which elution can be effected. A common technique is simply alteration of pH, usually from neutral to acidic, although basic conditions can be used. Basic conditions tend not to be used because they are generally harsher on the adsorbent and can promote leakage for certain immobilization chemistries as pointed out above. A common acidic elution buffer is 0.2 M glycine, pH 2.5.[35] Liptak and Tarnowski[1] reported using 0.2 M acetic acid, pH 2.7 for the purification of interferon and were able to operate their column for several hundred cycles. Another technique used is the alteration of ionic strength, where increasing ionic strength can disrupt coulombic binding forces.

Chaotropic salts such as KSCN or KCl have also been used as eluents.[10] Denaturants such as guanidine or urea have been used to reversibly denature proteins, thus destroying the binding interaction.[25] Organic solvents have also been employed including ethanol, dimethyl sulfoxide, and ethylene glycol.[10] Although the intent is to make the process reversible, there may be irreversible denaturation, which results in reduced capacity over many cycles.

When reduced affinity cannot be achieved without destroying adsorbent activity, electrophoretic-based transport has been used as an alternative.[61] This technique is much slower than conventional elution and therefore may be especially suited to eluting materials from membranes, which have shorter bed lengths.

As an alternative to these methods, the biological binding properties of the ligand and the product have been exploited to achieve elution. One example is biospecific elution where an enzyme is eluted by adding its substrate.[15] Other examples include co-factors, competitive inhibitors, and haptens. Metal-dependent binding phenomena, where binding of a metal ion either enhances or diminishes binding, have been used.[15] The metal may bind either the ligand or the product. Elution may be effected by either addition of the metal or a metal chelator depending on whether the metal binding enhances or diminishes the binding affinity.

Finally, physical manipulations can be used for elution. The primary techniques used have been thermal and pressure elution. Thermal elution is based on the fact that the affinity constant changes with temperature according to the van't Hoff equation:[62]

$$\left(\frac{\partial \ln K}{\partial T}\right)_P = \frac{\Delta H^\circ_{rxn}(T)}{RT^2} \tag{4.11}$$

where $\Delta H^\circ_{rxn}(T)$ is the standard molar heat of reaction at 1 atm. If $\Delta H^\circ_{rxn}(T)$ is negative, then the adsorption will be more favorable at low temperature than at high temperature, while the reverse is true for positive $\Delta H^\circ_{rxn}(T)$. The difficulty again is knowing what the value of $\Delta H^\circ_{rxn}(T)$ will be *a priori*. This technique is also limited by the thermal stability of the ligand and the product. With the development of more thermally stable proteins this may become a more attractive method in the future as thermally stable structural motifs may possibly be engineered into recombinant proteins.[63] Thermal elution is especially attractive since it requires no additional reagents.

Pressure elution has also been used as a milder alternative to the harsh chemical eluents frequently used. Olson *et al.*[13] found that 75% of reversibly bound bovine serum albumin could be recovered from an anti-albumin immunoadsorbent in one cycle of pressure treatment to 2000 atm and over 90% recovered by additional cycles. These immunoadsorbents were found to be subject to significantly reduced binding capacity upon treatment with glycine, pH 2.5.

Membranes. All of the above considerations apply to affinity membranes as well. However, membranes generally have better mass-transport properties and shorter bed heights. As a consequence, the time to complete one cycle of operation will be much shorter, although the amount processed will be smaller. An interesting alternative to convective transport for loading is the use of electrophoresis to drive the product into membrane. Olson *et al.*[56] found that electrophoretically driven loading could increase product yields from 6.6% to 67% while reducing adsorption times from 6 h to 45 min compared with crossflow adsorption.

4.2.4.2 Stirred tanks. Since most bioseparations are best suited to batch operations, continuous stirred tanks will not be considered. An important parameter in the analysis of a stirred tank for affinity adsorption is the equilibrium concentration of the final product in the liquid phase. By combining an isotherm with a mass balance on the product, the final concentration can be calculated.[6] Just as with the fixed bed, it is important that the capacity be greater than the amount fed and the feed concentration be greater than the inverse of the equilibrium constant.

Arnold *et al.*[58] developed a model to describe the adsorption in a stirred tank. The assumpions used in analyzing the fixed bed were retained and pore diffusion was assumed to be rate controlling. An expression was developed that gives the time required to reach a given bulk liquid concentration.

$$\frac{D_i(1-\alpha)t}{R^2\alpha} = \frac{1}{3}\ln\left|\frac{a^3+w^3}{a^3+1}\right| + \frac{1}{6a}\ln\left\{\frac{(a^2-a+1)(a+w)^2}{(a^2-aw+w^2)(a+1)^2}\right\}$$
$$+\frac{1}{a3^{1/2}}\left\{\tan^{-1}\left(\frac{2-a}{a3^{1/2}}\right) - \tan^{-1}\left(\frac{2w-a}{a3^{1/2}}\right)\right\} \quad (4.12)$$

where

$$\begin{aligned} a &= (\gamma-1)^{1/3} & \gamma > 1 \\ &= 0 & \gamma = 1 \\ &= -(1-\gamma)^{1/3} & \gamma < 1 \end{aligned}$$

$$w = \left\{1-\gamma\left(1-\frac{c}{c_0}\right)\right\}^{1/3} = \left(1-\frac{q}{q_0}\right)^{1/3}$$

$$\gamma = \frac{\alpha c_0}{q_0 \rho_b}$$

α = batch void fraction.

The model includes the parameters used for the fixed-bed analysis and defines a new void fraction, α. Again, the assumption that pore diffusion is limiting may not be valid for porous beads depending on the kinetics of the immobilization process and the chemistry used. For example, there may be a large fraction of ligands attached near the outer surface of the beads. Eveleigh and Levy[11] estimated that 80% of the ligands in their Sepharose™-based immunoadsorbent were 'readily accessible' as determined by kinetic analysis of both adsorption and desorption. In this case, fluid film mass transport may need to be considered as the rate-limiting step and traditional mass-transport correlations may be used.

4.3 Summary

Affinity adsorption has grown in prominence as a purification technique over the last 25 years and will probably continue to grow. Our ability to create better ligands, supports, and immobilization techniques is increasing. This development can become a boon to purification in general. Hopefully it will make ligands less expensive and therefore biological products less expensive.

References

1. Tarnowski, S.J. and Liptak, R.A. Jr. (1983). In *Advances in Biotechnological Processes, Vol. 2* Alan R. Liss, New York, pp. 271–287.
2. Kagel, R.A., Kagel, G.W. and Garg, V.K. (1989). *BioChromatogr.*, **4,** 246–252.
3. Pharmacia. (1984). *Gel Filtration: Theory and Practice*; Pharmacia Fine Chemicals, Uppsala, Sweden.
4. Porath, J., Janson, J.-C. and Laas, T. (1971). *J. Chromatogr.*, **60,** 167–177.
5. Nishikawa, A.H. (1983). In *Solid Phase Biochemistry, Vol. 66* (Ed. W.H. Scouten); John Wiley and Sons, New York, pp. 34–35.
6. Chase, H.A. (1983). *Chemical Engineering Science*, **39**, 1099–1125.
7. Haller, W. (1983). In *Solid Phase Biochemistry, Vol. 66* (Ed. W.H. Scouten); John Wiley, New York, pp. 535–597.
8. Regnier, F.E. and Noel, R. (1976). *J. Chromatogr. Sci.*, **14,** 316–320.
9. Coupek, J. (1982). In *Affinity Chromatography and Related Techniques* (Eds T.C.J. Gribnau, J. Visser and R.J.F. Nivard), Elsevier, Amsterdam, pp. 165–179.
10. Yarmush, M.L. and Colton, C.K. (1985). In *Comprehensive Biotechnology: the Principles of Biotechnology*. Pergamon, London.
11. Eveleigh, J.W. and Levy, D.E. (1977). *Solid-Phase Biochem.*, **2,** 45–78.
12. Goldberg, B., Nochumson, S. and Sloshberg, S. (1989). *Am. Biotechnol. Lab.*, **Nov/Dec,** 27–31.
13. Olson, W.C., Leung, S.K. and Yarmush, M.L. (1991). *BioTechnol.*, **7,** 369–373.
14. Regnier, F.E. (1991). *Nature*, **350,** 634–635.
15. Wilchek, M, Miron, T. and Kohn, J. (1984). In *Methods in Enzymology; Vol. 104*, Enzyme purification and related techniques (Ed. W.B. Jakoby), Harcourt Brace Jovanovich, New York, pp. 3–55.
16. Hiatt, A., Cafferkey, R. and Bowdish, K. (1989). *Nature*, **342,** 76–78.
17. Boss, M.A., Kenten, J.H., Wood, C.R. and Emtage, J.S. (1984). *Nucleic Acids Res.*, **12,** 3791–3806.
18. Better, M., Chang, C.P., Robinson, R.R. and Horwitz, A.H. (1988). *Science*, **240,** 1041–1043.
19. Axen, R., Porath, J. and Ernback, S. (1967). *Nature*, **214,** 1302–1304.
20. Kohn, J. and Wilchek, M. (1981). *Anal. Biochem.*, **115,** 375–382.
21. Tesser, G.I., Fisch, H.-U. and Schwyzer, R. (1972). *FEBS Lett.*, **23,** 56–58.
22. Tesser, G.I., Fisch, H.-U. and Schwyzer, R. (1974). *Helv. Chim. Act.*, **57,** 1718–1730.
23. Wilchek, M., Oka, T. and Topper, Y.J. (1975). *Proc. Nat. Acad. Sci. USA*, **72,** 1055–1058.
24. Nilsson, K. and Mosbach, K. (1981). *Biochem. Biophys. Res. Comm.*, **102,** 449–457.
25. Dean, P.D.G., Johnson, W.S. and Middle, F.A. (1985). *Affinity Chromatography: A Practical Approach*. IRL Press, Oxford.
26. Narayanan, S.R. and Crane, L.J. (1990). *Trends in Biotechnol.*, **8,** 12–16.
27. Cuatrecasas, P. and Parikh, I. (1972). *Biochem.*, **11,** 2291–2299.
28. Wilchek, M. and Miron, T. (1987). *Biochem.*, **26,** 2155–2161.
29. Coleman, P.L., Walker, M.M., Milbrath, D.S., Stauffer, D.M., Rasmussen, J.K., Krepski, L.R. *et al.* (1990). *J. Chromatogr.*, **512,** 345–363.
30. Chase, H.A. (1984). *J. Biotechnol.*, **1,** 67–80.
31. Sisson, T.H. and Castor, C.W. (1990). *J. Immunol. Meth.*, **127,** 215–220.
32. Hasemann, C.A. and Capra, J.D. (1989). In *Fundamental Immunology* (Ed. W.E. Paul), Raven, New York.
33. Comfort, A.R., Mullon, C.J.-P. and Langer, R. (1989). *Biotechnol. Bioeng.*, **32,** 554–563.
34. Schmer, G., Rastelli, N.L., Dennis, M.B., Detter, J.C., von Sengbusch, G. and Lehman, H.D. (1982). *Trans. Am. Soc. Artif. Org.*, **23,** 374–377.
35. Antonsen, K.P. (1989). PhD Thesis, Massachusetts Institute of Technology.
36. Office of Biologics Research and Review. (1985). *Points to Consider in the Production and Testing of New Drugs and Biologicals Produced by Recombinant DNA Technology*. Office of Biologics Research and Review.
37. Wilchek, M. (1973). *FEBS Lett.*, **33,** 70–72.
38. Gribnau, T.C.J. and Tesser, G.I. (1974). *Experientia*, **30,** 1228–1230.
39. Lasch, J. (1975). *Experientia*, **31,** 1125–1126.
40. Lasch, J. and Koelsch, R. (1978). *Eur. J. Biochem.*, **82,** 181–186.
41. Koch-Schmidt, A.-C. and Mosbach, K. (1977). *Biochem.*, **16,** 2105–2109.
42. Schnapp, J. and Shalitin, Y. (1976). *Biochem. Biophys. Res. Comm.*, **70,** 8–14.

43. Wilchek, M. and Miron, T. (1974). *Molec. Cell. Biochem.*, **4,** 181–187.
44. McManus, D.E. and Randerson, D.H. (1986). *Chem. Eng. J.*, **32,** B19–B28.
45. Pfeiffer, N.E., Wylie, D.E. and Schuster, S.M. (1987). *J. Immunol. Meth.*, **97,** 1–9.
46. Matson, R.S. and Little, M.C. (1988). *J. Chromatogr.*, **458,** 67–77.
47. Royer, G.P. and Uy, R. (1973). *J. Biol. Chem.*, **248,** 2627–2629.
48. Dennis, K.E., Clark, D.S., Bailey, J.E. *et al.* (1984). *Biotechnol. Bioeng.*, **26,** 892–900.
49. Hornby, W.E. (1968). *Biochem. J.*, **107,** 669–674.
50. Langmuir, I. (1918). *J. Am. Chem. Soc.*, **40,** 1361–1403.
51. Freundlich, H. (1926). *Colloid and Capillary Chemistry*. Dutton, New York.
52. Sips, R. (1948). *J. Chem. Phys.*, **16,** 490–495.
53. Steward, M.W. and Steensgard, J. (1982). *Antibody Affinity: Thermodynamic Aspects and Biological Significance*. CRC Press, Boca Raton, Florida.
54. Hill, A.V. (1913). *Biochem. J.*, **7,** 471–480.
55. Zubay, G. (1988). *Biochemistry*; 2nd ed. Macmillan, New York.
56. Olson, W.C., Colton, C.K. and Yarmush, M.L. (1991). *J. Membr. Sci.*, **56,** 247–278.
57. Yarmush, D., Personal communication.
58. Arnold, F.H., Blanch, H.W. and Wilke, C.R. (1985). *Chem. Eng. J.*, **30,** B9–B23.
59. Gribnau, T.C.J., Stumm, C. and Tesser, G.I. (1975). *FEBS Lett.*, **57,** 301–303.
60. Lasch, J. and Kuhnau, R. (1986). *Enzyme Microb. Technol.*, **8,** 115–119.
61. Dean P.D.G., Brown, P., Leyland, M.J., Watson, D.H. *et al.* (1977). *J. Biochem. Soc. Trans.*, **5,** 1111–1113.
62. Sandler, S.I. (1989). *Chemical and Engineering Thermodynamics*. John Wiley, New York.
63. Nosoh, Y. and Sekiguchi, T. (1990). *Trends Biotechnol.*, **8,** 16–20.

5 Membrane separations in downstream processing

W. EYKAMP

5.1 Background and need

In the living world, membranes keep things separated. They pass materials selectively. Discrimination is often very acute, permitting substances having very similar composition and structure to be either passed or excluded. They facilitate active separations with exquisite precision. The facility and ease with which very difficult selection in the passage of components is realized is testimony to the beneficial effects of millennia of natural selection.

Artificial separation membranes are new. First prepared in the 19th century, membranes useful for separations came into widespread use only in the 1970s. Compared with biological membranes, today's artificial separation membranes are crude but compared with the membranes of commerce of only 20 years ago, they are quite impressive. Today, the technology is sufficiently advanced that a large-scale industrial membrane process should be considered in every case where a membrane separation makes sense. The early days of many trials and frequent errors are now part of the membrane industry's history.

Membranes meet many downstream separation needs in pharmaceutical and biological processes. The streams treated successfully are usually aqueous, they are rarely at extremes of temperature, pH, or oxidation potential and they often contain mixtures easily separated by molecular size or charge. Particle sizes and types of filtration are shown in Figure 5.1. Membrane processing is evolving and the economics of the separation processes using membranes have steadily improved. There is now a wide selection of configurations, materials and skilled practitioners eager to fit their wares to new applications. Separations need not be only for recovery of prime products; membranes are also used to purify raw materials and to treat by-products and wastes.

Downstream separations needs may exceed the capabilities of conventional membrane processing. Often there is a requirement to pluck a specific molecule, even a specific enantiomer from a complex mixture in which the wanted component is a minor constituent. Conventional membranes cannot achieve exotic separations but some specialty membranes can.

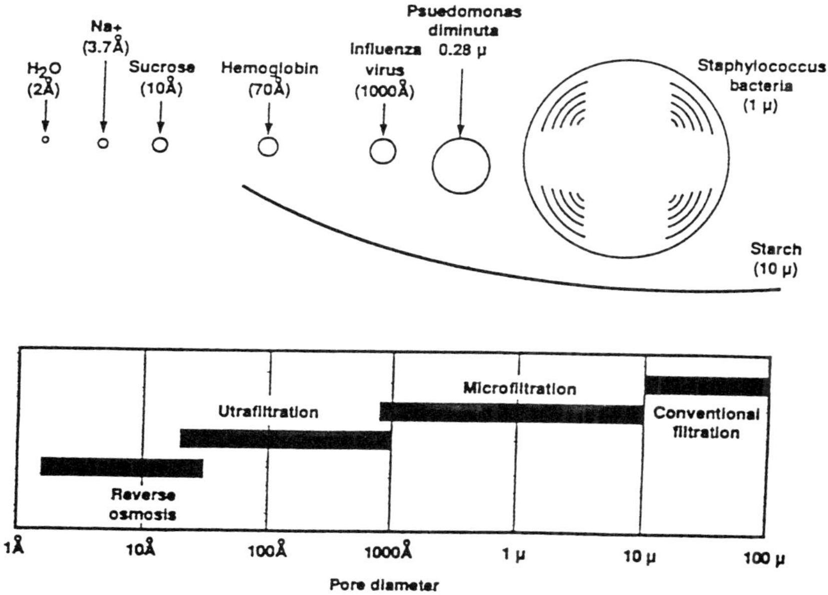

Figure 5.1 Sizes of things.

5.2 What membranes are

Membrane processes on a large scale developed after most technical personnel received their formal training Microfiltration membranes made as replacements for, and improvements over conventional dead-end flow filters have been available for limited applications since the 1930s but the total use of such membranes was quite small as late as the 1960s. Process applications for treating large volumes of liquids started on a very limited basis at the end of the 1960s. The general recognition of membrane separation as a reliable process tool occurred in one industry after another, and has yet to happen in a few.

Membrane separation processes use perm selective membranes, which are man-made structures fabricated to be selective in the passage of components. Membrane effective pore size covers a huge range. At the small-size end, gas membranes separate oxygen from nitrogen, carbon dioxide from methane and separate many other gas mixtures on the basis of size, solubility, or diffusivity. At the other extreme, microfiltration membranes are capable of passing large macromolecules while retaining bacteria. As a practical upper limit, the term membrane is not applied to devices to separate particles larger than 10 μm. Ion-selective membranes used in electrodialysis, and gas-separation membranes are essentially non-porous in the normal sense, the former function by the passage of ions and the latter by dissolution of the gas in the membrane material. Reverse-osmosis membranes are generally considered to be non-porous but there is controversy on that issue. Strathmann writes[1]

> In the most general sense, a synthetic membrane is a barrier that separates two phases and restricts the transport of various chemical species in a rather specific manner. A membrane can be homogeneous or heterogeneous, symmetric or asymmetric in structure; it may be solid or liquid; it may be neutral, may carry positive or negative charges, or may be bipolar. ... The electrical resistance may vary from several megohms to a fraction of an ohm. Mass transport through a membrane may be caused by convection or by diffusion of individual molecules, induced by an electric field, or a concentration, pressure, or temperature gradient.

5.3 Membrane history

The first membrane processes still commercially practised was dialysis. Then came microfiltration. The story began about 150 years ago but the important developments are quite recent.

Microfiltration (MF) grew out of the discovery of nitrocellulose in 1846. Cellulose nitrate membranes were reported by Frick in 1855. Early cellulose nitrate membranes were prepared by dipping a test tube in a collodion solution.[2] Surprisingly, some of these ancient materials are still used today.

Membrane development continued for decades, mostly in Germany. In 1906, Bechhold published means to vary pore size by varying polymer concentration. In 1918 Zsigmondy and Bachmann published means to make small quantities of 'parchment-like' MF membrane, including a description of how to vary pore size. Zsigmondy, in 1922, got a US patent on his invention, which was made by exposing a thin coating of a nitrocellulose solution to humid air.[3] Sartorius started producing membranes commercially in 1927.[4] Commercial development was very slow until shortly after World War II.

The discovery that propelled MF into large-scale commercial production was one attributed to Gertrude Mueller at the Hygiene Institute, University of Hamburg, that the micro flora from a large volume of water could be deposited intact on a small disk of microfiltration membrane. By culturing the membrane and counting the colonies, rapid and accurate determinations of the safety of drinking water could be made. The discovery was particularly important in post-war Germany, where much of the civil infrastructure was damaged.[5]

Microfiltration technology was investigated by the US Joint Intelligence Objectives Agency in their post-war assessment of German science and technology. The leading investigator was Alexander Goetz, who later elaborated on his findings and made improvements under contract with the US Army.[6]

In 1950, the Lovell Chemical Company won a contract to develop Goetz's work further. A series of reports ranging from November 1950 through April 1954 chronicles the difficult transition of MF membranes from the laboratory to semicommercial production. The report of September 1951 is less interesting because it despairs of ever making large sheets of MF membrane. Glass plates, it says, are 'too plastic' in large sizes to be kept flat. Membranes require a 'precisely flat surface'. Glass is subject to 'cold flow' and that

> Controls of great precision and nicety are so prerequisite to the making of a satisfactory MF that we strongly urge upon the Chemical Corps the concept that microfilters should not (and cannot) be a cheap item of manufacture comparable to filter paper. The day that expensive controls are abandoned for even normally careful standard production, the MF functional characteristics will be wholly undependable and the output will be rendered worthless as a reliable quantitative bacteriological tool.

The report was signed by John R. Bush, who later bought the technology from Lovell and founded Millipore Corporation, now the world leader in microfiltration.

Reverse osmosis (RO) and ultrafiltration (UF) came much later in time. Curiously, neither developed from microfiltration. Attempts to make UF membranes at microfiltration firms were unfruitful. UF is derived from RO in almost all important respects. The difference in the route of development of MF and UF continues to divide the two fields to this day; firms outstanding in one field have little presence in the other, even though the two disciplines are now closely related technically.

The major problem that prevented microfiltration membrane technology from being extended to smaller and smaller pores was the one of throughput. As pore size decreases, so does the amount of fluid that may be pushed through the membrane. The problem may be illustrated by considering the equation for flow through one cylindrical pore in a membrane.

$$q = \frac{\pi \Delta P d_{\mathrm{i}}^4}{128 \mu t} \tag{5.1}$$

where q is the volumetric flow rate through the pore (in m^3/s), ΔP is the pressure drop (in pascals), d is the diameter (in m), μ is the viscosity (in ns/m^2), π is the ratio of the circumference of a circle to its radius, and t is the membrane thickness (in m).

The output of a membrane, referred to as its flux, is the sum of the outputs of the individual pores in a given area. Flux is commonly reported in terms of volume per membrane area and time, such as in units of l/m^2-h or gal/ft^2-day. It is in fact a velocity (m/s) but is measured over the *entire* membrane area, not just the pores. So since the pores can make up only a small portion of the membrane surface, this fact must be kept in mind. So for a membrane:

$$J = q\frac{N}{A} \quad (5.2)$$

where N is the number of pores in area A m^2. From geometry, the number of uniform circular pores that will fit in a square is proportional to the inverse square of pore diameter.

$$\frac{N}{A} \propto d^{-2} \quad (5.3)$$

So, for constant pressure drop, viscosity and thickness, the flux of a membrane is proportional to the pore diameter squared.

The pore size suitable for UF is of magnitude 10 nm; for MF; it is of magnitude 200 nm. Thus, a microfiltration membrane with high throughput will, if made as a UF membrane, have a throughput less than 1 percent of the MF value.

The UF throughput problem was trivial compared with that for Reverse Osmosis. Insofar as it is reasonable to speak of 'pores' in an RO membrane, they would be smaller than those in a UF membrane by another order of magnitude. The problem was solved for RO with the invention of the skinned membrane (discussed later). The solution was quickly copied for UF, for it solved the UF low-flux problem. Thus, the RO membrane, not the MF membrane, is the direct predecessor of the UF membrane.[7]

The RO membrane breakthrough occurred in 1959, with the invention at UCLA of the Loeb–Sourirajan membrane. Working with cellulose acetate, which Reid[8] had reported as having superior salt rejection properties, the inventors found a way to make an asymmetric membrane — a very thin skin integrally attached to an otherwise porous backing. In effect, they found a way to make the factor t in equation (5.1) very small with no negative impact on mechanical toughness or durability. High flux membranes thereby became possible.

While the group at UCLA pursued RO membranes as their principal thrust, a group under Alan Michaels, first at MIT, later at Amicon Corporation went after ultrafiltration membranes. Although many materials were tried for UF membranes, grades of cellulose acetate similar to those useful in RO produced membranes with fairly good properties as ultrafilters. Since cellulosic polymers have chemical properties that limit their industrial suitability, Dorr-Oliver began a search for other polymers from which asymmetric UF membranes could be prepared — initially by sponsoring research at Amicon. The decade beginning in 1965 was one of explosive development activity in UF, both in membranes and in processes. Cellulose acetate membranes were soon displaced by a variety of non-cellulosic membranes invented independently by several investigators within a short time. Polyacrylonitrile UF membranes were discovered in a failed attempt to prepare an RO membrane, as described in a 1965 report. Polysulfone membranes may have had a similar origin. Polyvinylidene fluoride membranes were first made by accident.[9]

5.4 Membrane taxonomy

Based on the rather broad-membrane industry definitions, the following is a membrane taxonomy:

1. Membranes derived from physical solids:
   (a) Ceramics;
   (b) Sintered metal;
   (c) Sintered polymers;
   (d) Wire or fiber wound.
2. Membranes derived from homogeneous solid films:
   (a) Track-etched membranes;
   (b) Stretched polymers;
   (c) Aluminum derivatives;
   (d) Dense films.
3. Membranes derived from heterogeneous solid films:
   (a) Leached glasses;
   (b) Extracted polymers.
4. Symmetric membranes derived from solution:
   (a) Leached membranes;
   (b) Thermally inverted solutions.
5. Asymmetric structures derived from solution:
   (a) Loeb–Sourirajan membranes.
6. Asymmetric composite structures:
   (a) Dynamic membranes;
   (b) Thin-film composites;
   (c) Coated structures;
   (d) Self-assembled structures.
7. Liquid membranes:
   (a) Imbibed in capillaries;
   (b) Emulsions.

5.4.1 How they are made

Membranes may be made from micro particles by sintering or agglomeration. The pores are formed from the interstices between the solid particles. The simplest of this class of membrane is formed by sintering metal, metal oxide, graphite, ceramic or polymer.[10] Sintered membranes are used for MF and can be made to retain colloids with particle size of 0.1 μ. Silver, tungsten, stainless steel, glass, several ceramics and other materials are made into commercial membranes. Sintered metal may be coated by titanium oxide (TiO_2) or zicronium oxide to produce MF and UF membranes. Membranes may be made by the careful winding of microfibers or wires.

Porous media illustrate the issue of what is and what is not a membrane. Some media, such as diatomaceous earth deposited on a screen, serve as functional microfilters for larger particles. Diatomaceous earth filter media are not considered to be membranes under any definition. Dynamically formed membranes are generally regarded as true membranes. They are formed by depositing a material — hydrous zirconium oxide is the most common — onto a porous substrate in cross-flow, the normal means of operating a membrane in which the velocity across the membrane is far greater than the velocity through it. Why is the dynamically formed structure considered a membrane and the DE filter is not? The dynamically formed membrane does not use the forming material as a filter aid and, once the membrane is formed, the filtration operation proceeds, usually in cross-flow, very much the way any membrane operation is conducted. The fact that the membrane is deposited in cross-flow rather than in dead-end flow is a further distinction. Moreover, the ability to form dynamic membranes that have reasonable properties such as reverse-osmosis membranes confirmed the view that these devices were to be considered as membranes. In the final analysis, the definitions are a consensus of early workers in the field, and they are arbitrary.

Membranes derived from microparticles may be uncharged or charged, symmetric or asymmetric. The issue of charge is dominated by choice of material and by the usual rules of colloid chemistry. As illustrated in the case of ceramic membranes, surface charge may be a function of operating condition. Many micro-particle-derived membranes are symmetric; others have several orders of magnitude difference in effective pore size from the membrane surface to the support structure. Asymmetric porous membranes of this type are constructed of layers of particles of ever-decreasing diameter, each applied and stabilized in turn, and each acting as the support for the next finer layer. This is characteristically true in the case of ceramic membranes.

Membranes derived from microporous media are generally used for MF. Attempts to decrease pore size down to the ultrafiltration range are achieving some success, and membranes exhibiting reasonable UF properties are now made from alpha and gamma alumina, zicronia and other vitreous materials.

Membranes derived from homogeneous solid films are important in microfiltration. These are structures that contain pores or are a matrix whose openings are fixed. Given the incredible diversity of products fitting in this category, the sizes of the openings from one face to the other in this class of membrane are more constant than variable.

Stretched polymers form a major part of this class. Semicrystalline polymers, if stretched perpendicular to the axis of crystalline orientation, may fracture in such a way as to make reproducible micro channels. Best known are Goretex® produced from Teflon®, and Cellguard® produced from polyolefin. Stretched polymers have unusually large fractions of open space, giving them very high fluxes in certain applications, the microfiltration of gases, for example. Track-etched membranes have cylindrical pores. Originally, such membranes were made on a very small scale from mica. The membranes of commerce are made from polymers. A thin polymer film is first exposed to a collimated beam of radiation strong enough to break chemical bonds in the polymer chains. The film is then etched in a bath, which selectively attacks the damaged polymer. Successful implementation of the technique produces a film with photogenic pores, whose diameter may be varied by the intensity of the etching step. Commercially available membranes have a narrow pore-size distribution and are reportedly resistant to plugging. The membranes have low flux because it is possible to achieve high pore density without sacrificing uniformity of diameter. Using track-etched membranes, it is possible to prepare stunning photomicrographs of objects sitting on a well-defined membrane surface, and they are often pictured in that role.

A newer membrane form is prepared from a monolithic aluminum foil by electrolytic oxidation.[11] Asymmetric membranes with very uniform pores at a high area fraction are manufactured through careful control of the electrochemical formation process. The resulting films are unusual in many ways — they may be inorganic, asymmetric, high-flux and brittle.

Heterogeneous solid films may be extracted to form porous membranes with microfiltration properties. The most common are polymers extruded with high loadings of mineral or oil fillers, which are subsequently leached out. The common application for such materials is as battery separators but some are employed as membranes. Inorganic glasses may be extracted selectively to produce porous structures having a spectrum of pore sizes. Metals may be made into membranes by selectively dissolving one phase. Co-extrusion of two polymers followed by extraction of one is another technique.[12]

Symmetrical phase-inversion membranes are the most important commercial membranes produced today. They remain the mainstay of the microfiltration industry. These are two major variants in the method. The first, and most significant process involves preparing a concentrated solution of a polymer in a solvent. The solution is spread into a thin film, then precipitated through the addition of a non-solvent, usually water, sometimes from the vapor phase. The technique is impressively versatile, capable of producing fairly uniform membranes whose pore size may be varied within broad limits.[13]

The second process is thermal precipitation. A solution of polymer in poor solvent is prepared at elevated temperature. A sudden drop in solution temperature causes the polymer to precipitate. The solvent is then washed out. Membranes may be spun at high rates using thermal phase inversion. Asymmetric membranes derived from solution form the most important class of reverse-osmosis and ultrafiltration membranes; they are also important in microfiltration. These membranes, often referred to as 'skinned' membranes, divide two necessary functions of a membrane, allowing each to be optimized. First, there is the separating layer or skin, where separation is achieved, using a high concentration of uniform pores. Since the separation process is achieved at the surface, and resistance to flow through a pore is proportional to the pore length, the universal rule is 'thinner is better'.

The second functional part of the membrane is the support. Its job is to provide mechanical support for the skin, and to make the membrane able to withstand handling and processing. Desirable characteristics include minimal resistance to flow, adequate resistance to compression in service and chemical inertness at least equal to the skin. It must also resist plugging by any particle able to pass through the skin.

The discovery of the asymmetric membrane by Loeb and Sourirajan was a major breakthrough in membrane technology, and the techniques in use today are derived from their early work. In its simplest form, a polymer is dissolved at about 20% solids in a water-miscible solvent. The polymer solution is cast on a plate to form a thin film, which is then quenched in water. In normal practice, the skin forms on top and, immediately below it, the polymer forms a much more open porous support layer. Almost all commercial membranes are cast on fabric (sometimes called 'casting paper'), a non-woven polyester or polyolefin material. The membrane substrate must bond well to the fabric, which becomes a permanent part of the membrane structure. The skin is a small fraction of the finished membrane thickness.

Asymmetric *composite* structures are sometimes used for UF and MF. Lately, they have become the dominant type in reverse osmosis.

The oldest composite is the dynamically formed membrane mentioned above. Dynamic membranes do not require a membrane for a substrate. Microporous structures are commonly employed, as are microfiltration membranes. Hydrous zirconium oxide is the favorite material but others are cited in the literature. The major commercial embodiment of the technology uses porous carbon tubes for the substrate.

Thin composite membranes are relatively new and, as improvements and manufacturing experience accumulate, they are becoming the most important commercial membranes in reverse osmosis. The most successful of many membranes attempted is an interfacially polymerized fully aromatic polyamide formed *in situ* over a polymeric ultrafiltration membrane.[14]

A few UF membranes are prepared by coating a previously prepared organic membrane with a topcoat. Extra uniform pore size distribution is one goal.[15]

Self-assembled membranes are made from the natural membranes found on certain types of anachobacteria, primitive organisms recently discovered in very hot thermal vents. Microporous membranes may be coated with self-assembling fragments from these very unusual bacteria to form extremely uniform pore-size distribution membranes. The bacteria grow in an extremely aggressive chemical environment, and the assembled membranes maintain excellent chemical resistance.[16]

Liquid membranes are generally prepared from chemically active materials that do not exist in a solid phase. They are either emulsified using carefully chosen surfactants, or immobilized in a porous structure, such as a polymer or a ceramic. Although much studied, they are not widely used commercially.

5.4.2 *Membrane ratings*

Membranes are rated by the rate at which they produce permeate (*flux*), and the ability to discriminate between things they retain and things they pass. Three different words are used to describe what is held back: (i) retention; (ii) rejection; and (iii) reflection. For purposes of downstream processing, they are, practically speaking, synonymous. Reflection has a special meaning in certain exotic applications but the choice between 'rejection' and 'retention' is largely dependent on the author's background.

There is an arbitrary element in the definition. By convention, retention is defined as:

$$R_i \equiv 1 - \frac{c_i \text{ (permeate)}}{c_i \text{ (feed)}} \tag{5.4}$$

where c is the concentration (weight, volume, conductivity, etc.) of the ith species. The arbitrary element comes from the convention that concentration is measured in the *bulk* of the feed, well distant from the membrane. Physically, the concentration that matters is that at the membrane surface, and that can be quite different. The convention makes it much easier to perform calculations about important things like yield but it can be confusing for the unwary.

5.4.2.1 Reverse osmosis and nanofiltration. Reverse-osmosis membranes are usually rated by 'salt rejection', since their dominant use is for desalting water for potable or industrial use. Sodium ion is the usual concern, and membranes are normally tested using NaCl. A possible complication is the common practice of measuring RO membrane performance by measuring conductivity of the permeate. RO membranes commonly pass H^+ with far more alacrity than other cations. Since conductivity is influenced by H^+, the apparent retention becomes a function of pH. Similarly, membranes containing charge change the pH of water passing through them.

Nanofiltration is a new process, once referred to as 'loose RO'. Nanofiltration membranes usually retain larger polyvalent ions, but pass much of the smaller monovalent ions. This property is advantageous in some applications. The membranes are tested using many markers, usually including NaCl and $MgSO_4$.

5.4.2.2 Microfiltration. Membranes are rated by markers or pore size. (Markers are objects or molecules used to test membranes for their effective retention.) Only microfiltration membranes are easily tested by direct examination, as their pores can be observed by electron microscopy. Since the number of pores that may be observed directly by microscopy is so small, microscopic pore-size determination is useful mainly for membrane research and verification of other pore-size-determining methods.

Large areas of microfiltration membrane can be tested and verified by a bubble test. Pores of the membrane are filled with liquid, then a gas is forced against the face of the membrane. The Young–Laplace equation relates the pressure required to force a bubble through a pore to its radius and the interfacial surface tension between the penetrating gas and the liquid in the membrane pore.

$$\Delta P = \frac{2\gamma \cos \Theta}{r} \tag{5.5}$$

γ is the surface tension (in N/m) r is the pore radius (in m), and P is transmembrane pressure (in Pa). Θ is the liquid–solid contact angle. For a fluid wetting the membrane perfectly, $\cos \Theta = 1$. By raising the gas pressure on a wet membrane until the first bubble appears, the largest pore may be identified, and its size computed. This is a good test to run on a membrane apparatus used to sterilize a fluid, since bacteria larger than the identified largest pore (or leak) cannot readily penetrate the assembly. Pore-size distribution may also be run by bubble point but its most important function is to verify that all pores are smaller than a specific size. Bubble-point testing is particularly useful in assembled microfilters, since the membrane and all seals may be verified. Periodic testing ensures that the assembly retains its integrity.

Diffusional flow of gas is a complication in large microfiltration assemblies. It results from gas dissolving in pore liquid at the high-pressure side, and desorbing at the low-pressure side. It is so small that it is generally unnoticed in small-area membrane tests but it can be a perturbing factor in larger assemblies. If the number of pores and the average pore length are known, the effect can be computed. It is easily distinguished from gas flow at the bubble point, and special protocols are used to ensure that the apparatus meets the required level of bacterial reduction.

Membranes are further tested by challenge with microorganisms of known size; their ability to retain all the organisms is proof that all pores are smaller than the organism. The best known microorganism for pore-size determination is *Pseudomonas diminuta*, an asporogenous Gram-negative rod with a mean diameter of 0.3 μm. Membranes with pore size smaller than that are used to ensure sterility in many applications. In 1978, Leahy and Sullivan[17] provided details of validation procedures. Membranes may also be tested using latex particles.

There are several ASTM standard methods dealing with issues of membrane testing. The one describing the bubble point method is ASTM F316-86, 'Standard test method for pore size characteristics of membrane filters by bubble point and mean flow pore test'. It is a method for testing individual membrane disks, not an entire apparatus. The method includes corrections for diffusive flow and provides a good description of the technique. It also gives a method for determining pore-size distribution.

More detail on pore size values is found in the section on applications for sterile filtration (section 5.10.4). Fouling introduces further complications, described below following a discussion of ultrafiltration. Additional material is provided by Goel *et al.*[18]

5.4.2.3 Ultrafiltration. Ultrafiltration membranes cannot be tested by bubble point. The pores are too small, so other means are used. Direct microscopic observation of the surface is difficult and unreliable. Also, because the pores are very small, the pores usually close when samples are dried for the electron microscope. Equation (5.5) is of the same form as the expression for the force pulling a pore shut as it dries, and as r becomes very small, the force is enormous. Furthermore, its effect is greatest on the smallest pores. Critical-point drying of a membrane (replacing the water with a fluid that can be removed at its critical point) reduces γ to zero, and although not without complications of its own, it has been used to produce a *few* good pictures. They are scattered exceptions in many misleading photographs.

The best known method for UF membranes is molecular-weight cut-off. Unfortunately, it is widely misunderstood and has been the cause of much error.

Molecular-weight cut-off. The concept of molecular-weight cut-off is deceptively simple. Ultrafilters retain soluble molecules, so why not measure their porosity by seeing which molecules will pass through them?

The concept of a 'molecular-weight cut-off' was conceived and introduced into commerce by Amicon Corporation in the mid 1960s. Some of the complications arising from the name were foreseen, others were not. To persons totally unfamiliar with ultrafiltration, the name communicated a new concept. The convention set by Amicon in the early days, now generally but not universally followed, is to define molecular-weight cut-off as the *molecular weight of globular proteins that are 90% retained by the membrane.*

The first complication is that UF membranes have a distribution of pore sizes. In spite of decades of effort to narrow the distribution, most commercial membranes are not notably 'sharp'. The outstanding exception is the class of self-assembled membranes invented recently, and there are other 'sharp' membranes available. A graph of retention versus solute molecular weight is shown in Figure 5.2.

For membranes with a perfectly monodisperse pore-size distribution, there is the complication that membrane materials adsorb proteins, the adsorption

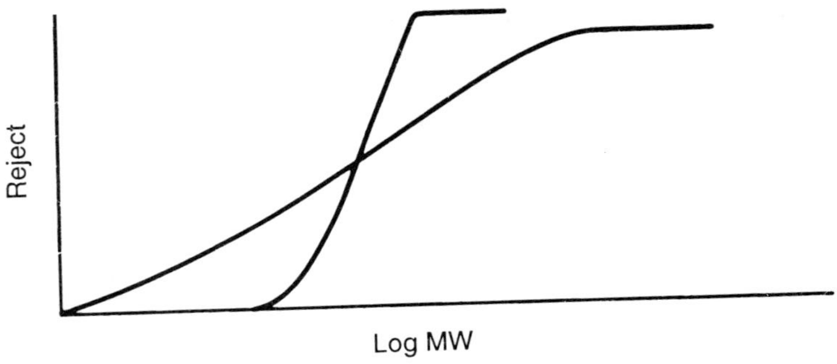

Figure 5.2 Retention (rejection) versus log of the molecular weight. The steep line reflects a sharper cut-of.

is material specific and is dependent on concentration, pH, ionic strength, temperature, etc. Adsorption has two consequences: (i) it changes the membrane pore size; and (ii) it removes protein from the permeate by adsorption in addition to that removed by 'sieving'. Porter[19] gives an old but illustrative table for adsorption of cytochrome C on materials used for ultrafiltration membranes, with values ranging from 1–25%.

For membranes with a normal pore-size distribution, the point where '90%' falls on the curve is perforce an approximation. The curves are often bending in a way that makes picking the '90%' point referred to in the molecular-weight cut-off definition arbitrary. What little is known about pore size distribution in commercial UF membranes suggests that the Poisson distribution is the most common; however, some, including ones of commercial importance, have distributions that are polydisperse!

The next level of complication, and of considerable confusion, is that of the marker. Markers for UF membranes are polymeric. Polymers of the same molecular weight can have very different molecular size. To further complicate the picture, molecular shape can change in the vicinity of a membrane. One well-known example is cited by Porter[19] who states that Dextram 250, a branched polysaccharide with molecular weight 250 kD passes through a 50 kD molecular-weight cut-off membrane. Linear molecules, such as polyacrylic acid, pass easily through membranes with molecular-weight cut-off values far below their molecular weight.

To get a reproducible number for molecular-weight cut-off, many factors must be held constant. To the producer of membranes, that is easy enough, for by standardizing the test, he can have some assurance that the same membrane is being made in successive lots. Such standardized testing is also useful to a user of membranes, who has some assurance that a constant material is being supplied.

To test the membrane using protein, in keeping with the definition of molecular-weight cut-off, it is necessary to keep the concentration in the feed very low to prevent polarization effects which result in 'auto filtration', the consequence of which would be to measure the boundary layer rather than the membrane itself. At low concentration, however, adsorption becomes important, perhaps dominant, in the retention measurement.

Owing to the severe difficulties in testing with globular proteins, most membrane manufacturers use surrogate probes. Materials selected are ones for which the complications are minimized, the probe is simple, fast and cheap to detect, does not readily biodegrade, and which gives results, whatever they are, that are reproducible. There is a good faith attempt to relate the findings back to globular protein but the frustrations of the protein test are sufficiently daunting to make the probe test the only test in many cases. And the convention adopted by manufacturer 'A' may well be quite different from that adopted by manufacturer 'B'.

Misunderstandings arise when membrane users assume that molecular-weight cut-off means what it says. The definition implies that a 50 kD membrane will separate a 25 kD material from a 75 kD material. Generally, it will not but there are special circumstances under which protein – protein separations are possible.

In an ideal world, membranes would contain a very high density of fully uniform cylindrical pores. It is perhaps instructive that, in spite of the very creative effort invested in 'sharp' membranes, their share of the overall membrane market is very small. Practically speaking, ordinary membranes have proven to be adequate for most separations.

Complications from fouling. Fouling is a major problem in all membrane operations. It causes significant problems in measuring and interpreting pore size in both microfiltration and ultrafiltration membranes. It has an effect on reverse-osmosis membranes as well, quite distinct from the effects in ultrafiltration and microfiltration.

Fouling is an irreversible process; it cannot be rectified by changing processing conditions such as flow or pressure. It can be reversed only by cleaning or replacing the membrane. It is often an adsorption involving significant binding energy.

Effects on pore dimension

Microfiltration and ultrafiltration membranes contain 'pores' and, for most membranes, they are not all of the same size. Fouling affects pores differently. Belfort[20] illustrated three cases affecting microfiltration membranes. A fourth case primarily affecting ultrafiltration, is added as shown in Figure 5.3.

In case A, adsorption causes all pores to become smaller, and may result in the smallest pores plugging. In the case of a protein probe present in dilute solution, this fouling error would cause the test to understate the size of all pores, and could truncate the distribution on the low pore-size end of the spectrum.

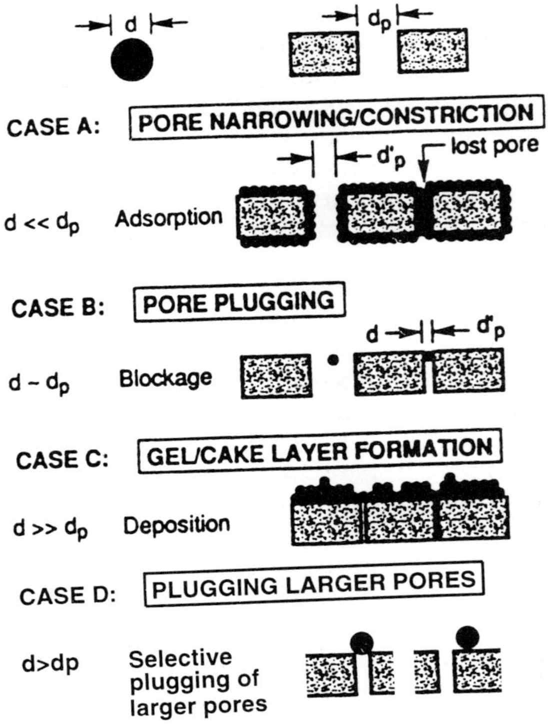

Figure 5.3 Fouling effects on pore dimensions (by courtesy of G. Belfort): Case A = adsorption; Case B = plugging smallest pores; Case C = gel/cake formation; Case D (added by author) = plugging larger pores.

Case B shows pore plugging. Any adsorption, means that particles may plug pores. In the example shown, smaller pores would be expected to suffer disproportionately. (Rember the drawing is an artistic simplification.) Pores are not created by precision drills in uniform materials. The entrance to a pore may be fairly sharp in a mica membrane exposed to radiation and etched by chemicals prepared by a careful chemist but, in the real world of membranes, pore openings are considerably less precise. Pore openings may be conical, they may have other shapes, and for many materials, 'pore' is a tenuous concept. Nonetheless, to some degree, the small openings are more susceptible to being plugged by small particles.

Case C represents the deposition of a material that supersedes the porous structure of the membrane. The membrane now becomes analogous to a precoat filter. Fouling is irreversible to the extent that the layer nearest the membrane is probably adsorbed onto it. Throughout the layer, the binding may or may not be irreversible. At the surface, quite a degree of dynamic reversibility remains. One may assume that in all cases, the effect is to shift the effective pore size downwards.

Case D is peculiar to membranes with small pores filtering particles much larger than the pores. An example is ultrafiltration membranes processing electrodeposition paint, all of whose particles are much larger than most membrane pores. The effect is to plug the large holes, leaving the smaller ones. It is case B in reverse. The effect is to reduce the average pore diameter by affecting only the high end of the distribution.

Case D is familiar to operators of industrial equipment who often find that macroscopic seal leaks are effectively plugged by the gross debris in a feed stream.

Effects on flux

Fouling affects flux dramatically. The pure-water flux through a virgin ultrafiltration membrane is commonly 10-fold greater than the water flux after the membrane has been exposed to protein.

Flow will be laminar through a cylindrical pore because of its size. To summarize equations (5.1, 5.2 and 5.3) flux through a given area containing N cylindrical pores each having diameter d:

$$J = \frac{\pi \Delta P}{128 \mu t} \sum_{0}^{\infty} \frac{N_i}{A} d_i^4 \tag{5.6}$$

When reviewing the four cases of pore narrowing, Case B, in which smaller pores are plugged and larger ones are unaffected will have the least impact on flux. Case A, narrowing of all pores and plugging of some of the smallest, will have a greater impact, because loss of some pore dimension is particularly important given the fourth power of diameter. Case C is a guess, as the porosity of a cake layer on the membrane can be anything. Case D results in a dramatic loss of throughput, because that form of plugging takes out the most productive pores.

Overall effect on retention

Passage of material through a pore obviously depends on how much is flowing and what that pore will pass. Big pores pass large quantities but their retention is different from smaller pores. As a membrane fouls, the retention characteristics worked out for the virgin membrane will change, often dramatically.

A few general rules apply. Fouling processes that plug only the very smallest pores have little effect on retention. In theory they would be expected to lower retention somewhat, as the most retentive pores are plugged leaving the larger less retentive ones but the contribution of the smallest pores to overall input is normally low anyway. When fouling lowers retention it is probably not because of this mechanism but because of the effect explained below under the heading 'Reverse osmosis and nanofiltration'.

Fouling by almost any other mechanism raises retention. Either it substitutes a cake layer on top of the membrane, or it narrows pores, or it selectively plugs larger ones. It is a poorly kept secret that many membranes do not do the job they are sold to do until they are a little bit fouled.

Reverse osmosis and nanofiltration

In membranes containing pores, fouling generally has the effect of making the pores smaller, thus raising the retention over what the membrane exhibits when it is unfouled. In reverse osmosis, the situation is quite different. A foulant layer at the surface of a reverse osmosis membrane has the dominant effect of lowering the diffusion of rejected solutes back into the feed. If the flux declines less rapidly than this impediment to back-diffusion, the concentration of retained salts at the surface of the membrane will rise. Thus, while the membrane seen in microcosm may exhibit constant retention, what we measure is the retention based on the mixed fluid distant from the membrane. So, by convention, according to practical experience, the retention of a fouled membrane declines. There may be an additional complication if the retention is measured by conductivity and the foulant is charged.

5.5 Large applications outside the field

In every field, there are successes, failures, and good ideas stuck in limbo. Membrane separation has its share of all three. It is like any other endeavor in the general reluctance to admit that many great ideas never come to fruition.

A few examples of areas related to the bioprocessing industry where membranes are significantly successful, provides perspective and a degree of insight into techniques and equipment that hold promise for newer biologically related processes.

5.5.1 *Hemodialysis*

Dialysis has a long history of use in a few industries, but the artificial kidney is by far the most pervasive and successful dialysis application ever. In fact, measured by either membrane area or by money, hemodialysis is more important than all other membrane applications *combined.* Hemodialysis began developing in the mid 1940s with hand-made experimental prototypes.

The manufacture of artificial kidney devices grew explosively in the 1970s. Within 20 years, the field progressed from an experimental one to a mature one, including a major 'shake-out' and a subsequent recovery. Companies in the artificial kidney membrane business make a variety of offerings but the dominant membrane design is a capillary having a 200 μm inside diameter. Cellulosic membranes are still the most common but are slowly giving ground to higher-performance materials. Newer designs feature greatly improved mass-transfer characteristics over the devices available 10 years ago.

Severe competition has lowered the market price dramatically. Ready-to-use sterilized cartridges can be purchased for US\$15/m^2, which is very much lower than the price of any other membrane assembly.

Kessler and Klein more recently published an excellent summary of dialysis membranes.[21]

5.5.2 *Wine microfiltration*

Most of the wine bottled for commercial sale passes through a membrane. Membrane microfiltration has virtually replaced heat and chemical treatment because it does not affect organoleptic properties. Some spoilage organisms will pass a 1 μm filter, so the preferred membrane pore rating is 0.45 μm.[22] These microfiltration membranes are run in dead-end flow, making them very easily plugged by the colloidal material present in wine. Pre-filtration is practised, occasionally as a built-in part of the membrane cartridge.

5.5.3 *Dairy industry applications*

Membranes are used for many large-scale applications in the dairy industry. One of the best illustrations is whey, the by-product of the manufacture of cheese from milk. Whey composition is a function of the type of cheese made but it is generally about 5% lactose, 1% ash (mainly sodium and calcium salts), 0.5% true protein, 0.2% non-protein nitrogen (NPN), and the balance water. pH is either 4.5 or 5.7, depending on the type of cheese. Roughly 9 kg of whey is produced for each 1 kg of cheese.

Reverse osmosis is often used to dewater whey in preparation for further processing. Good reverse-osmosis membranes produce permeate containing very little salt or lactose, giving a low BOD permeate. This water is often re-used in the dairy for washing or other non-critical applications, since it is a curious fact that dairies often lack a convenient source of good water. There is a natural limit to the concentration that may be achieved, since lactose starts to crystallize at concentrations above 25%. Owing to polarization (discussed in section 5.8), the practical limit is 20% total solids. Reverse osmosis competes with mechanical vapor recompression evaporation for this application: both are more economical than conventional evaporation.

Ultrafiltration is used to produce the high-value whey products, which range from 35% protein powder (a skim-milk replacement) to >80% protein products, used as high-value, high-functionality food ingredients. A large dairy ultrafilter operating on whey will contain 1800 m^2 of membrane, and have a whey intake of 1000 m^3/day.

Ultrafiltration selectively retains the proteins while passing everything else except fat. For this reason, residual fat must be separated before whey is ultrafiltered. The easiest way to conceptualize the ultrafiltration process is to recognize that the ultrafiltration membrane is practically 'transparent' to lactose, salts and NPN as well as to water. The concentration of these small molecules is not changed much by the membrane, but the concentration of protein on the feed side is. Everything is moved through the membrane except the protein, raising it relative to everything else. There are limits as to how high the protein concentration can rise without causing unacceptably low fluxes but the micro components can be washed out by diafiltration, introducing water to the protein rich feed when flux drops too low. Using diafiltration, the protein purity can be raised to almost any level desired.

In recent years, thin-channel membranes have become the standard for dairy applications. Spiral wound modules are reliable, available from multiple sources, and are relatively inexpensive. Properly made, they are cleanable, sanitizable, and composed of only FDA-approved materials.

5.5.4 *Juice filtration*

Apples are pressed to produce a juice containing pommace, fine solids and haze-producing components. Production of high-clarity juice requires de-pectinization and filtration. Conventional treatment is with pectinase, normally followed by diatomaceous earth filtration. Ultrafiltration membranes have largely displaced these diatomaceous earth filters. The colloidal material in the juice is easily separated by an ultrafiltration membrane, and any left over pectin is also captured. Tubular membranes are used because of the economic necessity of maximizing the juice yield. As the pommace content of the juice rises, it becomes liable to plug up the passages in a membrane system. Tubular membranes are less susceptible to plugging than thin-channel devices, so they are favored for this application. The usual way to increase solute yield, diafiltration, is inappropriate for apple juice, since it would dilute the product.

5.5.5 *Water filtration*

Membranes have numerous applications in the water field. The largest and most physical application is the production of potable water from saline sources. Brackish water desalination is the most important but sea water is also a major source of membrane-derived drinking water.

Process applications where the water product has very high standards are boiler feed, nuclear reactor coolant, pharmaceutical, and electronic microcircuit rinse water. New electronic chips have requirements for purity best described as fantastic in terms of absence of ions, microorganisms, and particulates. Ultra-high-purity water is produced by double passage through reverse-osmosis membranes, followed by non-membrane processes, followed then by ultrafiltration to remove any contaminants introduced downstream of the reverse-osmosis filter, followed by microfiltration at the point of use.[23]

Membranes are used in downstream water applications as well. Some waste treatment and water recovery from waste streams uses reverse osmosis. It is used for large-scale recovery of water from sewage for groundwater recharge at the Orange County Water District in California.[24] Very high purity boiler feed water is made from sewage, and even from sea water.

5.5.6 *Nanofiltration*

A new class of membranes, known for some time but with few applications are called *nanofiltration* membranes. The best current embodiment is made by techniques developed for the extremely high-rejection composite reverse-osmosis membranes. These membranes are used outside the downstream processing area for their ability to discriminate between monovalent and polyvalent cations. They show high retention of Ca^{2+} and Mg^{2+} but low retention of Na^{+}. They have high retentions for soluble organics and microorganisms, giving them use as water-softening and disinfection barriers in water systems. The advantage inherent in passing Na^{+} is that osmotic pressure is not raised and the process operates at lower pressure with higher flux. Naturally, the process is only applicable where Na^{+} levels do not require reduction.

5.6 Fluid management

The conventional way to operate a filter is called dead-end filtration. This is the familiar method used in everyday filtration (e.g. of water, or coffee or in a laboratory). When the filter medium is a membrane, it usually has the appearance of a piece of paper with a shiny coating on one side. The membrane may be in one of many different forms (flat sheet, pleated cartridge, capillary, tube, etc.) In dead-end membrane filtration, the fluid passes normal to and through the face of the membrane, and the particles or molecules retained by the membrane are held at its surface.

Dead-end filtration requires only the energy necessary to force the fluid through the filter. In the simplest applications, gravity, a laboratory vacuum or a simple pump provide enough motive force to drive the application at an acceptable rate. The ideal energy requirement, if rate is not critical, is negligibly low.

Another way is by cross-flow filtration. Here, the fluid to be filtered is pumped across the membrane parallel to its surface. Only the small fraction of fluid actually passing through the membrane flows normal to the filter. By maintaining a high velocity across the membrane, the retained material is swept off the membrane surface. Cross-flow filtration is preferred when significant quantities of material will be retained by the membrane, when the retentate is soluble, or when the solid retentate is compressible.

A principal difference in the operation of these two schemes is conversion per pass. In dead-end filtration, essentially all of the fluid entering the filter emerges as permeate, so the conversion is roughly 100% less the fluid entrained in the retained solids. For a cross-flow filter, far more of the feed passes by the membrane than passes through it, and conversion per pass is usually $<20\%$. Recycling permits the ultimate conversion to be much higher.

One important criterion of a feed stream is the level of solids that must be retained by the membrane. Streams containing high loadings of solids ($>0.5\%$) are likely to be processed by membrane filters operating in cross-flow. Feeds with medium loadings of solids ($<0.5\%$) are generally operated in dead-end flow. Fluids with low solids loadings ($<0.1\%$) are almost always filtered in dead-end flow, where the membrane acts as an absolute filter as fluid passes directly through it.

5.6.1 *Dead-end flow*

By far the largest use of dead-end membrane filters is in the quantitative removal of particulates in the size range 0.1–10 μm from aqueous liquids. This use is also likeliest to be of interest in downstream processing. The dominant design used for this purpose is the cartridge filter, usually pleated, sometimes fiber-wound.

Two immediate implications/limitations follow when cartridge filters are used: (i) the solids level is low; and (ii) the solids will not be recovered. A significant number of interesting applications fit within these restrictions, including purification, sterilization, and clarification.

In one form or another, the surface of the membrane used in dead-end filtration is often protected by a guard filter, which entrains most of the larger, easier-to-filter solids before they reach the membrane. The structure acts as a depth filter backed by a membrane filter. These devices may be quite sophisticated. Both pre-filter and membrane filter can be charged to give superior non-plugging characteristics. In some applications, these composite structural filters are comparable with the outstanding characteristics of asbestos filters.

A good source of information about dead-end microfiltration may be found in Goel *et al.*[25]

5.6.2 *Cross-flow*

Most membrane processes treating the products of biological activity will be run in cross-flow. It can be thought of as the industrial equivalent of a stirred cell filtration.

Cross-flow filtration as practised in industrial membrane applications derives from reverse osmosis. Its connection to conventional filtration is tenuous at best. Cross-flow processes grew out of the desalination of water, where, unlike conventional filtration, there was never any hope of producing a filter cake when starting with sea water. A reasonable goal was to remove a small fraction of the water, perhaps raising the salt content of the feed from 3.2% to 5% in the process. The resisting barrier to be overcome was not the dynamic passage of filtrate through a cake built up on a cloth but was the osmotic pressure of the salts dissolved in the feed. Reverse osmosis almost always operates in cross-flow. In general, that means fluid is moving parallel to the membrane *at least* 100 × its velocity component perpendicular to it. A comparison of cross-flow filtration and dead-end filtration is given in Figure 5.4.

Cross-flow is used to simulate stirring. The problem to be overcome arises from the fact that salt water is being convected to the membrane while only water (ideally) passes through it. The salt is left in the vicinity of the membrane, a condition called *concentration polarization.* To prevent several undesirable consequences, the polarization must be minimized. By pumping feed (sea water) past the membrane, concentration polarization may be reduced.

Diffusion helps reduce polarization, and for the small ions separated by reverse osmosis, diffusivities are relatively high. For larger entities, such as proteins and cells, unassisted diffusion is of little value.

Rate equations for cross-flow are very important. Their origin is in the observation that reverse osmosis is a steady-state process. Reverse osmosis membranes will operate at about the same rate for months without interruption. It is obvious that the membrane is operating at steady state since the few unsteady-state complications, including fouling and compaction of the membrane, arise slowly. It is clear that the concentration of the rejected salts near the membrane is time-invariant.

That salient feature is the basis of the process description. The fundamental concept is that salt is dragged towards the membrane with the water that passes through it. Salt is prevented from passing through the membrane (rejection). Since the concentration of salt in solution at a given point on the

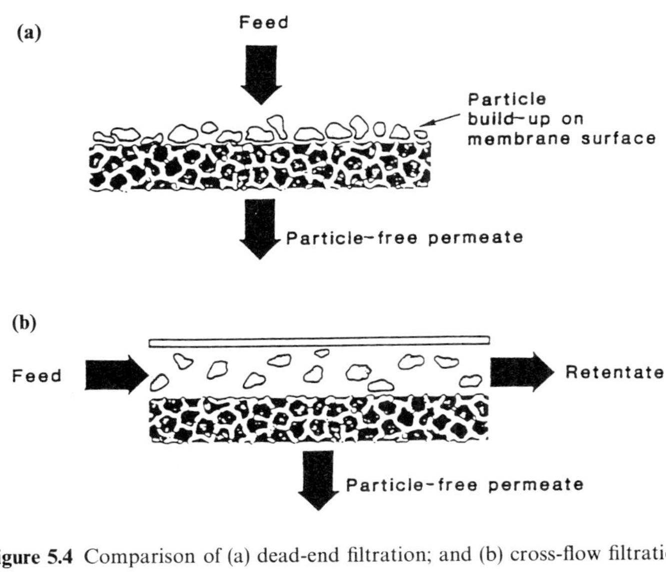

Figure 5.4 Comparison of (a) dead-end filtration; and (b) cross-flow filtration.

membrane is constant, the re-dispersion of salt into the feed must occur at precisely the rate at which salt arrives. By solving the re-dispersion kinetics, the equal-but-opposite kinetics of water passing through the membrane is solved. Movement of a solute along a concentration gradient is a much-studied engineering problem. Thus the approach is to apply the equations describing dispersion along a concentration gradient in a flowing system. There are three cases, turbulent flow, laminar flow, and no flow. The relevant dimensionless groups are:

$$Re = vd/\nu = HQ/\nu d; \quad \pi/4 \leqslant H \leqslant 1 \tag{5.7}$$

$$Sc = \nu/_D \tag{5.8}$$

$$Sh = kd/D \tag{5.9}$$

where Sh is the Sherwood number, Re is the Reynolds number, and Sc is the Schmidt number. D is diffusivity (in m^2/s), d is hydraulic diameter (in m), k is the mass-transfer coefficient (in m/s), v is the velocity in m/s, and ν is the kinematic viscosity in m^2/s. H is a constant, a function of the geometry of the conduit. It is $\pi/4$ for a circular conduit, and 1 for a square conduit. In all subsequent discussion, H will be taken as 1 as a convenient approximation.

Turbulent flow. Turbulent flow past the membrane (Reynolds number above about 2500) is by far the most common cross-flow case. While caution is appropriate in the range $2500 < Re < 10000$, the equations generally apply in that range. An equation variously attributed to Dittus-Boelter and Desalius is:

$$Sh = 0.023\ Re^{0.88} Sc^{0.33} \tag{5.10}$$

Solving for k:

$$k = 0.023 \frac{D^{0.67} V^{0.88}}{d^{0.12} \upsilon^{0.52}} \tag{5.11}$$

Since k is proportional to the rate at which solute leaves the membrane, it is also proportional to the rate at which water passes through the membrane. This simple but powerful fact takes a little getting used to but it follows inexorably from the logic of the steady-state process. This logic leads to the equation:

$$J \propto k \tag{5.12}$$

Substituting volumetric flow rate for velocity, since it is far easier to measure, the flux equation assumes a usable form. Incidentally, the values of the numerical coefficients in these equations are not particularly useful, since it is quite difficult to use them for *a priori* calculations of flux. For example, D

and v are rarely known. The relationship between flux and volumetric cross-flow flow rate is thus:

$$J = B\frac{Q^{0.88}D^{0.67}}{d^{0.88}v^{0.55}} \tag{5.13}$$

where B is an arbitrary constant. An important derivative of this equation, showing how flux is influenced by flow rate is:

$$\left(\frac{\partial \log J}{\partial \log Q}\right)_{v,d,D} = m \tag{5.14}$$

$$0 \leqslant m \leqslant 2 \tag{5.15}$$

The expected value of m is 0.88 or thereabouts, depending on which experimental study is used. In fact, m is an important experimental variable. While it is *often* the predicted value, it is quite variable from application to application. As a rule-of-thumb, true solutions exhibit values of m close to the value 0.88. Suspensions of fine particles raise m, and *fouling* lowers it. See the later discussion on fouling.

Laminar flow. For laminar flow, flow regimes up to a Reynolds number about 2200, an equation modified from Leveque's heat-transfer formulation shows that for practical situations.

$$Sh = 1.62\left(Re\,Sc\frac{d}{L}\right)^{0.33} \tag{5.16}$$

The mass transfer coefficient k contained in Sh is the rate constant for movement of solute along the concentration gradient. Solving for k:

$$k = 1.62\left(\frac{VD^2}{Ld}\right)^{0.33} \tag{5.17}$$

$$J \propto \left(\frac{QD^2}{Ld^3}\right)^{0.33} \tag{5.18}$$

where L is the length along the channel.

Equation (5.18) shows that flux is proportional to the inverse cube root of channel length, and the reciprocal of channel height. An attractive module design would thus have short, very small pores.

Practically, no economic way yet exists to manufacture such a design, and most membrane equipment suppliers furnish equipment designed to operate in turbulent flow. In higher viscosity applications, the same devices often operate in laminar flow but are unable to take advantage of the benefit of short path length.

Stagnant flow. In the absence of forced convection, simple diffusion is the predicted limit on the ability of a membrane to operate in steady state.

$$J = -\int \frac{D}{c}\frac{dc}{dx} = \frac{D}{\delta}\ln\frac{c_m}{c_b} \tag{5.19}$$

where c is the concentration of retained solute at m, the membrane or in b, bulk, and δ is the diffusive path length. Since the integration assumes D to be constant (not often true), the expression should be regarded as an approximation.

Equation (5.19) has two useful consequences. First its implication of an assumption underlying all steady-state mass-transfer calculations, namely that the concentration of retained solute at the membrane is constant. The solute at the membrane is sometimes referred to as a gel. While there is ample evidence that gels in the strict sense are unusual, near-constancy of concentration at the membrane is not. Second, equation (5.19) is a very useful expression for predicting the dependence of flux on concentration. If (D/δ) and c_m are treated as constants, the equation predicts flux as a function of concentration quite well. The usual technique is to plot log c versus J, with caution advised as the value $J = 0$ is approached.

Vibratory flow. No adequate theory is available to explain mass transfer in vibrating membrane systems. Summers[26] shows a correlation of mass transfer with the first power of shear rate over a narrow range of data and conditions. The first-order hypothesis is supported by other observations but it must be regarded as unverified and preliminary.

Taylor vortex flow. An elegant way to decouple polarization from driving force is by using a rotating filter device. When fluid flows around a curve in a duct, or when fluid is confined between differentially rotating cylinders, secondary flows called Taylor vortices are generated. Using these secondary flows to minimize polarization, provides a tool for membrane equipment design.

In a detailed analytic and experimental paper, Holeschovsky and Cooney[27] find a flux equation for this device analogous to equation (5.10):

$$Ta = \frac{\omega R g}{v}\sqrt{g/R} \tag{5.20}$$

$$Sh = C\,Ta^{0.5}Sc^{0.33} \tag{5.21}$$

where c is an experimental constant, R is the radius of the inner cylinder in m, g is the gap between inner and outer cylinders in m, and ω is the angular velocity of the rotating cylinder in radians/s. Ta is the Taylor number. For a fixed device on a given fluid, flux is predicted to be proportional to $\omega^{1/2}$.

5.7 Filtration

5.7.1 Comparison with membranes

Membrane processes sometimes operate in dead-end flow. When they do, they almost always operate as a batch, with either periodic membrane cartridge replacement or periodic backwash and cleaning. The rate equations for batch filtration are a little different from those used in cross-flow filtration but the difference is more superficial than real.

$$J = \frac{dV}{A\,d\Theta} = \frac{\Delta P}{\mu[\alpha(wV/A) + r]} \tag{5.22}$$

where A = area in m^2; V = volume of filtrate in m^3; ΔP is the pressure drop across the filter medium and cake in Pa; w is the mass of dry cake solids per volume of filtrate in kg/m^3; r is the filter cloth resistance (plus miscellaneous other constant resistances) in m^{-1}; α is the specific cake resistance in m/kg; μ is the viscosity of the filtrate in Pa s; and Θ is the time in s. The logic of ths expression is apparent. Regardless of the loading, more solids on the filter results in a lower rate of filtration. More pressure drop gives a higher rate, etc.

Alpha is constant only within an application. Part of α is determined by the size of the particles making up the cake. Part is proportional to P^s, s being a compressibility factor usually varying from 0 for rigid, incompressible, well-defined particles, to around 0.8 for highly-compressible cakes. This relationship is often expressed as:

$$\alpha = \alpha'\Delta P^s \quad \text{where } 0 \leqslant s \leqslant 0.8 \tag{5.23}$$

where J is the same flux given in the cross-flow equations, and it has the same units, m/s, gallons/ft^2-day (gfd) or l/m^2-h (lmh).

A similar formulation of the throughput equation as the sum of resistances is:

$$J = \Delta P/R \tag{5.24}$$

$$R = \mu R_r + R_c \tag{5.25}$$

$$R_c = \mu\alpha'\Delta P^s(wV/A) \tag{5.26}$$

R is the total resistance to flow, R_r is the resistance of the filter cloth and the apparatus, and R_c is the cake resistance. The total resistance is thus the sum of a constant part and a variable part. The cake resistance will always be the controlling resistance for most of a cycle.

For a filtration where $s = 0$, output may be kept constant by raising the pressure with time. If the cake is compressible, as it almost always is, raising ΔP uniformly with the volume processed, V, is not sufficient. In the extreme case for a completely compressible cake, $s = 1$ and the resistance is proportional to pressure. Therefore, in the flux equation, the ΔP term cancels. The rate of filtration is independent of pressure and cannot be maintained constant by manipulation of pressure. This case, flux independent of pressure, resembles the result of a cross-flow filtration.

A critical difference is that cross-flow membrane devices usually operate at constant output. Most are powered by centrifugal pumps, and are run at constant flow and pressure. The constancy in rate is in spite of the presumption that for many of the substances filtered, $s \cong 1$.

The practical methods of operation of batch type dead-end filtrations are to drive the system either with a positive displacement pump, and strive for constant output, or with a centrifugal pump, where the pressure rises slowly as flow decreases. Products of biological processes usually have high values of s, and when loadings are significant, constant output cannot be maintained.

5.8 Concentration polarization

Concentration polarization is simply the term used to describe the fact that a retained solute has a higher concentration at the membrane than it does in the bulk solution. So long as solvent is being forced through the membrane, the retained solute is 'polarized'. In turbulent cross-flow filtration, the equations describe behavior in the boundary layer, since the boundary layer, thin as it is, is by definition where 99% of the 'action' takes place. They describe a process driven by pressure but controlled by mass transfer. The behavior of materials within the polarized boundary layer are of particular significance to biological systems.

Figure 5.5 shows the experimental result of a typical cross-flow ultrafiltration experiment, plotting the flux as a function of pressure. The hydraulic resistance of the membrane itself is dominant for the lower left linear portion labeled 'region I'. In 'region III', mass transfer at the membrane is completely controlling, and the mass-transfer equations given above describe the operation. Almost all MF and UF operates in this region. 'Region II' is a transition zone, where some pressure dependence lingers and where mass-transfer limitations are becoming dominant but are not exlusively controlling.

Figure 5.5 can be described by the analogous equations (5.24) and (5.25) where R_c becomes R_{dy}, the resistance of the dynamic boundary layer.

As transmembrane pressure is increased, the driving force for permeation is increased but the flux is not increased at all. The resistance term is therefore rising exactly as fast as the pressure. This layer has significant autorejection. It operates in a fashion reminiscent of filter aid on a screen. When the solution at the membrane is diffuse, the retention properties of the system are those of the membrane. As R_{dy} rises, the boundary layer becomes the determinant not only of flux *but also of retention.* With a normal fully polarized membrane, microfilters can turn into ultrafilters, and ultrafilters start to reject microsolutes. Conditions causing high dynamic boundary layer resistance also promote

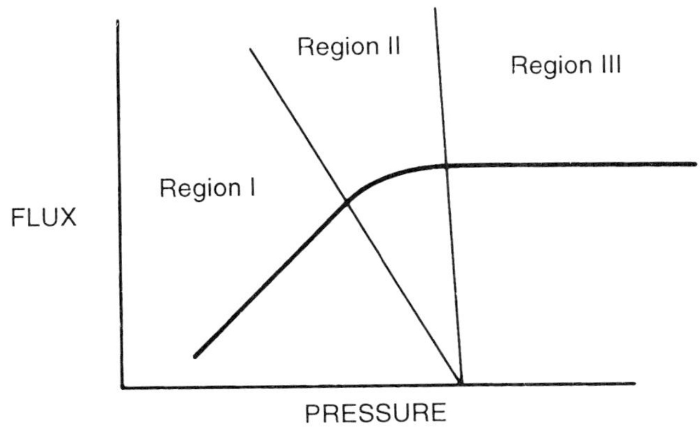

Figure 5.5 Graph of flux against pressure in a typical cross-flow ultrafiltration experiment.

fouling. Flux drops, and membrane retention increases permanently until the foulant is removed.

When the main objective is to separate cells and cell fragments from proteins, it is critical that microfilters do not operate in a fashion that significantly increases protein retention. For that reason, microfilters for cell-protein separation are designed not to become excessively polarized.

Most pumped systems are highly polarized for fundamental economic reasons. It is far easier and far cheaper to build a system where the membranes are connected in series than it is to construct one that is 'massively parallel'. According to Eykamp[28] the best massively parallel membrane system known, the kidney, consumes half of its entire energy input simply getting through the inlet and outlet manifolds. Membrane controlled selection is very important in the kidney, and R_{dy} is very low.

Equipment designers are not yet able to design an economic massively parallel system. Instead, thay have confronted the challenge using a variety of strategies. One is to design the system as a series array, then elevate the permeate pressure so that the transmembrane pressure at any point in the system is low. One example of that approach is a device for separating bacteria from milk[29] and from whey.[30] Several firms are trying innovative approaches that decouple the pressure driving force from the membrane depolarizing force, such as by the use of vibration to promote mass transfer,[31] spinning disks interleaved between membrane sheets,[32] or the use of Taylor vortices.[33] Limitations caused by the polarized boundary layer are exacerbated by two sharply conflicting needs in two of the more important microfiltration applications. Lysed cells, particularly *Escherichia coli*, are notoriously difficult to separate from proteins. They quickly form a highly rejecting dynamic layer, and they foul membranes. Both work against reasonable yields of protein. The usual approach is to operate the membrane at very high shear to keep polarization to the minimum. Mammalian cells, particularly those undergoing medium exchange to enhance protein expression, are intolerant of high shear. One solution will not work for both applications if it is based on conventional design pinciples.

5.9 Membrane components

5.9.1 Membranes

It will amaze many people struggling with separations problems just how easy it is to make a membrane. A Loeb–Sourirajan membrane can be prepared in a few minutes on a kitchen table. All that is needed is a solution of polymer in a water-miscible solvent, a flat surface, a device for leveling the solution on the flat surface, and a pan of water in which to immerse the coated flat surface. It then takes considerable skill and time to get a membrane with desirable flux and retention properties. *Reproducibility* is the most difficult and critical problem.

Polymeric membrane in capillary form is only a little more complicated. A novice guided only by the literature can make a passable membrane in a reasonably equipped laboratory in less than a week. Concentric hypodermic needles, the larger of which contains casting solution and the smaller of which contains water are the kind of simple equipment required.

Flat-sheet membranes, suitable for use in cells, or incorporation in spirals or cassettes, are made on casting machines with a considerable spectrum of sophistication. The primary design criterion is reproducibility and the reduction of statistical variance in the product produced.

If one learns to make reproducible membrane with acceptable attributes, and that can be the labor of a decade, not a day, there remains the problem of putting that membrane in a suitable module.

5.9.2 Modules

A ‘module’ is the term used to describe a replacable membrane-containing element. There is only general agreement in the industry regarding a strict definition of the term, so caution is indicated when comparing the exact use of the term between manufacturers. Modules currently offered are the survivors of a much larger universe of clever designs.

Tubes. Tubular membranes (Figure 5.6) are defined in the industry as membranes usually made singly as cylinders, customarily with diameter greater than 10 mm. Other configurations that are geometrically cylindrical are described under capillaries and ceramic monoliths.

The simplest, and oldest tubular membrane has a porous backing acting

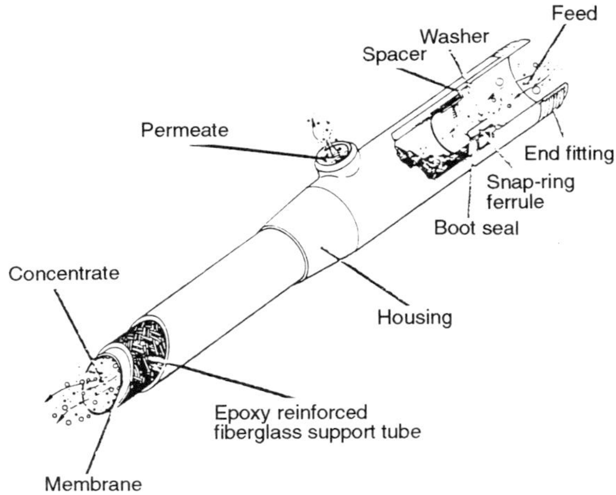

Figure 5.6 Example of a tubular membrane system — Abcor Tubular Membrane — in housing (by courtesy of Koch Membrane Systems).

as a pressure vessel, with the membrane cast directly on the porous backing. The entire assembly is leached, and since membranes want to contract as the solvent is removed, keeping everything where it is needed during the leaching step is an art. The major commercial embodiment is a 25 mm diameter tube 2.8 m long. It is incredibly rugged even to feed side insults, has a very efficient hydraulic design, and is expensive to buy and operate. A much cheaper tubular design uses polyester tapes to construct a cylinder in the manner of a soda straw, which is formed, welded or glued, and has a membrane coated on it in one continuous operation. By this process it is possible to prepare a tubular membrane of almost any length and diameter. Custom and economics, however, combine to produce typical modules with 12 mm tubes usually less than 3 m in length. The third tubular type separates the function of membrane and pressure vessel, and uses an insertable membrane in a cylindrical porous pipe. This is particularly popular in reverse osmosis, where the pressure is too high for the 'soda straw'-type backing to resist rupture, and membranes so made are subsequently inserted, usually into perforated stainless steel tubes.

Another variant is the porous carbon tube with an inorganic membrane formed dynamically inside it. Raised to prominence by Union Carbide in the 1970s, it was the by-product of the manufacture of carbon electrodes for cinema arc lights. Primarily an ultrafiltration membrane, it is currently offered by Rhône Poulenc.

Once produced, tubular membranes are connected in series/parallel arrays to manifolds using U bends. Smaller tubes are bundled into parallel arrays for better economy.

Tubular systems were very popular in the early days of reverse osmosis and ultrafiltration because of their great reliability and durability. As membrane technology became more familiar and accepted, the level of security that tubes provided became less important. Simultaneously, other designs proved adequate for many applications. Tubes remain superior for applications where operations at very high levels of suspended solids are required, and for unstable emulsions where 'stagnation points' on the membrane surface result in membrane malfunction.

Compared with any other design, an accurate generalization is that tubular membranes are tougher and more resistant to physical contamination and high concentrations of suspended solids but are much more expensive and require much more floorspace.

Parallel-plate designs. Historically, many membrane designs featured membrane arranged in parallel plate arrays. The three most successful were the parallel-plate cartridge, sold for many years by Dorr-Oliver, and now discontinued, the plate and frame devices most successfully offered by De Danske Sukkerfabriker (now Dow-DDS) and Rhône Poulenc, and cassette designs offered by Millipore and others.

Plate and frame membrane devices. Plate and frame membrane devices resemble plate and frame filter devices. The most prominent reverse-osmosis design is a tall stack of horizontal membrane plates surrounding a large bolt used to hold the device together (Figure 5.7). Ultrafiltration devices include large arrays of vertical plates of oval shape with circular inlet and outlet pores at the foci. Madsen gives a detailed rationale for the design.[34]

The other popular commercial design resembles a plate heat-exchanger with similar treatment of the fluid-management problems.

Cassettes. Cassette ultrafilters (Figure 5.8) and microfilters are popular in the pharmaceutical and biotechnical fields because they facilitate direct scale-up from laboratory equipment. It is at once their appeal and their limitation, since their fluid management design is inherently very limited and

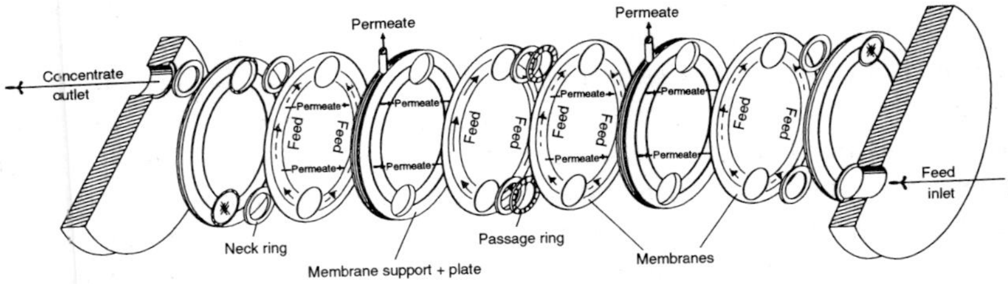

Figure 5.7 Exploded view of plate and frame system of a reverse-osmosis plate and frame membrane device depicting a fluid flow path (by courtesy of Dow Chemical Company).

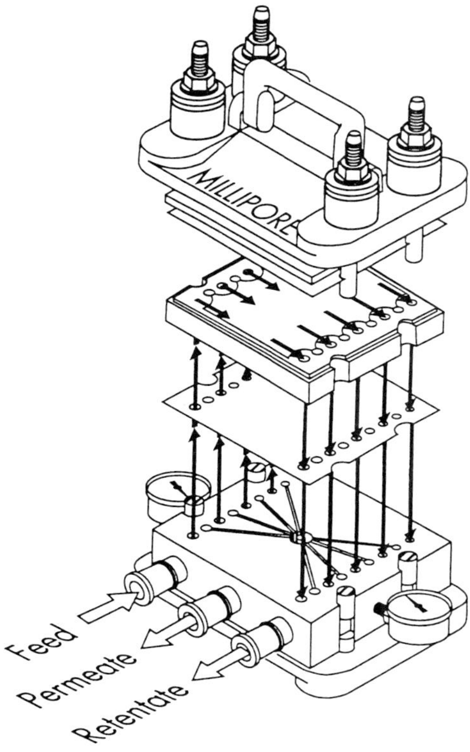

Figure 5.8 The Millipore Cassette (by courtesy of Millipore).

inefficient. These severe engineering limitations are sometimes outweighed by the simplicity inherent in taking the same membrane and equipment design to a larger scale of production.

Spirals. Spiral modules are made from flat-sheet membrane wound up like a jelly roll. Behind the membrane is a permeate carrier, usually a woven tricot cloth, which conducts the permeate from the back of the membrane to the central permeate collector, the tube at the middle of the spiral. The edges of the membrane with the tricot inside are sealed on three sides by a flexible curable sealant, most often polyurethane or epoxy. A polyolefin screen (usually Vexar™) is placed on the face of the membrane to provide space for feed to flow along the membrane sheet. As the membrane assembly is rolled up, membrane faces are separated by one layer of polyolefin screen, so feed passing along the screen is exposed to two membrane surfaces.

A diagram of a spiral is shown in Figure 5.9. There are many diagrams and many written explanations of the operation of the spiral but experience

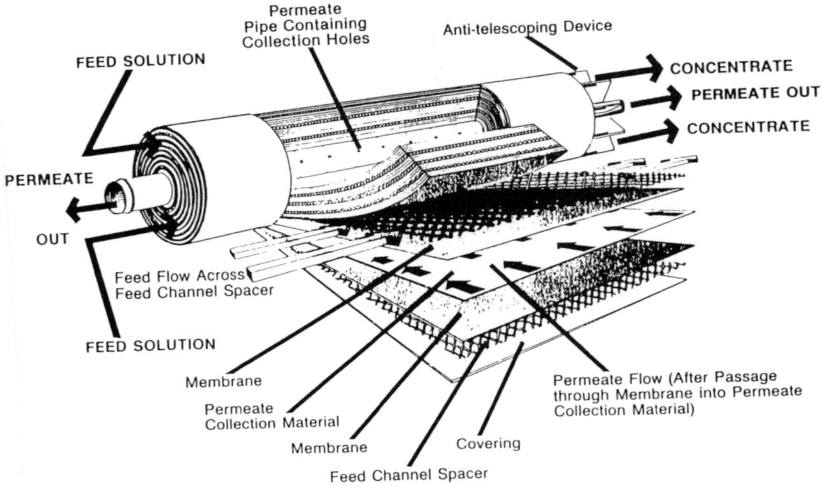

Figure 5.9 A spiral-wound module (by courtesy of Koch Membrane Systems).

shows that this device is hard to understand without seeing one disassembled. The most common misconception is that feed enters the central pipe. The core is part of the permeate envelope, and it must be completely sealed from the feed. Feed flows in one end of the cylinder and out the other. A gasket of some sort is placed around the outside diameter to keep this feed from bypassing around the annulus between the spiral module and the pressure vessel in which it is placed.

Spirals were originally developed for use in reverse osmosis.[35] Over the years, considerable investment and effort has expanded their application greatly. They look simple and 'low-tech', but that belies the great sophistication required to make a spiral that works well under conditions far removed from desalination. With knowledgeable process and product design, spiral modules operate well in a large number of disparate applications. Spirals are the dominant membrane module, based on their economy of manufacture, their rugged constitution in use, and their ability to put a lot of membrane area in a small factory area.

Capillaries. Capillary membrane devices (Figure 5.10) are much simpler to understand than spirals. They constitute bundles of membrane typically around 1 mm internal diameter. Hollow fibers are much smaller, but are manufactured in a similar way.

Hanks of capillary membranes are assembled in bundles of any length, but customarily around 1 m or less, and the ends are potted with typically epoxy or urethane sealant. After cure, the pot is sawed through exposing the ends of the capillary membranes. The potted bundle is sealed in an enclosing

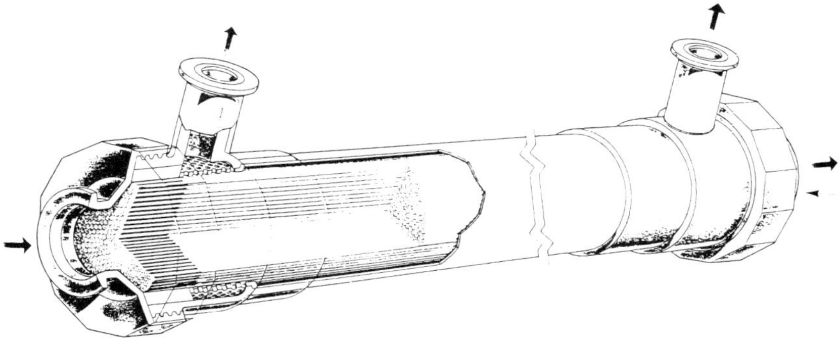

Figure 5.10 Conventional capillary module (by courtesy of Pall Corporation).

shell. Feed is typically passed through the inside diameter of the capillary under pressure. Permeate passing through the membrane is collected by the shell.

Capillary bundles, like spirals, underwent long evolution. Problems were more often with the bundle than with the membrane itself. Potting failures proved difficult to reduce to a low level. Dynamic effects, such as standing waves in the membrane which subject the membrane to great stress near the pot, were difficult problems. Unlike spirals, where a sealant failure tends to be closed by pressure, when a capillary breaks leakage is often excessive.

One firm uses capillary bundles in different modes (Figure 5.11), feeding on the outside of the membrane for microfiltration applications. The creeping flow and the distribution problems inherent in the shellside feed cause excessive polarization and build-up akin to a dead-end filter. Periodic pulsing by air from inside the capillary displaces accumulated material on the outside of the membranes and causes it to be returned to the concentrate steam.

Ceramic monoliths. Ceramic membranes (Figure 5.12) are made in multi-tube monoliths. The diameter of the channels is commonly about 2 mm, permitting 19 channels in a 21 mm diameter monolith. Large diameter monoliths are also made. Another manufacturer makes large monoliths with square channels but these are seldom used in downstream processing. The membrane formation process, described above, is achieved through the successive deposition of ever smaller spheres, followed by firing.

Ceramic monoliths certainly occupy the ultimate position in thermal stability, although polymeric membranes can be steam-sterilized at almost any required temperature. The ceramics also have the ultimate solvent resistance, and are generally good at low pH and under oxidizing environments. Caution is required at even moderately high pH but there is good resistance from some products. Cost is high but the life of ceramic monoliths is long, although finite, and is often unknown because of insufficient experience.

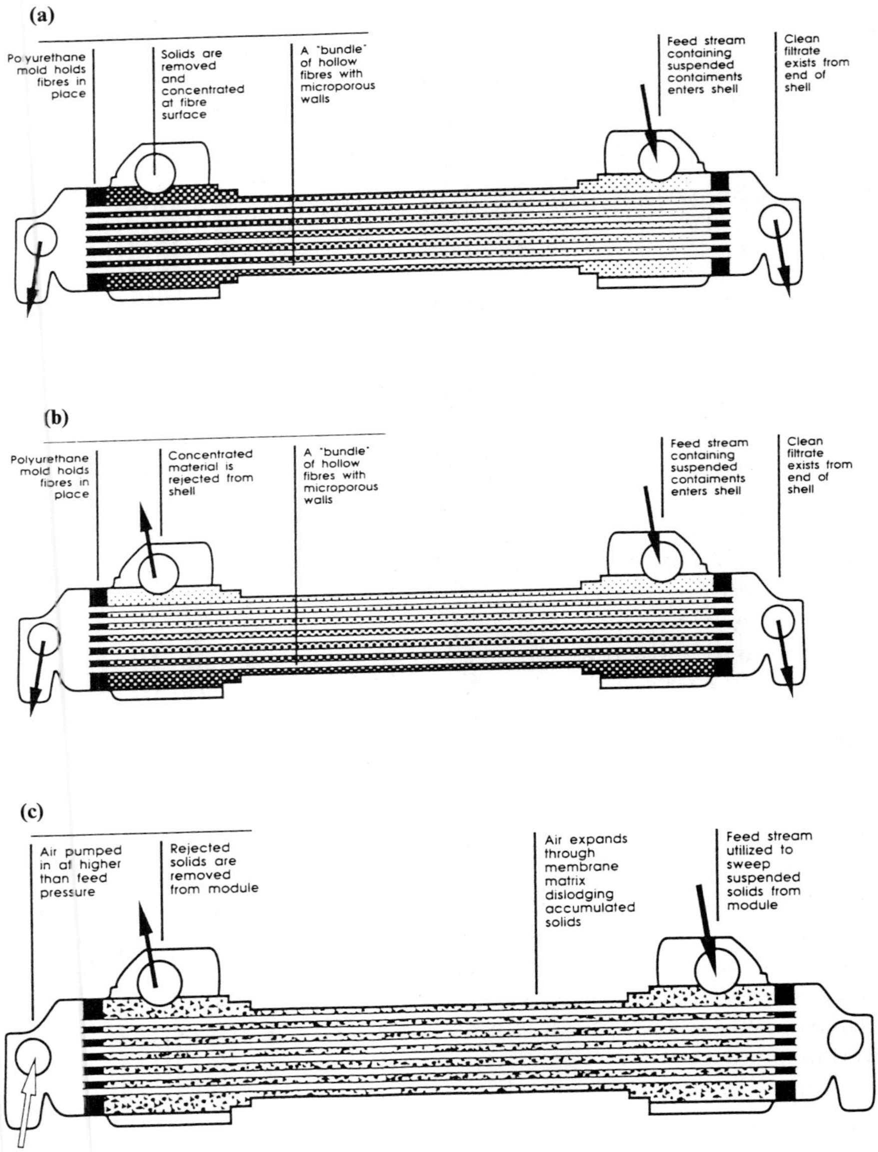

Figure 5.11 Operating modes: (a) dead-end; (b) cross-flow; and (c) back-wash mode (by courtesy of Memtec).

Figure 5.12 Example of ceramic membrane — the Alcoa Ceramic Module (by courtesy of U.S. Filter).

Hollow-fiber membranes. Hollow-fiber membranes (Figure 5.13) are so small (80 μm o.d., 20 μm i.d.) that it is hard to believe they are actually hollow. They are packed, by the million, into modules (an 8″ Permasep* module contains 2 750 000 M fibers). Feed flows to the outside of the fiber; permeate flows through the inside. Since the fibers are both fine and closely packed, they are excellent traps for particulates. Feed streams must therefore be extremely clean. Fine, hollow-fiber modules are used generally for reverse osmosis, and their special niche is for use at very high concentrations and thus at very high osmotic pressure.

Shake stacks. By mounting a plate and frame membrane assembly above a torsion bar spring, membranes may be depolarized by vibrating the stack at a resonant frequency (Figure 5.14). Unlike any other membrane device,

* Dupont product. Courtesy of Dupont.

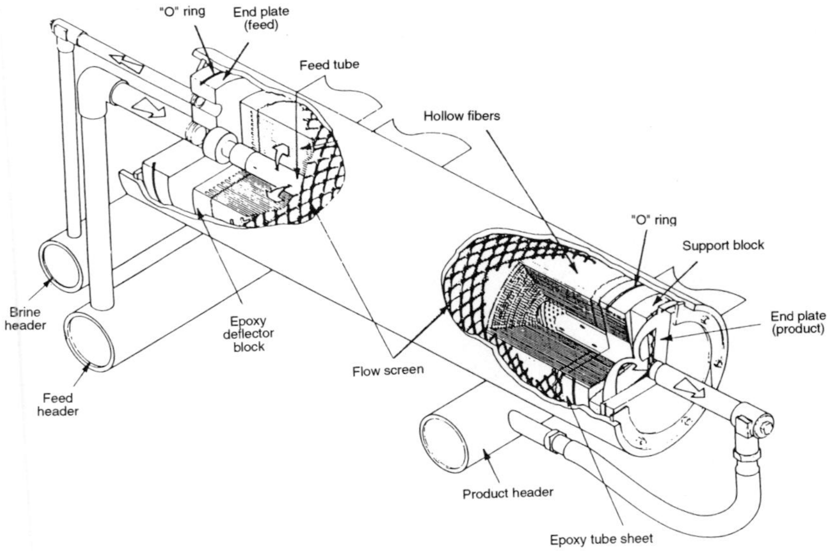

Figure 5.13 Example of a hollow-fiber membrane — the Dupont Hollow Fiber Module (by courtesy of Dupont).

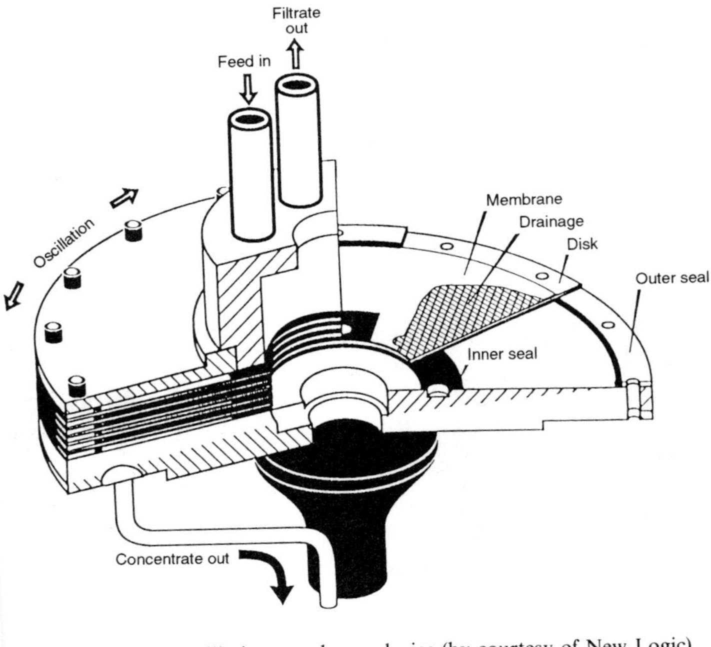

Figure 5.14 An oscillating membrane device (by courtesy of New Logic).

the membrane moves and the fluid is essentially stationary. At first thought to be ideal for and limited to solutions of very high viscosity or solids loading, the devices are now viewed as another economic competitor for a broader range of applications. The technique is still new, and it lacks large-scale long-term experience from use.

Rotary modules. Another way to enhance depolarization is to put a rotating device near the membrane, or rotate the membrane. Devices much more sophisticated than the simple laboratory bench-top cell and its magnetic stirrer are offered particularly for viscous downstream applications.

One scheme tends to be cylindrical, and uses a rotating concentric device to create Taylor vortices near the membrane. Another tends to be planar, although it is almost always stacked, featuring a rotating disk interposed between two flat membranes. High shear rates are readily obtained at the membrane.

Both rotary and shaking devices seek to decouple the depolarizing force from the pressure, permitting easy control over polarization, particularly for highly viscous lysates (Figures 5.15 and 5.16).

5.10 Applications to biotechnology

5.10.1 *Protein fractionation using ultrafiltration*

In the section on 'molecular-weight cut-off', various impediments to the use of ultrafiltration membranes as fractionation devices are discussed. These limitations are serious and real, and they have historically limited the use of ultrafiltration devices as protein fractionators.

With the growing need to fractionate proteins, and given the low concentration of proteins in many streams that would normally be fed to a

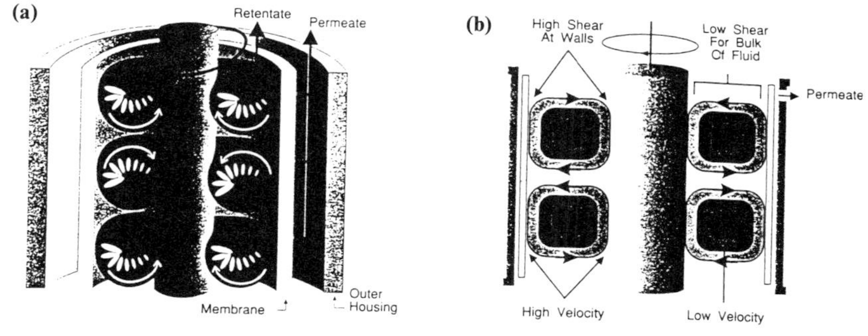

Figure 5.15 Taylor vortex module: (a) VFP scrubbing vortices; (b) VFP's sheer zone (by courtesy of Membrex).

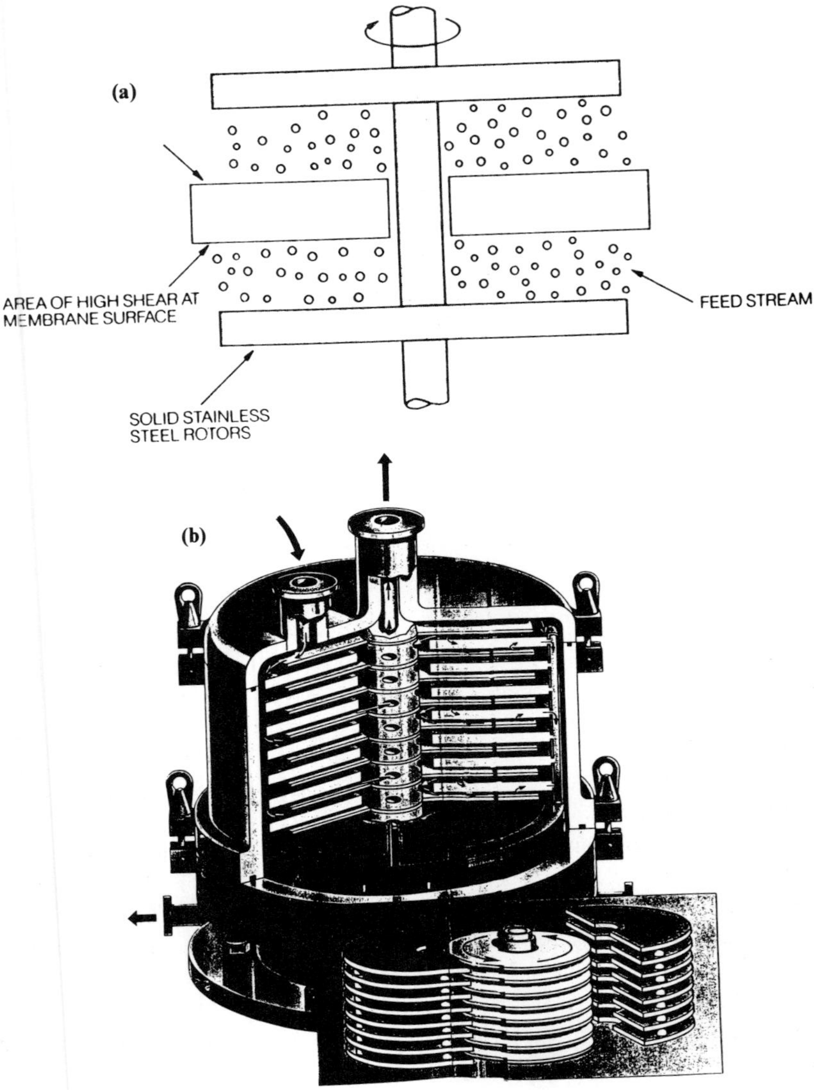

Figure 5.16 A rotary membrane unit (by courtesy of Pall Corporation).

chromatograph, renewed interest in using membranes has led to some successful applications.

When using a membrane for macromolecular fractionation, neither concentration polarization nor fouling may be extensive. These requirements imply operation on dilute solutions, at low transmembrane pressure, and for short time intervals.

(c)

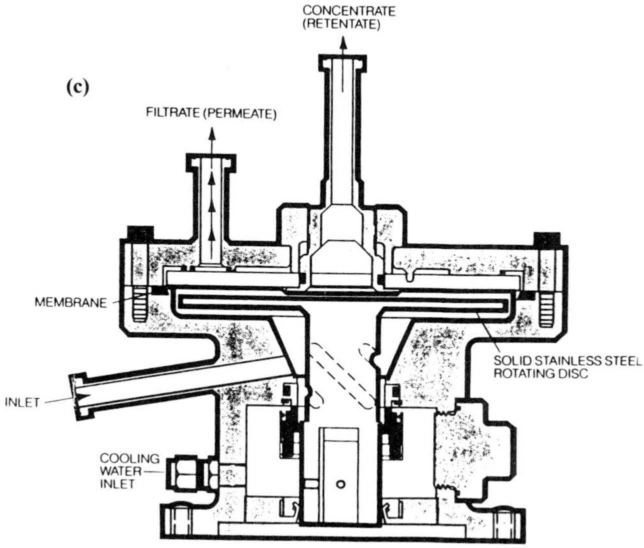

Figure 5.16 Continued.

Consider a tank containing a solution connected to a membrane. Liquid is pumped from the tank past the membrane. Some of it permeates. As the process progresses, the quantity of retained material remaining in the system decreases as a function of the amount permeated and the membrane retention. *Yield* of the retained species is the quantity remaining in the system divided by the amount at the start.

In membrane processing, there are two fundamentally different ways the operation is conducted, as a *batch concentration* in which a batch is concentrated while nothing is added, and as *diafiltration* (Figure 5.17), in which solvent containing no retained solute is added to the process stream at a rate that maintains overall volume constant. In both cases, good stirring is assumed. Although there are variants and combinations of these two operations, these two are the basic themes. For batch concentration, the yield equation is:

$$Y = \left(\frac{V_0}{V_f}\right)^{R-1} \tag{5.27}$$

Y is the fractional yield, V_0 is the volume of process fluid at the beginning of a batch run, and V_f at the end. The equation is valid only when R, retention, is constant.

$$Y = \exp\left(\frac{V_D}{V_0}\right)(R-1) \tag{5.28}$$

Table 5.1 Comparison of batch concentration and diafiltration methods. Batch run (columns 3 and 4) is expressed as batch yield for the material of the Rand volume ratio listed, and (in parentheses) the ratio of the final concentration of retentate to initial concentration. For the batch run as concentration (column 3) the number in parantheses will be between 1 and the volume ratio, that for the batch run as diafiltration the number will always be less than 1

Retention (R)	Volume ratio (V/V)	Batch run as concentration ($V_0 V_f$)	Batch run as diafiltration (V_D/V_0)
0.995	10	98.9% (9.89)	95.1% (0.951)
	20	98.5% (19.7)	90.5% (0.905)
	30	98.3% (29.5)	86.1% (0.861)
0.990	10	97.7% (9.77)	90.5% (0.905)
	20	97.0% (19.4)	81.9% (0.819)
	30	96.7% (29.0)	74.1% (0.741)
0.80	10	63.1% (6.31)	13.5% (0.135)
	20	54.9% (11.0)	1.83% (0.018)
	30	50.6% (15.2)	0.248% (0.002)
0.60	10	39.8% (3.98)	1.83% (0.018)
	20	30.2% (6.03)	0.034% (0.000)
	30	25.7% (7.70)	0.000% (0.000)

where V_D is the volume of diafiltration solvent (water) added, equal to the volume of permeate removed.

In Table 5.1 are given examples of the results of using the two processes to fractionate two proteins having different retentions. In a normal process, they would be run sequentially, diafiltration first.

Column 1 lists retention which is assumed to be constant; in practice, it might not be constant and would need to be determined experimentally especially in view of the caveats discussed in ultrafiltration membrane ratings. The caveats enumerated are very important, since fractionation implies the ability to run the batch with two or more materials, each with its own retention.

Column 3 lists the results of a concentration run. The first of the two numbers is the batch yield for the material at the R and volume ratio listed. The second (in parentheses) is the ratio of the final concentration of that retentate to its initial concentration. The number will be between 1 and the volume ratio.

Column 4 lists the results for a diafiltration run. The numbers are analogous to those in column 3. The number in parentheses, which is again the ratio of final concentration of the species to its initial concentration, will always be less than 1.

An example of this follows. A batch contains only two proteins, both with an initial concentration of 0.1%. Under the conditions of processing, one protein is retained at 0.990, and the other is retained at 0.80. Initially, the protein purity of the larger molecule is 50%. A 20× diafiltration is run. At the conclusion, the total batch volume is unchanged because diafiltration water was added to compensate for all permeate production. Working from column 4 in Table 5.1, 81.9% of the large protein remains in the batch but its concentration is reduced to (0.1%) (0.819) = 0.0819%. The yield of the small protein is 1.83%, and its concentration is now (0.1%) (0.018) = 0.0018%. The protein purity of the large protein is now 97.8%.

To continue the example, a 30× batch concentration is now run on the retentate. From column 3, the concentration of the large protein will be (0.0819%) (29) = 2.38%. The small protein will be (0.0018%) (15.2) = 0.027%. The protein purity of the large protein is 98.9%. The overall yield of the large protein since the beginning will be (74.1%) (96.7%) = 71.7%. The overall yield of the small protein will be (1.38%) (50.6%) = 0.698%. The yields in the permeate will be reversed.

The example is illustrative but optimistic. In practice, the retentions will not be constant; however, if a membrane has a high retention for one species and perhaps 0.2 retention units lower for another protein, it is possible *in principle* to make a significant purification of *either* protein, depending on which fraction is collected. It is possible *in principle* to achieve separations that are quite competitive with chromatography. When tanks and other equipment are considered, as well as the floorspace they occupy, the economics of membrane separation of proteins may be very attractive.

Fractionation requires a combination of these two types of operation. Since fractionation is possible only for very dilute solutions, and products are generally needed in concentrated form for subsequent processing, normal operation will begin with diafiltration and finish with batch concentration, as shown in the example.[36]

Major caveats need to be emphasized. Retention of individual species will not be the same as for that species in a mixture; the ideal separation is always better than the actual one. R is defined relative to the concentration in the well stirred bulk, not the concentration at the membrane where the separation actually takes place. Therefore, R is not actually constant for any real operating condition. If the membrane is ideal, R will actually rise a little as the solution becomes very dilute and is run with very low polarization, since the concentration at the membrane will be close to the concentration in the bulk. Conversely, in a typical batch concentration, as concentration and polarization increase, retention for the ideal membrane will decline. Finally, the ideal membranes exist only in books. Real membranes foul, and fouling increases retention. Some membranes function poorly on dilute solutions since they rely on a degree of fouling to 'plug the leaks'.

5.10.2 Membrane reactors

One way to simplify downstream treatment of biotechnical processes is to conduct them in membrane bioreactors. Heath and Belfort[38] published a brief review of the status and possibilities of using membranes as an integral part of the production sequence.

Immobilized whole-cell reactors have come in many types. Semipermeable microcapsules and capillary modules are perhaps the best publicized types but flat-sheet multi-layer devices are also reported.

Membrane reactors have long been used to contain enzymes for bioconversions. The enzyme may either be contained within a membrane-defined space, while reactants and products pass in and out, or the enzyme may be immobilized on the membrane, while the membrane performs both a separating function and serves as a fixed catalyst site. One intriguing possibility is for sequential reactions in which fast reaction times are not limited by diffusion, and back conversion is eliminated because the enzyme sequence is fixed.[39]

5.10.3 *Enzymes*

Enzyme concentration and purification was the first large application of membranes to a fermentation product. Ultrafiltration in particular proved to be far cheaper than conventional enzyme isolation techniques, and it is widely used for the concentration and purification of commodity enzymes such as proteases and pectinase.

At first there were concerns about things such as membrane stability in the presence of enzymes. Surprisingly, even cellulose acetate membranes proved reasonably stable in the presence of cellulases and, as membrane chemistry moved towards more durable polymers, concerns about membrane stability shifted towards the more familiar concern over their ability to withstand cleaning. Shear sensitivity of the enzyme products was another concern but, considering the shear in an aerated fermentor, even rudimentary engineering precautions in the design of a membrane system are adequate. More valid concern about membrane retention is a problem long-since solved and, typical of any commodity, the concern is now over cost. As such, the membrane configuration utilized has progressed from tubes to spirals and capillaries.

5.10.4 *Sterile filtration*

In the discussion on membrane ratings, the use of bacteria to determine the size of membrane pores is mentioned. Reference is given to ASTM tests for determining the suitability of microfiltration membranes for use as sterilizing filters. Sterile filtration using membranes is simple in principle. It may be obvious that a bacterium or a virus cannot go through a membrane rated smaller than its size but absolute ratings are difficult to achieve.

It has long been known that membrane thickness is a factor in microbial retention, the rationale being that membranes are not sheets of matter containing neat holes drilled in the surface. Martin[40] states that 0.22 μm membranes 'typically demonstrate surface openings as large as 1 μm or more'. Narrower restrictions are expected beneath the surface. Osumi *et al.*[41] show *Pseudomonas diminuta* organisms well beneath the surface of an 0.2 μm membrane but they do not observe monolayers of bacteria on the membrane surface, or plugging membrane pores, even at challenge levels of $10^7/cm^2$. Nonetheless, the bacteria did not pass into the permeate.

Meier[22] cites the use of microfiltration membranes in the wine industry. This application is a yeast removal and bacteria reduction process, the level of bacteria reduction being optimized with throughput and taste. Sterile processing is codified by the FDA, the *Guideline on Sterile Drug Products Produced by Aseptic Processing*[42] being one example.

DiLeo[15] reports the difficult process of removing viruses from biotherapeutics. Using an ultra thin ultrafiltration membrane formed as a composite above a microfiltration membrane to produce a very uniform pore size membrane. This membrane reportedly gives 3–8 log reduction in virus while passing up to 90% of a soluble protein, albeit in a very dilute solution.

5.10.5 *Process microfiltration*

Microfiltration can separate colloidal material such as bacteria, molds, and particulates from soluble material, such as proteins, peptides, and salts. The appeal of microfiltration to downstream processing is obvious. Specific means for realizing the potential are often not so obvious.

Microfiltration is practised both as dead-end flow filtration and as cross-flow filtration. Cross-flow filtration models come in all types and styles: Spiral, plate-and-frame, capillary, tubular, rotary and vibratory. Ceramic monoliths are a special type of capillary membrane. Each of these configurations has a market niche.

Typical applications. Processes are divided into '*clarification*' (Table 5.2) in which the main product is a clarified liquid, and '*solids recovery*', in which the major objective is to reclaim a solid component in a liquid stream.

Removal of suspended matter is a topic that covers numerous applications in downstream processing. Separating cells or their fragments from broth is the commonest application. Clarification of the broth in preparation for product reovery is the usual objective but recovery of cells is sometimes the primary goal. Cross-flow microfiltration competes well with centrifugation, conventional filtration by rotary vacuum filter or filter press and decantation. In return for higher capital costs, microfiltration delivers a supernatant that is cleaner than for any competing process, gives an uncontaminated cell product, which may be washed in the process, gives generally high yields, and has no filter-aid disposal problem.

Cell preparation may be practised by most membrane hardware

Table 5.2 Typical applications of process microfiltration

General application	Specific example	Process flow
Clarification		
Clarification of protein solutions	Plasma fraction purification	Cross-flow filtration
Clarification of bacterial lysates	*E. coli*	Cross-flow filtration
Clarification of yeast lysates	*Saccharomices cervesei*	Cross-flow filtration
Clarification of fermentation broths	Clarification of antibiotic broth	Cross-flow filtration
Analytical preparation	Pre-filtration for chromatography	Dead-end filtration
Sterile filtration	Aseptic packaging of final product	Dead-end filtration
Virus removal	Tissue-culture media	Dead-end filtration
Yield improvement	Mycoplasma removal from serum	Dead-end filtration
Solids recovery		
Continuous fermentation	Cell recycle	Cross-flow filtration
Harvesting bacterial cells	Cell washing	Cross-flow filtration
Harvesting yeast	Yeast concentration	Cross-flow filtration
Recovering mycelia	Mycelia concentration	Cross-flow filtration
Harvesting mammalian cells	Washing hybridomas	Cross-flow filtration
Analytical	Measuring particulates in water	Dead-end filtration

configurations. For historical reasons, plate and frame devices and cassettes are commonly used. Capillary membranes, particularly those from 1–3 mm i.d. are making inroads in newer applications. Larger tubes (12–25 mm i.d.) are available, and have large advantages in a few applications. For lowest cost, spirals are usually the entry to beat. Spiral use is concentrated in a few large applications, and their economic advantage is not decisive in the vast majority of smaller uses. Newer methods based on 'stirred' or vibrating membranes are making inroads in spite of their high cost.

In cell separation, the microfiltration process is generally separate from the reaction, although it need not be so. Continuous cell harvesting and broth clarification are often discussed and occasionally practised. Low shear devices are used in perfusion reactors to extend significantly the duration of a run.

When a broth containing cells is treated, cells may be lysed, separated intact, or treated by medium exchange to induce protein expression. The appropriate treatment conditions are quite different.

Lysed cells, especially *Escherichia coli*, are a challenge to membranes. Intracellular material fouls membranes, it polarizes readily at the membrane, and by polarizing forms a self-rejecting layer for protein, significantly inhibiting the separation of protein from cell debris. Shear rates in processing lysates are very high.

With animal cells, Maiorella[43] proposes a shear stress limit of 3000/s (average). For cells undergoing medium exchange, the maximum shear rate is significantly lower. Van Reis[44] reports successful operation at shear rates up to 8000/s. Conventional wisdom holds that there are many components that can affect cell viability and function, the membrane itself being merely the one most often studied. Cell deformation is also an issue, and restricting the transmembrane pressure is necessary.

For all applications, ancillary equipment is very important. Pumps are notorious for introducing problems — centrifugal pumps are particularly troublesome. Among many concerns, a pump may initially contain air that does not purge, or it may cavitate or draw air in through a seal. Pipe components are also suspect if running in turbulent flow with expansions and contractions in the circuit. Even the manifolding and internal fluid configuration of some membrane housings cause cells to lose viability or rupture.

In the case of cell recovery from aerated fermentation processes, it is hard for a membrane component to outdo the fermentor in shear or general abuse. If the cells are lysed, vigorous treatment is needed in the membrane filter to avoid excess polarization and fouling.

When cells accumulate at the membrane surface, Hodgson *et al.*[45] discovered that the inter-particle spacing that normally determines cake resistance was not the limiting factor. Rather, 99% of the resistance was caused by the thin layer of extracellular polymeric material surrounding the cells in the biomass.

5.10.6 Nanofiltration

For decades, reverse-osmosis membranes with very low rejection of sodium ions have found specialty applications in the pharmaceutical industry. At first they were called 'sugar' membranes. They would pass sodium chloride but retain about 50% of $MgSO_4$. Their sucrose retention was used as a measure of their 'tightness'. Far too costly and inefficient to realize the inventor's hopes in the sugar industry, this class of membranes found niche applications in the purification of pharmacologically active polypeptides.

Nanofiltration membranes are the modern successor, and nanofiltration is very useful in fields such as antibiotic purification, where product is concentrated before crystallization, and in synthetic polypeptides where sodium ion can be removed through diafiltration. Desalting and solvent exchange for reverse-phase chromatography are also practised.

5.11 Manufacturers

Companies involved in membrane separations are listed in Table 5.3.

Table 5.3 Manufacturers in the field of membrane separations

Manufacturer	Membrane specialty	Contact address
A/G Technology Corporation	Capillary MF, UF	101 Hampton Ave, Needham, MA 02194, USA Tel. (617) 449 5774
Berghof/America	Teflon sheet MF	125 Mason Circle, Concord, CA 94520, USA Tel. (510) 827 1868
Ceramem	Ceramic microfilters	12 Clematis St, Waltham, MA 02154, USA Tel. (617) 899 0467
Cuno Incorporated	Cartridge MF, Spiral UF	400 Research Parkway, Meriden, CT 06450, USA Tel. (203) 237 5541
Domnick Hunter	Microfiltration cartridge filters	6636 East W.T. Harris Blvd, Charlotte, NC 28215, USA Tel. (704) 568 8788
Dow/DDSS	Plate and frame UF, Spiral UF	Larkin Laboratory, Midland, MI 48674, USA Tel. (517) 636 6478
DSI	Spiral UF	1238 Simpson Ave, Escondido, CA 92029, USA Tel. (619) 746 8141
Filterite/Memtec	Capillary, flat-sheet, cartridge MF/UF	2033 Greenspring Drive, Timonium, MD 21093, USA Tel. (410) 560 3000
Filtron	Membranes, cassettes MF, UF	50 Bearfoot Road, Northborough, MA 01532, USA Tel. (508) 393 1800
Fluid Systems Corporation	Spiral UF	10054 Old Grove Rd, San Diego, CA 92131, USA Tel. (619) 695 3840
Fuji America	Porous metal sheet, tube MF	4044 Yancey Rd, Charlotte, NC 28217, USA Tel. (704) 527 3854
Gelman Sciences	Broad range MF firm	800 South Wagner Rd, Ann Arbor, MI 48106, USA Tel. (313) 665 0651
Gore Associates	Flat-sheet MF	Box 1550, Elkton, MD 21922, USA Tel. (410) 392 4440
Hoechst-Celanese	Fibres, flat-sheet, bundles	13800 South Lake Drive, Charlotte, NC 28217, USA Tel. (704) 588 5310
Hydranautics/ Nitto Denko	UF, MF spiral and other	401 Jones Rd., Oceanside, CA 92054, USA Tel. (619) 901 2500
Koch Membrane Systems	Tubes, capillaries, spirals, systems	850 Main Street, Wilmington, MA 01887, USA Tel. (508) 657 4250
Membrex	Vortex and rotating devices, proprietary membrane	155 Route 46 West, Fairfield, NJ 07004, USA Tel. (201) 575 8388
Microgon	UF and MF capillary	23152 Verdugo Drive, Laguna Hills, CA 92653, USA Tel. (714) 581 3880
Millipore/Norton Ceraflo	Biggest microfiltration firm	80 Ashby Road, Bedford, MA 01730, USA Tel. (617) 275 9200
Mott Metallurgical	Sintered metal microfilters	Farmington Industrial Park, Farmington, CT 06032, USA Tel. (860) 677 7311

Table 5.3 Continued

Manufacturer	Membrane specialty	Contact address
Nanosearch	Monodisperse pore size UF	Koppstrasse 103, A-1160 Wien, Austria Tel. 43-1-49 101 402 Fax. 43-1-49-101-100
New Logic	Vibratory MF, UF	1155 Park Avenue, Emeryville, CA 94608, USA Tel. (510) 655 7305
Osmonics	Spirals, systems	5961 Clearwater Drive, Minnetonka, MN 55343, USA Tel. (612) 933 2277
Pall/Asahi Kasei	Broad range esp. microfiltration	2200 Northern Boulevard, East Hills, NY 11548, USA Tel. (516) 484 5400
Sartorius	Cassettes, cartridges, low binding UF, MF	131 Heartland Blvd, Edgewood, NY 11717, USA Tel. (516) 254 4249
Schleicher & Schuell	Flat sheet MF	10 Optical Drive, Keene, NH 03431, USA Tel. (603) 352 3810
Sepracor, Inc.	Capillary MF, capillary affinity	33 Locke Rd, Marlborough, MA 01752, USA Tel. (508) 481 7683
US Filter/Alcoa	Ceramic membranes	181 Thorn Hill Rd, Warrendale, PA 15086, USA Tel. (412) 772 1427
W.R. Grace/ Amicon	Broad range of membrane products	72 Cherry Hill Drive, Danvers, MA 01923, USA Tel. (508) 777 3622
Zenon Environmental	Systems incorporating tubes, spirals, capillaries	845 Harrington Court, Buyrlington, Ontario, Canada L7N 3P3 Tel. (416) 639 6320

MF = microfiltration; UF = ultrafiltration.

Acknowledgement: Some of this work also appeared in *Membrane Separations Technology, Principles and Applications* (Eds R.D. Noble and S.A. Stern), Elsevier, Amsterdam (1995).

References

1. Strathmann, H. and Porter, M.C. (1990). In *Handbook of Industrial Membrane Technology*, Noyes Publications, Park Ridge, New Jersey, p. 2
2. Kesting, R.E. (1985). *Synthetic Polymeric Membranes*, New York, Wiley, p. 3.
3. Zsigmondy, R. and Bachmann, W. (1918). *Anorg. Allegm Chem.*, **103,** 119.
4. Peterson, John, Millipore Corp. Internal document.
5. Porter, M.C. (1990). *Handbook of Industrial Membrane Technology*, Noyes Publications, Park Ridge, New Jersey, p. 61.
6. Goetz, A. (1960). *Microporous Filter Film on a Solid Support*, US Patent No. 2, 926104.
7. Elford, W.J. (1930). *Roy. Soc. Proc. B*, **106,** 216. (As a historical fact, Elford apparently discovered asymmetric MF membranes far in advance of the discovery of skinned RO membranes.)
8. Reid, C.E. and Breton, E.J. (1959). *J. Appl. Polymer Sci.*, **1,** 133–43.

9. Sternberg Shmuel. Personal communication.
10. Bhave, R.R. (1991). *Inorganic Membranes: Synthesis, Characteristics and Applications*, New York, Van Nostrand Reinhold.
11. Rigby, W.R., Cowieson, D.R., Davies, N.C. and Ferneaux, R.C. (1990). An anodizing process for the production of inorganic microfiltration membranes. *Trans. Inst. Metal Finishing*, **68**(3), 95–98.
12. Birch, R.D. and Artus, K.J. (1990). *Manufacture of Tubular Membranes from Aromatic Polyetherketone-Aromatic Polyether Sulfone Blends*. Eur. Pat. Appl. No. 417908.
13. Strathmann, H. (1985). Production of microporous media by phase inversion processes. In *Materials Science of Synthetic Membranes* (Ed. D.R. Lloyd), *ACS Symposium Series* 269 American Chemical Society, Washington, pp. 165–195.
Kesting, R.E. (1985). Phase inversion membranes. In *Materials Science of Synthetic Membranes* (Ed. D.R. Lloyd), ACS Symposium Series 269, American Chemical Society, Washington. pp. 131–164.
Kamide, K. and Sei-Ichi Manabe, Role of microphase separation phenomena in the formation of porous polymeric membranes. In *Materials Science of Synthetic Membranes* (Ed. D.R. Lloyd), ACS Symposium Series 269, American Chemical Society, Washington. pp. 197–228.
Hiatt, W.C. *et al.* (1985). Microporous membranes via upper critical temperature phase separation, In *Materials Science of Synthetic Membranes* (Ed. D.R. Lloyd), ACS Symposium Series 269, American Chemical Society, Washington. pp. 229–244.
14. Cadotte, J. (1981). *Interfacially Synthesized Reverse Osmosis Membrane*. US Patent No. 4 277 344.
15. DiLeo, A.J. and Allegrezza, A.E. Jr. (1991). US Patent No. 5 017 292.
Dileo, A.J. and Allegrezza, A.E. Jr. (1991). Validatable virus removal from protein solutions, *Nature*, **351,** 420–421.
16. Pum, D., Sara, M. and Sleytr, U.B. (1989). Structure, surface charge, and self-assembly of the S-layer lattice from *Bacillus coagulans* E38-66. *J. Bacteriol.* **October** 5296–5303.
17. Leahy, T.J. and Sullivan, M.J. (1978). Validation of bacterial-retention capabilities of membrane filters. *Pharmaceut. Technol.* **2**(11), 65.
18. Goel, V., Accomazzo, M.A., DiLeo, A.J., Meier, P., Pitt, A. and Pluskal, M. (1992). In *Membrane Handbook* (Ed. W. Ho and K. Sirkar), Van Nostrand Reinhold, New York, pp. 506–527.
19. Porter, M.C. (1990). *Handbook of Industrial Membrane Technology*, Noyes Publications, Park Ridge, New Jersey, pp. 156–160.
20. Belfort, G., Pimbley, J.M., Greiner, A. and Chung, K.Y. (1993). Diagnosis of membrane fouling using a rotating annular filter. 1. *J. Memb. Sci.*, **77,** 1–22.
21. Kessler, S.B. and Klein, E. (1992). In *Membrane Handbook* (Ed. W. Ho and K. Sirkar), Van Nostrand Reinhold, New York, Ch. 11–15.
22. Meier, P.M. (1988). *Aseptic filling using membrane cartridge filtration*. Wine East Buyers' Guide, 15–17.
23. Eykamp, W. (1993). In *Separation Technology in Japan* (Ed. C.J. King), International Technology Research Institute at Loyola College, Baltimore, Ch. 3.
24. Nusbaum, I. and Argo, D.G. (1984). Design operation and maintenance of a 5-mdg wastewater reclamanation reverse-osmosis plant. In *Synthetic Membrane Processes: Fundamental and Water Applications* (Ed. G. Belfort), Academic Press, New York, pp. 377–436.
25. Goel, V., Accomozzo, M.A., DiLeo, A.J., Meier, P., Pitt, A. and Pluskal, M. (1992). In *Membrane Handbook* (Eds W. Ho and K. Sirkar), Van Nostrand Reinhold, New York, Ch. 34.
26. Summers, K.J. (1993). Effects of vibratory enhanced shear on membrane fouling. Unpublished report.
27. Holeschovsky, U.B. and Cooney, C.L. (1991). Quantitative description of ultrafiltration in a rotating filtration device. *AIChE J.* **37**(8), 1219–1226.
28. Eykamp, W. (1990). *Ultrafiltration and Microfiltration*. Plenary Lecture delivered at the International Congress on Membranes and Membrane Processes, Chicago.
29. Sandblom, R.M. (1978). *Filtering Process*. US Patent No 4 105 547.
30. Marshall, A.D., Hill, B.M., Smythe, B.W. and Deller, G.J. (1992). *The Removal of Thermo-tolerant Bacteria from Milk and Whey by Microfiltration*. IMSTEC '92 Abstracts, Centre for Membrane Science and Technology, University of New South Wales, Sydney, 269–271.
31. Culkin, B. (1991). Vibratory shear enhanced processing: an answer to membrane fouling? *Chem. Process.*, **January,** 42–46.

32. Kroner, K.H. and Nissinen, V. (1988). Dynamic filtration of microbial suspensions using an axially rotating filter. *J. Memb. Sci.* **36,** 85–100.
33. Kataoka, K., Doi, H. and Komai, T. (1977). Heat/mass transfer in Taylor vortex flow with constant axial flow rates. *J. Heat Mass Transfer*, **20,** 57–63.
34. Madsen, R.F. (1977). *Hyperfiltration and Ultrafiltration in Plate and Frame Systems*. Elsevier, New York.
35. Westmoreland, J.C. (1968). *Spirally Wrapped Reverse Osmosis Membrane Cell*. US Patent No. 3367504.
36. van Reis, R. (1993). *Tangential Flow Filtration Process and Apparatus*. US Patent No. 5256294.
37. Bray, D.T. (1968). *Reverse Osmosis Purification Apparatus*. US Patent No. 3 417 870.
38. Heath, C.A. and Belfort, G. (1992). Synthetic Membranes in Biotechnology: Realities and Possibilities *Adv. Biochem. Eng. Biotechnol.*, **47,** 45–88.
39. Michaels, A.S. (1990). *Desalination*, **77,** 5.
40. Martin, J. (1993). *Recent Developments in Bacteria Challenges of Sterilizing Reade Membrane Filters*. (Proceedings of the 1993 National Meeting of the American Filtration Society, Chicago.)
41. Osumi, M., Yamada, N. and Toya, M. (1991). Bacteria Retention Mechanisms of Membrane Filters, *Pharm. Tech. Japan*, **7,** 11. (English translation by Pall Corporation.)
42. FDA (1987). *Guideline on Sterile Drug Products Produced by Aseptic Processing*. Office of Compliance, Center for Drugs and Biologics, Food and Drug Administration, Rockville, Maryland, June 1987.
43. Maiorella, B., Dorin, G., Carion, A. and Harano, D. (1991). *Biotechnol. Bioeng.*, **37,** 121–126.
44. van Reis, R., Leonard, L.C., Hsu, C.C. and Builder, S.E. (1991). Industrial scale harvest of proteins from mammalian cell culture by tangential flow filtration. *Biotechnol. Bioeng.*, **38,** 413–422.
45. Hodgson, P.H., Leslie, G.L., Schneider, R.P., Fane, A.G. and Marshall, K.C. (1992). *Microfiltration of Bacterial Suspensions: The Effect of Suspension Ionic Strength and Bacterial Surface Modifying Agents on Cake Resistance*. IMSTEC '92 Abstracts, 272–4, Centre for Membrane Science and Technology, University of South Wales, Sydney.

6 Electrodialysis

T.A. DAVIS and D.A. GLASSNER

6.1 Introduction

Electrodialysis (ED) is a membrane process wherein ions are transported through ion-exchange membranes from one solution to another under the influence of an electrical potential. The electrical charges of ions allow them to be driven through solutions and water-swollen membranes when a voltage is applied across these media. Its ability to remove ionically charged components, electrolytes, from solutions without otherwise affecting the solution composition makes ED useful in the processing of food, pharmaceuticals and numerous products of biotechnology.

In this chapter, the concepts of ED will be introduced, together with the properties of the membranes that allow selective ion transport, some of the uses to which ED has been applied successfully and some considerations in selecting and designing an ED process. The limitation and potential problems associated with the use of ED will be examined, and some effective solutions to these problems will be described.

Some important concepts of electromembrane processes are presented in Figure 6.1. In this illustration, the feed solution is a fermentation broth containing acetic acid. The acetic acid-producing microorganisms produce more acetic acid at a neutral pH, so NaOH is added to the fermentation broth to control pH. The broth is circulated through the center compartment of a compartment of a three-compartment electrolytic cell containing two ion-exchange membranes.

The fermentation product exists in the broth as Na^+ cations and Ac^- anions. Application of an electric potential to the electrodes causes the ions to move through the solution at velocities proportional to the strength of the electric field. Average ionic velocities are surprisingly slow, usually in the order of 1 mm/min but the ions do not have to travel very far to reach the membranes because the solution compartments of ED stacks are typically about 1 mm thick. The combined motion of Na^+ to the right and Ac^- to the left carries the total electric current through the bulk solution (unless other ions are present). Electrode reactions transfer the current from the solution to the anode and cathode. O_2 gas and H^+ ions are generated at the anode while H_2 gas and OH^- ions are generated at the cathode. The anion- and cation-exchange membranes, designated as A and C in Figure 6.1, are selectively permeable to ions of a specific charge. The anion-exchange

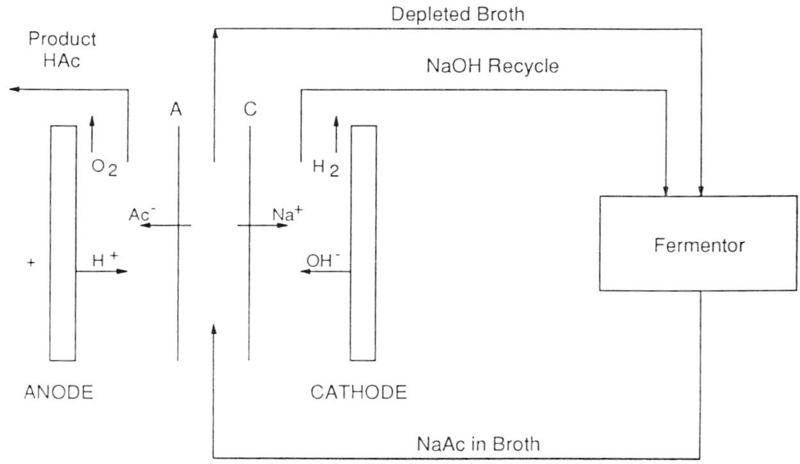

Figure 6.1 Simplified acetic acid recovery process.

membrane allows Ac^- ions to be carried by the electric potential out of the center compartment and into the anode compartment. Similarly, the cation-exchange membrane is permeable to Na^+ ions that enter the cathode compartment to form NaOH.

For reasons that will become evident later, the process illustrated in Figure 6.1 is not the most efficient way to recover a weak acid from a fermentor. One of the reasons is the relatively low concentration of the NaAc in the broth. The electrolytic cell would operate more efficiently if the NaAc were at a higher concentration than that normally produced by fermentation. Figure 6.2 shows a membrane arrangement that could be used to concentrate NaAc.

This array of alternating anion- and cation-exchange membranes is the membrane arrangement most commonly found in ED. The solutions between the membranes are alternately enriched in or depleted of NaAc when the electrodes are energized. The enriched and depleted solutions are withdrawn from their respective compartments to achieve useful changes in the electrolyte content of solutions without substantially affecting the non-electrolyte composition of the solutions. It is this selectivity for electrolytes that often makes ED the process of choice for certain separations, for example desalting of protein solutions or whey and recovery of salts of organic acids from fermentation broth.

6.2 Ion-exchange membranes

The ion-exchange membranes used in electrodialysis are essentially sheets of ion-exchange resins. Figure 6.3 illustrates the structure of a cation-exchange

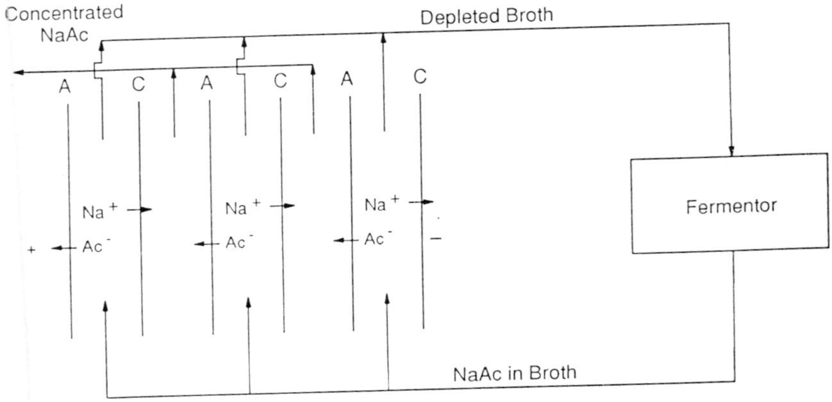

Figure 6.2 Ion transport in electrodialysis.

membrane, which has negatively charged sulfonic acid ($-SO_3^-$) groups chemically attached to the polystyrene chains. Ions with a charge opposite to the fixed charge (called counter ions) are freely exchanged at these sites.

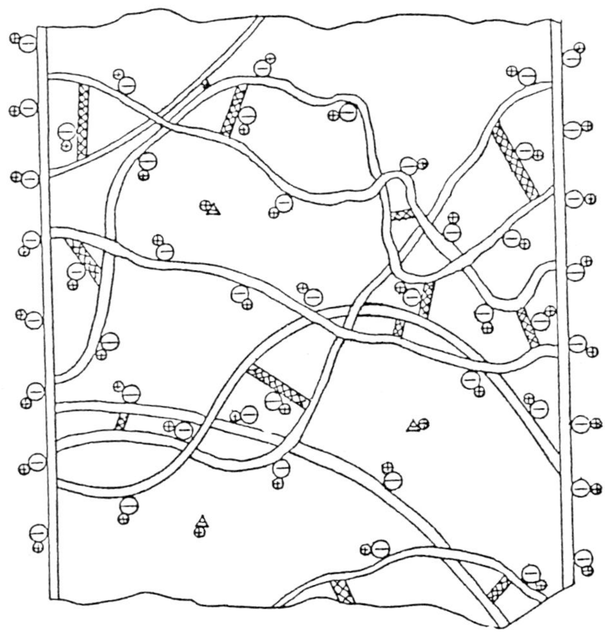

Figure 6.3 Structure of cation-exchange membrane: ⊖ fixed negatively charged exchange site, i.e. SO_3^-; ⊕ mobile positively charged exchangable cation, i.e. Na^+; △ mobile anion; ═ polystyrene chain; XX divinylbenzene crosslink.

Since there must be enough counter ions in the membrane to neutralize all of the fixed charges, the concentration of counter ions (e.g. Na^+) in the membranes is relatively much higher than the concentration of co-ions (e.g. Ac^-), therefore, counter ions carry most of the electric current through the membrane.

The exclusion of co-ions is described theoretically in terms of ionic activity and equilibria but it is more easily explained in terms of electrostatic repulsion. The fixed charges attached to the polymer chains repel ions of the same charge (co-ions), in this case the anions. This electrostatic replusion of co-ions is called Donnan exclusion. Since their concentration is very low, anions carry only a small fraction of the electric current through a cation-exchange membrane. Consequently, cation-exchange membranes selectively exclude anions and pass cations to transport electric current.

Attachment of positive fixed charges (e.g —NR_3^+) to the polymer chains forms anion-exchange membranes, which are selectively permeable to negative ions, because the fixed $-NR_3^+$ groups repel positive ions.

Since ion-exchange polymers (e.g. styrene sulfonic acid) are water soluble, cross-linking is needed to prevent dissolution of ion-exchange membranes. Divinylbenzene (DVB) is used to cross-link styrene. The degree of cross-linking and the fixed-charge density affect the membrane's properties in opposite ways. A higher degree of cross-linking improves selectivity and membrane stability, which reduces swelling but increases electrical resistance. High charge density reduces resistance and increases selectivity but it promotes swelling and thus necessitates higher cross-linking. A compromise among selectivity, electrical resistance, and dimensional stability is achieved by proper adjustment of cross-linking and fixed-charge densities.

6.2.1 Membrane types

There are two general types of commercially available ion-exchange membranes — heterogeneous and homogeneous. Both types usually contain a reinforcing fabric to increase tensile strength and improve dimensional stability. Heterogeneous membranes have two distinct polymer phases. They are made by dispersing powdered ion-exchange resin in a film-forming polymer (plasticized powder or solution) and pressing this mixture into a support fabric. Most commercially available membranes are of the homogeneous type with a continuous polymer phase containing ionic groups attached to the polymer chains. The most common methods of preparing homogeneous membranes are listed below.

1. Cast mixture of styrene (chloromethyl styrene) and DVB into support fabric and polymerize (ionics).
2. Mix plasticized PVC with styrene (or CMS) and DVB, cast on support fabric and polymerize (paste method, Japan).

3. Swell PE film with styrene and polymerize (AMF).
4. Irradiate PE or FEP film swollen with ionic monomer (Pall-RAI).
5. Treat PE film with SO_2 and Cl_2 (Israel).
6. Polymerize perfluoroalkyl vinyl ether sulfonyl fluoride co-monomer with TFE, extrude the polymer into film and hydrolyze (Dupont Nafion).

Many variations of the general methods described above are used by membrane manufacturers to produce membranes with the desired properties. In most cases, the membrane must be modified chemically to impart ion-exchange functionality. For example, in methods 1, 2 and 3 cation-exchange functionality is imparted by treatment of the styrene unit of the co-polymer membrane with a sulfonating agent such as sulfuric acid, sulfur trioxide or chlorosulfonic acid. In method 5 only hydrolysis is needed to generate the —-SO_3^- groups. Several years ago, anion-exchange functionality was imparted by treatment of the styrene unit with chloromethylether followed by treatment with an amine. Owing to its extreme carcinogenicity, chloromethylether has been largely abandoned. Alternative preparations use functional monomers such as chloromethyl styrene, which allows direct amination of the polymer, or vinyl pyridine that can be subsequently quaternized.

Since methods 1 and 2 account for most of the world's ion-exchange membrane production, some elaboration on these production methods is warranted. Method 1 is used by Ionics Inc. to produce thick, rigid membranes that are particularly suitable to the ED stacks made by Ionics. Ionics is reported to be making membranes by a continuous process.[1] A liquid comprising monomer, cross-linking agent, and a high-boiling, non-reactive diluent is applied to a fabric that is sandwiched between two sheets of polymer film. The sandwich passes between rollers that determine the membrane thickness and thence between heating elements that cause rapid polymerization. Then the containment sheets are removed, and the membrane is cut into sheets that are subjected to further processing, which can include further polymerization, chemical treatment, and leaching of the non-reactive diluent. Ionics reports the use of monomers that already have ionic functionality to avoid the need for chemical treatment.[1] Except for the inert and fully encapsulated support fabric, the Ionics membranes are an integral sheet of pure ion-exchange resin.

Method 2 is used by Tokuyama Corp. and another maker of ion-exchange membranes in Japan. This method is sometimes called the paste method because the monomers are mixed with plasticized PVC powder and a polymerization initiator before being applied as a paste to the PVC support fabric.[2] Both surfaces of the paste-impregnated fabric are covered with a polyester film, and this assembly is rolled up and cured for several hours to gel the PVC and co-polymerize the styrene and DVB. The PVC serves the dual role of a thickener for the monomers and a flexible component in the finished membrane. The monomers actualy permeate and swell the PVC powder and fabric before and during polymerization, so the PVC becomes an integral part of the polymer system. The cross-linked polystyrene is subsequently treated with chlorosulfonic acid to attach —SO_2Cl that is subsequently hydrolyzed to —SO_3^-.[3] In contrast to the thick, rigid Ionics membranes, the Tokuyama and Asahi Glass membranes are thin and flexible.

6.2.2 *Membrane properties*

Properties of ion-exchange membranes are usually measured by procedures that were adopted in the 1960s by the Office of Saline Water.[4] Manufacturers' data on the physical properties of some commercially available ion-exchange membranes are listed in Table 6.1. The diversity of membrane thickness is noteworthy. The membranes made by Ionics Inc. are thick and rigid because they must span rather wide spaces between support in the solution compartments without deflection. The thick Ionics membranes have relatively high resistance but these values are still low compared with the resistance of the brackish water they are designed to treat. RAI developed very thin membranes for

Table 6.1 Commercial ion-exchange membranes

Designation	Type	Thickness (μm)	Resistance[a] (ohm cm^2)	Counterion[b] transport (%)
Asahi Chemical, Tokyo, Japan				
K101	C	240	2.1	95.0
A111	A	210	2.5	73.0
Asahi Glass, Tokyo, Japan				
CMV	C	135	2.7	91.0
AMV	A	135	2.7	93.0
Sybron Chemical, Birmingham, NJ, USA				
MC-3470	C	343	9.6	83.0
MA-3475R	A	368	10.5	85.0
Ionics, Inc., Watertown, MA, USA				
CR-61 CZL-386	C	508	10.0	90.3
AR 103 QZL-386	A	508	12.5	86.8
AR 204 UZL-386	A	508	9.0	95.0
Rai Research Corp., Hauppage, NY, USA				
R-4010	C	114	1.0	86.0
R-4035	A	89	1.0	86.0
Tokuyama Soda, Tokyo, Japan				
C66-5T	C	155	1.4	94.0
AF-4T	A	175	2.0	96.0

[a] Resistance measured in 1.0 N KCl.
[b] Transport number measured between 0.5 and 1.0 N KCl.

battery separators and used the same radiation grafting techiques to make thin ED membranes with low resistance. The heterogeneous membranes made by Sybron are extremely tough and durable, and their anion-exchange membranes have found wide acceptance in electrocoat paint tanks as covers for anodes.

The transport number of the counter ion provides an indication of the permselectivity of the membrane. However, the conditions of measurement must be specified, because the counter-ion transport number drops as the concentration of the external solution increases. Techniques for measuring the transport number are described by Helfferich.[5]

The membrane properties reported in Table 6.1 are for new membranes. A long-term study by Kneifel and Hattenbach[6] revealed deterioration in some of these properties after prolonged exposure to 0.1 N solutions of NaCl, NaOH, and HNO_3. NaOH tended to be most destructive.

6.2.3 *Water transport*

The permeability of ion-exchange membranes to water is seldom tabulated as a membrane property because it varies so much with types and concentrations of ions. Although it is not often a significant factor in ED of dilute solutions, water transport determines the upper concentration limit of the enriching stream when ED is used to concentrate solutions. Each ion that is transported through the membranes by an electric potential drags along with it about five water molecules on average.[7] When solutions are dilute and membranes are loosely cross-linked, substantially more water transport occurs. (Membranes that are sufficiently cross-linked to restrict water transport in dilute solutions are referred to as 'tight' membranes.) Contact with concentrated solutions dehydrates the gel structure of ion-exchange membranes so that all membranes tend to transport less water when solution concentrations are high.

When a large amount of salt is being removed by ED, the volume of the depleting solution can be reduced substantially. Water removal from the depleting solution can be further enhanced by maintaining high salinity in the concentrate stream, which causes osmotically driven water transport through the membranes. Indeed, ED can be used to dehydrate organic solutions.

6.2.4 *Special membranes*

The ion-exchange membranes described above are those for typical salt-removal applications. In addition to these, there are several membranes made for special applications. The most widely used special membrane is Nafion, a fluorochemical-based cation-exchange membrane that was developed for chlor-alkali cells. Nafion is generally considered too expensive for use in ED stacks but its chemical stability often makes it the material of choice for electrolytic cells.

Composite membranes with special coatings are still considered a novelty to the casual ED user but they are used extensively in Japan for the recovery of salt from sea water. In fact, there are probably more composite membranes than conventional membranes installed in Japan. Most notable are the univalent cation-selective membranes that reject Ca^{2+} and Mg^{2+} but allow the passage of Na^+ and the univalent anion-selective membranes that reject SO_4^- while passing Cl^-. This selectivity is normally achieved by coating the membrane surface with a thin layer of ionic polymer of opposite charge.[8] This coating causes some repulsion of all counter ions but the repulsion is much stronger for multivalent counter ions.

Bipolar membranes are composites with the ability to reject virtually all ions. They essentially comprise a layer of anion-exchange polymer on one surface and a layer of cation-exchange polymer on the opposite surface.[9] The properties and use of bipolar membranes will be discussed extensively later in this chapter.

6.3 ED stacks

An ED stack consists of alternating anion- and cation-exchange membranes with solution compartments between them. The solution compartments are bounded by perimeter gaskets that are pressed tightly between the membranes to confine the solutions within the compartments. The compartments usually contain spacers that keep the membranes separated by a constant distance. The spacers also aid in distributing the solution velocity evenly throughout each compartment. (Poor flow distribution within and among the compartments severely limits the performance of an ED stack.) Each solution compartment has a means for introduction and removal of solutions. These features are often combined into a one-piece plastic device called a gasket-spacer. ED stacks can be designed to use identical gasket-spacers for enriching and depleting compartments.

An exploded view showing the components of an electrodialysis stack is presented in Figure 6.4. The end plate is typically made of rigid plastic. It contains an electrode and the connections for the solution that rinses the electrode. It also contains connections for solution streams into and out of the enriching and depleting compartments. Holes in the inside face of the end plate are aligned with holes in the membranes and gasket-spacers to form manifolds for solution flow through the ED stack. Large stacks have multiple manifold holes on both ends of the gasket-spacers to provide uniform flow over the width of the compartment. The two membranes and two gasket-spacers illustrated in Figure 6.4 constitute one cell pair. Commercial ED stacks typically contain hundreds of cell pairs between a single pair of electrodes.

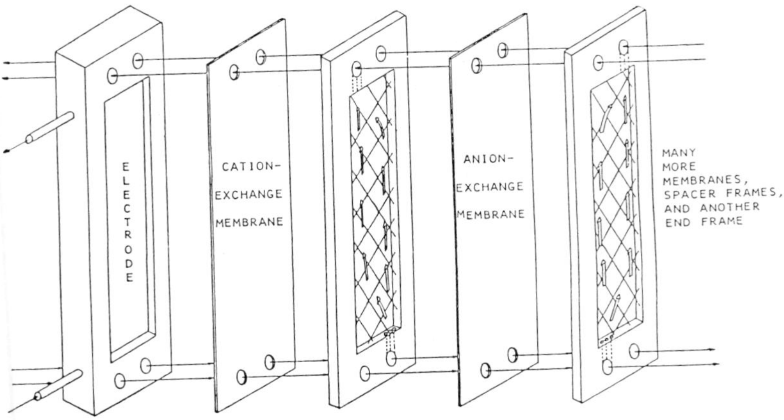

Figure 6.4 Solution flow in electrodialysis stack.

The ED stack in Figure 6.4 represents the sheet-flow type. The sheet-flow design is used by most ED manufacturers. (Ionics is the major exception.) The spacing between membranes of sheet-flow stacks is made as thin as practical to reduce electrical resistance and minimize the thickness of hydraulic boundary layers at the membrane surfaces. The thinness is ultimately limited by hydraulic resistance and by problems associated with getting solutions into and out of the solution compartments between the membranes. To maintain constant spacing between the membranes, it is necessary to support the membranes at close intervals, but the support material must not cover a large fraction of the membrane surface or cause stagnation in the solutions near the membrane surface. The Vexar non-woven polyolefin netting developed by DuPont has gained wide acceptance as a spacer that meets these criteria. The flow of solution in the plane of Vexar netting is illustrated in Figure 6.5.

The stack design developed by Ionics Inc. is quite different from the sheet-flow stacks previously described. The Ionics design utilizes the tortuous-path spacer illustrated in Figure 6.6 to direct solutions over a relatively long, open channel in the solution compartment. Superimposed on this channel is a spacer that supports the membrane and promotes turbulence in the solution compartment. These supports are more widely spaced than those of a sheet flow design, so a more rigid membrane is required to bridge the space without severe deflection. Therefore, the membranes made by Ionics are substantially thicker than those produced by most other manufacturers. The thick membranes and spacers result in higher electrical resistances than those found in sheet-flow stacks but the openness of the solution channels should make them less prone to fouling by suspending solids.

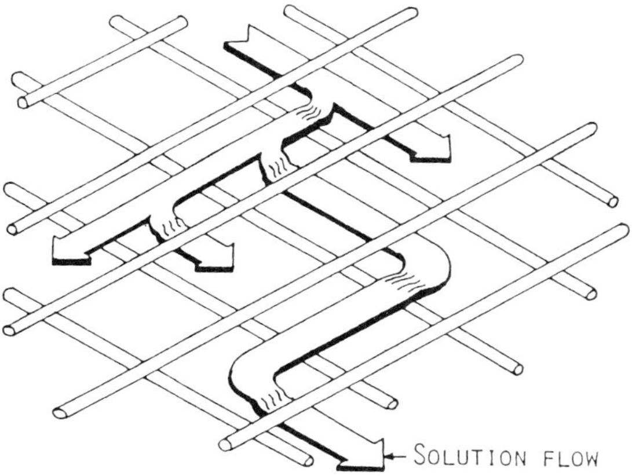

Figure 6.5 Solution flow through Vexar spacer.

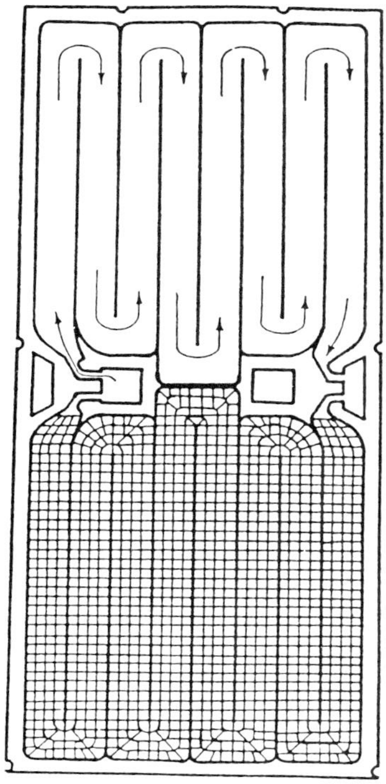

Figure 6.6 Tortuous-path spacer for electrodialysis.

6.4 Applications of ED

Outside of Japan, the largest application of ED has been in the desalination of brackish water. Most of the early research and development effort had the objective of improving marginal waters to make them potable. During the 1950s, the Netherlands National Research Organization, TNO, and the Office of Saline Water in the USA funded programs to improve membranes, equipment and processes to produce potable water more economically.

A recent growth area for ED is in the rough desalting of water that will be subjected to subsequent purification for use as boiler feed or rinse water in the electronics industry. Ion exchange has traditionally been used for preparing waters with low salinity but the cost of regenerants and the magnitude of the waste-disposal problem are proportional to the salinity of the feed water. The bulk of the dissolved solids can be removed more economically by ED or RO. These two processes are competitive in cost, and both offer the advantage that they do not contribute additional water pollutants.

ED desalting of sea water has also been demonstrated on a commercial scale but high-power consumption, 28 watt-h/gal with low resistance ion-exchange membranes,[10] makes ED uneconomical. Still, the greatest use of ED treatment of sea water has been for the recovery of NaCl. With no natural deposits of NaCl, the Japanese traditionally obtained the salt by solar evaporation of sea water. However, solar ponds occupied acreage that was needed to grow rice. Now the Japanese use ED to produce a brine concentration of about 20% TDS with less than 200 kWh of energy consumed per ton of salt recovered. Moreover, the use of membranes that are selective for univalent ions has improved salt purity to 97% and reduced precipitation problems. The approach has been so successful that essentially all the edible salt in Japan is prepared from ED enriched brines.[11]

Food processing provides many potential applications for ED because of its ability to separate electrolytes from non-electrolytes. The largest use of ED has been in whey deashing. Whey is the waste product from cheese making, and it contains useful quantities of proteins, lactose and lactic acid. However, its high mineral content makes whey unacceptable for human consumption and of marginal value as animal feed. Deashing by ED upgrades the whey so that subsequent processing can produce edible whey solids. The deashing of whey is discussed in more detail in the section on electrodialysis applications in biotechnology. Amino acid applications are also discussed in that section.

Cows' milk is more salty than milk from human mothers, and this limits its use in the preparation of infant formula. Desalting of cows' milk by ED allows larger quantities of cows' milk solids to be used for these purposes. Research has shown that desalting by ED to remove calcium improved the protein stability of frozen skim milk and its concentrates.[12] The lumpy texture of thawed frozen milk has been attributed to clumping of micellar casein, and calcium removal by ED led to the dispersion of casein from the micellar to the serum form.

The opposite effect was observed in blood ultrafiltrate. With the apparatus shown in Figure 6.7, Jain[13] used ED to desalt the blood plasma. Chilling aided in the precipitation of proteins, which were removed by ultrafiltration. The salt content of the plasma was restored by passing it through the concentrate compartments of the ED stack before returning it to the host. Such a process could be used to treat certain diseases where excessive amounts of undesirable proteins are present in a patient's blood or to obtain needed proteins from a donor.

Metal finishing processes offer numerous applications for ED in pollution control and material recovery. The rinse streams from such processes pose particularly troublesome pollution problems. They are usually too dilute for direct metal recovery and too concentrated for disposal. ED processing of a rinse stream from a nickel electroplating system[14] is illustrated in Figure 6.8. The used rinse water flows through the depleting compartments of the ED stack where the metal ions are transferred into the concentrate stream. The treated rinse water can then be re-used in the process. The concentrate stream can be recirculated to build up its metal content to a level that is useful for further recovery or direct return to the plating bath.

The discussion of ED applications will end on a sweet note with a description of a process to deacidify citrus juices. During the early part of the harvest season, autumn and early winter, citrus juices have high acidity levels that impart a tart flavor. In the spring, the acidity level decreases with a consequent improvement in flavor. In the early 1960s, AMF developed the ED process shown in Figure 6.9 to remove excess citrate ions. The ED stack contained only anion-exchange membranes. The citrus juice and KOH

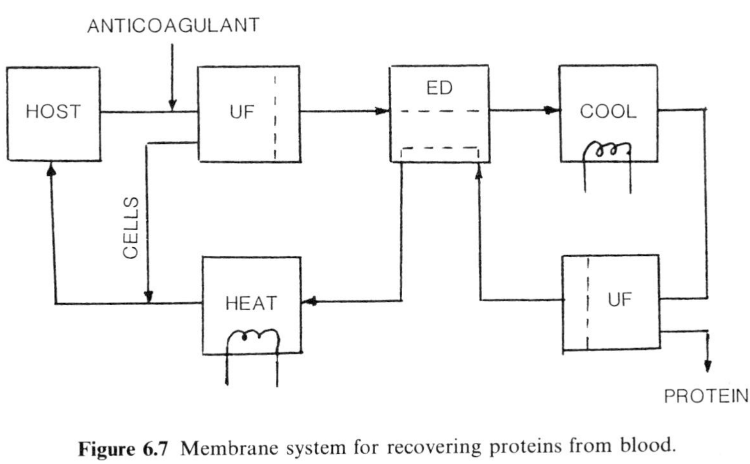

Figure 6.7 Membrane system for recovering proteins from blood.

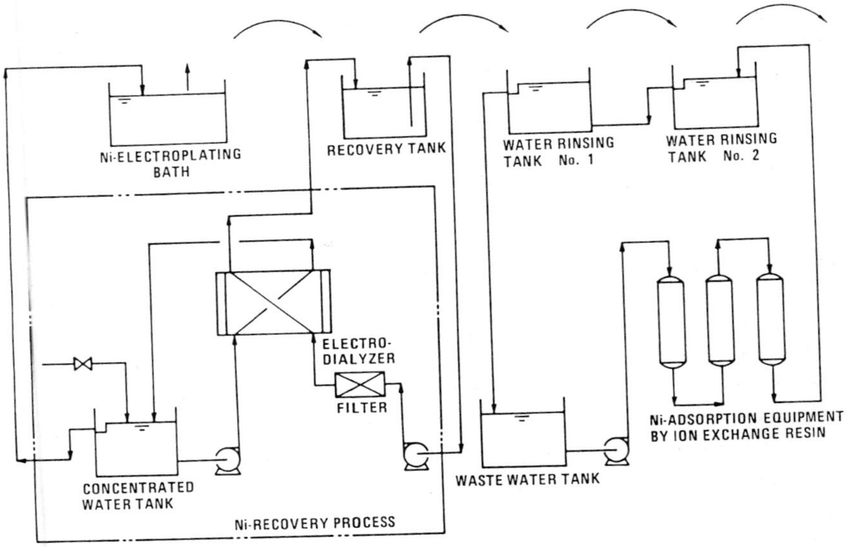

Figure 6.8 Flowchart of treatment of multi Ni-electroplating effluent.

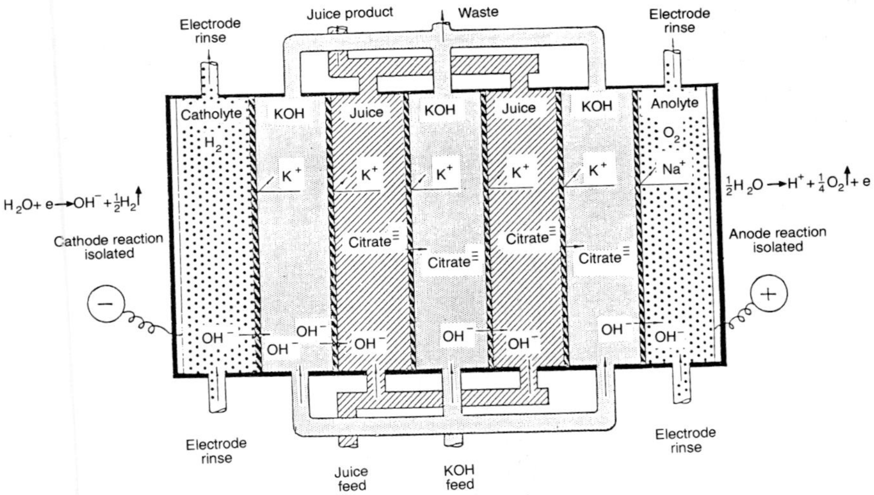

Figure 6.9 De-acidification of fruit juice by electrodialysis.

solution flowed through alternate compartments. The electric current carried the citrate ions out of the juice into the KOH solution, and OH^- ions moved through the opposite membranes to neutralize the H^+ ions. Owing to the symmetrical membrane arrangement, polarity reversal could be performed to remove pulp from the membrane surfaces without any disruption of solution flows. The process was demonstrated on a pilot scale to be effective in deacidifying both orange and grapefruit juices.[15]

6.5 Bipolar membranes to produce acids and alkalis

Bipolar membranes consist of an anion-exchange membrane and a cation-exchange membrane laminated together. Application of an electric potential to the bipolar membrane exposed to an electrolyte solution will cause current flow but the amount of current flow will be determined by the orientation of the membrane. If the cation-exchange membrane faces the anode, the current will be relatively high. If the anion-exchange membrane faces the anode, the current will quickly drop to a low value as the anions and cations are depleted from the membranes. Subsequent elevation of the potential in the latter case will increase the current, which is carried by the only available ions, H^+ and OH^- ions generated at the junction by water splitting. This results in the production of acidic and basic solutions at the surfaces of the bipolar membrane, as illustrated in Figure 6.10. Multiple bipolar membranes can be placed between a single pair of electrodes in an ED stack along with other ion-exchange membranes for the production of acid and base from a neutral salt.

The major benefit to the use of bipolar membranes is their ability to convert a salt into its constituent acid and base without generating unwanted by-products. Since there are no gases evolved at the bipolar membranes, the energy associated with gas evolution is saved, and the power consumption

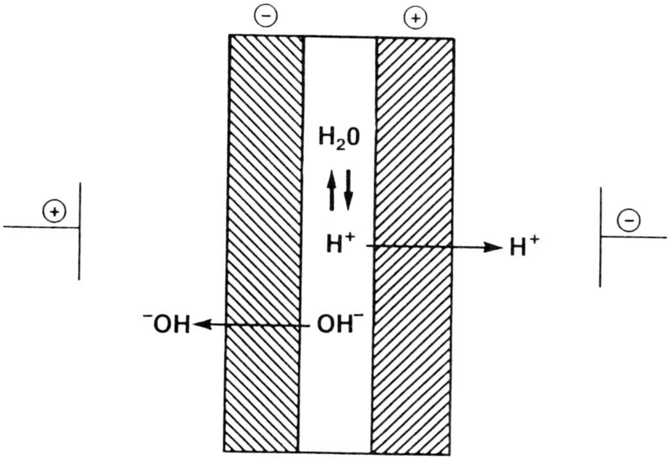

Figure 6.10 Water-splitting in bipolar membranes.

is about half that of electrolytic cells. Compared with the electrodes in conventional electrolytic cells, the bipolar membranes are inexpensive. Where dilute (up to 4 N) acids or bases are needed, bipolar membranes offer the prospect of low cost and minimum unwanted byproducts.

Figure 6.11 illustrates how HCl and NaOH are generated from NaCl. The Cl^- ions from the salt solution move through the anion-exchange membrane where they meet H^+ ions generated at the bipolar membrane to form HCl. Likewise, the Na^+ ions pass through the cation-exchange membrane and join the OH^- ions to form NaOH.

The acid/base production process is illustrated in Figure 6.12. Salt is dissolved in depleted brine, and the resulting saturated brine is softened with chelating ion-exchange resins. This softening is required to prevent the formation of hydroxide precipitates of Ca or Mg in the base compartments. The saturated brine is metered into the solution circulating through the bipolar membrane stack to maintain the salt concentration at about 10%. The dilute brine displaced by this addition is returned to the salt saturator. The concentrations of acid and base are allowed to build up to their respective recirculating streams until the desired product concentration is achieved. Then the products are discharged to their respective receiving tanks, and the volume is replaced with deionized water. This fill-and-drain mode of operation produces higher purities of acid and base than can be obtained with continuous addition of dilution water to the recirculating acid and base. Acid and base purities exceeding 99% were obtained with this method, and the only impurity was NaCl.[16]

The acid and base produced with bipolar membranes are rather dilute (3–4 N) compared with the concentrated acid and base sold commercially, so the process is not expected to make products for shipment. However, efforts to generate higher concentrations of acid and alkali are progressing.

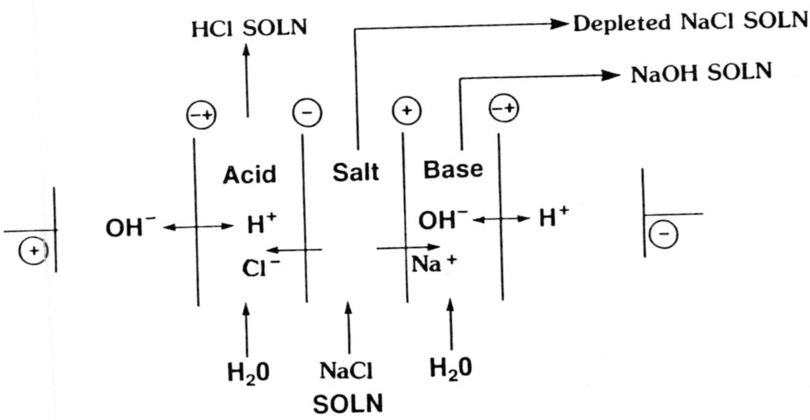

Figure 6.11 Salt splitting in bipolar membranes.

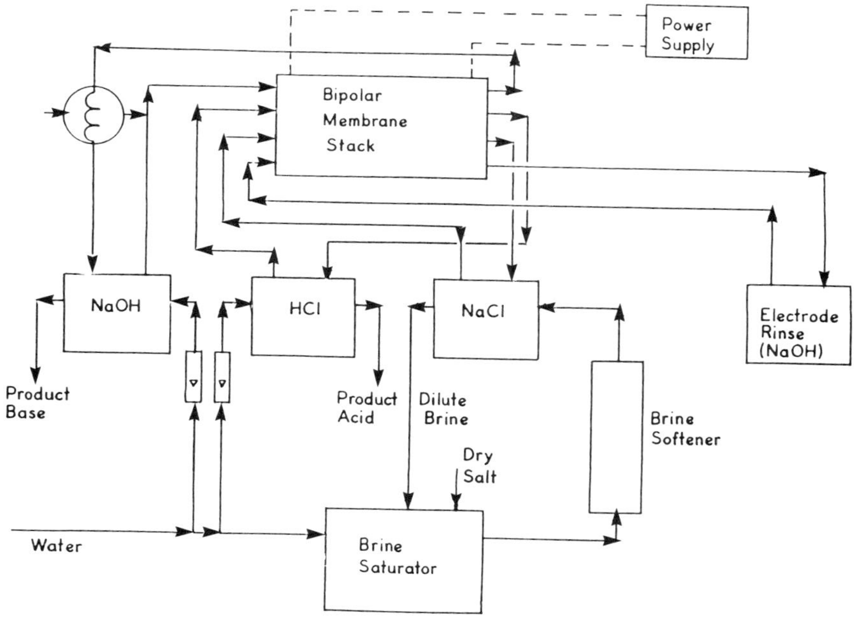

Figure 6.12 System for producing acid and base from salt.

When dilute acid and base, say up to about 3 or 4 N, are needed on site and a source of clean salt is available, then on-site generation with bipolar membranes is an alternative to purchasing them.

6.6 Membrane fouling

When the feed solutions to ED stacks are clean and relatively free of sparingly soluble materials, both desalting and water-splitting ED stacks can be operated for years with little concern about membrane fouling. However, most feed waters have constituents that can cause problems for ED stacks. Such problems are usually handled by reducing current densities, periodic current reversal (the EDR process), or pretreatment of the feed water.

The limiting current density (LCS) is the current density at the onset of concentration polarization. Concentration polarization is caused by low electrolyte concentration near the membrane surface. Since the electrolyte concentration is low, the voltage drop is increased. The limiting current density needs to be known to allow the operating current density to be chosen. In a later section, determination of the LCD is described. Typically, the electrodialysis stack is operated at 50–75% of the LCD to make sure concentration polarization is minimized during operation.

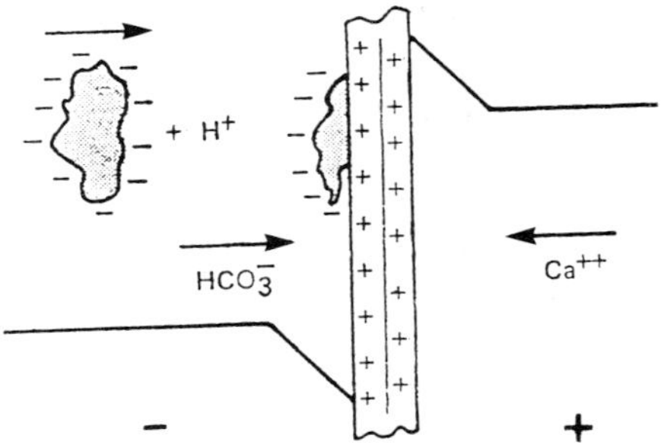

Figure 6.13 Fouling of anion-exchange membrane.

Figure 6.13 illustrates some of the fouling problems. Large organic molecules with ionizable groups (e.g. anionic surfactants and humic acids from decomposing vegetation) are troublesome foulants for ED membranes. Their negative charges allow them to migrate in the electric field but their large size prevents their passage through the membrane. Consequently, these large charged molecules tend to accumulate on the dilute side of anion-exchange membranes and cause a build-up of membrane resistance. Colloidal particles, which are also usually negatively charged, cause similar problems. These problems are worse when the electrodialysis stack is operated above the LCD and concentration polarization occurs.

Sparingly soluble salts such as $CaSO_4$ and $CaCO_3$ can precipitate on the concentrate side of the membrane, because their concentrations are highest there caused by concentration polarization. $CaCO_3$ precipitation occurs when the limiting current density is exceeded. The reaction of HCO_3^- ions with OH^- ions generated by water splitting forms CO_3^- ions that pass through the anion-exchange membrane and precipitate in the boundary where the concentration of Ca^{2+} ions is the highest.

Problems of membrane fouling are generally less severe when current densities are kept low to minimize concentration polarization. $CaSO_4$ precipitation is countered by limiting the concentration of $CaSO_4$ in the enriching stream below the saturation level and by addition of sodium hexametaphosphate to disrupt crystal formation. $CaCO_3$ precipitation can be prevented by acidification of the enriching-stream feed. Fouling of membranes by organic anions and colloidal material is best handled by pretreatment to eliminate the offending material. Alternatively, these membrane-fouling problems can be alleviated to some extent by suitable cleaning-in-place (CIP) procedures or by the use of the EDR process.

Bioprocessing fluids typically contain proteins and other non-ionic species. These species are very weakly charged and migrate slowly in an electric field. They can foul electrodialysis membranes over sufficiently long periods of time or when concentration polarization conditions exist. Suitable CIP procedures must be developed to solve bioprocess fluid fouling.

EDR is the electrodialysis-reversal process developed by Ionics Inc. to desalt brackish water.[17] The EDR process employs reversible electrodes and automatic valves to swap the flows of enriching and depleting product lines. The reversal is programmed to take place about four times per hour. This frequent reversal tends to dislodge the organic and particulate foulants from the membrane surfaces. Moreover, since current reversal causes concentrating boundary layers to become depleting boundary layers, freshly formed precipitates tend to dissolve. Immediately after the current reversal, the product water is of poor quality because of foulant discharge. Therefore, there is a period of a few minutes when the product water must be diverted to waste. The loss of feed water and the high costs associated with the automatic switching valves are the major drawbacks to EDR for desalination. Loss of product during the switchover could also preclude the use of EDR to desalt valuable feed streams.

6.7 Specific electrodialysis applications in biotechnology

This section will cover examples of specific recovery applications in biotechnology. Two broad categories of applications are discussed. The first is the removal of salts from bioproducts. In this case, the desired product is non-ionic and is recovered from the diluting compartment. The second type of application is the recovery of acids or salts from bioprocess streams. In this case, the product is the ionic species and is recovered in the concentrate stream.

6.7.1 Removal of salt from bioproducts

In bioprocessing, electrodialysis has found applications in the deashing or salt removal from product solutions containing the impurities. An example is the desalting of whey for use in ice-cream, cake, bread and baby food.[18] Whey desalting using electrodialysis is practised in all parts of the world. About 3.0 million tons of whey were processed using electrodialysis in 1984.[19] This represents the use of over 25 000 m^2 of installed membrane.

The inevitable accumulation of whey solids on the membranes and spacers requires routine cleaning-in-place and periodic disassembly of the stacks for mechanical cleaning. The apparatus for this application must be of sanitary design. Generally, the apparatus is cleaned and sterilized with dilute caustic and hydrochloric acid once a day.[19] These operations are now well established and can be applied to the treatment of other food products or biotechnology products.

Electrodialysis can be used to purify protein solutions.[19,20] Electrodialysis offers the following advantages for desalting proteins: (i) desalting low- or high-molecular-weight species efficiently with high yields; (ii) desalting without diluting the product; and (iii) precise control of the desalting process.[19] An example of protein desalting is the removal of 99% of the salt from an alpha-interferon solution. The interferon product concentration is also doubled during the process. Other products for which the electrodialytic desalting has been effective include Bovine Serum Albumin, Low Molecular Weight Polypeptide, Vasopressin and Angiotensin I-125.[19]

Low-molecular-weight peptides can be synthesized from amino acids using the appropriate enzymes. It is important to separate the product and reactants for downsteam processing of the product. Electrodialysis is used to remove the amino acids from the desired peptides.[21] The amino acids removed are recycled as reactants and the product is ready for further use.

Amino acid recovery using electrodialysis has taken two forms. One method is the direct recovery of the amino acid by electrodialysis. Another use of electrodialysis is to treat waste streams after the amino acid has been removed by ion exchange. The application of ion-exchange membranes and electrodialysis to the manufacture of amino acids is reviewed by Odagiri.[22]

Figure 6.14 shows the use of electrodialysis to treat the waste stream produced by the fermentation and ion-exchange recovery of lysine.[23] The lysine is collected by ion exchange and the spent broth is concentrated using reverse osmosis. Desalting electrodialysis produces a low-salt organic stream and a concentrated salt fraction. The concentrated salt fraction can now be recycled or reused. The organic stream can be used as a feedstuff.

Other separations of proteins and amino acids have been achieved by pH adjustment of the solution being treated. As in electrophoresis, a molecule at its isoelectric point will not migrate in an electric field. However, the same molecule migrates as an anion at a pH above its isoelectric point and as a cation below its isoelectric point. Therefore, if the pH of such a solution is maintained at the isoelectric point of a particular amphoteric solute, the solution can be desalted by ED with minimal loss of the subject solute.

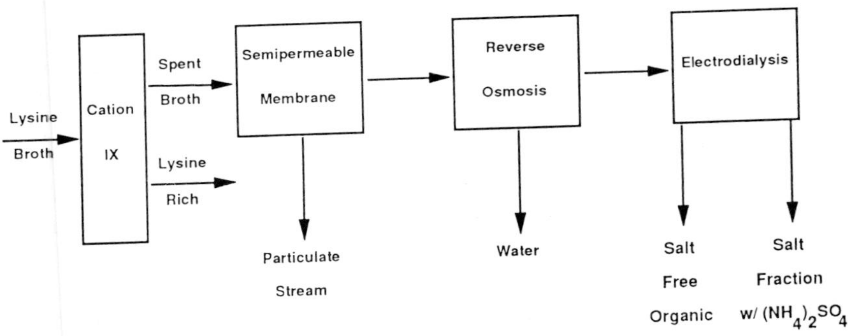

Figure 6.14 Electrodialysis processing of lysine waste stream.

Gavach studied the recovery of amino acids from the hydrolysis of poultry feathers.[24] The hydrolysate was neutralized and treated by ED to remove excess NaCl. During the desalting, it was observed that about 25% of the arginine, histidine and lysine, which are ionically charged at neutral pH due to amine groups, were lost through the membranes when 95% of the salt had been removed. Losses of other amino acids during desalting were much lower.

Soy sauce has a natural salt content of 16–18%. Electrodialysis can be used to reduce the salt content to low concentrations without adversely affecting the flavor, color or taste of the sauce. The low-salt soy sauce can then be used in low-salt diets and other applications requiring a reduced salt content.

The production of ethanol from lignocellulosic biomass is aided by the use of acid electrodialysis to recover sulfuric or hydrochloric acid used to hydrolyze cellulose.[25,26] Cellulosic material is subjected to a strong acid to produce the monomer glucose. The acid needs to be recovered and recycled for an economic process and acid electrodialysis can be used to accomplish this process. This process presents a potential large-scale use of electrodialysis because of the abundance of lignocellulose and the demand for liquid fuels.

6.7.2 Recovery of organic acids

The recovery of organic acids from fermentation broths using electrodialysis has been researched as long ago as 1954.[27] Early work was performed using Rohm and Haas 'Amberplex membranes' and Ionics Inc. 'Nepton membranes'. Although ion-exchange membranes have undergone many improvements over the years, these improvements have not yet resulted in major applications in the recovery of organic acids.

A major problem in the recovery of organic acids from fermentation broths has been the elimination of cations. Fermentations typically perform better in pH significantly above the pKa of the acid being formed. Fungal fermentations, like citric acid, are a notable exception. Not surprisingly, citric acid is the only organic acid produced by fermentation in reasonably large volumes.

Bipolar membranes offer a solution to the cation-elimination problem accompanying most organic acids. The bipolar membrane allows a salt to be split into the corresponding alkali and acid solutions. Recent ion-exchange membrane breakthroughs include the marketing of bipolar membrane systems by Aquatech Systems (Warren, New Jersey), WSI Technologies (St Louis, Missouri) and Graver Water (Union, New Jersey), which use membranes produced by Tokuyama Corporation (Tokyo, Japan).

The recovery of organic acids using bipolar membranes can be classified by the type of bipolar membrane-stack configuration used. Figure 6.15 shows the three bipolar stack configurations. The first described, in Figure 6.15a, is the three-compartment water-splitting membrane stack. The stack contains

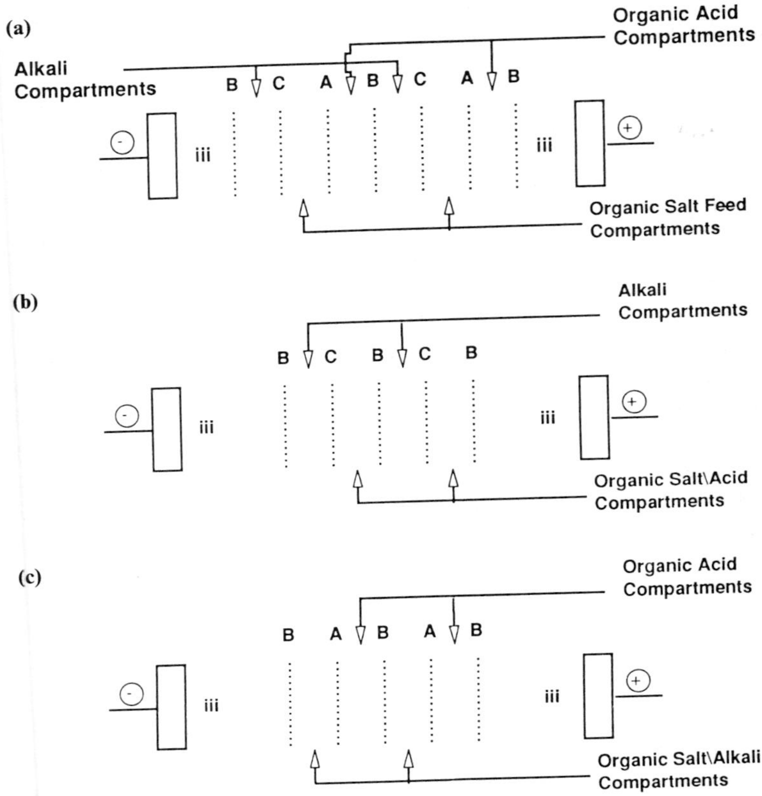

Figure 6.15 Bipolar membrane stack configurations for conversion of organic salts to organic acids and corresponding alkali: (a) three-compartment stack; (b) two-compartment cation exchange; (c) two-compartment anion-exchange stack.

a series of bipolar, cation and anion exchange membranes. The three compartments formed between the membranes contain the salt feed, the acid product and the alkali. The product acid can also be made using two-compartment membrane stacks consisting of bipolar membranes alternated with either anion- (Figure 6.15b) or cation- (Figure 6.15c) exchange membranes.

Several processes are reported to use each of the configurations shown in Figure 6.15. Mani and Johnson[28] describe a process that uses a three-compartment bipolar membrane stack as shown in Figure 6.15a. Van Nispen *et al.*[29] and Czytko *et al.*[30] recommend use of a two-compartment bipolar stack with cation exchange for organic acid recovery (Figure 6.15c). Glassner *et al.*[31] and Berglund *et al.*[32] describe processes using a recovery as shown in Figure 6.15b. All of the above systems have positive and negative features.

Development work on all three has been substantial in the past 5 years. The above systems are discussed more in the following paragraphs.

Van Nispen *et al.*[29] inserted an ultrafiltration unit prior to the bipolar membrane stack (two compartment-anion exchange, Figure 6.15c). The ultrafiltration units removed cell and other high-molecular-weight debris before the fermentation broth was processed by electrodialysis. The same configuration was described by Mani and Johnson.[28] They specified a three-compartment bipolar membrane stack as shown in Figure 6.15a.

Glassner *et al.*[31] and Berglund *et al.*[32] described a method for recovery of lactic and succinic acids that uses desalting electrodialysis and a two-compartment cation-exchange bipolar-membrane stack. The desalting electrodialysis step purified and concentrated the fermentation products. The purified and concentrated salts were then treated by water-splitting electrodialysis (two-compartment bipolar stack with cation exchange) to yield an alkali and the succinic acid or lactic acid product. The alkali was then recycled to the fermentation.

Despite all the interest in using ED to recover organic acids, no major installations were operating at the time of writing. With the emergence of bipolar membranes it seems likely that organic acid applications will soon appear.

6.8 Electrodialysis process design for bioprocessing

6.8.1 *Process selection*

The selection of an appropriate electrodialysis process or combination of processes to be used is determined by a careful definition of the desired product and consideration of how waste can be minimized. The product can be ionic or non-ionic. Typically, the non-ionic product would be produced in the diluting compartment of a desalting electrodialysis stack. Ionic products could be a salt, an acid, a base or an acid and a base. These applications require selection of the appropriate desalting, or bipolar electrodialysis operations. An appropriate solution may be a combination of electrodialysis operations or an appropriate design of membrane cell pairs. A number of examples of membrane cell stack and electrodialysis operations have been discussed earlier in this chapter.

Desalting electrodialysis. Electrodialysis in bioprocessing applications is used both for purifying products (non-ionic product) and for the recovery of products (ionic product). In a bioprocessing application, the biological stream with ionic and non-ionic components will flow through the electrodialysis diluting compartment. ED moves the desired quantity of ionic components to the concentrating compartment where the ionic product is a concentrated and purified salt. The non-ionic product is desalted to the desired degree and removed from the dilute compartment.

Water-splitting electrodialysis. Many bioprocesses produce acids that are desirable products. Usually, the bioprocess is sensitive to the concentration of acid present and an alkali must be added to neutralize the acid. Hence, the bioprocess would actually produce a salt. Water-splitting electrodialysis is capable of splitting the salt into the desired acid and the corresponding alkali. This alkali can be recycled to the bioprocess. Three configurations for water-splitting ED are shown in Figure 6.15. The use of three membranes (bipolar, anion and cation) in the repeating cell allows recovery of pure acid and base because the salt remains isolated from the product streams. However, weak acids or bases are not sufficiently ionized to carry the electric current through their respective product streams. In such cases, it is better to use only two membranes and allow some of the salt to remain in the product to maintain adequate conductivity.

6.8.2 Data required for design

The data required to design an electrodialysis unit depends on stack design, solution properties, hydrodynamics and the limiting current density (LCD). Additional data needed to design the electrodialysis system include the degree of desalting desired, the quantity of feed to be processed and the quantity of product desired. The composition of the feed should be known so that potential fouling or chemical reactions can be addressed.

The LCD must be determined using a representative sample of process fluid. This is especially true of bioprocess feeds, which are known to foul membranes.[33] There are two general methods to determine the LCD. One method is to measure the current as a function of the voltage.[34,35] The resulting data are plotted as V/I versus 1/I. This is called a Cowan plot,[36] which exhibits a point of inflection at the onset of polarization. This represents the LCD for the salt tested. A second method is based on the water-splitting activity that occurs during operation above the limiting current density.[37] The pH of the diluting compartment falls and the pH of the concentrating stream rises. A predetermined change in pH represents the beginning of concentration polarization. The current density at this pH change is the LCD. The same pH change must be used to yield good comparative results.

The flow compartments in a desalting electrodialysis stack are the anolyte, catholyte, concentrate and dilute. The flow systems should be configured to give co-current flow of the fluids through the electrodialysis stack. Counter-current flow results in an imbalance of pressures across the membranes that cause membrane damage and leakage of solution between the adjacent compartments. The system should also be designed with minimal hydraulic resistance in the piping downstream of the ED stack so that adjustments in feed rates do not disturb the pressure balance across the membranes.

6.8.3 *Special considerations for biotechnology applications*

The use of electrodialysis processes in biotechnology requires management of membrane fouling as do all electrodialysis processes. Czytko *et al.*[33] reported 30–100 h operation before membrane replacement was needed. To counter the problem, a microfiltration unit was used to recycle the cells and other debris. Management techniques applicable to bioprocesses have been developed for whey desalting[38] and are applicable with modification to processes described here.

Bioprocessing fluids will require special treatment both prior to processing and then to keep the stack and lines clean. It is important to remove particulate matter from process fluids prior to electrodialysis. Cell removal is widely documented as necessary to prevent electrodialysis membrane-stack fouling.[33,39] Microfiltration or ultrafiltration,[29] centrifugation and cell immobilization are some of the methods used to prevent cells from entering the electrodialysis stack. To maintain sanitary conditions and remove fouling materials, warm, dilute caustic and hydrochloric acid are used once a day to clean whey processing membrane stacks.[19,20] Strong caustic solution and citric acid were used every 24 h to keep an electrodialysis stack used for the recovery of lactate salt.[40] For new bioprocessing applications, the development of a new cleaning procedure may be required.

6.8.4 *Choice of electrodialysis membrane*

Since there are several commercial suppliers of anion- and cation-exchange membranes and equipment, a user has a wide variety of products from which to choose. Conversely, bipolar membrane choices are limited since they have been available commercially for only a short period of time. However, as applications of bipolar membranes increase, it is likely that more bipolar membrane choices will become available.

The cation- and anion-exchange membrane choices reflect material properties and permeability to cations and anions. As discussed earlier, selectivity to the transport or multivalent ions, protons and large molecules can be imparted to some membranes as well as mechanical strength and fouling resistance. Choice of these membranes is dependent on the application being considered. Bipolar membranes must be used when it is desired to generate alkali and acid. Bipolar membranes can be configured in two or three membrane cells. Choice of the membrane cell is dependent on the process application under consideration.

6.8.5 *Operating configurations*

Electrodialysis stacks can be operated in four different flow configurations. Continuous, batch, feed and bleed and current reversal are the four modes of operation.[34] Each has advantages for a particular application.

Continuous operation has the dilute and concentrate streams fed continuously from the feed source through each of the electrodialysis stages. Some recycle of dilute or concentrate streams from the last stage outlet to the first stage inlet is possible. When the feed rates and concentrations do not change, then flow rates, concentrations, voltages and current densities remain unchanged with time.

Batch operation is suitable for processing smaller quantities of material to a high degree of desalting. Even in ED stacks with a long solution path, less than half of the salt is removed in a single pass. With enough passes, however, even a small ED stack can eventually remove an appreciable amount of salt from a batch of feed. In practice, the dilute stream is cycled through the electrodialysis stack and returned to a dilute tank until the desired desalting is achieved.

Feed and bleed operation can be viewed as a combination of batch and continuous operating modes. The concentrate and dilute are recycled but feed is added continuously and product is bled off continuously. The desired salt concentrations in the concentrate and dilute streams are controlled by the recycle rates and the amount of salt removed by the stack.

Current reversal, commonly known as EDR, is an operating mode in which the direction that electric current flows through the stack is reversed periodically. The objective is to reduce membrane-fouling problems. Current reversal can be used with any of the operating schemes discussed previously.

In the current-reversal mode the polarity of the voltage applied to the ED stack is reversed on a periodic basis. The time between reversals might be 15–20 min. The stack is plumbed with three-way valves so that dilute and concentrate streams into and out of the stack can be reversed automatically. The electrodes must be constructed so that they can function as anode or cathode.

The EDR process has been successful in reducing fouling in water desalination where the concentration solution goes to waste. Current reversal would also be expected to remove membrane-fouling materials in bioprocessing as well but the utility of EDR might depend on the fate of the foulants dislodged by EDR. If such materials are unacceptable for return to the concentrate or dilute, current reversal could only be used when the ED stack is off-stream for cleaning.

6.8.6 *Process optimization*

Electrodialysis processes have three types of costs.[37] The first type is a cost that increases with current density. The cost of energy is the key component of this category. Increased current density causes increased voltage drop and power consumption. The second type of costs are those that decrease with increasing current density. The initial capital investment and, therefore, the depreciation charge is less, and the cost of membrane replacement is reduced. The third cost type are those that do not change with current density. Chemical costs, labor and maintenance are the costs in this category.

Since some costs increase with current density and some costs decrease with increasing current density, an optimum current density for minimizing costs can be projected. However, in most applications and particularly in bioprocessing applications the limiting current density is less than the economic optimal current density, because operation at a current density above the LCD increases the voltage drop greatly. This causes high power consumption and increases the electricity cost. Furthermore, in bioprocessing applications, as well as most other applications, operation of the electrodialysis stack at a fraction of the limiting current density is prudent. A typical operating current density is 50–75% of the limiting current density. Under this operating condition, the lowest operating costs possible are achieved.

6.9 Original equipment manufacturers

A list of the manufacturers of ion-exchange membranes for use in electrodialysis follows. These manufacturers or their vendors will help provide electrodialysis equipment and design for applications.

Asahi Chemical Co., Ltd., 3-2, Yako 1-Chome, Kawasaki-Ku, Kanagawa-Ken, 210, Japan.
Asahi Glass Co., Ltd, 1-2 Marunouchi, 2-Chome, Chiyoda-Ku, Tokyo 100, Japan.
Tokuyama Corp., Shibuya Konno Bldg., 3-1, Shibuya, 3-Chome, Shibuya-Ku, Tokyo 150, Japan.
Sybron Chemical Co., Birmingham, NJ 08011, USA.
Ionics, Inc., 65 Grove Street, Watertown, MA, USA.
Aqualytics, 7 Power Horn Drive, Warren, NJ, USA.
WSI Technologies, Inc., 2464 West Port Plaza Drive, St Louis, MO, USA.
Morgane, 6, rue du Gatinais, 77167, Bagneaux-sur-loing, France.
Pall RAI, 225 Marcus Blvd., Haupauge, L.I., NY 11787, USA.
Dupont Nafion, Wilmington, DE 19898, USA.

References

1. MacDonald, R.J., Hodgdon, R.B. and Alexander, S.S. (1993) US Patent No. 5264, 125.
2. Mizutani, Y. (1990). *J. Membr. Sci.* **49**, 121.
3. Sata, T. and Izuo, R. (1990). *J. Appl. Polym. Sci.* **41**, 2349.

4. Anon. (1964). Test Manual (Tentative) for Permselective Membranes, NTIS PB-181575.
5. Helfferich, F. (1962). In *Ion Exchange*. McGraw-Hill, New York.
6. Kneifel, K. and Hattenbach, K. (1980). *Desalination,* **34,** 77.
7. Trivijitkasem, P. and Osvold, T. (1980). *Electrochim. Acta* **25,** 271.
8. Mizutani, Y. (1990). *J. Membr. Sci.* **54,** 233.
9. Chlanda, F.P., Lee, L.T.C. and Liu, K.J. (1978). US Patent No. 4 116 889.
10. Tani, Y. (1982). *Tech. Annu. Conf. Int. Trade Fair Nat'l.* Water Supply Improv. Assoc. 9th., Paper Number 4.
11. Kawate, H., Seto, T., Komori, R. and Nagasato, Y. (1980). *Proc. Symp. 5th Proc., Salt* p. 316.
12. Lonergan, D.A., Fennema, O. and Amundson, C.H. (1982). *J. Food Sci.* **47,** 1429.
13. Jain, S.M. (1981). US Patent No 4 275 140.
14. Itoi, S., Nakamura, I. and Kawahara, T. (1980). *Desalination* **32,** 383.
15. Smith, R.N. and Lacey, R.E. (1982). Paper in Corn Refiners Association Inc., Scientific Conference, Lincolnshire, Illinois, June 16–18, 91–118.
16. Davis, T.A. and LaTerra, T. (1987). *Proc. Int. Water Conf. Eng. Soc., West. Pa.* 316.
17. Katz, W.E. (1983). *Proc. Int. Water Conf. Eng. Soc. West. Pa.* **44,** 51.
18. Reed, P.B. (1984). *Chem, Eng. Prog.* **47.**
19. Batchelder, B.T. (1993). US Patent 5 223 107.
20. Asahi Glass Co. Ltd. (1990). *Demineralization and Purification of Organic Substances by Electrodialysis,* Product Brochure.
21. Snedecor, B.R. and Hsu, C.C. (1991). US Patent No. 5 055 399.
22. Odagiri, H. (1983). *Shukuhin kogyo* **26**(18), 20.
23. Saeki, M., Kaneko, T., Tanaka, K. and Kawakita, T. (1983). *JP* **83,** 251036 (12-26-83).
24. Gavach, C., Sandeaux, R. and Sandeaux, J. (1990). *Proceedings of ICOM '90.* 870.
25. Goldstein, I.S. and Easter, J.M. (1992). *Tappi J.* **8,** 75.
26. Easter, J.M. (1993). World Patent Application WO9305186.
27. Lightfoot, E.N. and Friedman, I.J. (1954). *Ind. Eng. Chem.* **46,** (9).
28. Mani, K.N. and Johnson, W.L. (1980). Presentation at the Sixth Annual Membrane Technology/Planning Conference, Boston, Massachusetts, November 2, 1988.
29. Van Nispen, J.G.M., Zoom, B.O. and Jonker, R. (1991). US Patent No. 5 002 881.
30. Czytko, M., Ishii, K. and Kawai, K. (1989). US Patent No. 4 882 277.
31. Glassner, D.A. and Datta, R. (1990). European Patent Application, EP 393818.
32. Berglund, K.A., Elankovan, P. and Glassner, D.A., (1991). US Patent No. 5 034 105.
33. Czytko, M., Ishii, K. and Kawai, K. (1987). *Chem.-Ing.-Tech.* **59,** 952.
34. Leitz, F.B. (1986). *Desalination,* 381.
35. Yamabe, T. and Seno, M. (1967). *Desalination,* **2,** 148.
36. Cowan, D.A. and Brown, J.H. (1959). *Ind. Eng. Chem.* **51,** 1445.
37. Klein, E., Ward, R.A. and Lacey, R.E. (1987). In *Handbook of Separation Process Technology* (Ed. R. W. Rousseau), Chapter 21, Wiley-Interscience, Chichester.
38. Hiraoka, Y., Itoh, K. and Taneya, S. (1979). *Milchwissenschaft* **7,** 34.
39. Hongo, M., Numura, Y. and Iwahara, M. (1986). *Appl. Environ. Microbiol.* **52,** 314.
40. Kuppinger, F.F., Busch, C. and Eigenberger, G. (1992). *DECHEMA Biotechnol. Conf.* **5**(B) 671.

7 Large-scale column chromatography — a GMP manufacturing perspective

J. EDWARDS

7.1 Introduction

During the past decade the biotechnology industry has used process-scale chromatography operations as the primary method of protein purification. This critical unit operation continues to be the main focus of purification development teams as an ever increasing number of biopharmaceuticals are commercialized. Consequently, consideration of column chromatography scale-up and manufacturing issues is a critial element in the successful commercialization of therapeutic proteins.

To ensure a GMP purification process consistently produces a high-quality product at a cost that allows an acceptable return on investment, general manufacturing-related questions and issues need to be addressed at each stage of a developing purification process. The objective of this chapter will be to review the critical concerns of manufacturing during the process development, scale-up, and manufacturing start-up stages of chromatography-based purification processes.

7.2 Large-scale chromatography — concerns during the process-development stage

During the development of a protein purification process, many chromatography resins are screened for their ability to bind and elute a product of interest. The removal of gross protein contaminants is often addressed early in the development of a process, followed by evaluation of DNA, virus, and other trace contaminant removal. A given chromatography resin is often selected because it efficiently binds and elutes a desired product, while effectively removing a large portion of contaminants. As the purification process becomes more refined, various column wash and elution buffers are evaluated and optimized to provide maximum product yield and purity. With the focus on maximizing product recovery while effectively removing product contaminants, decisions that are made during this stage should take into consideration commercial GMP process concerns. Examples of commercial-scale problems include: (i) selection of chromatography resins that are difficult to sanitize because the ligand support structure is easily degraded (e.g. non-cross linked sepharose); (ii) the selection of relatively expensive buffers or scarce buffer components (e.g. HEPES* and methyl mannopyranoside,† respectively); and (iii) significant column equilibration or wash volumes are specified that when scaled-up require significant tankage volume (i.e. >5 column volumes of buffer for columns over 20 l requires ⩾100 l of a given buffer per batch). Viable alternatives, which could save a considerable amount of time and money during product manufacturing, often exist. Several common manufacturing concerns that should be considered early on during process development are summarized in Table 7.1

The lessons learned during large-scale GMP manufacturing have consistently proven that the issues outlined in Table 7.1 are likely to surface as problems if not considered prior to scale-up. Unfortunately several desired process characteristics can be overlooked. This includes: (i) the ability to wash resin with sodium hydroxide or similarly strong sanitizing agents; (ii) the need for reproducible chromatography resin performance; (iii) the desire to keep buffer volumes at or below 5 column volumes; (iv) collection of elution peaks using isocratic elution; and (v) collecting a single bulk fraction. The importance of each of these factors is discussed below.

7.2.1 *Resin compatibility with caustic and acid sanitization agents*

When operating a GMP production process, chromatography systems must be able to maintain a low bioburden level to ensure product quality is not compromised. To achieve this target, the use of resin sanitization agents can be very beneficial (e.g. 1.0 M NaOH applied between chromatography batches). The consequence of not periodically using a sanitizing agent is that once a system becomes contaminated the bioburden level will continue to increase until the point where product quality will likely be affected. The chromatography resin must then be discarded and the system disassembled and sanitized. This results in costly downtime, potential loss of product, and often replacement of expensive resin.

A well-conceived chromatography system design can reduce the risk of bioburden contamination. Despite the lack of harsh chemical sanitization, some large-scale chromatography columns have been kept free from detectable bioburden for entire production campaigns, which included over 100 cycles. None the less, the inability to use strong sanitization agents makes the

* In one clinical production process over $300 000/year in raw material costs was saved by switching from a HEPES-based buffer to an alternative and equally effective buffer component.

† A developing production process specified the use of methyl mannopyranoside. During the scale-up it became apparent that the world supply of this component would be required to meet annual demands for this one purification step. Furtunately, an alternative was found to replace the methyl mannopyranoside.

Table 7.1 Ideal characteristics of large-scale chromatography operations

Characteristic	Ideal	Rationale
Number of process steps	Minimum number (i.e $\leqslant 3$ steps)	Lower production costs Less opportunity for processing errors
Stability of resins	No ligand leakage	Longer resin life No product contamination from resin
	Resin can withstand 1 M NaOH and 0.1 M acetic acid for over 24 hours per batch cycle	Resins can be effectively cleaned to remove process contaminants
Resin life	At least 1 year life or 100 batches	Lower production costs
Resin vendor	Reproducible lot-to-lot resin performance	Ensure consistency of product
	Readily available resin at low cost	Lower production costs with less inventory required
Number of buffers per step	Minimum number (<5)	Lower process costs
Chromatography buffers	No flammable and hazardous solutions	Less expensive processing and facility equipment
	Readily available components	Avoid shutdown due to supply shortages
	Low-cost components	Lower production costs
	Low volumes required (5 CVs)	Minimize need for large buffer tanks and all the associated costs of operation
Process volume(s)	Reduce significantly at first step (e.g. 100-fold concentration)	Ease of handling Lower processing costs
Column operations	Ability to load, wash, and elute within a single manufacturing shift (8–12 h)	Less chance of batch failures during overnight operations Does not require additional production shifts
Column elution method	Isocratic elution	Simpler, less problematic equipment can be used
	Collect peak as a single bulk fraction	Faster processing times Less chance for processing errors

potential for process contamination greater, which should be avoided when possible. The benefit of chemical agents, such as sodium hydroxide or acetic acid, in removing non-specifically bound process-stream contaminants should not be ignored. In some cases the use of these agents can be even more important for the cleaning (regeneration) of resin, compared with its value as a sanitization agent.

7.2.2 *Chromatography resin lot consistency*

Variability in resin performance between lots can cause considerable problems during large-scale production. Often the development and scale-up of processes is carried out with only one or two lots of resin. Consequently, not until large-scale manufacturing is underway and new lots of resin are ordered to replace existing resin is there an assessment of resin consistency. Although vendors control the release of resin through their in-house quality-control tests, these tests do not necessarily encompass the range of relevant parameters for a specific biological purification process. Although a rare occurrence, lot-to-lot variation has been observed to affect the quality of the resin packing and ability to resolve contaminants. For example, this variability has been observed in certain large-scale reverse-phase chromatography columns.

A solution to the resin variability concern is to invest time during the process development phase to explore the effect of different lots of a chosen resin on product recovery, product purity, and other relevant processing parameters. Furthermore, the vendor's quality-control release criteria should be evaluated. In some instances deficiencies become apparent. If the vendor is unwilling to modify the testing procedures, new resin sources should be considered. Although this may appear drastic, the consequences of resin variability can be a substantial financial loss as a result of unreleasable product.

7.2.3 *Column buffer volumes required*

Occasionally a bench-scale process will use column equilibration and wash volumes that are set at a level well above necessary. For example, a column that can normally be equilibrated with 3–5 column volumes of buffer may be equlibriated with 10 column volumes. At the bench scale, this may not be given much consideration because it represents only a few extra milliliters of buffer and insignificant extra time. Unfortunately, by the time the process reaches full-scale production the few extra milliliters of buffer can translate into hundreds of liters (e.g. for a 50 l column each extra 5 CVs of wash requires 250 l of buffer). If the step is run several times per week, a considerable amount of buffer volume capacity is required. This results in high product cost from the increase in buffer volume, and higher capital costs because of the need for large buffer tanks. Consequently, the rationale for buffer-volume selection should be examined early in the development process. When possible buffer volumes should be reduced to levels that achieve the objective intent of the process, while ensuring that product quality is not affected. In practice, no more than 5 column volumes is typically required for column equilibration and washes.

7.2.4 *Gradient elution points to consider*

The removal of bound product from a chromatography resin can occur by either gradient or isocratic elution. While improved separation of product from contaminants typically occurs with gradient elution, this degree of separation may not always be necessary and with large-scale production can be operationally problematic.

The use of gradient elution requires specialized chromatography equipment. The development of large-scale chromatography equipment has come a long way during the past 10 years. However, failure of some commercial chromatography systems occurs at a rate of more than one incident per month, which is extremely costly when these systems are used in production. Although more reliable systems are available, gradient systems are generally problematic and users should beware that some systems are not suited for all applications. In particular, some gradient systems that have been demonstrated to work effectively at room temperature, fail to operate in a reproducible manner below 8°C. Finally, considerable capital outlays are required because of the high cost of these systems, which typically exceeds US$50 000 per system.

The drawbacks with performing gradient elution during large-scale production require that its application needs to be considered carefully during the development of a purification process. If a gradient is necessary for the successful purification of a protein, efforts should be made to ensure that the required flow rates at the large scale will not exceed 1 l/min, and ideally will be below 150 ml/min. These specifications are based upon the maximum flow rates of currently available field-tested equipment. In addition, prior to selection of a chromatography system, thorough testing of the unit using production buffers, flow rates, and temperature conditions should be conducted for up to 3 months. Experience has shown that some chromatograph systems operate accurately during a 2-week demonstration but fail during actual operation over several months.

7.2.5 *Drawbacks of fractionating product elution peaks*

An elution procedure for a chromatography column step will occasionally stipulate collecting product peak fractions, assaying fractions for purity, and then pooling these fractions based upon the purity data. When performed as part of a routine GMP process, fraction collection has several drawbacks. First, the collection of fractions typically must be performed in an open environment, which increases the probability of outside contamination. Second, the act of collecting and subsequently pooling fractions is more difficult than collecting a single bulk fraction. As a result, the process becomes more labor-intensive and is more prone to operational error. Third, the sampling and assaying of fractions is costly and time-consuming, often necessitating considerable delays in a process (i.e. typically > 12 h delay). For one process being scaled-up the use of fraction collection was required during two purification steps. A calculation of the number of assays required for these fractions indicated that each year over 5000 fractions would be sampled, submitted to quality-control, assayed, and then selected fractions pooled. In addition, each of these events requires thorough documentation. The costs in terms of both time and money is quite significant and could have potentially been avoided by using bulk-product collection. Therefore, the benefits of fractionating an elution peak should be weighed carefully against the increased possibility for product contamination, operational error, and the time and cost of collecting, assaying, and pooling fractions.

7.3 Scale-up of large-scale chromatography operations

After bench-scale purification steps have been developed, the next step is to begin scaling up the process. This will often occur in a series of steps; for example, bench scale to 1/10th of full-scale production, followed by a second scale-up to full-scale production. The two critical aspects to scale-up are; (i) technology transfer; and (ii) the methods used to scale-up a purifications step.

The speed and success of process scale-up is dependent upon the effective transfer of technology from one group to the next group in the process commercialization pathway (e.g. process development to the pilot plant). Besides the obvious need to increase the scale of a process to allow for adequate production capacity, the transfer of technology serves two critical purposes for GMP processes. First, documentation of chromatography procedures and developing process-control limits should be an integral part of any technology-transfer process. This documentation becomes a valuable record of the rationale and history behind the eventual production-scale GMP process. The Food and Drug Administration (FDA), as well as in-house groups such as quality assurance and regulatory affairs may request information to support many of the production procedures and control limits. The earlier that documentation of a process begins, the easier to compile more thoroughly the documentation package. Second, the transfer of information will not only be useful in highlighting what is known about a process but should also include information on what knowledge and data are still required to fully understand a process. If the 'unknowns' in a process are identified early in development, often simple concurrent experiments or process sampling that address these unanswered questions can be performed during scale-up. The cost and time of gathering information once a process is at the production scale can be significant, not to mention the cost related to an otherwise avoidable problem that occurs during production.

7.3.1 *Information collected and documented during process scale-up*

The information package that should be created as a chromatography process is developed and scaled-up and can be quite extensive. Relevant information needed includes the following:

1. Background information
   (a) Goal of purification step
   (b) Critical parameters to be monitored.
2. Resin information
   (a) Resin characteristics (e.g. particle size)
   (b) Resin-packing procedures
   (c) Evaluation of resin-ligand leaching
   (d) Resin-storage procedures
   (e) Procedure for removal of resin from storage
   (f) Resin life time
   (g) Lot-to-lot consistency
   (h) Stability of resin in the presence of harsh sanitization and regeneration agents.
3. Product information
   (a) Product solubility levels in process buffers
   (b) Acceptable product storage times and conditions during processing.
4. Product-resin data
   (a) Dynamic resin capacity
   (b) Product recovery range
   (c) Eluate purity range
   (d) Key impurities removed during step.
5. Chromatography operation
   (a) Column load preparation procedures
   (b) Maximum column load and eluate storage time and temperature
   (c) Volumes, flow rates, and pH/conductivity specifications for column load, washes, elution, and regeneration procedures and rationale for each
   (d) Expected product and contaminant peak shapes
   (e) Product collection and pooling procedures
   (f) Column operating temperature range
   (g) Evaluate for presence of product loss, aggregation, and precipitation during washes and elution of column when resin is loaded at capacity
   (h) Acceptable materials of construction for product contact surfaces.

This list does not include the extensive amount of process validation documentation that is also required. Unfortunately there are cases where important information that should be gathered is overlooked or considered unnecessary. Experience has proven, especially in the area of chromatography, that failure to fully understand a chromatography unit operation can result in costly difficulties.

7.3.2 *Process scale-up procedures*

Once an understanding of the bench-scale process is in hand the actual scale-up of a process can begin. This usually starts with a paper scale-up whereby column dimensions, flow rates, and volumes are calculated for the desired increase in scale. The mathematical procedures are straightforward and performed using the methods described in Table 7.2. The first set of challenges typically arise when the calculated column diameters, bed heights, or flow rates are not practically attainable (see Table 7.2 for details). At this point compromises must be made. For example, in order to achieve enough bed height so column channeling is less likely to occur, one may decide to increase bed height (with constant diameter) beyond what capacity requires. Although this will necessitate more resin and buffer, the alternative costs associated with repacking channeled columns and the expense of lost production batches would be significantly greater. As always, the end result of any process compromise should be an operationally more economical and efficient production process.

7.3.3 *Challenges encountered during scale-up*

Following the scale-up on paper, the process equipment is set-up and pilot production is initiated. Often the initial scale-up is the first time the process is run repeatedly more than three times. As a result, process problems are often uncovered. Recoveries may be inconsistent, contaminants are observed that previously were not seen, and resolution of peaks occasionally is decreased. For example, product elution volume may be greater than predicted by small-scale studies because of less-efficient flow of elution buffer through a large-scale column. This results in increased loading time on the subsequent column step. In some instances, this added time has adversely affected production schedules and methods to reduce the peak volume had to be developed. Slower elution flow rates, reverse elution, or narrower columns have proven effective in reducing the elution volumes.

During the scale-up phase one needs to approach each challenge carefully. In one process, small-scale column studies predicted that lower elution flow rates would not affect elution volume. However, at the large-scale a reduction in the flow rate sharpened the elution peak considerably, resulting in a 30% decrease in elution volume. This is probably because of the different fluid flow properties between small and large-scale columns (e.g. wall effects, flow-adapter configuration). This example illustrates how bench-scale data are not always directly scaleable.

Most process changes often affect more than one parameter. Therefore, changes should be evaluated for their potential effect on the entire process. For example, in one process the regeneration procedure was improved for the first chromatography step. This change enhanced the chromatographic

Table 7.2 Common practices and considerations for scale-up of large-scale chromatography operations

Parameter	Basis for scale-up	Considerations
Column size		
Diameter	Increase diameter to maintain constant product loading per unit of resin and constant bed height	Commerical columns are only available in discrete sizes; therefore, precise diameter scale-up is not usually possible Above 63 cm diameter columns packing procedures can be difficult
Bed height	Constant with increase in scale	Bed heights below 5 cm for columns above 35 cm in diameter can lead to channeling. Ideally, aspect ratio should be greater than 1:5 For chromatography columns above 35 cm in diameter, experience indicates dynamic resin capacity is reduced, especially columns below a 10 cm bed height
Flow rates		
Depyrogenation	Constant linear velocity, while maintaining equivalent contact times	
Load, washes, elution	Equivalent linear velocity	The linear flow rate may require significant product load times, so column diameter may need to be increased to allow for shorter processing times Column pressure limits and equipment flow rate limitations may restrict allowable flow rates
Regeneration	Equivalent linear velocity	Certain regeneration solutions are viscous and therefore limit the maximum allowable flow rates due to column pressure limit (e.g. acetic acid, ethanol)
Volumes		
Depyrogenation, load, washes, isocratic elution	Equivalent number of column volumes	
Gradient elution	Equivalent change in column volumes per unit of time for each buffer	

life and capacity of this step. The increase in column capacity, when combined with high bioreactor titers resulted in more product proceeding downstream. This, in turn, led to exceeding the capacity of an ion exchange column two steps later in the process. This was never observed in the past because the limited capacity of the first step resulted in loss of any additional product produced in the bioreactors.

7.3.4 *Compatibility of materials of construction with processing*

The compatibility of equipment and filter materials of construction with product and process fluids must be reviewed. In particular, studies to examine product binding to filters should be initiated. In one process, studies revealed that 20% of a product was binding to the harvest filters. This had not been observed previously because the variability in the product assay during the development phase could not identify a product-binding problem with sufficient statistical significance. As the assay's reliability was improved and as more data were collected during the large-scale operations, the loss of product became clearer. A switch to an alternative membrane material completely eliminated the product loss with no effect on product quality. In another case, a large-scale chromatography column was observed to over-pressurize during elution. This required a reduction in the elution buffer flow rate. Due to processing time constraints, the incremental elution time became a significant problem. The original hypothesis was that high protein concentrations during the elution resulted in the high back-pressure. After several attempts to solve the problem, the column frit was changed from polypropylene to titanium. This change completely eliminated the over-pressurization problem, ultimately revealing that the problem was related to the binding of a contaminant protein to the polypropylene. These examples underscore the need to examine carefully the compatibility of materials during the development and scale-up phases of a process.

7.3.5 *Environmental contamination control*

Finally, issues relating to producing a product under sanitary conditions should be addressed. This includes evaluating bioburden and endotoxin levels in the process stream. Given that biopharmaceuticals must have low bioburden and endotoxin levels, appropriate sanitization and depyrogenation procedures need to be defined and optimized for each process step. This is where the ability to flow sodium hydroxide or similar sanitizing agents over the chromatography resins becomes an important factor (see 7.2.1).

7.4 Preparations for large-scale GMP column chromatography operations

As the scale-up of a process nears completion several production issues need to be addressed, including:

- Equipment selection
- Evaluation of facility requirements
- Chromatography system design and construction
- Chromatography system qualification
- Writing necessary production documentation.

7.4.1 Equipment selection

Standard chromatography systems include a pumping device, chromatography column, and possibly a fraction collector (Figure 7.1). In addition, various sensors can be employed to monitor the operation and alert production personnel in the event of a process upset. Sensors include: (i) pH and conductivity cells to monitor changes in buffers, as well as buffer proportioning during elution gradients; (ii) UV detectors to monitor protein levels at the outlet of a column; and (iii) bubble sensors to detect the presence of air entering a column and shut off the pumping system if air is detected. Beyond the required operating specifications (e.g. flow rate, temperature, and pressure limits), the selection of process equipment is affected by several factors:

- Materials of construction that are compatible with the process-stream components.
- Sanitary design requirements
- Equipment that performs reliably under the operating conditions of the chromatography column (e.g. low temperatures).

Although commercially available equipment for large-scale chromatography is constructed of materials that are acceptable for most biological processes, occasionally incompatible materials are used. In one instance, a vendor upgraded equipment by switching to Viton® O-rings. Unfortunately Viton® was incompatible with the solvents used in a particular process for which this instrument was used. The system failed soon after the O-rings were installed and all batches exposed to the O-rings were discarded.

Another problem material can be stainless steel. If not properly electropolished and passivated, commercially available 316 L stainless-steel chromatography systems have been found to rust after only 6 months of exposure to commonly used chromatography buffers. Therefore, careful examination of compatible materials of construction for equipment must occur.

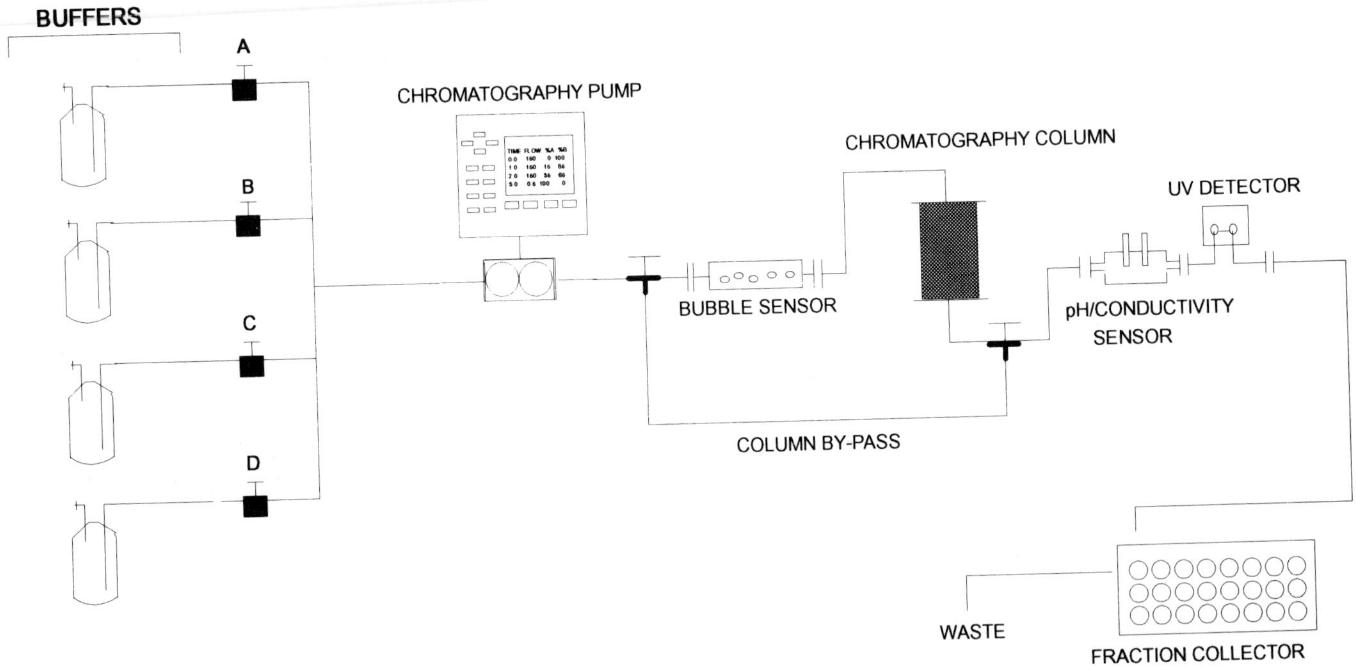

Figure 7.1 Typical chromatography system equipment layout.

With few exceptions, most equipment available for large-scale chromatography is designed for sanitary use. However, careful examination of system design should not be overlooked. In one instance, a process step bioburden problem was traced to a 10 inch (25 cm) piece of pipe that was not fully flushed during each batch. The removal of this pipe completely eliminated the bioburden problem. Interestingly, the system was designed and built by a major chromatography systems manufacturer. It is important not to take either an in-house design or a vendor's equipment design for granted.

Another pitfall to beware of during equipment selection is the reliability of a new system under the conditions typically encountered in biological purification processes. For example, below 8°C a commercially available chromatography system was found to proportion buffers incorrectly, whereas at room temperature it offered a high degree of operational consistency. This occurred in spite of the manufacturer's assurance that the system was tested and could operate properly at 2–8°C.

To help avoid chromatography system equipment problems, a customer should request references for equipment users that have considerable field experience with the equipment. Secondly, the proper qualification of equipment will identify problem areas prior to initiating an expensive production process (see section 7.4.4, System qualifications).

7.4.2 Evaluation of facility requirements

All GMP chromatography operations need to occur in a clean, sanitary, and controlled environment, which includes qualified heating, ventilation and air conditioning (HVAC) and other utility systems. Generally most equipment uses standard 120 V. To avoid system shut-down during power outages, uninterrupted power service is commonly used in production.

Beyond the standard use and GMP facility design requirements, certain chromatography operations require special systems or equipment. Reverse-phase HPLC operations typically use flammable organic solvents, which must be contained. This requires the use of walk-in fume hoods to contain waste drums, as well as product elution collection systems. Chromatography operations with open-fraction collection or exposure of product to the environment, should occur within laminar flow areas that provide added protection against environmental contamination (i.e. class 1000 or lower). However, one should avoid disrupting the laminar air flow with processing equipment. Only the equipment and piping that is exposed to the environment should be located within a laminar flow area. In one production process, the removal of ancillary process equipment from laminar-flow areas resulted directly in the decrease of in-process bioburden levels, from occasional 5–10 cfu/ml to consistent results of 0 cfu/ml.

7.4.3 *System design and construction*

Once the equipment is selected, the integration of all components into a working chromatography system can begin. A design criterion for any GMP purification unit operation should be a closed-system design that avoids exposure of the product and system to the environment. Individual pieces of equipment are typically connected using tubing of a diameter that is as small as possible without affecting pump performance or significantly increasing the system pressure. In general, the use of Teflon® or other similarly inert polymeric tubing should be used instead of stainless steel. Corrosion problems associated with stainless steel can be completely avoided by the use of Teflon®. In systems where steam sanitization is used (such as systems that cannot be periodically flushed with sodium hydroxide), 316 L stainless-steel piping will be necessary.

Additional features that are useful include column by-pass piping (see Figure 7.1) and column isolation valves. By-pass valves allow for flushing of the system up to the column, without having to disconnect the tubing entering the column and exposing the system to the environment. In one manufacturing facility, a portion of acetic acid was not fully primed through a small section of column piping leading to the column inlet. After the product was loaded onto the column, this section of piping was used for the elution. Since the acetic acid was not fully removed from the line, the product was exposed to the acetic acid and the entire production batch lost. An effective priming of the system would have completely avoided this problem.

An additional element of chromatography system design that should be considered is the degree of automation. The simplest forms of automation include automatic system safety shut-off devices. Monitoring devices such as pressure sensors and air sensors can easily be integrated into a chromatography system. If the system either over-pressurizes or if air enters the system, the pumps can be controlled to stop and prevent loss of the batch and/or the chromatography column. Over-pressurization of an expensive large-scale column for one firm resulted in a cracked column, loss of product, and the complete loss of expensive chromatography resin. This would have been avoided with a simple pressure transducer and controller.

Although a rare occurrence, air entering into a chromatography column can be a major problem, not only affecting the releasability of the batch of product but also affecting the integrity of the chromatography resin. The presence of air can occur for any number of reasons, from a leak in the system tubing or fittings, from a buffer line becoming detached, or the level of the chromatography feed solution falling below the inlet of the feed line. Given the constant daily operations in manufacturing, even a rare event like air entering a chromatography system has the potential for occurring several times during a year. Owing to the dire consequences, installation of an air-detection device is highly recommended.*

Automation can be taken a step further to include computer monitoring and control of processes, not to mention interfacing the computer with process documentation also within a computer system. The costs and time required to develop these systems can be significant; however, once developed the long-term improvement in efficiency and productivity of operations can be tremendous.

7.4.4 *System qualifications*

After a chromatography system has been designed and built, it needs to be qualified for GMP operation. Despite being required by cGMPs, the evaluation and documentation of a chromatography system's performance makes practical and financial sense. Generally three types of qualifications are performed, installation qualifications (IQ), operational qualifications (OQ), and process qualifications (PQ).†

IQs are used to ensure that all critical aspects of the equipment and/or chromatography system installation comply with the manufacturer's recommendations. In the case of a chromatography system comprised of many individual components, the IQ document will typically include, but is not limited to, the following information:

- Equipment and supplier information
- An overview of how each piece of equipment functions and will be used
- Materials of construction for all wetted parts
- Listing of installation and operation manuals for equipment
- Listing of any vendor equipment drawings
- Electrical and other utility requirements
- Description of how the system is to be assembled, with a corresponding check list
- Spare parts list
- Safety information

* A word of caution. Although air sensors can be extremely beneficial in a chromatography operation, certain air-detection devices have been found to be unreliable for chromatography applications. As with all equipment, request references and test equipment fully before putting into actual production. More than one US$60 000 commercial chromatography system was found to contain air detectors that failed during operation and, when fully tested, was confirmed to be highly unreliable.

† The three types of qualifications performed represent one common approach to systems and equipment qualification. The actual approach taken by each individual company will vary. Often in the case of a chromatography system, the IQ and OQ will be combined into a single system qualification document.

Most of the IQ contains reference information. However, both the materials of construction and utility requirements sections should contain a check to ensure compatibility with common process and facility systems. Potential problem areas should be noted.

The OQ is used to verify that the chromatography system and components perform properly over the desired operating ranges. The OQ document will typically contain the following information:

- System description (which includes a drawing)
- Listing of system components with associated calibration documentation information
- Commissioning procedures (e.g. cleaning, passivation)
- Test procedures to evaluate general operations of systems
- Test procedures to evaluate control/alarm systems of systems.

For chromatography systems where the individual components are not integrated, the evaluation of general operations will consist of testing each piece of equipment for satisfactory operational performance over pre-specified ranges. For some companies, these tests may have already been performed as part of the calibration procedures and can simply be referred to in the OQ document.

The PQ is used to ensure that a chromatography system will perform appropriately under the conditions encountered during the system's actual use in production. Beyond testing the equipment for general operational performance, the PQ can be used to evaluate whether the system operates in a satisfactory sanitary manner, as well as whether leaching of undesirable materials occurs during operation. The PQ is commonly performed using established standard operating procedures (SOPs) and other operational documentation. In this respect, the PQ serves as a useful pilot test of both the equipment and the documentation. PQs are commonly performed before the start-up of phase III processes.

7.4.5 Production documentation

The final key element in preparing for clinical production of any new process is to ensure that adequate documentation exists that allows the process to be operated safely and consistently and provides proper recording of all critical operational events. The documentaion package for a chromatography operation should include the following:

1. Raw material release specifications (e.g. buffer components and resins).
2. Standard operating procedures for equipment, product handling, and other process-specific procedures.
3. Batch records including
   (a) General process instructions

   (b) Alarm and/or action limits for critical operating parameters (e.g. flow rates, column equilibration pH levels)
   (c) Sampling and testing information
   (d) Process control parameters (e.g. step yields,* bioburden levels).
4. Logbooks that summarize:
   (a) Chromatography resin history
   (b) Chromatography column performance tests (e.g. theoretical plate analysis)
   (c) Chromatography equipment used to replace existing equipment.

7.5 Conclusions

The intent of this chapter was to highlight the manufacturing issues that should be addressed during purification development, scale-up, and pre-production planning. During purification development, an awareness of GMP manufacturing issues can have a significant positive impact on the project time line and eventual process productivity. Although the short-term consequence of addressing manufacturing concerns may delay the development effort or require increased resources, the long-term benefit is often justifiable. Note, however, that these comments above should be put into proper perspective. For some chromatography operations, the use of more difficult procedures, such as gradient elution and fraction collection, may be a critical element in the success of a purification unit operation. For these processes, seeking alternatives to improve manufacturing efficiency is not a viable option. However, for chromatography operations, the importance of difficult procedures is less certain. In these instances, time is warranted to evaluate manufacturing issues, such as column-elution procedures and buffer-volume requirements.

During process scale-up, a documentation package should be developed to support the process procedures and control limits. The actual scale-up of a process involves considerable process troubleshooting and optimization. To be thorough, one must be aware of the effect of any single change on the entire process. Upon completion of the process scale-up, most of the items listed on page 173 should have been addressed. Issues of process consistency, material of construction compatibility, and product-stream contaminant levels should be understood to the point where the process operates

*Regulatory agencies often desire specifications for chromatography column step yields. Unfortunately, one reality of chromatography operations is that, although a process can be operated consistently, any individual chromatography column on any given batch can have considerable yield variation from the average yield. Therefore, one needs to be careful when deciding to include column yield as an in-process control. This does not suggest that column yields should not be calculated and monitored very closely.

consistently and within reasonable control limits. At this point, the development or pilot plant group can begin to prepare for transfer of the process to manufacturing.

In preparation for GMP production, several equipment, facility, and documentation issues should be addressed; especially the importance of carefully examining and evaluating equipment performance and system design.

The overall message of this chapter is that, to avoid project delays and to maximize process productivity, scientists, engineers, and production personnel must work together at each stage in the development pathway, focusing on the requirements of cGMP production processes and potential trouble spots commonly associated with large-scale column chromatography.

Further reading

Courtney McGregor, W. (Ed.), (1986). *Membrane Separations in - Biotechnology*. Marcel Dekker, New York.

Deutscher, M. (Ed.), (1990). *Methods of Enzymology, Vol.* 182, *Guide for Protein Purification*, Academic Press, London.

Fiechter, A. (Ed.) (1982). *Advances in Biochemical Engineering*. Springer-Verlag, New York.

LeRoth, D., Shiloach, J. and Leahy, T. (1985). *Purification of Fermentation*. American Chemical Society.

Nauch, D. (1990). Industrial-scale downstream processing of biotechnology products. *BioPharm* **May**.

Scopes, R.J. (1987). *Protein Purification Principles and Practices, 2nd edn*. Springer-Verlag, New York.

Sofer, G.K. and Nystrom, L.E. (1989). *Process Chromatography: A Practical Guide*. Academic Press, San Diego, California.

Via, J.W., Costs, D.W. and Reed, R. (1991). "Select the right sensors for pharmaceutical processes", *Chem. Eng. Prog.*

8 Product recovery and purification via precipitation and crystallization

C.A. SCHALL and J.M. WIENCEK

8.1 Introduction

Precipitation of biomolecules such as antibiotics or proteins is a means of concentrating and, in some cases, purifying a desired product. Precipitation of a molecule results in an amorphous solid that is of low purity relative to a product obtained by crystallization. Precipitation is best utilized as a means of product isolation; whereas, crystallization may be an effective polishing step that provides a product of high purity. As in any unit operation, the design engineer must be concerned with equilibrium behavior, kinetic behavior and appropriate equipment selection. In all of these respects, both crystallization and precipitation are more similar than different. However, some differences exist and will be highlighted during the subsequent discussion. Although many of the examples presented in this chapter will focus on the application to protein molecules, the principles applied may be generalized to any system.

8.2 Equilibrium behavior in solid–liquid systems

8.2.1 Solid–liquid equilibria

Temperature effects. Classical thermodynamics may be used to describe the effects of controllable variables (e.g. temperature) on the solubility of a molecule in solution. For example, the effect of temperature on solubility is described by the van't Hoff equation:[1]

$$\ln x = (\Delta h_f/R)(1/T_m - 1/T) \qquad (8.1)$$

where x = mole fraction of solubilized solute; Δh_f = heat of fusion of solute; R = gas constant; T_m = melting point of pure solute; and T = temperature of interest. This equation suggests a semi-empirical equation of engineering use:[2]

$$\ln (c^*) = b - a/T \qquad (8.2)$$

where: c^* = saturation concentration of solute at temperature, T; a, b = experimentally determined parameters.

As shown in Figure 8.1, such dependency is displayed typically in protein systems that are not denatured. However, temperatures greater than 60°C will denature most protein molecules, leading to precipitation (i.e. amorphous solid). This observation is in opposition to the expected high solubility predicted by the above equation. Thus, thermodynamics will provide a useful tool for analyzing the lower temperature crystallization processes but will not be so effective in predicting high-temperature precipitation. As is usually the case, the design engineer must have at least an experimental laboratory knowledge of the solubility of the solute as a function of the temperature or amount of precipitating reagent employed.

The more usual method of causing solid formation (crystal or precipitate) is to employ a chemical precipitating agent. Common precipitants include neutral electrolytes, short-chain alcohols, non-ionic polymers and polyelectrolytes. The most commonly employed precipitant is neutral electrolyte.

Effect of electrolyte. Purification of proteins has long relied on precipitation via 'salting out'. The addition of neutral electrolyte causes the protein to precipitate under appropriate conditions, and the effect of electrolyte on biomolecule solubility is usually described in terms of 'salting-in' or 'salting-out'. Salting-in implies an increased solubility of a biomolecule with the addition of a neutral electrolyte. Conversely, salting-out is a decrease in solubility because of the addition of neutral electrolyte.

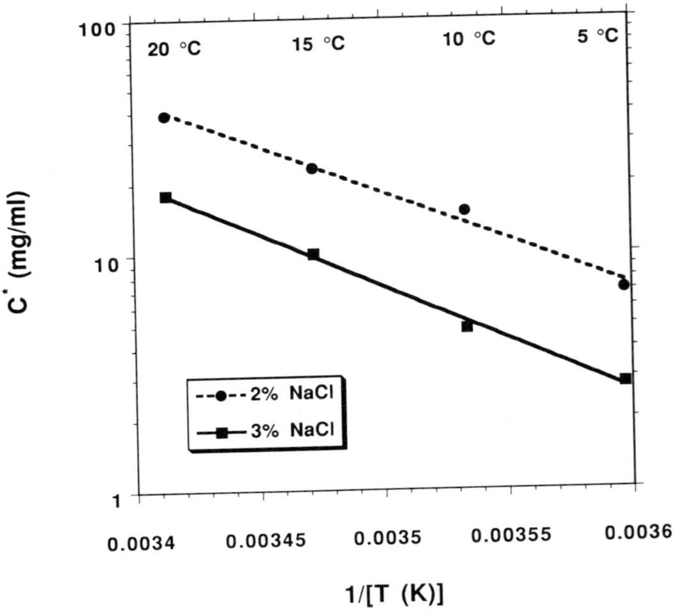

Figure 8.1 Van't Hoff plots of lysozyme at various pH values (data from Cole[31]).

Salting-in is actually predicted by the Debye-Huckel theory for ionic solutions.[3] This theory predicts the following solubility dependence upon the addition of neutral electrolyte:

$$\ln\left(\frac{S}{S_0}\right) = \frac{Z^2\varepsilon^2 N\kappa}{2DRT(1+\kappa a)} \tag{8.3}$$

where: S = solute solubility when electrolyte is present in the solution; S_0 = solute solubility when electrolyte is not present in the solution; Z = net charge on the solute; ε = charge of an electron; N = Avogadro's number; D = dielectric constant of solution; R = gas constant; T = temperature (absolute units); a = solute radii; and κ = inverse Debye length.

The Debye length ($1/\kappa$) is a quantitative measure of the distance over which electrostatic forces are important and is defined as:

$$\kappa = \sqrt{\frac{8\pi N\varepsilon^2 I}{1000DkT}} \tag{8.4}$$

where: I = ionic strength of the solution ($= 0.5\sum_{i=1}^{n} C_i Z_i^2$); k = Boltzmann constant; C_i = concentration of ionic electrolyte species, i; Z_i = valence of ionic electrolyte species, i; and n = number of ionic species present in solution.

Increasing electrolyte concentration causes a corresponding increase in κ. Thus, the Debye-Huckel theory predicts that solubility will increase with increasing electrolyte concentration. For many proteins, such effects are observed at low electrolyte concentrations. In addition, this theory also predicts that if the net charge (Z) falls to zero, the solubility will be minimal. Once again, this effect is often evident in protein systems (Figure 8.2). However, at higher electrolyte concentrations, the above equation fails to agree with observation. It is observed that the solubility often decreases at higher electrolyte concentrations where salting out occurs (Figure 8.3) and the solubility is not independent of electrolyte concentration at the isoelectric point ($Z = 0$).

Such observations are collectively referred to as salting-out. Salting-out is most easily viewed as a process wherein the electrolyte removes water of hydration from hydrophobic regions of the solute; thus, the solute solubility is diminished. The efficiency of various neutral electrolytes to salt-out proteins was documented by Hofmeister in 1887.[6] This ranking of electrolytes has come to be known as the lyotropic or Hofmeister series (Table 8.1) and is of importance to many colloidal and solution equilibrium phenomena. Setschenow proposed an empirical model to describe the salting-out of gas from aqueous solution in 1889:[7]

$$\ln\left(\frac{S}{S_0}\right) = k_s C_s \tag{8.5}$$

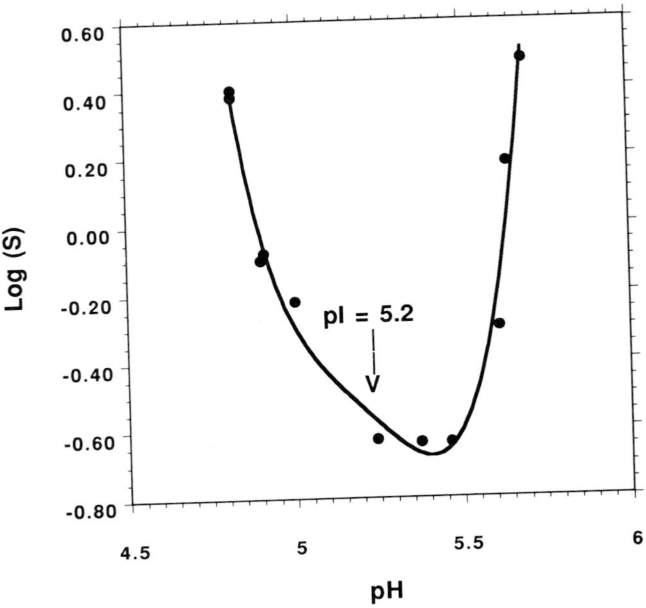

Figure 8.2 The effect of pH on solubility (S) for β-lactoglobulin (data from Gronwall[5]).

Where S and S_0 are defined as above, C_s is the concentration of the electrolyte, and k_s is the salting-out constant. Cohn [8] later generalized this equation to:

$$\ln(S) = \beta - k_s C_s \tag{8.6}$$

where β is an experimentally determined parameter. This modification allowed for the possibility shown in Figure 8.4; that is, a particular solute may have a solubility that is well described by the salting-out equation at high electrolyte concentration but salting-in effects predominate at low electrolyte concentrations. The effect of pH and other variables on the solute solubility are often incorporated into this β parameter. [10–12]

Effect of other precipitating agents. Other chemical reagents also have the ability to cause the formation of a solid precipitate. For example, many organic solvents that are water-miscible are quite effective precipitating agents. The cause of the precipitation is similar to that of salting-out. The water of hydration is removed from the surface of the solute, exposing

Table 8.1 The Hofmeister series: effectiveness for causing precipitation

Anion effectiveness: citrate > PO_4 > SO_4 > CH_3COO > Cl > NO_3
Cation effectiveness: NH_4 > K > Na

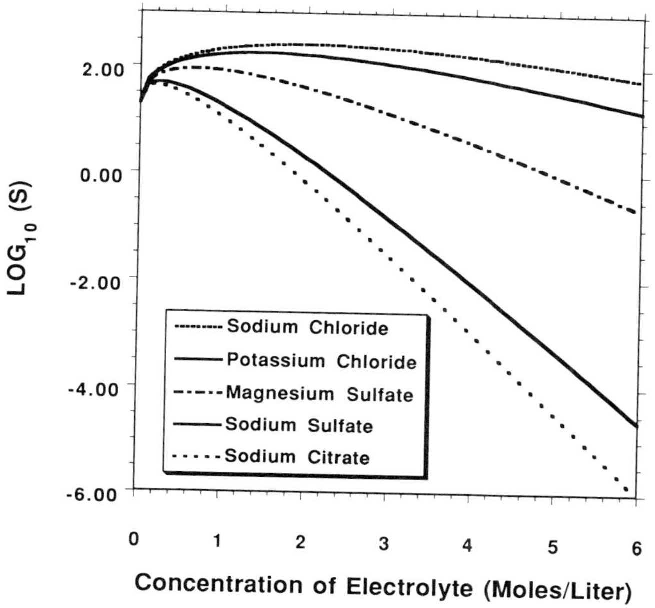

Figure 8.3 Effect of various electrolytes on horse carboxyhemoglobin solubility (from Green[9]). Chloride salts only 'salt in', whereas sulfates and citrates initially 'salt in' but the predominant effect is 'salting out' at high electrolyte concentrations. The symbol '*S*' refers to the hemoglobin solubility in g/l.

hydrophobic portions of the molecule.[13] In this case, the water of hydration on the protein surface may be replaced by the miscible organic material. In fact, the hydrophobic portions of the solute may become more soluble in the mixed solvent but other factors such as the ability of the solute's opposite charged regions to come together can lead to aggregation and phase-separation. The increased solvation of the hydrophobic regions can lead to denaturation of the solute (e.g. enzyme) and appropriate constraints must be placed on the operating conditions to minimize such effects (e.g. low operating temperatures). Models of protein solubility in systems employing organic solvents as precipitating reagents follow a form of the salting-out equation (replacing salt concentration with solvent concentration); however, the linear behavior is disrupted whenever conformational changes occur.[14] Hoare and co-workers [15] have suggested a form of a 'solventing-out' equation based on the dielectric constant of the solution:

$$\ln\left(\frac{S}{S_0}\right) = -\frac{k}{D_m^2} \tag{8.7}$$

where D_m is the dielectric constant of the mixture.

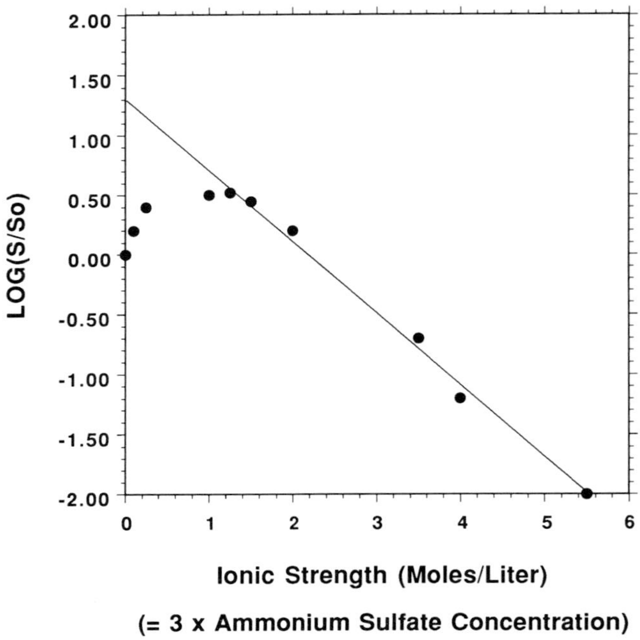

Figure 8.4 Cohn's equation is valid at high electrolyte concentration but does not predict salting-in at low ionic strength (data from Green[9]).

Other precipitating reagents include nonionic polymers, polyelectrolytes, and affinity ligands. The use of nonionic polymers follows the generalized salting-out equation. Polyelectrolytes require binding due to opposite charge and, thus, would require the precipitation be operated at an optimal pH, not at the isoelectric point, where both the solute and the polyelectrolyte display opposite net charge. Affinity methods require the design of molecules with highly specific binding properties for the protein of interest. Such methods are very selective but often require very expensive affinity ligands.

8.3 Theories of crystallization and precipitation kinetics

In order to design a crystallization (or precipitation) system, an understanding of the crystallization kinetics and the system properties (i.e. surface energy of the crystal, diffusion in the bulk phase and along the crystal surface, dislocation density, temperature) is a necessity. Nucleation and crystal growth kinetics are closely related and not entirely separable. A brief and quantitative review of the most important nucleation and crystal growth theories is presented here.

Most nucleation theories are centered on the concept of the formation of a critical-size nucleus, where the crystal bonding forces are in competition with the solvating bonding forces at the crystal-solvent interface.

In crystal growth, several processes are taking place simultaneously. Solute must be transported from the bulk phase to an adsorption layer at the growing crystal surface. There is surface diffusion of the solute molecules to a thermodynamically favorable site, or until a two-dimensional cluster of critical size is formed. Lastly, there is an incorporation step. Most crystal growth theories differ in the weighting given to these steps and in the assumptions as to the nature of the crystal surface.

8.3.1 *Homogeneous nucleation*

Homogeneous nucleation refers to the addition of monomers or clusters of solute to an ordered array of solute molecules, with the eventual formation of a cluster of a sufficient size where the free energy of formation of the crystal is large enough to cancel the effect of unsatisfied bonding interactions at the crystal surface. In other words, the crystal surface-to-volume ratio, the free energy of formation of the crystal, and the surface energy determine the critical-size nucleus. This 'activation barrier' is schematically described in Figure 8.5. Above the critical size, the crystal will tend to grow. Below the

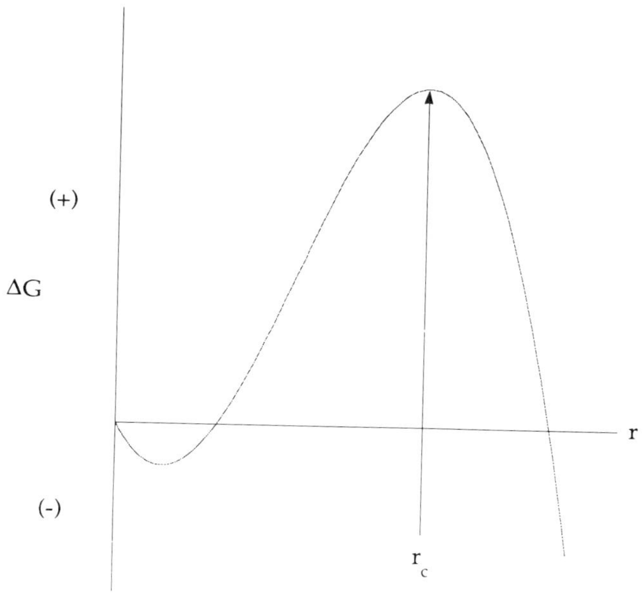

Figure 8.5 Plot of free energy change of cluster formation versus cluster radius: r_c denotes critical radius.

critical size, the nucleus will tend to dissolve. The Gibbs free-energy change for such a process is:

$$\Delta G = \Delta G_s + \Delta G_v \tag{8.8}$$

$$\Delta G = 4\pi r^2 \sigma - \frac{4\pi r^3 \Delta\mu}{3v} \tag{8.9}$$

where: σ = surface energy; r = radius of a spherical cluster; $\Delta\mu$ = change in the chemical potential of solute in cluster core; and v = molecular volume.

Clearly, this function will have a maximum at the critical radius. Differentiating with respect to the radius, and setting this expression equal to zero, yields the critical radius:

$$r_c = 2\sigma v / \Delta\mu \tag{8.10}$$

The chemical potential of the solid phase is equal to the chemical potential of the liquid phase in equilibrium with the solid phase. Therefore, the free-energy change can be written in terms of the solute activities:

$$\Delta\mu = kT \ln(a/a^*) \tag{8.11}$$

For ideal solutions, the activities can be replaced with concentrations.

$$\Delta\mu = kT \ln(c/c^*) \tag{8.12}$$

where c^* is the equilibrium solute concentration with respect to the crystalline solid phase. The ratio c/c^* is referred to as the relative supersaturation, 'S'. Substituting the above formula into the equation for the critical radius yields:

$$r_c = \frac{2\sigma v}{kT \ln S} \tag{8.13}$$

The Gibbs free energy at the critical nucleus size is often thought of as an activation energy for crystal formation and the value is substituted into a kinetic expression for nucleation:

$$\Delta G^{\#} = 16\pi\sigma^3 v^2 / 3(kT \ln S)^2 \tag{8.14}$$

$$B^0 = k_n \exp(-\Delta G^{\#}/kT) = k_n \exp(16\pi\sigma^3 v^2 / 3(kT)^3 (\ln S)^2) \tag{8.15}$$

The above expression is frequently invoked to model homogeneous nucleation kinetics where k_n is analogous to a specific rate constant.

The above development for classical homogeneous nucleation theory assumes that the interfacial energy is constant over the cluster radius, r. Experimental evidence indicates the existence of a population of clusters of a size smaller than the critical radius.[16,17] The theoretical basis of this phenomenon is the fact that the surface energy is a function of cluster size, which becomes important when the cluster radius is of the order of the interfacial thickness, 2δ. Larson and Garside[16] expressed the dependence of surface energy on cluster radius as an exponential function:

$$\sigma = \sigma_{\infty} \exp(-2\delta/r) \tag{8.16}$$

Substitution of this value into the expression for the free energy of nucleus formation, results in a prediction of clusters smaller than the critical radius and appears as a secondary minimum in the plot of ΔG versus r (see Figure 8.5). The secondary minimum implies that homogeneous nucleation can proceed by the addition of clusters instead of by addition of monomers. It is expected that the clusters would diffuse slowly because of their size. Agitation would be expected to greatly increase encounters of these clusters. High levels of agitation would result in the shearing of these clusters. Experimentally, an induction period is often encountered in industrial crystallization which could correspond to a cluster formation period. Also, nucleation often occurs in an initial burst when agitation is started, which could correspond to a greatly increased encounter frequency of clusters under agitation.

8.3.2 *Growth by surface nucleation*

Nyvlt *et al.*[18] have reviewed several growth theories. Among the surface nucleation theories, three theories are often invoked: (i) mononuclear growth; (ii) polynuclear growth; and (iii) birth and spread with surface diffusion.

8.3.2.1 Mononuclear growth. The primary assumption of this model is that growth occurs by formation of a two-dimensional nucleus of a critical size at the crystal surface and the nucleus then spreads at an infinite velocity compared with the rate of formation of the nucleus. The growth rate is then directly proportional to the area of the face, A, the height of the growth step, h, and the two-dimensional nucleation rate. This results in the following expression for the growth rate perpendicular to the surface.

$$G = k_1 h A (\ln(1 + s))^{1/2} \exp(-k_2/T^2 \ln(1 + s)) \tag{8.17}$$

where

$$s = (c - c^*)/c^* \tag{8.18}$$

the supersaturation; and c = the bulk concentration of solute. The symbol 's' is referred to as the supersaturation, as distinguished from the relative supersaturation, 'S'.

Equation (8.17) can be simplified at low supersaturations by substituting $\ln(1 + s) \approx s$, for small s, to give:

$$G = k_1 h A s^{1/2} \exp(-k_2/T^2 s) \tag{8.19}$$

An important consequence of this theory is that it predicts the growth to be proportional to the crystal surface area.

8.3.2.2 Polynuclear growth (PNG). The polynuclear growth theory differs from the mononuclear theory in that the growth of the two-dimensional critical nucleus is assumed to be essentially zero. Growth occurs by the addition of two-dimensional nuclei to the crystal surface.

$$G = J\pi r_c^2 \tag{8.20}$$

where J is the two-dimensional nucleation rate. The resulting growth dependence on supersaturation is:

$$G = \frac{k_3}{T^2(\ln(1+s))^{3/2}} \exp\left(\frac{-k_2}{T^2 \ln(1+s)}\right) \tag{8.21}$$

where k_3 is a function of adsorption rate of solute onto the crystal surface, the surface energy, the molecular volume, and the step height. At low supersaturations, this formula can be simplified to:

$$G = \frac{k_3}{T^2 s^{3/2}} \exp(-k_2/T^2 s) \tag{8.22}$$

8.3.2.3 Birth and spread model. The birth and spead model of crystal growth differs from the PNG model in that nuclei form randomly on the crystal surface but the spreading velocity of the nuclei is not taken to be zero. The nuclei can form anywhere on the crystal surface, including incomplete layers. The spreading velocity of the islands (the growing two-dimensional nuclei) is independent of the size of the island. Confusion can arise in reading the crystallization literature because this model is also referred to as the polynuclear growth model (PNG). The resulting equation for crystal growth is:

$$G = hJ^{1/3}(v^*)^{2/3} \tag{8.23}$$

where v^* is the nucleus spreading velocity and is assumed to be independent of the size of the nucleus and also to be equal to the rate of growth of a linear step.

Substituting in expressions for the two-dimensional nucleation rate and the spreading velocity with their dependencies upon supersaturation results in the following:

$$G = k_4 s^{2/3}[\ln(1+s)]^{1/6} \exp[-k_5/T^2 \ln(1+s)] \tag{8.24}$$

$$k_4 = 2h^{1/6} V_m^{5/6} (v_a/\pi)^{1/3} (n_1 D\beta\gamma s/x_s)^{2/3} \tag{8.25}$$

$$k_5 = \pi h \gamma^2 V_m/k^2 \tag{8.26}$$

where: h = height of a growth unit; V_m = the molecular volume; v_a = mean adsorption rate of the molecules on the surface; n_1 = equilibrium number of monomers on the surface per unit area; D = diffusion coefficient of solute in solution; β = the retardation factor for a linear step; γ = retardation factor for a kink; and x_s = particle mean free path.

At low supersaturations the above expression can be simplified:

$$G = k_4 s^{5/6} \exp(-k_5/T^2 s) \tag{8.27}$$

8.3.2.4 The BCF surface diffusion model. The surface nucleation theories predict lower rates of growth than seen experimentally for inorganic systems.[19] Burton *et al.* [20] proposed a model of crystal growth (the BCF model) where crystal growth steps originate at a dislocation in the crystal. Growth then proceeds as addition of units to the steps provided by the dislocation. Growth originating at an edge dislocation will appear as a spiral when observed from above. These dislocations are known to exist in crystals and growth spirals have been observed in micrographs of inorganic crystals. The growth rate of a crystal in this model is a function of the dislocation density on a crystal face.

The primary assumption of the BCF model is that diffusion from the bulk is not limiting. The growth rate is determined by surface diffusion and incorporation of a unit into a step. The resulting equation for this model is:

$$G = ks^2 \tan h(B/s) \tag{8.28}$$

At low supersaturations this equation (8.28) reduces to:

$$G = ks^2 \tag{8.29}$$

At high supersaturations the model reduces to:

$$G = ks \tag{8.30}$$

8.3.2.5 Mass transfer-limited growth. In the case of mass transfer-limited growth, the growth of the crystal is limited by the transfer of solute from the bulk to the crystal surface and not by the surface diffusion or incorporation steps. The rate of mass transfer is usually expressed in terms of the mass transfer coefficient k_1. The flux, J, which is the rate of mass transfer per unit time to an area, A, is:

$$J = k_1 A(c - c^*) = k_1 A c^* s \tag{8.31}$$

There exist empirical equations for k_1:[19]

$$Sh = 2 + 0.6\, Re^{1/2} Sc^{1/3} \tag{8.32}$$

where $Sh = k_1 d/D$, the Sherwood number; d = characteristic length of the crystal; D = solute bulk diffusion; $Re = dU/\eta$, the Reynolds number; U = relative velocity of the crystal and bulk; η = kinematic viscosity; and $Sc = \eta/D$, the Schmidt number.

At low relative velocities of crystal to bulk:

$$k_1 = 2D/d \tag{8.33}$$

At high relative velocities:

$$k_1 = \frac{0.6D^{2/3}v^{1/2}}{d^{1/2}\eta^{1/6}} \tag{8.34}$$

The growth rate can then be expressed as below, with the insertion of the appropriate approximation for k_1:

$$G = V_m k_l c^* s \tag{8.35}$$

where V_m is the solute molecular volume.

8.3.2.6 Secondary nucleation and empirical expression of crystal growth kinetics. Seconday nucleation refers to any process that results in the formation of nuclei that does not take place by the addition of monomers or clusters to a homogeneous cluster of solute below the critical nucleus size. Secondary nucleation implies the presence of a solid phase (not necessarily the crystalline form of the solute). In industrial crystallization, sources of secondary nucleation are initial breeding, attrition and fracture, and contact nucleation. Initial breeding refers to the crystalline dust adhering to seed crystals, which serve as nuclei. This is important only in seeded crystallization. Attrition and fracture result from collisions with crystallizer agitators or apparatus and could be important in mixed systems. The most important source of secondary nuclei in industrial and in microscale crystallization is probably contact nucleation. The adsorbed layer of solute at the growing crystal surface can serve as a source of secondary nuclei. As convective forces at the crystal surface increase (i.e. through agitation or density/temperature-driven convection) the adsorbed layer can be sheared from the growing crystal surface. In experiments where a growing crystal was slid along the bottom of a beaker,[21] nucleation was observed by new crystal formation along the beaker bottom. No visible damage to the crystal was observed. It has been observed that the crystal must be growing in order to induce contact nucleation.[21]

Secondary nucleation is expressed by an empirical equation that appears frequently in the industrial crystallization literature:

$$B^0 = k_n(T)w^k M^j \Delta c^i \tag{8.36}$$

where: k_n = nucleation rate constant with temperature dependence; M = suspension density, weight of crystals per unit volume or weight of solution; Δc = the concentration driving force, $c - c^*$; c^* = equilibrium concentration of solute; and w = a measure of the crystallizer hydrodynamics.

An empirical expression for crystal growth kinetics similar to the above expression for nucleation kinetics is:

$$G = k_g(T)\Delta c^g \tag{8.37}$$

The exponents in the above equations are normally determined by fitting experimental data and these equations are frequently used in industrial crystallizer design. The kinetic equations (8.36) and (8.37) are empirical.

8.4 Equipment design considerations

8.4.1 Population balances and crystal-size distribution

Crystallizers can be modeled using a population balance approach. This approach essentially sets up a material balance around the crystallizer, which includes terms for crystal nucleation and growth kinetics. The dependency of the kinetics on supersaturation combined with the dependency of the solute solubility on temperature and the material balances lead to a prediction of crystal-size distribution (CSD). The model can then be used to optimize the crystal-size distribution. Conversely, the crystal-size distribution and material balance can be used to determine the nucleation and growth kinetics.

The population balance and material balance equations below follow from the development of these equations expressed in the book by Randolf and Larson.[21]

The development of the population balance equation starts with a particle number continuity equation.

$$\text{Accumulation} = \text{Input} - \text{Output} + \text{Net generation} \tag{8.38}$$

With some manipulation the following equation represents the macroscopic population balance:

$$\partial n/\partial t + \partial Gn/\partial L + n(\mathrm{d} \ln V/dt) = B - D - \sum_{k=1}^{k} \frac{Q_k n_k}{V} \tag{8.39}$$

where: L = characteristic particle size; n = population density at size L (no./ml); $G = \partial L/\partial t$, linear growth rate of crystal in characteristic size L; V = crystallizer volume; B = particle birth function at size L; D = particle death function at size L; and Q_k = inlet and outlet volumetric flows (inlet negative).

The birth and death functions can be neglected when there is negligible attrition of the crystals.

Breakage of crystals by collisions with other crystals and the crystallizer equipment is the primary mechanism of birth and death of crystals from one size range to another. Birth and death functions will be assumed to be negligible for this development.

Crystals are frequently grown in a batch mode. In such a case, the inlet and outlet streams will be zero. Changes in crystallizer volume will be assumed negligible. Assuming crystal growth rates are independent of crystal size, equation (8.39) reduces to:

$$\partial n/\partial t + G\,\partial n/\partial L = 0 \tag{8.40}$$

The population density of nucleus size crystals (crystals of size zero are assumed to be nuclei), can be related to the population density:

$$B^0 = \text{nucleation rate} = (\mathrm{d}N/dt)_{L\to 0} = \mathrm{d}N^0/dt = \{(\mathrm{d}L/dt)(\mathrm{d}N/\mathrm{d}L)\}_{L\to 0} \tag{8.41}$$

where N^0 = the number of crystals of size zero per unit volume. Assuming that the growth rate, $(\mathrm{d}L/dt)$, is independent of size, and noting that n^0 is $\mathrm{d}N/dL$ as L goes to zero, the above equation reduces to:

$$B^0 = n^0 G \tag{8.42}$$

In population balances, moment analysis is frequently used. The zeroth through third moments of the population density are related to physical quantities. The integrated moments of the population density are defined below:

$$\mu_k = \int_0^\infty n(L,t)L^k\,\mathrm{d}L \quad \text{where } k = 0, 1, 2, 3. \tag{8.43}$$

$$\mu_0 = N_T = \int_0^\infty n(L,t)\,\mathrm{d}L \tag{8.44}$$

$$\mu_1 = L_T = \int_0^\infty n(L,t)L\,\mathrm{d}L \tag{8.45}$$

$$k_A\mu_2 = A_T = k_A\int_0^\infty n(L,t)L^2\,\mathrm{d}L \tag{8.46}$$

$$k_V\mu_3 = V_T = k_V\int_0^\infty n(L,t)L^3\,\mathrm{d}L \tag{8.47}$$

where: N_T, L_T, A_T, and V_T are total number of crystals, total end-to-end length of crystals, total area, and total volume of crystals per unit crystallizer volume respectively. k_A and k_V are the area and volume shape factors, respectively. These shape factors relate the characteristic length to the surface area and volume of one crystal.

The mean particle size, $\langle L\rangle$, and the coefficient of variance of mean particle length $(s/\langle L\rangle)$, c.v., can be expressed in terms of the moment equations.

$$\langle L\rangle = \mu_1/\mu_0 \tag{8.48}$$

$$\text{c.v.} = \left(\frac{\mu_0\mu_2}{\mu_1^2} - 1\right)^{1/2} \tag{8.49}$$

Other important relationships which can be expressed in terms of the moment equations are those for the nucleation rate, B^0, and the magma density, M_T.

$$B^0 = d\mu_0/dt \tag{8.50}$$

$$M_T = \mu_3 \rho_c k_V \tag{8.51}$$

$$\rho_c = \text{crystal density}$$

It should be noted that, in a batch reactor, all moments are a function of time.

In summary, the population balance, material balance, and kinetic expressions for a batch crystallizer are:

$$\frac{\partial n(L,t)}{\partial t} + G\frac{\partial n(L,t)}{\partial L} = 0 \tag{8.52}$$

$$\frac{dc}{dt} = -3\rho_c k_V G \int_0^\infty n(L,t)L^2\,\mathrm{d}L \tag{8.53}$$

$$G = k_g(T,w)(c - c^*)^g = k_g \Delta c^g \tag{8.54}$$

$$B^0 = k_n(T,w)M^j \Delta c^i \tag{8.55}$$

where k_n and k_g can be functions of temperature and hydrodynamics, and G is the linear growth rate, $\partial L/\partial t$.

The initial and boundary conditions for equations (8.52) to (8.55) are:

$$n(L, t = 0) = n_0(L) \tag{8.56}$$

$$n(L = 0, t) = \frac{B^0}{G} \tag{8.57}$$

$$c(t = 0) = c_0 \tag{8.58}$$

Note that equations (8.53) and (8.54) have no theoretical basis except when g is equal to 1. These equations are normally empirically fit to experimental growth and nucleation data.

8.4.2 Controlled cooling

In a reactor with changing temperature, the equilibrium concentration of solute, c^*, will be changing with time. The change in equilibrium concentration with time can be expressed as:

$$dc^*/dt = (dc^*/dT)(dT/dt) = F(t)(dc^*/dT) \tag{8.59}$$

where $F(t)$ is the cooling curve. dc^*/dT is known from the phase diagram information of the solute in equilibrium with the solute crystals, usually fit to a thermodynamic expression. For an isothermal reactor c^* is a constant.

A few studies have been performed on batch crystallizers with controlled cooling. Various cooling schemes and objectives have been applied. One objective has been to operate the crystallizer within the metastable zone of supersaturation. The metastable zone is not a strictly quantitative measure and varies with the definition, that is, the solution supersaturation in which nucleation is negligible but crystal growth proceeds, or the supersaturation in which nucleation induction times are large. Alternate objectives have been to operate the crystallizer at a constant nucleation rate or to maximize the terminal size of seed crystals.

In experimental work by Jones[22] and Jones and Mullin,[23] four cooling strategies were adopted for batch crystallization of an inorganic salt, potassium sulfate. Natural cooling was achieved by design of the cooling apparatus to maintain the same batch times as a constant-supersaturation controlled crystallizer and the same initial and final temperatures as all other cooling schemes. For constant supersaturation control, the supersaturation balance was solved numerically and a cooling curve calculated to maintain constant supersaturation (this is the same as assuming that the nucleation rate is constant if there is no dependency of nucleation on magma density). Metastable control was defined by making several simplifying assumptions on the constant supersaturation scheme. Nucleation was assumed negligible, solubility was assumed to vary linearly with temperature, and the growth rate was assumed to be independent of supersaturation. The supersaturation balance was then solved analytically for the cooling curve. The final strategy, size optimal control, used the Pontryagin Continuous Maximum Principle with the equations for the population and material balances to determine the operating policy (cooling curve) to maximize the terminal size of seed crystals. In all cases, the batch crystallizers were seeded. The results of these experiments are summarized below in Table 8.2. As expected, size optimal control yielded the largest crystals with the smallest coefficient of variation.

In earlier work by Mullin and Nyvlt,[24] two salts were crystallized under batch-seeded conditions with natural cooling and controlled cooling. The controlled-cooling curve was determined from the supersaturation material balance by maintaining a constant supersaturation. Nucleation was assumed to be negligible and solubility was assumed to vary linearly with temperature. Their results are summarized in Table 8.3.

Table 8.2 The effect of various control strategies on mean size and polydispersity of crystals.[22,23]

	Mean size (μm)		Coefficient of variation (%)	
	Theory	Experiment	Theory	Experiment
Natural cooling	594	792	75	58
Constant supersaturation	1236	970	39	53
Metastable control	1240	1031	38	53
Size optimal control	1291	1141	39	44

Table 8.3 A comparison of controlled to uncontrolled cooling. The values are mean crystal size expressed in millimeters.[24]

Substance	Uncontrolled cooling	Controlled cooling	Theoretical
K_2SO_4	1.0	1.4	1.5
$(NH_4)_2SO_4$	0.8	1.1	1.4

As can be seen from the experimental evidence, even very simplified models of crystal growth combined with a cooling strategy yield improved crystal size and in some cases smaller coefficients of variation.

8.5 Special considerations for precipitation processes

Crystallization is usually employed as a finishing step to produce very pure product. For this reason, the composition is not very complicated and more is known about the feed. Precipitation is usually an isolation technique employing a complex mixture, which usually poses special processing problems.

Grabenbauer and Glatz[25] have shown that precipitation at the isoelectric point may be viewed as a two-step process: (i) formation of primary particles (i.e. solid precipitate); followed by (ii) aggregation and growth of these primary particles into larger aggregates. Of course, there are complicating issues such as the effect of shear rate, which enhances the growth rate of the aggregates but also tends to break them up.[26] In addition, 'aging' the aggregates in a stirred vessel makes the aggregates more resistant to subsequent shear stress such as pumps or centrifuges.[27]

Once a precipitate is available, a device for solid–liquid separation must be chosen. It is wise to know the characteristics of the solid–liquid separation device and to design the precipitation device to suit the separation device. For example, if a dead-end filtration device will be used for the separation, the precipitator should, ideally, produce a solid with minimal compressibility and minimal cake resistance. For centrifugation, the density difference between the solid and liquid as well as particle size will control the effectiveness of the separation.

8.6 Summary

Both crystallization and precipitation are valuable techniques for product recovery. Although each has a separate role to play, both techniques employ the same phenomenon—phase separation from a liquid to a solid. This chapter has attempted to review the basic thermodynamics, kinetics and design equations for these processes.

References

1. Prausnitz, J.M. (1986). *Molecular Thermodynamics of Fluid-Phase Equilibria.* Prentice-Hall, Englewood Cliffs, New Jersey, p.415.
2. Wankat, P.C. (1990). *Rate-Controlled Separations.* Elsevier Applied Science, New York, New York, p.21.
3. Edsall, J.T. (1952). *Adv. Protein Chem.,* **3,** 383.
4. Cole, J.B., Bryan, M.L. and Bryan, W.P. (1969). *Arch. Biochem. Biophys.,* **130,** 86.
5. Gronwall, A. (1942). *Compt. Redn. Trav. Lab. Carlsberg,* **24,** 8.
6. Hofmeister, T. (1887). *Arch. Exptl Pathol. Pharmakol.* Naunyn-Schiedeberg's, **24,** 274.
7. Setschenow, J. (1889). *Z. Physik. Chem.,* **4,** 117.
8. Cohn, E.J. (1925). *Physiol. Rev.,* **5,** 349.
9. Green, A.A. (1932). *J. Biol. Chem.,* **95,** 47.
10. Sorenson, S.P.L. (1933). *Biochem. Z.,* **258,** 16.
11. Green, A.A. (1931). *J. Biol. Chem.,* **93,** 495.
12. Czok, R. and Bucher, Th. (1961). *Advances in Protein Chemistry,* **15,** 315.
13. Scopes, R.K. (1982). *Protein Purfication.* Springer-Verlag, New York.
14. Schubert, P.F. and Finn, R.K. (1981) *Biotechnol. Bioeng.,* **23,** 2569.
15. Hoare, M., Bell, D.J. and Dunnill, P. (1983). *Adv. Biochem. Eng. Biotechnol,* **26,** 1.
16. Larson, M.A. and Garside, J. (1986). *J. Crystal Growth,* **76,** 88.
17. Rusli, I.T. and Larson, M.A. (1987). *Industrial Crystallization, Vol. 87,* (Eds, J. Nyvlt and S. Zacek), Elsevier, Amsterdam, p.111.
18. Nyvlt, J., Sohnel, O., Matuchova, M. and Broul, M. (1985). *The Kinetics of Industrial Crystallization.* Elsevier, Amsterdam, p.154.
19. Ohara, M. and Reid, R.C. (1973). *Modeling Crystal Growth Rates from Solution.* Prentice-Hall, Englewood Cliffs, New Jersey. p.43.
20. Burton, W.K., Cabrera, N. and Frank, F.C. (1951). *Trans. Royal Soc. (London),* **A243,** 299.
21. Randolph, A.D. and Larson, M.A. (1988). *Theory of Particulate Processes.* Academic Press, San Diego, California.
22. Jones, A.G. (1974). *Chem. Eng. Sci.* **29,** 1075.
23. Jones, A.G. and Mullin J.W. (1974). *Chem. Eng. Sci.,* **29,** 105.
24. Mullin, J.W. and Nyvlt, J. (1971). *Chem. Eng. Sci.,* **26,** 369.
25. Grabenbauer, G.C. and Glatz, C.E. (1981). *Chem. Eng. Commun.,* **13,** 203.
26. Glatz, C.E., Hoare, M. and Landa-Vertiz, J. (1986). *AIChE J,* **32,** 1196.
27. Brown, D.L. and Glatz, C.E. (1987). *Chem. Eng. Sci.,* **42,** 1831.

9 Lyophilization

J.W. SNOWMAN

9.1 Introduction

Many labile, heat-sensitive products are stabilized by drying. Lyophilization, also called freeze-drying, avoids most of the problems of liquid-phase drying. A simple definition of lyophilization is that it is a means of drying, achieved by freezing the wet substance and causing the ice to sublime directly to vapor by exposing it to a low partial pressure of water vapor. In this context, 'low' means below about 6 mb (4.6 mmHg) — the triple point of water. Below this low pressure, by adding heat energy, ice can be sublimed directly into water vapor without first melting to liquid. Since the subliming ice crystals leave cavities, lyophilized dry material contains a myriad of interstices into which water can penetrate to give rapid and complete re-solution when needed. The name lyophilization is derived from the Greek for 'to make solvent-loving'. Constituents of the material are kept immobilized during sublimation drying by the ice crystals. The shape of the dry substance is broadly the same as that of the frozen, wet substance and migration to the surface to form a skin is reduced. Since drying takes place at a low temperature, damage is minimized and volatile components are retained.

Lyophilization is favored for parenteral products because the wet material can be sterile-filtered just before it is filled into final containers, reducing particulate and bacterial contamination. The principal advantages of lyophilization are:

- Minimum damage to, and loss of activity in, delicate heat-labile materials
- Creation of a porous, friable structure
- Possibility of accurate, clean dosing into final product containers, and
- Speed and completeness of rehydration.

The principal disadvantages are:

- High capital cost of equipment (about three times more than other methods)
- High energy costs (two to three times more than other methods), and
- Lengthy process time (typically 24–48 h/drying cycle).

Lyophilization is useful when the product meets one or more of the following criteria:

- It is unstable
- It is heat-labile

- Minimum particulates are required
- Accurate dosing is needed
- Quick and complete rehydration is required
- The product is of high value.

9.2 History of lyophilization

Lyophilization sometimes occurs in nature when material is kept frozen in a very dry atmosphere while the energy to create the phase change from solid to vapor is supplied. Such conditions can occur in cold, but sunny, weather. The prerequisite is to keep the partial pressure of water vapor low. However, providing a continuous supply of dry air is difficult, particularly if sterile conditions are needed. In practice, it is convenient to provide the necessary low partial pressure of water vapor by carrying out the process under vacuum.

Lyophilization was first reported in 1909 as a means of preserving small quantities of protein. Chemical desiccants were used to absorb water vapor sublimed from the frozen specimen. Later, a mechanically refrigerated ice trap was used to avoid expensive desiccants. The process first came into industrial use for the preservation of blood plasma; the first large-scale plant was described in 1940. Subsequently, one of the best-known applications was in the campaign that led to the worldwide eradication of smallpox by the 1970s. Between 1970 and 1980 lyophilization became a mature process. It was applied to a wide variety of products including diagnostics, hormones, liver extracts, vitamin preparations, antibiotics and blood fractions.

In the 1980s large-scale applications for new antibiotics, anti-cancer drugs and rDNA fermentation products led to the need for sophisticated control systems and automated material handling. Some of the new drugs were extremely potent and frequently toxic. The need to ensure operator safety led to the development of elaborate self-cleaning systems and isolation technology. Improvements in systems have also been necessary to meet regulatory requirements for good manufacturing practices (GMPs). During the 1990s, validation of production processes has been extended to include lyophilizers, resulting in requirements for much improved testing and documentation. In spite of its cost and complexity, lyophilization is now the preferred method for the stabilization of labile materials with short shelf-lives in solution.

9.3 Principles of lyophilization

A lyophilization process consists of three phases, indicated by the temperature/time and pressure/time diagrams of Figure 9.1; these are:

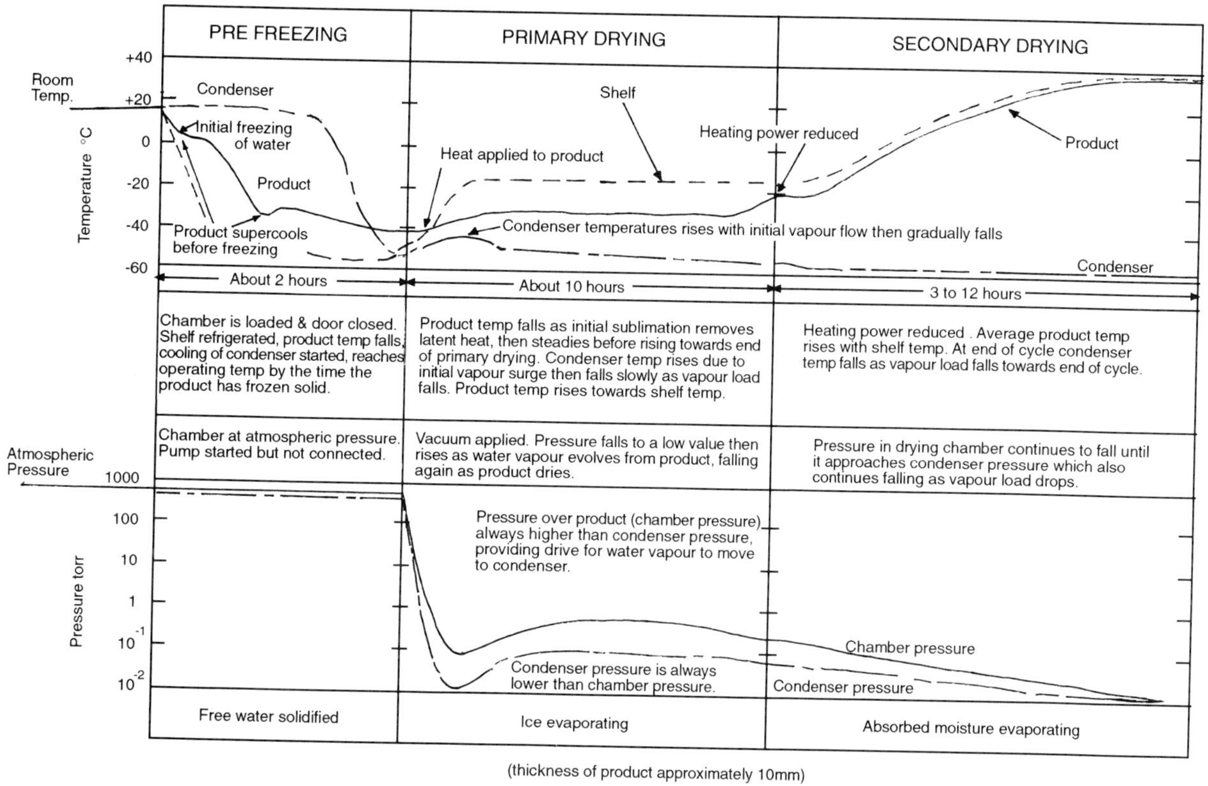

Figure 9.1 Graph of complete freeze-drying cycle.

- Freezing to solidify the material
- Sublimation drying to reduce moisture to around 4% w/w of the dry product and to leave an apparently dry cake of solid residue, which is substantially the same size and shape as the frozen mass, and
- Desorption or secondary drying to reduce bound moisture to the required final value (often below 1% w/w).

The three phases overlap to some extent in different parts of the drying material, usually called the 'cake'. For example, if low-melting-point materials such as ethanol are present in the material, it will not freeze completely at the temperatures available in practical lyophilizers. At the commencement of primary drying, such low-melting-point constituents will evaporate from the liquid phase before true freeze-drying of the product begins. It is also usual for some desorption drying to begin in the drier parts of the cake before sublimination drying is completed. During sublimation, vapor is generated from a distinct interface or freeze-drying front, which moves from the outer surface of the cake. Subliming vapor must pass through the interstices of the substantially dry layers. The shape of the interstices results from the ice crystals formed on freezing. If the crystals are small and discontinuous, the escape path for the vapor is limited; if large crystals are formed, escape is easy and the product can be dried more quickly. The method and rate of freezing are critical to the course of sublimation.

9.3.1 Freezing

In simple two-component systems, water and a solute may both crystallize. In this case freezing proceeds as follows. Cooling from room temperature to below 0°C eventually results in the nucleation of ice. The release of heat of crystallization raises the temperature toward 0°C. Crystallization of water then proceeds at a progressively falling temperature related to the equilibrium melting point of the concentrating solution as shown in Figure 9.2 which shows a typical freezing curve.

As the temperature falls the solution becomes closer to saturation so that crystals of solute are precipitated. Eventually a *eutectic point* is reached where the material becomes wholly crystalline. When a material contains more than one crystallizable solute, a similar situation exists but the eutectic point is lower than that of any two components combined.

In most practical cases, the solute does not actually crystallize but forms an amorphous glass, discussed in the section below. In such cases, there is no eutectic temperature as such but a maximum product temperature for freeze-drying with retention of microstructure. This is called the *collapse temperature.*

It should be appreciated that, during freezing, the concentration of the solute increases markedly as pure water separates into ice crystals. This

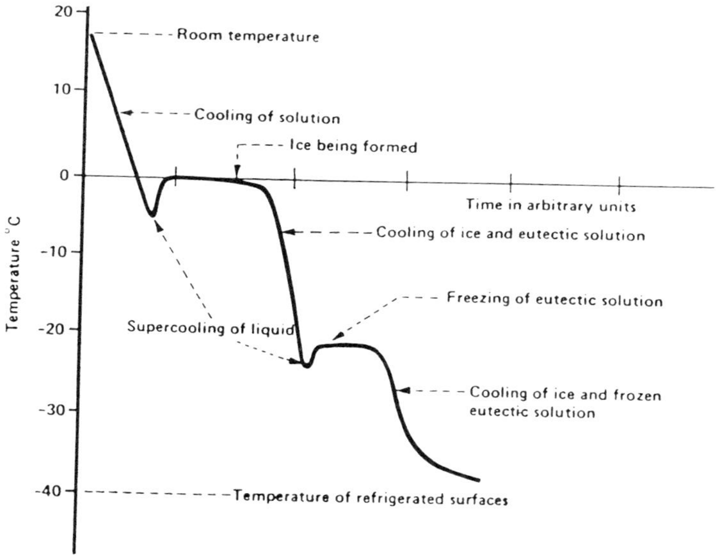

Figure 9.2 Typical freezing curve.

produces drastic changes in concentration and pH, which can have serious effects on the product.

9.3.2 Freezing methods

When liquids are to be lyophilized, the freezing and drying times can be reduced by disposing them in thin layers. A variety of means have been developed for achieving this in freezing baths, either by tilting containers to form wedges, rotating them slowly about horizontal axes or rapidly about vertical axes to form hollow cylinders. These methods are usually only applicable on a relatively small scale. For most aseptic production, it is usual to freeze on the shelves of the lyophilizer.

Consumable refrigerants, such as liquid nitrogen, are sometimes used to achieve fast-freezing. Such methods must be used with care because fast-freezing generates small ice crystals, which can retard vapor escape, resulting in either melt-back, collapse, or slower sublimation. It should also be noted that if there is direct contact with product, the consumable refrigerant itself must be sterile. This is difficult to achieve with liquid nitrogen, which is an excellent preservative for many organisms. It is more usual to use an indirect heat-exchange method. If small ice crystals cannot be avoided, a 'conditioning' or 'tempering' can be used to give the crystals of ice an opportunity to grow. This is performed by holding the temperature of the product just below the eutectic or collapse temperature for 1–2 h to encourage crystal growth, then re-cooling before starting sublimation.

9.3.3 *Sublimation*

After freezing, the partial pressure of water vapor must be reduced below the triple point pressure of water (about 6 mbar) to allow sublimation to take place. If there is a free escape path for water molecules, the speed of sublimation at the beginning of primary drying is limited largely by the rate at which heat can be transferred to the freeze-drying front.

If the lowest safe drying temperature of the material is exceeded during sublimation, then melting can occur. This gives rise to gross material faults, such as shrinking and puffing. These are sometimes confused with collapse, which can occur when one or more of the solutes does not crystallize but supersaturates to form an amorphous glass associated with unfrozen water. The glass is, in fact, supported by the pure water ice crystals. It appears to be rigid but, as the pure water ice crystals sublime away, the support is removed and apparently dry product collapses to form an impermeable mass. The phenomenon was explained and has been extensively researched by McKenzie,[1] who has measured the critical temperature that must not be exceeded during drying for several commonly used materials. This is called the *collapse temperature*, or Tc, for any particular material.

Since most of the heat energy to cause sublimation must be transferred through the frozen material, which usually has poor thermal conductivity, the need to avoid melting means that the thermal gradient is low and thus the heat conducted is small. Later in the process, the vapor flow from the freeze-drying front and hence the sublimation rate is restricted by the increasing depth of the dry part of the product cake. Figure 9.3 illustrates

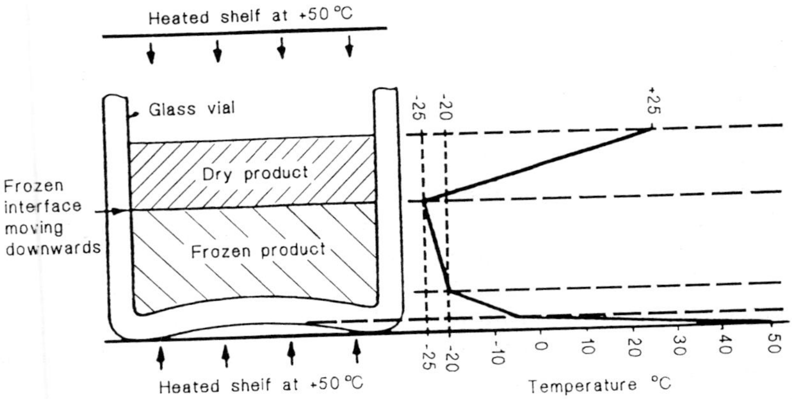

Figure 9.3 Temperature distribution in product.

the difficulty of transferring heat to the freeze-drying front without melting the product at the bottom of the vial, when drying at very low pressure. Assuming a maximum drying temperature around −20°C, the poor heat transfer under vacuum conditions from shelf to vial and through the glass bottom of the vial may require a shelf surface temperature of up to 40°C. Any restriction of the rate of sublimation can lead to disastrous melt-back or collapse and product destruction.

Heat transfer from shelf to product during sublimation can be increased by spoiling the vacuum, usually performed by bleeding air or dry nitrogen into the chamber or throttling the vacuum pump to maintain pressure between 0.1 mbar and 0.5 mbar. This improves convective heat transfer from the heating shelf to the bottom of the product container and also reduces the temperature difference required from shelf to product for a given drying rate. The techniques have been described by several workers; although treatments prepared by Nail[2] and Pikal[3] are helpful. For sterile products, air or nitrogen is bled into the chamber through a filter, usually a 0.2 μ steam sterilizable model.

Typical primary drying time for a 1 cm thickness cake of a 'simple' material is between 10 and 20 h, with favorable ice-crystal structure and optimized temperature and pressure conditions in the drying chamber. Drying time varies with thickness approximately by the relationship

$$\text{Time} = \text{K}\ (\text{Thickness})^{1.5} \tag{9.1}$$

where: K is a constant. This can also be expressed as:

$$\text{Time} \propto (\text{Thickness})^{1.5} \tag{9.2}$$

9.3.4 Secondary drying

This is the removal of bound moisture, which may be: (i) water of crystallization; (ii) randomly dispersed water in a glassy material; (iii) intracellular water; or (iv) absorbed water. Primary or sublimation drying accompanied by desorption removes the free water.

The remaining bound water usually amounts to about 4% w/w. It can be removed by heating the product under vacuum, typically to between 15°C and 30°C for about one-third of the primary drying time.

The residual moisture content of dry material is a critical factor in storage life. Many biological materials can be damaged by overdrying and retain their best titer when dried to between 1.5% and 2% final moisture content. Examples are vaccines such as BCG (1.5%), and live rubella, measles and others (2%). Other materials, such as chemotherapeutics and antibiotics, must be dried to as low as 0.1% residual moisture for best results. Most freeze-dried materials are hygroscopic and must be stored in sealed containers. In the early days of lyophilization, products were stored under vacuum.

Current practice is to store them under dry, inert gas — usually pure dry nitrogen. Other gases, such as argon and helium, have also been used. Some factors affecting storage life have been reported by Freyrichs and Herbert.[4]

9.4 Principles of equipment

Aqueous solutions can be frozen by placing them in a vacuum. The more energetic molecules escape and the temperature of the solution falls by evaporative cooling. Eventually it freezes. About 15% of the water is lost.

The simplest lyophilizer would consist of a vacuum chamber in which solutions could be placed together with a means of removing water vapor so as to freeze them by evaporative cooling and then maintain the water vapor pressure below the triple-point pressure. The temperature of the material would continue to fall below the freezing point, and sublimation would slow down until the rate of heat gain by conduction, convection and radiation to the material was equal to the rate of heat loss as the more energetic molecules sublimed away and were removed. This simple approach presents other difficulties. When a solution is frozen by evaporative cooling it froths as it boils. This can be suppressed by low-speed centrifugation but a more common alternative is to freeze the material before it is placed under vacuum. With small lyophilizers, material is usually prefrozen inside a flask. The flask is then attached to a manifold connected to the lyophilizer, which comprises an ice condenser and a vacuum pump.

For larger-scale equipment, it is usual to place the material on shelves inside the drying chamber (Figure 9.4), which can be cooled so that the material is frozen by conduction and convection at atmospheric pressure before the vacuum is created. Without a controlled heat input to the material, its temperature would fall during sublimation until drying was virtually at a standstill. It is usual to connect a heat supply to the product-support shelves so that, after their initial use for freezing, they can be used to provide heat to maintain the product at a constant low temperature.

At a typical lyophilization cycle pressure, 1 ml of ice produces more than 1 000 000 ml of water vapor. The vacuum pumps cannot handle this, so a refrigerated trap (called the ice condenser) is fitted between the lyophilization chamber and the vacuum pump. Figure 9.5 shows the major parts of an industrial freeze dryer. Some important modern refinements are shown in Figure 9.6 (a and b) and include:

- Separation of drying chamber and ice condenser to reduce cross-contamination, with an isolation valve between chamber and ice condenser to allow for end-point determination and simultaneous loading and defrosting
- Pressure-vessel rating for chamber and ice condenser to allow for sterilization by pressurized steam at 121°C or higher

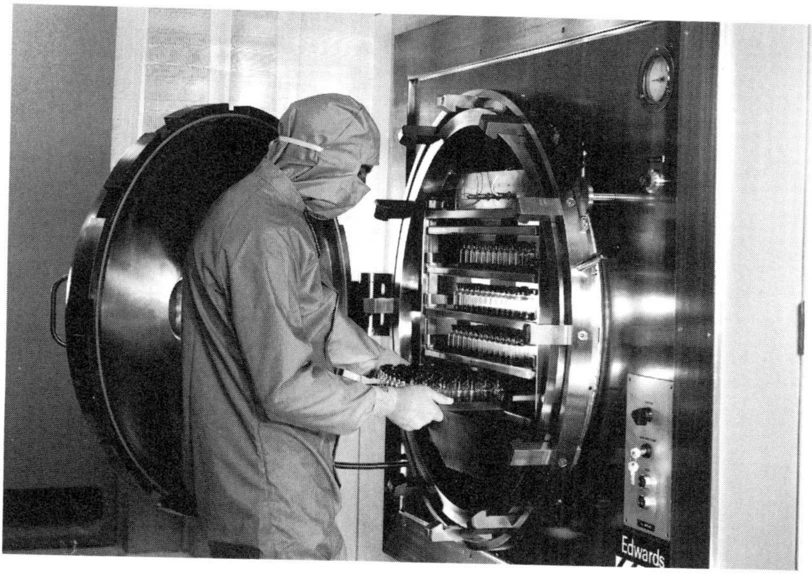

Figure 9.4 Loading vials onto shelves inside the drying chamber.

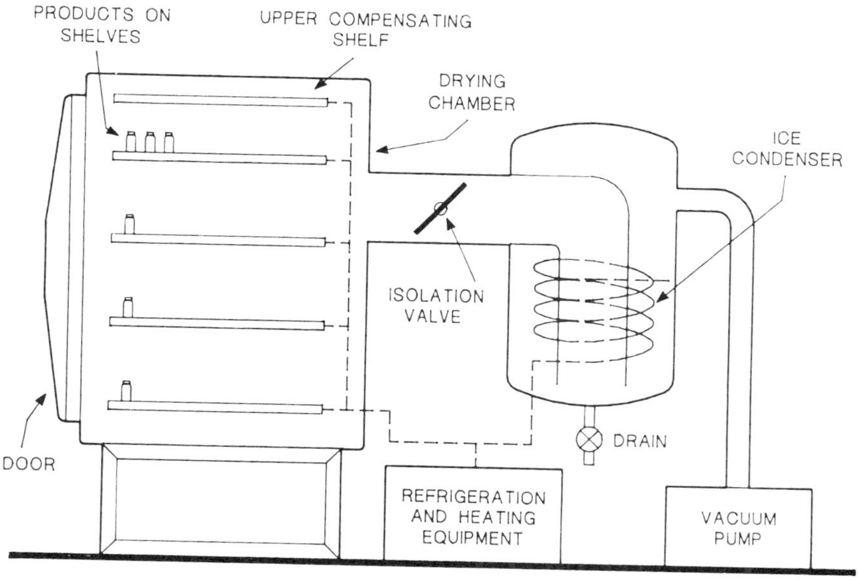

Figure 9.5 Basic industrial lyophilizer.

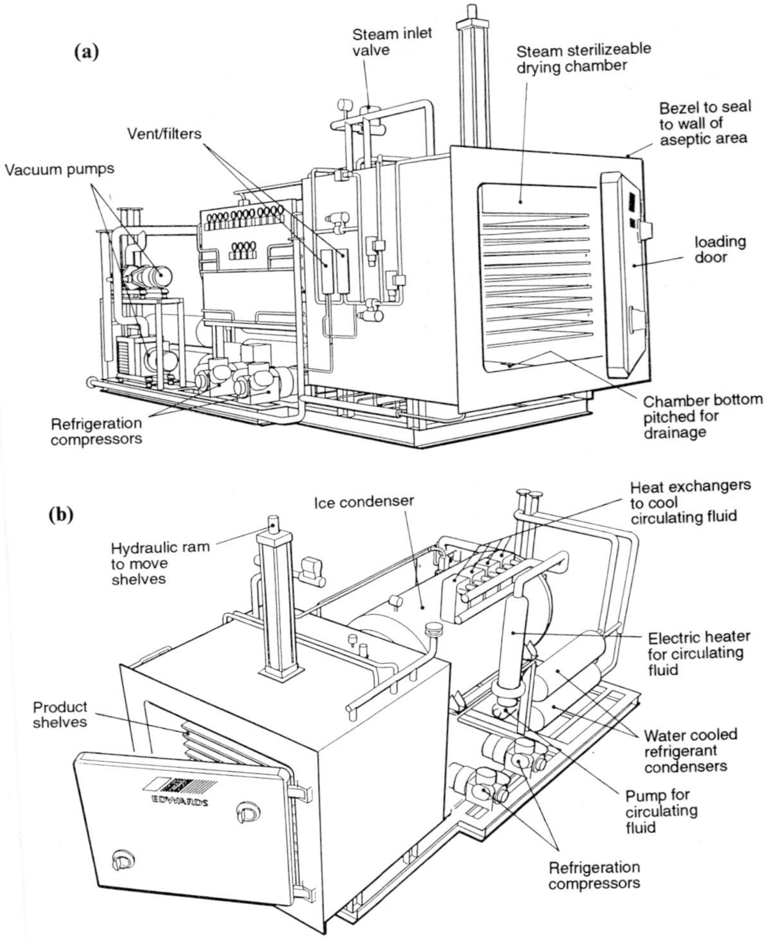

Figure 9.6 Typical industrial lyophilizer: (a) right-hand side; (b) left-hand side.

- Use of a circulating intermediate heat-exchange fluid to give even and accurate control of shelf temperature
- Movable product support shelves, which, at the end of freeze-drying, close the slotted bungs used in vials (Figure 9.7). Movable shelves can also facilitate cleaning and loading.
- Duplicated vacuum pumps, refrigeration systems and other moving parts to enable drying to continue without risking melt-back or collapse of the product in the event of breakdown
- Automatic control system with safety interlocks and alarms
- Additional instruments to control, monitor and record process variables.

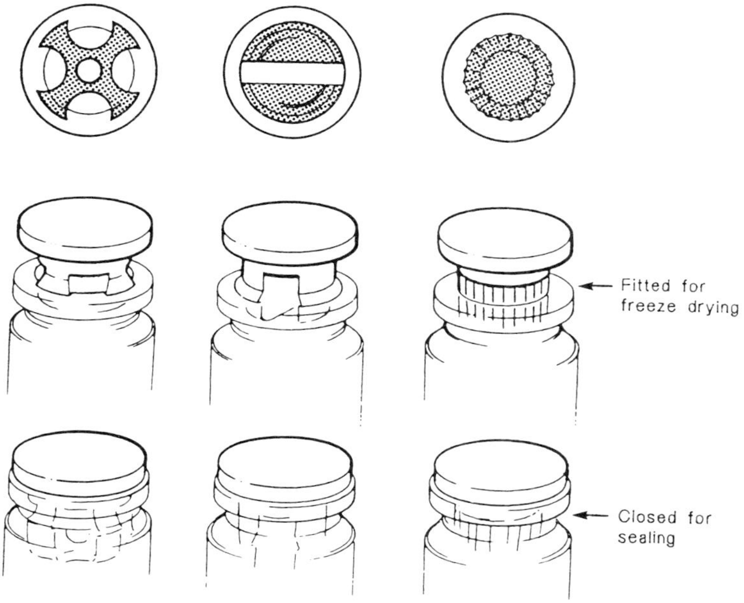

Figure 9.7 Bungs of various types.

For low-value products where safety and possible cross-contamination are not so important, the ice condenser is sometimes placed inside the drying chamber to reduce equipment costs. This is the normal method of construction for food freeze-dryers.

One batch of a pharmaceutical material in a freeze-dryer can often be worth millions of dollars. Deciding the optimum conditions for freeze-drying a complex product requires knowledge of its eutectic and collapse temperatures, and the yield of active product depends critically on the methodology adopted. An initial approach to formulating a product for lyophilization and subsequent cycle development was outlined by Snowman.[5] More extensive treatments are given by Pikal.[6,7] Owing to the risk and cost of investigating the process in a large production lyophilizer, small-scale pilot plants are needed with the characteristics and performance of industrial freeze-dryers. Also, where small-scale production is required, rather than optimization, the reliability and reproducibility of a scaled-up laboratory freeze-dryer are inappropriate. For these reasons, pilot lyophilizers with shelf areas up to 24 sq. ft. ($2\,m^2$) are built with characteristics similar to larger machines.

For scale-up work, if the refrigeration and vacuum systems are proportional, it is possible to scale-up by a ratio of 20:1. For larger scale-ups the chamber geometry must be taken into account because of the variations in drying rate that occur across large shelves. This was studied by Rowe *et al.*[8]

Lyophilizers with shelf areas between 2 m^2 (20 sq. ft.) and 20 m^2 (200 sq. ft.) are considered 'medium-scale'. They are often custom-built to suit the product, with features such as special refrigeration and vacuum-pumping performance, stand-by machinery and sophisticated control and data-acquisition systems to meet regulatory authority requirements for quality control.

Large-scale installations are considered to be from 20 m^2 (200 sq. ft.) shelf area upwards in size. In the pharmaceutical industry, the largest units in current use reach about 100 m^2 (1000 sq. ft.) total shelf area, whereas food-processing equipment is frequently built with shelf areas over 200 m^2 (2000 sq. ft.). Figure 9.8 shows a modern lyophilizer of 20 m^2 shelf area before installation in the wall of an aseptic area. It is large enough to process some 37 000 bottles of 23 mm diameter or 400 kg of bulk material.

Large capacities require special attention to loading and unloading procedures. Operator handling is the major source of contamination of sterile products and much attention is now focused on reducing or eliminating their presence by automation. Its high cost precludes the automation of small freeze-dryers under about 15 m^2 of shelf area, although an exception occurs with highly potent or toxic materials when operators must be protected from contact.

Where products are to be placed on the shelves — the normal practice in the pharmaceutical industry — automatic systems have been developed to place trays or bottles on the shelves without sliding contact (to avoid particle

Figure 9.8 Large (20 m^2) industrial lyophilizer.

generation). If the product to be lyophilized is adequately stable and can withstand a delay of some hours between being filled into its container and being frozen, it is usual to accumulate trays of product on a loading cart. When the loading cart is filled, it is placed in front of the lyophilizer and the trays are automatically pushed onto the shelves, via a roller system, in one operation. Figure 9.9 shows a fully automatic cart for loading vials onto shelves directly without trays. Cart systems improve the utilization of lyophilization capacity since it is possible for carts to be loaded ahead of time when the freeze-dryer is available for unloading and loading. This means that the loading operation can be carried out in a few minutes.

If the product is unstable and must be frozen within a short time of being filled into its container, the system of 'constant-height loading' is used. Product support shelves are lowered to the bottom of the drying chamber, then lifted one by one to a position in line with a small subdoor at the height of the filling machine conveyor as shown in Figure 9.10. Product is placed onto the pre-cooled shelves as soon as a tray load or shelf load has accumulated from the filling line. The disadvantage of this technique is that the loading time for the freeze-dryer is extended to the time taken to fill up the vials, which can be several hours for a large installation. The result is

Figure 9.9 Fully-automatic loading cart for direct loading of vials.

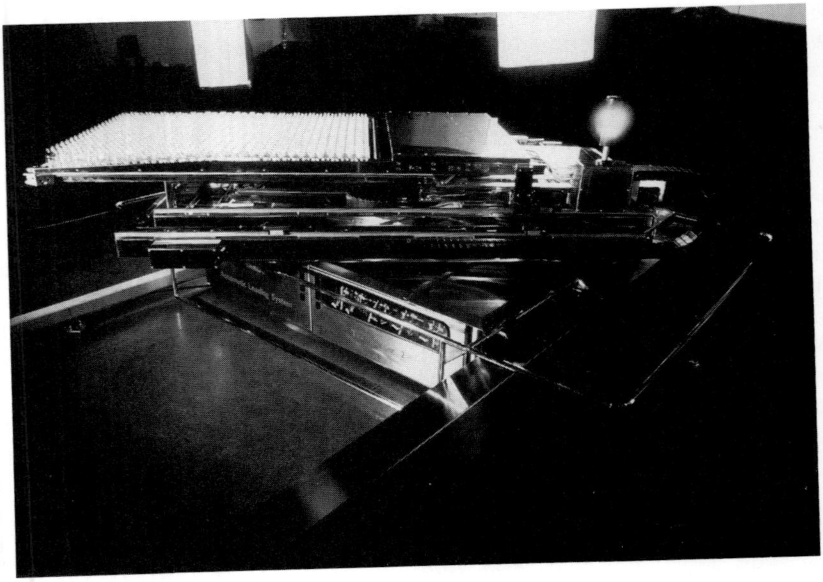

Figure 9.10 Automated constant-level loader (trayless).

productivity of the freeze-dryer can be reduced to as little as 80% of the theoretical production time available.

An alternative approach mostly used in food dryers involves loading trays of product onto a rack that can be moved into a separate pre-freezer and then into the lyophilizer. In this system, the trays do not come into direct contact with the shelves but are suspended between plates that supply heat by radiation and convection instead of conduction. External pre-freezing is also used for some diagnostic reagents, such as those printed onto paper sheets, and also for lyophilized tablets.

Continuous lyophilizers, which accept externally prefrozen granules or droplets, have been developed. Material enters the drying chamber through a vacuum lock and falls onto the highest of a series of moving belts, running in opposite directions. Material falls from the end of each belt onto the next lower one and finally into an exit vacuum lock. Twin ice traps are fitted so that they can be defrosted alternately for continuous operation.

Products to be continuously lyophilized should have the following characteristics:

- Low intrinsic value
- Constant process conditions
- Sterile conditions not required

- High eutectic point
- Non-friable product (to avoid dust creation).

9.5 Specification of freeze-dryers

Equipment is characterized by the useful shelf area available stated in square meters (m^2) or square feet (sq. ft.). Bulk freeze-drying in trays is unusual for finished products because the accuracy of powder filling is not usually appropriate for the final presentation. It is, however, used for intermediate products and, in this case, the freeze-dryer may be characterized by the ice-condenser capacity usually stated in kilograms of ice, which is virtually the same as liters of water deposited.

9.5.1 Sizing

9.5.1.1 Shelf area. Consider local work patterns, holidays and shift patterns. It is unwise to expect the lyophilizer to be run effectively for more than, say, 240 days/year. If the planned lyophilization cycle lasts for less than 22 h, possibly 6 cycles/week can be run. However, experience with many production units suggests that two or three cycles/week are more likely. Clearly longer cycles will give fewer cycles/week but also to be considered are the time to defrost and the time to cool down after steam sterilization, if appropriate.

The requirement is to define a realistic number of cycles per year (A). Added to the yearly production needed must be allowances for QC samples, rejects and packaging rejects. Adding these together gives the required yearly throughput (B). Dividing throughput by the number of cycles gives the number of vials per cycle (C), where:

$$\frac{B}{A} = C \tag{9.3}$$

To find the needed shelf area, it is only necessary to calculate the number of vials/m^2 (D) and then

$$\frac{C}{D} = E \tag{9.4}$$

where: E = shelf area in m^2.

At this point, a judgement must be made. The number of vials can be calculated using square formation or hexagon formation (Figure 9.11).

A hexagon formation allows about 12% more vials on an infinite field. However, space is lost at tray edges and for small trays, say 30×45 cm, the

(a)

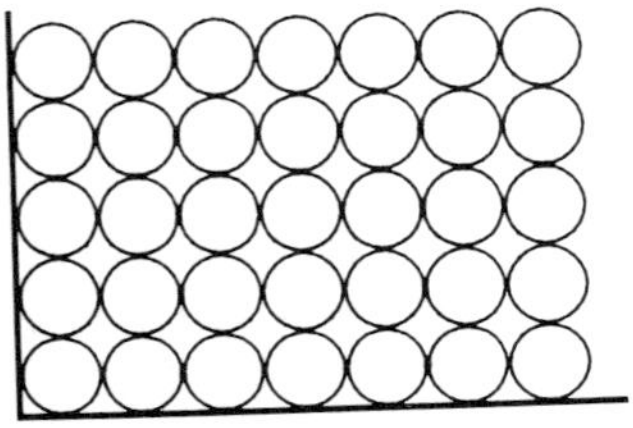

(b)

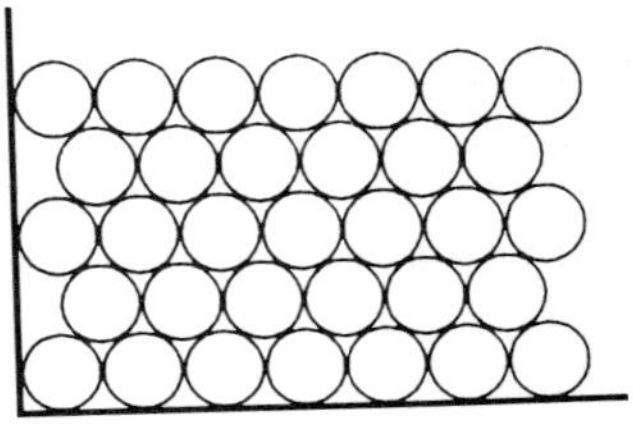

Figure 9.11 Possible vial formations: (a) square; (b) hexagonal.

square formation should be assumed. If a direct-loading trayless system is used, then a hexagon formation gives about 8% more vials on a given shelf than with the square formation.

Thus, the vials/m^2 are the number in the square formation D for small trays or $D \times 1.08$ for large shelves. For very large lyophilizers, the number of vials per tray or per shelf should be calculated accurately for the particular geometry in use.

9.5.1.2 Ice-condenser capacity. From the number of vials and the fill volume, the amount of water to be condensed every load can be calculated (remembering to subtract the solids content of the fill). It is recommended that the ice-condenser capacity at 10 mm thickness of ice be at least double the theoretical capacity needed. A further recommendation is that the ice-condensing surface be in the form of tubing, which is more tolerant of overloads than plates because the surface area of a tube condenser increases as ice is deposited. A tube condenser is also more resistant to back-migration of vacuum pump oil than a plate condenser.

9.5.1.3 Presence of solvents. If the product contains non-aqueous solvent, most will remain liquid even after the water in the product has frozen. The liquid solvent will evaporate quickly at the beginning of the drying cycle. Lyophilization in the presence of non-aqueous solvents was studied by Seager *et al.*[9] Up to 8% solvent, no special measures are necessary; if more than 8% solvent is present, it is useful to have a solvent trap that can be isolated and drained. It is also necessary to consider an explosion-proof installation.

9.5.2 Overall concept

The first decision to be made is on the method of loading to be used, which will also depend on whether vials are to be loaded on trays or in direct contact with the shelf. Loading may be:

- Manual
- Assisted manual at constant level

- Assisted manual by carts
- Fully automated at constant level
- Fully automated by carts.

9.5.2.1 Manual. This becomes difficult at shelf areas above 30 m^2 (300 sq. ft.) because of shelf height and depth. Increasingly, regulatory authorities prefer automatic methods. Shelf depth (front to back) should not be more than 2.1–2.4 m (7–8 ft).

9.5.2.2 Assisted manual constant level. This is a good compromise for vials on trays but has the disadvantages that the lyophilizer is out of production for the length of the filling run, and some shelf space is lost owing to the size of the filling arms. There is less restriction on shelf depth.

9.5.2.3 Assisted manual carts. This avoids the disadvantage that the lyophilizer is not operating for the length of the filling run, although some shelf space is still lost to the filling arms. There is no restriction on shelf depth. The weight of a cart may be up to 1.5 tons loaded but this has been found acceptable, especially if a powered wheel is fitted to drive the carts. It is suitable for vials in trays and direct vial-to-shelf contact.

9.5.2.4 Fully automated at constant level. This is a sophisticated and expensive approach. It is especially suitable for vials in direct shelf contact. Several installations have been made, mostly in Japan. The main impact on the freeze-dryer is that shelves are restricted to about 1.4 m (4′7″) deep and a large shelf interdistance is needed. This gives rise to tall drying chambers. Minimal shelf space is lost to loading arms.

9.5.2.5 Fully automated carts. The system uses automatic guided vehicles (AGVs) to move carts from the filling line to the lyophilizer. It can be used for direct vials to shelf contact and for vials on trays.

For automated loading operations, shelf flatness is important. Shelves should be flat at least to ± 0.5 mm/m. In addition, the positional accuracy of shelves is critical, especially in the vertical plane. By using active-feedback position-sensing systems, accuracy can typically be maintained at ± 0.25 mm.

9.5.3 Construction features

Lyophilizers not intended for steam sterilization are typically fabricated from stainless steel AISI 304 with a finish between 180 grit and 320 grit according to requirements. Detail points to be observed are that internal welds should be radiused to a minimum of 5 mm to facilitate cleaning, the floor of the chamber should be pitched to drain and additional ports should be fitted for validation purposes.

Where steam sterilization is required, it is usually specified for 121°C equivalent to a pressure of 1.05 bar gauge. Chamber and ice condenser must therefore be made to a recognized pressure vessel code. On occasion, higher temperatures and therefore pressures may be requested. Typical construction material will be stainless steel AISI 316L. All process contact pipework would also be in 316L, with inert gas purged full penetration orbital welds inspected by borescope. Passivation, if used, should be on weld areas rather than on all surfaces because of the deterioration in finish that it causes. Normally, individual steam condensate traps will be fitted to the chamber and ice condenser. Where a 0.22 μ hydrophobic filter is fitted, it should also have an individual trap to prevent steam condensate build-up. All pipework connections should use sanitary fittings and dead-leg pipe connections should have a maximum length:diameter ratio of 6:1.

One penalty of steam sterilization is the time required to cool the chamber and door. If cooling jackets are fitted, it is important that they cool the door as well as the chamber walls since the door is usually the heaviest section and without a cooling jacket might take 6–7 h to reach a manageable temperature.

9.5.4 *Stopper-closing system*

The half-inserted vial bungs are closed after drying by bringing the product shelves together in the stopper-closing operation. This is usually hydraulically operated. Screw jacks can be used but are best avoided because of possible particle generation.

Characteristics of the vials and bungs must be known since non-siliconized bungs require greater insertion force than is provided by the normal 1.0 kg/cm^2 maximum. Stoppering force should be adjustable. However, crushing of vials is a rare phenomenon owing to the high compression strength of glass. Small vials with non-siliconized bungs require special design of the stoppering system to prevent possible damage to the shelves and reaction plates. There is concern over the possibility of hydraulic oil being introduced into the drying chamber on the hydraulic stoppering ram. Several different solutions to this problem have been developed. Probably the simplest is to surround the ram with a special highly extensible bellows. Another is to use the saddle stoppering arrangement of Figure 9.12.

The saddle-stoppering and bottom-stoppering arrangements make it easy to change shelf spacing but cannot be used in a constant-level loading system where all shelves are moved to the bottom of the drying chamber, then lifted up one by one to a constant height for loading through a sub-door or 'pizza door'. Constant-level loading requires use of a top-mounted hydraulic ram. This needs to be fitted with a positive means to prevent downward creep of the shelves in operation and safety interlocks to prevent shelf movement when operators have the door open.

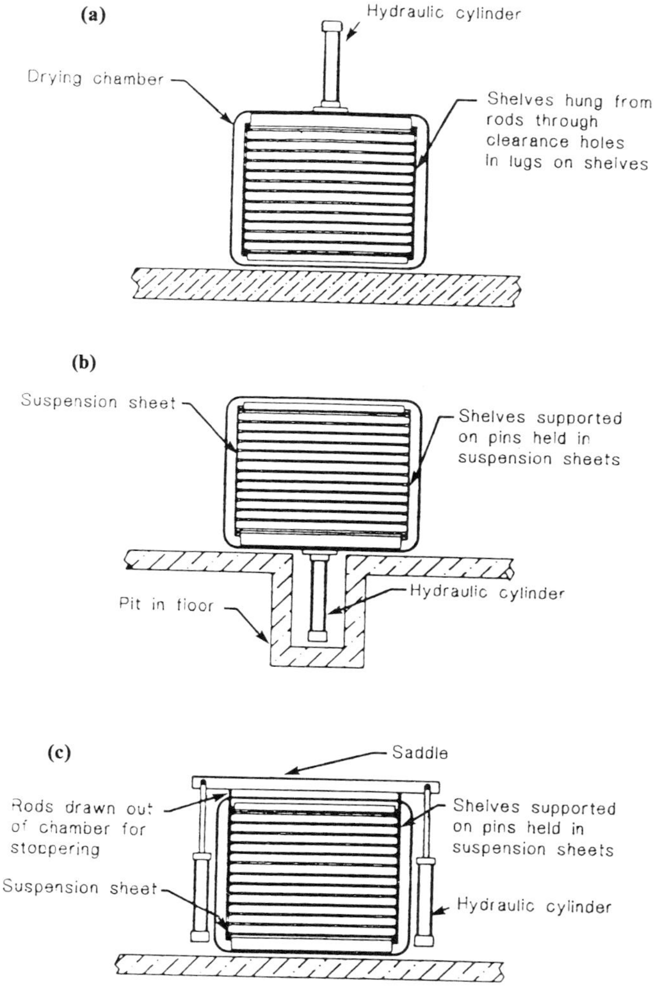

Figure 9.12 Common arrangements of the stoppering mechanism: (a) top stoppering, (b) bottom stoppering and (c) saddle stoppering.

9.5.5 Refrigeration system

Refrigeration compressors are used to cool the ice condenser and the shelves. The major function is the initial cooling of the shelves with product in place because the latent heat of crystallization must be removed to freeze water to ice. On basic machines, a single refrigeration system can be used to pre-cool shelves and product, the system then being switched to the condenser. However, this is only possible on very simple fast-drying materials. It is more usual to have a facility for the refrigeration system to maintain control cooling on the shelves during sublimation drying. The shelf cooling is usually carried out indirectly by a circulating fluid whose temperature is controlled by heat exchange with the refrigeration system and a heat source. The choice of circulating fluid is restricted as shown in Table 9.1.

Silicon oil is the material of choice for shelf temperatures down to −55°C. It can also be used to around −60°C shelf temperature but its increasing viscosity requires significant power input. Below −65°C it is necessary to use either Lexsol or TCE, both of which have good thermal properties that outweigh the other disadvantages of using them. Ice condensers are usually cooled by direct refrigeration. In the past, chlorofluorocarbons (CFCs) and halons have been popular choices of refrigerant. Halon R13b1 has been particularly useful because it allows the achievement of shelf temperatures down to −90°C without employing cascade systems with their associated difficulties in maintenance. Unfortunately, both CFCs and halons have high ozone-depletion potential (ODP), so they are rapidly being phased out in favor of hydrochlorofluorocarbons (HCFCs), which have small ODP and hydrofluorocarbons (HFCs), which have zero ODP. Environmentally friendly gases like R22 and HFC125 can give adequate refrigerating performance at low temperature.

Equipment using liquid nitrogen to replace conventional refrigeration systems is now available on the market. Operating costs are somewhat higher than with conventional refrigeration, especially in small lyophilizers under 15 m^2 shelf area.

The refrigeration system is typically sized to bring empty shelves to operating temperature in about 1 h. The ice condenser is of lower mass than the shelves so can be expected to cool to operating temperature in less than 30 min.

Table 9.1 Choice of circulating fluid

Fluid	Advantages	Disadvantages
TCE (trichloroethylene)	Best fluid for heat transfer Non-inflammable Low viscosity Cheap	Carcinogen Breaks down to form acid products
Lexsol (dewaxed aviation fuel)	Good heat transfer Low viscosity Cheap	Large fire risk
Silicon oil	Stable characteristics (can be exposed to +136°C without breakdown) Non-toxic Non-corrosive	Expensive Viscosity limits Shelf temperature to −65°C Low fire risk
Water/alcohol mixtures	Unsuitable because of instability and corrosive properties	

9.5.6 *Vacuum system*

The water vapor subliming from the drying product comprises the main load for the pumping group. This is essentially taken up by the refrigerated water vapor condenser. The duty of the vacuum pumps is to provide the initial pulldown of the chamber and condenser before sublimation starts and to provide pumping for permanent gases in the process such as air or nitrogen introduced for pressure control, and air emanating from leaks. Lyophilizers up to about 20 m^2 (200 sq. ft.) shelf area are often pumped by two-stage oil-sealed rotary pumps. Larger machines usually have an additional Roots blower to give a quick pull-down. Increasingly, non-oil rotary pumps are being specified to avoid any possibility of back-diffusion of hydrocarbon oil into the chamber and to allow filtration of the exhaust stream when dangerous materials are being dried. The vacuum pumping system is typically sized to achieve pumpdown to operating pressure in approximately 20 min.

9.6 Control systems

The temperature of the shelves and the pressure in the drying chamber control the lyophilization process. Many freeze-dryers operate with a fixed shelf temperature and a fixed chamber pressure during sublimation. In this situation, the only controls required are simple regulators for temperature and pressure.

Easy availability of automatic process controllers has led to their extensive use on lyophilizers where they give an opportunity to vary pressure and temperature throughout the drying operation in order to achieve the fastest possible cycle. Relatively simple process controllers are available for shelf temperature and chamber pressure, which include alarms for deviation from pre-set conditions. Batch records are invariably required, these are typically collected on a strip chart recorder.

Much more sophisticated electronic systems are used for multiple dryer installations. Frequently, several dryers are operated from a remote control room through a computer. Not only are the dryers operated but the computer is also used to display mimic diagrams, cycle status, alarm status and details of the pre-set program. Data logging is also achieved through the computer and, in some installations, electronic data logging has replaced the use of strip chart-recorders.

Considerable damage to the parts of a lyophilizer can be caused by incorrect operation. For this reason, there are many interlocks between different parts. The interlocks were traditionally made through hard-wired relays. With the advent of highly reliable PLCs, the logic is now usually controlled by the PLC on manually controlled as well as automatically controlled installations.

9.7 Good manufacturing practices

International regulatory agencies, requiring users to observe good manufacturing practices (GMPs), and calling for validation of processes and equipment before and during use, are responsible for many improvements in the equipment. These can be grouped loosely into three categories: (i) design and fabrication of equipment; (ii) calibration of instruments; and (iii) registration of control software.

Equipment must be oriented toward the elimination of potential dirt traps and the avoidance of rubbing contacts, which could generate particles. Trapped volumes can generate virtual leaks. Screw threads should be eliminated where possible. Some good and bad constructional features are shown in Figure 9.13.

Good cleaning access must be provided or clean-in-place (CIP) systems should be fitted. Figure 9.14 shows part of the piping and nozzles of a CIP system fitted to an 18 m^2 shelf area lyophilizer. Effective CIP systems require many nozzles. In the example shown, over 300 were used. Owing to the large flow rate of water for injection (WFI) required to operate the nozzles, the chamber was divided into zones, which were cleaned in sequence. The effectiveness of the CIP system can be determined by spraying a fluorescent product such as vitamin B1 around the chamber, allowing it to dry, and then running the CIP system. An ultraviolet lamp can then be used to look for remaining traces.

Surface finishes should be 240 grit or better to facilitate cleaning and avoid dirt traps of significant size. Where simple 'wipe-down' sterilization is employed, it is difficult to establish sterility, thus requiring media tests to be

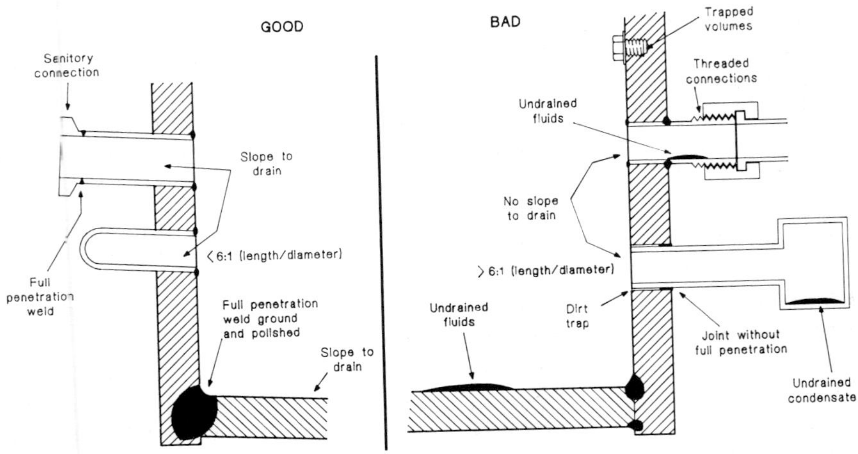

Figure 9.13 Good and bad constructional features.

Figure 9.14 Hoses and CIP system.

made. Gas sterilization using formaldehyde, ethylene oxide or mixtures of hydrogen peroxide is also less popular for similar reasons as well as the difficulty of penetrating pipe dead-legs. An additional problem with gas sterilization is the hazard presented to operators.

Steam sterilization is not a mandatory requirement at present but its use is becoming almost universal for parenteral production.

The use of clean steam at a temperature of 121°C or more is the preferred method. Once a steam sterilization cycle has been validated it is not necessary to carry out sterility testing each time it is performed. The procedure is to fit calibrated temperature sensors into the coldest parts of the chamber and condenser, normally the drains, and in any condensate drains from filters that are required to be sterilized *in situ* (e.g. absolute filters on the inlet gas lines for aeration and/or pressure control of the chamber). After pre-evacuation, steam is admitted until the drain sensors reach nearly 121°C. The steam is then pumped out with a water ring pump to flush out any remaining air. Steam is then readmitted until the drain sensors reach 121°C and that temperature is controlled there for 30 min. Subsequently, steam and condensate are discharged through the drain valve, and the interior of the unit is dried by pumping a rough vacuum with the water ring pump.

During validation, extensive temperature mapping on shelves and chamber walls must be undertaken to find if any cold spots exist. A typical specification is that all surfaces reach 121°C (plus 5°C minus 0°C). Independent measurements should be carried out to calibrate the temperature sensors in the drain.

Instrument calibration requires that suppliers of lyophilizers provide certificates of calibration traceable to known standards.

Shelf temperature control is critical for process reproducibility. In most modern lyophilizers, the shelf temperature is varied by varying the temperature of a fluid, often silicon oil, which is circulated through passages in the shelves. The uniformity of shelf temperature achieved throughout the shelf stack depends on the design of the passages, the fluid, and its speed of flow through the passages.

Since the rates of heating and cooling and the maximum and minimum temperature of the shelves during lyophilization depend on the nature and quantity of product loaded on them, operational qualification is usually carried out with an empty lyophilizer and the results compared with the lyophilizer specification. If a shelf-to-shelf uniformity test with simulated product on the shelves is required, it must be carried out after steady conditions have been reached.

The most suitable types of sensors are thermocouples for product and shelf surface-temperature measurements, and platinum resistance thermometers (RTDs) to measure the temperature of the circulating fluid. Calibration of sensors is carried out at the minimum and maximum operating temperatures defined in the specification by immersing them in a circulating temperature bath where the fluid temperature is monitored by an instrument traceable to a recognized standard. Thermocouple outputs and the monitoring instrument output are recorded by analogue or digital methods at temperature intervals throughout the specified operating range of the lyophilizer.

The minimum and maximum temperatures of the shelves can be established by running the shelf refrigeration of heating system at full capacity for the period defined in the lyophilizer specification to reach minimum or maximum temperature.

Shelf-to-shelf uniformity is measured at steady-state conditions at the maximum and minimum temperatures specified for the lyophilizer shelves and at 0°C. Typical shelf temperatures specified might be a minimum of −50°C, and a maximum of +70°C. In this case, the tests would be carried out after the shelves have been running for a period at each temperature.

Shelf-to-shelf uniformity and temperature drop across shelves are established with empty shelves and simulated product on the shelves by fitting calibrated temperature measuring devices to each shelf by a means that ensures good thermal contact with the shelf surface. A suitable method is to provide aluminum, brass, copper or other conductive blocks with a hole to receive the measuring device bored through the long dimension of the block and the measuring devices inserted in the holes with a conductive grease. The blocks are then placed on a film of conductive grease on the shelves. Self heating and cooling rate should also be established.

General levels of safety protection for operators are defined in national and company codes. As well as normal safety features to prevent electric shock or mechanical trapping, alarm and defence systems specific to lyophilization are fitted. They will generally fall into three types: (i) process protection against operator errors; (ii) protection against failure of utilities; and (iii) protection against lyophilizer failure.

9.7.1 *Protection against operator errors*

In manually-controlled lyophilizers, errors in the sequencing of operations can be protected against to some degree by the plant logic. However, mistakes in setting the critical variables such as shelf temperature and chamber pressure can still occur.

Automatic control systems should be protected by multiple password levels so that accidental interference with switches or set point controls cannot change cycle conditions. Passwords must be kept secure to the designated recipients and changed if they become known to others. They might be allocated as follows.

- Password 1 allows starting cycle, allocated to process operator
- Password 2 allows selecting cycle program, allocated to supervisor with authority to sign documents
- Password 3 allows modification to program, allocated to program preparer or an authorized and documented person who is competent in understanding the process, the control system, and the mode of operation of the lyophilizer.

9.7.2 *Protection against failure of utilities*

The following utilities are required for most lyophilizers:

- Electrical power
- Cooling water
- Compressed air (if used)
- Backfilling gas (typically nitrogen or sterile air)
- Water for defrosting the ice condenser
- WFI for clean-in-place systems
- Clean steam for sterilization.

The most potentially damaging utility failures occur when electrical power or cooling water supplies are interrupted during the first hour or two of sublimation, when an interruption of a couple of minutes could result in product damage. Later in the cycle the ice on the condenser may provide enough thermal mass to keep sublimation going for some time.

Complete power failure should be simulated during the following parts of the lyophilizer cycles.

- Pre-freezing
- Sublimation
- Secondary drying
- CIP cycle, if fitted
- Steam sterilization cycle, if fitted
- Ice-condenser defrost.

The simulation should consist of power interruptions lasting 5 s and 10 min. Interruptions of 5 s should result in the cycle restarting from previous condition, 10 min interruptions should restart shelf cooling at maximum rate.

When water supply is interrupted, refrigeration compressors will shut down after a period of a few minutes. The restarting of the cycle should follow the same pattern as for electric supply failures of 10 min.

In the case of failures of other utilities, depending on the part of the cycle in which the supply failure occurs, the importance or otherwise can only be assessed in the particular situation.

9.7.3 Protection against lyophilizer failure

Alarms fall into two categories, those directly concerned with critical process parameters and those concerned with lyophilizer operation.

Process alarms should be fitted for shelf temperature and chamber vacuum deviation from the set point. Also desirable are alarms for sterilization under-temperature and condenser over-temperature.

Moving parts fitted to the lyophilizer are also fitted with alarms showing malfunction.

Software validation essentially requires that contact-by-contact descriptions of control loops and alarms are available so that analysis of the effects of possible failures can be carried out, together with an algorithm for the system.

9.8 Application

The first requirement for efficient operation of lyophilizers is to have a formulation that takes into account the physics of the process. The first attempt at a universal medium for freeze-drying was Greaves' 'Misto Dessicans', which was developed for vaccinia and is still in use today for some biological agents. Numerous media have been described for particular applications. If possible, high sugar and solvent contents should be avoided. Most sugars form a surface skin preventing the escape of vapor. They also have very low collapse temperatures as tabulated by McKenzie.[1] Organic solvents do not freeze and can also produce impermeable surface skins of concentrated drug during freezing. In addition, if the solvent concentration is more than about 8%, it is necessary to take steps to trap the large volume of solvent, which vaporizes at the beginning of the drying in order to keep chamber pressure low enough for sublimation. This may require the fitting of specially refrigerated solvent traps, which can be valved off and drained during the sublimation of the ice. It may be necessary to use explosion-proof construction methods for the electrical machinery, which considerably increases equipment costs.

While the product is being formulated, the eutectic or collapse point should be determined using measurements based on product resistivity and differential thermal analysis during the freezing and warming of the product to identify phase changes. Knowledge of the eutectic or collapse point allows the product to be kept below it at all times. The vapor pressure of ice at this temperature is also a determining parameter in the cycle. Pilot drying tests can then be run to find the highest shelf temperature that can be employed safely without damage to the product. A good starting point for pressure is about 10^{-1} mbar, which will give reasonable heat transfer from shelf to product.

9.9 End point determination

Most industrial freeze-drying cycles operate on a time-historical basis developed from pilot testing. It is usual to take samples of material at the end of secondary drying and establish that the moisture content is within limits using a Karl Fischer titration. However, it is often helpful to investigate the moisture content during the course of drying. Some techniques commonly used are:

1. Monitoring shelf and product temperature. When they coincide it indicates that sublimation is complete.
2. Pressure-rise tests in which the main valve between chamber and condenser is closed while the rate of pressure increase is observed. This is useful during desorption drying. However, such tests are somewhat dangerous early in the sublimation drying because of the relatively high heat and mass-transfer rates. Any delay in re-opening the valve could cause severe product damage.
3. Extraction of samples from the chamber at various time intervals while the process continues using a 'sample thief', and then measuring the moisture that remains.
4. Application of a residual gas analyzer (RGA). The most commonly used RGA is a quadrupole mass spectrometer arranged to show water peaks and other peaks up to about mass 80 clearly. Results from this technique depend critically on the sample extraction methodology. For example, sample-extraction pipes must be heated to avoid the water vapor being absorbed

on the pipe walls before reaching the analysis chamber of the RGA. Up to the time of writing, this technology has remained experimental.

9.10 Leakage

The significance of leaks in a freeze-dryer is frequently not well understood since, during the drying process, there will always be vapor evolution from the product. Contaminants leaking into the freeze-dryer have a very high probability of being swept through the system to the ice condenser or vacuum pump. The only known cases of product failure from contamination through leaks have occurred when major mechanical breakdown has caused a loss of product through melt-back.

Real leaks from the atmosphere into an evacuated system may be by hydrodynamic flow through channels or gaps, or by diffusion through porous materials. Mass rates of flow through leaks are not significantly affected by variations in the system pressure when it is below some tens of millibar.

Real leaks are distinguished from virtual leaks, which are caused by desorption from surfaces, or from trapped volumes of porous materials within the system, or by the generation of vapors within the system. The last is important in freeze-dryers because it is difficult to dry them completely, and water vapor must be eliminated in order to obtain meaningful leak test results. The achievement of good results is assisted by cooling the ice-condenser during leak testing.

Vacuum leaks are expressed in millibar liters per second (mbar l/s). For a constant mass of a gas, the product of its pressure and its volume remains constant at constant temperature. Therefore, this measure holds for the rates of passage of gas entering and leaving a leakage pathway. Frequently, tests are based on 'leak-up' rates, which are pressure-rise tests with the chamber blanked off and which depend on the chamber volume. In this case, the units will be millibars per second (mbar/s).

For example, a user of vacuum equipment first obtained a pressure of 40×10^{-3} mbar in an empty dry chamber of volume 180 l. When the chamber was isolated, the pressure increased to 120×10^{-3} mbar over a period of an hour. It may be assumed that the pressure rise of 80×10^{-3} mbar was at a constant rate over the isolation period of an hour. Working from the volume of the chamber, this represents a leak rate of:

$$\frac{80 \times 180 \times 10^{-3}}{3600} = 4 \times 10^{-3}\,\text{mbar l/s} \qquad (9.3)$$

Typical leak rates required will be less than 0.001 mbar l/s on a clean, dry chamber with all its nozzles carefully blanked off and less than 0.01 mbar l/s on chambers in their final condition with all valves and connections in place.

It is important to analyze the possible sources of leaks and, hence, the likelihood of contaminants coming into contact with sterile products. The sources can be of three types: (i) leaks through the structure of the chamber and condenser; (ii) leaks around the seal between the door and the front of the chamber; and (iii) leaks through valves connected to the chamber.

Of these, leaks through the structure can be virtually eliminated by use of a helium mass spectrometer leak detector. These devices can detect leaks down to 10^{-7} mbar l/s. Assume that a structural leak takes the form of a capillary 3 mm long through a wall or between two sealing surfaces. Assume further that the dimensions of the leak are comparable with a path through a 0.2 μ absolute filter. The leakage rate through such a capillary would be about 1×10^{-5} mbar l/s. Thus, significant structural leaks can be detected and eliminated.

Leaks around the door seal of a freeze-dryer are likely to be large compared with leaks *through* the structure. However, leaks around the door seal will only introduce sterile room air into the chamber. Around the front of a freeze-dryer it is usual to have a class 100 area, and therefore contamination from this source is very unlikely.

It is ironic that the most significant potential leakage source is from the many penetrations through the chamber structure required to comply with GMPs as shown in Figure 9.15. These include:

- Large numbers of temperature sensors for validation (up to 40 in large freeze-dryer chambers)

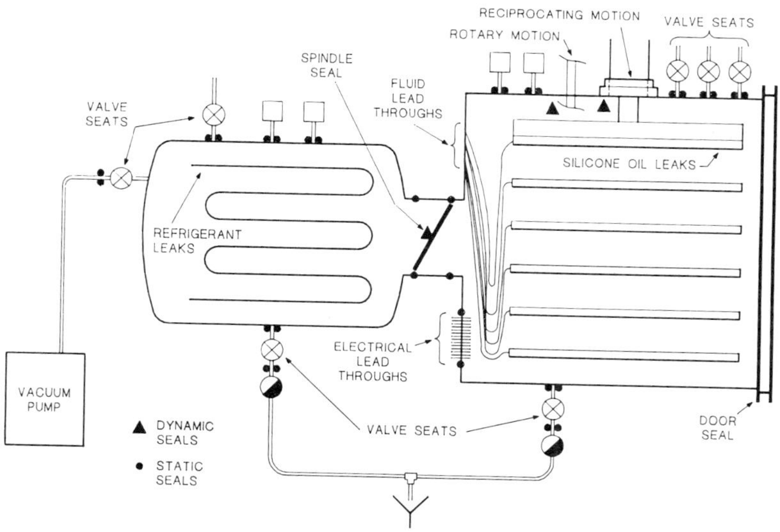

Figure 9.15 Potential leak sources into a lyophilizer.

- Steam inlet lines for sterilization
- Drain line(s)
- Lines for water for CIP
- Lines for sterile air or nitrogen admittance.

All lines are required to be closed by valves, which must remain leak-tight at the pressure and temperature required for steam sterilization and also must perform excellently as vacuum valves. Valves in process-contact lines must also be totally sterilizable. These mutually contradictory needs can be met with very high quality diaphragm valves. However, it is only necessary to deposit minute amounts of product, or glass debris, on a drain valve seat to create leaks that could easily admit organisms.

There is no real solution to this problem other than carrying out a leak test after each steam sterilization cycle. Such a test can be built into the automatic control system if one is fitted. However, it would not be realistic to place a limit on pressure rise so low that theoretically no bacterium could enter through a leak. If this were done, it would be necessary to carry out major servicing of the valves after every cycle. A typical compromise is to establish a leak rate when all valves have been cleaned and have new seats fitted. Leak-rate testing is then directed to looking for increases to this basic leak rate.

One 'rule of thumb' accepted in the pharmaceutical industry is that, after each sterilization, leak rate into the freeze-dryer should be tested by a pressure-rise test. The chamber should be pumped to about 70×10^{-3} mbar ($50\ \mu$) and then isolated. The rate of pressure rise should then not exceed 7×10^{-3} mbar (about $6\ \mu$) per minute over a period of 1 h.

9.11 The cost of lyophilization

Conventional drying from the liquid phase requires the addition of just over 500 calories of energy per gram of water evaporated. In freeze-drying, the energy flows are much greater. Allowing for typical efficiency of machinery, energy consumed will be as shown in Table 9.2.

Some energy can be recovered by using heat injected into the cooling water

Table 9.2 Typical energy consumed during lyophilization

Process	Energy consumed (cal/g)
Freezing product	400
Drying product	900
Condensing vapor	760
Total	2060

of the refrigeration compressors to heat the shelves during sublimation; however, this is only worthwhile on lyophilizers with shelf areas above about 20 m^2. Energy savings of 10–14% can be realized in practice. Assuming an energy cost per kilowatt hour of US\$0.06, the cost of evaporating 1 kg of water would be US\$0.14.

This energy cost, while high compared with the energy cost of conventional drying, is a minor factor in the overall cost of lyophilization. By far the greatest cost is the capital cost of equipment. Assume that a medium-sized industrial unit with a shelf area of 20 m^2, costing US\$700000 installed and commissioned, will be amortized over 10 years with average interest rates. Assume that 180 freeze-drying cycles are completed each year, and that 300 kg are sublimed in each cycle. Then the cost of sublimation of each kilogram of water is about US\$2.00. A more precise calculation would need to take into account the cost of cooling water for the refrigeration compressors, clean water for CIP, steam for sterilization and any other services. It should also consider the cost of loading and unloading operations, maintenance and record-keeping. Capital costs of freeze-dryers vary roughly in proportion to the capacity raised to the power 0.75.

9.12 Future considerations

There is no doubt the freeze-drying process is expensive. Although costs may be reduced by increasing the volume of production, even at the largest scale the added cost of freeze-drying is about US\$25/kg dry weight. It is also inconvenient. It is therefore always vulnerable to competing processes. For example, in the early days of antibiotics many were freeze-dried but, as spray drying developed, freeze-drying gave way to the simpler and cheaper processes. The use of freeze-drying in this area returned with the development of more complex and unstable antibiotics that could not withstand the stress of spray-drying. Also, freeze-drying is a cleaner process than spray-drying and allows the reduction of particulate contamination to presently acceptable levels. Another process that poses competition is reverse osmosis, which could be used to concentrate material and thus reduce the size of freeze-dryers. However, experience indicates that freeze-drying will continue to increase in importance and application.

Many new drugs are extremely potent and special means to protect operators are necessary, leading to the use of automated product handling systems on relatively small lyophilizers.

The medium-term future will see more complex freeze-dryers with elaborate safety and control systems to protect the expensive products they contain and the operators. Continuous freeze-drying is at present a 'solution looking for a problem' but real applications will develop in the downstream processing of biotechnology products.

References

1. Mackenzie, A.P. (1977). The physico-chemical basis for the freeze-drying process. International Symposium on Freeze Drying of Biological Products, Washington DC, 1976. *Develop. Biol. Studies*, **36**, 51–67.
2. Nail, S.L. (1980). The effect of chamber pressure on heat transfer in the freeze drying of parenteral solutions. *J. Parenteral Drug Assoc.*, **Sept/Oct**, 358.
3. Pikal, M.J., Roy, M.L. and Shah, S. (1984). Mass and heat transfer in vial freeze drying of pharmaceuticals: role of the vial. *J. Pharm. Sci.* **73**, 1224.
4. Freyrichs, C.C. and Herbert, C.N. (1974). Long-term stability studies in the international reference preparation of Newcastle Disease vaccine (live). *J. Biol. Standards* **2**(1), 59.
5. Snowman, J.W. (1993). Formulation and cycle development for lyophilization: first steps, *Pharmaceut. Eng.* **13**(6), 26–34.
6. Pikal, M.J. (1990). Freeze drying of proteins. Part I: process design. *BioPharm* **3**(8), 18–27.
7. Pikal, M.J. (1990). Freeze drying of proteins Part II: formulation selection. *BioPharm.* **3**(9), 26–30.
8. Rowe, T.W.G., Greiff, D. and Monroe, J. (1979). *Proceedings of 15th International Congress of Refrigeration*, Venice, 23–29 September.
9. Seager, H. *et al.* (1985). Structure of products produced by freeze-drying solutions containing organic solvents, *J. Parent. Sci. Technol.*, **39**, 161–179.

10 Drying in the pharmaceutical and biotechnology fields

A.S. MUJUMDAR and D.S. ALTERMAN

10.1 Introduction

This chapter provides a brief summary of the key dryer types used in the pharmaceutical and biotechnology industries, followed by a discussion of methods for removal of solvents from dryer exhausts. No discussion is included of typical equipment for the removal of solid particulates such as filters, cyclones, electrostatic precipitators or wet scrubbers from exhaust streams.

Drying of pharmaceutical and biotechnological products is used for a wide assortment of chemical and biochemical materials produced in quantities from large tonnages to very small batches. In general, the production requirements are small but the quality constraints understandably very stringent. Any form of potential risk of degradation from microbial attack, thermal instability, oxidation, contamination with metallic compounds, residual organic solvents etc. must be avoided. Dryer construction must be polished stainless steel or enamelled iron. In view of the limited space available, this chapter will focus mainly on the most commonly used dryers for synthetic pharmaceutical as well as biotechnological products. For details about the various types of dryers, their classification and selection, the interested reader is referred to the *Handbook of Industrial Drying*.[1] Further, it should be noted that many of the dryers can also function as granulators; this aspect is beyond the scope of this chapter.

10.2 Dryers for pharmaceutical and biotechnology products

Drying of pharmaceutical products is characterized by the following:

- Highly heat-sensitive (thermolabile) products
- Very-high-value products
- Generally low production capacity requirements
- No tolerance for contamination
- Frequent presence of organic solvent (or mixtures)
- High susceptibility to oxidative damage (requiring drying in inert atmosphere) or risk of explosion/fire caused by the presence of either organic solvents or explosive dusts
- High sensitivity to moisture content.

Production of products in solid form (e.g. powders, tablets etc.) involves drying at different stages of production *viz.* synthesis of intermediate products, synthesis of drug formulations and manufacture of various dosage forms. Besides conventional drying this often involves granulation, tabletting, coating, application of multiple-layer coatings for controlled drug release etc. Selection of production equipment and operating conditions for all these operations must take into account the thermal sensitivity, dry kinetics, quality etc. Cost of equipment and energy costs are generally of secondary importance. Also, products are often produced and dewatered (using filters, centrifuges, etc.) in a batch mode, while dryers used may only operate continuously (e.g. spray dryers).

Drying of products of microbiological origin must be carried out with extreme care to maintain their chemical and biochemical activity. Examples of such products are various vitamins, enzymes, yeasts, molds, pesticides, proteins, amino acids etc. Drying of, for example, antibiotics, enzymes, amino acids, require that the product retains a high level of activity after drying. In the drying of yeasts, it is important to minimize destruction of the microorganisms during thermal drying. Strumillo, Markowski and Adamiec have reviewed the main quality constraints in terms of the chemical, biochemical and physical changes in bioproducts during drying.[2] Although beyond the scope of this chapter, it is important not to consider the drying operation in isolation. Non-thermal dehydration processes such as filtration, centrifugation, ultrafiltration etc. should be considered in conjunction with drying.

Fermentation broths are suspensions of biomass with some components partly dissolved as a solution or gel. Most bioproducts are biopolymers consisting of proteins or protein-like compounds. Their properties depend upon how water binds to the molecules. Strumillo and co-workers group bioproducts into two categories:[2]

- Group 1: water acts as life medium for microorganisms (e.g. yeasts, bacteria), or as reactant (e.g. enzymes)
- Group 2: water is a solvent (e.g. vitamins, antibiotics, steroids etc.).

For group 1 products, the final moisture content must be such as to achieve the state of anabiosis, in other words a reversible form of life of dried microorganisms, which depends on the final moisture content. This must be determined experimentally.

Another characteristic of bioproducts is their xerostability in addition to their thermostability. Xerostability refers to their sensitivity to moisture (water) content. Both are determined experimentally in order to select the drying conditions to maintain viability of the product. Generally, products of group 1 are very thermolabile and a maximum product temperature of 40°C is suggested on the basis of data from scientific literature. For very short residence times, higher temperatures may be acceptable. Xerostability is closely connected to the way water is bound to the bioproduct. It appears that biochemical activity of most bioproducts is preserved fully if the water is in an unbound state.

Product temperatures and exposure times in dryers must be selected carefully to avoid undesirable changes during drying. Table 10.1 summarizes the types of changes and their possible effects during the drying of bioproducts.

10.2.1 Basic principles

Drying involves removal of solvent(s) (aqueous or nonaqueous) from a wet solid, paste, suspension or solution by application of heat. The selected dryer must be able to handle the material to be dried as the first requirement. Heat may be supplied to the wet product in dryer by one or more of the following mechanisms:

- Convection (i.e. by direct contact with drying gas), which supplies heat and also removes moisture
- Conduction (i.e. by heat conduction through metallic heat transfer surfaces); evaporated moisture is removed by gentle gas flow or by application of vacuum
- Radiation (with gentle gas flow or vacuum)
- Dielectric heating *viz.* microwave or radio-frequency (RF) fields.

Depending on the mechanism of heat supply, the mechanisms of drying may differ. In general, moisture is removed by transformation from liquid to vapor phase. However, for highly heat-sensitive products it may be necessary to employ the freeze-drying process, which involves operation of the dryer well below the triple point of water so that moisture removal is by sublimation via a solid-to-vapor phase transformation. Heat of sublimation may be supplied by conduction, radiation, or in some more recent dryers, by microwave energy sources. The vapor is removed from the drying chamber by application of vacuum.

Depending on the physical form of the product to be dried and the amount of moisture contained in it, the drying kinetics (i.e. rate of drying, w_D, as a

Table 10.1 Changes in bioproducts during drying

Type of change	Parameter affected	Examples
Biochemical	Cell survival	Yeasts, molds
Enzymatic reaction	Loss of activity	Vitamins, enzymes
Chemical reaction	Loss of nutrition value or activity	Proteins, antibiotics, amino acids, steroids, etc.
Physical changes	Rehdryation, solubility, shrinkage, aroma loss	Various products

function of moisture content, X) generally displays an initial constant-rate period, which is followed by one or more so-called falling rate periods. This transition takes place at the critical moisture content, X_{cr}, which, in general, is a function of the drying rate itself. The constant rate period depends only on the external heat and mass-transfer rates, while the falling rate depends primarily on the rates of internal heat and mass transfer within the product. Moisture transfer within the solid may occur by one or more of the following mechanisms depending on the product morphology, properties of the solvent, temperature field and so on:

- Liquid diffusion
- Capillary transport
- Vapor diffusion
- Evaporation/condensation.

Mechanisms of moisture transfer may change during a single drying process for a given material. Liquid diffusivity may vary over up to two orders of magnitude from the initial to the final moisture content during drying. Since the transport parameters governing the moisture transfer by one or more of the aforementioned mechanisms cannot be predicted, the design and scale-up of dryers must be based on careful experimentation at laboratory and pilot stages. The dwell time in the dryer must at least equal the drying time required to achieve the final moisture content from the initial. Note that the drying rate curves (w_D versus X) shown in Figure 10.1 are obtained under constant drying conditions, while the drying conditions in a typical dryer vary with

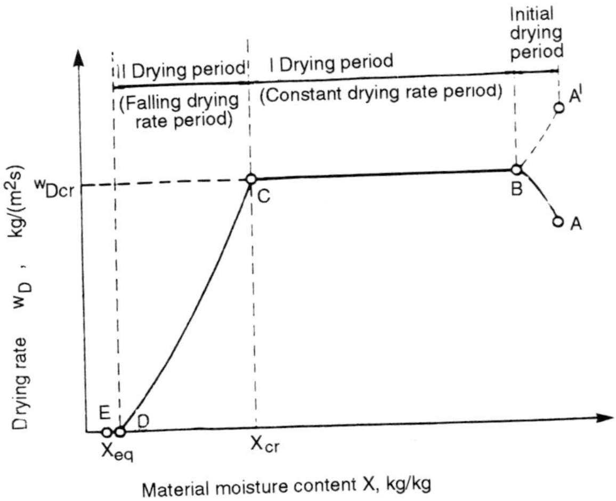

Figure 10.1 Typical drying curve under constant conditions: X_{cr} = critical moisture content; X_{eq} = equilibrium moisture content.

time and space. Hence, care is required to extrapolate or interpolate drying kinetic data based on few experiments. Scale-up of most dryer types is difficult and needs empirical information or prior experience.

Aside from the drying rate data, which determine the dwell time needed within the dryer and hence the size of the dryer for a given throughput, it is necessary to know the equilibrium desorption moisture isotherm data for the wet solid. If the solid contains a mixture of solvents, such data must be obtained experimentally for conditions under which the dryer is operated. Note that only moisture in excess of the equilibrium value participates in the transport process. With a gas at a given temperature and a given solvent vapor content, the wet solid can be dried only to its equilibrium moisture content, even with infinitely long contact time. Thus, the equilibrium moisture content under ambient conditions is the minimum level to which the product needs to be dried in the dryer. Figure 10.2 shows schematically typical sorption isotherms; in drying one is interested in the desorption isotherms. Liquid moisture in the solid, which exerts a pressure that is lower than the equilibrium vapor pressure of the liquid at a given temperature is called 'bound moisture'; it is bound, physically or chemically, to the solid matrix. This lowers its vapor pressure. Indeed, to remove bound moisture it is necessary to supply the so-called energy of desorption in addition to the latent heat of vaporization. The lower the moisture content, the greater the heat of desorption. As can be seen from Figure 10.2, the equilibrium moisture content at a given temperature depends on the moisture content of the gas phase in equilibrium with the solid.

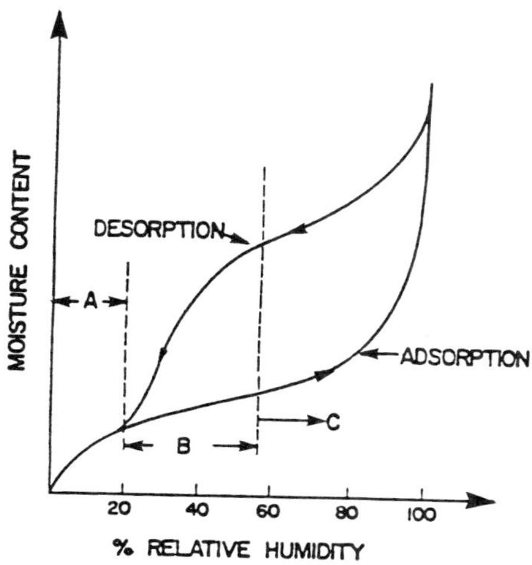

Figure 10.2 Typical sorption isotherm; A: tightly bound water, B: more loosely bound water, C: water in large capillaries.

10.2.2 Classification of dryers

Most products of interest in the pharmaceutical and biotechnology industry may be categorized as:

- Granular (or particulate) solids (particle size 0.05–5 mm)
- Paste-like solids (free-flowing pastes; particle size 0.1–50 μm)
- Solutions/suspensions (fine = 10–50 μm ultrafine = 0.1–10 μm; colloidal = <0.1 μm).

For granular solids, the most common dryers in use are:

- Direct type: tray; continuous-tray; fluid-bed (and modifications)
- Indirect type: vacuum contact dryers (double-cone, paddle dryers; conical dryers with helical mixer etc.).

For 'pasty' solids, one may use batch or continuous-tray dryers, drum dryers, band dryers, or spin-flash dryers. If the paste is pumpable, a spray-dryer may be used (if not, additional water may be added). For some biotechnology products, one may consider fluid-bed or spouted-bed dryers of inert materials into which the suspension or solution may be sprayed. Also, more recently units, which combine both filtration and drying (so-called Nutsch filter-dryers), have demonstrated successful applications in the pharmaceutical industry. These have the process advantages of not requiring a second step in moving and exposing the wet solids of uncertain flow parameters.

For solutions, a spray-dryer is the preferred choice, although a drum-dryer may also be used for this purpose. For highly heat-sensitive products, a drum-dryer may be operated under vacuum. Also, for very thermolabile products, freeze-drying may be used if the production rates are small. While freeze- and drum-dryers operate in a batch mode, spray-dryers can be operated only in a continuous mode.

10.2.3 Properties of pharmaceutical products

Most pharmaceutical products are produced by crystallization. Their particle size is generally no larger than 1 mm. However, it is possible to have polydispersed materials with wide particle-size distributions. In some cases (e.g. ASA or sodium *p*-aminosalicylate) the crystals are nearly monodispersed or they may have a bimodal distribution (e.g. codeine, streptomycin or sulfanilamide). Polydispersed materials are difficult to handle in dispersion (e.g. fluid-bed) dryers because of their inherent tendency towards aerodynamic classification. A vibrated fluid bed may be used to handle such materials because of its ability to pseudo-fluidize the solids at low aeration rates with appropriate choice of vibration frequency and amplitude; smaller particles can remain in the bed with larger particles, since the aeration velocity is below their terminal velocity. Vibrated beds operate at air flow-rates, which are small compared with flow-rates used in fluid beds. However, for aseptic duty, they may pose some operational or maintenance problems.

Thermal decomposition of solids is a critical characteristic of pharmaceutical and biotechnology products. Commonly a first-order kinetic Arrhenius-type temperature dependence of the rate constant is accepted to represent the decomposition kinetics of such products. The applicable equations are:

$$C = C_0 \mathrm{e}^{-kt} \tag{10.1}$$

and

$$k = A \mathrm{e}^{-E/RT} \tag{10.2}$$

where C = concentration of active material at time t; C_0 = initial concentration of active material; k = first-order rate constant in 1/s; t = time in s; A = empirical constant; E = activation energy in J/mol; R = gas constant in J/mol K; and T = temperature in K.

Note that the thermal decomposition kinetics also depend on moisture content. These relationships must be obtained empirically. It is easy to see that the products must go through a rigid thermal history to ensure desired quality.

10.2.4 Commonly used dryer types

Only the most commonly used dryers are described briefly in the following. For details, the interested reader may refer to the literature cited as well as to the several recent text and reference books on drying principles and technology.[3–7]

It is impossible to overstress the need to maintain aseptic processing conditions in all stages of drug production. High sterility standards are needed for pharmaceuticals of biologic orgin. For heat-sensitive materials thermal sterilization is not possible; other means of sterilization such as ultraviolet, chemical or radioactive (where permitted) sterilization must be applied. To avoid microbial contamination, high-efficiency air filters and laminar flow units are used in packaging rooms. When drying with low-temperature air, sterilization of air with an antibacterial liquid may be used.

Care must be taken in the design, fabrication and operation of dryers for aseptic operations. For example, rotating or sliding parts, corners and so on should be avoided in dryers. Dryer housings, ductwork and so on may be sterilized by heating them to 200–300°C for 2–4 h. Samples of selected bacteria in petri dishes may be placed at strategic locations in the dryer housing and ductwork to check if the sterilization is complete. Note that high-efficiency air filters that remove over 99% of particulates over 0.3 μm are available. This includes most bacteria but not viruses. Hence air filtration is necessary but often insufficient to ensure aseptic conditions.

Table 10.2 lists some industrial applications for different dryer types in the pharmaceutical industry. Table 10.3 lists limits on some of the residual values of commonly found solvents used in the production of various pharmaceutical products according to the *European Pharmacopoeia*. Table 10.4 is a similar listing for several solvents. The dryer operating conditions must be such as to achieve residual solvent values below the limits specified. These limits may be different in different parts of the world. Also, regulations may be revised.

10.2.5 *Some selected applications*

In 1987, Pakowski and Mujumdar tabulated several applications in the pharmaceutical industry.[9] It should be noted that several different dryers

Table 10.2 Examples of applications of dryers in the pharmaceutical industry

Dryer type	General application	Examples
Direct type		
Ovens	Small batches	Various products
Band-dryers	Pre-formed (pelletized) pastes	Various products
Turbo-shelf dryers	Coarse, granular products with long drying times	Penicillin, ascorbic acid, riboflavin, sorbitol
Flash-dryers	Removal of surface moisture; very short residence times	Tetracycline, ASA, aminopyrine
Cyclone (vortex) dryers	Longer residence times than flash-dryers	Folic acid, nicotinic acid, thiamine bromide
Vibrated beds (aerated)	Suited for polydispersed granular solids	Ascorbic acid, lactose
Fluid-bed dryers	Fluidizable granular solids (also used for granulation, agglomeration, coating, etc.); one of the most common dryers and can be operated under vacuum for heat-sensitive materials	Vitamins, gluciose, sodium glutamate, ASA, various antibiotics, etc.
Spray-dryers	Pastes, suspensions, solutions; extensively used to produce powders	Dextran, enzymes, gamma globulin, blood serum, tetracycline, casein hydrosylate, penicillin, vitamins, etc.
Indirect type		
Drum-dryers	Thin pastes, slurries	Streptomycin sulphate, calcium pantothenate
Vacuum-tumbler dryers	Heat-sensitive, granular products	Various products
Freeze-dryers	For ultra-heat-sensitive granules, suspensions, slurries; very expensive	Antibiotics, vitamins, chemotherapy dryers, etc.
Agitated-pan dryers		Synthetic pharmaceutical agents

Table 10.3 Recommended limits on residual solvent contents according to the *European Pharmacopoeia*

Compound	Solvent	Limit
Ampicillin sodium	Methylene chloride	0.2% m^3/m^3
Cefaloridine	All solvents	2.5% m^3/m^3
Doxycycline hydrate	Methanol	0.2% m^3/m^3
Gentamycin sulphate	Ethanol	1% wt.
Lactic acid	Methanol/methylesters	500 ppm
Streptomycin sulphate	Methanol	0.3% m^3/m^3
Tobramycin	2-methyl propanol	1% wt.

Table 10.4 Limits on residual solvent content in various pharmaceutical substances according to the *European Pharmacopoeia**

Solvent	Limit (ppm)
Ethylene oxide	10
Acetonitrite/chloroform	50
Benzene/dioxan/trichloroethylene	100
Butanol/2-methylpropanol/methylene chloride	500
Methanol/toluene/formamide/dimethylformamide	1000

* Lower limits than above may be applicable in some parts of the world.

can, in general, be used to dry a given material. Cost figures are rarely published and so it is difficult to generalize regarding dryer selection for a given product. The interested reader is referred to the aforementioned reference for description of the dryer types mentioned in Table 10.2, for example, flash, convex, cyclone, spin-flash, vibrated-fluid beds, drum-dryers etc. Batch-tray dryers are commonly used for small batches and are by far the simplest dryers. Continuous-tray dryers (e.g. turbo dryers) are recommended when larger capacities are needed.

Among the more common dryers in the pharmaceutical and biotechnology industries are: fluidized-bed dryers, vacuum dryers, freeze-dryers, drum-dryers, continuous-tray dryers, filter–dryers, mixer-dryers, etc. These are described briefly in this chapter.

10.2.6 Combined filtration/drying

Recently, combined filter–dryers have been shown to be especially advantageous since the product to be dehydrated does not have to be transferred from one vessel to another, thus minimizing the possibility of contamination. These so-called Nutsch filter–dryers, in fact, permit several operations to be carried out in one vessel, for example, reaction, extraction, filtration, washing, drying, deodorization, mixing and granulation. Furthermore, they can be operated under pressure or vacuum and are claimed to have lower space requirements, and lower capital and installation costs. They eliminate the need of transferring (and possibly exposing) the wet filter cake, which has a behavior of flow that is difficult to predict.

Heat for drying is supplied indirectly by heating the vessel walls, bottoms and internal rotors, which also act as mixers. Nitrogen may be used for drying if needed. Figure 10.3 shows the Nutsch filter–dryer made by Krauss-Maffei for a sterile application.

Among indirect dryers suitable for drying of pharmaceutical and biotechnological products one may cite the plate-dryer (Figure 10.4) and the mixer–dryer (Figure 10.5) as typical examples. The plate-dryer (or its variant turbo-dryer) is suitable for granular materials. In the plate-dryer, heating is by conduction through the metal walls while, in the turbo-dryer, forced air is commonly used. The product is fed onto a circular shelf at the top of the dryer housing. The shelves rotate slowly while stationary blades level the product layers. All blades are staggered to produce new layers continuously. After one rotation the material is wiped off and falls to the lower shelf through a slit. The dryer operates continuously and may have 10–30 circular trays.

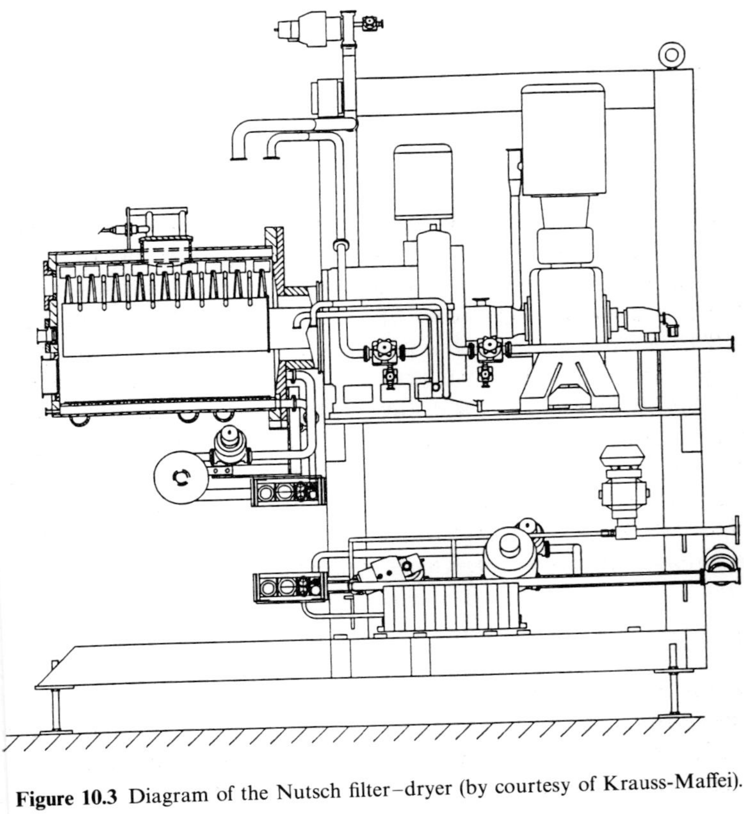

Figure 10.3 Diagram of the Nutsch filter–dryer (by courtesy of Krauss-Maffei).

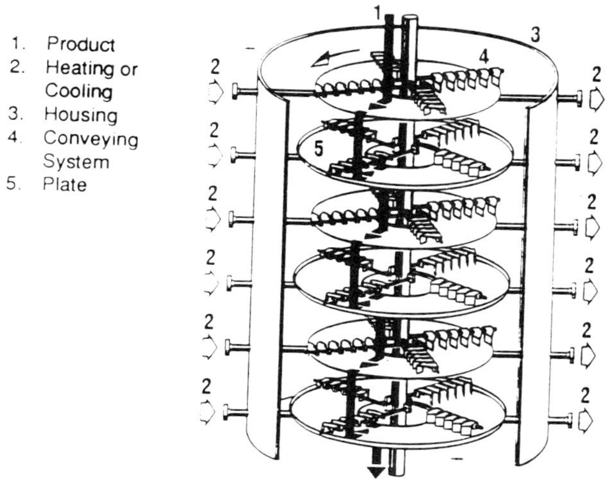

Figure 10.4 Diagram showing a plate-dryer (by courtesy of Krauss-Maffei): 1 = product; 2 = heating or cooling; 3 = housing; 4 = conveying system; 5 = plate.

Hermetic sealing of the dryer is possible so that solvent recovery is easy to achieve.

10.2.7 Flash-dryers

A flash-dryer may be used to convey and dry cakes produced by centrifuges. For manufacture of products, such as vitamins, penicillin, tetracycline etc., which do not tolerate contamination, a closed-loop system using nitrogen as the conveying gas has been used commercially. Preheated nitrogen (to 50–70°C) is used as the carrier gas, which flash-dries the product during conveying from the centrifuge to the mixer/hopper. The drying time is about 3 s (to 0.1% solvent) in a flash-drying section that is 3–5 m long. Solvents (e.g. alcohols, methylene chloride) are condensed and the nitrogen is drawn into the system under pressure control. Oxygen content is monitored and nitrogen added as required to make up the remainder. The mixer–hopper doubles as a cyclone separator and storage hopper.[9] Figure 10.6 shows a diagram of a closed-cycle flash-dryer in conjunction with a centrifuge.

10.2.8 Drum dryers

Used for thin pastes and slurries that stick to metal surfaces, these dryers are designed to apply a thin layer of the wet feed onto a slowly rotating steam-heated drum. The dry product is removed by scalpels at the end of drying. The contact time on the drum is designed to be enough to dry the

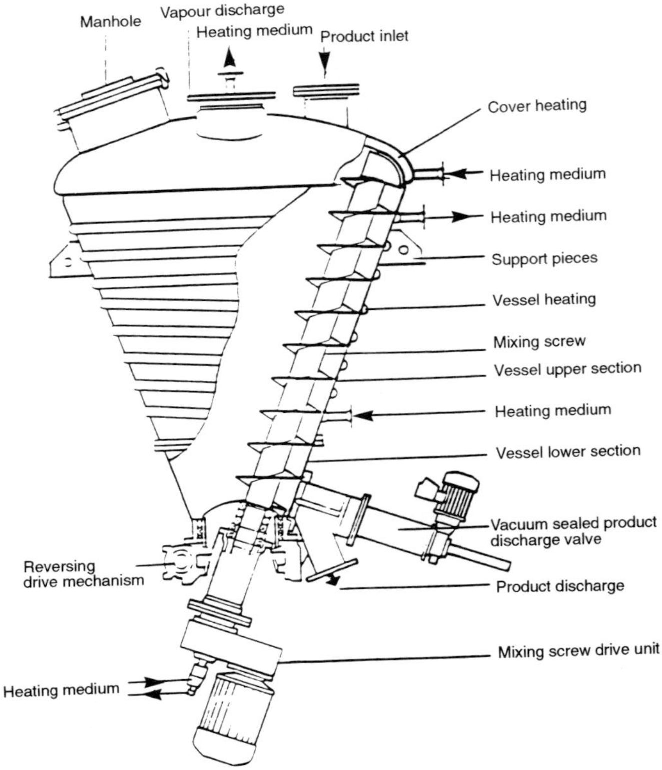

Figure 10.5 Mixer-dryer for granular solids (by courtesy of Krauss-Maffei).

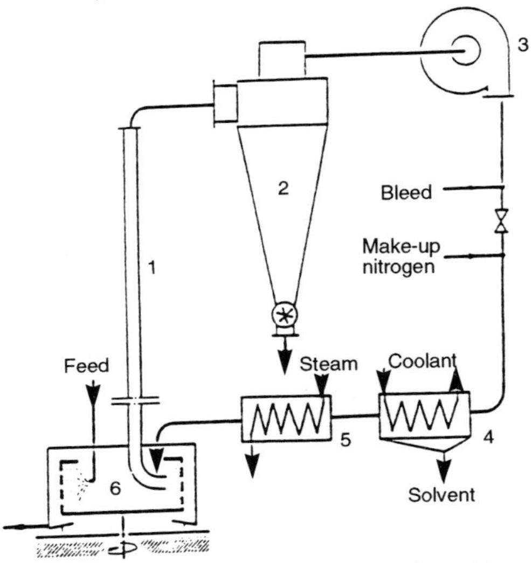

Figure 10.6 Closed cycle pneumatic (flash) dryer in conjunction with a centrifuge: 1 = dryer tube; 2 = cyclone; 3 = blower; 4 = condenser; 5 = heater; 6 = centrifuge.

thin layer fully. Numerous feeding arrangements are possible. The drums may be enclosed fully and maintained under vacuum for heat-sensitive materials or for solvent recovery. Drying rates up to $10\,kg/m^2h$ can be obtained with drum dryers. Drying areas of up to $5\,m^2$ are available commercially.[1]

10.2.9 Fluidized-bed dryers

Beds of granular products (200 μm–2 mm average size) can be fluidized by passage of a gas (generally air) through a low open area distributor (Figure 10.7). If there is excessive surface moisture or the product is sticky, fluidization without mechanical assistance (e.g. vibration or agitation) may not be possible. High relative velocities between the fluidized particles and gas give rise to high efficiency (i.e. result in high heat- and mass-transfer rates). Compared with tray drying (which may require up to 24 h or more), fluid-bed drying can reduce the time by an order of magnitude or more with no adverse effect of quality. Vacuum fluid-bed processors are now available commercially for highly heat-sensitive materials. Some new designs incorporate microwave heating with vacuum-drying.

By incorporating a spray nozzle within the fluid bed it is possible to achieve granulation or agglomeration. Controlled-release dosage forms require one or more layers of coatings. Fluid-bed processors are now being used as a pelletizing technique and for application of water-based or solvent-based films ranging from small particles to tablets. Mehta[10,11] and Jones[12] among others have discussed the fluid-bed processing technique in detail. In particular, scale-up considerations in production of controlled-release products are presented and discussed in the light of their influence on product quality.

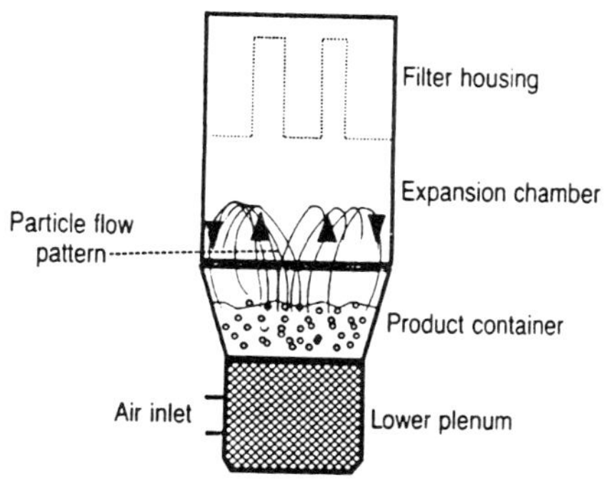

Figure 10.7 Diagram of a fluid-bed dryer.

Advantages and disadvantages of fluid bed coating using different locations of spray nozzle within the fluid bed are discussed by Mehta.[11] Figure 10.8 shows a typical schematic of a vibrated fluid-bed dryer.

10.2.10 Mixer–dryers

A mixer–dryer for granular, free-flowing solids is shown in Figure 10.5. Heat is supplied indirectly through a heated jacket. Along the inside of a conical vessel a vertical rotating screw is set, which stirs the product well to ensure uniform drying. Such dryers claim high energy efficiency.

10.2.11 Vacuum dryers

Various types are available, such as double-cone blender/dryer, paddle dryer, vacuum band dryers etc. The double-cone dryer is a batch unit with capacities of up to 10 m^3 and with drying rates of up to 2–5 kg/m^2h of heated area. The double-conical vessel is steam-jacketed and rotates at low speed to continuously blend the product and bring fresh material in contact with the heated surfaces (Figure 10.9).

10.2.12 Paddle dryers

Continuous-vacuum paddle dryers are used when longer residence times are needed. They are suitable for pasty and granular solids. A horizontal rotating shaft scrapes the product from the walls and mixes and transports it along

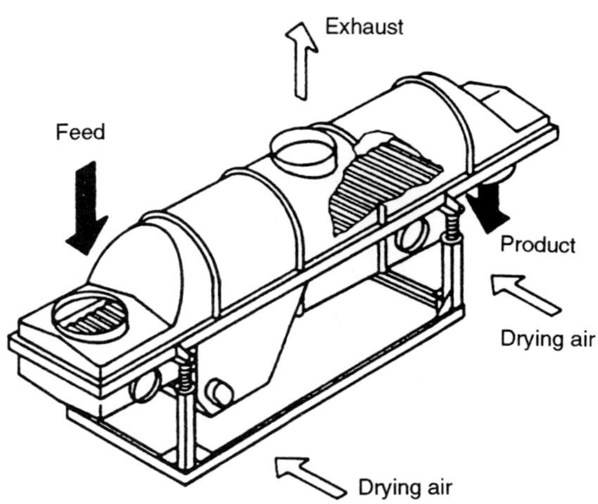

Figure 10.8 Diagram showing a vibrated fluid-bed dryer.

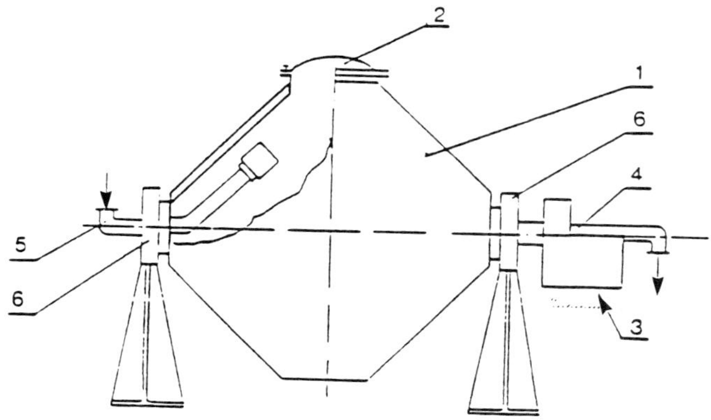

Figure 10.9 Double-cone batch vacuum dryer for particulate solids: 1 = steam jacket; 2 = hatch; 3 = drive; 4 = vacuum ducts; 5 = steam supply; 6 = bearings.

the dryer length. Solvents can be recovered with suitable condensers. Heat is supplied indirectly by a steam jacket around the body of the dryer (Figure 10.10).

10.2.13 Vacuum band dryers

Here the wet pasty product is spread mechanically onto a moving metallic band. Heat is supplied by radiation or by heating the band. In principle, it

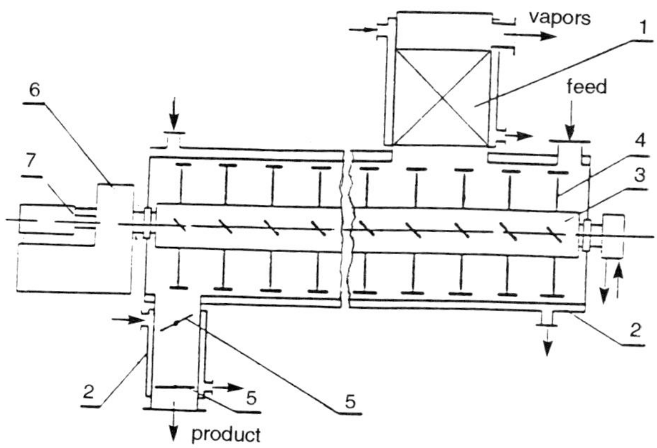

Figure 10.10 Continuous-vacuum paddle dryer: 1 = vapor filter; 2 = steam jacket; 3 = shift with paddles; 4 = paddles; 5 = valves; 6 = shaft drive; 7 = shaft oscillator.

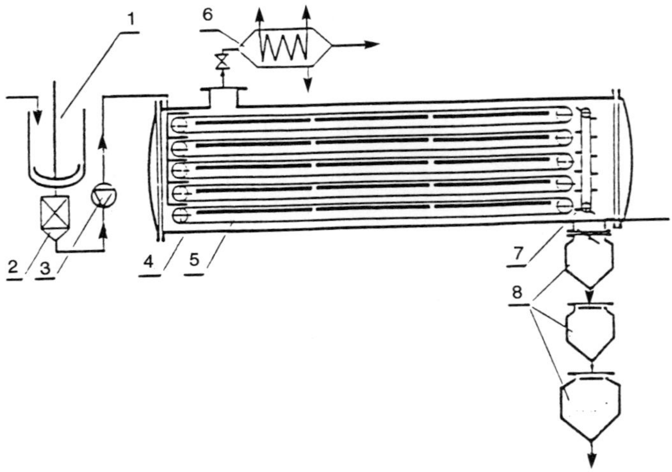

Figure 10.11 Vacuum band dryer installation: 1 = feed mixer; 2 = filter; 3 = feed pump; 4 = band; 5 = heating panels; 6 = vapor condenser; 7 = scraper; 8 = product collector system.

is similar to the drum dryer operated under vacuum. Figure 10.11 is a sketch of a typical vacuum band dryer installation.

10.2.14 Freeze dryers

Highly labile or readily oxidizable materials can be best dried at very low temperatures and at low vacuums by sublimation of the moisture from solid to vapor. As the moisture sublimes, a porous matrix remains that is easily reconstituted. Products that are freeze-dried industrially include antibiotics, certain vitamins and chemotherapy drugs. Owing to the high value of the product, freeze dryers for such products often have standby vacuum pumps, refrigeration units, power supply etc. Table 10.5 lists some of the parameter ranges found in pharmaceutical freeze dryers. Figure 10.12 shows key parts

Table 10.5 Some parameter ranges for batch freeze dryers

Parameter	Range
Chamber pressure (μm Hg)	
Peak	100–200 µm Hg
Final	20–50 µm Hg
Shelf temperature (°C)	
Minimum	−40
Maximum	70
Shelf heat-up rate from −40° to 70°C (min)	90
Evaporation rate (kg/m²h)	1–1.5

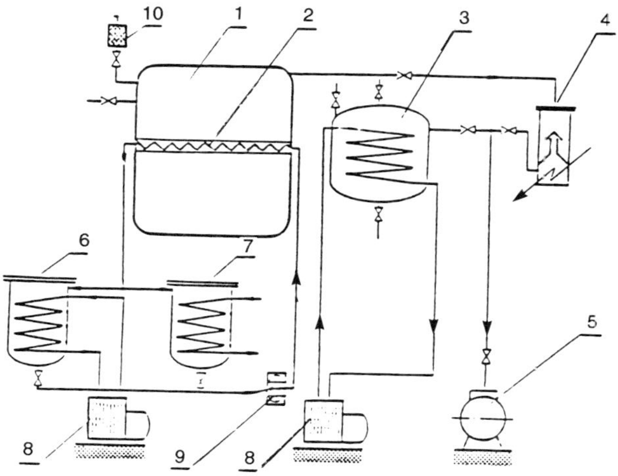

Figure 10.12 Freeze-drying system: 1 = drying chamber; 2 = heated or cooled shelf; 3 = ice-condenser; 4 = diffusion vacuum pump; 5 = rotary vacuum pump; 6 = tank for cooling shelf circulating liquid; 7 = tank for heating shelf circulating liquid; 8 = freezing aggregate; 9 = circulating pump; 10 = air filter.

of the freeze-drying system. Most freeze dryers operate in a batch mode. Although continuous freeze-drying is possible it is an expensive process and is rarely applied in practice. Freeze-drying offers the following advantages:[13]

- No concentration effects (i.e. no salting-out of proteins) so distribution of components remains unchanged upon drying
- Very low final moisture levels can be attained
- Vacuum drying ensures no oxidative denaturation
- Product has a porous structure that permits instant reconstitution.

It is not a suitable process, however, for materials that supercool to glasses that form an impermeable surface upon cooling. Some products (e.g. eukaryotes) can retain viability on freezing only in the presence of certain additives. Following is a list of the types of materials suitable for freeze-drying:

- Non-biological materials as they are highly heat-labile
- Non-living bioproducts such as enzymes, hormones, antibodies, vitamins, antibiotics etc.

Suspensions of bacteria in complex mixtures of salts, proteins and sugars have several eutectic temperatures. Prefrozen products for freeze-drying must be at least 20°C below the lowest eutectic point. Liquid bacterial concentrates have eutectic temperatures in the range −10°C to −40°C. High sugar contents (2–20%) provide good cryoprotection. Various bacteria dried in the presence of different additives show different degrees of stability and cryoprotection.

Table 10.6 Some cryoprotective additives for freeze-drying of bacterial products

Product type	Examples
Polymers	Skim-milk powder, PVC, dextrans, gelatin, algenates, maltodextrins
Carboxyl/carbonyl fixers	Sodium glutamate, malic acid
Dryness buffers	Lactose, sodium glutamate, sucrose
Antioxidants	Ascorbic acid, thiourea
Buffers and salts	$MgSO_4$, $MnSO_4$, $Ca(CO_3)_2$

Table 10.6 shows a typical list of protective additives for freeze-drying of bacterial products.

It is noteworthy that Gram-positive bacteria are more resistant to freeze-drying than Gram-negative ones. Mutations and DNA damage have been reported for Gram-negative bacteria upon freeze-drying. The rate of freeze-drying also has an influence on the quality of the bioproduct. This must be ascertained empirically.

Freeze-drying is five to ten times more expensive per unit amount of moisture removed than conventional drying in fluid beds, drum-drying or spray-drying. It is used only when the product cannot be dried conventionally or the quality demand is stringent.

10.2.15 Spray dryers

Spray-drying is one of the most common downstream operations for bioproduct dehydration. Vacuum evaporators are used to concentrate solutions to be atomized. Biological cells are collected by filtration or centrifugation to solids contents of 15–25% by weight. Small bacteria (~1 μm) may be concentrated by membrane filtration or flocculation, while mycelial fungal cultures can be filtered (e.g. using a rotary filter). Extracellular products may be recovered by precipitation, ion exchange, gel chromatography, ultrafiltration or reverse osmosis.

Masters[14] has presented spray-dryer data for several applications, for example, single cell yeast protein, autolysed yeast extracts, microalgae, spore-forming bacteria, dextrans, antibiotics, enzymes, protease, lysozyme, spent fermentation broth etc. Spray-drying can be used to produce high-energy drugs with high solubility compared with those made from precipitation procedures. For example, spray-drying of phenolbarbitol and hydroflumethiazide forms high-energy polymorphic or amorphous forms with apparent solubilities 1.25 to 2.5 times that of the crystalline form.

When drying spent fermentation broths from bacterial or fungal cultures for feed supplements, or in silage aids as a source of hydrolytic enzymes, low oxygen or closed-cycle systems are suggested to avoid danger of dust explosion or emission of odor.

Spray dryers of various construction are used extensively to produce powders from pumpable slurries. The suspension is atomized with a pressure nozzle (or rotating-disk atomizers for high capacities) and dried by direct contact with hot gases into which the droplets are entrained. A wide assortment of atomizers, spray-dryer chamber designs and flow patterns are used for various reasons. If the feed contains a solvent that is either flammable, toxic or valuable, a closed-cycle dryer is used; the solvent is recovered from the exhaust by a condenser and the gas re-heated and recycled to the dryer. If an inert gas is used as the drying medium (e.g. N_2 if the product is flammable or oxidizable) a closed-cycle operation is used so that the inert gas is not lost. Besides drying, closed-cycle spray dryers are also used for microencapsulation *viz.* coating of powders with polymeric coating material dissolved in the feed. This is an important operation in the pharmaceutical industry to produce controlled-release dosage forms. Figure 10.13 shows a layout for an asceptic spray dryer and the packaging area.

10.3 Drying in the pharmaceutical and biotechnology fields

Treatment of dryer exhaust in the pharmaceutical or biotechnology industries is concerned with removal of solid particles and solvent vapor from the carrier gas. The solid particles consist of product while the solvent vapor could consist of combinations of water and/or multiple organic compounds.

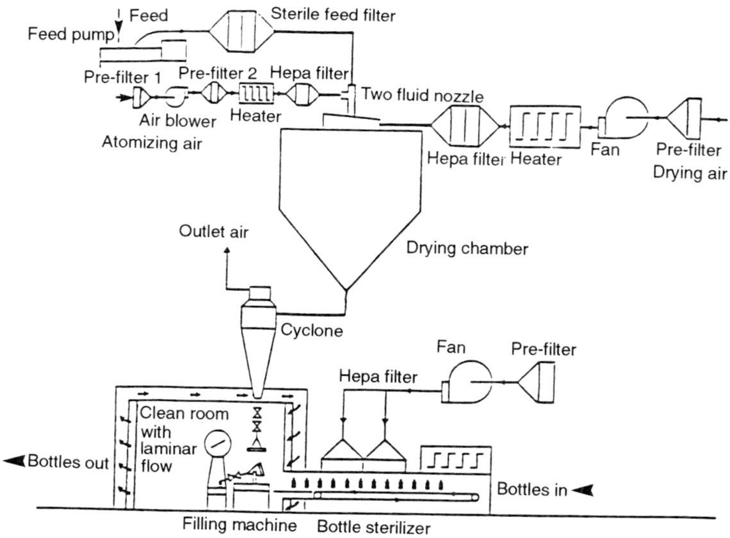

Figure 10.13 Aseptic spray dryer and packaging area.

The carrier gas is usually air or nitrogen. The discussion following will concentrate on methods of solvent removal.

10.3.1 *Removal of solvent vapor*

Solvents discussed in this section will be organic in nature. There are three basic reasons for removing solvents from the exhaust gas stream. These are:

- To satisfy the pollution laws
- To recover the solvent for reuse thereby saving the cost of make-up solvent, and
- To eliminate a potentially hazardous condition of high concentrations of solvent-laden air near the manufacturing plant, which can be dangerous because of flammability and toxicity.

The most common methods of removing vapor from a gas are: (i) condensation; (ii) carbon-bed adsorption; (iii) incineration; and (iv) biofiltration. Incineration and biofiltration are used when it becomes uneconomical to recover solvent for reuse or to satisfy the pollution requirements. At times, incineration and biofiltration are used as an adjunct to the condensation and carbon-bed adsorption methods. Very often, an economic analysis is required to determine the optimum method of solvent removal from the dryer-exhaust gases.

The present cost of common solvents from the *Chemical Marketing Reporter* on 12 February 1996 are shown in Table 10.7.

Inspection of the cost table for these solvents indicates that, while methanol might be a candidate for incineration, THF should be able to support an extensive solvent recovery investment.

10.3.2 *Condensation*

Conceptually, removal of solvent vapor by condensation is the simplest method to use. The recovered solvent has no contamination and the basic equipment consists of a heat-exchanger and a collection tank to collect the solvent.

However, the solvent vapor is always in the carrier gas above its dew point.

Table 10.7 Cost of common solvents*

Solvent	Cost of bulk delivery in the USA (US$/lb)
Methanol	0.064
Toluene	0.097
Methylene chloride	0.40
Isopropanol (IPA)	0.32
Tetrahydrofuran (THF)	1.36

* Data from *Chemical Marketing Reporter*, 12 February 1996.

As a result, it can require extremely cold temperatures to condense a significant percentage of the incoming solvent. This percentage is a function of the solvent, the solvent concentration in the incoming gas and the exit temperature of the cooled gas.

With some solvents or with low concentrations of solvents in the inlet vapor, mechanical refrigeration temperatures must be used to remove sufficient solvent to satisfy the pollution laws. Consider the following examples.

Example 1. Assume that a gas mixture from the dryer contains 25% by volume of IPA in air at 1 atmosphere initial pressure and the total gas flowrate is 1000 cubic feet/min (cfm) at a temperature of 100°C.

As the gas mixture cools, the IPA will condense when its vapor pressure is reduced to the partial pressure of the IPA in the entering gas stream. This initial condensation temperature is called the dew point of the gas mixture. (Vapor pressure is an intensive property and is a function only of temperature.)

Assuming ideality of the gas, the partial pressures are:

$$PP_{IPA} = 0.25 \times 760\,\text{mm Hg} = 190\,\text{mm Hg} \tag{10.3}$$

$$PP_{\text{dry air}} = 0.75 \times 760\,\text{mm Hg} = 570\,\text{mm Hg} \tag{10.4}$$

where the dew point of IPA in the mixture is 52°C. Assume that cooling water is available at 30°C, chilled water at 2°C and brine at −20°C. Assume countercurrent condensing in a 1:1 heat exchanger, a maximum allowable rise of 5°C in the cooling water and an approach temperature of the exit gas of 5°C so that:

$$\text{Rate of IPA vapor inflow} = 0.25 \times 1000\,\text{cfm} = 250\,\text{cfm} \tag{10.5}$$

$$\text{Rate of dry air inflow} = 0.75 \times 1000\,\text{cfm} = 750\,\text{cfm} \tag{10.6}$$

$$\text{IPA inflow} = 0.25 \times 1000\frac{\text{ft}^3}{\text{min}} \times \frac{1\,\text{lb mole}}{359\,\text{cfs}} \times \frac{273}{273+100} \times 60\frac{\text{lbs}}{\text{lb mole}} = 30.6\frac{\text{lbs}}{\text{min}} \tag{10.7}$$

$$\text{Air inflow} = 0.75 \times 1000 \times \frac{1}{359} \times \frac{273}{373} \times 29\frac{\text{lbs}}{\text{lb mole}} = 44.3\frac{\text{lbs}}{\text{min}} \tag{10.8}$$

The exit gas stream is saturated with IPA vapor at 35°C, the vapor pressure of IPA at 35°C is 80 mm Hg, and the partial pressure of IPA in the exit gas is 80 mm Hg.

$$\text{Mole fraction of IPA in exit gas}\ \frac{80}{80+680} = 0.105 \tag{10.9}$$

$$\text{Mole ratio of IPA/air in exit gas}\ \frac{80}{760-80} = 0.118 \tag{10.10}$$

$$\text{Dry air outflow} = 44.3\frac{\text{lbs}}{\text{min}} \times 1\frac{\text{lb mole}}{29\,\text{lbs}} = 1.52\frac{\text{lb moles}}{\text{min}} \tag{10.11}$$

$$\text{IPA leaving in gas stream} = 0.118\frac{\text{moles IPA}}{\text{mole air}} \times 1.52\frac{\text{lb moles dry air}}{\text{min}} \tag{10.12}$$

or

$$\text{IPA leaving as vapor} = 0.179\frac{\text{moles}}{\text{min}} \times 60\frac{\text{lbs}}{\text{mole}} = 10.8\frac{\text{lbs}}{\text{min}} \tag{10.13}$$

$$\text{IPA condensed} = 30.6\frac{\text{lbs in}}{\text{min}} - 10.8\frac{\text{lbs out}}{\text{min}}\ \text{(as vapor)} \tag{10.14}$$

$$\text{IPA condensed} = 19.8\frac{\text{lbs}}{\text{min}} \tag{10.15}$$

$$\%\ \text{recovery of IPA (as liquid)} = \frac{19.8}{30.6} = 65\% \tag{10.16}$$

Similar calculations using chilled water as the cooling media at 2°C and brine at −20°C yield recoveries of IPA of 94% and 98.6%, respectively.

Even with 98.6% recovery of IPA by condensation, the discharge rate of IPA from the condenser is still approximately 0.42 lbs/min. Since the present pollution laws in Puerto Rico allow only 3 lbs/h or 15 lbs/day of volatile organics to be discharged into the atmosphere from a point source, the calculated discharge of (0.42 × 60) or 25 lbs/h is well above the legal limit. The remaining IPA in the gas stream could be removed by incineration, biofiltration or carbon-bed adsorption. With an inlet concentration less than 25%, the dew point of the gas stream is less than 52°C and it will require lower coolant temperatures than −20°C to yield recoveries in the 90% or more range.

Example 2. As the next example, consider methylene chloride (dichloromethane) as a solvent vapor in air. Methylene chloride provides the dual advantages of being non-flammable and possessing a low boiling point, 40.7°C. It has use as a solvent where the drying temperature must be kept low to prevent thermal degradation of the product. The low boiling point, however, has the disadvantage of making it difficult to condense.

Calculations similar to those with IPA indicate that the dew point of a 25% volume mixture of methylene chloride in a fixed gas at one atmosphere is 7°C. Hence use of 30°C or 2°C coolant will not condense any methylene chloride.

Use of coolant at −20°C will condense 74% of the incoming methylene chloride. In order to condense 99.6% of the methylene chloride, its vapor pressure would have to be reduced to 1 mm Hg, which occurs at −70°C.

Hence, refrigerant temperatures in the −75°C to −100°C range would be required.

A potential problem in using coolant termperatures less than 0°C is the freezing of condensed water in the heat exchange equipment. If the carrier gas is dry (dew point ⩽60°C), there is little chance for freezing of water. If the carrier gas is 'wet' a method is available to circumvent this problem. This would entail the use of two heat exchangers in parallel; one would condense vapor while the other 'regenerates' by melting the ice formed in the exchanger. The timing cycle would have to be estimated based on the particular case in question. Using a solvent that is infinitely soluble in water (e.g. IPA) is not a problem since the condensed solvent would effectively dissolve the frozen water.

10.3.3 Carbon-bed adsorption

Carbon-bed adsorption is usually used when it is desired to recover the solvent, and the inlet vapor concentration is so low as to make condensation impractical. It makes use of the principle of selective affinity of organic vapors for porous carbon. It is useful for solvent recovery as well as odor removal from a dryer exhaust stream. In the recovery of solvent from a dryer exhaust stream, it is merely necessary to pass the gas through a bed of activated carbon to remove most (usually 95–99%) of the solvent.

For small installations or very low gas flows (<150 SCFM), the carbon can be purchased in a chamber designed for easy replacement, usually equipped with duct-transition pieces for easy mounting. For a continuous process, it is advantageous to use multi-chambers in parallel. The used carbon is returned to the supplier who removes the solvent and returns fresh carbon. This type of installation requires a minimum of capital cost but has high operating cost because of the frequent carbon replacement.

If the air flow or solvent flow from the dryer is of such magnitude that the use of replacement carbon is impractical or it is desired to recover the solvent, a regenerable carbon bed design must be used. In this case, the activated carbon is located on trays within a vessel in order to allow the gas stream to contact the carbon without excessive pressure drop. Typical gas flow through the carbon bed is 24.4–30.4 m/min (80–100 ft/min). The carbon will eventually become saturated with organic vapor to the point that it will no longer be able to remove the organic. At that point it must be regenerated.

A complete cycle would consist of the following steps:

1. Passage of solvent laden gas through the bed.
2. Regeneration of the bed with low-pressure steam. Note that the steam temperature must be greater than the solvent boiling point, and that regeneration is best performed countercurrent to the flow during the adsorption step.

3. Passage of hot, dry gas through the bed to evaporate water.
4. Passage of cool, dry gas through the bed to cool the carbon, thereby increasing its adsorption capacity during the next cycle.

A permanent installation such as this would usually have automatic controls and operate on a strict time cycle. For a continuous operation, dual beds operating in tandem would be required.

Activated carbon beds are not 100% efficient and previous comments apply on removing the residual solvent from the carbon-bed exhaust stream. Neither do carbon beds last for ever, and the carbon must be completely changed every 2–3 years, depending on the solvent collected and the rate of usage.

Another variable must be considered when using a carbon bed for solvent recovery. When regenerating the bed with steam approximately 1.8 kg (4 lbs) steam will be used for every 0.45 kg (1 lb) of organic material; hence the condensed vapor is only 20% organic in nature. Evaluation must be made of the method of recovery of the solvent from its water mixture. A combination of decantation and distillation is usually used. Very often, simple distillation is insufficient to separate the organic from water and azeotropic distillation must be used.

The engineer, during economic evaluation, must pay considerable attention to the cost and ease of operability of the solvent recovery system after the carbon-bed adsorption system. Very often the recovery system after the carbon bed will cost more and be more difficult to operate than the carbon-bed system itself.

Carbon-bed adsorption has its limitations because of the chemical compounds suitable for collection by this method. Compounds such as organic acids, aldehydes, some ketones, phenols, glycols and amines are considered unsuitable for carbon-bed adsorption. This is an important consideration when choosing the method of removal of solvents from the dryer exhaust.

10.3.3.1 Vapor-separation membranes. In the past 4 years, a vapor-separation technique making use of composite polymeric membranes to separate volatile organic compounds from a carrier gas such as air or nitrogen has been developed. The organic passes through a membrane yielding a concentrate permeate from which the organic can be condensed for reuse. The membranes are fabricated as flat sheets and incorporated into spiral-wound elements. These elements are housed in stainless steel pressure vessels.

To date, there have been approximately 25 commercial installations. The largest system has been able to handle total gas flows of 100 SCFM. Volatiles handled have consisted of vinyl chloride, chlorinated-fluorinated hydrocarbons, aldehydes, ketones, aromatics and alcohols.

The minimum inlet concentration that can be handled is 1000 ppm. The membrane is not regenerated but is replaced: the expected membrane life is 3 years.

With regard to a cost analysis of alternative methods of dryer-exhaust treatment, the following should be considered.

Capital costs comprise:

- Purchased equipment cost
- Cost of equipment installation
- Cost of equipment tie-ins (if an existing plant)
- Cost of utility generating equipment
- Cost of utility equipment installation
- Cost of utility tie-ins
- Cost of engineering and construction
- General contractor's fee
- Contingency.

Operating costs comprise:

- Amortization cost
- Cost of capital for equipment purchase
- Cost of raw material (e.g. water, steam, carbon)
- Cost of utilities operating
- Cost of direct operating labor
- Cost of supervision
- Cost of maintenance and repairs
- Overhead charges.

References

1. Mujumdar, A.S. (Ed.) (1987). *Handbook of Industrial Drying.* Marcel Dekker, New York.
2. Strumillo, C., Markowski, A.S. and Adamiec, J. (1991). Selected aspects of drying biotechnological products in *Drying '91* (Ed. A.S. Mujumdar), Elsevier, Amsterdam, pp. 36–55.
3. Strumillo, C. and Kudra, T. (1987). *Drying: Principles, Applications and Design.* Gordon and Breech, New York, p. 466.
4. Keey, R.B. (1978). *Introduction to Industrial Drying Operations.* Pergamon Press, New York, p. 376.
5. Keey, R.B. (1992). *Drying of Loose and Particulate Materials.* Hemisphere/Taylor Francis, New York.
6. Cook, E.M. and Dumont, H.D. (1991). *Process Drying Practice.* McGraw-Hill., New York, p. 254.
7. Van't Land, C.M. (1991). *Industrial Drying Equipment.* Marcel Dekker, New York, p. 362.
8. Pakowski, Z. and Mujumdar, A.S. (1995). Drying of Pharmaceuticals in *Handbook of Industrial Drying*, 2nd edition, vol. 2 (Ed. A.S. Mujumdar), Marcel Dekker, New York, pp. 743–774.
9. Anon. (1981). *Chem. Eng.* **November,** 80–81.
10. Mehta, A.M. (1986). Factors in the development of oral controlled-release dosage forms. *Pharmaceut. Manufact.,* **January**.
11. Mehta, A.M. (1988). Scale-up considerations in the fluid-bed process for controlled-release products, *Pharmaceut. Technol.* **February**.
12. Jones, D.M. (1985). Factors to consider in fluid-bed processing. *Pharmaceut. Technol.* **April**.
13. Adam, G.D.J. (1991). Freeze drying of biological materials. *Drying Technol.* **9**(4), 891–925.
14. Masters, K. (1985). *Spray-Drying Handbook.* Gordon Goodwin, London.

Further reading

Kadam, K.L. (Ed.) (1991). *Granulation Technology for Bioproducts*. CRC Press, Boca Raton, Florida, USA. 336 pp.
Mellor, J.D. (1978). *Fundamentals of Freeze Drying*. Academic Press, London.
Membrane Technology and Research Inc. (1994). Technical Bulletin and personal communication.
Mujumdar, A.S. (Ed.) (1992). *Drying '92*, Part 1 and Part 2. Elsevier, Amsterdam.
Pakowski, Z. (1992). Drying of solids containing multicomponent moisture. In *Advances in Drying*, Vol. 5 (Ed. A.S. Mujumdar), Hemisphere, New York.
Regulation for the Control of Atmospheric Pollution of the Environmental Quality Board, Rules 417, 419 (EPA Equivalent in Puerto Rico).
Treybal, R.E. (1955). *Mass Transfer Operations*. McGraw-Hill, New York.
VIC Manufacturing Company (1992). *Carbon Adsorption*/Emission Control.

11 Sterilization in the pharmaceutical and biotechnology industry

P.M. ARMENANTE and A.C. KIRPEKAR

11.1 Definitions and classification of sterilization processes

Sterilization can be defined as the process through which all forms of life are destroyed, removed, or permanently inactivated. Although, according to this definition, sterility is an absolute concept, in industrial practice sterilization is often referred to as the process through which the probability of survival of undesired organisms is reduced to an arbitrarily low level. Typically this level is so low that the chance that even one organism will survive the sterilization process can be taken to be negligible. Such a strict requirement is justified, especially in those situations in which the potential health effects (e.g. in the preparation of parenteral drugs) or the validity of the process (e.g. in antibiotic fermentation) are such that no level of microbial contamination can be considered acceptable. This probabilistic approach to sterility does not imply that the sterility requirements of the process can be relaxed. Instead, it reflects the fact that especially large-scale industrial sterilization processes are affected by many variables (e.g. the number of initial organisms, the intrinsic variability of the resistance of each individual microorganism to heat, the difficulty to regulate precisely the temperature in every part of a given equipment) that cannot be controlled absolutely. In any case, sterilization processes should not be confused with other less-stringent processes, such as pasteurization or disinfection, whose aim is the reduction of the microbial count so that the probability of spoilage of a product over its useful lifetime is minimized (e.g. pasteurization of milk), or other specific operations can be carried out without interference (e.g. disinfection of 'sterile' rooms).

In the pharmaceutical and biotechnology industries, sterilization processes are applied to materials, equipment and products of different chemical composition and physical characteristics. A list of the most common applications is provided in Table 11.1. The corresponding, most appropriate sterilization methods will depend on such properties.

Of the three methods mentioned to eliminate life (i.e. destruction, physical removal and permanent inactivation) destruction is the least industrially significant since it requires the establishment of such extreme conditions (such as the attainment of incineration temperatures or the use of concentrated acids) to be impractical, except in laboratory practices (e.g. flaming of

Table 11.1 Typical industrial applications of sterilization processes*

Material, equipment or product to be sterilized	Type of material, equipment or product	Physical state of material to be sterilized
Fermentation vessels	Permanent equipment	Solid
Continuous sterilizers	Permanent equipment	Solid
Liquid dispensers and other pharmaceutical equipment	Permanent equipment	Solid
Sterile rooms	Permanent equipment	Solid
Microbiological equipment and apparatus	Permanent or disposable equipment	Solid
Syringes	Disposable supplies	Solid
Sterile room clothing	Disposable supplies	Solid
Disposable filters	Disposable equipment	Solid
Vials	Disposable supplies	Solid
Injectable solid drugs	Product	Solid
Fermentation media	Nutrient	Liquid
Fermentation nutrient oils	Nutrient	Liquid
Silicon oils	Supply	Liquid
Antifoam	Nutrient	Liquid
Fats	Nutrient	Liquid
Water for injection	Product/supply	Liquid
Sterile process water	Supply	Liquid
Fermentation air	Supply	Gas
Sterile room air	Supply	Gas
Laminar flow air	Supply	Gas
Nitrogen blanket in sterile area	Supply	Gas

*From reference 1.

inoculation loops). The remaining two approaches, namely removal or inactivation of microorganisms, are both used in the industrial practice, including large-scale operations. Accordingly, all sterilization processes used currently in industry fall into one of the following categories, listed in order of importance:

- Thermal sterilization
- Filtration
- Chemical inactivation
- Irradiation.

Thermal sterilization is the oldest, most dependable and most common sterilization method, relying on the proven effect of temperature on the chemistry of life. In addition, it is practically the only method used industrially to sterilize liquid fermentation media. For this reason, this method will be examined in the greatest detail in the following sections.

Filtration relies on the physical removal of microorganisms from a stream passing through the filter. It is used extensively for gas sterilization (especially fermentation air) because of the large volumes involved, low initial bacterial count, and the lower energy consumption and costs associated with it. Liquid filtration is not so popular because of its cost and the stringent demands imposed on the filtering element (e.g. the ability to withstand high pressure differences across it, high filtration area, homogenous and consistently small pore size to prevent microorganism to pass through it). In addition, the absolute proof of filtration effectiveness is hard to establish in spite of the many improvements made in the fabrication of the filters (e.g. well-defined pore size distribution, predictable pressure differentials, improved chemical compatibility). Therefore, liquid filtration is not well-accepted by governmental regulatory agencies as a sterilization process.

Chemical-inactivation processes use liquid or gaseous chemicals to produce microbial death. The most popular of such chemicals are chlorine, formaldehyde, glutaraldehyde, quaternary ammonium salts, and ethylene oxide. They are used primarily to sterilize equipment or medical supplies. Their main disadvantages are that they have a relative low penetration effectiveness, are not particularly effective against spores, and their residuals may interfere with subsequent processes, such as fermentation. Furthermore, their chemical reactivity makes them a potential health hazard for the operators, and special precautions must be taken to avoid or minimize human exposure. This, in turn, introduces handling problems. For all these reasons, they are typically out of favor in most sterilization processes. Conversely, these and other chemicals are used extensively as disinfectants in laboratories, clean rooms, and other fixed installations in which the primary aim is a significant reduction of the microbial count (disinfection) rather than the complete inactivation of all microorganisms (sterilization).

Ionizing radiations also have limited but important uses in sterilization processes. They are used primarily in maintaining asepsis (i.e. the continuous state of exclusion of harmful organisms) in particular areas such as laboratories, sterile cabinets, water-for-injection loops, and equipment. Ultraviolet (UV) radiation is especially useful for this purpose. Owing to its limited penetration power, UV light is used especially for maintaining sterility in air, on walls, and on other flat surfaces without creating severely hazardous working conditions. More penetrating ionizing radiation, such as gamma (γ) rays, has a much more significant sterilizing effect; γ rays are used primarily in all those cases in which other sterilization methods are unusable because of the thermal lability of the products that have to be sterilized.

A description of all the sterilization methods mentioned above is provided in the following sections.

11.2 Thermal sterilization

Thermal sterilization is the most commonly used sterilization method in the pharmaceutical and biotechnology industry, especially in large-scale processes, such as fermentation and in the thermal sterilization of vials. The method relies on the exposure of the material to be sterilized to a sufficiently high temperature and for enough time to produce the desired level of sterility. This produces a loss of viability in the microorganisms that are initially present in the material, without necessarily producing their physical destruction. When spore-forming organisms are involved, the achievement of sterile conditions also implies the inactivation of spores. This is typically a much tougher requirement.

All life forms, including microorganism spores (the most heat-resistant of all), are temperature sensitive and can be inactivated permanently if exposed to a high enough temperature. However, other factors such as pH, chemical composition of the medium, and way of administering heat can also have a significant impact on the rate of thermal microbial inactivation. For example, it is a well-known fact that moist heat, typically provided by the use of saturated steam, is much more effective than dry heat, such as air heating. This is probably caused by the different mechanisms involved in microbial inactivation. In fact, although all sterilization processes appear to be fairly well described by first-order kinetics (indicating that microbial death is the result of nuclear mutations) dry heat sterilization has been associated with an oxidation process within the cell or spore, whereas steam sterilization is believed to result from protein denaturation processes.

Other practical considerations point to the enhanced effectiveness of steam sterilization. Condensing steam releases a significant amount of energy because of latent heat. In addition, the heat-transfer coefficient of condensing steam is among the highest achievable. This maximizes the chances that heat will penetrate and be released into all parts of the material undergoing sterilization. Conversely, air heating can only rely on sensible heat, using a medium (air) having little thermal capacity, and whose heat-transfer coefficient is notoriously low.

The main disadvantage of steam sterilization is that water boils at only 100°C. This temperature is too low to be of any practical use for sterilization since it would result in unacceptably long exposure times to be effective. Therefore, all steam-sterilization processes are carried out under pressure in autoclaves or pressure vessels.

Thermal sterilization is invariably associated with the partial deterioration of the material, such as a nutrient, that was originally intended to be protected by the sterilization process. Fortunately, most of the nutrients used in fermentation are degraded at a rate that is much slower than the inactivation rate of even the most heat-resistant organisms. In those cases in which this is not true, as with thermally-labile products, thermal sterilization cannot be used advantageously and other sterilization methods must be employed.

11.2.1 Kinetics of microbial thermal death and thermal degradation of media

Thermal sterilization is based on the well-known fact that the enzymes necessary to sustain life in microorganisms are inactivated at high temperature.

This enzyme-degradation process is not instantaneous but can be assumed to proceed, like any other chemical reaction, at a rate that is a function of several parameters, such as temperature and activation energy. Depending on the assumptions that one makes for the mechanics responsible for thermal death, several models are possible — two types are used commonly: (i) one based on a simple first-order kinetic process, typically used for most micro-organisms (the logarithmic microbial thermal-death process); and (ii) a more complex one used for microbial populations containing spores (the non-logarithmic microbial thermal-death process). In many cases, however, the simpler model is assumed to be valid also for spore-producing organisms in order to simplify the calculations.

11.2.1.1 Logarithmic microbial thermal-death process. According to this model, the permanent microbial inactivation process (microbial death) is assumed to be produced by the inactivation of a single enzyme or class of enzymes within the organism. In addition, the process is assumed to follow a first-order kinetics, that is, the microbial death rate is assumed to proceed at a rate proportional to the number concentration of organisms present at any given time in the system. In quantitative terms this implies that:[2,3]

$$r_N = K_d N \tag{11.1}$$

where: r_N = microbial death rate in $(\text{ml} \cdot \text{s})^{-1}$; N = number concentration of viable microorganisms per ml; and K_d = proportionality parameter in s^{-1}. The parameter K_d is the specific kinetic rate of microbial thermal death and can be expected from kinetic theory to be a function of temperature according to the Arrhenius equation:

$$K_d = K_{d0}\, e^{-E_d/R_G T} \tag{11.2}$$

where: K_{d0} = Arrhenius constant of the thermal-death kinetic process in s^{-1}; E_d = activation energy of the thermal-death kinetic process in cal/g mol; R_G = ideal gas constant in cal/g mol K; and T = temperature in K.

This equation shows that any increase in sterilization temperature produces a reduction in the absolute value of the argument in the exponential term, thus resulting in an increase in the value of K_d and a faster sterilization process. In the same equation, the activation energy term, E_d, determines how rapidly the specific sterilization rate, K_d, varies as a function of temperature. For example, if the activation energy is high, then K_d will increase very rapidly when the temperature is increased even just a few degrees, thus resulting in a faster sterilization process.

The numerical values for K_{d0} and E_d depend on the type of microorganisms and their intrinsic capability to withstand high temperatures. However, spore-forming organisms can produce spores with significant greater ability to resist high temperatures than the same organisms in the vegetative state. For most microorganisms and their spores the value of the activation energy is usually quite high, typically in the range 50000 – 100000 cal/mol.[4] Some typical values for these constants are given in Table 11.2. This table shows that *Bacillus stearothermophilus* is one of the most heat-resistant organisms commonly encountered. Therefore, sterilization calculations are often made using the death-kinetic parameters for this organism as a 'worst-case scenario'.

The first-order kinetic model for microbial death given above can be used to determine the degree of sterility that can be achieved during a simple sterilization process in which the organisms are exposed to a constant high temperature. Quantitatively, for a given microbial system occupying a constant volume, V, a batch sterilization process can be described by the following microbial balance:

$$V \cdot \frac{dN}{dt} = -K_d V N \tag{11.3}$$

which states that the change in the viable number of organisms in the system (left-hand side of the equation) is equal to the rate at which the organisms are 'removed' as a result of thermal death (right-hand side). If the sterilization temperature is kept constant (thus implying a constant value of K_d) the simple integration of this equation yields:

$$\ln \frac{N}{N_0} = -K_d t \tag{11.4}$$

where N_0 = initial number of viable organisms per ml.

The ratio N/N_0 is the fraction of viable organisms that have survived sterilization at time t. The equation shows that a semi-logarithmic plot of N versus t results in a straight line whose slope is a function of temperature. This is indeed the case for a number of microorganisms in the vegetative state. An example of the semi-logarithmic relationship between the fraction of surviving organisms N/N_0 and time is shown in Figure 11.1.

Table 11.2 Numerical values of the Arrhenius constants and activation energies for the thermal death of different types of microorganism

Microorganism	K_{d0} (min^{-1})	E_d (cal/mol)
Bacillus stearothermophilus	7.9×10^{38} *	68 700 *
Bacillus stearothermophilus	9.5×10^{37} †	67 700 †
Bacillus stearothermophilus	9.8×10^{37} ‡	68 700 *
Bacillus subtilis	3.6×10^{42}–5.2×10^{42} §	76 000 ‖
Clostridium botulinum	—	82 000 ‖
Escherichia coli	1.5×10^{84} *	127 000 *

* After Aiba *et al.*[4]

† After Deindoerfer and Humphrey.[5]

‡ Extracted from Moss and Smith[6] using Deindoerfer and Humphrey's[5] value for the activation energy.

§ Extracted from Moss and Smith[5] using value for the activation energy from Wang *et al.*[3]

‖ After Wang *et al.*[3]

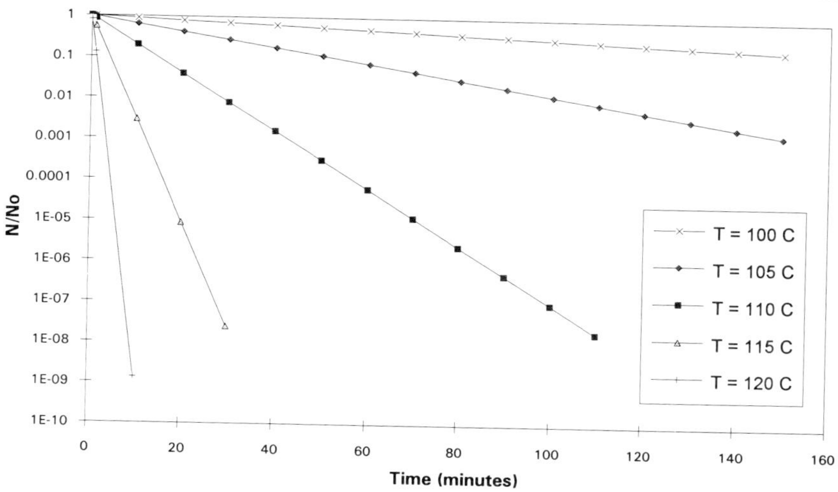

Figure 11.1 Typical death-rate plot for *Bacillus subtilis.*

11.2.1.2 Non-logarithmic microbial thermal-death process. The equations reported above and the first-order kinetic death rate on which they are based are typically used in sterilization equations. Nevertheless, these equations are only approximately correct to represent the thermal-death process if spores are present. Different models have been proposed to interpret the experimental death-kinetics data of spores. An often quoted model is that based on the assumption that inactivation of spores is a two-step process. Accordingly, the critical enzymes in the spores first undergo a reaction in which a labile intermediate is produced. Subsequently, this intermediate is inactivated resulting in thermal death of the spore. For a kinetic point of view this process can be represented by the following scheme:[7]

$$\text{Viable spores} \xrightarrow{K_{d1}} \text{Intermediate viable spores} \xrightarrow{K_{d2}} \text{Inactivated spores} \quad (11.5)$$

where: K_{d1} = rate of production of intermediate viable spores per s; and K_{d2} = rate of production of inactivated spores per s. As in the previous case, K_{d1} and K_{d2} are assumed to be functions of temperature according to the Arrhenius equation (equation (11.2)). A mass balance for the viable spores and intermediate viable spores yields

$$V \cdot \frac{dN_v}{dt} = -K_{d1} V N_v \quad (11.6)$$

$$V \cdot \frac{dN_{\text{int}}}{dt} = K_{d1} V N_v - K_{d2} V N_{\text{int}} \quad (11.7)$$

where: N_v = number of viable spores per ml; N_{int} = number of intermediate viable spores per ml.

For the case in which the sterilization temperature is maintained constant and the process is conducted in batch, these two equations can be integrated to produce:

$$\frac{N}{N_0} = \frac{K_{d1}}{K_{d1} - K_{d2}}\left(e^{-K_{d2}t} - \frac{K_{d2}}{K_{d1}}e^{-K_{d1}t}\right) \tag{11.8}$$

where: N = total number concentration of viable organisms or spores = $N_v + N_{int}$ (ml^{-1}); N_0 = initial number of viable organisms per ml. The corresponding plot for the case of *B. stearothermophilus* is given in Figure 11.2. This plot shows that the sterilization curves deviate significantly from the semi-logarithmic case for relatively low sterilization temperatures. However, this deviation becomes much less pronounced as the sterilization temperatures approach the values typically used in most commercial applications ($T_{ster} > 121°C$). This justifies the use of the simpler first-order kinetics described in the previous section even when spores are present.

11.2.1.3 Empirical correlations for microbial death kinetics. The expressions given above for the kinetics of microbial thermal death have a sound thermodynamic and kinetic theory basis. However, there are several other definitions and empirical correlations often encountered in the industrial practice to describe sterilization processes. These definitions have been developed over time to convey quickly the idea of the effectiveness of a

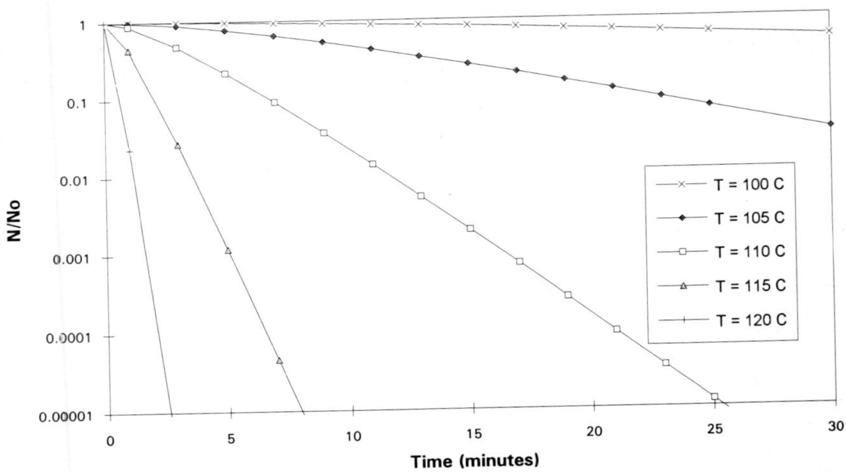

Figure 11.2 Typical death-rate plot for *Bacillus stearothermophilus*.

particular sterilization process. In some cases, however, this aspect has resulted in oversimplifications that may produce incorrect prediction of sterilization effectiveness.

The *decimal reduction time*, *D*, is defined as the sterilization time required to produce a reduction of one order of magnitude in the microbial population concentration for a given sterilization temperature.[8,9] The value of *D* is directly and univocally related to K_d and is therefore a measure of the kinetics of the sterilization process. However, unlike K_d, the *D* value and its unit (typically min) immediately convey the idea of how fast the sterilization process proceeds at any given temperature: the greater the *D* value the more resistant the organism will be to sterilization, and the slower the sterilization process will be. For this reason the *D* value and its multiple described below, the *F* value, are extremely popular in industrial practice.

From the definition of *D* one can easily establish the relationship between the *D* value and K_d using equation (11.4). The resulting equation is:

$$\ln \frac{10N}{N} = -K_d D \tag{11.9}$$

in which it is assumed that the sterilization process is conducted in batch at constant temperature. From this equation it follows that:

$$D = \frac{\ln(10)}{K_d} = \frac{2.303}{K_d} \tag{11.10}$$

Table 11.3 gives the values for K_d and *D* for different organisms at 115 and 121°C. This table shows that most heat-resistant organisms have *D* values of the order of minutes at temperatures near the common sterilization temperature of 121°C. The same table also shows that organisms such as *Escherichia coli* are much more easily inactivated than heat resistant organisms such as *Bacillus stearothermophilus*.

Another parameter often encountered in sterilization calculations is the thermal death time, or *F* value, defined as the time required to produce a specific decrease in the number of surviving organisms (typically by a 10^{-6} factor, i.e. six orders of magnitude). Owing to its definition, the *F* value is

Table 11.3 K_d, *D* values, and *Z* values for different organisms at two different temperatures (partially readapted from Wallhäusser[10])

Microorganism	K_d (min^{-1}) at $T = 121°C$	*D* value (min) at $T = 121°C$	*D* value (min) at $T = 115°C$	*Z* value (°C)
Bacillus stearothermophilus	0.4–1.5	1.5–6.0	10 – 24	6–7
Clostridium botulinum	1.6–2.9	0.8–1.4	2.8–3.6	—
Bacillus subtilis	3.3–5.8	0.4–0.7	2.2	8–13
Escherichia coli	5.3×10^{13}	4.3×10^{-14}	5×10^{-13}	—

just a multiple of the D value. For example, if a reduction of 10^{-n} is desired, then from equation (11.4) it is:

$$\ln\frac{N}{10^{-n}N} = \ln\frac{10^{n}N}{N} = -K_d F \tag{11.11}$$

from which:

$$F = \frac{\ln(10^n)}{K_d} = \frac{2.303n}{K_d} = nD \tag{11.12}$$

Owing to their definitions the D value and the F value are implicitly dependent on the temperature. Therefore, another parameter, the Z value, is often encountered to describe the susceptibility of sterilization to temperature effects. The Z value is defined as:[8,11]

$$Z = \frac{T_2 - T_1}{\log_{10} D_1 - \log_{10} D_2} \tag{11.13}$$

where D_1 and D_2 are the values of D in correspondence of the temperatures T_1 and T_2 for which $D_1 = 10D_2$. The Z value represents the temperature increase corresponding to a 10-fold reduction in sterilization time because of its definition. Typical values for the Z value are given in Table 11.3.

It is often assumed that $\log_{10} D$ is a linear function of the temperature T, as shown in Figure 11.3. If this were truly valid then the Z value would be a constant for each type of organism. However, this assumption is not really correct, as one can verify by substituting the expression for D from equation

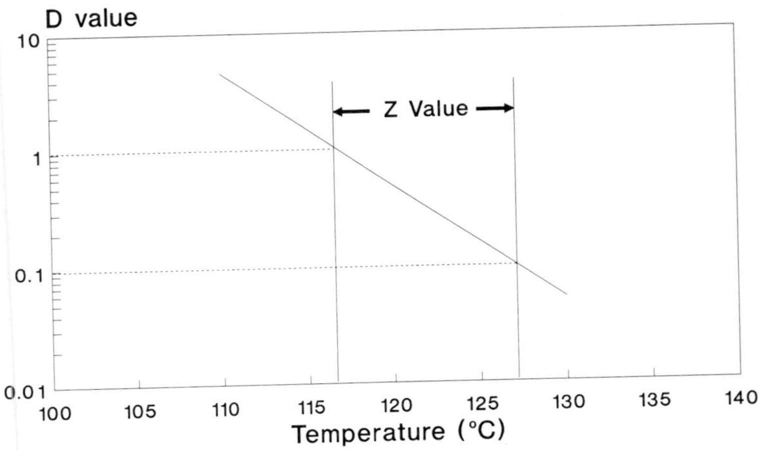

Figure 11.3 Decimal reduction (D value) and Z value for *B. stearothermophilus*.

(11.10) into equation (11.13). After some simplifications, one can easily obtain Z as a function of the generic temperature T from the equation:

$$Z = \frac{2.303 R_G}{E_d} T^2 \tag{11.14}$$

(Whereas, according to its definition, the Z value can be expressed in any kind of temperature scale, absolute temperatures must be used in equation (11.14).) Equation (11.14) is clearly a function of temperature and not a constant as often stated. Therefore, caution should be exerted when Z values are used to predict the behavior of the system at temperatures much higher than those used to determine the Z value experimentally. Conversely, if the Z value is used to predict the change in sterilization effectiveness resulting from small temperature changes within the typical temperature range used in commercial sterilization applications, then the term T^2 in equation (11.14) is approximately constant, and so is the Z value. In quantitative terms, if the Z value at $T = 121°\text{C}$ is known and is assumed to be equal to Z_{121} then the Z value at 140°C, 150°C, 160°C and 170°C will be 9.9%, 15.3%, 20.8% and 26.4%, respectively, greater than Z_{121}. These limited variations of the Z value (at least in the common sterilization-temperature range) partially justify the popular use of the Z value in the sterilization calculations.

Once the D value and the Z value (assuming it is a constant) have been determined at a given temperature, it is possible to estimate the effect of both time and the actual sterilization temperature on the reduction in microbial population. To do this the initial F value, F_{ref} (calculated at the reference temperature, T_{ref}, for which the data are available), is first determined using equation (11.12), after an initial microbial reduction level has been selected (e.g. 10^{-6}). This will give the time required for the sterilization process. Then the effect of temperature can be obtained from equation (11.13), which can be rewritten in terms of F instead of D, and used to determine the F value at a generic temperature, T. The resulting equation in the final reduction in microbial population is:

$$F = F_{\text{ref}} \cdot 10^{(T_{\text{ref}} - T)/Z} \tag{11.15}$$

Extensive tabulations of the F value (calculated through equation (11.15)) as a function of the Z value, T_{ref}, and T are available.[12]

Since the Z value is not really a constant but a function of temperature (as shown) equation (11.15) represents only an approximation. If the variability of the Z value with temperature is accounted for then a slightly different but more correct expression can be obtained. Such an expression is:[13]

$$F_T = F_{T\text{ref}} \cdot 10^{[(T_{\text{ref}} - T) || T)/(Z_{\text{ref}} || T)} \tag{11.16}$$

which is identical to equation (11.15) except for the term T_{ref}/T appearing in the exponent. Tabulations of the F value calculated with this equation are available elsewhere.[13]

11.2.1.4 Kinetics of thermal degradation of media. During the thermal sterilization of fermentation media, microorganisms are not the only ones to be affected by high temperature. The organic nutrient components of the media also undergo chemical degradation, which can result in the partial loss of some of the nutritive characteristics of the medium. This is especially true for the proteins contained in the medium since they are likely to be denatured by the temperature, just as the proteins that make up the microorganisms.

For most nutrients the rate of inactivation is given by a first-order expression similar to that used before to describe the microbial death process. The expression is

$$r_m = K_m C_m \tag{11.17}$$

where: r_m = rate of inactivation of nutrient in the media in g/l s; C_m = concentration of nutrient in the medium in g/l; and K_m = proportionality parameter (s^{-1}).

The parameter K_m is the specific kinetic rate of nutrient degradation, and can be expressed as:

$$K_m = K_{m0}\, e^{-E_m/R_G T} \tag{11.18}$$

where: K_{m0} = Arrhenius constant (frequency factor) of nutrient degradation process (s^{-1}); E_m = activation energy of nutrient degradation process in cal/g mol; and R_G = ideal gas constant in cal/g mol K.

Fortunately, most of the components of fermentation media have activation energies that are typically much lower than those associated with microbial thermal inactivation. This implies that as the sterilization temperature is increased the kinetics of the microbial death process increases much more rapidly than the kinetics of nutrient degradation. This results in a 'safe' sterilization process in which the nutritive content of the sterilized medium is nearly intact.

As an example, one may consider the case of thiamine degradation during heat sterilization, assuming that the medium must be sterilized against *B. stearothermophilus*.[3,14] The specific kinetic rates for thermal death and thiamine inactivation, respectively, are reported in Figure 11.4 as a function of the inverse of the sterilization temperature. The slope of each of these curves represents the activation energy of the process. If the sterilization took place at 100°C then the time required to reduce the concentration of the organisms by 12 orders of magnitude would be about 1300 min. Even discounting the impracticality of such a long sterilization process, more than 75% of the thiamine would be degraded during this period. However, if the sterilization is conducted at 140°C then the same degree of sterility would

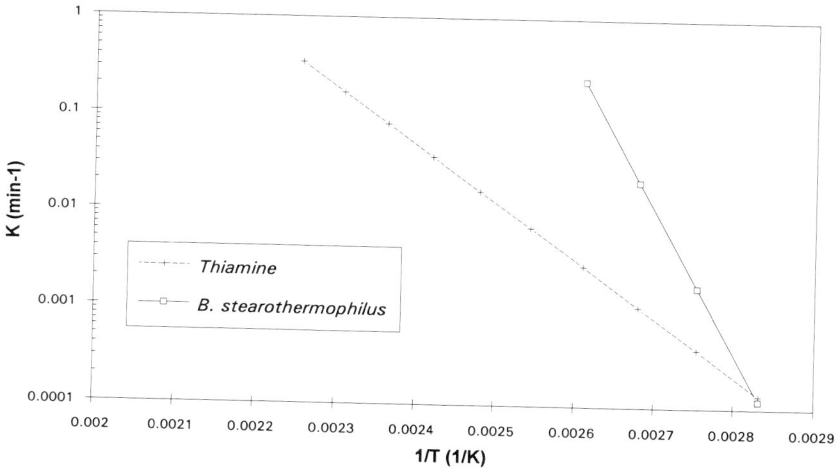

Figure 11.4 Specific kinetic rates for thermal death of *B. stearothermophilus* and thiamine inactivation.

be achieved in less than 12 s, and the loss of thiamine would be only about 0.7%. This is one of the reasons (in addition to the considerably shorter operating time and reduced overhead) why sterilization processes are conducted at high temperatures and for a short period of time rather than *vice versa*.

11.2.2 Sterilization requirements

The approach used above to describe the kinetics of the microbial thermal process is essentially deterministic. However, because of variability to heat resistance within a microbial population, this approach is only an approximation of an actual probabilistic process taking place during sterilization. This is especially true for the case in which the reduction in microbial population concentration is so significant that the expected number of surviving organisms is only a small fraction of unity. In any case, it can be shown[15] that equations, such as equation (11.14), which predict the fraction of remaining organisms after sterilization still hold if one now assumes that N/N_0 is the probability of survival of the average organism after sterilization. This implies that it is impossible, at least in principle, to produce with absolute certainty the destruction of all organisms through sterilization. What is achievable, however, is to design a process in which the probability of complete final sterility is as close to unity as one desires.

In practical terms, in order to determine the level of microorganism reduction to be achieved by a sterilization process, one needs to establish what probability of contamination of the final product is acceptable, what the initial concentration of organisms in the medium is (the so-called 'bioburden'), and how large the batch is. Then, assuming that even the presence of a single organism in the final batch can produce contamination, it is possible to determine the required value of the ratio N/N_0 to be achieved through sterilization. For example, a large-scale fermentation operator may decide that an acceptable level of risk of batch contamination is one contaminated batch in a thousand (if each fermentation lasted 10 days, this would imply that, on average, only one batch would be contaminated every 28 years). If each batch is 100 m^3 and the initial concentration of microorganisms in the medium is assumed to be 10^6/ml then the total number of initial organisms in the batch will be 10^{14}. A population reduction by such a factor would still leave, on average, one viable organism per batch. Since the risk accepted by the operator is only 10^{-3} the reduction to be effected by the sterilization process must be of 10^{-17}.

From this example one can derive a general rule for the sterile level to be achieved through sterilization:

$$\text{Required microorganism reduction ratio} = \frac{N}{N_0}$$

$$= \frac{\text{acceptable probability of contamination}}{\text{batch size} \times \text{initial microorganism concentration}} \quad (11.19)$$

11.2.3 Batch sterilization

11.2.3.1 Batch sterilization of liquid media. Batch sterilization of liquid media is most commonly encountered in large-scale fermentation. Concentrated, non-sterile medium is first prepared in batching tanks and then charged to the fermenters, which are typically provided with an external jacket and/or internal coils for heating and cooling. The fermenters are heated up to the sterilization temperature by passing steam in the jackets and coils while sparging steam directly through the air sparger. The live steam directly added to the fermenters produces a dilution effect that must be taken into account when preparing the medium. Once the sterilization temperature has been achieved, the steam supply is shut off and the fermenters are allowed to remain at that temperature for the time period required for sterilization to take place (holding period). Then, cooling water and/or brine are used to cool down the fermenters and their content prior to inoculation.

A typical temperature–time profile for batch sterilization is shown in Figure 11.5. This figure shows that the time required by the overall sterilization process can be quite significant because of the presence of long heating and cooling cycles. This is especially true for large fermenters because of the reduced ratio of heat transfer area to fermenter volume.

During batch sterilization the rate of microbial death can be calculated using the equations given in the previous section. More specifically, by

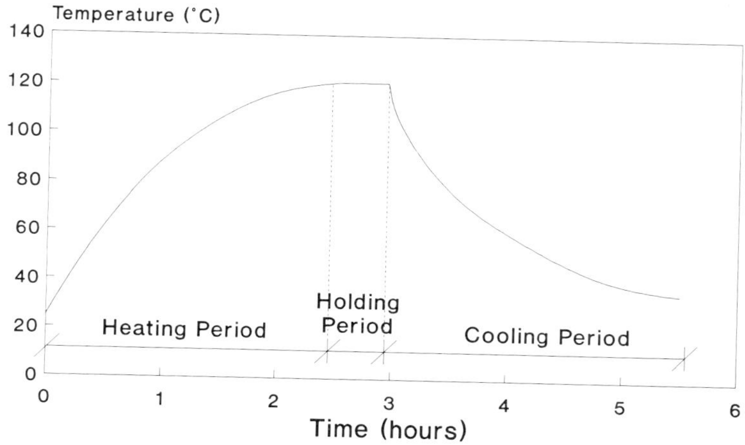

Figure 11.5 Typical time–temperature profile for batch sterilization in large-scale fermenters.

combining equations (11.2) and (11.3) one obtains:

$$\frac{dN}{dt} = -K_d N = -K_{d0} \exp(-E_d/R_G T) N \tag{11.20}$$

This equation can be applied to the heating, holding, and cooling steps, respectively, in order to calculate the microbial reduction produced by each one of them. The result is

$$\ln \frac{N_0}{N_1} = K_{d0} \int_0^{t_1} \exp(-E_d/R_G T_h(t)) dt \tag{11.21}$$

for the heating cycle;

$$\ln \frac{N_1}{N_2} = K_{d0} \int_{t_1}^{t_2} \exp(-E_d/R_G T_q(t)) dt \tag{11.22}$$

for the holding cycle; and

$$\ln \frac{N_2}{N} = K_{d0} \int_{t_2}^{t} \exp(-E_d/R_G T_c(t)) dt \tag{11.23}$$

for the cooling cycle.

In these equations N_1 and N_2 are the microbial concentrations at the end of the heating and holding cycles, respectively, and $T_h(t)$, $T_q(t)$, and $T_c(t)$ are the temperature profiles for the heating, holding and cooling cycle respectively. Notice that $T_q(t)$ is maintained constant in the vast majority of cases. Therefore, $T_q(t)$ in such cases is just equal to the sterilization temperature, and integration of equation (11.22) becomes very simple.

From these results one can see that all parts of the cycle and not only the 'sterilizing period' (i.e. the holding cycle) contribute to the overall sterilization. However, the integration of these equations requires the knowledge of the initial condition for each step (i.e. the residual microbial concentration at the beginning of each step) as well as the temperature change as a function of time during each step. This, in turn, depends on the way in which the heating or cooling cycles are conducted, in other words, on how quickly heat is supplied or removed.

In Table 11.4 the equations for the most common temperature profiles for the heating and cooling cycles are presented. These functions can be substituted in the previous equations to obtain N_0/N_1, N_1/N_2, and N_2/N. The integrated expression for each one of these functions is given elsewhere.[16] Then, the final ratio of microbial concentration reduction is given by:

$$\ln\frac{N}{N_0} = -\left[\ln\frac{N_0}{N_1} + \ln\frac{N_1}{N_2} + \ln\frac{N_2}{N}\right] \tag{11.24}$$

In order to achieve an adequate sterility level this value for N/N_0 must be lower than the value set through equation (11.19).

11.2.3.2 Batch sterilization of solids. There are several situations in which a solid or a stagnant liquid placed in a container must be heat-sterilized. In such a case, the main route by which heat is transported within the solid (or liquid) is conduction rather than convection since convection is a notoriously slow heat-transfer mechanism. This implies that, in convection, even if the temperature around the solid is quickly raised to the sterilization temperature it will still take some time for the heat to travel to the core of the solid and sterilize its content. If the solid is large and/or the sterilization time too short

Table 11.4 Temperature profiles in batch sterilization of media (after Bailey and Ollis[16])

Type of heat transfer	Equation for temperature profile of medium	Type of function
Electric heating	$T(t) = T_o + \frac{q}{V\rho_l C_{pl}} t$	Linear
Steam heating	$T(t) = T_w - (T_w - T_0)\exp\left(-\frac{UA}{V\rho_l C_{pl}} t\right)$	Exponential
Water cooling	$T(t) = T_{co} + (T_0 - T_{co})\exp\left[\frac{Q_c C_{pc}}{V\rho_l C_{pl}}\left(1 - \exp\left(-\frac{UA}{Q_c C_{pc}}\right)\right)\right]$	Exponential

T_0 = initial temperature of medium in °C; q = rate of heat transfer in kcal/s; V = volume of medium to be sterilized in m^3; ρ_l = density of medium in kg/m^3; C_{pl} = heat capacity of medium in kcal/kg °C; T_w = steam temperature in °C; U = heat transfer coefficient in kcal/m^2 s °C; A = heat transfer area in m^2; T_{co} = initial temperature of cooling water in °C; Q_c = flow rate of cooling water in m^3/s; and C_{pc} = heat capacity of cooling water in kcal/kg °C.

it may well be that the solid center never reaches the sterilization temperature, or the core is maintained at the sterilization temperature for too short a time period to ensure adequate sterility at the end of the process.

These problems can be analyzed using unsteady-state heat-transfer equations. For the case in which the solid is a slab or a parallelepiped (i.e. a prism whose bases are rectangles) the unsteady-state heat-transfer equation in three dimensions can be written as:[16]

$$\rho_s C_{ps} \frac{\partial T_s}{\partial t} = k_s \left(\frac{\partial^2 T_s}{\partial x^2} + \frac{\partial^2 T_s}{\partial y^2} + \frac{\partial^2 T_s}{\partial z^2} \right) \tag{11.25}$$

where: T_s = temperature inside the solid in °C; ρ_s = density of the solid in kg/m^3; C_{ps} = heat capacity of the solid in kcal/kg°C; and k_s = heat conductivity of the solid in kcal/m h °C.

For the simpler case in which heat transfer occurs only (or primarily) in one direction, as in a long (semi-infinite) slab of fixed thickness heated from both sides and having a uniform initial temperature, the previous equation becomes:

$$\rho_s C_{ps} \frac{\partial T_s}{\partial t} = k_s \frac{\partial^2 T_s}{\partial x^2} \tag{11.26}$$

with initial condition of $t = 0$ and $T = T_{s0}$; and boundary condition of $x = \pm b$ and $T = T_{s1}$; and where: b = 1/2 of the thickness of the slab in m; T_{s0} = initial temperature of the slab in °C; T_{s1} = temperature at the surface of the slab in °C; and x = axial distance in m.

The solution of this equation is:[16,17]

$$\frac{T_{s1} - T_s}{T_{s1} - T_{s0}} = 2 \sum_{n=0}^{\infty} \frac{(-1)^n}{(n + \frac{1}{2})\pi} \exp\left(-(n + \tfrac{1}{2})^2 \pi^2 \frac{\alpha t}{b^2} \right) \cos\left[\left(n + \frac{1}{2} \right) \frac{\pi x}{b} \right] \tag{11.27}$$

where:

$$\alpha = k_s / \rho_s C_{ps} \tag{11.28}$$

Although this equation may look complicated because of the presence of the infinite summation, the solution typically converges very rapidly using only the first few terms (sometimes even only one) of the summation. The temperature at the center of the slab can be obtained by imposing that $x = 0$. The solution for this equation has been tabulated in graphs for convenient use.[18] In addition, if a finite parallelepiped is considered the solution can still be obtained as the product of three solutions such as that represented by equation (11.27) for the three dimensions.[17,19]

If the solid is a sphere with homogeneous initial temperature and internal composition, and is heated by keeping its external surface at constant temperature the following equation applies:

$$\rho_s C_{ps} \frac{\partial T_r}{\partial t} = k_s \left(\frac{\partial^2 T_r}{\partial r^2} + \frac{2}{r} \frac{\partial T_r}{\partial r} \right) \tag{11.29}$$

where: r = radial distance from the center of the sphere in m; and T_r = temperature in the solid at a position r inside the particle in °C; with boundary and initial conditions:

$$\left.\frac{\partial T_r}{\partial r}\right|_{r=0} = 0 \tag{11.30}$$

$$T_r\ (r = R,t) = T_{r1} \tag{11.31}$$

$$T_r\ (r,t = 0) = T_{r0} \tag{11.32}$$

where: R = radius of the sphere in m; T_{r0} = initial temperature of the particle in °C; and T_{r1} = temperature at the surface of the partical in °C.

The solution of this equation takes the general form:[17]

$$T_r(r,t) = T_{r1} + \frac{2R(T_{r1} - T_{r0})}{\pi r} \sum_{n=0}^{\infty} \frac{(-1)^n}{n} e^{-(n\pi/R)^2 \alpha t} \sin\frac{n\pi r}{R} \tag{11.33}$$

In particular, at the center of the sphere this equation reduces to:

$$T_r\ (r = 0,t) = T_{r1} + 2(T_{r1} - T_{r0}) \sum_{n=1}^{\infty} (-1)^n e^{-(n\pi/R)^2 \alpha t} \tag{11.34}$$

Once the temperature profile is known as a function of time and position within the solid it is possible to calculate the fraction of residual organisms by integrating over the entire volume of the solid and for the time period of sterilization. For the case of a sphere in which the initial concentration of organisms (bioburden) is constant and equal to N_0 this integral takes the form:[20]

$$\frac{N(t)}{N_0} = \frac{3}{4\pi R^3} \int_0^R \exp\left[\int_0^t k_{d0} \exp\left(\frac{E_d}{R_g T_r(r,t)}\right) dt\right] 4\pi r^2 dr \tag{11.35}$$

Owing to the presence of the temperature function $T(r,t)$ given by equation (11.33) this integral can only be integrated numerically.

11.2.3.3 Autoclaving. Autoclaves are pressure vessels commonly used to sterilize equipment, materials, or supplies. They are typically used in a variety of sterilization tasks but are employed primarily to sterilize solids or stagnant liquids in small- or medium-scale operations. Sterilization of laboratory equipment, medical instruments, or final solid products are typical applications.

Most autoclaves rely on steam as the sterilizing medium. However, small-scale sterilizers also exist in which a sterilizing gas, such as ethylene oxide, is used. A steam autoclave consists of a pressurizable chamber provided with a removable lid or door. The material to be sterilized, typically stacked on trays, is charged to the chamber. Then, the door is closed and the steam is admitted to the chamber. The autoclave must be designed to allow the air initially present in the chamber to be removed prior to sterilization. Failure to do so will result in the presence of air pockets in which moist heat will not penetrate, thus resulting in inadequate sterilization.

Two methods are typically used to remove air. The first relies on gravity displacement of air by steam as the steam enters the chamber from the top. Hot steam does not mix readily with the air in the chamber. Hence, air can be removed through an opening at the bottom of the autoclave. Alternatively, the air can be removed using a vacuum pump prior to sterilization. Indeed, most autoclaves use repeated vacuum cycles to eliminate the air initially contained within the autoclave.

After the air-purging cycle is completed the actual sterilization step begins. This consists of closing all venting valves and admitting steam to bring the autoclave to the sterilization temperature. Steam condensate is removed at the bottom of the autoclave through a steam trap. Enough time must be allowed to enable all parts of the autoclave to reach the sterilization temperature. Then, this temperature is maintained for the required time period (typically 20 min at a sterilization temperature of 121°C).

Microprocessors are commonly used to monitor and control the entire autoclaving cycle. The control strategy relies on the use of sensors placed in different parts of the autoclave, most typically the drain. If solids or stagnant liquids are present, sensors should be placed inside them in order to ensure that the sterilization temperature is reached even in the solid core or in the bulk of the liquid before the actual sterilization cycle begins.

Validation of the sterilization effectiveness involves the achievement of the required sterilization temperature and its maintenance for the required sterilization time for a given load of material to be sterilized. If a different material is to be sterilized in the same autoclave, validation of the sterilization procedure must be repeated unless it is possible to prove that the new material is clearly more easily sterilized than the material for which the original validation was conducted (e.g. if sterilization in an autoclave has been validated for a certain number of vials of given size, the validation process must not be repeated if a new batch of smaller vials is to be sterilized).

Chemical or biological indicators can be used to test periodically the effectiveness of the sterilization process. Different chemical indicators exist. The most common types rely on a change in color to indicate their exposure to a predetermined temperature for a predetermined time. Biological indicators use strains of pure cultures of known resistance to temperature (typically spores that are the most heat-resistant) to determine the sterilization effectiveness. Strips containing these organisms can be placed in the autoclave, retrieved after the sterilization is complete, and then cultured to determine the rate of survival and hence the effectiveness of the sterilization process.

11.2.4 Continuous sterilization of media

Continuous sterilization is a very common and extremely valid alternative to batch sterilization of fermentation media. We have seen previously how the kinetics of the sterilization process increases dramatically with temperature.

As a result, a sterilization process, which would normally require 20–30 min to complete at 121°C requires only a few seconds at temperature as high as 145°C (processes using high sterilization temperatures are typically labeled very-high temperature (VHT) processes. Such short sterilization times can only be achieved in continuous sterilizers in which the temperature of the medium is raised rapidly in a heat exchanger, maintained at such a high sterilization temperature as the medium flows, unheated, in a holding tube, and then rapidly lowered in a cooler.

Continuous sterilization has several advantages over batch sterilization. Since short sterilization times are required, continuous sterilization can be achieved when the medium is charged to pre-sterilized fermenters. This drastically reduces the downtime of the fermenters, especially in large-scale operations. Energy savings can also be significant since the thermal energy content of the just-sterilized, high-temperature medium can be partially used to preheat the incoming unsterile medium. In addition, the surge capacity for steam is reduced. Furthermore, the process can be easily monitored, controlled, and provided with a number of safeguards to avoid charging unsterile medium to the fermenters.

In the previous sections it has been seen that exposing the medium to high temperatures results not only in sterilization but also in the partial degradation of nutrients. The most significant improvement of continuous sterilization over batch sterilization is that associated with the reduced level of nutrient degradation of the medium. Owing to the lower activation energies of the nutrient degradation reactions versus those of the microbial inactivation reaction, an increase in the operating temperature reduces the level of nutrient degradation for the same degree of sterilization required. This is precisely what can be achieved with high-temperature, short-time processes such as continuous sterilization. This improvement is not only predicted theoretically but can also be verified experimentally. In fact, data on the actual performance of fermentation processes using either batch of continuous sterilization have been reported in the literature.[3] For example, the production of riboflavin has been shown to improve by more than a factor of 7.4 in one case and 70 in another when the process was switched from batch sterilization (at 121°C for 25 or 45 min) to continuous sterilization (at 135°C for 5 min). Similar improvements (from 50% to 1200%) were observed in the fermentation of vitamin B_{12}. Such improvements are the direct result of preservation of the nutrient content of the medium produced by continuous sterilization. For all these reasons, continuous sterilization is now a commonly employed sterilization method.

11.2.4.1 Continuous sterilization processes, equipment and operation. Continuous sterilization can be accomplished using two basic types of continuous sterilizers. The first type, shown in Figure 11.6, relies on the continuous direct addition of steam to the feed stream of concentrated unsterile medium in

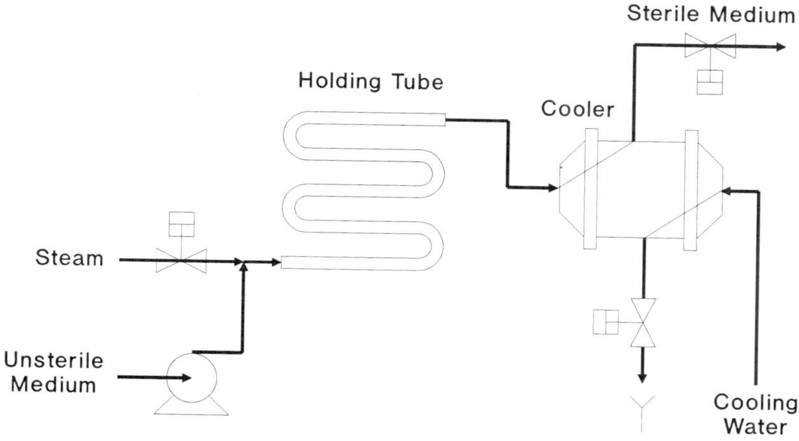

Figure 11.6 Direct steam continuous-sterilization system.

order to rapidly bring the temperature to that required for sterilization. The dilution effect produced by the steam addition must be taken into account when designing the system. The medium is then circulated in a holding tube having the appropriate residence time for sterilization to occur. The holding tube is just a pipe whose length is such to ensure that the medium flowing through it will spend a predetermined period of time to produce the required level of sterility. The sterilized medium is then passed through a cooling system before being charged to the fermenter.[21–23]

This type of continuous sterilizer has the advantage of simplicity since no contact heat-exchanger is required to heat up the medium. However, the energy requirements of such a sterilizer are significant since no preheating of the medium takes place. This implies that all the energy that has been supplied by the steam must then be removed in the cooling section and cannot be easily recovered or used.

For these reasons, *indirect* heat-transfer continuous-sterilizers are typically used in large-scale industrial processes. A diagram of such a continuous sterilization process is shown in Figure 11.7. Unsterile medium from the batching area is pumped from the holding tank through a heat-recovery exchanger where it is pre-heated at the expense of the hot medium that has just undergone sterilization. The pre-heated medium is then brought to its sterilization temperature in a steam heater, and then passed through the holding tube. As before, the holding tube is just a pipe whose length is such to ensure that the medium flowing through it will spend a predetermined period of time to produce the required level of sterilization. After the holding tube, a fraction of the energy content of the sterilized medium is recovered to pre-heat the incoming medium, as already described. The sterile medium

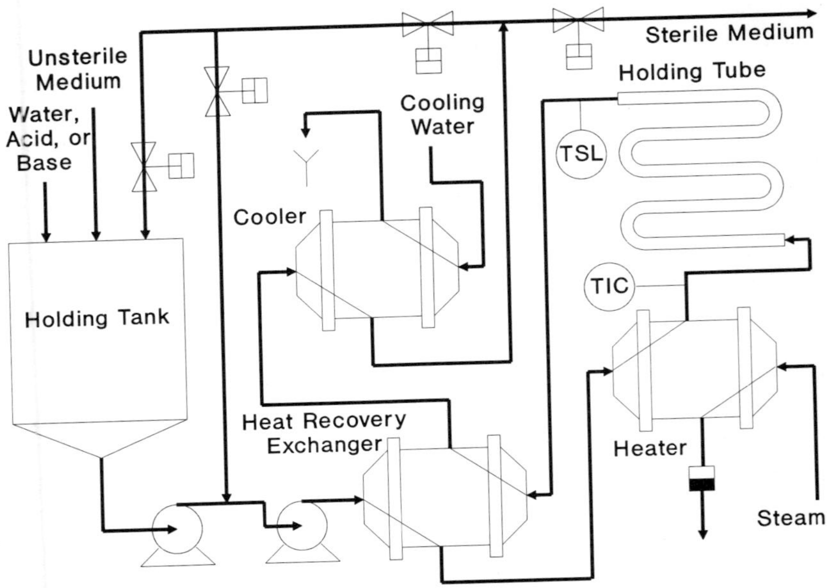

Figure 11.7 Typical continuous sterilization system (after Armenante and Li[24]).

is then cooled down in a cooler exchanger and charged to the pre-sterilized fermenters.

Different types of heat exchangers are found in continuous sterilizers. The traditional shell- and tube-heat exchangers common in the chemical industry are not typically encountered in continuous strerilization (except perhaps in antifoam sterilization) because they are difficult to clean and sterilize, and because of their tendency to foul. Sanitary construction, single-tube pass or double-tube pass (hairpin), shell-and-tube heat exchangers are available, however. They can only be used if the process fluid (e.g. a parenteral solution to be sterilized) flows on the tube side of the exchanger and the service fluid on the shell side.

Concentric double-pipe heat exchangers can be used advantageously, at least as heaters and coolers, since they are cleaned easily, and are relatively easy to operate. These exchangers tend to be expensive and occupy a significant amount of space. However, because of their sanitary construction, they are often encountered in continuous sterilizers in the pharmaceutical industry where they constitute the industry standard.

Plate-and-frame exchangers typically have high film-transfer coefficients, and moderate pressure drops (at least if no solids are present). However, plate-and-frame exchangers have the greatest length of gasket material per exchange area. This results in a high potential for leaks, and can make gasket replacement a time-consuming and labor-intensive operation. In addition, plate-and-frame exchangers force the fluids to change flow direction at each pass through a plate. This results in the accumulation of solid particles between the plates, thus increasing the chances of fouling and the possibility of producing an unsterile product. The tendency to build up solid deposits is also increased by the fact that the velocity profile is non-uniform as the fluid moves through the exchanger. For this reason, they are not encountered in the terminal sterilization of pharmaceutical liquids, such as parenterals, for which sanitary construction of the sterilizer is required. Nevertheless, plate-and-frame exchangers are often encountered in large-scale sterilizers for fermentation media because of their compact size, relatively low cost, and flexibility.

Spiral heat exchangers are also quite often encountered in large-scale fermentation medium sterilization. One advantage of this type of heat exchanger is that the fluids flowing between the spirals of the exchanger experience a constant cross-sectional area since these exchangers are built by rolling over each other two thin metal sheets maintained at a constant distance from each other by metal studs. The fluids flow in the constant cross-sectional area gaps between the two metal sheets, thus reducing the risk of solid sedimentation and fouling. This also makes the exchangers easy to clean via periodical chemical cleaning. Spiral heat exchangers have other advantages similar to those listed for the plate-and-frame exchangers but have a much smaller gasket area, with obvious benefits. As a result, spiral heat exchangers are often encountered in large-scale continuous sterilizers. Additional details on these types of heat exchanger can be found in the literature.[26 30]

The operation of a continuous sterilizer, such as that shown in Figure 11.7, begins with the start-up operations. Initially, process water is charged to the holding tank and circulated through the entire system by two pumps (the pressurization pump and the circulation pump). During this phase all valves, including the bypass valves, are opened to ensure complete flushing of all parts of the sterilizer. The system is initially operated in the recirculation mode in which the process water leaving the cooler is recycled back to the holding tank, and the isolation valve on the line to the fermenters is closed. Meanwhile, steam is admitted into the heat exchanger, and the different parts of the system are allowed to reach their operating temperatures. Those pipe sections that may contain water trapped between two valves should be provided with relief valves.

Once the holding tube has achieved the sterilization temperature, the entire sterilizer is maintained for a set period of time in the recirculation mode to ensure proper sterilization of the hardware. Then, unsterile medium is continuously fed to the holding tank, while the appropriate valves are switched to put the sterilizer in the forward-operation mode in which sterile material is fed to the fermenters. The system is operated at steady-state until the fermenters are charged.

In the case of a system upset, the temperature in the holding tube can drop below the set temperature value meaning that the sterility level of the medium is considered unacceptable. The control system is designed to switch the sterilizer into recirculation mode when this happens. Depending on the control strategy and the length of the 'unsterile' operation period, the medium may be recirculated until the sterile conditions are met again. Then the system may be switched to the forward-operation mode. If the system upset is too long, the control system can be designed to discharge the recirculating medium to the drain in order to prevent excessive fouling of the sterilizer. This occurs when the nutrients contained in the medium are maintained at too high a temperature for too long a period of time. When this condition occurs, the system will have to be flushed with water and re-sterilized as during start-up operations. The optimal control strategy will depend on the composition of the medium and its tendency to form fouling deposits.

Once the fermenter-charging operations are completed, the sterilizer proceeds to shut itself down by switching the incoming medium with water until the system is flushed. Periodic cleaning, either at the end of a full charging cycle or after a specific level of fouling has been achieved (e.g. as detected by a drop in heat-recovery efficiency) is common. Depending on the composition of the medium an acidic or basic cleaning solution (or both, sequentially) can be used. The cleaning cycle can also include partially heating the cleaning and rinsing solutions.

11.2.4.2 Continuous sterilization of liquid media. Sterilization of homogeneous liquid media is one of the most common applications of continuous sterilization. One of the key assumptions made in the development of the design equations for such a system is that no solid particles are present in the medium or that they have been removed prior to feeding the medium into the sterilizer.

In general, each section of the sterilizer will partially contribute to the achievement of sterility. A simple but very conservative design of a continuous sterilizer can be obtained assuming that the desired level of sterility must be achieved in the holding tube alone, thus neglecting any contribution of the other sections.[31] In addition, one can further assume that the medium moves in piston flow in the holding tube (this assumption will be removed later to account for axial dispersion). As a result, each slug of medium flowing through the holding section will not be affected by the presence of other slugs of material preceding or following it. Since the medium enters the holding section at the desired sterilization temperature, and leaves it at the same temperature after a time, t_{res}, which is equal to the residence time in the holding tube, the level of sterility can be calculated using an equation similar to equation (11.4) as follows:

$$\ln \frac{N}{N_0} = -K_0 \exp(-E_d/RT_{ster})\, t_{res} \tag{11.36}$$

where:

$$t_{res} = \frac{L}{v} = \frac{LS'}{Q} \qquad (11.37)$$

with: t_{res} = residence time of medium in holding tube in s; L = length of holding tube in m; v = velocity of medium in holding tube in m/s; S' = cross-sectional area of holding tube in m^2; and Q = volumetric flow rate of medium in m^3/s.

In the more general case in which the contributions of all the sections of the sterilizing system are considered, the equations to calculate the rate of microbial reduction produced by each section can be obtained using a method similar to that used in the batch-sterilization process case. The only difference is that the temperature–time profiles will now be obtained by making the appropriate energy balance in each heat exchanger so as to obtain the temperature of each infinitely thin slug of medium as it travels through it. The resulting temperature-time profiles are reported in Table 11.5[5a,b,20]. These temperature profiles can then be inserted in the following equation:

$$\ln\left(\frac{N_0}{N}\right)_{\text{exchanger}} = K_{d0}\int_0^t \exp(-E_d/R_G T(t))\,dt \qquad (11.38)$$

in which $T(t)$ is one of the expressions given in Table 11.5 to obtain the rate of microbial reduction produced by each exchanger. The final microbial reduction can be obtained by an equation similar to equation (11.23) in which all contributions are accounted for.

Table 11.5 Temperature–time profiles in continuous sterilization systems

Type of heat transfer	Equation for temperature profile	Type of function
Electric heating	$T(t) = T_o + \dfrac{q}{S\rho_l C_{pl} L}t$	Linear
Steam heating	$T(t) = T_w - (T_w - T_0)\exp\left(-\dfrac{UA}{S\rho_l C_{pl} L}t\right)$	Exponential
Countercurrent heat exchange	$T(t) = T_{c1} + \dfrac{(T_0 - T_{c1})}{a-1}\left\{1 - \exp\left[(a-1)\dfrac{UA}{S\rho_l C_{pl} L}t\right]\right\}$	Exponential
Countercurrent heat exchange with	$T(t) = T_o + \dfrac{UA}{S\rho_l C_{pl} L}(T_{c1} - T_o)t$	Linear

$a = \dfrac{Q\rho_l C_{pl}}{Q_c \rho_c C_{pc}} = 1$; q = rate of heat transfer in kcal/s; ρ_l = density of medium in kg/m^3; C_{pl} = heat capacity of medium in kcal/kg °C; L = length of exchanger in m; T_w = steam temperature in °C; U = heat transfer coefficient in kcal/m^2 s °C; A = heat transfer area in m^2; T_{c1} = exit temperature of service fluid in °C; Q and Q_c = flow rate of medium and service fluid, respectively, in m^3/s; C_p and C_{pc} = heat capacity of medium and service fluid, respectively, in kcal/kg °C.

11.2.4.3 Effect of axial dispersion on continuous sterilization of liquid media. In the previous section the assumption was made that the medium flows through all parts of the continuous sterilizer in piston flow. In practice, this is quite difficult to accomplish. Firstly, the medium is always subjected to several flow expansions, contractions and direction changes (e.g. when it flows through a plate-and-frame heat-exchanger), which introduce some degree of back-mixing. Secondly, even assuming that the flow occurs in a straight circular pipe, the mean velocity of the fluid is different from the velocities of the single fluid elements.[32,33]

The type of flow in a circular pipe depends on the Reynolds number, *Re*, defined as $vD\rho/\mu$, where v = average velocity; D = pipe diameter; ρ = fluid density; and μ = fluid viscosity.[18] For $Re < 2100$ the flow in the pipe will be laminar, in other words the fluid elements will all move in the same direction of the flow. For $Re > 3000$ the flow will be turbulent and will be characterized by the presence of turbulent eddies superimposed on the average longitudinal flow. If the flow is laminar, the distribution of velocities will be parabolic with respect to the radial position. This implies that the fluid elements near the pipe wall will travel much more slowly than, say, the fluid element in the center of the pipe (having a velocity twice the average velocity). If the flow is turbulent, radial mixing will also occur, and the maximum velocity will typically be approximately 25% greater than the average velocity. Therefore, piston flow constitutes only the limit case of a fluid flowing under very turbulent flow conditions.

The existence of velocity distributions implies that if the average velocity is used in the design of the sterilizer then those fluid elements traveling at higher velocities will be associated with a lower degree of sterility. A possible remedy is to determine the velocity of the fastest moving element and use that in the design calculation instead of the average velocity. This is a conservative approach since all fluid elements except the fastest one will be over-sterilized.

A more sound approach is to model the phenomenon by assuming that an axial diffusion effect (produced by the flow turbulence and back-mixing) is present. This effect can be superimposed on the axial translation effect produced by the average fluid velocity. Then, a steady-state microbial balance can be written for an infinitely thin element of medium maintained at a constant sterilization temperature.[34] In the derivation of the following equations it is assumed that the temperature is constant. Therefore, this approach can only be used for the design of the holding tube. Then, the microorganism balance takes the form:[4,15]

$$v\frac{dN}{dx} - D_x\frac{d^2N}{dx^2} = -K_dN \qquad (11.39)$$

where: D_x = axial dispersion coefficient of the microorganisms in m^2/s; and

x = axial coordinate in m. In this equation, the first term represents the net number of organisms entering the element because of the average fluid flow, the second term represents the contribution to the mass balance of axial dispersion, and the last term the microbial reduction due to the thermal-death process.

The boundary conditions for this equation can be established at the beginning and at the end of the holding section. They are:

$$vN_0 = \left(vN - D_x \frac{dN}{dx}\right)\Bigg|_{x=0} \tag{11.40}$$

and:

$$\frac{dN}{dx}\Bigg|_{x=L'} = 0 \tag{11.41}$$

where: L' = length of the holding tube in m. The integration of this equation gives:

$$\frac{N(x=L')}{N_0} = \frac{4\zeta e^{Pe/2}}{(1+\zeta)^2 \exp\left[\frac{Pe\zeta}{2}\right] - (1-\zeta)^2 \exp\left[-\frac{Pe\zeta}{2}\right]} \tag{11.42}$$

where:

$$\zeta = \sqrt{1 + \frac{4Da}{Pe}} \tag{11.43}$$

This solution contains two important non-dimensional numbers, the Peclet number, Pe, and the Damköhler number, Da, defined respectively as:

$$Pe = \frac{vL'}{D_x} \tag{11.44}$$

and:

$$Da = \frac{K_d L'}{v} \tag{11.45}$$

The solutions of equation (11.42) are graphically displayed elsewhere.[15,16] For large values of Pe (i.e. for small deviation from piston flow) equation (11.42) can be rewritten as:[16]

$$\frac{N\ (x=L')}{N_0} \cong \exp\left[-Da + \frac{Da^2}{Pe}\right] \tag{11.46}$$

Once again, the equations derived in this section are valid only for isothermal systems. Therefore, unlike the equation obtained in the previous section, they can only be used for the design of the holding section and not for the design of the heat-exchangers.

11.2.4.4 Continuous sterilization of liquid media containing suspended solids. The presence of solid particles in industrial liquid media used for large-scale fermentation is quite common. However, suspended solids can present some special problems during medium sterilization since heat transfer through solid particles proceeds at a slower rate than that in the liquid in which the particles are immersed. As a result, the temperature in the core of the particles may be quite lower than that of the liquid during a transient operation such as continuous sterilization. This implies that those microorganisms that are lodged deep inside the particles may undergo a less-effective sterilization process than those in the liquid bulk.[35,36]

This effect is typically negligible in most batch processes since the long sterilization time (typically 20 min) and long heating and cooling times usually provide ample time for small particles to thermally equilibrate with the surrounding liquid, thus ensuring a similar level of microbial reduction in both the particles and the liquid. However, the same is not true for continuous processes because of the much shorter sterilization time. In such cases, large temperature differences between the solid core and the surrounding liquid can be predicted for short sterilization times, with significant consequences for sterilization effectiveness. For this reason, solid particles should be excluded, whenever possible, from the medium prior to continuous sterilization. This is the case for most pharmaceutical formulations, which have very stringent requirements on the maximum size of the solids (typically of the order of μm). However, most commercial fermentation media have a high concentration of suspended solids.[25] In such cases, the impact of the presence of solids on sterilization effectiveness should be accounted for in the design of the sterilization process.

A design approach to continuous sterilization in the presence of solids has more recently been described in the literature.[20,24] Several assumptions were made in the development of the design model. The most important are that the fluid (and the particles suspended in it) moves in piston flow, the particles are homogeneously distributed in the fluid, are neutrally buoyant, and spherical in shape. In addition, the concentration of the particles is assumed to be sufficiently low to play a negligible role in the energy balance of the fluid. The latter assumption implies that the energy balance for the fluid moving in a given heat exchanger is the same as that obtained previously for the case of liquid media and given in Table 11.5. The equations in this table then provide the temperature change of the liquid as a function of time. The temperature inside a solid particle moving along with the liquid can be obtained from an energy balance in an infinitely thin shell inside the particle itself according to the following equation:[17]

$$\frac{\partial T_r}{\partial t} = \frac{k_s}{\rho_s C_{ps}} \left(\frac{\partial^2 T_r}{\partial r^2} + \frac{2}{r} \frac{\partial T_r}{\partial r} \right) \tag{11.47}$$

with boundary conditions

$$-k_s \frac{\partial T_r}{\partial r}\bigg|_{r=R} = h_r[T_r(r,t)|_{r=R} - T(t)] \tag{11.48}$$

and:

$$-k_s \frac{\partial T_r}{\partial r}\bigg|_{r=0} = 0 \tag{11.49}$$

and initial condition:

$$T_r\ (r, t = 0) = T_0 \tag{11.50}$$

since the initial temperature of the particles has been assumed to be constant and equal to T_0. In the above equation: T_r = temperature in the solid at a position r inside the particle in °C; k_s = thermal conductivity of the particle, kcal/m s °C; ρ_s = density of the particle in kg/m^3; C_{ps} = heat capacity of the particle in kcal/kg °C; r = radial position within the spherical particle in m; R = radius of the spherical particle in m; and h_s = particle-liquid heat-transfer coefficient in kcal/m^2 s °C.

Equation (11.47) can be integrated once the temperature of the liquid is known as a function of time. For example, if the medium first enters the heat-recovery exchanger, as shown in Figure 11.7, then the liquid temperature on the cold side of the exchanger can be obtained from Table 11.5, and integration of equation (11.47) gives:

$$T_r(r,t) = T_0 + \frac{AU}{S\rho_s C_{ps} L}(T_1' - T_0) + \sum_{n=1}^{\infty} \left\{ \sin\left(\beta_n \frac{r}{R}\right) \frac{2AU}{S\rho_s C_{ps} L}(T_1' - T_0) \right.$$

$$\left. \times \frac{R^2\{\cos(\beta_n) - [\sin(\beta_n)/\beta_n]\}}{\dfrac{k_s}{\rho_s C_{pt}} \dfrac{r}{R}[\beta_n^3 - \beta_n^2 \sin(\beta_n)\cos(\beta_n)]} \left[1 - \exp\left(-\frac{k_s}{\rho_s C_{ps}} \frac{\beta_n^2}{R^2} t\right)\right] \right\} \tag{11.51}$$

where: A = surface area of heat exchanger in m^2; U = overall heat transfer coefficient in heat-recovery exchanger in kcal/m^2 s °C; S = cross-sectional area available for fluid flow in heat exchanger in m^2; T_1' = hot side bulk fluid temperature at the exit of heat-recovery exchanger in °C; and L= length of the heat-recovery exchanger in m; and in which the eigenvalues β_n are the solution of the equation:

$$\beta_n \cot \beta_n + \frac{h_s R}{k_s} - 1 = 0 \tag{11.52}$$

Equation (11.51) can be used to calculate the temperature in the solid as a function of time and position inside the particle, as shown in Figure 11.8.

This figure shows a plot of the temperature in the core of a 6mm radius particle made of flour (a typical component of media containing corn meal) as it moves through the entire heat-exchanger train constituting the continuous sterilizer. The temperature of the surrounding liquid is also reported. One can see clearly that a significant temperature difference exists between the liquid temperature and the core temperature. Equations such as equation (11.51) can be derived for particles flowing in any type of heat exchanger by using the equations contained in Table 11.5 to get the temperature of the fluid as it moves through the heat exchanger.

To calculate the level of microbial reduction inside the particles one can then use equation (11.35). In this equation, the temperature distribution inside the particle must be known, and can be obtained from equation (11.51) or similar equations. Owing to the complexity of the resulting equation integration can only be performed numerically. Figure 11.9 shows that the level of sterility in the particles (given as $\log_{10}(N_0/N)$) is much lower than that of the corresponding level in the liquid.

The effect of several operating parameters on the sterilization of media containing suspended solids is shown in Figures 11.10 and 11.11. The first of these two figures shows that the sterilization temperature has a positive impact on the overall degree of sterility (in this figure the steam temperature was maintained constant at 150°C). The reason for this is that as the sterilization temperature increases the time the medium has to spend in the heat exchanger increases too (since the steam temperature is constant). This produces a longer residence time in the heater and a higher sterility level. In addition, the longer residence time also gives more time to the particles to

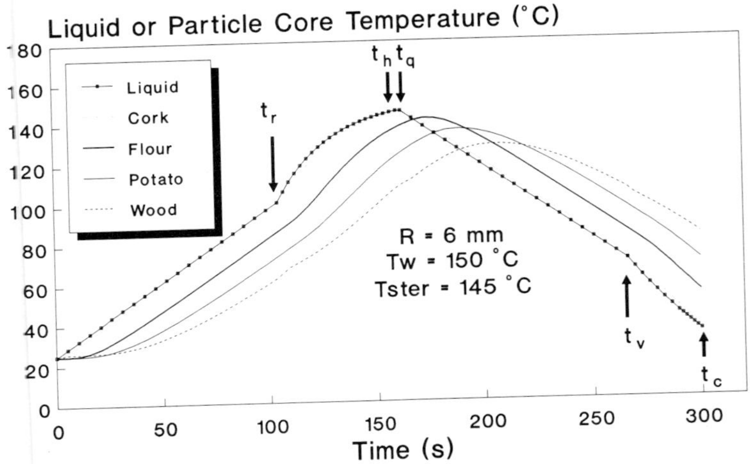

Figure 11.8 Temperature–time profiles in a continuous sterilizer for media containing suspended particles (after Armenante and Li[24]).

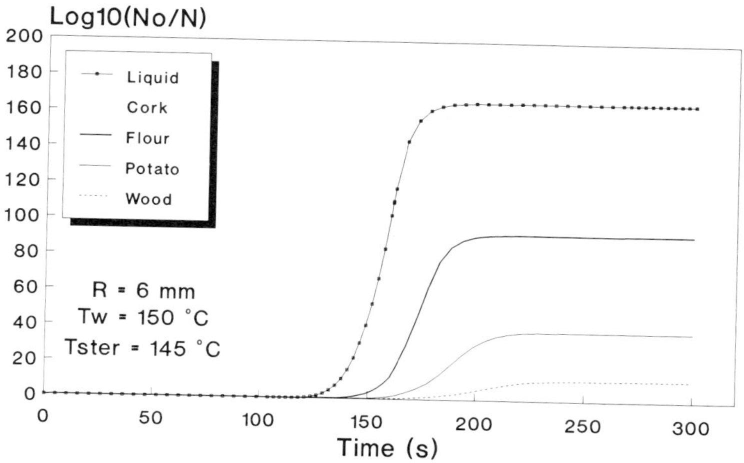

Figure 11.9 Sterilization level in a continuous sterilizer for media containing suspended particles (after Armenante and Li[24]).

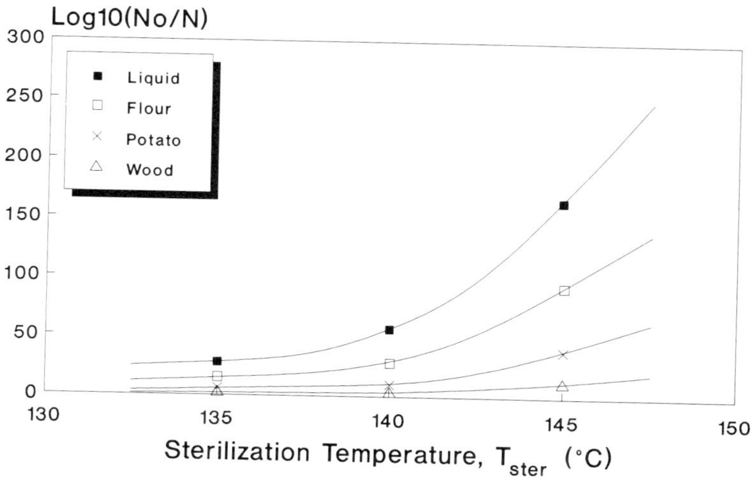

Figure 11.10 Sterilization effectiveness in continuous sterilizers for media containing suspended particles (after Armenante and Li[24]).

thermally adjust to the temperature of the surrounding liquid, thus producing an increased level of sterility in the particles. In any case, a significant difference exists between the degree of sterility in the liquid and that in the particles. This gap widens as the sterilization temperature increases.

Figure 11.11. shows the effect of particle size on the degree of sterility as

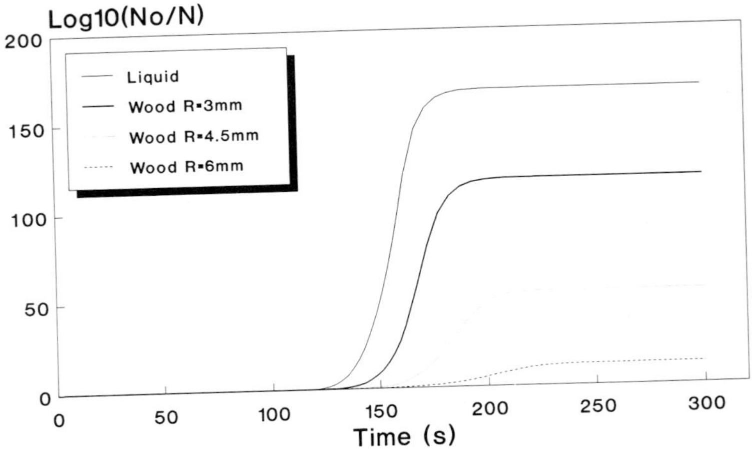

Figure 11.11 Effect of particle size on sterility in continuous sterilizers for media containing suspended particles (after Armenante and Li[24]).

the medium flows through the heat exchanger and the holding tube. Once again, one can see that larger particles result in a lower degree of sterility.

Once the sterility levels of the particles and the liquid have been calculated the overall sterility level for the medium as a whole can be obtained from the following equation:

$$\left(\frac{N}{N_0}\right)_{\text{medium}} = \xi_s\left(\frac{N}{N_0}\right)_{\text{solids}} + (1 - \xi_s)\left(\frac{N}{N_0}\right)_{\text{liquid}} \tag{11.53}$$

where: ξ_s = volumetric fraction of solids in the medium.

Additional information on this subject can be found elsewhere.[20,24]

11.2.5 Dry-heat sterilization of non-aqueous media and pharmaceutical products

In the biotechnology industry, dry-heat sterilization is primarily carried out when the medium is non-aqueous and water-immiscible, and therefore cannot be exposed to steam without forming another phase. This situation is encountered when materials such as antifoam oil, nutrient oils, or fats have to be sterilized independently from the main fermentation medium. In addition, dry heat is used in the pharmaceutical industry to sterilize glassware (e.g. vials and ampuls used as sealed containers for drugs), non-volatile thermostable powders (e.g. zinc oxide, talc, sulfamerazine), or oily solvents and drugs (e.g. corn oil, cottonseed oil, sesame oil, desoxicorticosterone).

Dry-heat sterilization is considered separately from steam sterilization or autoclaving because the rate of microbial death under dry conditions is much slower than when the medium is aqueous or when sterilization is conducted in the presence of steam. If temperature were the only factor affecting thermal sterilization this should not be the case. However, microorganisms typically live in aqueous environments. Therefore, they may not be wetted so easily when exposed to non-aqueous media. This can have an effect on the heat transfer and its penetration within the cell wall. In addition, it appears that the mechanism of temperature inactivation is different when the organism is not immersed in an aqueous medium because of the lack of water penetrating the cell. Dry heat produces microbial death as a result of dehydration followed by protein oxidation. Death by moist heat is caused by denaturation and coagulation of essential cell proteins.[37]

In any case, sterilization parameters such as *D* values and *Z* values are significantly higher for dry heat sterilization rather than for steam sterilization, as shown in Table 11.6. All other factors being equal, this results in higher sterilization temperatures and/or longer sterilization times, since the procedure to determine the degree of sterilization in such a process is the same as that outlined for the general thermal sterilization case.

A mitigating factor in the design of dry-heat sterilization processes for oily media is the fact that the initial bioburden of oil is typically much lower than that of nutrient aqueous media, and is about 100–200 organisms/ml (as opposed to 10^4–10^9 organisms in aqueous media).[10] In addition, the volumes to be sterilized are typically much smaller.

In the sterile formulation of drugs, vials and ampuls are typically washed with de-ionized water and a final rinse of pyrogen-free water, and then dried, sterilized, and de-pyrogenated in hot-air ovens using dry heat. If the process is carried out in batch the vials are typically kept at 170°C for 2 h.[37] Alternatively, vial drying and de-pyrogenation is conducted in continuously operated dry-heat tunnels connecting the non-sterile area of the plant, where the clean but non-sterile vials are charged, to the sterile area.[25]

Table 11.6 *D* values and *Z* values for microorganisms during dry-heat sterilization[10]

Microorganism	*D* value (min)	*Z* value (°C)
Bacillus subtilis	30 (at 120°C) 2.7 (at 140°C) 2 (at 150°C) 1.3 (at 160°C) 0.5 (at 180°C)	18.3–23.3
Bacillus stearothermophilus	0.3 (at 160°C) 0.15 (at 180°C)	14.4–24.4
Clostridium sporogens	6 (at 150°C) 2 (at 160°C) 0.25 (at 180°C)	18–33

11.2.6 Sterilization validation

A product or a process is considered sterile if the probability of the survival of undesired microorganisms is extremely low. To have a high sterility assurance level (for example less than 1 in a million sterility failure) the sterilization processes must be validated.

In the fermentation industry, validation of a process for sterility assurance involves developing and testing sterilize-in-place (SIP) protocols for equipment and sterility testing of the 'sterile' product (the latter point may not apply to final formulation products, such as parenterals, for which the FDA has different requirements). The protocols for SIP systems involve:

- Leak testing under vacuum
- Demonstration of consistent operation via measurement and documentation of temperatures, pressures and other parameters
- Microbial challenge tests using biological indicators such as *Bacillus stearothermophilus* spore strips.

Leak testing under vacuum indicates whether or not non-sterile material can enter the underpressurized process equipment. Multipoint temperature and pressure measurements are carried out to ensure there are no 'cold' spots. For biological challenge tests spore strips inoculated with about a million spores are used. The expected *D* value (resistance to steam sterilization) should be at least 1.0 but not more than 2.5 min. Upon completion of the test, the spore strip is placed in a suitable growth medium and incubated. A positive control for this test is the incubation of an unsterilized spore strip in the same medium using the same conditions. A negative control for the test is incubation of the medium without inoculating with any spore strips.

The United States Pharmacopoeia (USP) has developed a reference protocol for sterility testing of medical products.[38] Each manufacturer of sterile products uses his own protocol, and the Food and Drug Administration (FDA) regulates and approves such practices. Protocols vary from product to product. The protocols describe the method of sampling, and the environment, equipment and personnel required. Once the SIP validation of equipment is complete and a protocol for sterility testing of the product is established, a request for certification for routine manufacturing can be initiated.

11.3 Sterilization via filtration

Sterilization of raw materials and products can be carried out by either destroying or removing microorganisms. Filtration, a means of segregating suspended particulate matter from fluids, can be used to remove microorganisms and obtain a sterile filtrate. Compared with thermal destruction of microorganisms, filtration is an expensive method for medium sterilization and cannot be used for sterilizing particulate matter. In the biotechnology industry, filters are used to sterilize fermentation air and heat-labile liquids. The objective of this section is to describe the basic principles of filtration applicable to the removal of microorganisms, and to review the filtration equipment and procedures used in the biotechnology industry. For additional information the reader is referred to books Aiba *et al.*[15] and Bailey and Ollis (*Biochemical Engineering Fundamentals*, 2nd edition, 1986, McGraw-Hill, New York).

11.3.1 Filtration mechanisms and types of filters

The history of filtration dates far back: clarification of fruit juices and wine using fabrics is probably the earliest known application. For centuries people have used fibrous material such as paper, cloth, asbestos, granular solids that include sand, charcoal, metal meshes, sintered powders and ceramics to remove suspended material. Curiously, what was meant to remove visible matter often prevented the beverages from turning sour for an extended period of time. Only after the advent of the microscope was it recognized that filtration not only removed coarse particles but also contributed to remove microparticles, that is, microorganisms. It was precisely the removal of microorganisms that reduced the rate of the food-spoilage process.

In order to understand better how filters can be used to achieve sterility in fluids one needs to look at the basic mechanism of microorganism removal via filtration. Size plays a very important role. Fungi, that are 5–7 μm in diameter and several millimeters long, are the easiest to exclude. Single-cell microorganisms, ranging from 0.5–2 μm in diameter, are the hardest to remove. The problem of separation becomes formidable when one has to filter submicroscopic entities such as viruses and bacteriophages that are only 0.04–0.1 μm in size.

Microorganisms are not rigid. Moreover, their size varies according to growth stage. For this reason, most sterilizing filters are designed to exclude particles larger than 0.22 μm, that is much smaller than most microorganisms. Electrical charge and surface-adhesion characteristics of the microorganisms can also be effectively used in the design of a filter.

Filters are broadly classified in two categories: (i) depth filters; and (ii) membrane filters. This classification is based on the type of filter medium (i.e. the material that produces the separation of particles larger than a certain size) and how the medium is packed. Typically, depth filters do not have a homogeneous pore size and are therefore identified by their normal pore size. Conversely, membrane filters associated with an absolute pore size.

11.3.1.1 Depth filters. Depth filters are composed of fibrous, granular or sintered materials. This material is loosely pressed, wound, or packed in a casing so that a particle-bearing fluid passing through it has to take a tortuous path, thus producing the removal of solid particles as they become retained in the filter media. The removal effectiveness depends upon the fluid velocity, particle size, and filter bed porosity and depth. Particle retention results from impaction, interception, diffusion, and electrostatic interaction.

Large particles are retained on the surface by direct impaction. Moderately large particles that pass through the surface are trapped in the interstices via interception or impaction. High velocities result in impaction, while moderate velocities lead to entrapment of particles as they are intercepted by filter media and stick to it. Small particles, present in the fluid boundary layer, diffuse to the filter media and stick to it. Diffusion is a slow process and is facilitated by low fluid velocities. The particles (microorganisms) adhere to the filter media because of sedimentation or electrostatic interaction. A detailed mathematical analysis of this phenomenon can be found elsewhere.[15]

The particle retention ability of depth filters is also a function of fluid flow velocity, temperature, pressure, viscosity, and the presence of impurities in the fluid to be filtered, such as water droplets in air or immiscible organics in aqueous solutions. Additionally, the operating characteristics of depth filters change with usage. Owing to the probabilistic nature of the mechanisms responsible for depth filtration, it is impossible to rate the filtration capacity of a depth filter in absolute numbers. Thus, the particle retention capability of depth filters is expressed in normalized values indicating what percentage of a certain particle size could be retained in the filter under given conditions of flow, temperature, pressure and viscosity.

11.3.1.2 Membrane filters. These filters are essentially sieves. Recent advances in membrane technology have facilitated the production of rigid, uniformly thick membranes with evenly distributed pores having a very narrow size distribution. These filters are called absolute because they do not allow the passage of particles larger than their pore size. In addition, because of bridging (i.e. the accumulation of particles around pores) even particles smaller than the pore size are likely to be removed from the fluid passing through the filter. Membrane filters with 0.22 μm pore size are used in the final step of sterile product preparations.

Membrane filters are made up of metals, cellulose-derived compounds or organic polymers. Microporous and macroporous membrane filters made up of PVA, PVDF or Teflon® are most commonly used in the fermentation industry for a wide variety of applications. An extension of membrane filtration is ultrafiltration and nanofiltration, where molecules can be separated on the basis of their molecular weight.

The particle retention ability of membrane filters depends on temperature, pressure and the presence of impurities in media. Repeated use, and fluctuations in temperature and pressure can cause stretching and elongation of the pores. This, coupled with chemical degradation of the membrane material, can lead to loss of filter integrity. Vendors of filter for large-scale fermentation typically indicate the number of cycles that the filters can tolerate without losing their integrity.

A relatively new development in filter membranes is sintered ceramic membranes for air/gas sterilization. These membranes are claimed to have a narrower pore size distribution than the conventional polymeric membranes, and have a higher resistance to pressure, thermal and chemical breakdown.

11.3.1.3 Depth filters compared with membrane filters. The pharmaceutical and biotechnology industries rely on filtration for many production operations, from fermentation to final sterile product formulation. Although expensive, liquid filtration is an essential operation in this fermentation industry because a wide variety of the raw material and products that have to be sterilized are heat-labile. In any given application, a combination of depth and membrane filters is used to get the desired results at a minimum cost. Table 11.7 provides a comparison of these two types of filters. Nevertheless, in the final sterilization of parenterals, the use of such filters is uncommon.

Depending on the content of impurities, microbial load and the characteristics of the fluid undergoing filtration, depth filters are used as pre-filters where most of the large particles and other impurities are removed while the membrane filters are used for the final sterilization to obtain a sterile raw material or product.

11.3.2 Filtration of fermentation air and other gases

Aerobic fermentations require a large quantity of sterile air. The air intake ranges from 0.1 to 1 unit volume of air per unit volume of fermentation broth per minute (vvm). The operating volume for industrial fermenters is typically 10000 to 100000 liters. Thus air flow rates as high as 100000 liters/min may be required.

Air typically contains 1000 to 10 000 microbes/m^3.[15] This implies that at a 1 vvm air flow rate as many as 1 million microbes could be introduced in a 100 000 l/min batch fermenter.

However, the problem is not so formidable as it may sound. The air to be introduced in a large fermenter has to be compressed, and the heat of compression raises the air temperature to such an extent (around 150°C) that microbial destruction of several orders of magnitude occurs in a short time. Good air filtration can further reduce the concentration of microorganisms by a factor of 10^6.

Depth filters are used widely for air sterilization during fermentation. For large fermenters, these filters come in cartridges. BIO-X brand depth filters manufactured by Domnick Hunter Micro-filtration Incorporated, can be found in fermentation plants built in the 1960s and are still operational today.

For shaker flasks and bench-top-stirred fermenters, cotton plugs are used for air filtration. The filtration efficiency of cotton plugs is extremely sensitive

Table 11.7 Comparison of the main characteristics of depth filters and membrane filters

Depth filter	Membrane filter
Exhibits a greater capacity for handling solid impurities (due to the large interstitial space)	Retains solids on surface, cannot handle large loads of solid impurities
Retains particles smaller than the rating as a result of diffusion and electrostatic interactions	Retains particles smaller than the rating as a result of 'bridging'
Pressure build-up is gradual	Pressure build-up can be rapid; particles smaller than the pores that escape bridging may deposit in the interstices and increase the pressure drop
In the long run, the solids that are retained on the non-sterile side may migrate to the sterile side	Surface retention excludes any possible migration
Microorganisms retained in the filter can grow/sporulate and enter the sterile side	Surface growth and sporulation of microorganisms is possible; however, passage to the sterile side is unlikely
Product hold-up is high; filter is typically several inches deep	Product hold-up is low; filter thickness is typically about 150 μm
Relatively inexpensive	Relatively expensive

to moisture. Cotton is replaced with fiberglass when the air is humid or high air flow rates are required, and when the fermentation is sensitive to airborne contaminants.

The microorganism retention efficiency of depth filters decreases with the moisture content. While a relative humidity (RH) of 95% may be enough to retain bacteria (0.5–2 μm in diameter) an RH value of less than 60% may be necessary to eliminate phage (0.04 – 0.1 μm) from air. The integrity of depth filters can be tested by microbial challenge and aerosol tests similar to MIL STD 282 DOP test. The retention capacity of depth filters depends on the manufactures and is typically $\geqslant$ 99.99%.[39] However, different manufacturers may have different retention capacities for their filters.

Membrane filters are appropriate where a high level of retention ($\geqslant$ 99.999999%) is required. Furthermore, strict FDA guidelines on fermentation-based pharmaceuticals is forcing the industry to use 'absolute' filters to reduce the risk of contamination. Hydrophobic membrane filters are nowadays being used to filter fermentation off-gas before venting it to the atmosphere. In the fermentation industry, validation of absolute membranes/filters can be carried out using the microbial challenge, aerosol or bubble point tests. Every major filter manufacturer has their own version of such tests.[39–41]

Recent advances in polymer chemistry have led to strong and durable membrane air filters for large-scale fermentation that may withstand 150 steam sterilization cycles in the forward direction, or 50 cycles in the reverse direction. Typically, the validation of these claims is carried out by the manufacturer. Pleated-membrane filters are compact and provide a large filtration area. These filters also come in cartridges. Filters are sized on the basis of the required sterile air flow rate required and the allowable pressure drop. A typical filter housing may contain one or more cartridges.

Fermentation air supply pressure ranges from 35–75 psi. A pressure drop of 1–3 psi across the filter is considered acceptable. The valves on a fermentation sterile-air header are designed in such a way that filters do not have to be sterilized each time a fermenter is sterilized. A positive air pressure, with a slow bleed on the sterile side, keeps the filters sterile. This technique extends the life of filters. Loss of air pressure, contamination, and periodic maintenance are the primary reasons for which filters are sterilized. To ensure proper filtration, fermentation operations personnel monitor routinely the pressure drop across the filters, as well as air humidity and temperatures.

Sintered ceramic cartridge filters are likely to replace the polymeric membranes in the future. These strong and rigid filters can be washed with acid/alkali solutions and are expected to have a very extensive lifetime.

The above discussion may lead the readers to think that use of depth filters for fermentation air sterilization is rapidly diminishing. However, this is not true. A major application of depth filters is in providing clean air for several industrial uses. In the fermentation area all manual sterile transfers such as inoculating shaker flasks, sampling and culture plating have to be carried out in a laminar flow hood that provides air obtained via filtration through high efficiency particulate-arresting (HEPA) filters. These filters consist of banks of a filter medium made up of fibrous material, separated by corrugated paper or metallic sheets. The metal grill that holds the filter helps sustain the laminar air flow. To maintain laminar flow, air velocity of 230 ± 90 cm/min through the filter is recommended. The integrity of HEPA filters and installation is checked using the dioctyl phthalate (DOP) smoke test.

In hospitals, laminar flow hoods are recommended for extemporaneous preparation and manipulation of sterile dosages. In the pharmaceutical industry, where bulk sterile products are manufactured, processed and packaged, clean rooms with the classification of 100 000 ($\leqslant 100\,000$ particles/cubic foot) or better (class 10 000) are required. Depth filters are used for this application. Class 100 is typically used for aseptic applications.

11.3.3 Filtration of fermentation media and liquid products

As mentioned previously, thermal deactivation of microorganisms in fermentation media is the most cost-effective method of sterilization. However, during sterilization, interaction between different medium components can generate compounds inhibitory to the biosynthesis of the desired product. Sometimes the medium components are heat-labile. Strict FDA guidelines for all parenteral products require these to be sterile and free of large particles. Under these conditions, radiation sterilization or filtration sterilization (for liquids) are the only possible alternatives. Filtration sterilization is the most economical, effective and validatable method of sterilization for heat-labile fluids.

Filtration of liquids is much more difficult than that of air or gases. Owing to the relatively high viscosity of liquids, small particle retention produced by diffusion through the boundary layer, is almost non-existent. In addition, the particle-to-filter medium interactions are less intensive in liquids. Thus, depth filters for liquid filtration can only be used as pre-filters to remove large particles, while membrane filters are used to perform the final 'polishing' step.[42] To maximize the absolute filter life, usually a large pre-filter (approximately, 10 × the filtration area) is commonly used.

A wide variety of polymeric membrane filters are used in the biotechnology industry. For aqueous systems, polycarbonate, cellulose, PTFE (Teflon), PVDF, and polypropylene filters are used, while for organic systems solvent-resistant nylon or fluoropolymeric membranes, such as Teflon, are used. A major problem with the sterile filtration of organics/solvents is contamination with extractable materials, which are released by the membrane itself when exposed to solvents. These extractables are chemicals used to bond the membrane to the support elements on the filter cartridges. Filter manufacturers and users pre-wash the filters in organic solvents to minimize leaching of such extractables materials. Nevertheless, Teflon and polypropylene membrane filters are not affected by this problem since they are heat-sealed.

Dead-end filtration is the most common mode of filtration if the volumes to be processed are not very large and the liquids do not contain an excess of particulate matter. For rapid processing of large volumes of liquids, or to eliminate the bulk of particulate matter, tangential flow filtration or vortex flow filtration methods can be implemented. Typical applications of such modes of filtration (and/or ultrafiltration) are separation and concentration of proteins, viruses, antibodies and other biologic materials. In the fermentation industry, special hydrophilic membranes made up of polysulfones and cross-linked polyacetonitrile are used to reduce the loss of valuable biological materials by retention in the membrane structure.

11.3.4 Validation and integrity testing

Two issues are involved in the validation of filtration systems. One pertains to the performance of the filter medium or membrane, while the other to its installation and usage. Improper installation and operation can compromise the integrity of a validated filter. Guidance for validation is available from the American Society of Testing Materials (ASTM),[43] the Food and Drug Administration (FDA),[44] the Health Industry Manufacturers Association (HIMA)[45] and the Pharmaceutical Manufacturers Association. Tests developed by filter manufacturers that allow non-destructive, *in situ* testing, and validation of filtration systems can also be obtained.

Most of the filters used in the fermentation industry today undergo several routine sterility tests. The most notable ones are: (i) aerosol testing using fine droplets of oil or dioctyl phthalate (DOP); (ii) bubble point and diffusion testing; (iii) microbial challenge testing using greater than 1 million/cm^2 filter area of *Pseudomonas diminuta* (American-type culture collection, ATCC No. 19146); and (iv) viral challenge testing using greater than 10^8 plaque-forming units (phage particles).

Tests (i) and (ii) are non-destructive and are performed routinely *in situ* or as needed. The bacterial and viral challenge tests are mainly carried out by the filter manufacturers and sometimes for validation at the site and at the time of filter and other equipment installation. A brief description of these tests is given below.

11.3.4.1 Aerosol test. Fine droplets of vegetable oil or dioctyl phthalate, in the size range of 0.3 to 3 μm, can be produced by an aerosol generator. These droplets are introduced into the air passing through the filter to be tested for integrity. On the downstream side of the filter, the concentration of the droplets is measured with the help of an aerosol photometer. Dioctyl phthalate (DOP) forms a smoke and can be used to determine 'visually' the integrity of filters. Only HEPA filters are integrity tested using DOP. The aerosol test applies to depth as well as membrane filters.

11.3.4.2 Bubble point and diffusion test. This test is limited to membrane filters. An enlarged view of a membrane filter will show that it is formed by a sponge-type matrix, with small pores on each face connected via capillaries. If a gas (e.g. nitrogen) were to pass through such a membrane filled with a fluid (e.g. an organic solvent for the case of a hydrophobic membrane), it would have to overcome the capillary forces. The critical pressure essential for the gas to pass through the membrane is a function of pore size and the wetting liquid surface tension.

Experimentally, the critical pressure is determined to be the gas pressure at which a gas bubble appears at the downstream side of the filter. Thus, using a wetting liquid of known surface tension, one can calculate the effective pore size based on the experimentally determined bubble point. In practice, for a given filter membrane and a wetting liquid, manufacturers usually specify the bubble point required for the filter to pass the integrity test. The bubble point is inversely proportional to the pore size. Thus, a bubble point lower than the specified bubble point indicates integrity failure.

The diffusion test, also suitable for large filters, is a slight variation of the bubble point test. To perform a diffusion test, the test filtration system is wetted with a specified liquid and a gas pressure 80% of the bubble point is applied. At this pressure, no convective flow of gas occurs through the membrane, and the passage of gas to the downstream end of the filter results from the following three events occurring in series. First, the gas dissolves in the wetting liquid at the upstream (high-pressure) end of the filter. Then the gas diffuses through the liquid, as given by Fick's law. At last, the gas evaporates at the downstream (low-pressure) end of the filter. For a given wetting liquid and gas combination, and specified operating temperatures and pressures, the diffusion rate depends on the membrane thickness, frontal area and porosity. In practice, the flow rate of a gas is determined under specified conditions, and compared with the manufacturers specifications. A flow higher than that specified indicates loss of filter integrity.

11.3.4.3 Microbial challenge test. *Bacillus stearothermophilus* is used as a biological indicator for thermal sterilization because of its well-characterized resistance to heat. Similarly, *Pseudomonas diminuta* is used as a biological indicator for filter sterilization because of its size. Using the appropriate strain (ATCC No. 19146) and a specified cultivation medium, rapidly growing and non-aggregating cells can be obtained. This microorganism has been accepted by the ASTM,[43] the FDA,[44] and the HIMA.[45] Most filter manufacturers use this microbe for testing. Some manufacturers also have a correlation between this microbial challenge test and the diffusion/bubble point test.

11.3.4.4 Viral challenge test. This test is similar to the microbial challenge test. However, phages (which are considerably smaller than bacteria having a size between 0.04 and 0.1 μm) are used. Filters designed to retain bacteria are also found to remove phage particles. Most of the major manufacturers of filters for fermentation applications test their filters for phage challenge with a procedure similar to that for the bacterial challenge test. Details of phage testing can be obtained from the manufacturers of air/gas filters.

11.4 Chemical sterilization

Thermal destruction is by far the most common and efficient mode of sterilization. However, use of dry or moist heat for sterilization is not always feasible. A common example is sterilization of plastic/disposable devices. A wide variety of chemicals in the form of gases, liquids or solids are known to affect the capability of microorganisms to survive and reproduce. The antimicrobial activity of such chemicals ranges from simple 'cleansing' to complete destruction or sterilization. In the biopharmaceutical industry, such chemicals are used for disinfection/sanitation of clean rooms, biological safety hoods, glove boxes, centrifuges and other bulk sterile processing equipment. The main problem with disinfecting chemicals is their toxic effect on living beings and the environment.

11.4.1 *Mode of action of antimicrobial chemicals*

It is not always easy to determine the mechanism of microbial death for microorganisms exposed to a given chemical compound since more than one cell constituent may be affected by the chemical in use. The methodology typically used to elucidate the fundamental process of the antimicrobial action consists of applying the chemical at a very low concentration to avoid multiple effects on an axenic microbial population at a given growth stage.

11.4.2 *Sterilizing and disinfecting agents*

Several sterilizing and disinfecting agents are used industrially in a variety of applications. In order to minimize the development of strains resistant to a particular agent it is sometimes convenient to cycle between sterilizing or disinfecting agents. The most important of these agents are now reviewed.

11.4.2.1 Alcohols. Ethanol is the only industrially significant alcohol used for disinfection purposes. Ethanol is commonly used in hospitals for surface cleaning and disinfecting surgical devices. Generally speaking, alcohols have limited biocidal activity and do not affect spores. Their mechanism of action appears to involve protein denaturation.

In the fermentation industry, ethanol is used to disinfect small glassware and metallic devices. The procedure consists in immersing these items in ethanol and flaming them. Sterility is attained as a result of rapid incineration of all surface-attached biomaterials. In biopharmaceutical manufacturing (e.g. protein, vaccine and antibody preparation), chromatography columns are kept in ethanol[46] and sterile buffers to avoid contamination.

11.4.2.2 Aldehydes. For centuries formaldehyde has been used as a preservative and as a disinfectant in hospitals. In the pharmaceutical industry, it is used as a surface sterilizing agent for centrifuges, glove boxes, biosafety hoods and in the form of mist of vapors (low-temperature steam with formaldehyde) for clean rooms. The mechanism of cell destruction by formaldehyde is assumed to be a reaction with cell protein and DNA or RNA.[47] Owing to the toxic nature of formaldehyde, its use as a disinfectant is gradually diminishing.

In the early 1960s, glutaraldehyde, a 5-carbon dialdehyde, was found to have an extremely broad biocidal activity, and is now considered a 'chemosterilant'.[48] It is very effective against bacteria, spores, mycobacteria, fungi and viruses. As with other dialdehydes, glutaraldehyde readily reacts with proteins, specifically with the amide group of amino acids. Currently it is used in the fermentation industry as a surface disinfectant, particularly when sterile needles are used to pierce through rubber stoppers to add nutrients/inoculum to sterile fermenters. In spite of its toxicity, glutaraldehyde has found a broad range of applications, and new areas of application are being found.[49]

11.4.2.3 Ethylene oxide. Implementation of ethylene oxide sterilization has made possible the use of sterile, low-cost, disposable thermoplastic devices for medical and pharmaceutical applications. Ethylene oxide is soluble in water at 10°C and reacts with alcohols, organic acids and amines. The vapors are flammable and explosive.[50] Ethylene oxide is effective against bacteria, molds, yeasts as well as viruses. The most probable mode of such activity is alkylation of nucleic acids and other cellular components.[51]

11.4.2.4 Peroxides. Liquid-phase hydrogen peroxide is a widely used surface disinfectant for aseptic food and dairy packaging and processing equipment. It is considered safe and is an approved food preservative in several countries.[52] Hydrogen peroxide kills a wide variety of organisms,[53,54] including spores[55] and viruses.[56] Bacteria contain enzymes capable of reacting with oxygen free-radical, for example superoxide dismutase, that produces hydrogen peroxide. Most microorganisms do not accumulate significant quantities of lethal hydrogen peroxide, as it is broken down into water and oxygen by catalase.[57] A hydrogen peroxide concentration greater than 3% is sufficient to kill most microorganisms.[53]

A new development in the application of hydrogen peroxide is to use its vapors as a sterilizing agent. Vapor-phase hydrogen peroxide (VPHP) can be used as an effective sterilizing agent.[58,59] Hydrogen peroxide is non-toxic, easily degradable to oxygen and water, and is effective at low concentration ($\sim$10 mg/l VPHP). It is possible to sterilize biological safety hoods, glove boxes, centrifuges, fermenters and clean rooms using this technology.

11.4.2.5 Other antimicrobial chemicals. In addition to those discussed above, there are many chemicals used for disinfection/sanitation and sometimes for sterilization. Owing to their limited industrial applications, only some of the compounds are described here.

Chlorine and its oxides, and ozone. These gases are used mainly for water disinfection. Ozone can be used as a sterilizing agent because of its high oxidizing capability. Attempts have been made to use ozone to deodorize fermentation vent gases. Other gases such as propylene oxide (compare with ethylene oxide) and β-propiolactone (compare with formaldehyde) have had restricted use because of their toxicity.

Peracetic acid. This is a peroxide of acetic acid. It has all the advantages of hydrogen peroxide and is unaffected by catalase and peroxidases. However, its utility as a vapor-phase sterilizing agent is limited because of the high boiling point and high flammability.[60]

Quaternary ammonium salts. These compounds can be used in combination with detergents to clean, disinfect and deodorize floors, walls and surgical instruments. In the fermentation and biopharmaceutical industries, such compounds are used with extreme caution to clean, wipe and disinfect surfaces that do not come in direct contact with the raw materials and products but which may harbor contaminating microorganisms and viruses.

11.5 Ultraviolet light and radiation sterilization

When neither heat nor chemicals can be used to sterilize a given product (usually in the particulate form) or a piece of equipment, radiation is considered as an option. The dosage and type of radiation sterilization is a function of the type and amount of the microbial flora that has to be removed from a given product or equipment.

11.5.1 Basic physics

Radiations can be ionizing or non-ionizing. The energy of incident ionizing rays results in the ejection of electrons in biologic targets. Subsequent free-radical or activated molecular reactions cause microbial death. Ionizing radiations can be further classified into corpuscular (i.e. particulates such as streams of atoms, electrons or protons) and electromagnetic waves, such as X rays or γ rays. The non-ionizing radiations include electromagnetic waves such as infrared, visible, and ultraviolet (UV) light. Such radiations do not create ionization in the biologic targets.

Electromagnetic radiation is characterized by wavelength and penetrating power or intensity. Wavelengths range from 0.00001–400 nanometer (nm) for electromagnetic radiation. The energy of radiation is inversely proportional to its wavelength. The penetrating power is a function of the energy and the material. X rays and γ rays have a high penetration power. Radiation dosage, either particulate or electromagnetic, is expressed in 'rad' units, to indicate the amount of energy absorbed by a unit amount of cells. Sterilization is generally expressed in megarads (10^6 rads).

11.5.2 Ultraviolet (UV) radiation

The UV spectrum ranges from 400 nm (violet in the visible range) to 1 nm (X rays). In the fermentation industry, the traditional random mutagenesis programs for microbial productivity improvement use UV radiation in the 230–260 nm range. The presumed mechanism of mutagenesis is pyrimidine dimerization.[61] Use of excessive UV radiation causes cell destruction and thus becomes a means of sterilization. UV radiation is considered non-ionic and has a low penetration power. The main application of UV rays is in destruction of infectious airborne particulate matter, especially for hospital HVAC systems, and in the production of water for injection (WFI). Owing to its low penetration power, UV light can only be used to inactivate bacteria present on surfaces or floating on liquid media.

11.5.3 Ionizing radiations

Electrons, β-particles, γ rays and X rays are all used to achieve sterilization through irradiation. Microbial inactivation by such rays probably is produced by the formation of free radicals and activated molecules. The main application of such radiation is in the food industry for pasteurization (partial destruction of microorganisms) or sterilization (complete destruction). Hospital supplies, antibiotics, vitamins, plastic devices and sutures have been sterilized using γ rays.[62,63]

References

1. Richards, J.W. (1968). *Introduction to Industrial Sterilization*. Academic Press, New York.
2. Deindoerfer, F.H. (1957). Calculation of heat sterilization times for fermentation media. *Appl. Microbiol.* **5,** 221–228.
3. Wang, D.I.C., Cooney, C.L., Demain, A.L., *et al.* (1979). *Fermentation and Enzyme Technology*. Wiley and Sons, New York.
4. Aiba, S. and Sonoyama, T. (1965). Residence time distribution of a microbial suspension in a straight pipe, *J. Ferment. Technol.* **43,** 534–539.
5. Deindoerfer, F.H. and Humphrey, A.E. (1959). Principles in the design of continuous sterilizers, *Appl. Microbiol.* **7,** 265–270.
6. Moss, M.O. and Smith, J.E. (1977). *Industrial Applications of Microbiology*. Surrey University Press, Guildford.
7. Prokop, A. and Humphrey, A.E. (1970). Kinetics of disinfection, in *Disinfection* (Ed. M.A. Bernardo). Marcell Dekker, New York, pp. 61–84.
8. Blackebrough, N. (1968). *Biochemical and Biological Engineering Science*, Vol. 2, Academic Press, London.
9. Charm, S.E. (1971). *The Fundamental of Food Engineering*, 2nd ed. AVI Publishing Company, Westport, Connecticut.
10. Wallhäusser, K.H. (1985). Sterilization in biotechnology. *J. Chem. Technol. Biotech.* **2,** 699–723.
11. Mark, Herman F. *et al.* (1982). *Kirk-Othmer Encyclopedia of Chemical Technology*, Vol. 21, Wiley and Sons, New York.
12. Pflug, I.J. (1972). Heat sterilization, in *Industrial Sterilization*. (Eds G.B. Phillips and W.S. Miller), Duke University Press, Durham, NC.
13. Armenante, P.M. (1994). Derivation of a rigorous equation for the calculation of the *F*-value in isothermal sterilization processes. *J. Pharm. Sci.* **83,** 668–673.
14. Feliciotti, E. and Esselen, W.B. (1957). Thermal destruction rates of thiamine pureed meats and vegetables, *Food Technol.*, **11,** 77–84.
15. Aiba, S., Humphrey, A.E. and Millis, N.F. (1973). *Biochemical Engineering*, 2nd edn, Academic Press, New York.
16. Carslaw, H.S. and Jaeger, J.C. (1959). *Conduction of Heat in Solids*, 2nd edn, Clarendon Press, Oxford.
17. Bird, R.B., Stewart, W.E. and Lightfoot, E.N. (1960). *Transport Phenomena*. Wiley and Sons, New York.
18. Perry, R.H. and Green, D.W. (1984). *Perry's Chemical Engineers' Handbook*, 6th edn, McGraw-Hill, New York.

19. Geankoplis, C.J. (1983). *Transport Processes and Unit Operations*, 2nd edn, Allen and Bacon, Boston.
20. Armenante, P.M. and Leskowicz, M.A. (1990). Design of continuous sterilization systems for fermentation media containing suspended solids, *Biotech. Progress*, **6**(4), 292–306.
21. Lin, S.H. (1975). A theoretical analysis of thermal sterilization in a continuous sterilizer. *J. Ferment. Technol.* **53,** 92–98.
22. Lin, S.H. (1976). Continuous high-temperature/short-time sterilization of liquid foods with steam-injection heating, *Chem. Eng. Sci.* **31,** 77–82.
23. Lin, S.H. (1979). Residence time distribution of flow in a continuous sterilization process, *Process Biochem.*, **14**(7), 23–27.
24. Armenante, P.M. and Li, Y.-S. (1993). Complete design analysis of a continuous sterilizer for fermentation media containing suspended solids. *Biotechnol. Bioeng.*, **41,** 900–913.
25. Vogel, H.C. (Ed.) (1983). *Fermentation and Biochemical Engineering Handbook*, Noyes Publications, Park Ridge, New Jersey.
26. Baird, M.H.I., McCrae, W., Rumford, F. and Slesser, C.G.M. (1957). Some considerations on heat-transfer in spiral-plate heat exchangers, *Chem. Eng. Sci.*, **7,** 112–115.
27. Hargis, A.M., Beckman, A.T. and Loiancono, J.J. (1967). *Application of Spiral-Plate Heat Exchangers*. AIChE Houston Meeting, February, 1967.
28. Lamb, B.R. (1962). The Rosenblad spiral heat exchanger, *Chem. Eng.*, **June,** A-108-111.
29. Tangri, N.N. and Jayaraman, R. (1962). Heat transfer on a spiral-plate heat exchanger, *Trans. Instn Chem. Eng.* **40,** 161–167.
30. Rohsenow, W.M., Hartnett, J.P. and Ganiz, E.N. (Eds) (1985). *Handbook of Heat Transfer with Applications*, 2nd edn, McGraw-Hill, New York.
31. Ashley, M.H.J. (1982). Continuous sterilization of media, *Chem. Eng.*, **February,** 54–58.
32. Levenspiel, O. (1958). Longitudinal mixing of fluids flowing in circular pipes. *Ind. Eng. Chem.* **50,** 343–346.
33. Brenner, H. (1962). The diffusion model of longitudinal mixing in beds of finite length: numerical values. *Chem. Eng. Sci.*, **17**, 229–244.
34. Levenspiel, O. (1971). *Chemical Reaction Engineering*, 2nd edn. Wiley and Sons, New York.
35. de Ruyter, P.W. and Brunet, R. (1973). Estimation of process conditions for continuous sterilization of food containing particles. *Food Technol.*, **July**, 44–51.
36. Manson, J.E. and Cullen, J.F. (1974). Thermal process simulation for aseptic processing of food containing discrete particulate matter. *J. Food Sci.*, **39**, 1084–1089.
37. Parrott, E.L. (1970). *Pharmaceutical Technology*, Burgess Publishing Co., Minneapolis,
38. *United States Pharmacopoeia, National Formulary, 1985*, 16th edn. The USP Convention Inc., Rockville, Maryland.
39. Domnick Hunter Filters (1988). In *Validation Document for Hyflo Tetpor™ Air Sterilization Grade Filter Cartridges*. Publication Ref. No. 41/USA/2/85.
40. Conway, R.S. (1984). State of the art in fermentation air filtration. *Biotechnol. Bioeng.* **286**(6), 844–847.
41. Millipore (1990). Aerex™-FC cartridge filters, in *Millipore Validation Guide*. Lit. No. VG 011, Millipore, Bedford, MA.
42. Meltzer, T.H. (1989). Filtration: the practice of prefiltration and its related considerations. *Ultrapure Water*, **6**(3), 24–34.
43. American Society of Testing Materials (ASTM) (1986). Determining the bacterial retention of membrane filters utilized for liquid filtration, in 1986 *Annual Book of Standards*, Vol. 14.02, F838-83, pp. 938–944.
44. Food and Drug Administration (FDA) (1987). *Guideline on Sterile Drug Produced by Aseptic Processing*. Center of Drugs and Biologics and Office of Regulatory Affairs, Washington, DC (June, 1987).
45. Health Industry Manufacturers Association (HIMA), (1982). *Microbiological Evaluation of Filters for Sterilizing Liquids*. HIMA document No. 3, Volume 4.
46. Angal, S. and Dean, P.D.G. (1989). In *Protein Purification Methods, A Practical Approach*, (Eds E.L.V. Harris and S. Angal), Oxford Unversity Press, pp. 254–261.
47. Russell, A.D. (1976). Inactivation of nonsporing bacteria by gases. *Soc. Appl. Bacteriol.* **5,** 61–68.
48. Borick, P.M. (1969). Chemical sterilizers (chemosterilizers). *Adv. Appl. Microbial.* **10,** 291–312.
49. Scott, E.M. and Gorman, S.P. (1991). In *Disinfection, Sterilization and Preservation*, 4th edn, (Ed. S.S. Block), Lea and Febiger, London, pp. 596–614.

50. *Merck Index, 1989*. 11th edn, Merck & Co., Rahway, New Jersey.
51. Phillips, C.R. (1977). Gaseous sterilization, in *Disinfection, Sterilization and Preservation*, 2nd edn (Ed. S.S. Block), Lea and Febiger, London.
52. Schumb, W.C., Satterfield, C.N. and Wentworth, R.I. (1955). In *Hydrogen Peroxide*, Van Nostrand Rheinhold, New York, pp. 813–816.
53. Wardle, M.D. and Renninger, G.M. (1975). Bacterial effect of hydrogen peroxide on spacecraft isolates. *Appl. Microbiol.* **30,** 710–711.
54. Wang, J. and Toledo, R.T. (1986). Inactivation of microorganisms on polyethylene exposed to hydrogen peroxide vapors in air at various temperatures. *Curr. Technol. Flexible Packaging.*
55. Ito, K.A. (1973). Resistance of bacterial spores to hydrogen peroxide. *Food Technol.*, **11,** 59–66.
56. Mentel, R. and Schmidt, J. (1973). Investigation on rhinovirus inactivation by hydrogen peroxide. *Acta Virol.*, **17,** 351–354.
57. Stanier, R.Y., Adelberg, E.A. and Ingraham, J. (1976). Effect of environment on microbial growth. In *The Microbial World*, 4th edn Prentice Hall, New York.
58. Moore, F.C. and Perkinson, L.R. (1979). *Cold Gas Sterilization Using* H_2O_2. US Patent No. 4 366 125.
59. Forstrom, R.J. and Wardle, M.D. (1979). *Low Temperature Sterilization Using* H_2O_2 *at Reduced Pressure.* US Patent No. 4 169 124.
60. Block, S.S. (1991). In *Disinfection, Sterilization and Preservation*, 4th edn (Ed. S.S. Block), Lea and Febiger, pp. 167–181.
61. Carlton, C.B. and Brown, B.J. (1981). Gene mutation, in *Manual of Methods of General Bacteriology* (Ed. P. Gerdhardt), American Society for Microbiology, Washington, DC.
62. Alexander, P. and Bacq, Z.M. (Eds) (1961). In *Fundamentals of Radiology*, Vol. 5. Pergamon Press and Macmillan Co., New York.
63. Gaughran, E.R.L. and Lloyd, R.S. (Eds) (1974). In *Technical Developments and Prospects of Sterilization by Ionizing Radiation*, Vol. 1. MultiScience Publication, Montreal.

12 Pharmaceutical Packaging Operations

W.C. CZANDER and R.V. LEUNG

12.1 Introduction

This chapter addresses non-sterile liquid and solid dosage form product packaging lines.

The packaging for liquid and solid dosage form products for over-the-counter consumer consumption is often designed as a marketing vehicle for increasing the product's salability, even though the efficacy and therapeutic attributes of the product are known. Despite the aesthetics associated with packaging design, all pharmaceutical packaging is required by the Food and Drug Administration (FDA) to state clearly the product's active and inactive ingredients, recommended dosage, warnings and other information required to ensure the product's intended and safe use by the consumer. This vital information is contained on the product's labeling, insert leaflets and individual folding cartons.

The packaging associated with sterile products is similar to non-sterile products in that the same type of information as found on non-sterile product packaging materials is also required on sterile product labeling, insert leaflets and individual folding cartons. Also, after the sterile product's container has been filled and sealed, downstream packaging operations can be performed in a non-aseptic environment similar to non-sterile packaging operations.

The basic difference between sterile and non-sterile packaging operations is that the former require:

- Sterilized container components or the means to sterilize these components if they are received in a non-sterilized condition
- Aseptic environment for handling, filling and sealing of containers
- Off-line, post-filling and post-sealing operations such as lyophilization, inspection, leak testing, terminal sterilization, etc.
- Packaging of sterile products in a 'batch' operating mode because of off-line operations
- Recording of materials used in filling and sealing in a non-aseptic environment to minimize particulate generation in the aseptic environment if this information is recorded on paper.

During packaging operations, information in code form is applied onto printed materials to enable the manufacturer and the FDA to trace the manufacturing steps associated with the product's manufacture in the unlikely event of a product recall by manufacturer or the FDA. The coding information is specific to all products manufactured from the same batch and is the link to the product's manufacturing history from the initial process step where the ingredients are combined until the finished product is packaged. The FDA also requires the drug manufacturer to control the storage, issuance and coding of printed packaging materials to prevent their potential misuse or mixup with other products.

The following discussion focuses on the operational and procedural controls on a packaging line during labeling and final packaging that are required for adherence to current good manufacturing practices (cGMPs) and FDA guidelines. Although there are a variety of methods and equipment used to perform these tasks, from hand or semi-automatic application of labels to high-speed equipment that labels containers, places containers and literature into cartons, detects missing or incorrect components, and packages into shipping cases at high speeds, at the end of each packaging run, all printed materials have to be accounted for! The operational reliability of a packaging line is generally more successful with a highly mechanized line rather than one that requires human involvement, particularly when the mechanized line has been validated successfully by undergoing installation qualification (IQ), debugging, and operational qualification (OQ) procedures once the equipment is installed. Regardless of the level of automation, reconciliation of all materials used during production still applies.

12.2 Pharmaceutical filling and packaging systems

12.2.1 Liquids packaging

The filling of liquid pharmaceutical products into containers requires an infinite variety of packaging container shapes, materials and sizes. The packaging machinery industry has met the challenge in the design of machinery to fill containers of all shapes, sizes and materials at a wide range of filling speeds.

For the purposes of this discussion, a hypothetical packaging line will be used to illustrate the controls and procedures required in the packaging of liquid pharmaceutical products to meet current cGMPs and guidelines set forth by the FDA.

12.2.1.1 Bottle unscrambler. Released cases of bottles (plastic or glass) are brought to the packaging line for delivery to the downstream filling equipment. This can be performed manually, by forklift truck or by conveyor. The bottle manufacturer's identity and lot number are recorded on the batch record associated with the production run. The batch record provides an audit trail for the FDA to track all the materials that comprise the finished product. This information is usually recorded on paper, although many pharmaceutical firms are now progressing to paperless or electronic batch records. After the bottles have been 'unscrambled' and oriented in a single lane, they are conveyed through a bottle washer or cleaner on their way to the filler. The conveyor on which the bottles are conveyed between the cleaner and filler is covered with a plexiglass cover to ensure the bottles' cleanliness.

12.2.1.2 Filler/capper. The empty bottles are metered through the filler by timing screws and starwheels. The liquid product is delivered to the filler's reservoir via gravity from holding tanks located a level above the filler or pumped from portable or fixed tanks located at the same level as the filler. The product name, lot number, and volume of liquid product used during filling are recorded in batch records. Other controls include recording the volume of product filled, gathering samples required for QC testing and estimation of process losses. Immediately after filling, the filled containers are capped in a capping machine. Caps are usually fed to the capper unit via a bulk hopper-feed system.

12.2.1.3 Tamper-evident seal. The capped container is conveyed to a tamper-evident seal machine, which applies a seal or plastic band around the cap of the container to provide evidence of tampering if the seal or band is broken.

12.2.1.4 Labeler. Filled containers are labeled either manually or automatically. The labels are either pre-cut or are fed from a roll, typically from highly mechanized labelling equipment. Regardless of label type, all the information on the product is contained on the label. To identify the bulk product in the filled container, the batch number of the bulk product is identified on the label in the form of a bar code or printed code. This information is applied to the labels in a label room prior to issue, or during the labeling operation on the packaging line. As the completion of the packaging operation, all the labels issued must be accounted for, reconciling the quantity remaining of labels at start of labeling, quantity used, quantity damaged and quantity remaining at the completion of packaging.

12.2.1.5 Cartoning. The labeled containers are manually or automatically inserted into individual folded cartons (IFC) together with folded leaflets (literature). Both the literature and IFCs contain printed information about the product and have to be reconciled similarly to the printed labels. Labels, leaflets and IFCs are stored and dispensed from a label room where strict control over their storage, issue and use is required.

12.2.1.6 Bundling. The individual cartons of product are then conveyed to a bundling machine, which collates and applies plastic film over-wrap around a specified number of cartons. The individual bundles of cartons are usually not coded with the product's lot number or other product specific information, since each individual product carton already contains this information.

12.2.1.7 Case-packing. The bundled or individual cartons of product are packed into corrugated shipper cases, either manually or automatically on higher speed lines and sealed with tape or hot melt adhesive. The shipping cases are identified by in-line carton coders or by hand-applied case labels with the product's name and product batch numbers. Preprinted case labels are accounted for and reconciled at the end of the packaging run. Rubber stamps or mats used to print product information on the shipper cases are retained for the batch records. Samples of the printed impressions also become part of the batch record.

12.2.1.8 Case palletizing. The finished cases of product are then palletized at the end of the packaging line or are conveyed to a remotely located case-palletizing operation. At the completion of the case-palletizing operation, the total case count and number of palletized loads of product are recorded for the batch record. Hand-printed labels are usually affixed to each pallet load of finished product. One label identifies the product; the other label identifies the product as 'unreleased' pending final QC assay testing. The palletized product is moved to the 'unreleased' area of the warehouse.

After QC clearance of release, the pallet loads become picking stock, available for inclusion in outgoing shipments.

12.2.2 Solids packaging

The filling and packaging of solid dosage form products (e.g. powders, capsules and tablets) into bottles are performed on packaging lines similar in design to a liquids packaging line.

Depending on filling speed requirements, tablets are filled into bottles with a slat filler or a high-speed rotary filler. After filling and capping, the downstream labeling, cartoning, case packing and case-palletizing operations are similar to a liquid packaging line.

Capsules and tablets are also packaged into blister packs. The major components of an automated blister pack line are as follows.

12.2.2.1 Thermoformer. The thermoformer is a form-fill-seal packaging device. This equipment forms rigid pockets from a roll of PVC/PVDC film, which is then filled with tablets or capsules fed from the equipment's overhead hopper.

Aluminum foil fed from a roll seals the filled pockets of product and code information is printed directly onto the aluminum foil as it is drawn off the roll. Samples of printed foil and print mats are retained at the completion of the packaging operation for the batch record. The thermoformer punches out individual blister cards, which are conveyed to downstream operations.

12.2.2.2 Cartoner. An automatic cartoner will fill the blister cards, leaflets or other literature into an individual folded carton. Individual leaflets can also be roll-fed. The cartoner cuts and folds the leaflets and inserts them. Coding of the cartons can also be performed by the cartoner. At the completion of the packaging operation, the samples of leaflets and coded cartons are retained for the batch record.

12.2.2.3 Checkweigher. An in-line checkweigher located after the cartoner weighs each individually filled carton and will reject a carton that is underweight because of missing leaflet, blister pack or product in the blister pack.

12.2.2.4 Bundling, case-packing, case-palletizing. Downstream bundling, case-packing and case-palletizing operations are similar to those described for the liquids packaging line.

12.3 Packaging-line clearance and set-up

At the completion of a packaging operation and before the start of the next one, certain procedures are taken to ensure prevention of product mix-up between packaging operations and to set up the packaging line for the next packaging order.

12.3.1 Packaging-line clearance

Packaging line clearance procedures involve the physical checking of the line areas and equipment for the presence of leftover bulk product, packaging components, line-signage or labeling, equipment components and paperwork.

In most instances, the degree of line clearance between packaging orders is dependent upon the previous and next scheduled order to be run on the particular line and the degree of necessary line changes between packaging orders. This is often dictated by:

- Bulk product lot number and packaging lot number
- Pack size, type and market
- Equipment changeover.

After the line has been cleared, the paperwork authorizing the transfer of materials to the line identifies:

- Line number
- Product name
- Work order number
- Description of components required.

All the components on the line are checked by supervision against the packaging order to verify:

- Component identity
- Lot and quantity.

Appropriate signatures are recorded on the packaging order and 'label-reconciliation' sheet to verify material checks.

12.3.2 Packaging-line set-up

Authorization paperwork for line changeover includes the following information:

- Line number
- Work order number
- Product name
- Pack size
- Machines requiring changeovers
- Required components
- Packaging date.

After the line changeover has been completed, the line is set-up and operated as follows:

- Setting the batch code on various machines
- Operator sets machine parameters for the production run and runs product samples
- Line supervisor checks samples off the line against the packaging order to see that the equipment was properly set up and machine parameters were attained.

The line set-up also includes checking and calibration of weigh scales and checkweighers for compliance with actual pack weights recorded on the equipment set-up sheet. The QA supervisor then approves the completion of the line for start-up by signature and date in the appropriate section of the line-clearance form.

12.4 Packaging line start-up

With approval to start up the packaging line, the required line operators are sent to the line and their presence is recorded on the time sheets.

12.4.1 In-process checks

Immediately on start-up, a specific number of bottles are checked to ensure the fill is within specific set limits. Additional fill checks are performed throughout the production run by:

- Operators, on a regular basis
- QA inspectors and/or line supervisors, on a random basis; all fill checks are recorded for inclusion in the batch record with dates, times, results and action taken as necessary for each fill check.

Random packaging checks can also be performed by operator, supervision and QA to check the following attributes:

- Correct, legible coding of relevant components
- Presence of all correct components in a pack
- Elegance of the packs.

When a new label reel is required on the line, the operator affixes the identity sticker from each label roll used to the back of the label-reconciliation sheet along with a sample of the label from the roll. The operator who changed the label roll records the date, time and his/her initials on the samples.

On blister packaging lines, operators record the following attributes at start up:

- Blister-integrity vacuum dye test
- Checkweigher operation test
- Filler-pin-detector operation test (where applicable).

Supervision and QA personnel perform these checks at regular and irregular intervals. These checks are recorded on relevant check sheets, which are included in the batch record.

12.4.2 Line stoppages

Packaging line mechanics assigned to the packaging line will tend to mechanical related line stoppages. Other line stoppages can be attributed to a packaging line running out of product or packaging components. The operator or supervisor records the following information on line-stoppage sheet:

- Description of the problem
- Duration that line was down
- Corrective action taken.

12.5 Packaging-line completion and clean-out

At the completion of the packaging operation, the line is stripped down to facilitate cleaning. Operators clean down the line to the extent required dependent upon the nature of the next job on the line.

Supervisor completes the appropriate paperwork for the return of unused components to the warehouse or label room.

- Unused components are recorded on the packaging order.
- Unused labels are recorded on the label-reconciliation form.

Materials handlers return any components remaining on the line to their respective warehouse areas. Returned components are checked by the warehouse operators by weighing or counting. The supervisor completes the following paperwork, as required to closeout the packaging order:

- Completes label reconciliation.
- Checks materials and quantities used.
- Checks yield of the job.
- Checks for operator signatures after each process step and compliance with all packaging and work order standard operating procedures (SOPs)
- Attaches samples of all required components to the packaging order
- Fills packaging line statistics form identifying:
  - —Performance and line-efficiency statistics
  - —Line operating time
  - —Line downtime, delineated by categories
  - —Number of operators

The QA inspector performs the following paperwork at the completion of the packaging operation:

- Physically checks the yield of the job
- Checks final paperwork in the packaging order
- Approves final paperwork by signing relevant section of the packaging order.

12.6 Packaging trends

12.6.1 Pre-qualified suppliers

Pharmaceutical companies are now using pre-qualified suppliers of packaging materials, giving them the advantages of not having to perform in-house 100% inspection, and minimizing the need for large inventories.

The first step for a pre-qualified supplier program is to publish detailed specifications for the packaging material that include tolerances that can meet the requirements of packaging equipment. The specifications also comply with government regulations and marketing requirements of print size and color. In addition, the specifications should include how the material is to be packaged, stored and handled.

The supplier must then show that its quality control and inspection methods, procedures and systems are equal to or better than the pharmaceutical companies' counting and strobing inspection systems and QA and inspection procedures.

After the supplier is qualified, a sampling program of the incoming material is conducted by the user. This sampling program requires that samples from the start of the supplier's production run, along with samples from the middle and the end of the run, be taken and sent to the pharmaceutical company's packaging engineers for inspection and approval before shipment.

12.6.2 Bar-coding

The use of bar codes is a trend to reduce further the possibility of human error in packaging. The bar-code label can be applied to the equipment and scanned by the operator or supervisor to give real-time status of the availability of the equipment. The scanning of the label relays information electronically, such as its cleaning status, change-over time, readiness for production and date of last maintenance. The bar code on the containers of materials to be supplied to the packaging line is used to ensure that there are no mix-ups of wrong material being introduced to the packaging run, especially the wrong printed material and/or the product contained in the wrong package.

12.6.3 Equipment

The trend of equipment suppliers is to use more mono-block systems that integrate more than one operation into a single piece of equipment, such as filling, stoppering, sampling and testing. The advantages of these systems are that they perform more self-diagnostics at higher speeds and more self-adjustments without shutting down.

12.6.4. Computer control

Programmable logic controllers (PLCs) and personal computers (PCs) are being used more frequently to give operators more real-time information on the status of equipment during operations, such as speed, accuracy of fill, down-time and its causes. In addition, these systems make it easier for change-overs, ensuring through sensors that the maintenance people and operators are setting up the machine properly for the next run of a different size package.

13 Clean-in-place and sterilize-in-place systems

J.A. COVEY and J. BROWN

13.1 CIP basics

13.1.1 Historical background

Thirty-five years ago the concept of cleaning in place (CIP) was unknown. In all industries manual cleaning of equipment was universally accepted as the best way of achieving acceptable standards of cleanliness. However, there were many disadvantages to manual cleaning:

- Cleaning time was excessive and extended production-cycle time resulting in larger equipment sizes
- The standard of cleanliness achieved was low and not repeatable
- Labor, energy and chemical cost was high
- The need for manual cleaning limited the development of new processes and ideas.

In the 1950s, the dairy industry pioneered the concept of CIP, a name used to stress that pipes no longer needed to be disassembled. The dairy industry started with the spray-cleaning of milk tanks. It has since been used in all industries where sanitation is important. It now includes the cleaning of virtually all types of production equipment. The idea was simple, and even with manually controlled CIP systems the advantages were obvious. The cleaning time, cost of cleaning and the standards achieved were improved significantly. The capital investment required was relatively small. Reduction in cleaning time resulted in shorter production cycle, leading to smaller production equipment sizes.

CIP gave plant designers freedom to change traditional production techniques. Horizontal tanks used to be very common because they were easier to clean manually. Vertical tanks are used now because they are easier to clean using CIP. Automation of production and the use of welded pipes are examples of technologies that could never have been introduced without the development of CIP techniques.

13.1.2 Cleaning using detergents

Cleaning and disinfection are the subject of many ongoing research and development programs. Chemists are developing increasingly more effective detergent chemicals. The choice of applicable detergent chemicals requires a high level of expert knowledge, because the detergent should have:

- ability to hold salts in solution
- high disinfectant effect
- low foaming quality
- compatible with equipment surfaces
- safe to handle
- non-polluting and biodegradable.

No chemical substance exists that combines all these desirable properties. Commercially available detergents are a combination of different chemicals, each contributing one or more of the desirable properties. Most detergents contain alkalis, polyphosphates, surfactants and chelating agents of some kind, which are capable of dissolving and dispersing limescale deposits.

Alkalis used in detergents are caustic soda (NaOH), sodium carbonate and sodium metasilicate. Caustic soda is commonly used because it has a good dissolving effect of organic dirt, has a good fat-dissolving property at high temperature, is an effective bactericide and is cheap.

Polyphosphates are excellent emulsifying and dispersing agents. The most common are sodium triphosphate and complex phosphate compounds. They also are corrosive inhibitors.

Surfactants (wetting agents) are of several types: (i) anionic; (ii) non-ionizing; (iii) ampholytic; and (iv) cationic. Anionic surfactants are alkyl sulfates or alkyl aryl sulphonates. Cationic surfactants are mainly quaternary ammonium bases. Anionic and non-ionizing surfactants are good detergents. Chelating agents prevent salts from forming insoluble compounds. Several types of chelating agents are available. The choice is determined by the pH of the detergent solution. Polyphosphates are suitable chelating agents for weakly alkali solutions. The most common used types are EDTA (ethylene tetra-acetic acid) and NTA (nitrilo-tri-acetic acid). Gluconic acid and citric acid are also used.

13.1.3. Disinfection and sterilization

Disinfection means the destruction of microorganisms that may affect product quality. This may be accomplished by heating the equipment with 90°C water or by using chemical agents. If total destruction of bacteria is required, the equipment should be sterilized.

Chemical disinfectants are commonly used in the food and dairy industry. They can be acid, basic or neutral pH. Neutral disinfectants are the most commonly used, and include quaternary ammonium compounds and several halogen derivatives.

In the pharmaceutical industry, sterility is usually required. This is normally accomplished by steaming equipment at a minimum temperature of 121°C for at least 20 min.

13.1.4 Cleaning procedures

The required degree of cleanliness may be achieved by conducting the cleaning operations strictly according to a carefully worked out procedure. This requires that the sequence of operations must be exactly the same every time. A typical cleaning cycle is:

- Recovery of product residues by drainage, displacement with water or blow down with compressed air
- Water pre-rinsing to remove dirt
- Detergent wash
- Water post-rinsing
- Disinfection/sterilization with chemical agents or by steaming if sterilization is required.

Each step requires a certain length of time to achieve the desired result.

13.1.5 CIP/SIP definitions

Manual cleaning of equipment was a common practice many years ago. This method has now been replaced by automated cleaning. The method is named CIP (cleaning-in-place), which means that rinsing water and cleaning solutions are circulated through tanks, piping and process lines without dismantling the equipment.

CIP is described as the pumping of cleaning fluid through equipment and piping interconnected to form a cleaning circuit. The high-velocity flow of liquids in the pipe generates turbulence, which removes dirt. This applies to pipes, heat-exchangers, pumps, valves, etc. Tanks are cleaned by spraying the detergent on the upper surfaces and then allowing the liquid to drain down the walls by gravity.

SIP is a term meaning 'sterilize in place'. In the pharmaceutical industry, small pieces of equipment are dismantled and sterilized in an autoclave. This is sometimes called 'sterilize out of place'. Large equipment is not practical to disassemble, so to sterilize this equipment steam is injected into the unit. The complete system vessel, pipes, filters and other associated units are sterilized all together without disassembly. SIP may be incorporated as the final cleaning sequence in a CIP operation.

13.2 System design

13.2.1 The 3A standards

The dairy product industry established standards for system design and operation of CIP systems. In a joint effort, three organizations, the International Association of Milk, Food and Environmental Sanitarian, the United States

Public Health Service, and the Dairy Industry Committee published *'3A Accepted Practices for Permanently Installed Sanitary Product-Pipelines and Cleaning Systems'*. These documents form the basis for the physical construction and cleaning procedures in the dairy, food, pharmaceutical and biotechnology industries. These standards are summarized below.

- Product contact surfaces are limited to the following materials: American Iron and Steel Institute's 300 series stainless steel or the corresponding Alloy Casting Institute's types, equally corrosion-resistant metal that is non-toxic and non-adsorbent, heat-resistant, glass piping, and rubber or rubber-like sealing materials
- Product contact surfaces of CIP sanitary piping shall have a minimum number 4 mill finish on stainless-steel sheets
- Product lines and equipment shall have sanitary fittings or welded joints
- Equipment and liquid lines shall have sanitary fittings or welded joints
- Welded joints shall be smooth and free from pits, cracks, inclusions, or other defects
- Removable fittings shall be designed to form flush interior joints
- Lines shall be self-draining, and pitched to drain points
- The minimum average circulation velocity through each pipe or fitting in the circuit shall not be less than 5 ft/s.
- The temperature of cleaning solution and rinses shall be controlled automatically within ± 5°F range.
- The return solution temperature shall be measured and recorded
- The tungsten-shielded arc method or equivalent welding method shall be used for welding of all lines
- A proven or tested rinsing, cleaning and sanitizing procedure shall be used.

13.2.2 Mechanical design

CIP may be manual, semi-automatic or fully automated. The degree of automation depends on the number of pieces of equipment and the availability of operators.

To CIP a complete plant, equipment and lines are grouped together into what is called 'cleaning circuits or loops'. If possible, the objective is to have one circuit. In most plants this is not feasible. In many cases, some equipment is being used, while other equipment and piping is being cleaned. In this case, the plant is divided into several sectors that can be cleaned separately using CIP.

13.2.2.1 CIP lines. Pipelines for CIP should slope toward return lines to ensure drainage. The minimum acceptable slope is 1 inch every 10 feet (2.5 cm every 305 cm). The slope should be as great as possible.

Pipelines should be properly supported to eliminate sagging, which may result in liquid traps. Spacing and type of support depends on plant/piping layout. In general, support is recommended every 10 to 15 feet (300–400 cm approx.) for straight runs. Supports are also usually required around automatic valves and instrument measuring devices like flow meters. Thermal expansion and contraction should also be considered in the design since CIP solution is usually hot.

13.2.2.2 Pressure drop and velocities. For CIP a minimum velocity is required to guarantee turbulence or good mixing in the pipe. The recommended velocity is 5 ft/s (150 cm/s).

The guidelines listed in Table 13.1 may be used to obtain the recommended velocity in several sizes of piping.

Higher velocities result in effective cleaning. It also results in higher pressure drop and higher power requirement for the pump.

When piping of different diameters are in the same circuit, the flow rate is determined by the larger size pipe. Maintaining this velocity will result in higher pressure drops in smaller pipes.

To determine the volume of CIP solution it is necessary to calculate the capacity of the pipe circuit; Table 13.2 may be used as a guide.

Table 13.1 Guidelines for obtaining recommended velocity for piping of different sizes

External pipe diameter inches (cm)	GPM required to obtain velocity of 5 ft/s (152.4 cm/s)
1 (2.54)	10
$1\frac{1}{2}$ (3.81)	25
2 (5.08)	45
$2\frac{1}{2}$ (6.35)	70
3 (7.62)	100
4 (10.16)	180

Table 13.2 Pipe size and capacity of pipe (approximate guide)

External diameter inches (cm)	Capacity gallons/ft
1 (2.54)	0.0332
$1\frac{1}{2}$ (3.81)	0.08
2 (5.08)	0.142
$2\frac{1}{2}$ (6.35)	0.230
3 (7.62)	0.334
4 (10.16)	0.600

13.2.2.3 Fittings and valves. The selection of the proper fittings and valves is very important. The piping and equipment should have approved sanitary fittings or pipes must be welded. The most common type of fittings are the sanitary clamp and the bevel seat.

13.2.2.4 Storage and process tanks. The most effective and efficient method of cleaning tanks is through spray balls located in the tanks. The spray balls may be mounted permanently, or they may be removable.

Process vessels cleaned using spray balls require careful attention at the design stage. Vessel connections such as thermowells, sight glasses, agitators, inlet/outlet must be self-draining. Purchase specifications must state that the 3A code must be followed. Information relating to CIP must be supplied with vessel drawings. Sanitary design and cleanability factors must be planned before fabrication.

13.2.2.5 Design considerations. The first step in CIP design is examining the supply line to the spray ball and the return line from the tank. It is important for the return line to have an adequate slope. Air in the return line must be purged. If the air is not eliminated the pump will vapor lock. A slope of 1″ in 20 feet (2.54 cm in 609.6 cm) will allow air to leave the line. To allow for upset, a slope of 1″ in 10 feet (2.45 cm in 304.8 cm) is recommended. An air vent may also be installed in the return pump casing or in the suction line.

Determining the flow rate of the solution is the next step. There are two different rules-of-thumb for static spray balls. Cleaning solutions should be applied at a rate of 0.10–0.5 gpm/ft^2.[1] In general, 0.2 gpm/ft^2 may be used as a first estimate. Fluids can be applied at a rate that results in a falling film, Reynolds number of 2000 or more. The formula is:[2]

$$NRE = 5952\, M/Wu \tag{13.1}$$

NRE = Reynolds number; M = mass per time in lbm/s; W = surface width in feet; u = liquid viscosity in CP.

It is very important that the CIP feed line provides adequate flow and pressure to the spray ball. Return lines also must be sized correctly to avoid vapor locking the return pump.

An inductor type of return system may be used. This method eliminates sloping, venting and draining problems by applying vacuum in the return line.

To eliminate the possibility of the tank collapsing, vessels should be designed for full vacuum or suitably-sized air vents should be provided. Factors to consider when sizing these air vents are the construction and capacity of the tank or vessel, flow rate of solutions, maximum temperature of solution and minimum temperature of the cold water. Vessels that are sterilized shall be designed for full vacuum. If this is not economical, vent and filter sizing should consider the effect of the change in vapor density when steam is cooled.

13.2.2.6 CIP key pieces. Protection must be provided when cleaning a tank and its associated piping where adjacent tanks and piping may contain product or the adjacent tanks and piping are clean or sterile. There are two methods to prevent contamination of the clean side. The most common method is the use of key pieces. These are sections of pipe, which, by virtue of their length and connections, fit only two positions in the system—in the product line alone or in combination with other key pieces to complete the CIP circuit. The second method uses a block-and-bleed arrangement of shut-off valves. Figure 13.1 shows examples of each method.

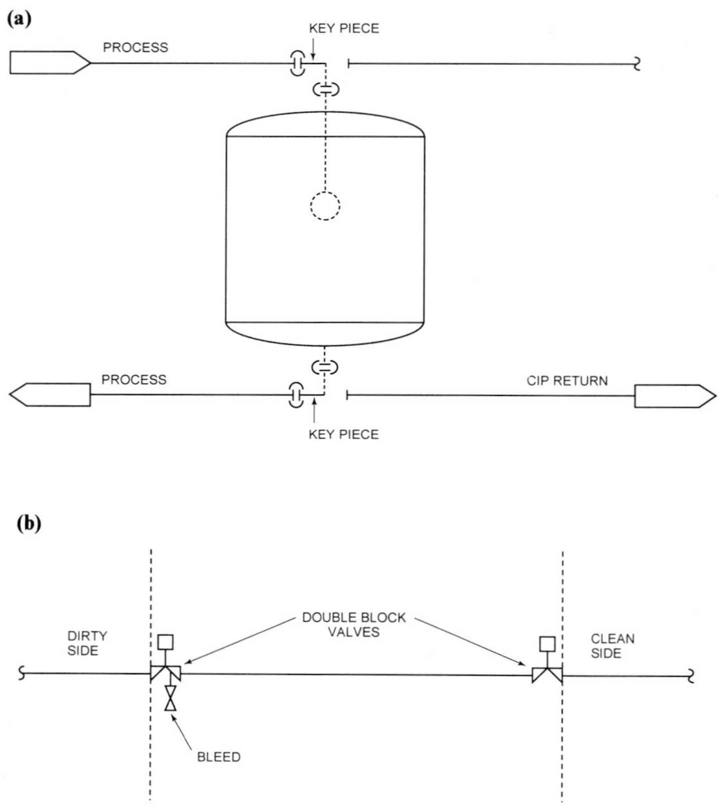

Figure 13.1 CIP key pieces: (a) in combination with other key pieces to complete the CIP circuit; (b) a block-and-bleed arrangement.

13.2.3 SIP design considerations

The important factors in SIP design are:

- Venting of air
- Draining of condensate
- Adequate supply of steam
- Automation versus SOP
- Maintaining sterility.

The first step in a steam sterilization cycle is removal of air. If excessive air is trapped in the piping, vessel cold spot will result and steam penetration or heat up will be slow. Air is normally removed by venting at the high point and through the steam trap. Most steam traps have an air-venting capacity of about 2–15 SCFM at 20 psig. Traps will readily allow air to bleed out and keep steam inside the system. Figure 13.2 shows the air capacity of typical thermostatic traps. Air at the high points should be removed by providing vent valves. Figures 13.3 and 13.4 illustrate typical vessel and filter connections for SIP.

Condensates should be removed as fast as they accumulate. All low points in the system including piping should have a drain. These drains may be either a steam trap, regulating valve or an orifice. Condensate is detrimental to the sterilization process. The condensate will have a lower temperature than steam and can remain cold enough to prevent adequate sterilization.

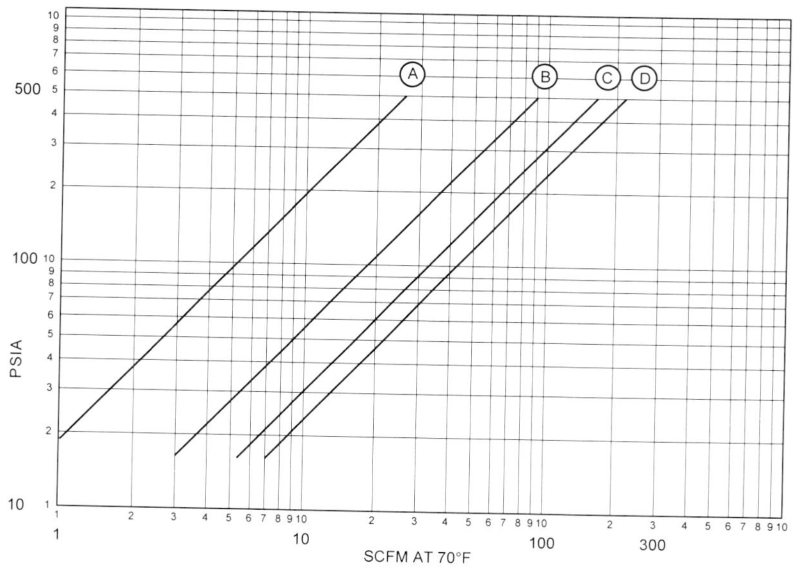

Figure 13.2 Air capacity of various size thermostatic traps.

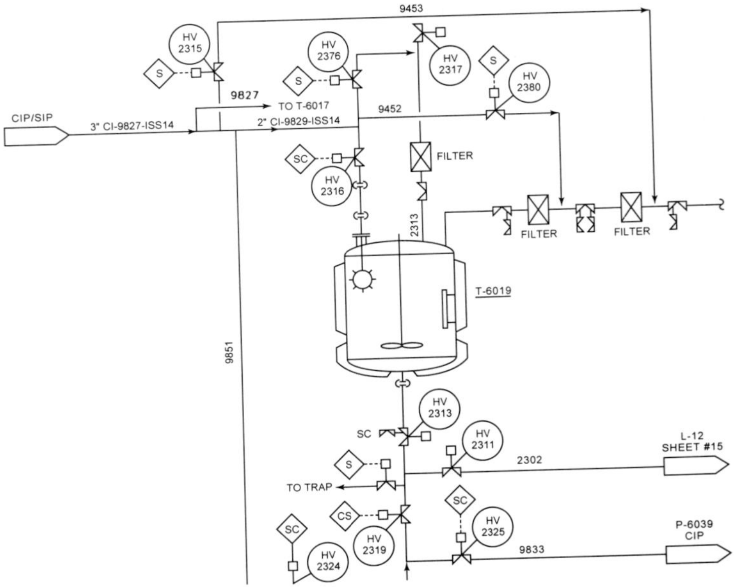

Figure 13.3 A typical vessel and filters connected for SIP and CIP.

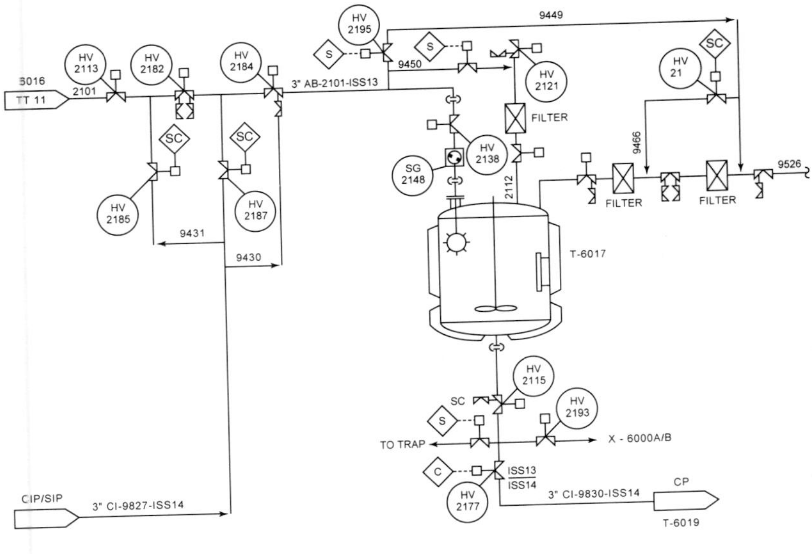

Figure 13.4 Another example of a typical vessel and filters connected for SIP and CIP.

Adequate supply of steam must be provided to the system. The steam unit should be sized to meet the heat up requirement of the largest vessel or equipment. An uninsulated vessel requires about 0.066 kg/h per liter.

A SIP may be accomplished using a control system. A computer system will provide an increase in process reliability compared to a manual system. Most SIP, particularly in pharmaceutical applications, are manual. For this type of system a very detailed standard operating procedure (SOP) is required to guarantee a successful sterilization.

Maintaining sterility after completing the sterilization process is attained by blowing or charging a pressurized gas (air or nitrogen) into the vessel or equipment through a sterile filter. The pressurized gas is used to purge the system of residual steam and condensate. The pressurized gas may also be used to dry the system. This may be an important consideration for a non-aqueous system.

13.2.3.1 Steam quality. The quality of steam used for sterilization is very important. For pharmaceutical use, pyrogen-free steam is specified for sterilization. Pyrogen-free steam is often referred to as pure steam and is used for applications such as the sterilization of mammalian cell-culture systems where the presence of pyrogens will severely retard cell growth. Other uses are found in parenteral drug manufacture, where any pyrogens remaining in the process system could find their way into the end product. Pure steam is generated by a special clean-steam unit designed to remove pyrogens. Water for injection (WFI) is usually the feed water, and is pyrogen-free.

Clean steam is produced in several ways. The most common methods are:

- Filtrations of plant steam using a high efficiency filter. A typical specification will call for the removal of 95% of particles larger than 5 μ.
- Direct generation from di-ionized or distilled water in an electric or steam unit.

Clean steam is used for sterilization in non-critical areas in the food and drug industry where the presence of pyrogens is not an important issue.

13.3 Types of CIP systems

There are several types of CIP systems that can be designed to meet the requirements of a specific cleaning application. The CIP unit consists of a tank or tanks for the supply of cleaning fluids, a supply pump to recirculate the solution through the vessel or line, valves to direct the flows at the unit, and instrumentation to verify that cleaning conditions have been achieved. The common types of systems used today are:

- Re-use system
- Single-use system
- Solution recovery.

13.3.1 Re-use system

Re-use CIP units (Figures 13.5 and 13.6) are designed to re-use the cleaning fluids for several CIP cycles. The units usually have two tanks for holding water and cleaning solution. More tanks may be added as required. This system configuration is used where soil loads are minimal, and the unit cleans the same type of equipment repeatedly. The advantages of this configuration are: (i) lower detergent usage, since chemicals are re-used for several cycles;

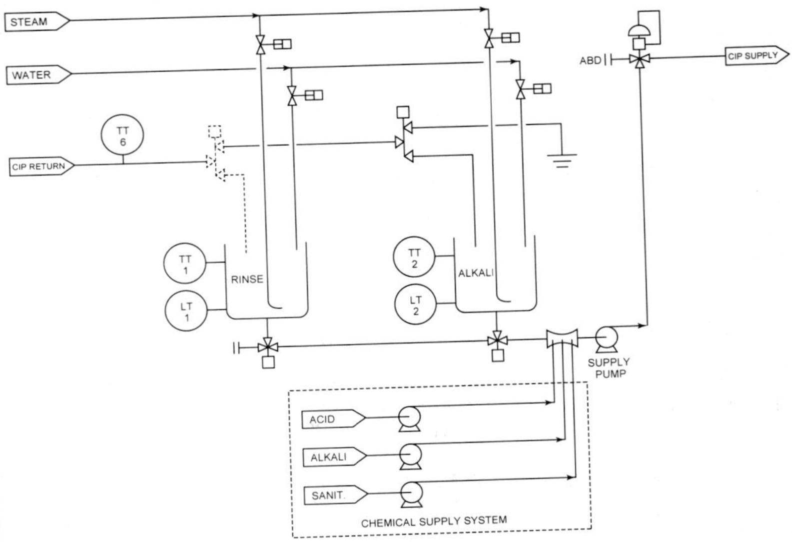

Figure 13.5 Two-tank alkali re-use CIP system: TT = temperature transmitter; LT = level transmitter.

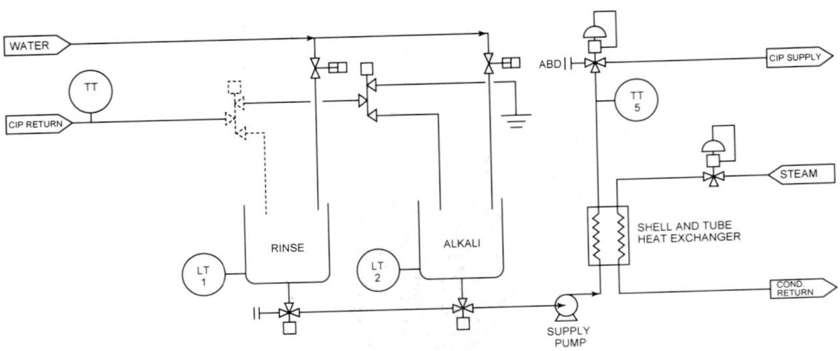

Figure 13.6 Two-tank alkali reclaim CIP system with heat exchanger. TT = temperature transmitter; LT = level transmitter.

and (ii) savings in energy cost since hot cleaning solution is re-used in the next cycle. The disadvantages are: (i) higher initial equipment cost; (ii) larger tanks and more floor space are required for unit; and (iii) a longer pre-rinse cycle may be required to reduce soil load.

13.3.2 Single-use units

Single-use CIP systems are based upon the single use of chemicals which are dumped to drain at the end of each cleaning task. These systems are designed to use the minimum quantity of solutions required to maintain recirculation for each CIP circuit. Single use units have one tank which is sized to provide sufficient liquid for line circuits and water supply surge during pump down. The single use system is applicable when cross contamination is an issue because all spent solution is discharged to drain after use. The advantages of this system are: (i) very flexible, (ii) lower initial investment, (iii) eliminates cross-contamination, (iv) smaller space requirement, and (v) solution concentration is easier to control. The disadvantages are: (i) higher water and chemical consumption, and (ii) higher energy cost since hot water and chemical are drained.

13.3.3 Solution-recovery units

Solution-recovery CIP systems (Figures 13.7–13.10) are units that save the caustic wash solution and post-rinse water for pre-rinse in the next cycle.

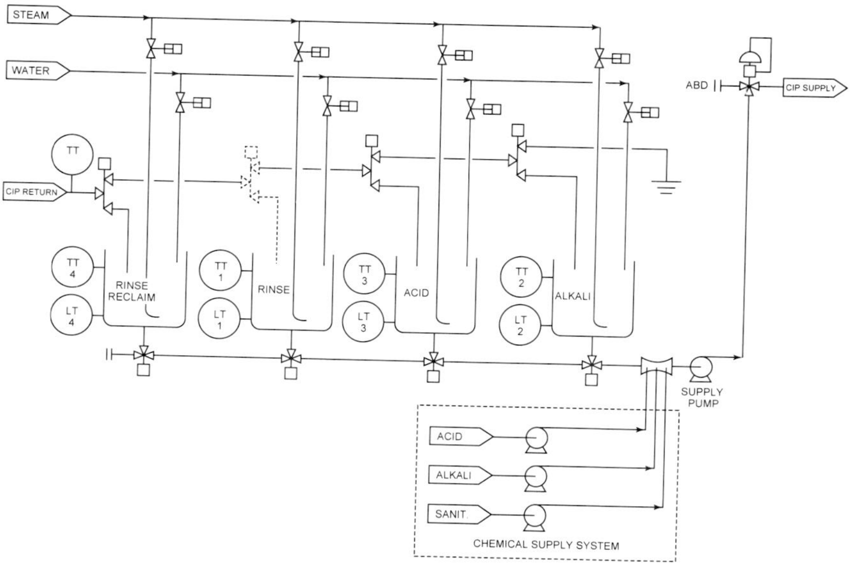

Figure 13.7 Three-tank CIP system. TT = temperature transmitter; LT = level transmitter.

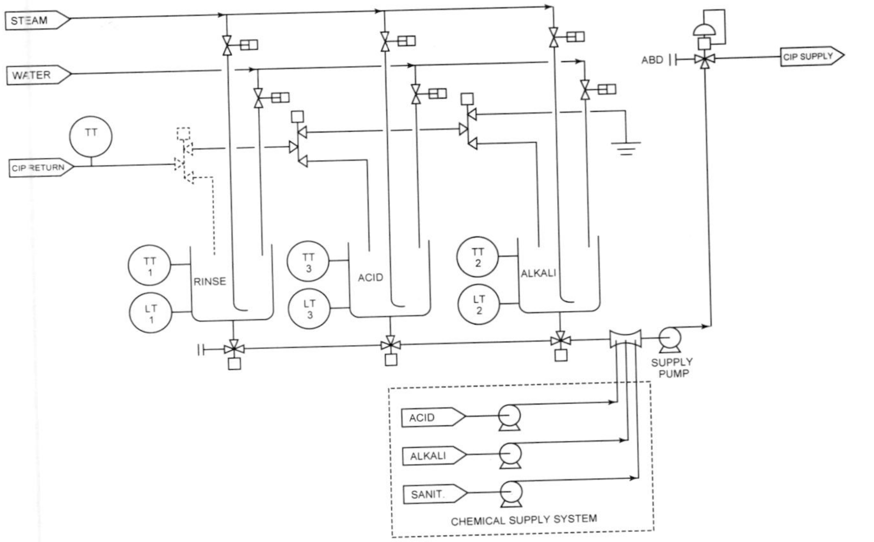

Figure 13.8 Four-tank CIP system with alkali/acid and rinse reclaim. TT = temperature transmitter; LT = level transmitter.

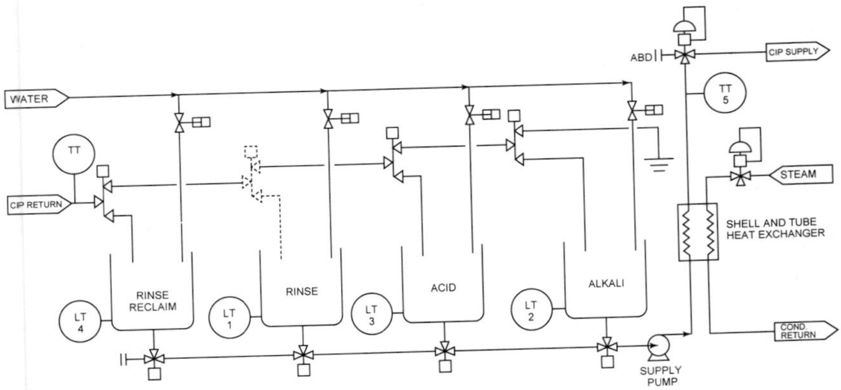

Figure 13.9 Four-tank CIP system with alkali/acid rinse reclaim and heat-exchanger.

The system operates as a single-use system, except that cleaning fluid and last rinse are pumped to the recovery tank. The advantages of this unit are that it: (i) minimizes water and chemical comsumption; (ii) reduces energy cost. The disadvantages are: (i) higher capital investment; (ii) potential cross-contamination; (iii) a larger space requirement; and (iv) stagnant water and solution in recovery tank may be a potential source of bacterial contamination.

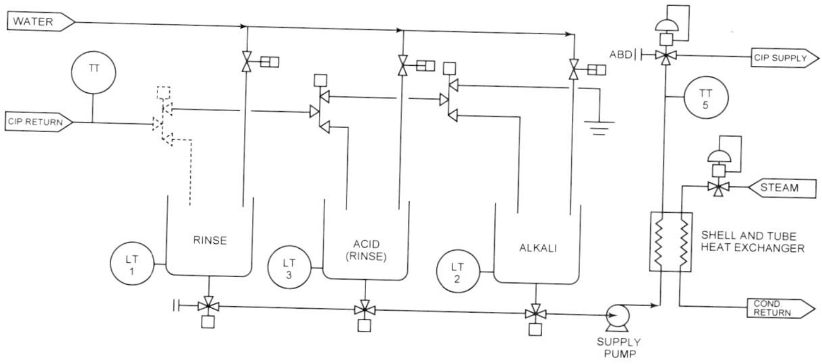

Figure 13.10 Three-tank CIP system with heat-exchanger.

CIP units may also be classified by the methods of returning the rinse water or cleaning fluid. The three most common methods of return are: (i) gravity return; (ii) return pumps; and (iii) eductor return.

Gravity return may be used when the equipment being cleaned is located at a higher elevation than the return and drain to the CIP return. The return line must be sized carefully to allow adequate flow to supply pump. Advantages of this design are: (i) low capital cost; and (ii) minimal space requirement. The disadvantage is that the CIP unit must be closed for equipment to be cleaned.

The use of a pump to return CIP solution is the most common method, as they can operate at higher flow rates and longer distances. Return pumps are flexible, their advantages are: (i) high flow-rate capability; (ii) that they can clean remote locations; and (iii) a multiple return. The disadvantages are: (i) the flow in the system must be balanced; (ii) the pump will cavitate if vapor is trapped in the system.

The educator return system uses a vacuum to return CIP solutions. The unit includes a motive pump, which recirculates water from a motive tank through an eductor. Advantages of this system are: (i) complete evacution of equipment and lines; (ii) lower equipment cost as a result of eliminating multiple pumps from multiple areas. Disadvantages are: (i) larger return lines; (ii) return lines must be sealed tightly to prevent leaks of outside air; and (iii) that it is limited by the length of the return line.

13.4 CIP cleaning process

The CIP cleaning cycle is determined by process requirements, nature of soil and the type of equipment but the cycle steps are similar. The typical steps or cycles are:

1. First pre-rinse
2. Second pre-rinse
3. Wash cycle
4. Post-rinse
5. Acid wash
6. Final rinse
7. Steaming (for equipment that requires sterilization).

The first pre-rinse with water removes gross soil. It is typically a 15–45 s burst. The rinse water is pumped to drain. The second pre-rinse with water, if required, may use recovered water from previous cycle. A second pre-rinse is required for heavy soil load. In the wash cycle, detergent is recirculated at a specified temperature and concentration for 5–45 min depending on soil load. Washed solution may be drained to sewer or pumped to a recovery tank for use in the next cleaning cycle. The post-rinse with fresh water flushes detergent from the system. The rinse water may go to the sewer or be recovered. The acid wash, if required, recirculates acid solution at a specified concentration and ambient temperature for 5–10 min. Acid wash may be required when thermal incrustations still remain on the cleaned surface after treatment with detergent. In the final rinse, the system is rinsed with purified water repeatedly until rinse water reaches an acceptable final-rinse conductivity.

In the pharmaceutical industry, sterilization of the system may be required. The best time to steam the equipment is immediately after the final rinse step of the CIP. Pure steam (15–20 psig) is applied to pipe and equipment for a minimum of 20–30 min. A minimum temperature of 121°C is required to achieve sterilization. Condensates are collected at low-point drains and are usually drained to the sewer.

13.5 CIP control system and documentation

The type of controls used for CIP systems depend on the capital budget, plant requirement, and availability of in-house technical expertise. The basic philosophy in CIP control is to provide for repeatability of operations. Once a cleaning program has been proven, the program sequence is ‘locked-in’ so that it can be repeated as required with predictable results. There are many types of CIP controls that can provide a broad spectrum of requirements from simple systems to complex multiple systems. Many units are still running using push-button/mechanical time, cam timer/relay-sequencing or drum-programmer, which are considered obsolete today. Presently the most common control systems used for CIP are electronic sequencers, PLCs, microprocessors, and PC-based controllers.

If a CIP unit has a few loops, a simple electronic sequencer could be used. These devices provide no logic. It provides stepper controls that energize outputs for a programmed time before advancing to the next sequence steps. Digital displays can provide status information. Documentation is provided as a separate recorder.

Programmable logic controllers are also used to control CIP systems. They are reliable and have the ability to handle many control tasks simultaneously. Their disadvantage is the complex programming. Maintenance personnel must receive training for this type of equipment.

A dedicated microprocessor is an approach to CIP control based on providing an economical control system that is user-programmable, easily maintained, and with the reliability of current electronics hardware. Many units using this technology are currently available, with pre-defined cycle operations and user definable operating parameters. The various possible flow configurations are entered in the order desired, along with set points, and cycle status is displayed. Programs can be changed easily by a plant engineer. Cycle flows are generally limited to the pre-defined configurations in the systems.

PC-based controllers are commercially available with dedicated software for CIP. The cycle steps can be chosen from pre-defined menus or can be defined by the user. Floppy disks can be used for storing programs. Program steps can be printed out for validation and documentation. They can be used for process and SIP control. CRTs are used for display. Historical data can be collected and stored to electronic data storage. Modems can be used to transmit programs for troubleshooting and revision.

13.5.1 Documentation and monitoring

The success of a CIP cleaning cycle is dependent on time, temperature, chemical concentrations and flow. The following parameters determine the cleaning performance of a CIP system.

1. Pressure: a pressure transmitter at the discharge of the supply pump will provide documentation of hydraulic performance. An adequate pressure reading is an assurance that the spray balls provide coverage.
2. Supply temperature: temperature measurement will operate the steam valve. For safety, temperature should not exceed limitations of the system.
3. Return temperature: this temperature must be recorded to confirm that the CIP cycle maintained minimum acceptable temperature at the coldest point in the system.
4. Flow rate: a flow-measuring device at the discharge of the supply pump should verify that the minimum flow rates have been achieved. Flow rates should be documented. A low flow may be an indication of clogged spray balls or line blockage.
5. Chemical composition: conductivity and pH are measures of composition. The sensing elements are installed in the return line. These measurements are used to control the addition of feed chemicals.

6. Rinse conductivity: a good indicator that the system is clean is the final rinse conductivity.

References

1. Adams and Agarnal (1990). "CIP System Design and Installation"; *Pharmaceut. Eng.* **10**(6).
2. Adams and Agarnal (1989). Clean-in-place system design. *BioPharm;* **June.**

14 Controls and automation for biotechnology and pharmaceutical industries

P.W. YANG

14.1 Introduction

For better quality products and more efficient operations, the process control industry had been seeking continuously new technologies to automate the manufacturing process. In the 1940s and early 1950s, control systems were predominantly mechanical and pneumatic, with large case instruments that measured about 36 × 46 cm (14″ × 18″ approx.) in size. These instruments were often mounted directly at the sensing point scattered locally throughout the plant. The advance of pneumatic transmission technology gradually brought the instruments into a central control room. With all instruments in a centralized location, the productivity and efficiency of plant operations were greatly improved. However, the pneumatic control panel was enormous with sizes of 9 m (30 ft approx) or longer, a common occurrence, and operators had to move around in order to see all process variables. Additions or change to pneumatic control loops were extremely difficult and expensive.

With the advance of solid-state technology in the late 1950s, the 7.6 × 15.2 cm (3 × 6 inch) miniature electronic instruments were employed and electronic transmission was gradually taking over from pneumatic transmission. The miniature electronic instruments greatly increased the flexibility of control systems. The control panel and control room were smaller with fewer operators. In the early 1960s, computers were added to the control systems. A typical process control computer has a mean time between failure (MTBF) of around 1000 h. With the advancement of digital integrated circuit technology, the introduction of microprocessors to control the process became a reality. 'Computerization' had a great impact on the process controls. During the 1980s and 1990s single-loop controllers, programmable logic controllers and microcomputers (PC) were all employed by process-control systems in various industries. The MTBF for a microprocessor-based process controller had also greatly improved to about 50 000 h or better. To ensure control reliability, redundancy designs of central processing units, data highways and input/output modules were also incorporated into the control systems.

To meet the Food and Drug Administration's (FDA's) process validation requirement an automated computer-controlled system is indispensable for the biotechnology and pharmaceutical industries. Automation systems cannot only improve product yield, product quality and operational efficiency but also give the plant and corporate management a great deal of manufacturing information vital to the plant and corporate decision-making. The relation of different levels of control automation can be illustrated as shown in Figure 14.1.

Refer to Figure 14.1, the first level is the process control level, which includes the continuous and the batch controls. The process control level interfaces with field-sensing instruments and sends a control signal to the field-control element through a pre-programmed algorithm resident in the process controller. The second level is the process supervision level, which provides the operator and the process control system interface. The process management level involves process batch data management. The process batch data management provides current and historical process data information, material and batch information, statistic quality control (SQC) and/or statistic process control (SPC) information. The batch data management can also provide a customized process data report, such as batch material report and shift operation report. The next two levels involve plant management and corporate management for the plant or corporate planning, scheduling and the plant operation optimization. The plant or corporate planning of future production output is based on the collected product market demand information. The scheduling is to meet the production demand based on the analysis of the past production information. The optimization allows for the determination of the process settings based on the process operation criteria formulated from the collected process data. The operations of planning,

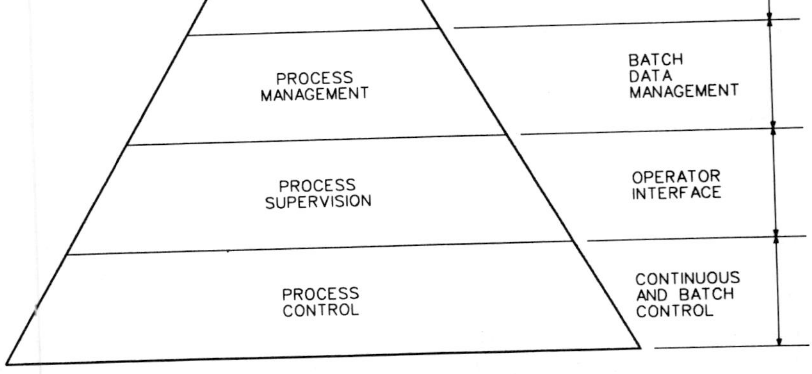

Figure 14.1 Levels of control automation.

scheduling and optimization are most often handled by a dedicated host computer with customized high-level language programming and/or a specialized process management software from a software developer.

14.2 Biotechnology control system

14.2.1 Process equipment

The biotechnology process may vary from product to product but, from an equipment point of view, it can be characterized as follows.

1. *Media preparations.* Typical media preparations consist of one or two rooms, each containing media preparation tanks and portable tanks.
2. *Inoculum preparations.* The inoculum preparation equipment consists of water batch tank, centrifuge unit, incubator and biosafety cabinet. In general, there will be one inoculum preparation room for each fermentation suite.
3. *Bioreactor suites.* Most biotechnology plants have three or four bioreactor suites. Each suite may have three to four bioreactors, harvest tanks, buffer storage tanks and one kill tank.
4. *Recovery systems.* The recovery system is also called the cold room. It has several portable, wheeled, hold tanks, buffer tanks and two or three sets of columns and fraction collection systems connected in tandem. The cold room is to be kept around 7°C, and there is usually one cold room for each fermentation suite.
5. *Purification systems.* Purification of the product is achieved by a series of liquid columns. Usually, there will be one purification system for each fermentation suite.
6. *Product packaging.* The product packaging system involves a sterilizing washer, depyrogenation oven, scale, lyophilizer, product container filling station, label and package station.
7. *Utility systems.* The utility systems include water for injection (WFI), clean steam, clean-in-place (CIP) controls and valving, carbon dioxide supply, oxygen supply, de-ionized water supply, chilled and tempered glycol supply, vacuum system, steam boiler, air compressor system, cooling water system and wastewater treatment system, etc.

14.2.2 Process operations

The process equipment arrangement of a typical biotechnology plant is shown in Figure 14.2, and the typical operation sequences are shown in Figure 14.3.

From Figures 14.2 and 14.3 it is evident that the biotechnology system is a multi-train system with single or multi-products automated by the recipe of media and inoculum preparation. However, each train is operationally independent. It is a common practice that most of the biotechnology

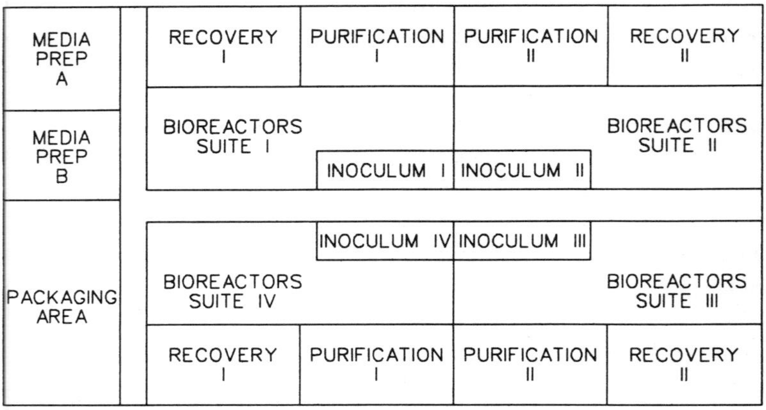

Figure 14.2 Typical biotechnology plant-equipment arrangement.

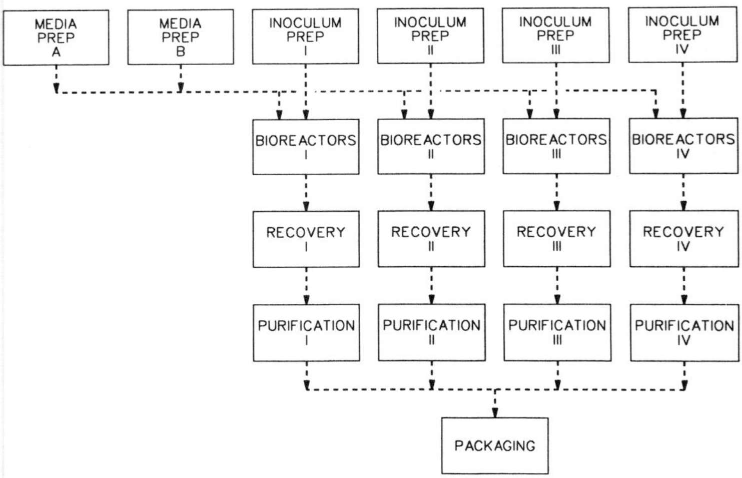

Figure 14.3 Typical biotechnology operation sequences.

equipment is supplied by vendor package systems. The utility systems are generally self-contained control systems and only process date monitoring is required. Media preparations, bioreactors, kill tanks and waste-water treatment systems will be controlled by an automated control system.

14.3.3 *Process controls*

Most biotechnology facilities require FDA validation. This validation requirement results in a large amount of process data-acquisition and historical data storage. Since the biotechnology process is distributed and independent, there is, in general, no centralized control room to monitor the entire plant. The industry practice is to have a small control room in each of the bioreactor suites, media preparation rooms and the utility room. In the control room, there is at least one operator station, one graphics printer, one archiving device and controller cabinets for controlling and monitoring field devices. This distributed control system is shown in Figure 14.4. The controller cabinets can also be located in the field near the controlled equipment and a single cable from the controller cabinet runs back to the operator station to save substantial field cable-installation costs. The operator stations and the supervisory computer are connected through a local area network (LAN) and each operator station can be programmed to view the data of any of the other stations.

The graphic printer is used to print out alarm status, event logging, and reports, as well as graphics displays. The archiving device is used to store process historical data, which can be retrieved at a later date.

The supervisory computer is used for the plant-management system. This management system will collect data from every operator station and furnish a customized data report to the management and the regulatory agencies.

The operation of the biotechnology system is usually a mix of manual, semi-automatic and automatic operations. For instance, raw materials are very often measured manually and added to the media-preparation reactors. The transport of media to the bioreactor and the final product packaging are mostly in semi-automatic operation. However, the controls of the

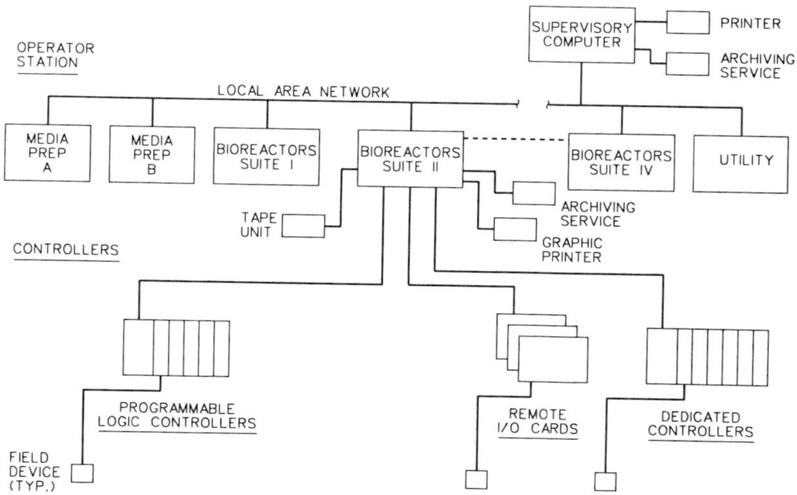

Figure 14.4 Typical biotechnology plant-control system, with choice of different type of controllers (only one suite shown).

bioreactor such as agitation, temperature and pressure, pH and dissolved oxygen controls are performed in fully automatic mode.

The biotechnology control system as shown in Figure 14.4 can be implemented with a programmable logic controller (PLC) computer or a true distributed control system (DCS).

14.3 Bulk pharmaceutical control system

14.3.1 Process equipment

The bulk pharmaceutical process is typically a batch process with a complex sequential control of operation. A typical process block diagram for the bulk pharmaceutical process is shown in Figure 14.5. The process and equipment involved are as follows:

1. *Reagent preparations.* These involve truck unloading, storage tanks, transfer pumps and mixing tanks with agitation.
2. *Raw material preparations.* These involve material weighing system, material transfer system, and material mixing and holding tanks.

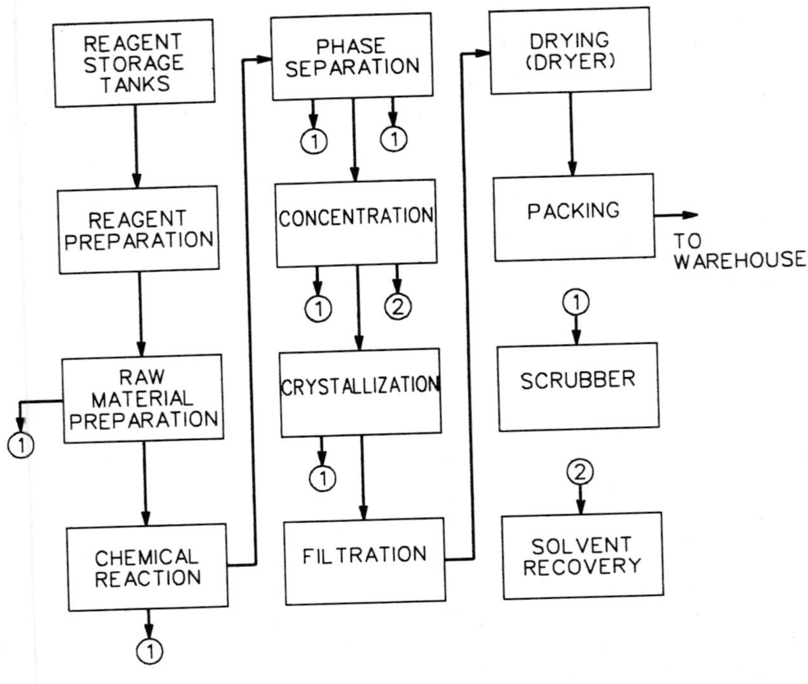

Figure 14.5 Processing block diagram for bulk pharmaceutical process.

3. *Chemical reactions.* One or more reactors with variable speed agitators are employed for chemical reaction process. Reactors can be glass-lined, hastelloy C, or stainless steel as the process dictates.
4. *Separations.* Separation process can be in single or multistages. The equipment involved may be one or more tanks for each separation process as follows:
   (a) Phase separations
   (b) Concentrations (evaporation and resolution)
   (c) Crystallizations.
5. *Filtration.* Filtration or dewatering is usually performed using a centrifuge. The centrifuge unit is normally supplied as a package unit with its own control systems. Operation and safety interlocks are required between the centrifuge unit and other process units.
6. *Drying.* A single or multiple dryer may be used to dry the products. Dryers can be furnished as a complete package unit or furnished as drying drums with separate purging, vacuum, and heating and cooling units. If separate units are furnished, a substantial amount of control integration work will be required.
7. *Packaging.* Each container or drum is to be filled to a pre-determined weight and measured by a weighing scale. The final product is then transported to a refrigerated storage warehouse.
8. *Auxiliary systems.* These are usually gaseous absorption scrubber systems and liquid solvent distillation systems. Both systems can be supplied as package units either with or without control systems. The scrubber and distillation systems are normally a continuous control system with feedback closed-loop controls.
9. *Utility systems.* The utility systems usually required are chiller, vacuum, de-ionized (DI) water systems, clean-in-place (CIP) and steam-in-place (SIP) systems, water-for-injection (WFI) system, cleaning and process steam systems, cooling water, nitrogen and compressor air systems. All of the utility systems can be furnished as package units either with or without control systems.

14.3.2 Process operations

A typical sequence of operation of a bulk pharmaceutical plant is shown in Figure 14.5. The reagent is fed from the reagent storage tank to the reagent preparation tank where the dilutant is added to reach desired concentration. The diluted reagent, solvent and raw materials are then fed to the raw materials preparation tank according to a pre-determined recipe. The prepared material is then pumped to a reaction tank and reactants are added based on product formula. After reaction is completed, the phase separation is accomplished, if separation is required. Otherwise, the medium is pumped to the concentration tank of the process directly. The concentration is normally achieved by evaporation and aqueous resolution repeatedly until the desired concentration is achieved. The medium is then crystallized and pumped to a batch or continuous centrifuge unit for de-watering.

The mother liquor from the centrifuge unit is concentrated and then fed back to a centrifuge to maximize the yield. The dewatered medium is transferred to a dryer, which can be either a batch or continuous dryer. The drying process is normally accomplished by applying in combination a heating and vacuum system. The dried product is weighed, filled into drums and transported to a warehouse for storage. Scrubber and solvent distillation systems are used to protect the environment by ensuring that no toxic materials are released into the air or sewer.

14.3.3 Process controls

The bulk pharmaceutical process is primarily a batching process with multiple recipes. The process can be single-train or multiple-train. Owing to the availability of batching time-sequences systems, multiple trains are commonly adopted for most pharmaceutical plants. Recipe handling, complex and flexible control strategies, quality controls, validation requirements, report-generation and data-archiving are major factors when considering which type of control system should be used for the pharmaceutical process.

The batching process performs not only sequential controls but also many analog loop controls. In addition, the number of input and output points for a bulk pharmaceutical process is relatively large, typically around 1500 points. Therefore a distributed control system (DCS) is commonly chosen for process controls. A typical DCS architecture is shown in Figure 14.6; the engineering work station is used to perform database configuration, programming, testing, diagnostic and application program generation. A minimum of two operator stations should be used for operator interface. Through the operator station, the operator can view the status of each step of process control and decide the necessary action to be taken so as to respond to changes in the process requirements. The historical trend unit is essentially for process data archiving, which is essential for validation purposes. Depending on the DCS vendor, the historical trending function may also be performed by the engineering work station or operator station.

The unit controller is a microprocessor-based dedicated controller. The input/output cards are placed in slots of the input/out card chassis, which is connected to the unit controller through a twisted pair wire acting as a local communication cable.

Data-highway redundancy is desirable for establishing reliable data communications. The data highway provides process data transmission among all devices connected to the data highway.

The engineering workstation, operator stations and historical trend unit are normally located at the central control room in a controlled environment.

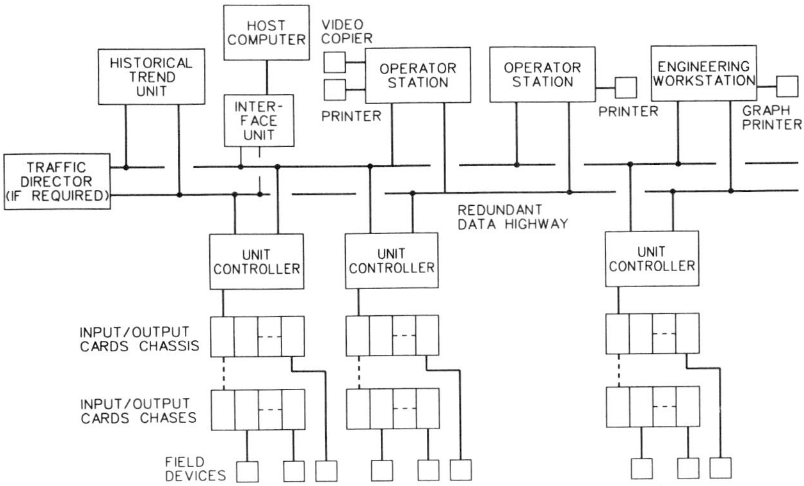

Figure 14.6 Typical distributed control system.

The unit controllers and input/output (I/O) chassis are housed inside the factory assembled cabinets. These I/O cabinets can be either centrally located in the equipment room or distributed in the field near the process equipment.

The host computer is used for the management control system, which includes planning, scheduling and process optimization.

The control of the pharmaceutical process is, in general, designed for fully automatic operation. The need of operator interface or manual operation should be minimized.

14.4 Control hardware and software

14.4.1 General considerations

The hardware and software requirements are application-dependent. Especially in software, more sophisticated packages may be required for one application and less sophisticated for other applications. However, both modular designed hardware and software are currently available and should be selected for all applications. The existence of large number of hardware and software systems, makes the selection of the right system sometimes difficult. Often a software demonstration disk is available from the vendor for evaluation but this 'demo-disk' only highlights what it can do and not what it cannot. The best way to ensure that the end user will have the right hardware and software is to hire a qualified consultancy firm or consultant to evaluate the system before placing the purchase order. The discussions in the following sections may be helpful in selecting and evaluating the system.

14.4.2 Operator station

The operator station is a device for operators to communicate with the control process. Through an operator station, the process operations can be viewed, process information can be collected and commands to the process controller can be made. The operator station is available in two types of computers, namely, industrial computer and industrial workstation. The industrial computer is generally a rugged personal-computer (PC)-based operator station running on a disk operating system (DOS), operating system 2 (OS/2), UNIX, Windows, Windows NT, Windows '95 or other operating system. The industrial workstation is a proprietary operator station with a proprietary operating system. The specifications of operator stations can be summarized as follows:

14.4.2.1 General requirements. The operator station should be at least a 66 MHz, 32-bit microprocessor that is based with either a proprietary designed computer or personal computer (PC) such as Apple Computer, IBM PC/AT, IBM Microchannel Computer. More powerful and higher-speed computers are also available such as DEC (Digital Equipment) Alpha system with clock speed of 150–200 MHz. A higher performance and higher-speed power PC is also available.

The mounting of operator station should be available in desktop console or rack mount. The construction of operator station should be available in NEMA 1, 4 and 12 for general purpose area classification and ISA type Z purge construction for hazardous area Class 1, Division 2 classification and ISA type X purge construction for Class 1, Division 1 hazardous area classification.

The operator station should meet FCC Part 15 Class A emission limits and shall also have EMI and RFI immunity. In addition, operator stations should be able to operate under the following environmental conditions:

- Temperature: 0–40°C
- Humidity: 10–90% non-condensing
- Vibration: 1 G max. 60–500 cycles

Power supply for operator stations should be within 108–125 VAC at 58–62 Hz or 220–250 VAC at 48–52 Hz.

Operator stations should have, at least, an operator interface software with the following basic capabilities:

- Interface with process controller
- Password security operation
- Priority alarms and alarm direct access

- Recipe handling and scheduling
- Dynamic graphics, control loops, current and historical trending
- Displays
- Free-formatted report-generation
- System diagnosis and testing.

A typical block diagram of the operator station is shown in Figure 14.7

14.4.2.2 Monitor. The monitor screen should be a minimum of 17 inches (43.18 cm) with high resolution (680 × 480 pixels) minimum (a pixel is the smallest element or dot that can be displayed on the monitor screen). The display should be in color with mouse and/or touch-screen operation. The monitor display update time should be no less than 1 s to facilitate operator interface and reaction.

14.4.2.3 Keyboard. The engineering keyboard is used for configuration and programming, and the operator keyboard is used for process operation. The engineering keyboard should be of the full-stroke type. The operating keyboard should be of the membrane type. Keys for the operator's keyboard should be functionally grouped together and have user-configurable function keys for direct functional access, for example, function key 'F2' may be configured for an emergency stop function. Once the key F2 is pushed, the operator station will automatically access to the emergency stop program and safely shut down the process operation.

14.4.2.4 Memory. The capacity of memory should be at least 8 MB (mega-bytes) of RAM (random access memory) with expansion to 16 MB (a

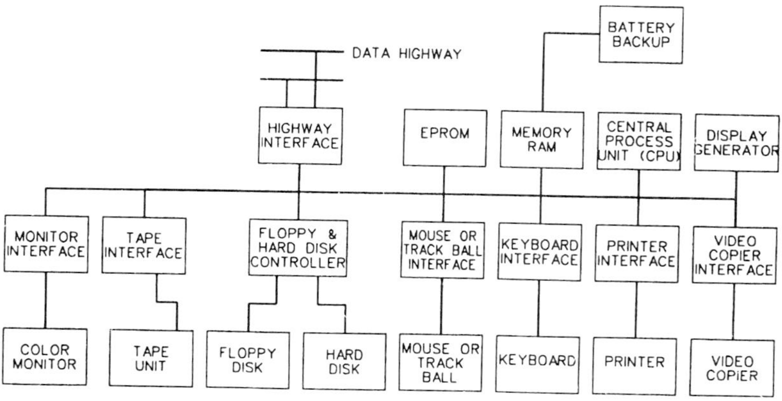

Figure 14.7 Typical block diagram of operator station.

byte consists of 8 bits and a bit is a basic memory element storing binary information of 1 for true, or 0 for false). For process data archiving, a 150 MB hard disk memory with option up to 300 MB should be considered. An electrical erasable programmable read-only memory (EEPROM) is also desirable to back up the software in the event of loss of power for long periods of time. The EEPROM will automatically download the software once the power is restored.

14.4.2.5 Periphery. Peripheral support should include, but not be limited to, the following devices:

- One monitor with option for up to three or more monitors
- Graphic printer
- Video copier
- Tape unit
- Operator keyboard and engineering keyboard
- Internal hard disk
- External hard disk
- Two floppy disk units
- Mouse or trackball-controlled cursor.

14.4.2.6 Communication. The operator station should be able to communicate directly to a redundant data highway system. The secondary communication port must operate automatically in the event of failure of the primary communication port.

14.4.2.7 Battery back-up. It is imperative that the operator station have battery back-up to retain the data and operating programs in RAM after power loss.
A typical block diagram of the operator station is shown in Figure 14.7.

14.4.2.8 Operating system. The operating system for the operator station must be a real-time, multi-tasking and multi-user system. A real-time operating system means the computer must provide a quick event response within a pre-defined time frame. The execution of tasks is according to a time scheduling, which is normally based on the task priority with the highest priority task executed first. This time scheduling of suspending the lower priority tasks and executing the higher priority tasks is called priority-based pre-emptive scheduling. The computer with real-time operating system must also be able to respond quickly to an external event through an interrupt (request). For multi-tasking, a computer executes on more than one task at a time. There are several types of multi-tasking, namely, context-switching, cooperative multi-tasking and time-slice multi-tasking. In context-switching, two or more tasks or applications are loaded to the computer at the same time but only the foreground (on-screen) task is executed and the task in background (not on screen) can be activated to the foreground for execution through an interrupt. The cooperative multi-tasking executes the foreground task and executes the background task only during the idle time of the foreground task. The time-slice multi-tasking is to execute each task in sequence or in rotating priority with equal time slice. A multi-user system allows two or more terminals (people) to have access to the computer.

It is also desirable for the operating system to have multi-processing capability; under multi-processing, multiple processors can be running concurrently. For example, one processor may handle the control system database, another processor manages the analyzing and calculation and a third processor may handle the graphical database to the screen. A memory-management program should be part of the operating system so that random memory access will not be limited.

The major operating systems available are disk-operating system (DOS), IBM operating system 2 (OS/2), UNIX, Windows NT, Windows '95 and proprietary operating system such as System 7 from Apple Computer and VMS from Digital Equipment. DOS is possibly the most widely used operating system but, in general, it is not a real time, multi-user and multi-processing operating system. The OS/2, UNIX, Windows NT, Windows '95 and VMS are operating systems with real-time, multi-user and multi-processing characteristics but the availability of the operator interface software is very limited at the present time except for a few vendors starting to offer UNIX and NT based systems.

14.4.2.9 Operator-station software. The software for operator station should be designed so that the operator in the control room is able to monitor process operations and batch-sequence control operations. This operator station software is also commonly called operator interface (O/S), data acquisition and control (DAC) or supervisory control and data acquisition (SCADA). The DCS vendor usually furnishes a well-designed proprietary operator interface package and most PLC or PC control systems depend on a third-party designed industrial DAC or SCADA software package. The DAC or SCADA is usually PC based software running on either DOS, OS/2 or the UNIX operating system. The capabilities and features of the DAC or SCADA may differ from vendor to vendor. A careful evaluation of the DAC or SCADA software is required to ensure that software meets the operating requirements.

In general, monitoring operations should comprise at least the following:

- Overview display to give the operator indications of all process points
- Group display provides a faceplate to display process information for a related number of points
- Detail display of process information for a single point

- Graphic display of the process should be a dynamic and interactive picture of current status
- Current trend display provides real-time trending of process variables
- Historical trend display gives data versus time plots of process variables messages from the system
- Message display provides user defined messages and error messages from the system
- Alarm displays should indicate off, normal, or critical process conditions with time and engineering units on a first-in basis
- Real-time operator log should document changes made by the operator
- Report printing to print out the pre-defined shift reports and batch reports
- Security should include multi-level password access
- Process data-archiving must store process data to facilitate trending.

Process operations should include at least the following:

- Set point change (e.g. either an increase or decrease)
- Final control device position change (e.g. on or off)
- Control parameters adjustment (e.g. ratio, bias, gain, integral, derivative, etc.)
- Control modes changes from manual to auto or auto to manual.

Batch and sequence control operations should incorporate the requirements of the batch data-management. The methodology should consist of a high-level control language and recipes with batch data access. Once a production activity is defined, batch procedures should be written with the high-level control language. The procedure coordinates the unit models that operate all field devices. The human interface with batch and sequence controls should:

- Start activities
- Select batch procedures
- Select recipes
- Assign batch and operator identification number
- Determine the number of batches to be made
- Enter batch time and date.

14.4.3 Engineering workstation

14.4.3.1 General requirements. If the operator workstation is not designed to perform process data configuration and control scheme programming, then an engineering workstation is required.

The engineering workstation must be complete with a full-stroke engineering keyboard including the required keys for special configuration functions, a high-resolution (640 × 480 pixel or better) color monitor, dual floppy disk drives, a hard disk for storage of the configuration files and a printer. The printer must have the capability of printing graphic displays.

The engineering workstation electronics, hard disk and floppy disk drives usually are mounted in its own free-standing enclosure. The monitor and keyboard may be mounted on the desk top, with the printer mounted on its own stand.

The engineering workstation should communicate with other highway devices through the data highway and have a real-time, multi-tasking, multi-user and multi-processing system as described under the operating system of the operator station. A multi-port connector should be available for multi-terminal use.

The engineering workstation should have at least one high-level language compiler, preferably 'C' or other commonly used language for developing special user programs.

A batch-information management database may reside within the engineering workstation.

The engineering workstation should also be able to accept external American Standard Code for Information Interchange (ASCII) files. Most configuration and programming files are in ASCII format, which is a 7-bit code for representing characters including letters, digits, punctuation and control instructions.

14.4.3.2 Control scheme programming. To design a batch control system, process control information has to be made available to the control engineer. This information must be properly documented, and the documentation should become part of validation packages. Based on Process Flow Diagram (PFDs) and P&IDs (Piping and Instrument Diagram), batch-control schemes can be documented using:

- Sequence logic diagrams
- Sequence descriptions
- Sequence matrices.

Sequence logic diagrams should be based on ISA Standards S5.2—*Binary Logic Diagram for Process Operations*. Sequence descriptions are step-by-step English-language descriptions of control sequences. Sequence matrices are tabular descriptions of control sequences. At each step, the matrices clearly define normal sequences and failure conditions of each control device. A typical sequence matrices control diagram is shown in Figure 14.8. Once the process control information is available, the control engineer starts to formulate the batch control strategy. The ISA committee SP-88 is currently working on the standards of process batch controls, which may be made available as a design guide in the near future. However, the control system vendor usually has some batch control literature helpful in designing the batch process control.

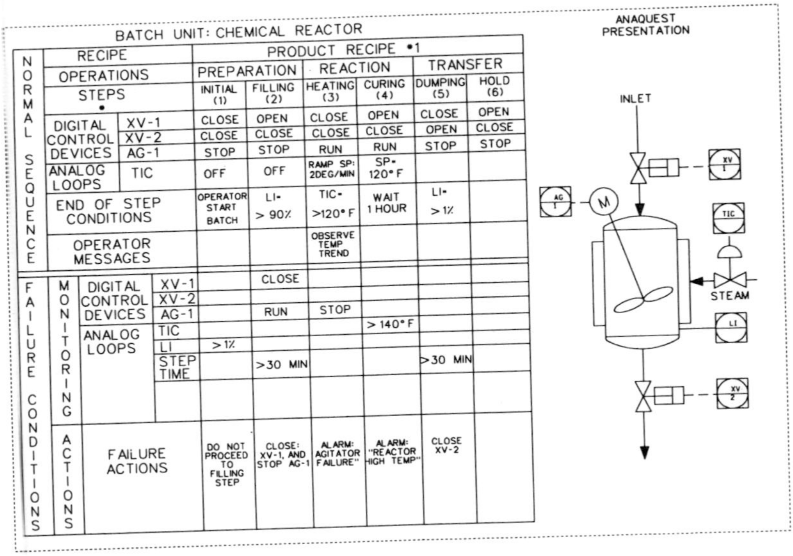

Figure 14.8 Typical sequence matrices control diagram.

The most common way to design a batch process control is to break down the process into units, which is a group of equipment operations or instructions for that unit. The batch procedure (program) then operates on the selected units based on the requirements of making a product. For example, a batch procedure may operate six units in sequence as follows:

1. Charge Material A
2. Charge Material B
3. Mixing
4. Heating
5. Cooling
6. Dumping.

Each of the 6 units above is a set of individual instructions designed for the intended process operation.

14.4.3.3 Batch-control programming languages. Batch controls are usually accomplished through programming. Ladder-diagram programming and English-type statements are most often used by PLC control system and high-level language progamming is employed for the DCS control system. The high-level language is, in most cases, a vendor's version of Basic, Fortran or 'C' language. For sequence interlocks, some DCS vendors use 'assembly' language. Training, in general, is required before performing any custom programming.

14.4.3.4 Database configurations. Process data configurations are generally performed with fill-the-blank block diagrams or pre-defined masks. Block diagrams and masks are relatively user-friendly and, with a few days of training, a proper system configuration is possible.

14.4.4 Historical trend unit

If the operator station or the engineering workstation is not designed for historical process-data trends, a historical trend unit may be used. Process data is stored on a periodic basis and the data are transmitted to the operator stations through the data highway for graphic trending display. The unit should also have the capability of printing out data trends in a numerical format.

The historical trend unit must have an external high-capacity storage device for storing trend data permanently. Storing of data is selectable and programmable-based on a pre-determined time basis or on demand.

14.4.5 Controller and input/output cards

The selection of controller for a control system is dependent on the size and complexity of the process control and the available budget for the process project. In general, the controller can be classified into the following three categories:

- Direct digital control (DDC)
- Programmable logic controller (PLC)
- Dedicated controller (DC)

14.4.5.1 Direct digital control. DDC is a computer control system. For a small control system, the I/O signals are connected to the I/O boards, which are plugged into the computer chassis. For a larger system, the I/O signals are generally connected to the I/O cards, which are plugged into an I/O card chassis. This chassis also contains a communication card, which passes the input signal information to the control computer or receives the output signal information from the control computer.

The control computer takes the input (measured process variable) information and sends out the desired output (controlled process variable) signal, based on the designed algorithm (control program), which resides in the central process unit (CPU) of the control computer.

The control computer is not only performing the continuous analog control but also the sequential (digital) control. However, most of the control computers used for process control are PC-based computers with only single microprocessor. This single microprocessor performs the external controls and internal housekeeping, such as internal checks and memory updating, in an orderly and sequential fashion.

The recommended sampling time (frequency of taking inputs) for the most common process variables is:

- Flow 0.5–2 s
- Pressure 0.5–2 s
- Level 5–10 s
- Temperature 10–20 s

The control-loop response time should be no longer than 0.5 s. The control algorithm for DDC is usually programmed in a high-level language such as Basic or C. In order to perform programming, careful planning and extensive training are required.

14.4.5.2 Progammable logic controller. The PLC is a single microprocessor controller and is very similar to DDC described above, but PLC is a well-designed vendor package system including all necessary hardware and software.

The PLC accepts data from field-input devices, performs logical decisions or algorithm determined by a program stored in memory, and provides output control to the process control device. The PLC executes the designed program in an orderly and sequential fashion from first to last instruction. For advanced PLC, instructions have been added to allow subrouting calling, interrupt routings and bypassing or jumping certain instructions. Advanced control functions such as piping and instrument diagrams (P&ID), ratio and cascade controls have also been added to make the PLC suitable for batching as well as continuous controls. In recent years, some PLC vendors also developed an integrated operator interface and process data-management software and make the PLC a one vendor supplied control system.

The PLC is a sequential control device; the capability of continuous loop controls is dependent upon the update time required by the control loops. The longer the update time and the less sophisticated the loop control, the more loops can be controlled by the PLC. In general, a single PLC unit can control approximately 30–60 × 1 s P&ID loops.

There are several types of languages currently used by PLC:

- Ladder diagrams
- Function blocks
- English statements.

The type most often used is a ladder diagram.

14.4.5.3 Dedicated controller. The DC is a specially designed controller with multiple microprocessors for the distributed control system. The controller generally has one microprocessor to perform data highway communication interface, one microprocessor to perform I/O device interface and one microprocessor to perform the controls functions. A typical DC architecture is shown in Figure 14.9. The dual date highway interface is

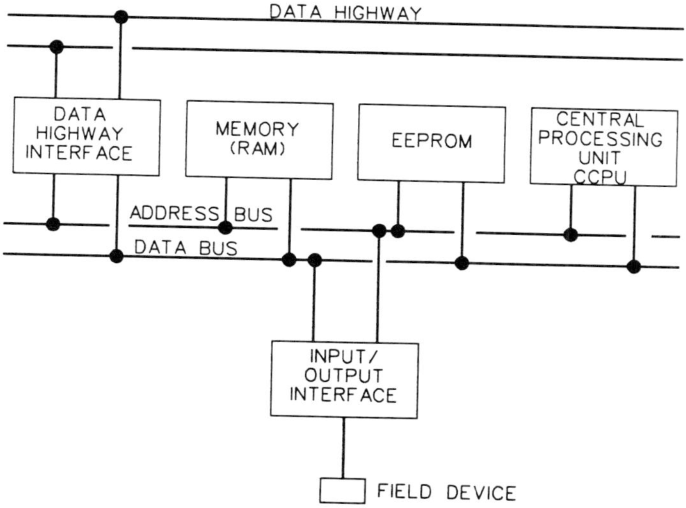

Figure 14.9 Typical dedicated controller architecture.

required for redundant communications. The microprocessor for the central processing unit (CPU) should be 32-bit with 16 MHz or higher clock speed. The electrical erasable programmable read-only memory (EEPROM) is used to store the program and constant tables and the RAM is used to store the database and the control programs. All field devices are controlled by the controller through the I/O interface unit.

The controller operating system should be a real-time, multi-tasking, multiprocessing system. The data-highway interface and CPU should have redundant capability. Switching over must be automatic in the event of failure. Battery back-up should be provided for memory retention so that all data and programs will be retained in the event of power failure.

The controllers should perform, at least, the following functions:

- Analog (continuous) controls
- Digital (batching) controls
- Functions such as square root, linearization for analog inputs
- Digital filtering for analog inputs to eliminate signal noise
- Binary and binary-coded decimal (BCD) I/O conversion
- Alarm detection for analog inputs
- Engineering unit conversion for analog inputs and outputs.

The output response for any analog loop should be no longer than 0.5 s. The I/O scan rate should be selectable from 0.1 s to 60 s maximum, and the I/O points should have optical isolation if so desired.

Contact input cards should be designed to eliminate contact bounce because, as a contact input is commanded to close, it will bounce in between close and open for several cycles before it finally becomes closed. The analog input cards must have a filter to eliminate electrical noise.

Every channel of the I/O cards should be supplied with an internal fuse protection. The digital output card should have the capability to start and stop a size 4 motor starter directly. The analog input and output cards should have at least a 12-bit resolution and be able to drive 1200 ohms loading at 20 mA, with 24 V direct current power supply.

Controllers and I/O cards must be immune to both electromagnetic and radio-frequency interference.

14.4.6 Data-highway communication

14.4.6.1 Introduction. The data-highway system provides communication among devices that are connected to the data highway. Data-highway systems vary in size and complexity, according to the process control system requirements. The traffic director shown in Figure 14.6 may or may not be required depending on the system used. Redundant communication interface is, in general, preferred to increase the system reliability. At present, there are three major data-highway systems adopted by most of the data-acquisition and control or DCS suppliers. These three systems are:

- CSMA/CD (carrier-sense multiple-access with collision detection)
- Token-passing bus
- Token-passing ring

14.4.6.2 CSMA/CD. The CSMA/CD data highway (network) is defined by IEEE Standard 802.3. The protocol of CSMA/CD allows all devices on the data highway to send their data at any time, so long as the highway is not used. Sometimes two or more devices find the highway is not used and start transmitting data at the same time. In this instance, it is said that a collision has occurred. After a collision is detected, all devices will stop transmitting immediately. To avoid another collision, all of the retransmitting devices on the data highway will then wait for differing random amount of time until the data are sent without a collision. Since the time of the data transmission cannot be precisely determined, it is said that the CSMA/CD system is non-deterministic. In practice, however, most collisions are resolved on the first attempt. The Ethernet developed by Xerox in the 1970s is one of the leading CSMA/CD systems, operating at 10 MBPS (megabits per second) over a maximum highway cable distance of 7500 feet (229 m approx.). The data highway cable can be coaxial, twisted pair or fiber-optic cable. The maximum number of devices that can be connected to the CSMA/CD highway is 1024.

14.4.6.3 Token-passing bus. The token-passing bus (data highway) system is defined by IEEE Standard 802.4. This system allows a highway device to transmit data only when it receives the token that regulates the traffic on the bus. The token is systematically passed from device to device in address order.

When a device receives the token, it transmits data to a destination device and waits for its acknowledgement, and then passes the token to the next higher address device. Finally, the highest-address device passes the token back to the lowest-address device, and the cycle repeats. However, if a device on the highway is powered off, it is automatically dropped from the token-passing bus. Since the transmission of data is predictable, it is said that the token-passing bus system is deterministic.

The Arcnet developed by Datapoint in 1977 is a token-passing bus system operating at 2.5 MBPS (mega bits per second) over a maximum highway cable distance of 4 miles (6.4 km). Arcnet can accommodate up to 255 highway devices. The data highway cable can be coaxial, twisted pair or fiber-optic cables.

14.4.6.4 Token-passing ring. The token-passing ring system is defined by IEEE Standard 802.5 and is shown in Figure 14.10. From Figure 14.10, it is evident that the token-passing ring is a string of hubs. Each hub forms a ring (closed loop) and can connect up to eight devices.

When a device receives the token and has no data to send, it passes the token to the next device immediately. However, if this device has data to send, it replaces the token with the data, and the data are passed around the ring. When the data reach the destination device, it reads, then re-transmits the data, with the addition of 'acknowledgement bits'. The data along with 'acknowledgement bits' are returned to the original sending device. The 'acknowledgement bits' indicate that the data was received and the original

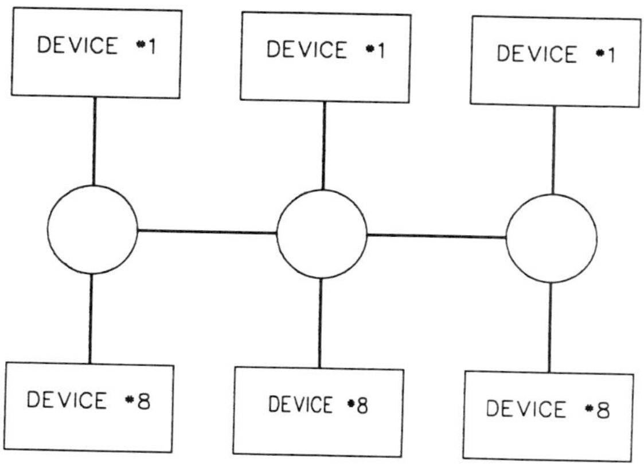

Figure 14.10 Token-passing ring network.

sender removes the attached data, and the freed token is then passed to the next device.

The token-passing ring system is also a deterministic system. The IBM token-ring network is a token-passing ring system that operates at 4 MBPS (mega bits per second) over a twisted pair or a fiber-optic cable with a maximum of 260 highway device and 1.25 miles (2 km) of cable.

15 Agitation in fermenters and bioreactors

R.J. McDONOUGH

15.1 Mechanically stirred fermenters

15.1.1 Introduction

Agitators have been used in fermenters for many years. The functions of the mixer in fermentation processes requiring dispersion of air or oxygen in a broth are that they:

- Provide uniform dispersion of gas bubbles
- Produce small gas bubbles by shearing the inlet gas with fluid velocity gradients
- Maximize retention time of the gas in the broth by driving the gas bubbles to the bottom of the tank
- Produce good bulk velocity and top-to-bottom turnover to:
  - —create the driving force for driving bubbles to tank bottom
  - —enhance heat transfer
  - —reincorporate nutrients from the surface froth back into the tank contents
  - —provide uniform nutrient and broth concentrations and temperature profile throughout the broth.

The biological products of fermentation are living organisms with a cell wall of limited resistance to fluid shear stress damage. The sensitivity of microbial cells to shear varies greatly. There has been great concern over the effect that fluid shear has on yield and productivity of various fermented products. When in doubt as to the threshold of shear sensitivity of a particular microbial organism, the fermenter mixer should be designed to minimize fluid shear. Shear rate in mechanically agitated fermenters will be discussed in section 15.1.2.

Animal cell fermentations are encountered in biotechnical processes, and the fermenter is often called a bioreactor. Animal cells frequently must attach themselves to a surface for growth and reproduction. Inert solids have been used successfully to increase the surface area on which these cells can attach and grow. Depending on the size of the cell and surface irregularities in the inert carrier (microcarrier), the animal cells may grow either internally or exterior to the surface of the inert solid.

The cultivation of animal cells that must be attached to the surface has revealed an intriguing set of problems. In addition, there are very sensitive cells made using genetic engineering techniques that have more diverse requirements for cultivation. This has led to the proliferation of many mixing devices other than impeller mixers (i.e. airlifts and bubble columns), in part because many animal cell cultures require 7–20 weeks for processing rather than the 5–7 days that is common to antibiotic fermentations. This requires that attention be given to developing equipment that can be maintained aseptic for long periods of time. Airlifts and bubble-column fermenters will be presented later in this chapter.

Anaerobic and aerobic processes are involved in fermentation. Anaerobic processes require low oxygen uptake rates (low mass-transfer rates of gas to liquid). The mixer design is controlled by mass transfer across the liquid–solid slurry boundary, possible effects of fluid shear on the organisms, and overall blending or bulk mixing of nutrients in the tank. Mammalian cell fermentation is an example of an anaerobic process.

Aerobic fermentation processes, on the other hand, require moderate to high oxygen rates between gas and liquid phases. The controlling factor in mixer design is mass transfer across the liquid–gas boundary, usually accommodated by uniform physical dispersion of gas in the fermented broth or medium. Products of aerobic fermentations are not usually shear sensitive and heat transfer can often be a major consideration in agitator design. Unlike mammalian cell bioreactors that are typically small in scale (10 liter to perhaps as large as 1000 gallon (4537 l) in capacity), aerobic fermenters are much larger with commercial scales typically ranging from 5000 up to 50 000 gallons (22 685–226 850 l). Bacterial fermentation of *Escherichia coli* strains and yeasts, and mycelial fermentations to produce antibiotics such as penicillin, are examples of aerobic fermentations.

Process and agitation requirements for various fermentation processes are tabulated in Table 15.1.

15.1.2 Fluid dynamics—flow and shear

Impeller-type mixers are essentially pumps (albeit not very efficient ones) having many of the same characteristics. The most fundamental of these is that the mixer-impeller power consumption is proportional to its flow and head development.

$$P \propto QH \tag{15.1}$$

where: P = power; Q = flow rate; and H = head. All the power supplied from the agitator to the fermenter media produces flow, velocity, head, or shear. Usually, impeller flow is expressed as the pumping capacity normal to the discharge plane of a turbine.

The discharge area of a typical constant pitch axial flow turbine is a conical

Table 15.1 Comparisons of fermentations and agitation requirements[1]

Fermentation type	Oxygen transfer rate	Shear sensitivity	Agitation requirements	Impeller types
Anaerobic				
Mammalian cell	Low	High (need low fluid shear)	Mild agitation Little or no gas dispersion Little or no heat transfer Liquid–solid mass transfer controlling Low viscosity Small-scale Shear sensitive	Low-shear propellers or hydrofoils running at slow speed
Aerobic				
Escherichia coli bacteria (yeast)	Moderate to high	Low (need high shear for gas dispersion)	Vigorous agitation Uniform gas dispersion Heat transfer required Gas–liquid mass transfer controlling Low viscosity Large-scale Not shear sensitive	Radial flow turbines or wide-blade hydrofoils (or a combination) at moderate to high speed
Mycelial				
Antibiotics	Moderate to high	Low (need high shear for gas dispersion)	Not shear sensitive but viscosity can be low to moderate	Radial-flow turbines or wide-blade hydrofoils (or a combination) at moderate to high speed)

surface, because the flow has both axial and radial components. The flow from streamlined airfoil impellers of variable-blade pitch tends to be more completely axial, so its discharge plane approximates a horizontal disc surface. An impeller in a fermenter emits a jet stream. This stream has a certain velocity, flow, and cross-sectional area. As the jet expands within the confines of the tank, the area, flow, and velocity change markedly and additional fluid is entrained. Fluid momentum, however, is conserved. The momentum of the stream leaving the impeller equals the momentum of fluid in the expanded jet. The momentum of the impeller, which manifests itself as velocity of motion in the tank, is a measurement of mixing or process result.

Impeller flow refers only to the flow directly produced by the impeller. There is also entrained flow, which is that fluid set into motion by the turbulence of the impeller stream. Entrained flow is often a major portion of the total flow as illustrated in Figure 15.1.

Equation (15.1) may be expressed in terms of impeller rotation speed, N, and impeller diameter, D, as:

$$P \propto N^3 D^5 \tag{15.2}$$

for turbulent flow or:

$$P \propto N^2 D^3 \tag{15.3}$$

for laminar flow. Other important relationships are:

$$Q \propto ND^3 \tag{15.4}$$

$$H \propto (ND)^2 \tag{15.5}$$

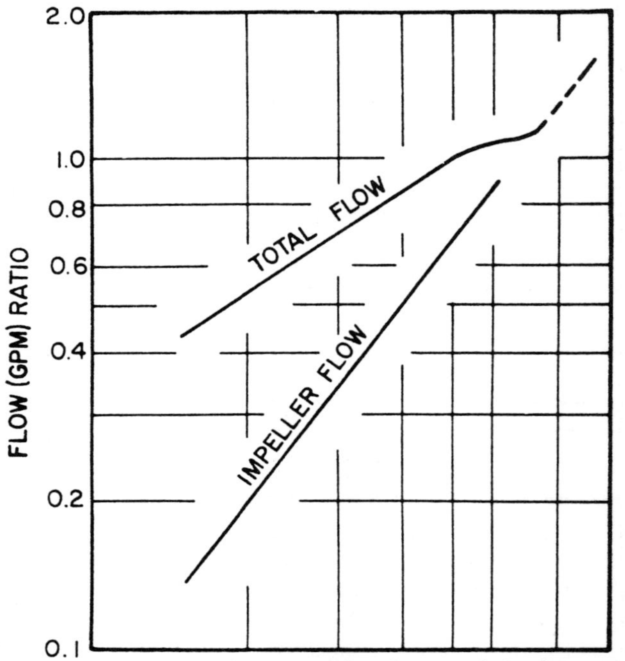

Figure 15.1 Total flow versus impeller flow. (Mixing equipment, Inc.)

$$Q_P \propto D^{4/3} \tag{15.6}$$

$$N_P \propto D^{-5/3} \tag{15.7}$$

Equations (15.6) and (15.7) are derived from equation (15.1) and express Q and N at constant power (denoted by subscript P) in terms of D. As impeller diameter increases, the impeller flow increases and the speed decreases rapidly at constant power. It follows that greater mixing or process results predicated on flow (i.e. flow-controlled fermentation) can be obtained at constant power by using a larger diameter impeller running at a slower speed.

Another consideration is the effect that a larger diameter impeller operating at reduced speed and at constant power has on torque. Torque is related to power and speed as follows:

$$\text{Torque} \propto P/N \tag{15.8}$$

As diameter increases and speed decreases at constant power to obtain greater process results, in flow-controlled fermentation, the torque rises rapidly (see Figure 15.2).

As shown above, there are numerous combinations of power and torque accomplishing the same mixing results for flow controlled fermentations. The

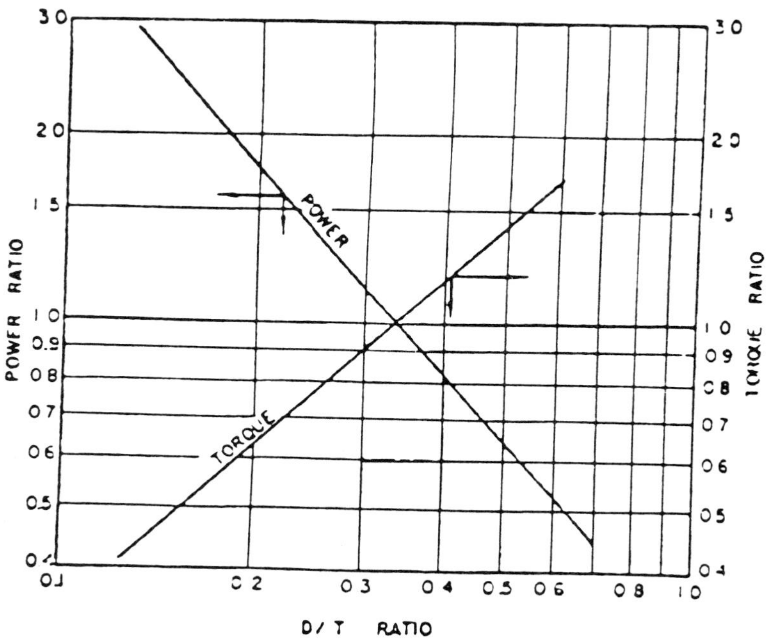

Figure 15.2 Power–torque relationship for constant process results at constant power input. (Mixing Equipment, Inc.)

remaining question is which selection is optimum for the individual. The answer lies in economics. When buying an agitator, both cost (i.e. purchase price of the machine) and operating cost (i.e. power requirement) must be taken into account. It must be remembered that:

$$\text{Capital cost} \propto \text{torque} \tag{15.9}$$

$$\text{Operating cost} \propto \text{power} \tag{15.10}$$

Economic situation will dictate which selection is right. Inevitably the trade-off between higher capital cost for lower operating cost or vice versa must be evaluated to select the optimum agitator.

A rotating mixer impeller will produce a flow velocity profile along the length of impeller blade. The profile may be nearly uniform or variable from point-to-point along the blade. The profile consists of velocity vectors with magnitude and direction. The direction will be either predominantly axial or radial (i.e. parallel or perpendicular to the axis of rotation) depending on the impeller geometry. There is a center of rotation about which all the flow can be considered to pass.

When the flow is turbulent, velocity fluctuations are significant. These can be analysed mathematically to give one value of average velocity at a given point in a stirred fermenter. When the flow is laminar, fluctuating velocity components are not present. In qualitative terms, the shear rate between adjacent layers of fluid, based on average velocity at any profile point, is used to develop the concept of macroscale shear rate. The shear operates on particles that are too large to respond to the high-speed velocity fluctuations.

Using either average velocity in turbulent flow or actual velocity in laminar flow, a velocity profile may be established for a particular impeller geometry of given diameter and speed. Measuring the slope of the velocity gradient along the blade length gives the shear rate, in reciprocal seconds, at any point on the profile. The maximum shear rate and average shear rate around the impeller can be calculated.

The product of shear rate at a given point and viscosity gives fluid shear stress (typically measured in lb/in^2 (6.9 kPa)). It is shear stress that performs the work necessary, and is ultimately responsible for, the shear-stress phenomena in the fluid. Although viscosity may have little effect on power consumption in the turbulent (high Reynolds number) flow regime, it is a direct multiplier on shear rate from the impeller, yielding the shear stress (i.e. viscosity increase from 1 to 10 cP produces a tenfold increase in shear stress from the same shear rate around the impeller).

For a realistic concept of an agitated fermenter, at least four different shear-rate values must be known: the maximum and average impeller-zone shear rates occurring within the periphery of the impeller, the average shear rate throughout the fermenter, and the minimum shear rate in the most remote, lowest velocity region of the fluid flow pattern.

From equations (15.2) and (15.4) in turbulent flow regimes, the flow-to-head ratio at constant power is:

$$(Q/H)_P \propto D^{8/3} \tag{15.11}$$

This relationship says that a large impeller operating at slow speed produces high flow and low impeller head. Impeller head, being related to the square root of fluid shear rate, is a measure of the flow-to-fluid shear rate around the impeller. Hence it follows that a small impeller turning at high speed develops a high shear rate and low pumping capacity. Every fermentation mixing application has an optimum balance of flow and fluid shear.

For a given impeller geometry, maximum shear rate is proportional to impeller tip speed,

$$\text{Maximum shear rate} \propto \Pi DN \tag{15.12}$$

whereas average shear rate is proportional to speed,

$$\text{Average shear rate} \propto N \tag{15.13}$$

Fluid discharge from an impeller can be measured with a device that has a high-frequency response, allowing velocity as a function of time to be determined. At any point in time, the fluid velocity can be expressed as an average velocity plus some fluctuating velocity component. Integration of the average velocities across the discharge of the impeller allows for the calculation of the impeller or primary pumping capacity normal to the discharge plane. This plane is bounded by the impeller blade diameter and height.

Velocity gradients between the average velocities operate only on larger particles (typically greater than 1000 μm in size). This phenomenon is defined as macroscale mixing. The fluctuating velocity gradients have an effect on smaller particles. In the turbulent mixing regime, these fluctuations are attributed to a finite number of impeller blades passing a finite number of tank baffles. For particle sizes typically less than 100 μm in size, turbulent properties of the fluid become an important consideration. Microscale mixing occurs in this regime.

As previously noted, all the power applied by an agitator to the fluid through the impeller produces flow, velocity head, or shear. Through viscous shear, power is converted to heat at the rate of approximately 2500 BTU/h per hp. Viscous shear, created only in the turbulent flow regime, is present at the microscale mixing level. Hence mixer horsepower per unit volume is the dominant component in microscale agitation. At very small particle sizes (less than say 1 μm), impeller geometry and type is of no consequence to viscous shear development and its effect on the particles, only the application of mixer power to generate shear is important. Experiments have shown that the power per unit volume in the impeller zone is about 100 times greater than in the rest of the tank.

Mixers are not specified for industrial fermentation applications to meet fluid mechanics parameters. Processes are so complex and integrated that it is impossible to isolate and define the effect of fluid mechanics on fermenter process results and mixer requirements. Pilot plant testing and experimentation affords a study of the sensitivity of the fermentation process to macroscale mixing variables (as a function of power, pumping capacity, impeller geometry, diameter, tip speed, and shear rate) as well as microscale mixing parameters (those related to power per unit volume, velocity fluctuations, etc.).

Excessive impeller shear rate can be detrimental in shear sensitive mammalian cell fermentations. The bond between the mammalian cells and the microcarrier can be broken, exposing the cell to shear-stress damage in the free liquid medium. In addition, the metabolism of cells may be inhibited, which reduces their ability to produce (reproduce) new cells. Studies to date have shown that the cell metabolism is not easily harmed by increasing the power level of a mixer. At some point, however, the cells will break away from their attachment to the microcarrier surface. The amount of dissociated cells can be measured.

Depending upon the scale being practiced, mixing systems (devices) other than impeller mixers may be practical. This includes placing the microcarriers in essentially two phase bioreactors through which nutrient streams pass or, perhaps, diffuse through the equipment boundaries. Other techniques use diffused gas bubble columns, airlift-type motions, or fluid jets to minimize fluid shear. These alternative approaches to mechanically stirred fermenters are presented later in this chapter.

Fluidfoil or hydrofoil impellers have been successfully applied in many mammalian cell fermenters. Care must be taken when evaluating hydrofoil performance in small scale fermenters and anticipating similar results on the larger commercial scale. Required blend time and appropriate shear rate witnessed on the small scale is difficult, although not impossible, to obtain on the larger scale. Appropriate scale-up considerations must be taken. These are addressed later in this chapter.

15.1.3 Gas–liquid dispersions

There are two distinct areas of gas dispersion requirements for fermentation; physical dispersion and dispersion for mass transfer. Radial flow impellers and wide-blade, high solidity axial flow hydrofoils are used because they have lower flow-to-head ratios (and consequently greater head-to-flow balance) than other impellers. For gas dispersion applications, not only is flow important, but also shear development by the mixing turbine. Flat bladed radial flow turbines and wide-blade axial flow hydrofoils provide the proper balance of flow and shear required in fermenters. Figure 15.3 shows that these impellers are best suited for dispersing gas in a liquid.

Flow is essential in gas–liquid dispersion requirements because it provides

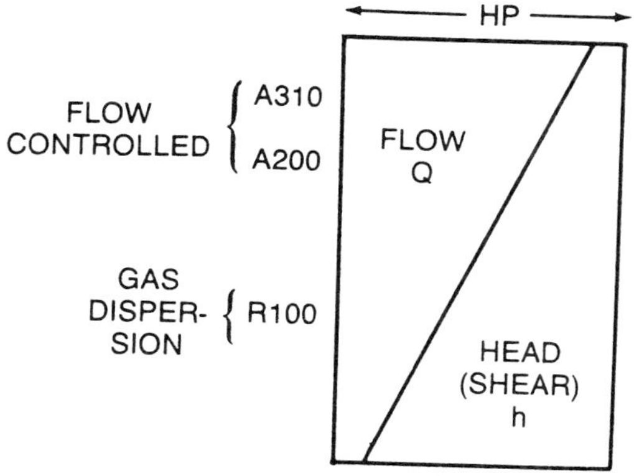

Figure 15.3 Flow–shear spectrum for mixing applications. (Mixing Equipment, Inc.)

the driving force (reflected in bulk fluid velocity) for shear forces and physical dispersion of the gas. Flow also provides gas–liquid contacting when the solubility of gas in the liquid is great enough, such that the controlling step is physical dispersion and not mass transfer of the gas into the liquid phase.

A mixer is required in a gas–liquid system to:

- increase the interfacial area of gas bubbles (i.e. shear the gas into small bubbles)
- maximize the residence or retention time of gas in the liquid
- provide bulk fluid velocity significant enough to shear the gas into small bubbles and drive the bubbles of gas to the tank bottom.

Shear is actually the velocity gradient of flow across the mixing impeller blade. This differential in velocity profile must be adequate enough in gas–liquid applications to produce smaller gas bubbles from larger ones to increase surface contact area between gas and liquid. This will inherently decrease gas–liquid reaction time while simultaneously increasing system efficiency due to increased gas hold-up time. Increased gas hold-up time means a longer residence time and, therefore, higher efficiency when adequate shear is applied to the fluid.

15.1.3.1 Physical dispersion. In order for gas to be dispersed by an agitator, the applied impeller horsepower (shaft horsepower at the impeller) must be equal or greater than the isothermal gas expansion horsepower. As shaft power is increased, physical dispersion patterns of gas in the liquid are intensified. Without adequate shaft horsepower to overcome the gas expansion horsepower, 'geysering' would occur. This type of flow pattern is shown in Figure 15.4.

When shaft horsepower at the mixing impeller is greater than gas expansion power, gas dispersion will occur and the degree of dispersion will be a function

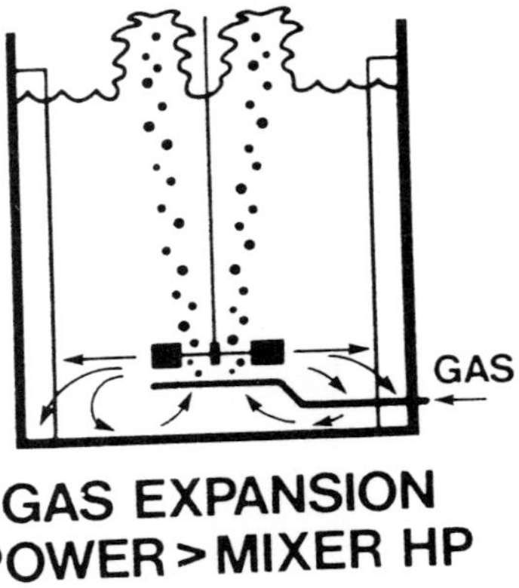

Figure 15.4 'Geysering' flow pattern. (Mixing Equipment, Inc.)

of the applied shaft horsepower. The three basic degrees of physical gas dispersion are the minimum, the intimate and the uniform. Figures 15.5 to 15.7 inclusive illustrate these basic degrees of gas dispersion.

In order to achieve the proper levels of dispersion for the basic degrees of physical gas dispersion, the following shaft power levels at the turbine are required:

- for minimum gas dispersion, 1.0 times the gas expansion hp
- for intimate gas dispersion, between 1.0 and 3.0 times the gas expansion hp
- for uniform gas dispersion, 3.0 times the gas expansion hp.

To calculate the isothermal gas expansion horsepower, one must start with

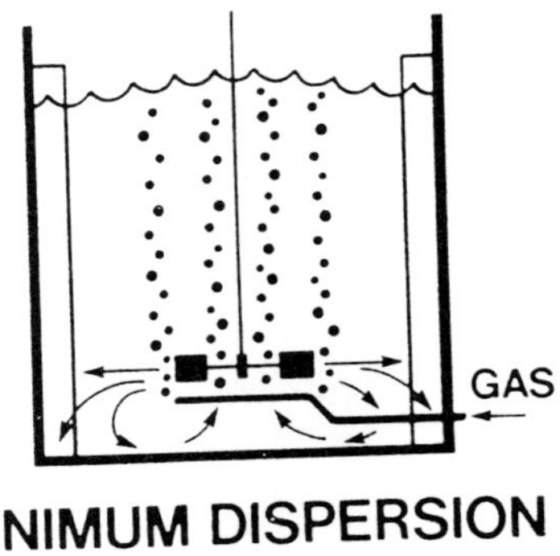

Figure 15.5 Minimum dispersion. (Mixing Equipment, Inc.)

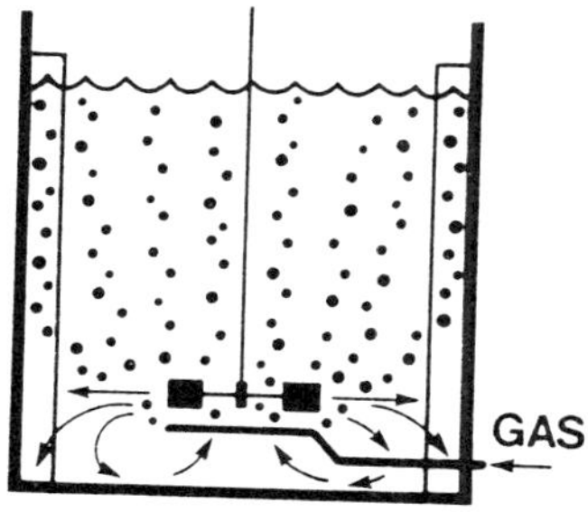

Figure 15.6 Intimate dispersion. (Mixing Equipment, Inc.)

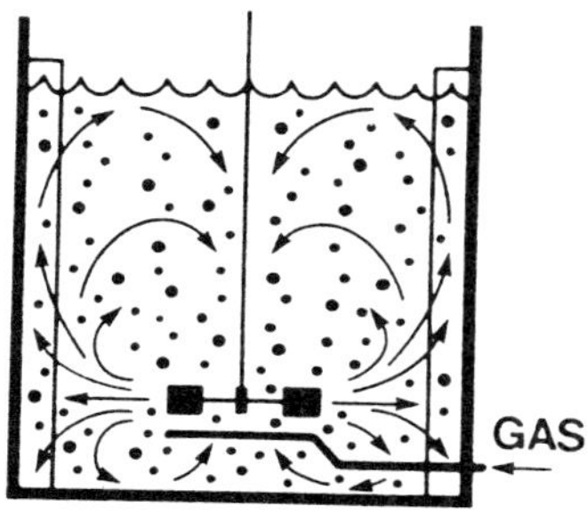

Figure 15.7 Uniform dispersion. (Mixing Equipment, Inc.)

the general equation for work, since work is what is accomplished by the gas:

$$W = P_1 V_1 \ln(P_1/P_2) \tag{15.14}$$

$$HP/Vol = \frac{(4.051 \times 10^{-3})(P_2 + 62.4\,Z\rho)\ln((P_2 + 62.4\,Z\rho)/P_2)F_1}{Z} \tag{15.15}$$

where W is the work performed (ft-lb/s (J/s^1)), P_1 is the pressure at tank bottom (lb/ft^2 (kg/m^2)), V_1 is the volumetric flow rate of gas (ft^3/s (m^3/s)), P_2 is the pressure at tank top (lb/ft^2 (kg/m^2)), Z is the batch height or depth (ft(m)), Vol is the batch volume (ungassed) (gallons (l)), ρ is the ungassed liquid specific gravity, and F_1 is the superficial gas velocity at gas inlet (ft/min (m/min)).

$$F_1 = Q_1/A \tag{15.16}$$

where Q_1 is the volumetric gas flow rate at the gas inlet location, actual cubic feet per minute (ACFM) and A is the cross-sectional area of the tank (ft^2 (m^2)).

Once the gas expansion horsepower is calculated, the agitator can be designed to accomplish the various degrees of physical dispersion. However, agitator horsepower is not the only important parameter for the required dispersion. Impeller to tank diameter ratio (D/T) is also a significant variable. The most effective ratios to use for the horsepower and gas rates required are depicted in the matrix in Table 15.2.

It may seem that designing agitators for physical gas dispersion is the only criterion for consideration. However, a check must be made to be sure that the mixing impeller is not flooded by the gas.

In a gas–liquid mixing system, the mixer and gas compete to gain control of the flow pattern. When the gas is the controlling factor, the mixer loses control of the flow pattern and the system flow pattern becomes gas controlled. The result is loss of dispersion, poor mixing and loss of gas hold-up and retention time. In essence a 'geysering' flow pattern similar to that shown in Figure 15.4 takes over. The impeller is therefore said to be flooded by the gas.

When the agitator is controlling the flow pattern, gas is dispersed in such a way that one of the patterns depicted in Figures 15.5 to 15.7 develops. Gas is dispersed and hold-up of gas in the liquid phase can actually be measured (by measuring the depth of liquid expansion under gassed conditions and calculating the fluid volume increase).

There is a maximum gas rate that a specific impeller type, of certain diameter and running speed, can disperse without being flooded. If this gas rate is exceeded, the impeller will become flooded and the system flow pattern will be gas controlled creating a 'geysering' effect. Some impellers such as axial flow flat bladed turbines, narrow bladed axial flow hydrofoils, and propellers can disperse gas at only very low rates without flooding and as such are impractical for consideration in gas dispersion applications, at even modest rates of gas introduction. Radial flow flat bladed turbines and wide blade axial flow hydrofoils can disperse modest to high gas rates without flooding and are considered the ideal impeller geometry for gas dispersion mixing requirements. All impellers regardless of type of geometry, diameter and/or speed will have a threshold of gas rate before flooding occurs. Certain impellers have low thresholds while others are better suited for dispersion at higher gas rates without flooding of the impeller.

Table 15.2 Optimum D/T ratios for gas dispersion

	Hp	
Gas rate	High	Low
High	0.15–0.25	0.33–0.50
Low	0.15–0.50	0.15–0.20

15.1.4 *Mass transfer*

Dispersion requirements for mass transfer of gas into a liquid or liquid–solid slurry are either uniform dispersion or at even greater power level (i.e. more than 3 times the isothermal gas power). For example, in fermenter applications power levels are much greater than that required for uniform physical dispersion. In fact, most gas–liquid applications are controlled by mass transfer, not physical dispersion. The absolute value of the overall mass transfer coefficient (K_{Ga}) cannot be calculated or even predicted with any accuracy without running a pilot plant program. Mass transfer coefficient is a function of the physical properties of the materials involved. The overall mass transfer relationship may be expressed as:

$$K_{Ga} \propto (\mathrm{Hp}/V)^{x}(F)^{y}(\mu)^{z} \tag{15.17}$$

where K_{Ga} is the gas-to-liquid mass transfer coefficient (kg mol/h/m^3/atm), or (lb mol/h/ft^3/atm) Hp is the horsepower, V is the volume, F is the superficial gas velocity $= Q/A$ (ft/min (m/min)), Q is the volumetric gas flow rate (ft^3/min (m^3/min)), A is the cross-sectional area of tank (ft^2 (m^2)), x, y, z are exponents and are functions of the process variables, and μ is the viscosity of liquid. Mass transfer is often thought of in terms of the the following relationship:

$$\text{Mass transfer rate} = K_{La}(\Delta C) \tag{15.18}$$

where K_{La} is the mass transfer coefficient (h^{-1}) and ΔC is the concentration gradient. It is not always possible to calculate the concentration driving force, ΔC, in given liquids, since solubility data and reaction rate data are often needed.

A pilot plant program run with a variable speed mixer at various gas rates, with instrumentation to measure dissolved gas concentration in the liquid and gas concentration in the off gas, affords one the opportunity to identify the exponents x and y and K_{La} for scale-up criteria, etc.

Most liquid–gas mixing applications involve mass transfer. A frequent requirement is the dispersion of a certain quantity of gas per unit time. However, this is not usually the only mixing requirement. A more basic one is for a required rate of mass transfer, which should normally be the design basis for the mixer. Basic measurement in terms of mass transfer is the mass transfer rate, usually expressed as the product of the volumetric coefficient and the average driving force of concentration.

There are numerous complexities in determining the proper mass transfer driving force. Henry's law relates the partial pressure in the gas phase with the concentration in the liquid phase. The problem, however, is that Henry's law constants are not known for many gas–liquid mixtures. In such cases a decision must be made about whether to use a gas concentration or liquid concentration driving force.

It follows that the liquid mass transfer coefficient K_{La} is more suitable for liquid film controlled processes, while the gas mass transfer coefficient, K_{Ga} is more suitable for gas film controlled processes. In concept this may be valid, but practically, with a Henry's law relationship, the only difference between K_{La} and K_{Ga} is the choice of concentration units.

In the periphery of the mixing impeller, the complex mechanism of mass transfer is not known. In this example, there is no reaction, or other mass transfer activity, acting to remove the dissolved gas from the process. If a large part of the mass transfer occurs in the impeller zone, the dissolved gas concentration may approach saturation and any further increase in power may not improve process results.

Two driving forces can be defined. In the first, the gas phase and the liquid phase are both well mixed, the concentration driving force will be the difference between the partial pressure of the exiting gas and the concentration of dissolved gas in the exiting liquid. In the second, if the gas is unmixed and flowing through the liquid in plug flow, there will be a concentration driving force at the bottom of the tank and another at the top. The log-mean value of the two driving forces should be reasonably valid and accurate. There is no way to prove that one is better than the other without frequent sampling of the gas phase at various depths of the tank. Experience has indicated that the log-mean concentration driving force is more consistent and practical for full size tanks.

15.1.4.1 Effect of gas rate on mass transfer. As gas rate increases at a given power level, pumping capacity becomes more important than fluid shear rate to optimize mass transfer. At low power levels, pumping capacity becomes still more important, while at high power levels, fluid shear rate is still the predominant factor affecting mass transfer. These findings are summarized in Figure 15.8. Here we show the optimum D/T ratio (shaded area) for various ratios of gas flow to mixer power.

At high gas rates and lower mixer power levels typical of a gas controlled regime, pumping capacity is more important than fluid shear rate for optimizing mass transfer. This is reflected in the high D/T ratios. At higher power levels typical of the mixer controlled regime, fluid shear becomes much more important, and the D/T approaches relatively low values of 0.1 to 0.2. At extremely high power levels, the energy from the mixer is so great that the way it is distributed makes no difference and there is no optimum D/T ratio.

Typical optimum D/T ratios in fermentation further illustrate that not everything in a mixing application can be optimized. The high shear rates produced by the optimum D/T ratios of 0.15 to 0.2 would be detrimental to biological organisms. Although D/T ratios from 0.35 to 0.50 are necessary for blending and heat transfer requirements in fermenters, as well as to prevent shear damage to the organisms, this range is not ideal or most effective for gas–liquid mass transfer.

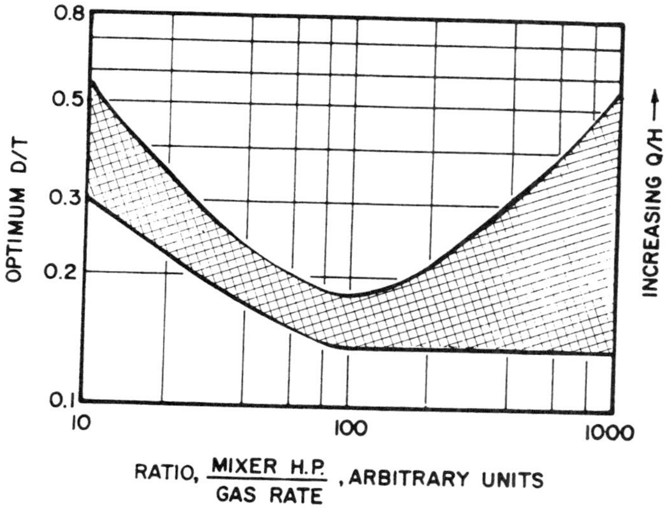

Figure 15.8 Optimum D/T ratios at various power to gas rates. (Mixing Equipment, Inc.)

15.1.4.2 Mass transfer in fermenters. Sufficient oxygen mass transfer in aerobic fermenters is not usually a problem in a small scale pilot demonstration. Regardless of scale, comparable rates of mass transfer per unit volume are achievable on small and large capacity fermenters, assuming the proper mixer scaling criteria are identified and applied in scale-up or scale-down procedures.

A problem often encountered in scaling mixer performance relative to mass transfer rate comes from the blending phenomena. A large tank has a much longer blend time associated with it than the smaller tank and oxygen deficiencies often result in remote portions of the tank contents. Mass transfer zones should be made a part of the overall flow pattern in the entire system.

A small mixing tank has a much lower maximum shear rate in the impeller zone relative to large tanks. It might be required to use a non-geometric impeller and mixing tank on a small scale to duplicate maximum shear rate in the impeller zone and performance of proper variables in the larger scale.

In terms of minimum fluid shear rates required for some mammalian cell fermentations and other types of shear sensitive cells, a different approach to mixing requirement will be needed, i.e. it will be necessary to determine a threshold of acceptable shear rates with a particular impeller type on a demonstration scale, and duplicate that shear rate on the required scale, with perhaps a geometrically dissimilar impeller.

Flat-bladed radial flow turbines with disks have been the workhorse in fermenters over the years. As a rule of thumb, 10 hp (K W) per 1000 gallon (4537 l) of ungassed broth with $D/T = 0.33$ flat-bladed radial flow turbines with disks achieved good results in large scale (10 000–50 000 gallon (45 370–226 850 l)) aerobic fermenters for bacterial (yeast) and mycelial (antibiotics) fermentations of low viscosity. However, the need for bulk velocity and shear in proper balance for optimum process results, the introduction of shear-sensitive cell fermentations in biotechnology, and the increasing cost of energy in the late 1970s paved the way for the use of hydrofoils in fermenters.

It has been shown in pilot plant demonstrations and full scale fermenters that hydrofoils afford a better flow/shear balance, at reduced power requirements than the traditional workhorse flat-bladed radial turbines in fermenters. Various configurations of hydrofoils have been installed, primarily on a retrofit basis, whereby a hydrofoil(s) replaces a radial flow turbine(s). Usually in a retrofit campaign the mixer speed does not change, and the impeller diameter varies either to draw the same power and produce greater mass transfer or to produce the same mass transfer at reduced power draw (typically 40–50% less power required). It has been shown that wide-bladed hydrofoils with 4–6 blades of high solidity ratios are optimum design for the dispersion of gas, shearing of gas bubbles and driving of the bubbles to the tank bottom—better than the flat-bladed radial flow turbine in most cases—and there is less shear, reducing damage to shear sensitive cultures.

Upper impellers in fermenters should be hydrofoils of narrow-bladed design for optimum results. These impellers need not be designed with gas dispersion in mind (the bottom impeller must be designed with gas dispersion in mind), but rather top-to-bottom turnover and bulk velocity to optimize blending, suspension, and heat transfer. Some fermentations require a combination assembly of impellers, the lower one being a high-shear radial flow turbine for greater shear while the upper ones can be hydrofoils of narrow blade design. Others are optimized with all impellers being hydrofoils (the bottom one a wide blade and the uppers of narrow blade design). Thus experimentation with various geometries of impellers, number of impellers, agitator speed, etc., should be conducted in fermenters to optimize mixing and process results and obtain initial scale-up parameters.

Oxygen is known to be a growth limiting nutrient for submerged aerobic fermentations. Oxygen is obtained by microorganisms through mechanical agitation and gas dispersion aeration of the fermentation medium or broth. Therefore it is very important to understand the mechanism of the transfer of oxygen for aerobic fermentation to optimize the design and scale-up of mixers in fermentation processes.

Aerobic fermentations are an intricate and complex number of different chemical constituents. Of importance is the need to understand oxygen transfer in the systems containing these individual components, before the highly complicated fermentation systems can be studied. The oxygen mass transfer process in fermentation is pictorially presented in Figure 15.9.

The interest in hydrofoil impellers for gas–liquid processes and fermentation was precipitated from studies directed at retrofitting existing fermenter tanks

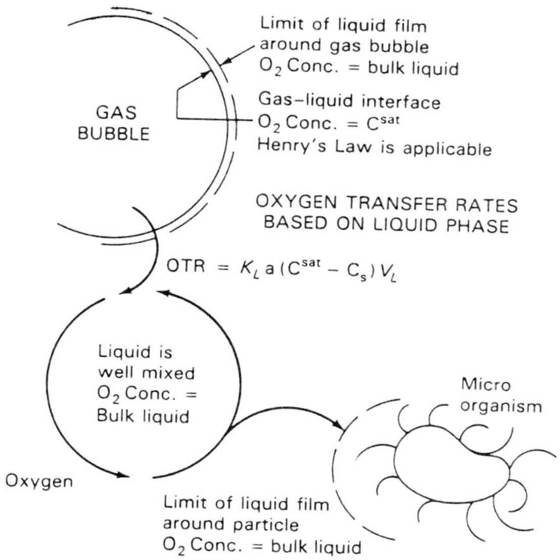

Figure 15.9 Oxygen gas–liquid mass transfer process. (Prochem Mixing Equipment, Inc.)

and mixers. Typical power levels were 5–20 hp per 1 000 gallons (10 hp per 1000 gallons was an average). Multiple radial flow disk impellers of $DT = 0.33$ were used. Superficial gas velocities were typically 0.1–0.5 ft/s.

The hydrofoil impeller provides better mixing and process results (i.e. mass transfer/heat transfer). Radial flow impellers can create a staged mixing pattern with non-uniform oxygen and nutrient concentrations throughout the fermenter. In addition, damage to the microorganisms may occur due to high energy levels going into macro- and micro-scale shear rates from the radial flow turbines. Replacing one or two of the radial flow impellers with hydrofoils, narrow bladed ones at the top of the vessel and wide bladed versions at the bottom, has proved to be the optimum in many fermentation systems. Hydrofoil impellers with wide blades and high solidity ratios (greater than 80%) have proved to be comparatively advantageous in most cases to radial flow impellers for gas dispersion and gas–liquid mass transfer.

Studies continue pertaining to mixer performance and process results with wide-bladed hydrofoils in fermenters. Several points can be made. First, the slope of the mass transfer coefficient as a function of power and gas rate has a marked break point and 'knee' in the curve for a radial flow impeller. This characteristic is not shown by wide-bladed, high solidity hydrofoils. The slope is constant. Hence the hydrofoil is more effective, particularly at higher power input and/or greater mass transfer requirements. The relative performance of both impellers is shown in Figure 15.10(a) and (b).

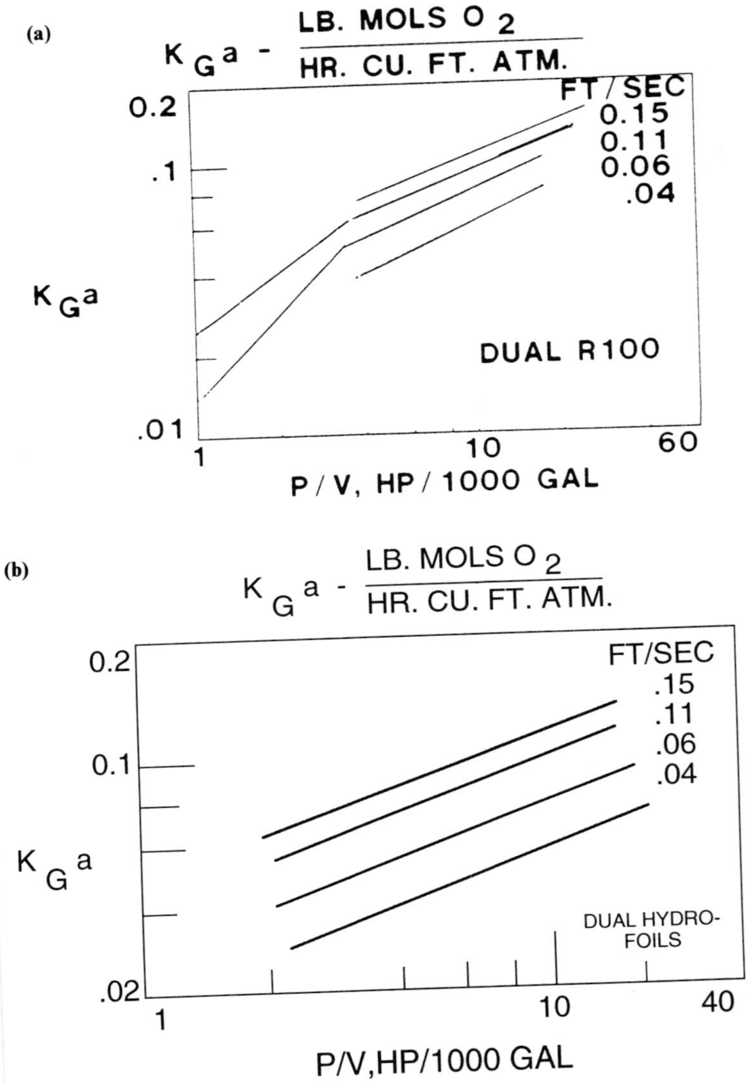

Figure 15.10 Typical plot of mass transfer coefficient K_Ga versus power per unit volume for various impellers and gas rates; (a) radial flow impellers, (b) wide-bladed, high solidity hydrofoils. (Mixing Equipment, Inc.)

Secondly, retrofitting existing agitators is practiced to optimize mixing and process results. Typically this involves keeping mixer power and speed constant while retrofitting the impeller. Hydrofoils replacing radial flow impellers will always be of larger diameter in this situation. As a result, essentially constant torque is maintained on the drive assembly and shafting. When the ratio of D/T for a radial flow turbine is 0.33–0.4, the gas–liquid mass transfer characteristics of the wide-bladed hydrofoil will be better than those of the radial flow impeller. Mass transfer, blending, and heat transfer will be improved. Improved flow pattern will also result with the hydrofoils in fermentation service. This is depicted in Figure 15.11. Typical improvements in mass transfer are shown in Figure 15.12.

15.1.5 Impeller types, geometries, and location

Radial flow impellers and axial flow hydrofoils with wide blades provide the best flow-to-head ratios for gas dispersion. Open impellers such as flat bladed paddle-type impellers and narrow bladed hydrofoils should not be considered for gas dispersion. The solidity ratios are too low with these impellers. There is too much open space between blades and gas bypasses the high-shear zones for gas dispersion. To compensate for this tendency for gas to bypass the blades of these open impellers, much more horsepower is required to disperse the gas. In addition, the flow patterns of these types of turbines are not conducive to good dispersion. The natural tendency of the gas bubbles to rise in the liquid competes with the open impeller flow pattern. A gas bubble of a certain diameter will rise through a liquid (or slurry) of a given viscosity at a particular rise velocity.

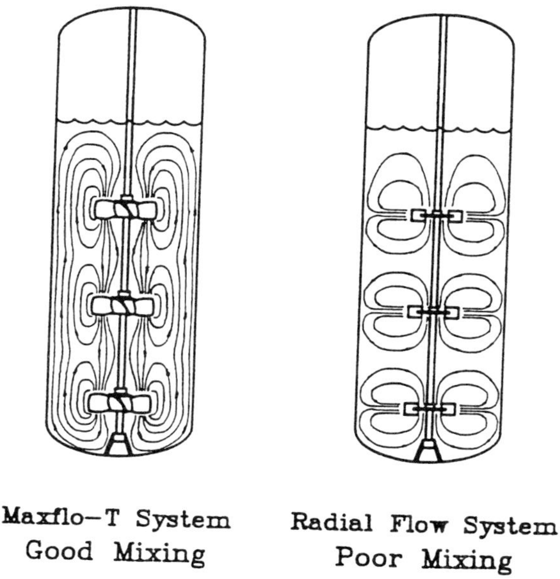

Figure 15.11 Flow patterns of hydrofoil and radial flow impellers in fermenters. (Prochem Mixing Equipment, Inc.)

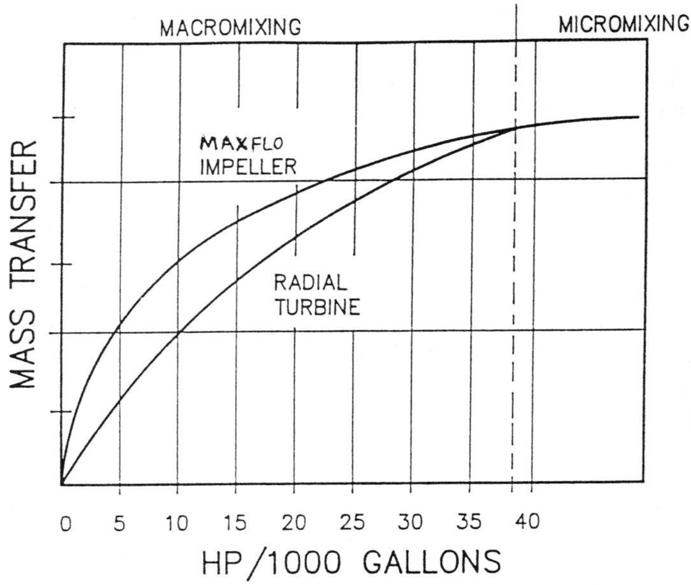

Figure 15.12 Relative improvement in mass transfer with hydrofoils at various power levels. (Prochem Mixing Equipment, Inc.)

Table 15.3 lists bubble diameters for terminal rise velocities ranging from 0.3 to 0.0005 m/s. As can be seen, as viscosity increases, bubbles must be considerably larger to flow upward at a given rise velocity. Impeller blade width should be a minimum of two to three times the size of the gas bubble being dispersed. Therefore, larger impellers are required for gas dispersion in higher viscosity liquids.

Retrofitting of existing fermenter mixers is common. Typically, this involves keeping the mixer power and mixer speed the same while changing the impellers. As a result, essentially constant torque is being maintained on the equipment. In order to do this, the wide-bladed hydrofoil impeller must be

Table 15.3 Bubble diameter required to give various terminal rise velocities

	Bubble diameter (mm) at terminal rise velocities of			
Viscosity (Pa/s)	0.3 m/s	0.05 m/s	0.005 m/s	0.0005 m/s
0.001	1.0	0.5	0.13	0.05
0.01	2.3	1.0	0.25	0.10
0.1	5	2.5	0.75	0.25
1.0	15	7.5	2	0.75
10.0	50	15	5	1.6
100.0	100	50	10	5

somewhat larger because its power number is lower than the radial flow turbine.

When the ratio of D/T for a radial flow impeller is of the order of 0.35 to 0.4, the gas–liquid mass transfer characteristics of the wide bladed hydrofoil impeller will be better than those of the radial flow impeller. Mass transfer and, in general, the overall blending flow pattern will be improved. There are, however, situations where this combination does not improve the process. In 5–20% of cases studied, no improvement or even a decrease in performance was observed when replacing radials with hydrofoils.

Some ask, "In going to the hydrofoil impeller, is it better to replace only the upper one or two impellers, maintaining the lower one or two impellers of the radial flow-type, or should the entire assembly be hydrofoil impellers?" We don't know. Quantitative data gathering is still going on, and the question is an open one.

15.1.5.1 Number of impellers. Gas dispersed by one impeller at the tank bottom tends to coalesce into larger bubbles. Multiple turbines will maintain small bubble diameter, particularly for a relatively tall fermenter. Spacing can be farther apart than for flow controlled mixing applications, typically 1–2 D apart, depending on tank geometry and number of impellers. Multiple impellers provide multiple high shear zones in the mixing tank to prevent coalescence of bubbles.

One method of dispersion with multiple impellers is to design the mixer such that most of the power is dedicated to the lower turbine for gas dispersion, while using less power in the impeller(s) above the lowest one for pumping, circulation, blending, solid suspension and heat transfer. Experimenting with various impeller combinations and configurations will achieve the proper balance of flow, shear, and power distribution.

A wide bladed hydrofoil of high solidity will not have its flow pattern destroyed by the rising gas bubbles and hence gas dispersion and flow controlled applications can be achieved simultaneously. These mixing results can be achieved in open tank and draft tube circulation arrangements. Hence, since the flow from the turbine is confined within the tube, enough power is added by the mixer to overcome the upflow drive of the gas. A net downflow of gas bubbles to the bottom of the tank results, with effective mass transfer in most cases.

With a radial flow turbine that has no disk (i.e. blades emanating from a hub on the shaft) and is of 'open' geometry, the gas can enter the low shear zone around the hub and pass through the impeller without being dispersed.

15.1.5.2 Impeller location. At a given gas rate and constant D/T, various arrangements of impellers, positioning of impellers and liquid levels show very little effect on gas–liquid mass transfer coefficient. This indifference is observed if the upper impeller (or only turbine in a single impeller installation) has at least 0.5 *D* coverage of ungassed liquid level above it. At coverage less than 0.5 *D*, the upper impeller will begin to splash surface liquid, simulating a surface aerator, and mass transfer will in fact be enhanced. A surface splashing impeller will not transfer oxygen as well as a submerged turbine with air introduced beneath it. In addition, it will pump very little.

In an unbaffled or partially baffled tank, vortex formation on the surface of the mixing contents will incorporate gas from the vapor space above the liquid back down into the liquid. Re-entrainment of gas in this manner will also enhance the mass transfer coefficient. However, the negative aspects of a vortexing impeller (and surface splashing impeller) are first that greater hydraulic loads are imparted on the mixer shaft and drive train and the mechanical design of the agitator must compensate accordingly, and secondly that there is poor pumping and a terrible flow pattern for top to bottom circulation.

Even though there are some minor differences in the flow to fluid shear-rate ratios of multiple impellers versus a single turbine, the overall effect of impeller arrangement, position, and liquid level on mass transfer is minimal. This illustrates that, even though the instantaneous mass transfer coefficient around the impeller is much higher than it is throughout the batch, moving to different spots in the tank does not significantly change the integrated average mass transfer throughout the entire volume.

Therefore, even though Figure 15.8 shows that D/T is very critical, the choice of multiple or single impellers, spacing, etc., is not critical. Impellers may be located to achieve other mixing results, such as solid suspension or blending, and still have effective gas–liquid mass transfer.

15.1.6 *Power requirements and effect of tank shape on power*

Should the fermenter mixer be designed for proper power loading in the gassed or ungassed state?

Power consumption of a flat-bladed radial flow turbine (with a disk and blades) in a gas–liquid system is less than the power consumed when the gas is off. In the gassed scenario the impeller 'sees' and operates in a region of a gas–liquid mixture, with lower specific gravity than that of the ungassed liquid. The ratio of gassed to ungassed power draw, P_g/P_0, is called the *K* factor of the impeller. It is a function of the type and geometry of impeller (wide-bladed axial flow hydrofoils have higher *K* factors, even approaching unity under certain conditions, than radial flow disk turbines with flat blades at similar gas rates and diameters). Other variables affecting *K* factors are gas rate and impeller speed. Figure 15.13 shows the relationships for a radial flow turbine with disk and flat blades in an air–water gassed system where P_g is the gassed shaft horsepower (kW), P_0 is the ungassed shaft horsepower (kW), Q is the air flow rate, actual cubic feet per minute (ACFM) (m^3/min),

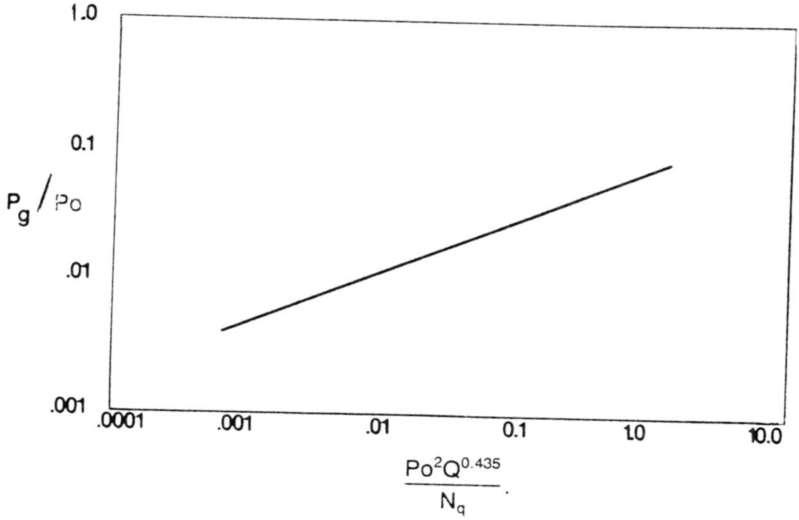

Figure 15.13 Relationship of K factor and gas flow for air–water agitated systems. (Mixing Equipment, Inc.)

N_Q is Q/ND^3 the dimensionless flow number, N is the impeller speed (rpm), and D is the impeller diameter (ft (m_0)).

Assume that for a given level of gas dispersion that $P_g = 13.5$ hp and $P_o = 27$ hp for a K factor of 0.5. If the mixer is sized on the ungassed basis, a larger mixer would be required than one designed on a gassed basis. However, we would be assured of operating safely, without concern for power overload, in either gassed or ungassed scenarios.

On the other hand, if we sized the agitator on the basis of gassed power draw, an overload condition might exist in a reduced gas flow or most certainly in an ungassed situation. Does this mean that the agitator is limited to operate only in the gassed condition? The answer is 'yes' with a single speed motor and 'no' with a two-speed motor. If a two-speed motor is supplied, both modes of operation (i.e. gassed and ungassed) could be pursued without oversizing the agitator, while still satisfying the process results. For an agitator with a design premise of power draw based on operating in a fully gassed regime, the higher motor speed (and hence greater mixer impeller speed) can be chosen. Upon reduction of gas rate, including complete shutting off of the gas, the motor speed can be switched to the low speed to compensate for the higher power draw at reduced gas rate, without jeopardizing on overload of power on the agitator. The change in motor speed can be accommodated automatically by the use of a gas flow switch interlocked with the motor to reduce speed based on reduced (or complete loss of) gas flow rate.

Infinitely adjustable motor speeds utilizing frequency controllers and/or hydraulic motors can accommodate subtle changes in gas rate by subtle adjustments to motor speed. Loading a mixer based on the gassed horsepower draw and using a two-speed, variable speed motor or gas flow-motor speed interlock offers these advantages: (i) smaller mixer size due to lower power and torque requirement, therefore lower initial purchase and installation cost; (ii) lower operating cost due to reduced power draw during reduced gas conditions; (iii) reduced spare parts cost (and faster delivery of smaller sized mixer components); (iv) easier and less expensive maintenance cost; (v) smaller, less expensive starter box.

15.1.6.1 Tank shape. For a given volume of sparged air, the power required by both the air and the mixer is less with tall, thin fermenter tanks than with short, squat tanks of the same volume. Table 15.4 depicts this relationship. The greater pressure and high superficial gas velocity in deep fermenters provide process improvements that more than offset the extra horsepower required to introduce the air.

In general, axial flow impellers are more effective for solid–liquid suspension whereas radial flow disk turbines are more effective for mass transfer. It is not uncommon to install both radial and axial flow turbines on the same mixer shaft to optimize the balance of flow-to-shear ratio. The axial flow pattern is completely distributed when gas is sparged into the tank. Only wide-bladed, high solidity hydrofoils have shown a propensity to disperse gas, while at the same time providing enough flow in the tank for flow controlled applications, such as solids suspension of heat transfer. When an open axial flow turbine (a wide-bladed hydrofoil like a flat-bladed disk radial turbine, are considered closed impellers) is subjected to gas for dispersion purposes, the open turbine is rendered ineffective for either suspension, heat transfer, or mass transfer. The gas chokes the open impeller and prevents it from forming adequate flow velocity for any of the traditional flow controlled applications. For a combination gas–liquid–solid system, the radial flow and wide-bladed hydrofoil provide the proper flow to head (or shear) balance for most process results.

Table 15.4 Power requirement for gas dispersion with various tank shapes (Mixing Equipment, Inc.)

Tank height (in)	*F* (m/s)	Air power (kW)	Mixer power (kW)	Total power (kW)
3	0.0015	0.4	1.6	2.0
6	0.0030	0.75	0.75	1.5
12	0.0045	1.0	0.3	1.3

15.1.7 *Gas hold-up and flooding*

Many published studies have been done on gas–liquid mixing in fermenters under various gas rates. At low gas rates of less than 1 volume of gas per volume of liquid per minute (VVM), mixing is fairly straightforward. In industrial scale fermenters, gas rates are usually much higher, often exceeding 3 VVM, and impeller flooding and cavitation can be a problem. When this occurs, impeller discharge flow diminishes and the distribution of gas in the fermenter can change dramatically from uniform to geysering. As impeller discharge rates decrease, impeller flow can no longer entrain gas and carry it to all regions in the fermenter. The region immediately below the impeller near the tank bottom is usually the first to become void of bubbles. Under flooded conditions, the impeller can often only distribute gas radially to regions above it. The '*K*' factor or ratio of gassed to ungassed power draw falls to less then 0.5 in flooded situations.

Gas hold-up values are determined through measurements of the change in liquid height in dispersions with aeration in the agitated liquid.[2]

$$H = \frac{h_a - h_o}{h_a} \tag{15.19}$$

where H is the gas hold-up, h_a is the liquid height with aeration and agitation, and h_o is the liquid height without aeration and agitation. Gas hold-up in a fermenter can be correlated with gassed power per unit volume, P_g/V, and superficial gas velocity, V_s, since both affect the amount of gas in the gas–liquid dispersion. A correlation has been developed such that:

$$H \propto (P_g/V)^{0.45} (V_s)^{0.61} \tag{15.20}$$

Gas hold-up is a strong function of (P_g/V) and V_s. At constant V_s and with increased (P_g/V), gas bubbles become more dispersed predicting a longer residence time and thus leading to greater gas hold-up. On the other hand, at constant (P_g/V) and with increased V_s, more gas is introduced into the dispersion resulting in greater gas hold-up.

Mechanical differences between agitation systems and tank internal configurations account for variations between actual and predicted gas hold-up. Discrepancies in impeller design, baffle placement, sparger placement, etc., will affect gassed power per unit volume and superficial gas velocity with respect to gas hold-up in fermenters.

Solids in fermentation broths will reduce the rise velocity of bubbles and liquid surface tension in the aerated broth will vary over the fermentation cycle. These variables will affect gas hold-up during fermentation.

Addition of antifoam agents can reduce gas retention from as high as 20% down to 3% in a novobiocin fermentation under constant gas rate and power level. In practice, it is difficult to distinguish between percentage hold-up and actual foaming. It is likely that hold-up and foaming are similar phenomena, although it is unlikely that they both fit a single correlation such as equation (15.20).

During normal fermenter operation, it is not necessary to predict percentage hold-up, as it is controlled by the addition of an antifoam agent. This requires some method of measuring percentage hold-up, which is difficult in large equipment. At best, it can be estimated from physical observation of the fermenter and is limited in the extreme by the use of antifoam probes. The typical range observed for percentage hold-up is between 5 and 15%.

It is difficult to determine the effect of percentage hold-up on the mass transfer performance of a fermenter. Low levels of entrained gas may indicate poor gas dispersion and low bubble surface area for mass transfer. Very high levels of gas hold-up lead to decreased agitator efficiency, wasted tank volume, and may or may not indicate high bubble area. Any effect of hold-up on bubble surface area is taken care of separately in the K_{La} relationship.

Figure 15.14 shows a model prediction of the effect of increasing the volume expansion on a 150 000 liter fermenter from 10 000 to 40 000 liter (6.67 to 26.67% hold-up). Only minor changes in the axial DO profile were predicted. If, however, the increase in entrainment substantially increased bubble area

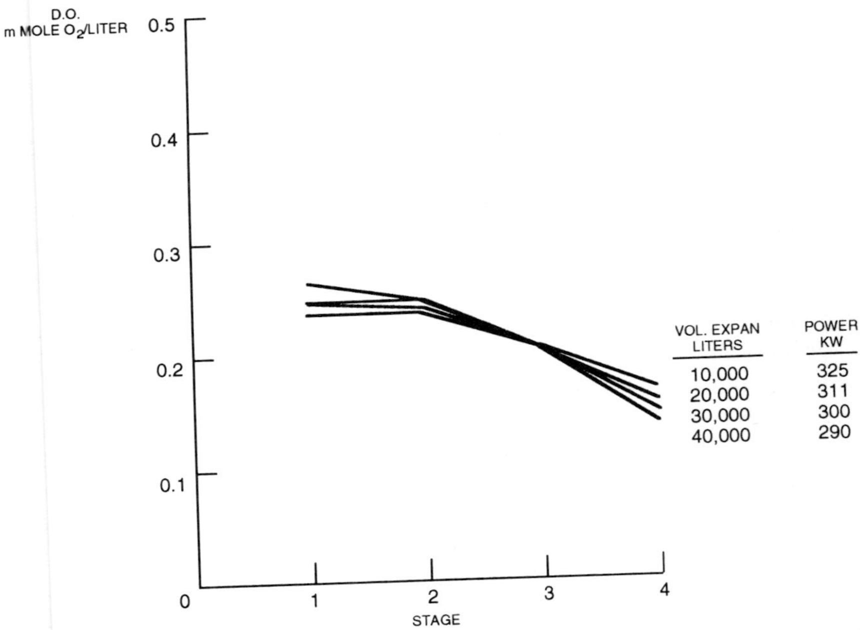

Figure 15.14 Effect of volume expansion or percentage hold-up on the D.O. profile for a hypothetical fermenter.[3]

and K_{La}, the effect would be larger. This will be discussed in more detail in section 15.1.7.1.

Certain recommendations can be made related to future work on percentage hold-up. First, a simple on-line method of measuring it would be useful both for control and future modeling work. Second, more work is needed to correlate H with variables such as percentage solids, surface tension, and viscosity. Third, studies are needed to understand better the effect of gas hold-up on agitator and mass transfer performance. Fourth, the effect of antifoam on hold-up needs to be studied. Finally, the potential for using surface tension reducing agents to increase hold-up and improve mass transfer may be worth investigating.

15.1.7.1 Bubble diameter. Bubble diameter also affects gas hold-up in fermenters. Smaller gas bubbles are less buoyant and rise more slowly than larger ones in the dispersion. Thus residence time of the gas is increased and gas hold-up is greater.

The interfacial area of the gas bubble increases with decreasing mean bubble diameter. The volumetric oxygen transfer coefficient (K_{La}) is the product of the oxygen transfer coefficient (K_L) and interfacial bubble area per unit dispersion (a). Therefore an increase in interfacial area will inherently increase the volumetric oxygen transfer coefficient in a fermenter. Interfacial bubble area can be calculated from measurements of gas hold-up and mean bubble diameter as follows:

$$a = 6H/d_m \tag{15.21}$$

where a is the interfacial bubble area per unit dispersion (cm^2/cm^3) and d_m is the mean gas bubble diameter (cm).

Interfacial bubble area has been correlated with gassed power per unit volume and superficial gas velocity as:

$$a \propto (P_g/V)^{0.62}(V_s)^{0.45} \tag{15.22}$$

Interfacial area (like gas hold-up) is a strong function of (P_g/V) and V_s. At constant V_s and with increased (P_g/V), gas bubbles experience greater shear resulting in smaller bubble diameter and greater interfacial area. On the other hand, at constant (P_g/V) and with increased V_s, more gas is introduced into the dispersion thereby increasing the number of gas bubbles (assuming uniform dispersion of gas still occurs), and inherently the overall interfacial bubble area per unit dispersion increases.

Differences between actual and predicted interfacial bubble area are most likely to be due to variations in the operating systems. Different types of fermenter mixers may take more (or less) power to disperse the gas phase in the liquid phase. Impeller design and other tank dimensions will also have an effect on power draw and will further affect both (P_g/V) and V_s with respect to the relationship to interfacial area.

15.1.8 *Sparging devices*

Gas introduction into the fermenter broth may be accomplished in many ways. The manner in which gas is introduced is not arbitrary, however. Three common sparge designs are illustrated in Figures 15.15, 15.16 and 15.17.

A sparge ring (Figure 15.15) is best suited for clean gas and liquids and non-slurries. One must be careful to avoid plugging of the small sparge ring holes with dirt, fouling, or settled solids from a slurry. Mass transfer is optimized with the sparge ring due to the gas discharging from the ring as uniformly distributed small diameter bubbles.

The sparge ring diameter should be 80% of the impeller diameter and should be located no closer than 2 in (5.08 cm) running clearance between the bottom of the impeller blade and the top of the sparge ring. The number and size of the gas outlet holes are predicated on gas rate and allowable

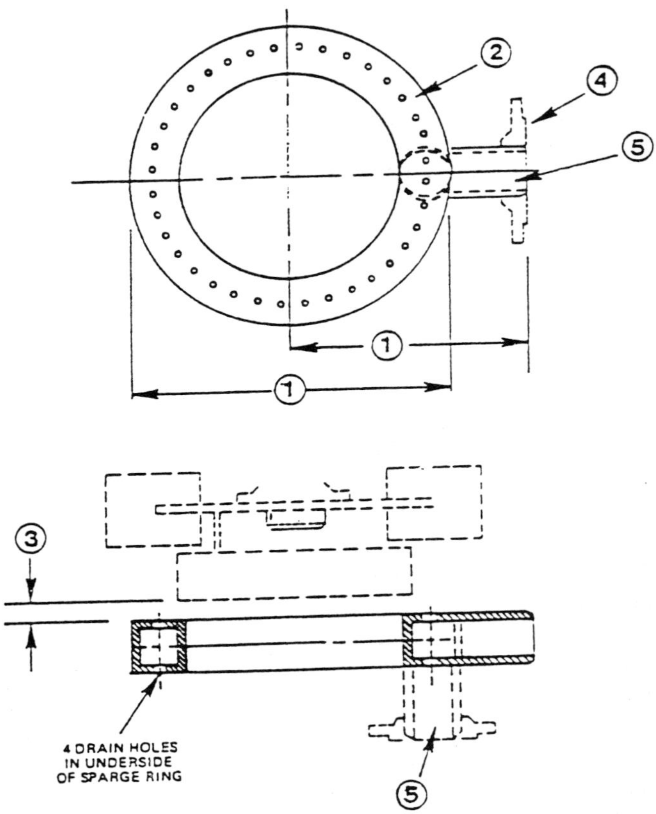

Figure 15.15 Gas inlet sparge ring design. (Mixing Equipment, Inc.)

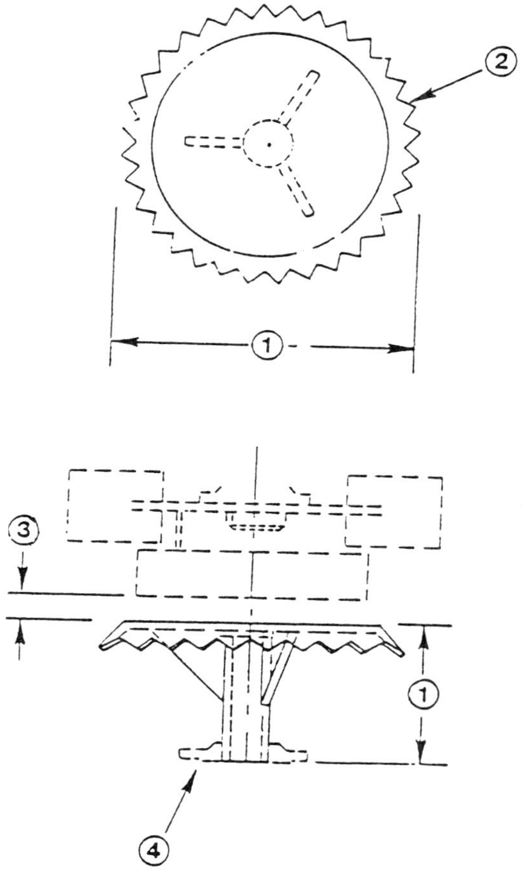

Figure 15.16 Gas inlet sparge plate assembly. (Mixing Equipment, Inc.)

pressure drop. Pressure drop through the holes should be limited to 2–5 lb/in^2 to pass the desired volumetric flow rate of gas, and avoid plugging of holes of small diameter.

Figure 15.18 shows the effect of sparge ring diameter on mass transfer coefficient, all other things being equal. The optimum ratio of sparge ring to impeller diameter is of the order of 0.8. This ratio can provide an increase of 10–15% in mass transfer rate, at a given gas rate and power level.

In the area where the mixer flow pattern prevails, there is very little effective difference due to the size and orientation of holes in the sparge ring. Figure 15.19 shows that, regardless of whether the holes are pointed upward or downward, or are of any particular diameter, there is little effect on mass

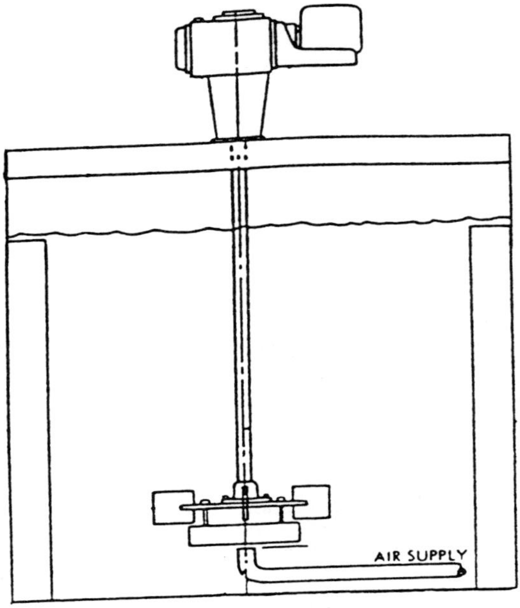

Figure 15.17 Open pipe sparging system. (Mixing Equipment, Inc.)

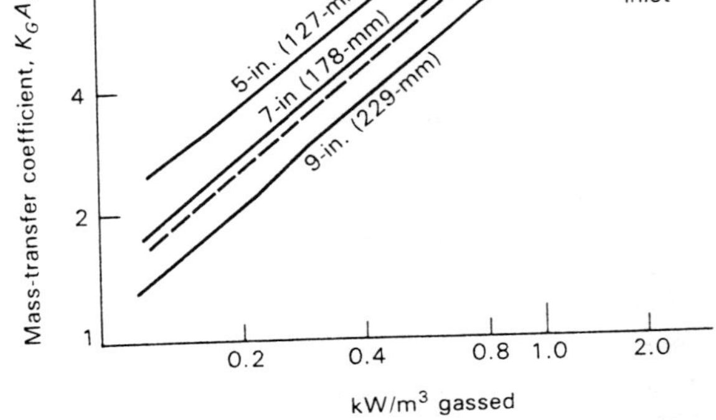

Figure 15.18 Coefficients for various sparge ring diameters. (Mixing Equipment, Inc.)

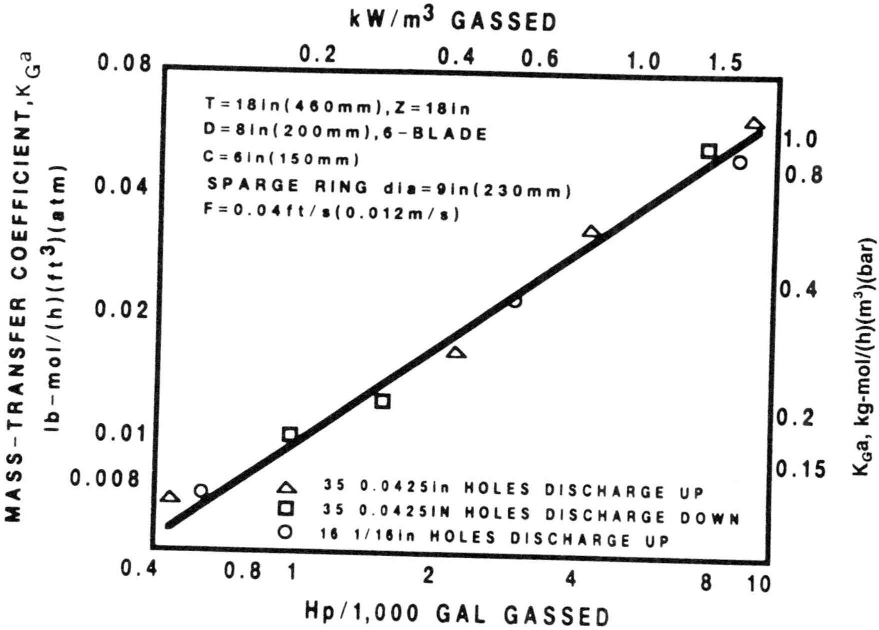

Figure 15.19 Coefficient with various hole diameters and number of holes. (Mixing Equipment, Inc.)

transfer coefficients. This is because the impeller, and not the gas bubbles produced by the sparge ring, controls the gas dispersion.

In designing a sparge ring, the objective is to get uniform flow distribution out of all the holes in the ring. Usually the diameter of the ring and pipe diameter are given and the design focuses on the number and diameter of the holes. A rule-of-thumb for the design is that the ratios for kinetic energy in the inlet stream to pressure drop across the outlet holes and of friction loss in the ring to pressure drop across the outlet holes should be equal to or less than one-tenth. If this rule is followed, the expected maldistribution of flow is less than 5%. Drain holes should be drilled on the bottom side of the sparge ring to promote self-cleaning and prevent plugging of the ring.

The sparge plate, shown in Figure 15.16, is less likely to plug than the sparge ring, and is much easier to clean in the event that it does plug. Like the sparge ring, the sparge plate diameter should be 80% of the turbine diameter and no closer than 2 in (5.08 cm) running clearance to the impeller blade bottom. Plate serrations promote uniform distribution of relatively small diameter bubbles of gas to enhance mass transfer.

The open pipe discharge of gas (Figure 15.17) is ideally suited for slurries and/or liquids that are dirty. Of all the gas discharge assemblies the open pipe is least likely to plug up, and if it does plug can be unplugged by rodding out the pipe or unclogging with high pressure gas or liquid. The open pipe discharges large agglomerated gas bubbles that must be sheared into smaller bubbles and distributed uniformly by the gas–liquid mixing impeller. As such, mass transfer is relatively poor in comparison with sparge plates and rings.

Gas inlet pipe design should be based on minimizing gas side pressure drop to ensure proper presentation of volumetric flow rate of gas. Each type of gas inlet assembly increases the fluid forces on the impeller due to the gas impinging on the impeller blades. This must be taken into account in the mechanical design of the agitator impeller, shaft, and drive train. Non-uniform gas loading on the impeller can occur if the impeller and gas inlet device are not installed in symmetry or if the inlet device is not designed and located properly or if the gas rate varies frequently creating pulsing on the turbine. Not only can these situations increase mechanical load on the impeller and mixer, but process results can be adversely affected by poor gas distribution.

15.1.9 Heat transfer

Heat transfer requirements in fermenters require flow controlled mixing design criteria, as opposed to shear controlling considerations for mass transfer. Applications include agitation to provide heating, or cooling, or merely to maintain a uniform temperature of the tank contents. Maintaining temperature uniformity is the easiest mixing application. Heating is the next easiest application, followed by cooling as the most difficult.

In processing, heat is transfered by conduction from the wall of the vessel or surface of the internal heating elements (i.e. coils) to the contents of the vessels. As the contents are agitated, heat is transferred into the mass by convection. Mixing thereby speeds up the conduction of heat for the heat transfer surfaces and helps promote the transfer of heat by forced convection.

Heating and cooling applications are best handled by providing adequate flow velocities of fluid across the heat transfer surfaces, with appropriate measures taken to promote good top-to-bottom circulation and turnover of tank contents. This flow pattern will ensure good exchange of fluid from areas remote to the heat transfer surface to the heat transfer element (i.e. between fluid in the center of the tank and the jacketed tank wall). In this manner reasonably uniform temperature profiles will result.

Heat transfer mixing applications may be considered similar to blending requirements. Consequently, for low viscosity fermentations, conventional propeller, impeller, and state-of-the-art hydrofoils are generally used. One qualification on horsepower levels for low viscosity applications is that unlike blending, a point of diminishing returns can be reached for heat transfer where an increase in input horsepower has an incrementally diminishing effect on process result.

For higher viscosity fermentations, larger diameter, slower speed impellers are usually required to obtain the required flow patterns. Often multiple large diameter, slow speed turbines are needed with accompanying increases in horsepower similar to that required in high viscosity blending applications.

Complete data on all contributing factors in a heat transfer agitator application are often difficult to obtain. As much data that are available should be furnished, including the following:

- Tank dimensions and details of heat transfer surface (i.e. jacket-type-half pipe, dimple, etc.; surface area; coil pipe or tube size and wall thickness; number of banks of coils and diameter of each bank; coil spacing, etc.)
- Whether heating, cooling, or temperature uniformity is the required process result
- Time available for heating and/or cooling
- Are solids present so that solid suspension must be addressed? If so, details must be provided as in any solid suspension problem to design a mixer to meet this criterion.
- Viscosity and specific gravity of fluids at the temperatures to be encountered
- Details of heating or cooling media and temperature of batch at the start and end of the cycle
- Specific heats and thermal conductivities of tank contents
- Fouling characteristics of the heating or cooling medium and whether the process material is heat sensitive (i.e. will it decompose at the temperature of the heating medium?)

Heat transfer in a fermenter is optimized with open impellers running at some distance from the tank wall. Examples of open impellers are propellers, radial and axial flow turbines of constant pitch, and hydrofoils of variable pitch. The impellers provide a greater balance of flow than shear to enhance heat transfer.

The transfer of heat to and from an agitated fermenter depends on the temperature difference driving force (ΔT) between the heat transfer medium and process fluid, the heat transfer area (A_o) and the overall heat transfer coefficient (U_o):

$$Q = U_o A_o \Delta T \tag{15.23}$$

where Q is the heat transfer rate (BTU/h (J/h)), U_o is the overall heat transfer coefficient (BTU/h-ft^2/°F (J/h/m/°C)), A_o is the heat transfer area (ft^2 (m^2)), and ΔT is the temperature difference between heat transfer media and tank contents (°F (°C)).

The calculation of ΔT and A_o is straightforward, but U_o is a combination of several resistances to heat transfer. Thus,

$$\frac{1}{U_o} = \frac{1}{h_o} + \frac{l}{K} + \frac{1}{h_i}\frac{A_o}{A_i} + ff \tag{15.24}$$

The subscript o refers to the process side or mixer side and the subscript i to the heat transfer medium side. The overall heat-transfer resistance ($1/U_o$) is the sum of the individual resistances due to the outside film coefficient on the mixer side of the heat transfer surface ($1/h_o$), the wall material and thickness (l/k), the inside film coefficient of the heat transfer medium ($1/h_i$) corrected to the common heat transfer area basis, and a fouling factor (ff) to account for corrosion, dirt films, and scale formation on either side of the heat transfer surface.

The agitator can only affect the mixer-side heat transfer film coefficient (h_o). In many cases, the mixer side film coefficient will be the governing resistance, and therefore the effect of a mixer on the overall heat transfer coefficient will usually be quite significant.

To predict heat transfer rates in agitated fermenters, the mixer-side heat transfer film coefficient (h_o) must be estimated. By means of dimensional analysis of heat-flow and energy-balance equations, the Nusselt number (which contains the coefficient h_o) can be expressed as a function of Reynolds number and Prandtl number:

$$N_{Nu} = f(N_{Re}, N_{Pr}) \tag{15.25}$$

where N_{Nu} is the Nusselt number (dimensionless), N_{Re} is the Reynolds number (dimensionless), and N_{Pr} is the Prandtl number (dimensionless).

Equation (15.25) can be expanded into a form similar to the heat transfer relationship dependent on impeller characteristics, geometrical properties of the tank, and fluid properties:

$$N_{Nu} = \frac{h_o D}{K} = f\left(\frac{D^2 N \rho}{\mu}, \frac{C_p \mu}{K}, \mu/\mu_s, D/T, d/T\right) \tag{15.26}$$

where D is the impeller diameter, K is the thermal conductivity of broth, N is the impeller speed, ρ is the density of broth, μ is the viscosity of broth, μ_s is the viscosity of broth at the temperature of heating or cooling medium (i.e. at the heat transfer surface), C_p is the specific heat of broth, d is the diameter of coil, pipe or tubing, and T is the tank inside diameter.

The relationship expressed in equation (15.26) includes many independent variables in the form of dimensionless groups (i.e. $N_{Re} = D^2 N\rho/\mu$ and $N_{Pr} = C_p\mu/K$) but only a portion of those that affect the flow field and heat transfer coefficient.

All the variables can be broken down into three categories, geometrical properties, impeller characteristics, and fluid properties. For a vertical cylindrical fermenter, the following are the principal variables for these three classes:

1. Geometrical properties:
   Tank diameter, T
   Liquid depth, Z

Bottom geometry (flat or dished, etc.)
With or without baffles (number of baffles = η)
Ratio of impeller to tank diameter (D/T)
Jacketed, helical coil, vertical panelcoil, or vertical tube heat transfer surface
Coil or tubing spacing, S
Number of coils or tubes

2. Impeller characteristics
   Type (axial or flow turbine, hydrofoil, etc.)
   Diameter, D
   Number of blades
   Blade geometry (height, width)
   Impeller off-bottom and spacing (if multiple turbines)
   Impeller blade pitch
   Rotational speed, N
   Placement of impeller with respect to coils
3. Fluid properties
   Newtonian or non-Newtonian characteristics
   Viscosity as a function of temperature and shear rate
   Thermal conductivity, K
   Thermal heat capacity, C_p
   Density, ρ

The three most common types of heat transfer surfaces in fermenters are jacketing, helical coils, and vertical tubes. These are illustrated in Figure 15.20. For each type, experimental correlations are presented for estimating

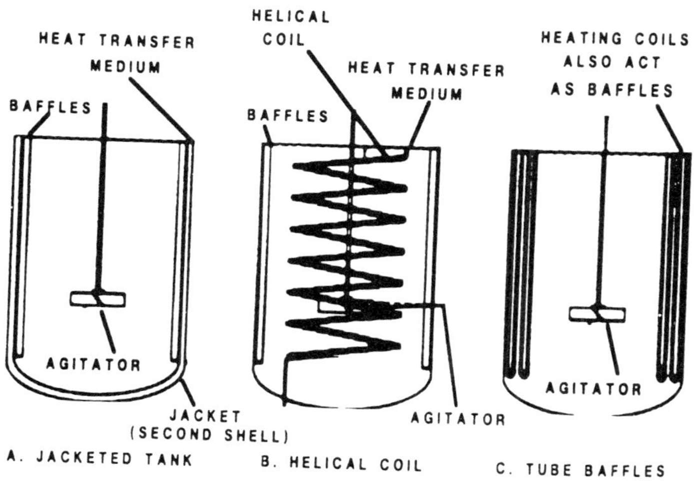

Figure 15.20 Three constant heat transfer surfaces with agitated tanks. (Mixing Equipment, Inc.)

the process fluid film coefficient. The calculated coefficient can only be estimated, because many of the variables affecting heat transfer in agitated tanks are not included in the correlations. All the correlations are based on fully or substantially baffled tanks, since the industrial use of unbaffled tanks in fermenters is quite limited.

15.1.9.1 Jacketed tanks. The jacket surface, which has the fewest variables to analyse, offers advantages with respect to initial cost and operating economy. Its principal limitations are that the heat transfer surface area is small for a given tank volume, in a commercial scale application. Another drawback to jackets is that adjustment of the jacket area after the tank has been constructed is virtually impossible.

Equation (15.27) shows the relationship of the process-side (mixer-side) heat transfer film coefficient, h_o to geometrical and fluid properties, and impeller characteristics, for jacketed heat transfer, with a flat-blade turbine, covering a range of low to moderate viscosities:

$$\frac{h_o T}{K} = 0.74 \left(\frac{D^2 N \rho}{\mu}\right)^{2/3} \left(\frac{C_p \mu}{K}\right)^{1/3} \left(\frac{\mu}{\mu_s}\right)^{1/7} \tag{15.27}$$

The exponents on Reynolds number, $D^2 N\rho/\mu$ and Prandtl number $C_p\mu/K$, have been well documented for paddle, axial flow turbines and hydrofoils, and flat-bladed radial flow impellers. The exponent on viscosity ratio, μ/μ_s, may range from 1/7 to 1/4, but unless the temperature differential is extremely great between the process-side film on the tank wall and the tank contents, the ratio is likely to be close to unity. In such a case, the value of the exponent is inconsequential. When calculating the process-side heat transfer film coefficient, one should assume the viscosity ratio to be unity.

More recent experimental work seems to support the following correlation:

$$\frac{h_o T}{K} = 0.85 \left(\frac{D^2 N \rho}{\mu}\right)^{2/3} \left(\frac{C_p \mu}{K}\right)^{1/3} \left(\frac{\mu}{\mu_s}\right)^{1/3} \left(\frac{Z}{T}\right)^{-5/11} \left(\frac{D}{T}\right)^{1/8} \tag{15.28}$$

15.1.9.2 Helical coils. A significantly greater amount of heat transfer surface can be obtained with helical coils in a fermenter for a given volume of process fluid than with jacketed tanks. However, coils are usually more costly to fabricate and more difficult to maintain. The following general correlation for helical coils in baffled tanks is based on experimentation and thorough investigation.

$$\frac{h_o d}{K} = 0.17 \left(\frac{D^2 N \rho}{\mu}\right)^{2/3} \left(\frac{C_p \mu}{K}\right)^{3/8} \left(\frac{D}{T}\right)^{1/10} \left(\frac{d}{T}\right)^{1/2} \left(\frac{\mu}{\mu_s}\right)^{m} \tag{15.29}$$

As can be seen, the exponents on the dimensionless groups are nearly identical to those of jacketed tanks, and are consistent with earlier helical-coil heat transfer investigations. An additional variable for helical coils compared to jacketed tanks is tube diameter d. Equation (15.29) is based on work covering the practical and commercially available range of tube diameters. Tube spacings of between two and four tube diameters (centerline to centerline) are required for consistent results. Spacing tighter than two tube diameters will not allow for good flow and velocity of fluid across the tube surfaces. Spacing greater than four tube diameters apart minimizes coil surface area, and may require multiple banks of coils or outside surface area to compensate.

A correlation obtained for exponent m, for the viscosity correction factor (μ/μ_s) results in values ranging from 0.1 to 1.0 for m:

$$m = 0.1\,(\mu 8621 \times 10^{-5})^{-1/5} \tag{15.30}$$

A plot of the exponent m versus viscosity for a typical fluid is shown in Figure 15.21.

15.1.9.3 Vertical tubes. For this style of heat transfer surface, the tubes are areas of high turbulence, improving heat transfer. The best work on vertical tubes as a surface for heat transfer has resulted in the following relationship:

$$\frac{h_o d}{K} = 0.09\left[\left(\frac{D^2 N\rho}{\mu}\right)^{2/3}\left(\frac{C_p\mu}{K}\right)^{1/3}\left(\frac{D}{T}\right)^{1/3}\left(\frac{2}{\eta_B}\right)^{1/5}\left(\frac{\mu}{\mu_s}\right)^{1/7}\right] \tag{15.31}$$

where η_B is the number of vertical tubes acting as baffles.

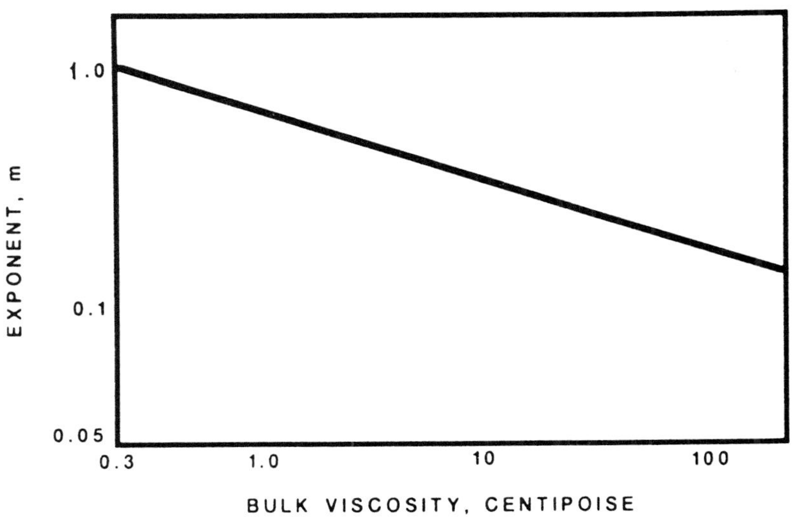

Figure 15.21 Exponent m versus bulk velocity. (Mixing Equipment, Inc.)

Previous work indicates that with six tube baffles (i.e. $\eta = 6$), each consisting of three tubes of size $d/T = 0.031$, the vertical tubes account for 75% of the power consumption of an impeller in a tank having standard wall baffles. Thus, tube baffles provide reasonable, but not full, baffling.

15.1.9.4 Vertical plate or panel coil. Plate or panel coils are variations of vertical tubes and are easier to install and maintain. The baffling effect with plate or panel coils is greater than with vertical tubes. However, they restrict fluid flow more around the heat transfer surface and hence spacing is critical.

The following correlations have been proposed for heat transfer with vertical plate coils.

For $N_{Re} < 1400$:

$$\frac{h_o L}{K} = 0.1788 \left(\frac{D^2 N \rho}{\mu}\right)^{0.488} \left(\frac{C_p \mu}{K}\right)^{1/3} \left(\frac{\mu}{\mu_f}\right)^{1/2} \tag{15.32}$$

for $N_{Re} > 4000$:

$$\frac{h_o L}{K} = 0.0317 \left(\frac{D^2 N \rho}{\mu}\right)^{2/3} \left(\frac{C_p \mu}{K}\right)^{1/3} \left(\frac{\mu}{\mu_f}\right)^{1/2} \tag{15.33}$$

where L is the plate width per plate of panel coil and μ_f is the viscosity of fluid at mean film temperature along the plate coil surface. The angle at which the plate coil intersects the tank radius has been found to have no significant effect on heat transfer or power consumption. This means that plate coils can be angled for maximum heat transfer surface area within a tank. However, 90° is best to minimize stagnant flow.

Panel coils should be equally spaced around the inside tank wall and they should be spaced no closer than 10° apart from each other to allow fluid to flow between them.

Overcrowding the inside of fermenters with panel coils, helical coils, or vertical tubes spaced too close together can result in the agitator performing in essence in an unbaffled tank. Tight spacing of the transfer surface will not provide flow through and around the surface and the flow pattern will take on a swirling regime. Inadequate velocities across the surfaces will not achieve the anticipated mixer-side heat transfer film coefficient.

Theoretically, the best heat transfer coefficient for the least mixer horsepower can be obtained if swirling due to lack of baffles is allowed. However, unbaffled systems are limited to relatively narrow ranges of Reynolds number because of the limiting condition of totally rotational motion. Since baffled systems are capable of much higher Reynolds numbers, they are ultimately capable of higher heat transfer coefficients.

Although agitation is necessary to provide the driving force for good forced-convection heat transfer, sizing an agitator to achieve a desired heat transfer coefficient may be impractical. As a typical case, equation (15.29) can be expanded to show the effects of each variable on the heat-transfer coefficient.

$$h_o \propto \mu^{-0.30} D^{1.44} N^{0.67} T^{-0.6} K^{0.63} d^{-0.5} \rho^{0.67} C_p^{0.37} \left(\frac{\mu}{\mu_s}\right)^m \quad (15.34)$$

The energy consumption of a mixing impeller is transformed into heat input into the tank contents. An input of 2550 BTU/h additional heat load is transferred into the batch for each mixer hp. In fermentations requiring high power levels, the heat load due to the mixing impeller may be a significant factor. For example, the temperature rise in a batch due to mixing alone will typically be 0.2–0.3°F at a power level of 1 hp (745.7 W) per 1000 gallons (4537 l) of fluid.

Usually axial flow turbines are best suited for heat transfer applications, as they are for any other flow controlled mixing requirement. They produce the greatest amount of flow per unit horsepower with reasonable capital cost implications. Along this line of thought, a hydrofoil impeller is the optimum choice of impeller geometry, since it affords the highest flow efficiency of the family of axial flow turbines. When internal space is a problem, a high power number flat-bladed radial turbine is recommended. In some cases, batch depth may predicate that multiple impellers be installed to ensure forced convection throughout the fermenter contents.

In general, hydrofoils optimize heat transfer rates because these impellers produce the greatest flow per unit power. Figure 15.22 is a typical plot showing the effect of agitator speed on measured heat transfer coefficient. In the middle speed range is a zone where the physics of heat transfer by forced convection predominates. In this range there is a straight line relationship with a slope of 0.67, which is the exponent of Reynolds number in the heat transfer correlation.

Below point A, fluid turbulence decreases to where natural convection forces begin to influence heat transfer. At an agitator speed of zero, natural convection alone provides heat transfer, and the two curves converge.

At agitator speeds above point B, fluid motion is so turbulent that further increases in such motion provide little driving force for additional heat transfer. At these speeds, the surface motion is likely to be so intense that air will be entrained into the liquid, further complicating heat transfer relationships. Most well baffled mixing systems operate below point B.

The greatest incremental gains in heat transfer efficiency are achieved by installing an agitator in a tank previously limited to natural convection heat transfer. For example, an agitator operated at 80 rpm (Figure 15.22) increased the mixer-side heat transfer film coefficient by nearly 100% over the coefficient at 0 rpm. Yet to double the heat transfer coefficient again would require a speed increase of over 300% (or over 27 times as much power), due to diminishing returns for heat transfer with power in this speed range.

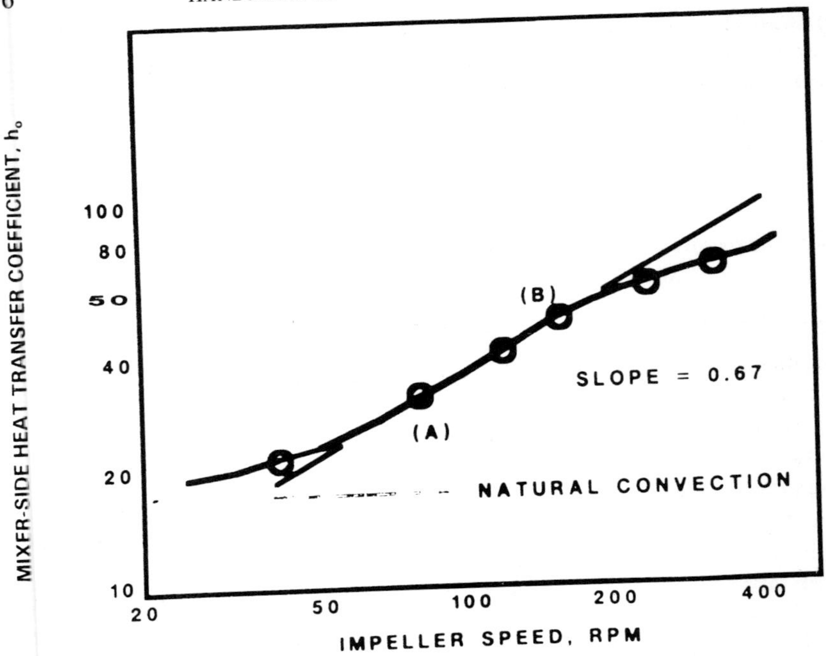

Figure 15.22 Typical logarithmic plot of heat transfer coefficient versus impeller speed. (Mixing Equipment, Inc.)

It is imperative to operate a fermenter mixer in the forced convection zone of Figure 15.22. Experimental data may be plotted to generate the curve or if such data are not available, a fast selection may be made without excessive cost by using twice the minimum power required for continuous blending. This is also a good design method to follow in making preliminary budget estimates of mixer size and cost for heat transfer applications. Keep in mind that mass transfer and/or dispersion requirements may be the limiting factor in fermenter mixer design.

Empirical relationships have been developed to determine minimum agitator speed to ensure forced convection. For example, the following relationship has been presented to be used with equation (15.31).

$$N \geqslant 43\left(\frac{T}{D}\right)^{2.5}\left(\frac{B}{2}\right)^{0.3}(N\upsilon)^{0.2} \tag{15.35}$$

where N is the impeller speed (rpm), T is the tank diameter, D is the impeller diameter, B is the baffle width, and υ is the kinematic viscosity. As long as the inequality is satisfied and the mixing is not too severe, equation (15.35) defines the heat transfer characteristics of the system tested. However, such empirical relationships tend to be particular to the system studied. Visual observation is usually the only test necessary to confirm if a mixing system is one of forced circulation in the fermenter. Because of their higher thermal conductivities, the coefficient of aqueous fermenter broths will be three to four times greater than those of the organic liquids in Figure 15.23.

15.2 Pneumatically agitated fermenters

15.2.1 Introduction to bubble columns and airlifts

Since its discovery in 1954, airlift fermentation has increasingly been explored as an important and potentially superior alternative to tank fermentation. The special feature that distinguishes an airlift fermenter from the more common stirred tank reactor is that the pressurized gas which is furnished for mass transfer also provides the necessary circulatory driving force. The main advantages of this design are the capacity for satisfying the high oxygen demand of certain fermentations without mechanical moving parts, low energy input, especially in large configurations, and the possibility of easier removal of the heat generated in the fermentation process.

Pneumatically agitated bioreactors (bubble columns and airlifts) are relatively new types of devices (compared to the classical stirred vessels) being

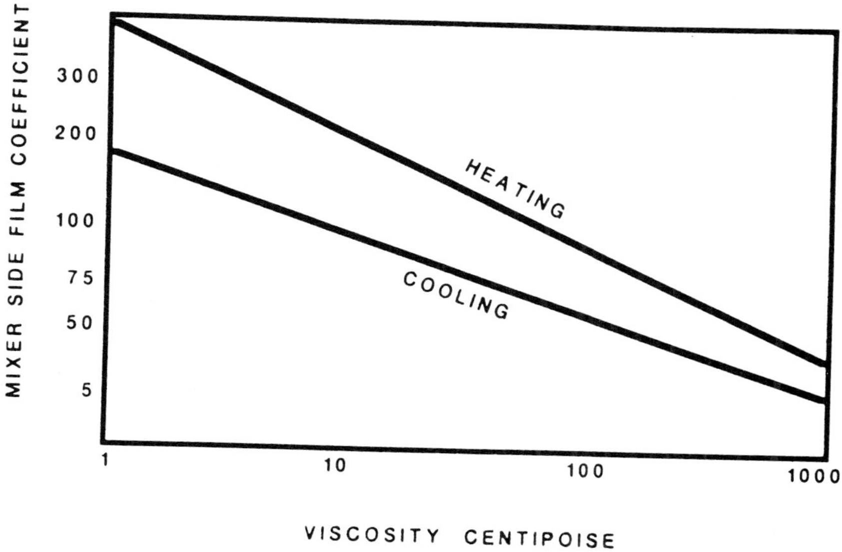

Figure 15.23 Typical mixer-side film coefficients for heating and cooling organic chemicals (thermal effectiveness 1.54). (Mixing Equipment, Inc.)

applied to aerobic fermentations. Despite its limitations, the stirred tank reactor remains the workhorse for the fermentation industry. However, the relative simplicity of the design of pneumatic devices, which leads to mechanical robustness, ease of monoseptic operation over extended periods, low shear fields, and low power consumption makes it attractive for new or improved processes in commercial practice. For highly shear-sensitive fermentations such as with some mammalian and plant cells where applications are increasing, there may be no choice other than the use of pneumatic reactors.

Airlift fermenters are noted for high mass transfer and good mixing with relatively low shear and low power output. Due to these features, airlift reactors are currently employed in certain large scale industrial applications, most notably for single cell protein fermentation. Recently, researchers have utilized the airlift design for applications which develop non-Newtonian broths, such as fermentation of filamentous fungi and plant cell culture. Other possible applications, including culture of animal cells and production of citric acid and xanthan gum, have been cited.

While the construction and operation of airlift reactors are particularly simple, the complexity of recirculating two-phase flow hampers design optimization and scale-up. The complexity is, of course, compounded when non-Newtonian effects are included. Over the last decade a number of hydrodynamic airlift models have been generated based upon a macroscopic balance approach that are limited in their practical applicability owing to their rheological simplicity, restriction to zero gas recirculation, and/or dependence upon highly specific correlations.

A large (1000 m^3) airlift aerobic reactor has been operated for a production (as opposed to waste treatment) fermentation, for the manufacture of single cell proteins. However, process engineering information for the design of airlift devices remains limited. Design correlations for the effects of various geometric and operating parameters on reactor performance are not well-established. This uncertainty about performance appears to be the principal factor which restricts wider use of airlift reactor devices in biotechnology industries.

Airlift fermenters can be designed in one of two basic configurations. Either the upflowing gassed liquid and the downflowing liquid are separated by a draft tube or a baffle plate suspended inside the reactor (internal loop), or the two streams flow in separate pipes connected at the top and bottom (external loop). It has been shown that, because of the introduction of gas, a strong liquid circulation pattern is developed. The liquid circulation rate is considered to be one of the key parameters in the design and scale-up of the airlift reactor. It affects mixing characteristics of the reactor, gas hold-up, volumetric mass and heat transfer coefficients and uniformity of reaction temperature and cell concentration; thus, it affects the overall performance of the reactor.

One of the earliest studies of the draft-tube airlift column predicted the liquid circulation rates in a Pachuca tank using airlift pump theory. In this model the energy transferred as the air expands in rising through the vessel is corrected for energy dissipation due to bubble slip and the remainder gives rise to velocity heads associated with liquid flow in the vessel.

15.2.2 Theoretical and actual performance

Based on studies and theoretical modeling, liquid circulation rates can be predicted based on a macroscopic energy balance and airlift pump theory in draft tube bubble column fermenters. There is evidence of close agreement between actual and theoretical liquid circulation velocities based on the energy balance method. The basis of the energy balance method is to equate the rate of energy input to the fermenter by the gas to the rate of energy dissipation by the fluid motion.

For buoyancy-driven two phase flow, the supply of energy results from introduction of the dispersed phase. The energy input rate for gas–liquid dispersion is given by equation (15.36):

$$E_i = QP \ln(1 + \rho_L g H_L / P) \tag{15.36}$$

where E_i is the energy input rate, Q is the volumetric gas flow rate, P is the pressure at the top of the column, ρ_L is the density of the liquid phase, H_L is the ungassed liquid height, and g is the acceleration due to gravity.

The energy dissipation in an airlift column is due to the following factors:

1. Energy dissipation in the upflow zone due to wakes behind the bubbles, E_{d1}.
2. Energy dissipation in the top and bottom regions due to liquid turnaround and other resistance to liquid recirculation in the column, E_{d2}.
3. Energy dissipation in the downflow zone due to the upflow motion of bubbles with respect to the liquid, E_{d3}.
4. Energy dissipation due to the viscous drag at the walls of the column, E_{d4}.

An order of magnitude calculation indicates that the dissipation due to viscous drag, E_{d4} makes up only 1 to 2% of overall energy input for liquid viscosities up to 20 mPa. Therefore, the energy dissipation due to this mode is negligible.

The energy dissipation rate, E_{d1}, in the bubble wakes can be evaluated by considering a bubble cloud with uniform fractional hold-up E_{G1}, ungassed liquid height H_L, liquid density ρ_L, horizontal cross-sectional area in the upflow zone, A_1, and liquid and gas superficial velocities U_{L1} and U_G. E_{d1} is the result of three terms, namely the rate of loss of pressure energy of the liquid $= A_1 \rho_L g H_L U_{L1}$, the rate of loss of pressure energy of the gas $= A_1 \rho_L g H_L U_G$, and the rate of gain of potential energy of the liquid $= A_1 \rho_L g H_L U_{L1}/(1 - E_{G1})$, giving

$$E_{d1} = A_1 \rho_L g H_L E_{G1} (U_G/E_{G1} - U_{L1}/(1 - E_{G1})) \tag{15.37}$$

The slip velocity, the relative velocity of the bubbles with respect to the liquid, is equal to $U_G/E_{G1} - U_{L1}/(1 - E_{G1})$. Assuming that slip velocity is equal to the terminal velocity of bubbles rising in a stagnant liquid, V_b, gives

$$E_{d1} = A_1 \rho_L g H_L E_{G1} V_b \tag{15.38}$$

As might be expected, gas and liquid flow rates do not appear in this equation; the energy dissipation due to wakes behind the bubbles is mainly determined by the gas hold-up.

Evaluation of the energy dissipation E_{d2} in the airlift column gives

$$E_{d2} = \sum_{1}^{2} \tfrac{1}{2} \rho_L V_i^3 e_i A_i (1 - E_{Gi}) \tag{15.39}$$

where e_i is the friction loss factor due to liquid turnaround and other resistance to liquid recirculation in the column, V_i is the interstitial liquid velocity $= U_{Li}/(1 - E_{Gi})$, A_i is the horizontal cross-sectional area, and 1, 2 represents the upflow zone and downflow zone, respectively.

The energy dissipation E_{d3} can be evaluated by considering the macroscopic energy changes of the liquid between the top and bottom regions of the downflow section. If we neglect the change of kinetic energy, E_{d3} is the result of two terms, namely the rate of loss of potential energy $= A_2 \rho_L g H_L V_2$ and the rate of gain of pressure energy $= A_2 \rho_L g H_L (1 - E_{G2}) V_2$, giving

$$E_{d3} = A_2 \rho_L g H_L E_{G2} V_2 \tag{15.40}$$

The energy balance, therefore, takes the following form:

$$E_i = E_{d1} + E_{d2} + E_{d3} \tag{15.41}$$

If we substitute the expressions of E_{d1}, E_{d2}, and E_{d3} into equation (15.41) we get

$$E_i = A_1 \rho_L g H_L E_{G1} V_b + \sum_{1}^{2} \tfrac{1}{2} \rho_L V_i^3 e_i A_i (1 - E_{Gi}) + A_2 \rho_L g H_L E_{G2} V_2 \tag{15.42}$$

Since there is no net liquid flow from the column, the primary liquid volumetric flow rates in upflow and downflow zones are equal, and so from continuity:

$$V_1 A_1 (1 - E_{G1}) = V_2 A_2 (1 - E_{G2}) \tag{15.43}$$

Hence, the interstitial liquid velocities V_1 and V_2 in a given geometry of airlift column may now be predicted from equations (15.42) and (15.43) for a given volumetric gas flow rate, Q, and ungassed liquid height, H_L, provided the friction loss factors, e_1 and e_2, are known and the characteristic bubble rise velocity, V_b, and gas hold-ups, E_{G1} and E_{G2} can be either estimated from appropriate correlations or measured.

15.2.2.1 Bubble rise velocity. The bubble rise velocity, V_b, can be predicted from the expression:

$$V_b = (2\sigma/\rho_L d_e + g d_e/2)^{0.5} \quad (15.44)$$

where σ is the surface tension and d_e is the equivalent bubble diameter in the gas–liquid system.

The influence of equivalent bubble diameter and surface tension on the terminal rise velocity of bubbles has been evaluated. The bubble rise velocity increases gradually with surface tension, the incremental difference diminishing with increasing bubble diameter. So, the influence of surface tension on bubble rise velocity can be neglected unless the broth contains significant amounts of impurities. The bubble rise velocity does not change too much in the range of equivalent bubble diameter from 0.002 m to 0.01 m. Since under normal operating conditions the diameters of most of the bubbles generated in an airlift column lie in this range, it can be assumed that the terminal rise velocity of bubbles in the range of interest is independent of bubble size and approximately equal to 0.23 m/s.

15.2.2.2 Friction loss factor. The friction loss factors, e_1 and e_2, can be determined by one of two methods, simultaneous solution of the macroscopic balances or experimental measurement. Theoretically, these values will vary with flow condition and column configuration. However, due to the lack of information of the experimental data for e_1 it will be assumed that the friction loss factors are constants and roughly equal to two to four depending on the resistance to liquid recirculation in the system.

15.2.2.3 Gas hold-ups. Previous testing in draft tube airlift columns has revealed the relationship between fractional gas hold-up in the upflow zone, E_{G1}, and downflow zone, E_{G2}, to be

$$E_{G2} = 0.89\, E_{G1} \quad (15.45)$$

Therefore, measurement or calculation of the gas hold-up in the upflow zone will lead to a prediction of same in the downflow zone as in equation (15.45).

The energy balance model described above was solved numerically using Newton–Raphson iteration and was tested by comparing the liquid circulation velocities predicted by this model with the experimental data reported from previous studies of three different types of airlift columns (see Figure 15.24).

15.2.2.4 Draft-tube airlift column. Data were generated from a draft tube airlift column when the liquid used was fermentation medium. Liquid circulation velocities predicted by the energy balance model were very close to the experimental data, the difference being within 7% for superficial gas velocities below 0.2 m/s. However, the difference between the predicted values and the experimental data increased with increasing superficial gas velocities

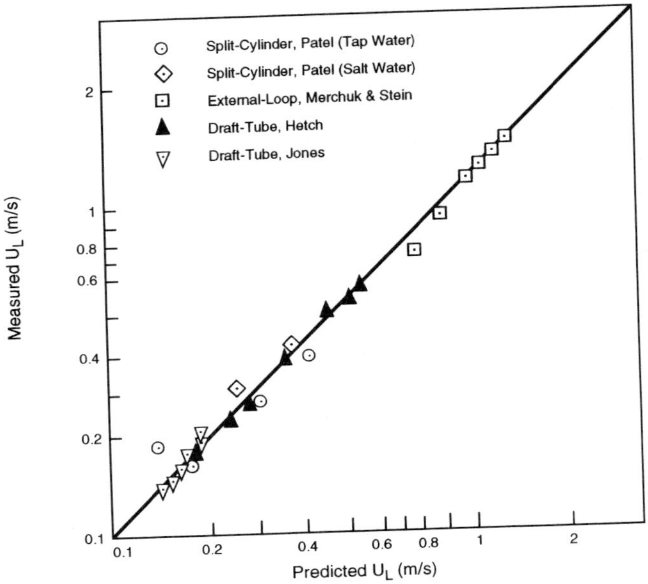

Figure 15.24 Comparison of liquid circulation velocities predicted by the energy balance model with the experimental data obtained in three different types of airlift columns.[4]

above 0.2 m/s. This probably resulted from the transition in flow regimes from bubbly to slug flow.

In the bubbly flow regime, which is of most practical importance in the operation of airlift fermenters, the influence of bubble size on the terminal rise velocity of bubbles can be neglected. Consequently, the energy dissipation due to the bubble wakes is independent of bubble size and directly proportional to the gas hold-up in the upflow zone. Thus, the liquid circulation velocity can be predicted from the measurement of gas hold-up in the upflow and downflow zones, respectively. In the slug flow regime, where the bubble diameter is much larger than that in bubble flow, the terminal rise velocity of bubbles is strongly affected by the bubble size. Therefore, the prediction of liquid circulation velocity is only possible when the characteristic bubble size for a given operating condition can be measured or estimated beforehand.

15.2.3 Fluid dynamics

Traditional methods are not suited to diagnosis and analysis of important effects of transient and localized velocity vectors experienced by cells in the turbulent two-phase flows of conventional bioreactors. There is presently little in the way of fundamental information and understanding of how the physics and dynamics of the fluid mechanisms contribute to the accumulation of biomass and recovery of products from cell cultures, including such deleterious aspects as clustering of suspended cells and their attachment to walls and accumulation in foams, or the possible damaging effects of encounters with bubbles and energetic eddies.

A shortcoming of mechanically agitated bioreactors (fermenters) is that a biologically unfavorable environment often exists within the zones of high shear caused by rapidly rotating impeller blades. Bioyields of shear sensitive cells can be adversely affected. In light of these findings, favor has shifted increasingly over the past decade to a 'stirrer free' airlift bioreactor for shear sensitive fermentations. Caution should be employed over the potential deleterious shear effects of 'overgassing' in airlift fermenters.

15.2.3.1 Circulation time in airlift fermenters. Circulation times around the flow loop of draft tube airlift fermenters decrease with increasing height of the liquid column and increase with increasing elevation of the bubble sparger above the base of the vessel. Both these results compare favorably with predictions on global energy fluxes of bubble buoyancy and frictional dissipation.

Circulation times are independent of geometrical details of the bubble sparger. The ratio of diameters of the draft tube and bioreactor have been investigated with respect to their effect on circulation time. The shortest circulation times are realized when the diameter ratio takes a value of about 0.6, whereas 0.8 to 0.9 gives rise to the shortest mixing times. Circulation time increases significantly with increased viscosity of the biomass medium, despite the fact that bubble sizes are larger in viscous solutions.

For a given draft tube airlift bioreactor geometry, there is a minimum liquid volume (and therefore liquid depth) associated with optimized circulation time, irrespective of gas flow. The physical explanation for this is as follows.

Consider first what happens with a reduction in the volume of liquid in the vessel, or equivalently, a lowering of the free surface within the disengagement chamber. Eventually this gives rise to weir flow plunging over the top of the draft tube, the perimeter of which then imposes a hydraulically critical control section. Ultimately, of course, the flows entering the downcomer are entirely suppressed. At low gas rates, the incremental increase in circulation time is significant as one deviates from the optimum volume (depth).

The increase is comparatively greater than that observed at higher gas rates. Increased hold-up of bubbles and hence increased elevation of the free surface accounts for a smaller critical volume realized at higher air rates.

Conversely, increasing the volume of liquid and hence elevation of the free surface eventually serves only to enlarge the volume of overlying 'dead-liquid' in the disengagement chamber. Passages of the rising bubbles through this zone then contribute only localized, recirculatory patterns of pluming and eddying motions, without appreciably affecting conditions in the riser and

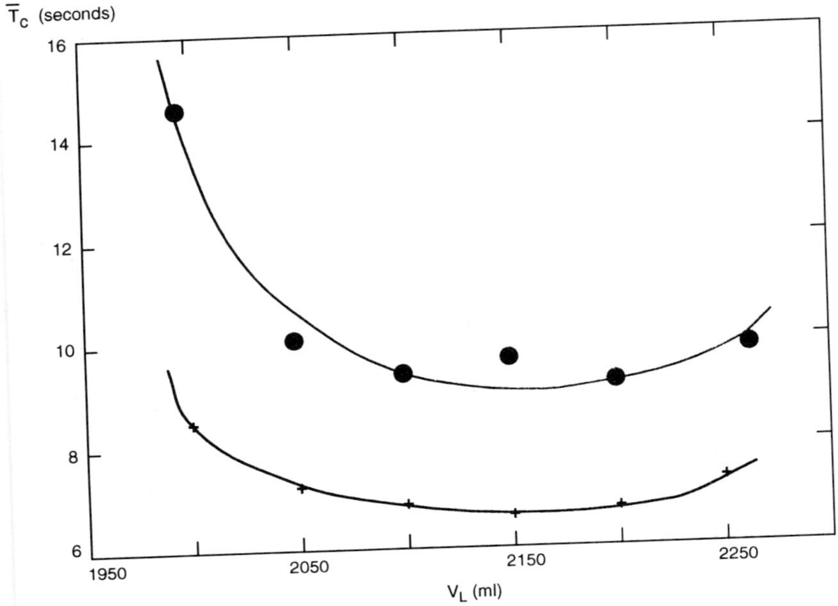

Figure 15.25 Mean-value circulation times, $\overline{T}_c$, versus liquid volume, V_L at air-flow rates (ml/min) of ●, 400; + 1000.[5]

downcomer. Figure 15.25 reflects test data observed in a nominal 2 liter draft tube airlift with bubble disengagement chamber. Circulation times were evaluated at various liquid volumes (depths) for two air rates.

Circulation times decrease with increasing gas flow rates. The reduction is significant initially as gas rate is increased. Circulation times then level off and approach a constant value, independent of further increases in gas rate. At higher gas rates, circulation times are not so strongly affected by volume (depth) of liquid. Figure 15.26 supports these findings based on testwork conducted with the previously described 2 liter draft tube airlift.

At effective viscosities of about 400 cP, demonstrated by the glycerol and carboxymethylcellulose (CMC) solutions in Figure 15.26, the mean value circulation time, $\overline{T}_c$, is of the order of ten times that of water. Bubble sizes are substantially bigger in both the glycerol and CMC solutions, compared to water.

The glycerol and CMC solutions closely resemble viscosity and shear rate characteristics of non-Newtonian pseudoplastic fermentation broths, with biomass loadings typically observed in commercial applications.

With glycerol and CMC there was no visible motion near the walls of the vessel in the disengagement chamber. This would extend residence time of biological cells within the bioreactor and possibly lead to surface encrustation

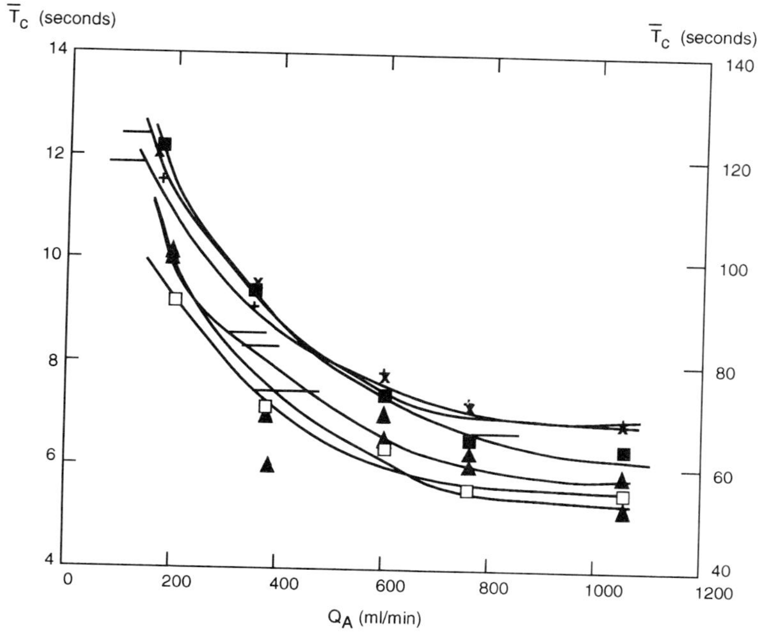

Figure 15.26 Mean-value circulation times $\bar{T}_c$ versus air-flow rate Q_A with $V_L = 2100, 2200$ ml respectively of ×+, water; ▲▲ 96% glycerol; □■ 2.5% carboxymethylcellulose (CMC).[5]

problems associated with concentrated broths. This phenomenon could eliminate the airlift fermenters from consideration for concentrated broths, while favoring mechanically mixed systems.

15.2.4 Mass transfer

The liquid film control of mass transfer of a sparingly soluble gas from the gas to liquid phase (or in the opposite direction) is well known and the transport rate is given by

$$\frac{dC_L}{dt} = k_L a_L (C^* - C_L) \tag{15.46}$$

where k_{LaL}, C^*, and C_L are the overall volumetric mass transfer coefficient, the saturation concentration of oxygen in the liquid and the liquid phase oxygen concentration at any time, t, respectively. Integration of equation (15.46) between the limits $C_L = C_0$ at $t = 0$, $C_L = C_L$ at $t = t$

$$\ln \frac{(C^* - C_0)}{(C^* - C_L)} = k_L a_L t \tag{15.47}$$

which may be utilized for k_{La_L} determination based on the transient gassing-in method. For point measurements, equation (15.47) is valid only when the reactor is fully backmixed.

Apart from $k_L a_L$, another important parameter in bioreactor design is the fractional gas holdup, ε. The residence time (t_G) of the gas in liquid, the total dispersion volume (V_D), the gas–liquid interfacial area for mass transfer (a_L) and, in airlift reactors, the circulation of liquid, all depend on gas hold-up. Thus,

$$a_L = \frac{6\varepsilon}{d_B(1-\varepsilon)} \tag{15.48}$$

$$V_D = \frac{V_L}{1-\varepsilon} \tag{15.49}$$

and

$$t_G = \frac{V_L\,\varepsilon}{Q_G(1-\varepsilon)} \tag{15.50}$$

where Q_G is volumetric flow rate of gas and V_L is volume of liquid.

15.2.4.1 Gas–liquid mass transfer. Based on experimental work in several different bioreactors (including bubble column, internal, and external loop airlifts), and with solutions to simulate typical fungal fermentation media, the following observations were seen. The overall mass transfer coefficient data showed trends very similar to those observed for gas hold-up in the pneumatic reactors. For suspensions in the bubble column the following correlation was obtained:

$$K_L a_L = (0.317 \pm 0.047)\,C_s^{-(1.781 \pm 0.183)}\,U_{sg}^{(0.637 \pm 0.114)} \tag{15.51}$$

where C_s is the concentration of solids and U_{sg} is the superficial gas velocity.

Equation (15.51) applied to suspensions only ($C_s = 1$ to 3 wt/vol%) for $0.016 \leqslant U_{sg}\,(\text{m/s}) \leqslant 0.35$ and it correlated the data within about 20% (Figure 15.27). The very strong negative effect of solids on mass transfer is clear. It is mainly due to gas hold-up reduction which occurred in the presence of solids because of their turbulence dampening and bubble coalescing effects. There was some evidence that k_L is also influenced by the presence of solids. Solutions of solka floc (SF) cellulose fibers in dilute saltwater were used to simulate mycelial fermentation media.

For external loop airlift reactors:

$$k_L a_L = \left(1 + \frac{A_d}{A_r}\right)^{-1} (0.349 - 0.102\,C_s)\,U_{sg}^{0.837 \pm 0.062} \tag{15.52}$$

where A_d is the cross-sectional area of downcomer and A_r is the cross-sectional area of riser.

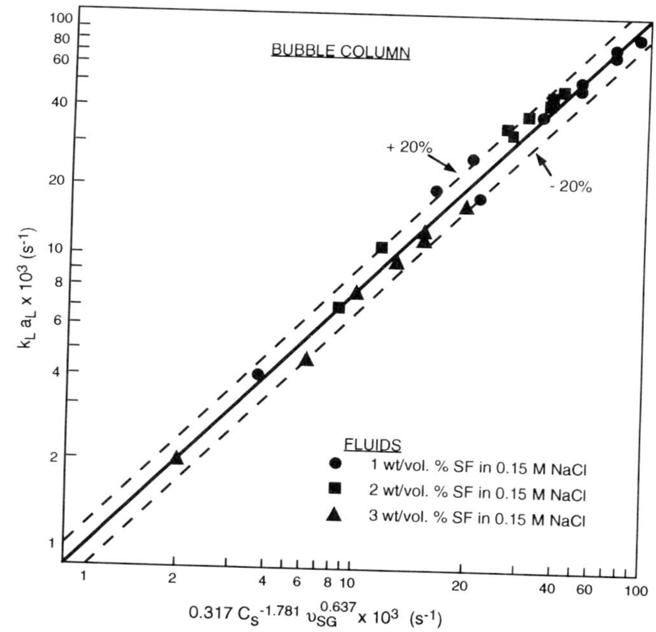

Figure 15.27 Overall mass transfer coefficient for suspensions in the bubble column.[6] For fluids, ●, 1 wt/vol% SF in 0.15 M NaCl; ■, 2 wt/vol% SF in 0.15 M NaCl; ▲, 3 wt/vol% in 0.15 M NaCl.

Figure 15.28 is a graphical representation of equation (15.52). The influence of solids on overall mass transfer coefficient depends on the type of gas–liquid contacting device and it was found to be stronger in the bubble column than in the external loop geometry. This is due to the direction of bubble motion being less chaotic in the airlift reactors than in bubble columns and in the former there was less bubble–bubble interaction. Consequently, the effects of such properties as solid content, which enhance coalescence, were less pronounced in the airlift vessels. However, the absolute value of $K_L a_L$ was always substantially lower in the external loops compared to the bubble column and the internal loop. This may be readily explained in terms of the different gas hold-up characteristics of these reactors.

15.3 Scale-up considerations

15.3.1 Macro- and micro-scale mixing

A major concern pertaining to fluid mixing processes revolves around scale-up. The role of macro- and microscale agitation is better understood

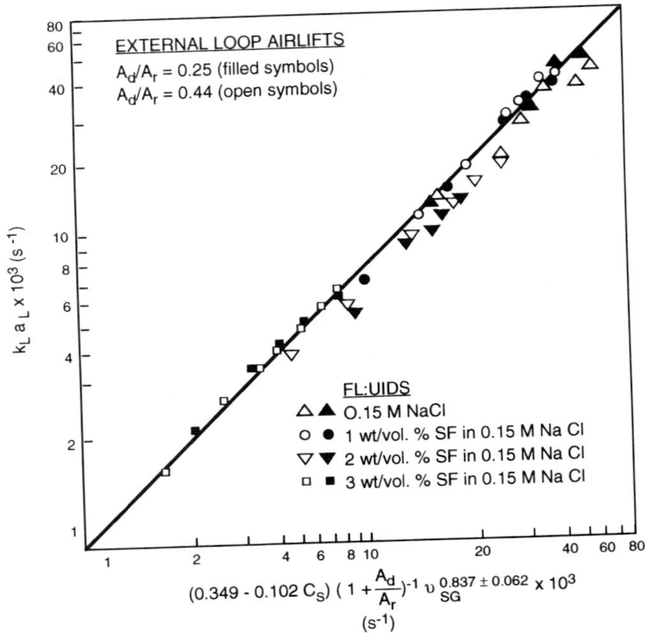

Figure 15.28 Overall mass transfer coefficient in external loop reactors.[6] For fluids, △▲, 0.15 M NaCl; ○●, 1 wt/vol% SF in 0.15 M NaCl; ▽▼, 2 wt/vol% SF in 0.15 M NaCl; □■, 3 wt/vol% SF in 0.15 M NaCl. External loop airlifts, $A_d/A_r = 0.25$ (filled symbols), $A_d/A_r = 0.44$ (open symbols).

today and continues to evolve. Geometry, and possibly geometric similarity, and the roles they play in selecting pilot and commercial scale mixing systems, with similar mixing process characteristics, are also better understood today. Study in this area as well as work on numerical fluid mechanics, with an understanding of the role of energy spectrum, has led to successful modeling and scale-up of reactors.

All of the power transmitted by a mixer to the tank contents through the impeller appears as heat ultimately, by way of viscous dissipation. Power is converted to heat through viscous shear at the rate of 2500 BTU/h/hp. Viscous shear is only present in turbulent flow regimes and exists at the microscale level. Consequently, power per unit volume is a major component of microscale mixing. The boundary limit for microscale mixing occurs when particle or droplet size is mush less than 100 μm. At 1 μm particle size, in fact, it does not matter what specific impeller design is chosen to transmit the power, for the flow regime will be turbulent and microscale mixing will predominate.[7]

Power per unit volume within the periphery of the impeller (which is typically about 5% of the total volume) is about 100 times higher than the power per unit volume in the balance of the vessel. The root mean square velocity gradient in the turbine zone is roughly 5–10 times greater than that in the rest of the tank. This has been confirmed by experimental work with sophisticated laser velocity measurement instrumentation.

For many open impellers (i.e. constant pitch and variable pitch hydrofoil axial impeller, propellers, etc.), the ratio of the root mean square velocity fluctuation to the average velocity in the mixing turbine zone is approximately 50%. However, looking at the ratio of root mean square velocity gradient to the average velocity in the rest of the vessel indicates that this figure is of the order of 5–10%. Coincidentally, this is also the ratio of root mean square velocity fluctuation to the mean velocity in pipeline flow. Mixing tanks with mechanical agitation can produce microscale mixing phenomena not encountered in pipeline reactors. The process requirements will dictate whether this is good or bad.

Figure 15.29 depicts velocity versus time gradients for three different types of impellers: narrow bladed hydrofoils, constant pitch axial flow turbines, and radial flow disk turbines. As can be seen, the differences between these impeller geometries are quite significant and can be important factors in reactor process results.

There are many fluid mechanics spectra applicable to process design and mixing results; however, it is impossible to specify and/or quantify these and apply them to mixer design. Industrial mixing processes are too complex to define required process results in terms of these fluid mechanics parameters.

Without question, the most practical aspet of fluid mechanics studies has been the ability to design pilot plant experiments and programs (and in some

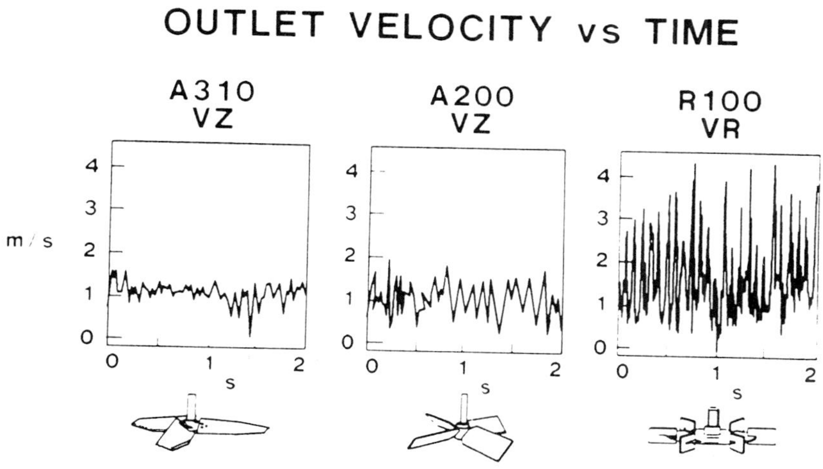

Figure 15.29 Comparison of velocity versus time for various impeller geometries at constant total impeller pumping capacity. (Mixing Equipment, Inc.)

cases full scale experiments) to establish process sensitivity to macroscale mixing variables (i.e. power, pumping, capacity, impeller diameter and speeds, macroshear shear rates, etc.) in comparison with microscale mixing parameters (i.e. power per unit volume, root mean square velocity gradients, and an estimation of microscale eddy size). As previously stated, using geometric similarity, the macroscale variables can be summarized as follows:[8]

- Blend and circulation times will be much longer in the full scale than small scale (pilot) tank.
- Maximum impeller zone shear rate will be greater on the full scale, while average shear rate throughout the tank contents will be less than the corresponding values on the small scale. Therefore, a much greater variation in shear rates exists on the full scale.
- Reynolds numbers in the larger tank will be higher, typically running 5 to 25 times greater than those on the small scale. Thus, the flow regime is more likely to be turbulent on the full scale even though transitional or laminar on the small scale.
- Large tank flow patterns tend to be recirculatory from the impeller through the tank back to the impeller. This simulates a number of individual tanks in series. The mean circulation time is increased over that predicted from the impeller pumping capacity. Likewise, the standard deviation of the circulation times around the mean also increases.
- Sufficient heat transfer is usually more difficult to achieve on the full scale. In-tank surface area for heat transfer in the form of helical coils, vertical tubes, panel coils, etc., can create resistances to flow and dead spots of low or no circulation. This is why external heat transfer surface is often used.
- In gas–liquid mass transfer applications, gas superficial velocity usually increases upon scale-up, if constant stoichiometric proportions of gas rate to liquid volume (i.e. equal VVM) are maintained. This increased resistance to flow will further increase circulation time on the larger scale.

Microscale phenomena depend largely on the energy dissipation per unit volume, although energy spectra are also of concern. As mentioned before, the energy dissipation per unit volume in the mixing zone 'seen' by the impeller is approximately 100 times greater than that occurring in the rest of the tank. The result is a ratio of root mean square velocity gradient to average velocity on the other of 10:1 between impeller mixing periphery and the rest of the tank.

The approach to pilot plant testing and scale-up is predicated by the specific process and required process results. No set of rules can be laid out to guide one through the myriad of possible test campaigns. A few guidelines that can be followed in carrying out a successful pilot plant program are:

- For a given process look qualitatively at the possible role of fluid shear rate and shear stress. Evaluate the effect, if any, fluid shear rate and shear stress have on process results. If there is no effect, then these extremely complex fluid mechanics can be dismissed and the testing can focus on process design based on macroscale variables such as uniformity, circulation time, blend time, or velocity variables. Usually flow controlled mixing applications fall into the scenario (i.e. blending, suspension of solids, and heat transfer), unless shear rate/shear stress is critical in addition to flow controlling criteria.

- Assuming fluid shear stresses are significant factors in obtaining a successful process result, one must qualitatively evaluate various scales (volumes) against which shear rate/shear stress influence results. If particles, bubbles, droplets, or fluid clumps are on the order of 1000 μm or larger, the mixing variables will be macroscale and average velocities (not root mean square gradients) at any given point in the tank contents will dominate the mixer design and process results.

Every geometric design variable can affect the role of shear stresses when macroscale mixing variables are involved. Power, impeller speed and diameter, blade shape and width or height, thickness of material used to make the impeller, number of impeller blades, turbine location, baffle location and number of impellers are examples of macroscale mixing variables that should be investigated on the small scale to determine their effect on shear stress and process results.

Microscale mixing variables will dominate process results when particles, droplets, bubbles, or fluid clumps are on the order of 100 μm or less in size. In this case the critical parameters usually include power per unit volume (energy dissipation per unit volume), distribution of power per unit volume between the impeller and the rest of the tank contents, root mean square velocity fluctuations, energy spectra, eddy size in the wake of the impeller blade passage for a particular power level, and fluid viscosity.

In order to make the pilot plant simulation more like a commercial scale unit in macroscale characteristics, the pilot plant mixing impeller must be designed such that the blend time is lengthened and maximum shear rate is increased over that traditionally observed. This will yield a greater range of shear rates than normally found in the pilot plant, and will closely resemble the macroscale considerations to be obtained on the larger scale.

These conditions can be met using smaller D/T ratios and narrower blade widths than normally used in traditional pilot plant impellers. Care must be exercised when using the same impeller type in both the pilot and full scale, since it may not be possible to come close to the long blend time to be experienced in the commercial scale on the small scale. Going to a different impeller geometry in this case may be the answer. For example, radial flow impellers can be an excellent model in a pilot plant for axial flow turbines on the full scale.

Flow efficient hydrofoil impellers will produce a shorter blend time than other impeller types. The outstanding performance of hydrofoils may not stand out in pilot plant testwork, primarily because a pilot unit usually is already too good a blending mixer compared to what really happens on the larger scale.

A pilot plant should be at least of minimum size to measure accurately data and quantify process results. From the viewpoint of minimum blade width based on particle size, consider the following. Assume for example that the blade width is about 1 cm. If the maximum impeller zone shear rate at the boundary of the discharge stream has a value of $10\,s^{-1}$, then the shear rate across 1.25 mm would be 9.5, across 2.5 mm it is 7.0, and across 6.25 mm it would be $5.0\,s^{-1}$. A shear rate of 5.0 is also roughly the average impeller zone shear rate.

The shear rate across a 1 cm particle would be zero, since the fluid velocity on both sides of the impeller blade is the same and shear rate is defined as $\Delta V/\Delta r$ or velocity difference divided by incremental blade width difference. This also indicates that a small particle would see a shear rate of 10 (maximum impeller zone shear rate), while a big particle of say 1 cm size would see a shear rate of zero. A general and very practical rule of thumb is that the impeller blade height or width of the impeller discharge stream should be at least four times bigger than the largest particle to be mixed in a process.

Remember that no mixing application is purely macroscale or microscale mixing controlled. By the same token, no impeller design or configuration will produce purely macroscale or microscale mixing.

Examples of macroscale mixing controlled processes are:

- Blending and storage
- Solid suspension
- Heat transfer
- Crystallization
- Fermentation.

Examples of microscale mixing controlled processes are:

- Mass transfer
- Liquid–liquid extractions
- Dispersions
- Gas dispersions (exclusive of fermentation).

Macroscale mixing of the bulk contents of a tank can be broken down into components of microscale agitation. This is illustrated in Figure 15.30.

Momentum theory stipulates that equal bulk fluid velocity and velocities equal at corresponding points in different scale vessels, guarantees similar macroscale mixing results based on scale-up criteria. Macromixing is deeply rooted in bulk velocity as the prime scale-up criterion and is a yardstick to quantify process results.

The average or bulk velocity in an agitated tank is directly proportional

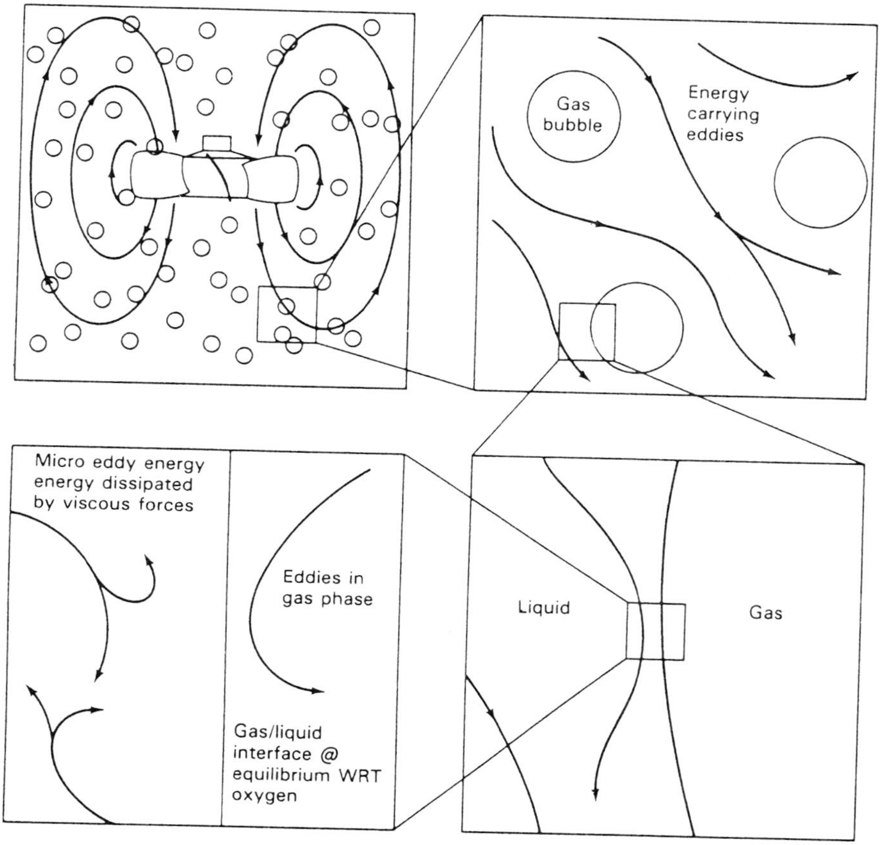

Figure 15.30 Macroscale mixing in a fermenter broken down into microscale components. (Prochem Mixing Equipment, Inc.)

to the momentum (QV) of the fluid developed by the impeller. It has been shown that:

$$V_{\text{bulk}} \propto \sqrt{\frac{QV}{(volume)^{2/3}}} \tag{15.53}$$

where V_{bulk} is the average of bulk fluid velocity, Q is the flow rate, V is velocity (QV is momentum), and *volume* is the volume of fluid in the tank. Equal momentum levels at various scales will yield equal bulk velocities and equal velocities of fluid at corresponding points in different size vessels. Equal velocities in two different scales will produce similar mixing results for macroscale controlled applications.

15.3.2 *Mass transfer scale-up*

As an example of mass transfer scale-up, consider aerobic fermentation. The general equation for mass transfer rate is:

$$R = K_{la}\,\Delta C \tag{15.54}$$

where R is the mass transfer rate (lb mol/ft^3 s (kg mol/m^3 s)), K_{la} is the mass transfer rate coefficient (s^{-1}), and ΔC is the concentration driving force (lb mol/ft^3 (kg mol/m^3)). The only variable that the mixer controls is K_{la}. A pilot plant demonstration program should be run with various types of impellers at various operating speeds, and with various gas rates or superficial velocities. Instrumentation should be incorporated to measure dissolved oxygen in the tank contents and gas analysis equipment to measure incoming or outlet gas composition. With this setup, a matrix of experiments should be conducted to evaluate mass transfer as a function of mixer horsepower and superficial gas velocity.

Mass transfer coefficient, K_{la}, can be calculated from equation (15.54) based on measured concentrations of oxygen in the liquid and off-gas and based on a mass balance calculation to determine mass transfer rate R.

Superficial gas velocity can be calculated based on measured gas rate and cross-sectional area of the tank. Horsepower can either be measured with a torque cell or dynamometer or calculated based on measured speed.

A plot of K_{la} versus superficial gas velocity F and K_{la} versus horsepower per unit volume should produce a straight line plot on log–log paper, the slope of which defines an exponent for scale-up. Figures 15.31 and 15.32

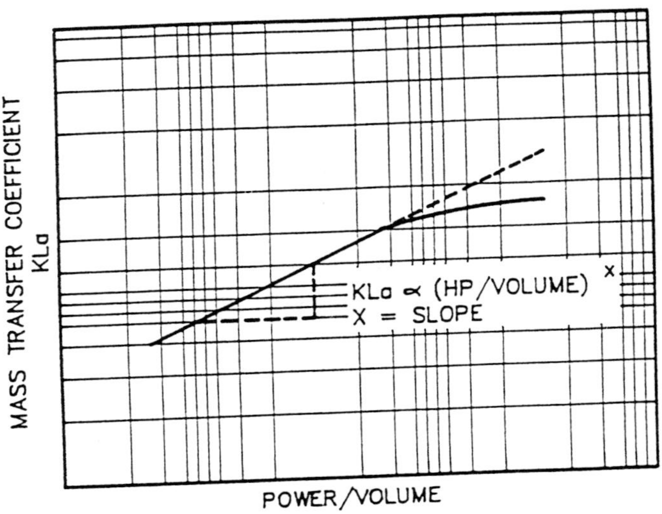

Figure 15.31 Mass transfer as a function of power. (Prochem Mixing Equipment, Inc.)

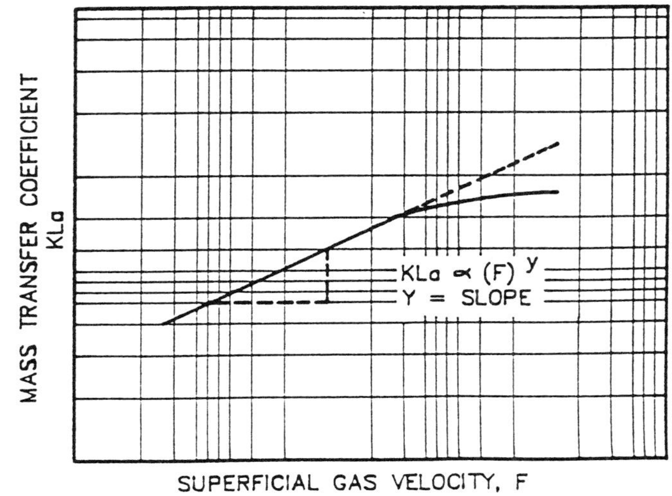

Figure 15.32 Mass transfer as a function of gas velocity. (Prochem Mixing Equipment, Inc.)

show typical plots of mass transfer coefficient as a function of power and gas velocity.

Assuming a desirable K_{la} is achieved in the pilot plant, one can achieve the same mass transfer coefficient on the larger scale by using the following ratio analysis:

$$K_{la1} \propto (hp/volume)_1^x (F)_1^y \tag{15.55}$$

$$K_{la2} \propto (hp/volume)_2^x (F)_2^y \tag{15.56}$$

Subscript 1 is for the pilot plant and subscript 2 is for the larger scale. Exponents x and y are slopes of straight line plots of K_{la} versus *hp/volume* and K_{la} versus F, respectively, on log–log paper.

Since the desired full scale mass transfer coefficient K_{la} is the same as that achieved in a successful pilot program, and provided that *hp/volume* and F on the pilot scale and F on the full scale are known, the following ratio analysis can be used to calculate the full scale power per unit volume to achieve the desired process result (mass transfer coefficient).

$$\frac{K_{la1} \propto (hp/volume)_1^x (F_1)^y}{K_{la2} \propto (hp/volume)_2^x (F_2)^y} \tag{15.57}$$

$$1 \propto \frac{(hp/volume)_1^x (F_1)^y}{(hp/volume)_2^x (F_2)^y} \tag{15.58}$$

Values of $(hp/volume)_1$ and F_1 are known based on a successful pilot test and the exponents x and y are the same for both scales. Provided the fullscale gas velocity F_2 is known, the only unknown to be solved is $(hp/volume)_2$ or fullscale power per unit volume. Finally, fullscale mixing power can be calculated by multiplying $(hp/volume)_2$ by fullscale ungassed volume.

The exponents x and y should be determined based on pilot testing. Literature values can be used without testing to confirm the exponents, but the risk is great. Overestimated exponents can result in an oversized (overpowered) agitator to achieve desired process results. Underestimated exponents can result in too small a mixer (underpowered) to achieve the desired mass transfer. An overpowered agitator will not be cost effective from an operating and capital cost perspective. An underpowered agitator simply will not produce the desired mass transfer process results on the large scale.

References

1. McDonough, R.J. (1992). *Mixing for the Process Industries*, Van Nostrand Reinhold, New York, p. 123.
2. Ho, C.S. and Oldshue, J.Y. (1987). *Biotechnology Processes Scale-up and Mixing*, American Institute of Chemical Engineers, New York.
3. Bader, F.G. (1987). Improvements in multi-turbine mass transfer models, in *Biotechnology Processes Scale-up and Mixing*, (Eds. Ho, C. and Oldshue, J.Y.), American Institute of Chemical Engineers, New York, p. 97.
4. Lee, C.H., Glasgow, L.A., Erickson, L.E. and Patel, S.A. (1987). Liquid circulation in airlift fermenters, in *Biotechnology Processes Scale-up and Mixing*, American Institute of Chemical Engineers, New York, p. 58.
5. Thomas, N.H. and Janes, D.A. (1987). Fluid dynamic considerations in airlift bioreactors, in *Biotechnology Processes Scale-up and Mixing*, American Institute of Chemical Engineers, New York, p. 65.
6. Chisti, M.Y., Fujimoto, K. and Moo-Young, M. (1987). Hydrodynamic and oxygen mass transfer studies in bubble columns and airlift bioreactors, in *Biotechnology Processes Scale-up and Mixing*, (Eds Ho, C. and Oldshue, J.Y.), American Institute of Chemical Engineers, New York, p. 79.
7. Oldshue, J.Y. (1987). Current situation in fluid mixing for fermentation processes, in *Biotechnology Processes Scale-up and Mixing*, American Institute of Chemical Engineers, New York, p. 4.
8. Oldshue, J.Y. (1989). Fluid mixing in 1989, *Chemical Engineering Progress*, 38.

16 Distillation in the pharmaceutical industry

R.F. WILCOX

16.1 Introduction

16.1.1 Overview

In the pharmaceutical industry separation is one of the most widely used unit operations. Its main uses include the recovery of reactants and solvents for recycling, the purification of products for sale, and the processing of byproduct streams for waste minimization. Though many separation techniques exist, the one most often used is distillation. Distillation is simple, has few moving parts, is well understood, can be simulated readily, and often has a low capital cost. On the negative side is the high operating cost associated with high energy requirements to vaporize and condense.

Most distillation simulations are done on the computer now and many excellent references exist (see the Bibliography) that give detailed accounts of the mathematics. The emphasis of this chapter therefore will be on the various uses for distillation to purify a product, break an azeotrope, and perform batch distillation. Equilibrium thermodynamics will be discussed in such a way that the engineer can have a feel for the behavior of a given system, if only in a general sense. After reviewing this chapter the engineer will understand why a system behaves as it does, for example, why the phase is split between water and butanol or why more water will dissolve in benzene than hexane, and will be enabled to make initial decisions about systems quickly without detailed simulations.

16.1.2 Molecular interactions

Understanding how a system behaves is more important now than in previous eras. With the advent of the computer, distillation design is being pushed closer and closer to the theoretical limit. In the past a distillation column required less front end engineering, but included more design allowances to account for uncertainties. This gave rise to distillation units that greatly exceeded their nameplate capacity. This meant higher capital cost than required for the original scope of a project. Now with computers and improved correlations, the design allowances have been cut to reduce capital outlays, thus requiring accurate predictions. But we as engineers must remember that the computer does not simulate the real world but only represents reality by using a mathematical model. In this way the engineer will need to know which mathematical model is best for the system, and just as importantly, when a model has performed poorly. This can be done with a thorough understanding of the working of the molecules in a system. Why does the system of benzene and water separate into two phases? Using common sense an engineer would think that more energy and less entropy is required to keep the benzene layer floating on the water. As seen in equation (16.1) minimizing the Gibbs free energy involves two terms, the entropy and the enthalpy of a system.

$$G = H - TS \tag{16.1}$$

where G = Gibbs free energy, H = enthalpy, S = entropy, and T = temperature.

From thermodynamics we know that mixing components increases entropy, i.e. the randomness of a system, and thus reduces Gibbs or internal energy of a system. In some cases this entropy change does not offset the increase in enthalpy. As we all know, benzene and water do form separate phases and this is because the Gibbs free energy is at a minimum when the phases are separated. But does this really tell us why? Using equation (16.1), could one look at the system of methanol and heptane and tell if it is miscible or not? Probably not, but with an understanding of the way molecules interact, a good guess can be made. This sort of understanding can be invaluable in reviewing the output from a computer distillation run. The engineer may find that the wrong model for thermodynamic equilibrium was used or better still would choose the right one first.

16.1.3 Phase equilibria

16.1.3.1 Understanding intermolecular interactions. Molecular material can exist as either a solid, liquid, or vapor over specific temperature and pressure ranges. These specific ranges exist because interactions between molecules cause either the solid, the liquid or vapor phases to be more stable. In other words there may be an energy saving in having molecules closely packed in a somewhat rigid arrangement rather then loosely and randomly flying about. All molecular material, whether a single component or a mixture, will seek to have the lowest potential energy possible[1]. For thermodynamic equilibrium to exist between phases in a system, the temperature T and the chemical potential $\grave{I}$ of each component must be equal in all phases. The chemical potential is defined as the molar Gibbs free energy of a component. Generally the pressure is also equal in all phases but this is not a necessity. For example when phases are separated by a semi-permeable membrane, thermodynamic equilibrium exists but both phases are not at the same pressure.[1]

The most stable form for a system will be that which minimizes the Gibbs free energy (see equation 16.1) and therefore the chemical potential. At lower temperatures the Gibbs free energy will be dominated by the enthalpy term H, which means that a structured system with greater molecular interactions and order is the most stable, i.e. the solid phase. At high temperature the Gibbs free energy is dominated by the entropy term TS. In order to have higher entropy the system must exist in a more random state which would correspond to a gas. The liquid phase is a balance of entropy and enthalpy effects which minimize the Gibbs free energy. Figure 16.1 depicts the change in chemical potential with rising temperature for each phase. One can see that the melting point T_m marks the point where the chemical potential would be minimized by following the liquid line instead of the solid and likewise the boiling point T_b is where the vapor takes the path to minimizing the chemical potential. Also note that thermodynamic equilibrium exists where the temperature and the chemical potential in each phase are equal: both at the melting point and the boiling point.

16.1.3.2 Deviation from ideal. An ideal gas is a gas in which no molecules interact with any other molecule. The molecules that make up the gas will exert no attractive or repulsive forces. Ideal gases and ideal solutions are only concepts. No real molecules are truly ideal. The concept is applied to molecular interaction to give a basis for comparison. Real molecules can be described by how much they deviate from the imaginary ideal state.

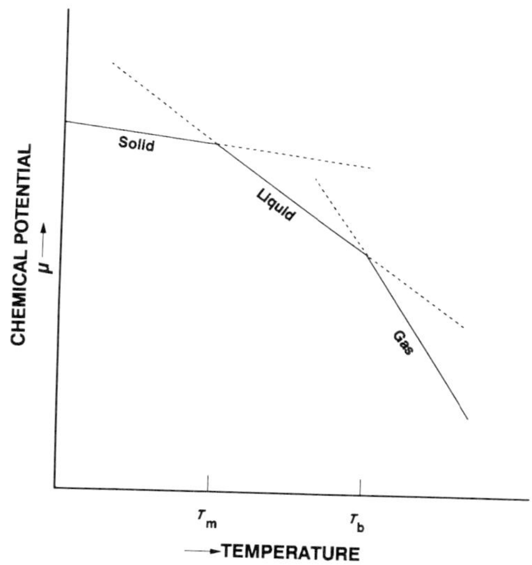

Figure 16.1 Representation of the chemical potential of a system as temperature increases.

An ideal gas has the following properties[1]:

- No interaction between like molecules
- No interaction between unlike molecules
- The equation $PV = nRT$ holds true for all components.

Likewise when mixing molecules to form an ideal solution the following properties will need to exist[1,2]:

- No volume change occurs when an ideal solution is produced
- The heat of mixing is zero
- The entropy change is equal to that which would be expected from the extra degree of freedom. In other words the mixture is perfectly random in concentration throughout
- The Gibbs free energy change is only provided by the entropy gain
- Ideal solutions are always miscible. Only very non-ideal solutions form multiple liquid phases.

All molecules deviate from ideality but some as we know have greater deviations than others. Why is this? The answer is provided by molecular interactions. In an ideal solution, a molecule does not care who its neighbors are, but in reality molecules do discriminate against their neighbors, some more so than others. This discrimination is caused by a molecule being drawn towards or repelled by a like molecular neighbor more than an unlike neighbor. When a molecule has a greater attraction for like molecules in a mixture the system is said to deviate positively from ideal. If on the other hand a molecule has more attraction for unlike molecules, then the deviation from ideal is negative. For example, a negative deviation from ideal is seen in the mixture of acetone–chloroform[1]. This deviation is attributed to hydrogen bonding between the two unlike molecules. Evidence is seen to support this in the increase in intensity of the C–H stretching band of chloroform which occurs on addition of acetone. As the new component acetone is added the acetone's oxygen exerts an attraction towards the chloroform's hydrogen, thus distorting the bond length slightly. There is also a change in the carbonyl stretching band of the acetone. This interaction gives rise to a complex with one acetone and two chloroform molecules as shown in Figure 16.2[1]. These data were obtained from the infrared spectrum of the mixture.

Positive deviation from ideality is more commonly seen, such as when a highly polar molecule such as water is mixed with a less polar substance. Water molecules have a great attraction for one another and water associates into an ordered liquid with many hydrogen bonds between the molecules (see Figure 16.3). When a diluent is added it will tend to break apart many of the hydrogen bonds. The energy of the hydrogen bond in water is 5.7 kcal/mol which means the heat of mixing will be non-zero, since energy

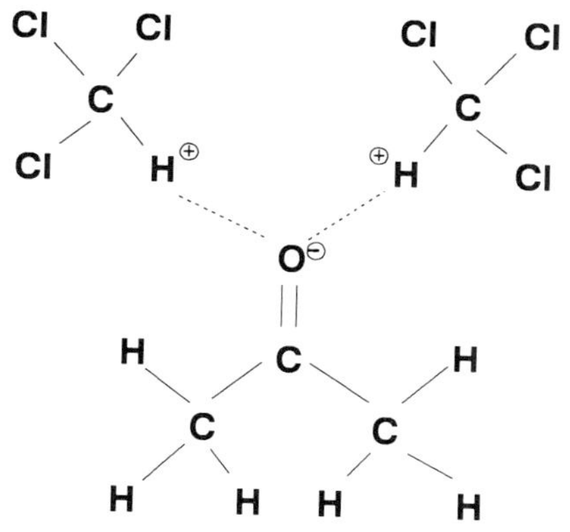

Figure 16.2 Acetone–chloroform interaction.

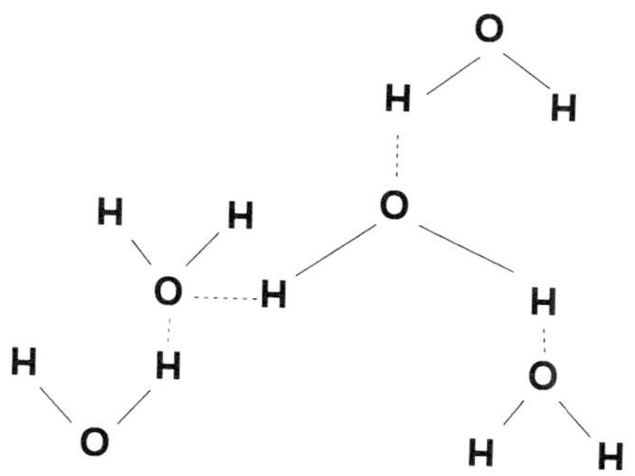

Figure 16.3 Possible hydrogen bonding sites for water molecules. The hydrogen bond is shown as a dashed line.

must be expended to break these hydrogen bonds and create a cavity between the water molecules[1,3]. If the diluent is small and can replace some of the broken hydrogen bonds with hydrogen bonds of its own then less energy is required for it to go into solution with water. But if the cavity required is large requiring the breaking of many hydrogen bonds, the energy increase required by the enthalpy term in equation (16.1) will exceed the reduction in free energy obtained from the entropy term by increasing the randomness. In this case the system's minimum potential energy is achieved by maintaining separate phases.

The only class of mixtures that closely resembles ideal is called a regular solution[2]. These mixtures obey the ideal mixing law for entropy but the heat of mixing is not zero although it is small. The number of solutions that meet these requirements is very limited.

16.1.3.3 Intermolecular forces. Intermolecular attractive forces are caused by various factors such as molecular shape, size, electron charge density, and ionic effects to name a few. Because of the inherent short range over which these attractive forces are felt the effect is more pronounced in liquids and solids than in vapors[1,2,4–6]. A greater deviation from ideality can be seen in a liquid mixture of water and heptane than in the vapor mixture of these same components.

The first attractive force, which is exhibited by all molecules whether polar or not is the dispersion forces also known as the London forces[1,6]. These dispersions forces are caused by an instantaneous fluctuation in the distribution of the negative electron cloud over the positive nucleus resulting in the appearance of dipoles. Although the collective instantaneous dipoles must average to zero the resulting local attractive forces do not. If the electron cloud shifts in the other direction (see Figure 16.4) the dipole will reverse which will reverse the induced dipoles shown. Even though the average over time of the instantaneous dipole is zero, as it must be for molecules without permanent dipole moments, the attractive forces it is associated with will not average to zero. The dispersion force is proportional to R^{-6}, the distance separating the molecular centers, which accounts for the small distance over which the force is effective. In most non-poplar molecules, dispersion forces will be the only attractive forces to cause deviation from ideality.

Molecules with permanent dipoles can induce a dipole in a neighboring molecule by distorting the negative electron cloud in such a way that the affected molecule will also have a dipole, whether polar or not. The induced and permanent dipoles will be mutually attracted (see Figure 16.5). This attractive force is known as the dipole-induced dipole or Debye interaction[1,6]. The dipole-induced dipole attraction can be large depending on the size of the permanent dipole moment, for example with ions, or small for nearly spherical molecules like CH_4, or even zero for inert gas atoms which have no permanent dipole moments. Since the induced moment will readjust itself nearly instantaneously, regardless of the molecular rotation of the permanent

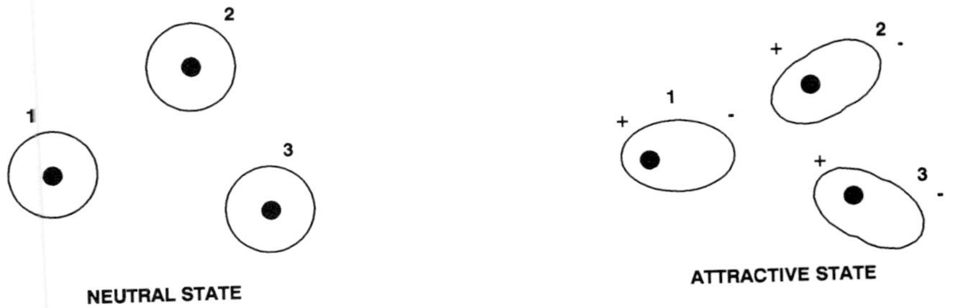

Figure 16.4 A shift in the electron cloud of molecule 1 induces a shift in molecules 2 and 3 causing attractive forces to set up.

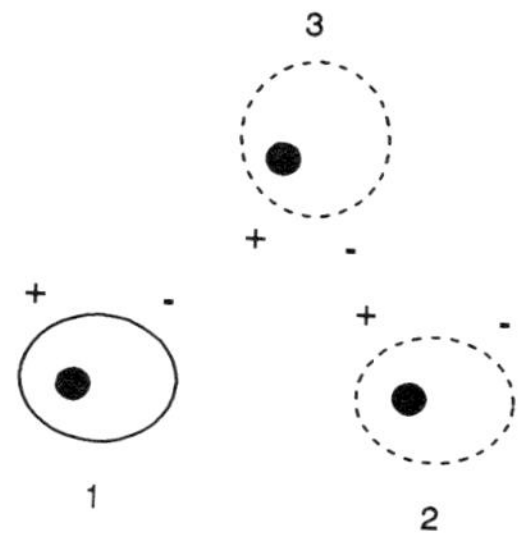

Figure 16.5 The electron cloud from the permanent dipole of molecule 1 induces a shift in molecules 2 and 3 causing attractive forces to be set up.

dipole, the effect is thus independent of temperature as long as the mixture density and therefore molecular spacing is not affected. The attractive force is proportional to R^{-6}, the distance separating the molecular centers.

Dipole–dipole or Keesom interactions occur between molecules in which both have permanent dipole moments[1,6]. This attraction is proportional to the strength of the contributing dipole moments. As shown in Figure 16.6, if molecule A rotates the associated molecules 1–5 must also rotate, but there is an inherent time lag in this rotation. This time lag will decrease the attractive force, proportional to the amount of time that the optimum alignment is not achieved. Since the force is an orientation effect it is temperature dependent. As the temperature rises, all molecules will have more random motion and the optimum orientation will not be achieved as often. Dipole–dipole interactions have contributions from all associated molecules. The force is proportional to R^{-3}, the molecular separation, which means this effect is felt over a greater range than the previous forces mentioned.

Electron donor–acceptor interactions are of great importance in explaining non-ideality. These interactions include Lewis acid–base attractions and the well-known phenomenon of hydrogen bonding which is a Bronstead acid–base interaction[1,4 8]. These interactions cause greater departures from ideality because the attractions are much stronger than all the others mentioned

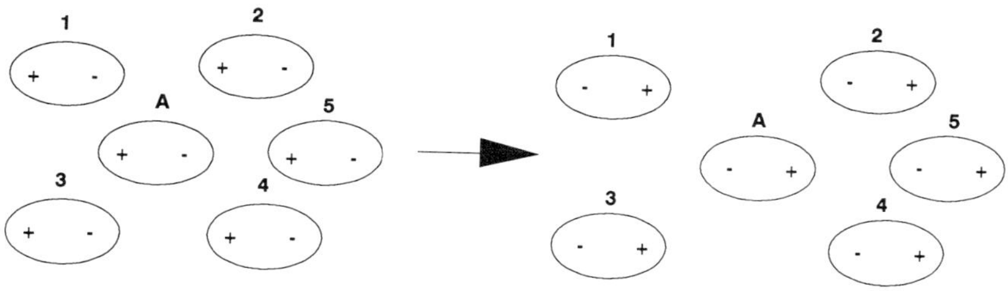

Figure 16.6 Rotation of dipole molecule A and its effect on dipole molecules 1 to 5.

previously. The attractions come in two forms. Association, in which like molecules are attracted, and complexing, where unlike molecules have the greater attraction. These molecules form quasi-polymeric systems with defined structures. Water forms a three-dimensional structure while alcohols will give rise to either cyclic or linear quasi-polymers of indefinite length. The strength of hydrogen bonding (though greater than dipole effects) is still less than regular bonds. The energy of the C–H bonds is 87 kcal/mol, while typical hydrogen bond energies vary from only 2–8 kcal/mol depending on the electronegativity of the atom bonded to hydrogen. Hydrogen bonding is favored by atoms with higher electronegativity than hydrogen and the strength of the hydrogen bond increases in the following order: $C < I < S < Br < Cl < N < O < F$.

Any mixture of molecules with hydrogen bonding capabilities will deviate greatly from ideal. If there is an increase in hydrogen bonding due to mixing then miscibility is enhanced, but if hydrogen bonding is broken miscibility is reduced as with water and oil. A special case exists where the hydrogen bonding molecules fit together in complete oligomer packages. Examples are dimers such as carboxylic acids and some glycol ethers, and hydrogen fluoride which forms a hexamer, H_6F_6.

Figure 16.7 shows an example of the acetic acid dimer. This results in diverse solubility properties such as the ability of acetic acid to dissolve in water and heptane. In water, acetic acid can open up and interact with water both through its polar properties and hydrogen bonding. But in non-polar heptane, acetic acid is a self-contained dimer with no remaining site to hydrogen bond, which causes it to exhibit non-polar tendencies. This ability to assume the property of the surrounding environment is termed chameleonic after the famous camouflaging reptile[6]. Since these dimerizing hydrogen bonding molecules form compact packets they are freer to move about at higher temperature than the larger structures seen in the quasi-polymers which tend to break hydrogen bonds as the temperature is increased. Thus these molecules will stay as dimers even in the vapor phase. This effect must be accounted for in designing distillation systems involving oligomerizing molecules.

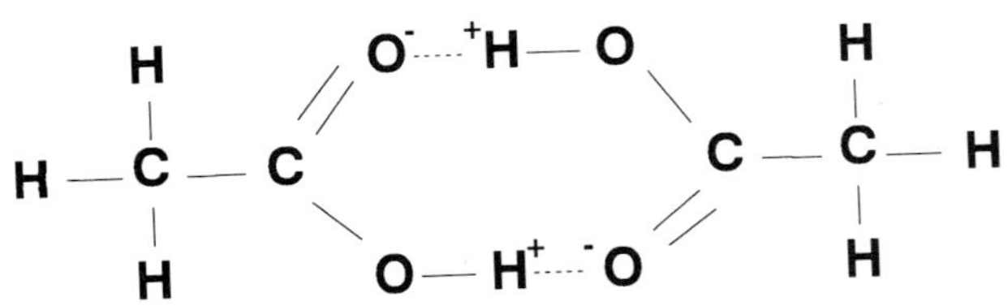

Figure 16.7 Acetic acid dimer with hydrogen bonds indicated by the dashed line.

The last effect to be discussed is often overlooked. This is the effect of molecular size. Three molecules, as shown in Figure 16.8, are mixed in equal molar concentrations, one polar and the other two non-polar; with the only difference between the non-polar molecules being size. The polar molecule will interact with the larger molecule in proportion to the size difference in non-polar molecules. This is based on simple molecular collision statistics. It is easier and therefore more probable to hit a larger target. But as discussed earlier the larger molecule will have to create a bigger cavity in the polar solvent, so these two phenomena will counteract each other. With a highly polar molecule 'A' that forms a quasi-polymeric network, the size difference will hinder creation of a cavity in the polar network more than it will facilitate an increase in collisions. But conversely if molecule 'A' is only slightly polar or forms packets of dimers and trimers, then collision frequency with the larger molecule will overshadow its resistance to forming a cavity.

Given this basic understanding it should be possible to make certain assumptions about the behavior of mixtures. If we take the local composition model, which says that the effect of a molecule is only felt over a short range, to the nearest neighbor at the least and marginally to the next row (see Figure 16.9), and assume no concentration gradients exist throughout the mixture, the interaction between a few molecules can be applied to the whole mixture. These assumptions will be general and may not apply to mixtures with many different molecules. Also some factors work for and some against reducing the system potential energy, as seen with molecular size.

Ewell *et al.* did this for the relationship between hydrogen bonding and azeotrope formation[1,8,9]. They concluded that hydrogen can coordinate between two molecules at the O_2, N_2, or F site, or even at C if a number of highly electronegative atoms are attached to the carbon atom. They suggested the following classifications of hydrogen bonds as strong or weak.

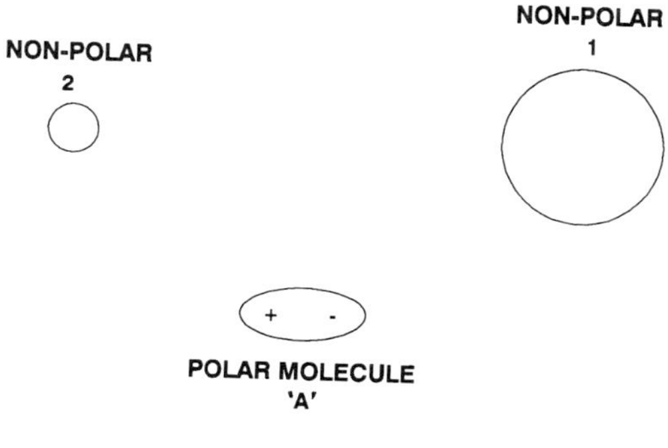

Figure 16.8 Effect of size on molecular interaction.

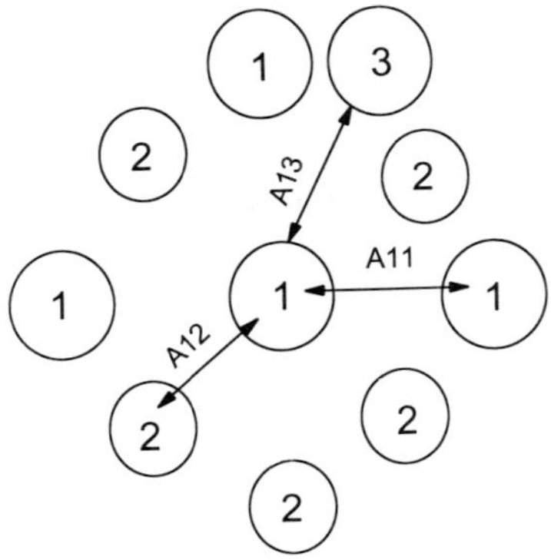

Figure 16.9 Local composition model with interactions A_{12} and A_{11} and very weak interaction A13.

- Strong bonds
  O–HO
  N–HO
  O–HN
- Weak bonds
  N–HN

$$\begin{matrix} \text{O} \\ \text{N} \end{matrix} \left\{ \begin{matrix} \text{HCCl} \\ \text{HCCl}_2\text{–CCl} \\ \text{HCNO} \\ \text{HCCN}_2 \end{matrix} \right.$$

Similarly liquid mixtures can be classified by functional group, which help determine the deviation from ideality (8). See Tables 16.1 and 16.2.

16.2 Design of a distillation system

It is not the intent of this chapter to go into the details of the mathematics involved in the calculations for the various distillation methods. Some key equations will be given but it is assumed that the reader has access to the many good references listed in either the bibliography or the reference list. Furthermore most readers will utilize a computer to perform these calculations.

16.2.1 Distillation methods

16.2.1.1 Simple distillation. Distillation accomplishes separation by maintaining a composition gradient between the vapor and liquid phases caused by difference in volatilities of the components. The separating agent

Table 16.1 Classification of liquid mixtures

Class I	Liquids capable of forming three-dimensional networks of H bonds: water, glycol, glycerol, amino acids, hydroxylamine, hydroxy acids, polyphenols, amides, etc. These are 'abnormal' or 'associated' liquids having high dielectric constants and are water soluble.
Class II	Other liquids containing active hydrogen atoms and other donor atoms: acids, phenols, alcohols, primary and secondary amines, oxides, nitro compounds and nitriles with H atoms, NH_3, hydrazine, HF, HCN, etc. The characteristics of class II liquids are the same as class I.
Class III	Liquids composed of molecules containing donor (O, N, F) atoms but not active H atoms: ethers, ketones, aldehydes, esters, tertiary amines. These liquids are also water soluble.
Class IV	Liquids composed of molecules containing active H atoms but no donor atoms: $CHCl_3$, CH_2Cl_2, CH_3CHCl_2, $CH_2Cl{-}CHCl_2$, etc., which show only slight water solubility.
Class V	All other liquids, i.e. compounds having no hydrogn bond-forming capabilities: hydrocarbons, CS_2, RSH, non-metallic elements, etc. This class of materials shows essentially no solubility in water.

Table 16.2 Deviations of class combinations from Raoult's law

Classes	Deviations	H bonding
I + V II + V	Always positive with I + V frequently showing limited solubility	H bonds broken only
III + IV	Always negative	H bonds formed only
I + IV II + IV	Always positive, I + IV limited solubility	H bonds both formed and broken (dissociation common)
I + I I + II I + III	Usually positive Some negative, giving maximum azeotropes	H bonds both broken and formed boiling
II + II II + III III + III III + V IV + IV IV + V V + V	Quasi-ideal always positive or ideal Minimum azeotropes Minimum azeotropes (if any)	No H bonds involved

is the heat energy suuplied. Simple distillation is usually a continuous process with the lighter or more volatile fraction leaving the top as the distillate and the less volatile or heavy fraction recovered as the bottoms product. A portion of the overhead vapor is condensed and returned as reflux to enrich the vapor rising through the column. The section above the feed where the vapor is enriched by reflux thus condensing the heavy portion is called the rectifying section. The section of the distillation column below the feed where the liquid is stripped of light components by the rising vapor from heat input is called the stripping section (see Figure 16.10).

16.2.1.2 Azeotropic distillation. For special cases where in a mixture the differences in effective volatilities are very small, these systems are known as close boiling mixtures. In some cases the difference is zero, called an azeotropic mixture. These two systems require special handling to be separated. In close boiling mixtures simple distillation is prohibitively expensive and an azeotrope is impossible to separate. Azeotropic distillation or extractive distillation could be an option to make the separation possible.

In azeotropic distillation (see Figure 16.11) an additional component is added which will azeotrope with one or more of the components in the original mixture. This new azeotrope must be easily recoverable and recycled for reuse or else one separation problem has been traded for another. Selection of the azeotrope agent is of great importance and since there are a plethora

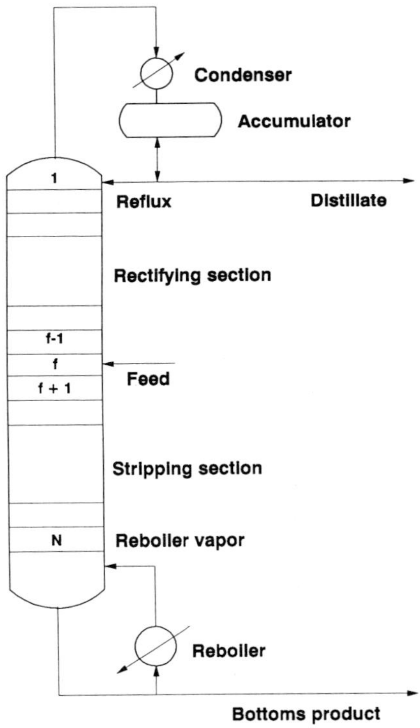

Figure 16.10 Simple distillation column.

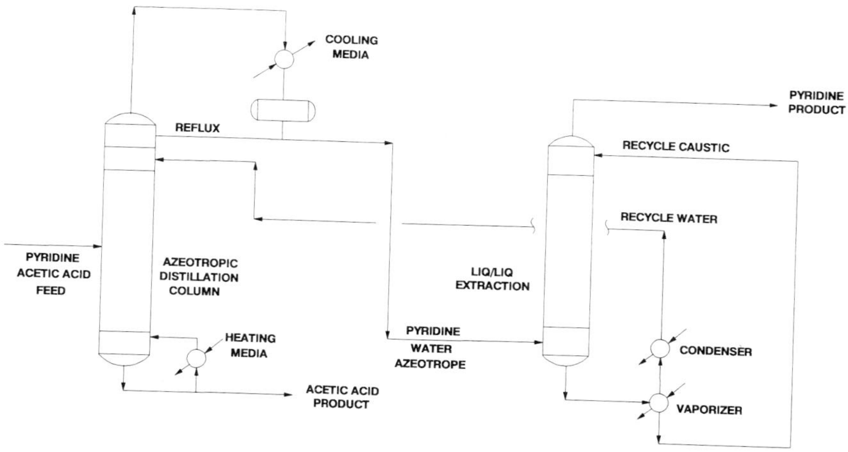

Figure 16.11 Azeotropic distillation of pyridine from acetic acid using water and subsequent dehydration of pyridine using caustic soda. (Courtesy of Jacobs Engineering.)

of compounds to choose from, a selection technique is necessary to winnow down the choices.

The azeotropic agent when added to the initial binary azeotrope must form a minimum boiling azeotrope with one of the components in order to separate the compounds[10]. If a maximum azeotrope is formed then the original azeotrope will rise overhead with no benefit obtained. There are two types of azeotropes. Those separating into two phases when condensed, which are known as heterogeneous azeotropes, while those remaining as a single phase are homogeneous azeotropes. If the agent added forms a heterogeneous azeotrope then the two phases can be separated by decantation in the reflux drum with the azeotroping agent-rich phase returning to the column as reflux while the recovered component-bearing phase is sent on for further purification if necessary. An example of this type is the dehydration of ethanol using benzene. Benzene forms a heterogeneous azeotrope with water[10]. The water/benzene azeotrope is condensed and decants into two phases. The water-rich phase is sent for further distillation to remove any residual benzene. The benzene is recycled as reflux. The bottoms product from the azeotropic distillation column is anhydrous ethanol. When using an azeotroping agent which forms a homogeneous azeotrope, breaking the new azeotrope must be dealt with. A few examples of possible methods to use for this new azeotrope are pressure swing distillation, extractive distillation, or liquid–liquid extraction. There should, however be an overriding reason to replace one azeotrope by another. These reasons could be, for example, because of the nature of the materials of construction or the toxicity of the component removed in the azeotrope column, or simply because no other method could be found. A homogeneous azeotropic distillation scheme requires a minimum of three distillation columns, perhaps more depending on product purity, whereas the heterogeneous azeotropic scheme could use a single column if high purity of the overhead product is not required. The azeotropic distillation scheme is best used with agents that form heterogeneous azeotropes and the smaller the mutual solubility the better this scheme becomes.

Table 16.4 shows a survey used to make a first pass selection of an azeotroping agent. The example is pyridine recovery from acetic acid. This system forms a fairly strong acid–base complex. Because of this there is negative deviation from Raoult's law caused by the greater attraction of acetic acid for pyridine than for itself. This system forms a maximum boiling azeotrope (see Table 16.3)[11]. Table 16.1 can also be used to predict the possibility of a maximum boiling azeotrope. Acetic acid is a class I liquid and pyridine is a class III liquid which indicates the possibility of forming a maximum boiling azeotrope (confirmed by Horsley[11]).

Suitable agents are ones which will azeotrope with either acetic acid or pyridine not only below the boiling point of the acetic acid–pyridine azeotrope but also below the boiling point of the pure component which is not being azeotroped. Finally, the boiling point must be less than that of an azeotrope that might be formed with any other components present. For this example, we require an agent which will have a positive deviation from ideal to give a minimum boiling azeotrope. From Table 16.1 the likely candidates would be as follows:

- Acetic acid (Class II) — Class I — Water
  Class IV — Carbon tetrachloride
  Class V — Hexane or xylene
- Pyridine (Class III) — Class I — Water
  Class V — Hexane or xylene

From the above candidates, the survey (Table 16.4) is then completed. The first item to be filled in is if the candidate azeotropes with either of the components and at what temperature and concentration. A good place to find this information is Horsley [11]. Of the candidates only xylene does not work as an azeotroping agent because the boiling point of the azeotrope is higher than the normal boiling point for both acetic acid and pyridine. For all other candidates the remaining survey information can be completed, which also has an impact on the design such as the materials of construction,

Table 16.3 Azeotrope conditions for acetic acid and pyridine

	Normal boiling point (°C)	Azeotrope composition (wt%)	Azeotrope boiling point (°C)
Acetic acid	118.5	51.1	138.4
Pyridine	115	48.9	

Table 16.4 Azeotropic distillation agent survey

Azeotropic agent	Formula	Mole wt	Normal B.P. (°C)	Stability (Rx) (DCMP) [a,b]	Toxic	Mat'l of constr[c]	Feed comp to be azeotr'd	Azeo w/other feeds	Wt% agent in azeo	Type of azeo[d]	Ease of separ from azeo comp	B.P. azeo with agent (°C)	Wt% agent in azeo comp[e]	Wt% azeo comp in agent[e]	Cost rel to first agent	Azeo agent rank	Comments
Carbon tetrachloride	CCl_4	153.8	77	N/A	Y	CS	2		98.5	0	Hard	76				3	Too much agent required
Water	H_2O	18	100	N/A	N	CS	1		41	0	Hard	94				1	Best choice
Heptane	C_7H_{16}	100.2	98	N/A	N	CS	2	1	67, 13	0	Hard	92, 95				2	Azeotropes boil too close
Xylene	CHM_{10}	106.2	140	N/A	Prob	CS	2		55	0		122				N/A	B.P. of azeotrope is > other components B.P.
Feed Comp #1	C_5H_5N	79.1	115	N/A		CS											Pyridine: Azeo comp 1-2, 138°C
Feed comp #2	$C_2H_4O_2$	60.1	118.5	N/A	N	304L											Acetic acid

[a] Rx indicates whether the components will react at operating conditions. [b] DCMP indicates whether the component will decompose at operating conditions. [c] Material of construction to handle listed chemical. [d] Type of azeotrope—0 homogeneous, 1 heterogeneous. [e] For heterogeneous azeotropes at reflux drum temperature.

toxicity, cost of the agent, and the ease of breaking the new azeotrope formed. This is why it is usually best to look for agents which form heterogeneous azeotropes since they are easier to separate. One can see from the completed survey that the best candidate for further testing is water because it is readily and inexpensively available and is safe to handle and also a low weight percent of agent is required to azeotrope.

Typically more than one candidate meets the requirement for further study. These candidates are set-up for testing, if required, or if the data are available, the system can be simulated and optimized to obtain preliminary equipment and operating costs as a basis to select which agent is the best.

16.2.1.3 Extractive distillation. Extractive distillation is commonly used to separate close boiling compounds or azeotropes.. It takes advantage of a relatively non-volatile component which has a greater affinity for one or more components in the original mixture[10] (see Figure 16.12). When this extractive agent is added to the column this affinity pulls these components down to the bottom of the column thus allowing the remaining components to rise overhead very readily. The agent is added a few trays below the top so the effect can be seen throughout the column. The top few trays are required to knock back any agent that might be carried up the column.

The extract agent is chosen to induce a greater relative volatility between the compounds to be separated, but by the same token it must not have so great an affinity that it becomes immiscibile. Once the agent separates it has no effect on the remaining extractive compound in the other phase.

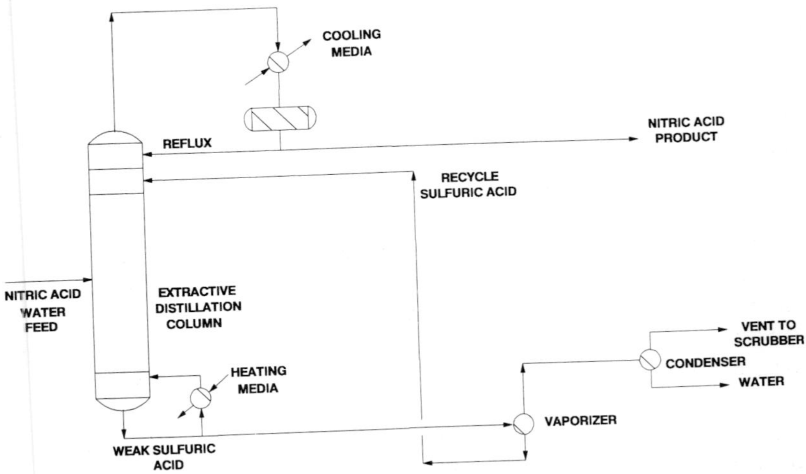

Figure 16.12 Extractive distillation of nitric acid from water using sulfuric acid and subsequent reconcentration of sulfuric acid. (Courtesy of Jacobs Engineering.)

The selection of an extractive agent is not a straightforward process. The selected agent must increase the relative volatility of the compounds without requiring a large ratio of agent to feed. As the relative volatility increases the ease of separation increases which reduces the number of theoretical stages required, but the smaller the ratio of agent to feed the smaller the column diameter. These two factors along with the following assessments, such as how easy is it to recover and reuse the extract agent, what materials of construction are required, is it a toxic substance, is it expensive, does it react at the operating conditions , and what are the operating costs, are used to make the evaluation. As with azeotropic distillation, using a survey form allows the engineer to compare all these variables when searching for the best entrainer without drawing up a complete design for all candidates (see Table 16.5).

Various methods of evaluation can be applied to select the extractive agent. One quick method is the Scheibel method in which another member of the homologous series of either of the components is used [12]. For example, the search for an extract agent for the separation of the methanol/acetone azeotrope should start by evaluating ketones such as methylethyl ketone, methyl n-propyl ketone, or methyl isobutyl ketone. Similarly for the alcohol series the search starts with ethanol, propanol etc. The logic behind this approach is to find an agent which behaves as moderately ideal with one compound and non-ideally with the other. A similar method but not limited to homologs is to use solubility parameters to select agents which are much more non-ideal in behavior toward one compound than another, since the greater the difference in solubility parameter the greater the deviation from ideality. The best method though is to use actual infinite dilution data and determine the selectivity. The selectivity is defined as the ratio of the infinite dilution activity coefficients of each species in the extract agent [13].

$$S_{ij} = \gamma_i^{\infty} / \gamma_j^{\infty} \qquad (16.2)$$

where S is the selectivity and γ^{∞} is the infinite dilution activity coefficient for components i and j.

Various sources of infinite dilution data are available and are listed in Table 16.10. It is very easy to obtain experimental infinite dilution data from various techniques such as differential ebulliometry and chromatography[14]. If no actual data can be found then the infinite dilution can be determined using either the modified separation of cohesive energy density model (MOSCED) or UNIFAC [15,16]. Before an agent can be finalized it must be proved in the laboratory or simulated using actual binary data.

16.2.1.4 Pressure swing distillation The methods mentioned previously have all required the addition of another component to permit the separation of an azeotropic mixture. However, another method which has received little study is pressure swing distillation. The mechanism behind pressure swing

Table 16.5 Extractive distillation agent survey

Extractive agent	Formula	Mole wt	Normal B.P.	Stability (Rx) (DCMP)	Toxic	Matl of constr[c]	Key comp to be extractd	Soluble in key comp	Ratio entrainer to feed	Solubility-parameter	Ease of separ. from key comp	Latent heat	Infinite dilution act coeff LK comp	Infinite dilution act coeff HK comp	Maximum entrn-factor[c]	Cost rel to first comp	Extract agent rank	Comments
LK feed comp[d]																		
HK feed comp[e]																		

[a] Rx indicates whether the components will react at operating conditions. [b] DCMP indicates whether the component will decompose at operating conditions. [c] Maximum entrainment factor = (inf dil alpha LK)* (vapor press LK)/, inf dil alpha HK)* (vapor press HK). [d] LK is the key feed comp which goes overhead. [e] HK is the key feed comp which is extracted.

distillation is the effect of pressure on an azeotrope composition. Figure 16.14 shows that at 150 mm the azeotrope point of acetonitrile and water is 92.8 wt% acetonitrile. If the pressure is raised to 760 mm then the azeotrope is 85.8 wt% acetonitrile. With this information a separation system can be designed (Figure 16.13). The primary feed and the recycled azeotrope from the higher pressure column are fed to column 1. The overhead from column 1 is the minimum boiling azeotrope at 150 mm, while the bottom is nearly

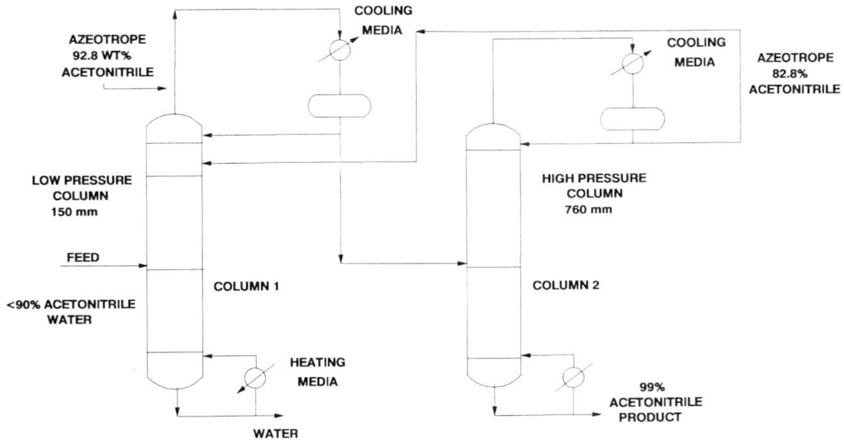

Figure 16.13 Pressure swing distillation scheme to purify acetonitrile.

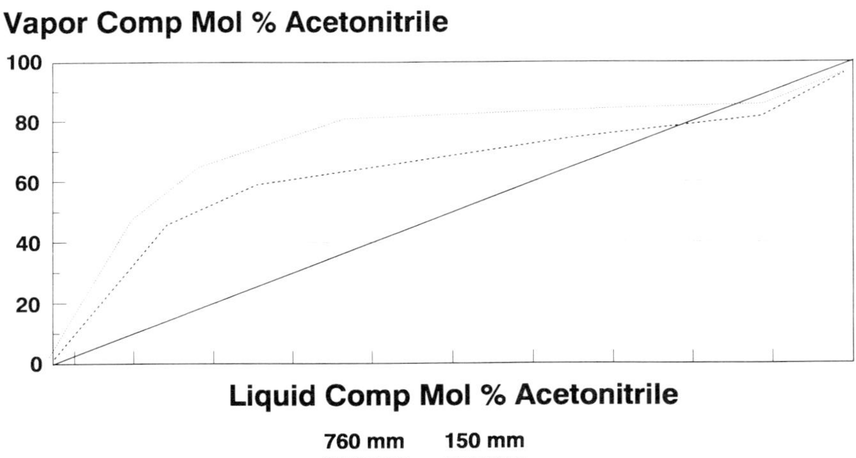

Figure 16.14 Vapor–liquid equilibrium for water and acetonitrile at (– – –) 760 mm and (· · · ·) 150 mm.

pure water. Column 1 distillate is fed to column 2 which operates at 760 mm. This change in pressure means the feed to column 2 is now on the other side of the azeotrope point. Thus the distillate is the azeotrope at 760 mm which is recycled to the first column while the bottom component is nearly pure acetonitrile. Obviously this technique is only appropriate for azeotropes that show large enough deviation with changes in pressures. If the change in azeotrope composition with pressure is small, then the recycle can become enormous or the pressure change required can be so great as to make the cost of this scheme prohibitive. A good rule of thumb for preliminary screening is that the change in azeotrope composition should be at least 7–10 mol%. The vapor–liquid diagram must also be checked for pinch points as the concentration approaches the azeotrope. It does little good if the azeotrope composition changes by 10 mol% if a pinch is formed which makes it nearly impossible to reach the azeotrope. Once a system has been assessed to be a candidate for pressure swing distillation the optimum pressure levels need to be determined.

The number of pressure levels to choose from is endless without some constraints. These constraints are listed below:

1. At least one column should be at a pressure to allow air or water condensing at a temperature approximately 20 to 40°F (10–25°C) above the cooling medium temperature. This column will be the anchor point column.
2. Avoid vacuum operation if possible. The increase in equipment size and need for vacuum equipment makes it costly.
3. For a first trial choose a pressure change in the second or swing column which gives a 7–10% change in the azeotrope molar composition.
4. Choose a pressure for the high pressure column which will allow the overhead to reboil in the low pressure column if possible.

A cost estimate is done at selected pressures and then this is compared to the cost of systems where the pressure of the swing column is 1.5 times higher and 1.5 times lower. If the selected pressure scheme does not give the minimum cost then take the minimum and bracket it with a higher and lower swing pressure at 1.3 times the minimum swing column pressure. Continue until a local minimum is achieved. This will correspond to the optimum pressure levels for a pressure swing distillation system.

16.2.1.5 Batch distillation. Batch distillation (see Figure 16.15) is similar in theory of separation to simple distillation but does not run continuously. Two types of batch distillation are normally encountered. The first and most common is batch rectification in which the feed is charged to the column and boiled up through a refluxing column with distillate product being drawn off. Conversely batch stripping, also known as the inverted batch column, has the feed charged into the reflux drum and impurities stripped from the

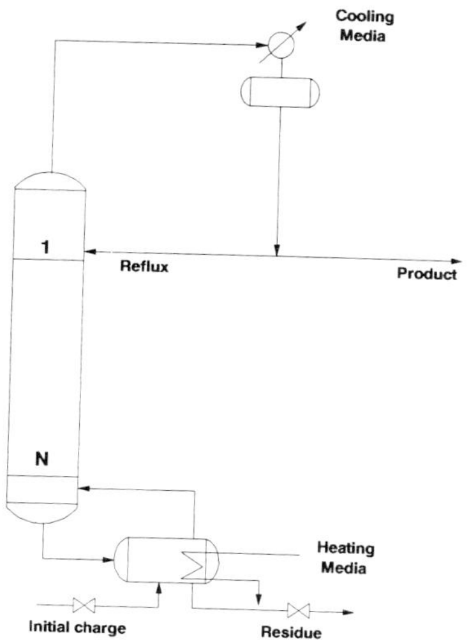

Figure 16.15 Batch rectification distillation.

falling liquid producing a bottoms products. Batch distillation is a non-steady state phenomenon with the composition changing throughout the column with time as material is drawn off.

Batch distillation is preferred over continuous distillation under certain circumstances. The following criteria can be used to determine if batch distillation is the separation option of choice.

1. Excellent for small quantities of materials handled at irregularly scheduled periods.
2. Excellent for dampening the effects of widely varying feed compositional changes.
3. Lower capital cost for producing multiple products than continuous operation. The batch option requires only a single column while the continuous operation requires $N - 1$ columns for N products.
4. Tolerant of uncertainties in VLE.
5. Excellent for handling residues which have a tendency to foul column internals and heat transfer surfaces.
6. Excellent for pharmaceutical and biotechnological process because the equipment can be cleaned and sterilized between each batch run.

On the down side, batch distillation is more expensive per kilogram of product and requires more operator attention and instrumentation control due to the variation of operation with time. After each batch charge the system must be shut down, drained, possibly cleaned, and then recharged with feed causing a loss of on-stream time.

The batch distillation can be operated in two modes: It either operates at a constant reflux rate which will produce a variable product composition high in the lightest material at the initial starting time and progressively decrease in purity, making the product composition an average of the composition throughout the product draw. Alternatively the reflux can be varied to produce a constant product composition until prevented by material balance limitation[17]. This method will require constant analysis of the product to be fed back to the reflux controller to make the changes in the rate required to maintain product purity. Though these are the two most common operating methods they are not necessarily the optimum ones [18].

When making pure products regardlesss of the reflux mode there will be a period of time when the amount of the light component is too small to meet the product specification but still too great to allow the next product to reach its purity specification. Therefore a slop cut is taken which removes the last of the light component and some of the next component. These slop cuts between products are dependent on product purity. Higher purity requires larger slop cut. The slop cuts are normally recycled back with the next charge. Redistillation of these components is expensive so the slop cuts should be minimized by specifying the minimum product purity needed and proper operation of the column to obtain the greatest separation possible. The method of recycling the slops can be optimized to reduce operating costs [19]. The slops can all be combined and fed with the next feed charge or they can each be fed continuously at the appropriate time into the column at an optimum location. Another option is to collect the slop cuts over a number of batch cycles and distil each slop cut when enough material has been collected.

A key variable which is encountered in batch distillation but not in continuous distillation is the effect of hold-up. The size of the hold-up in the charge drum can have a major impact on the sharpness of separation[20]. A batch column will achieve a sharper separation if the amount of material in the charge drum is small at the time of transition from one product to another. This can be reflected as the larger the charge drum liquid the larger the slop cut between products for sharp separations. This presents an inherent problem in scaling up batch distillation from laboratory and pilot sized equipment since there will be a difference in hold-up which must be accounted for.

16.2.1.6 Molecular distillation. Molecular distillation differs from ordinary distillation in a number of ways. First since very low pressure is used, as low as 0.001 mbar, the temperature the system experiences is moderate and with the inherent short residence times, heat sensitive high molecular weight materials can be separated effectively. Secondly the vapor phase is produced by surface vaporization not boiling. The maximum evaporation rate can be assumed to be a function of the pressure, molecular weight, and the temperature[21,22].

$$V = 5.83 \times 10^{-2} p(MW/T) \tag{16.3}$$

where V is in g/s-cm^2, p is in mmHg, MW is the molecular weight, and T is in °K. This rate is based on the provison that none of the vapor leaving the liquid surface returns. In other words, the molecules that evaporate must not encounter any other molecules and collide, thus returning molecules to the liquid. To meet this requirement, a very low vacuum is required to minimize the molecules per unit volume along with a short travel time to the condenser which minimizes the time available for collisions. In conventional vacuum distillation, a molecule on average will be vaporized about a thousand times before reaching the condenser; for true molecular distillation, every molecule that escapes the liquid reaches the condenser. In most cases to obtain this close spacing the condensing and evaporating surfaces are parallel to each other and separated by a distance of less than the mean free path of the evaporating molecule. The mean free path is the average distance a molecule travels between collisions and like the maximum evaporation rate is a function of pressure and molecular weight.

In molecular distillation boiling does not occur to aid liquid mixing and both high molecular weight and high viscosity tend to hinder diffusion. Efficient molecular distillation requires mechanical mixing to eliminate composition gradients caused by low natural mixing. This is done in three ways:

1. Agitating the bulk liquid
2. Utilizing a falling film
3. Mechanically wiping the film along the wall.

Separation occurs not because of boiling point differences but by the difference in the evaporation rate[21,22] (see equation 16.3). This driving force is set only by the molecular weight difference since the temperature and pressure experienced by each component will be identical.

The vacuum in a typical molecular distillation can be as low as 0.001 mbar to increase the mean free path to an acceptable value. The mean free path of a molecular distillation system may be approximately 2–3 cm[21]. To accomplish this a vacuum system must be designed to reach these low pressures. As a consequence, if there are any volatiles, provision must be made to degas the feed prior to entering the molecular distillation unit. Volatiles at these low pressures include many components one would not normally consider, such as water and even light oils if a wax is being processed. Care must be taken to allow degassing of the feed. With the low pressures required it does not take much volatile to overwhelm the vacuum system. Similarly if care is taken to minimize the chances of collisions in the vapor space before reaching the condenser one must also make sure that the molecule that reaches the condenser does in fact condense. This can usually be achieved by providing a temperature difference of 100°C between the evaporating surface and the condensing surface.

Short path distillation differs from molecular distillation by the distance between the vaporizing and condensing surface. Where this distance is the molecular mean free path in molecular distillation, it may be twice this distance for short path distillation. This allows a larger unit to be more easily and cheaply built for production of up to 10 000 kg/h. The drawback is that more collisions between vapor molecules will occur meaning only 40 to 60% of the molecules evaporated will reach the condenser without returning to the liquid, unlike the 90% obtained in molecular distillation.

Molecular or short path distillation is used for viscous heat sensitive materials which cannot be separated by more conventional methods. Product degradation can occur with some materials if maintained at high temperature for an extended period of time. A commercial short path unit will operate at the lowest theoretically possible temperature and at the shortest possible residence time presently obtainable. By comparison, the hold-up time for a batch column is of the order of 400 times longer than a short path unit[22]. This means a minimum of product degradation.

16.2.1.7 Supercritical fluid extraction. Supercritical fluid extraction (SFE) is not a new phenomenon as it was first reported in 1879[23]. However, little motivation until recently existed for its development due to cheap energy and few if any health regulations. Many studies were carried out to explain why a system near its critical point vaporizes when the pressure is raised while the converse, raising temperature, causes condensation. This underlying phenomenon gives rise to the benefit of supercritical fluid extraction. To appreciate this effect take the example of supercritical ethylene and naphthalene. For this system at 100 atm and 12°C the observed solubility of naphthalene in ethylene is more than 15 000 times greater than if ethylene were an ideal gas[24].

The supercritical phenomenon can be explained if it is understood that under these conditions the compound is no longer a gas but has become a dense fluid whose density resembles that of a liquid more than a gas. This explains the seemingly abnormal behavior of SFE; with an increase in pressure we see an increase in solubility as the supercritical fluid becomes more dense, but as temperature is increased, the density decreases therefore the solubility drops and material leaves the solution. Figure 16.16 shows the change in density for a pure component near its critical point. While other properties such as the transport properties, diffusivity and viscosity remain gas-like above the critical point and surface tension is zero. This allows excellent penetration by a high density fluid into microporous materials for better extraction from solid substances.

As mentioned previously the solubility of one substance in another is based on minimizing the free energy of a system. A substance will only dissolve up to the point where the system's free energy is a minimum and then no more

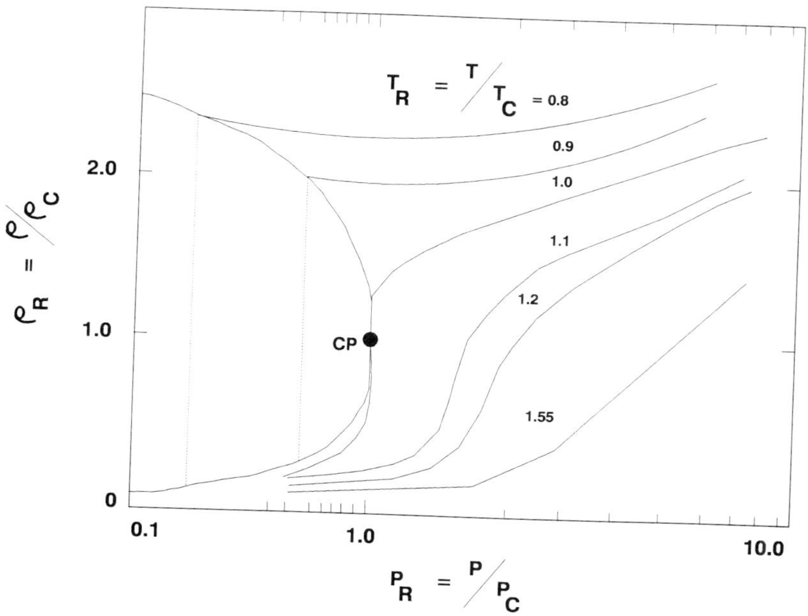

Figure 16.16 Behavior of a typical pure component near its critical point.

can dissolve unless the system changes. Factors affecting the system's free energy are the intermolecular forces between the like and the unlike molecules. The intermolecular attractive effects are inversely proportional to the molar volume which means that as the pressure increases the density increases and therefore the molar volume decreases. This results in the solubility increasing nearly exponentially with the density.

The trend in SFE is to use compounds which are non-toxic and whose critical pressure is usually less than 100 atm and critical temperature is close to ambient. The main workhorses of the SFE field have been carbon dioxide and light hydrocarbons. The properties of carbon dioxide, being non-toxic, non-hazardous, non-corrosive if dry, inexpensive, readily available, and the most studied compound in the supercritical region, make it the choice for the food and pharmaceutical industry to produce non-contaminated products. But both carbon dioxide and the light hydrocarbons are non-polar, non-hydrogen bonding and therefore can rely only on dispersion as the attractive force to cause a separation. Many compounds to be separated are polar, form hydrogen bonds, or are electron donor–acceptor complexes which the above SFE compounds cannot exploit. To take advantage of these effects, a cosolvent can be added to the original supercritical fluid to enhance its attractive forces. The cosolvent cannot usually be used by itself because of its high critical pressure, which would increase costs to the system, or its high critical temperature which may cause product degradation. An example is the extraction of benzoic acid by carbon dioxide at 35°C and 200 atm which can be more than doubled by adding as little as 3.5 mol% of a hydrogen bonding compound like methanol or acetone[25].

Little commercial work had been done until recently. The new motivation for SFE is the following:

- Increase in governmental requirements for removal of toxic, harmful, and/or unfavorable components without damage to the product
- Desire for higher purity heavy molecular weight products
- Heat sensitive materials
- More stringent pollution control
- No other separation method is viable
- Rise in energy costs making distillation more expensive.

Some industrial applications in commercial use include caffeine removal from coffee, Kerr–McGee's ROSE process, cholesterol removal from butter, and oilseed extraction[24].

New work in the field consists of both commercial applications and trying to understand and develop models for supercritical fluid extraction. Work to date has normally centered around an initial screening of conventional separation methods. If none of these methods, i.e. distillation, liquid–liquid extraction, vacuum distillation, proves viable, then some of the more exotic methods like SFE can be investigated. But to determine the viability of SFE a feasibility bench scale test and then pilot tests are required, which are time consuming and expensive. If a proven simulation were available then the initial SFE screening and economical evaluation could be done cheaper and faster saving the pilot test for verification and optimization purposes. Numerous SFE models have been proposed but none has achieved industry-wide acceptance.

The advantages of SFE are that it can be used to separate heat sensitive materials of high molecular weight with much better purity than other commercial methods and with little or no residual solvents. Also SFE is a function of temperature, pressure, and choice of solvent so these three parameters can be varied to achieve the desired product. The solvent can be a safe, non-toxic compound for food and pharmaceutical use. The higher transport properties and zero surface tension allow excellent penetration into solids enhancing extraction. With the addition of cosolvents even better extraction capabilities can be attained. A major disadvantage is the absence of readily available modeling techniques thus requiring expensive tests. Though SFE may reduce energy requirements, the capital outlay is high because of the high pressure equipment. The SFE process may add 15 cents per pound (33 cents per kg) to the cost of decaffeinated coffee[26].

16.2.2 Distillation design criteria

16.2.2.1 Thermodynamic equilibria prediction. There is a plethora of methods to predict thermodynamic equilibrium. Table 16.7 lists many, with the most commonly used emboldened, and their applicability.

The engineer responsible for simulating a distillation separation now must decide which model to use. There are expert systems available such as CONPHYDE and Simulation Science has one incorporated in its simulating program. Likewise there are flow charts which help in selecting the right model[27]. Some simple conditions to take into consideration prior to selecting a model are:

1. What are the key components to be separated, and based on their vapor pressures, what is the relative volatility?
2. Are there any minor components which may affect the key component separation?
3. Do experimental data exist for the composition, pressure and temperature range of interest?
4. Are there any significant polar interactions?

Table 16.6 Methods for predicting thermodynamic equilibrium

Model	Applicability
Raoult's law	For nearly ideal non-polar systems at low pressure and high relative volatility
Equation of state (EOS) **Soave–Redlich–Kwong (SRK)** **Peng–Robinson (PR)** **Benedict–Webb–Rubin (BWR)** **Chao–Seader (CS)** **Grayson** and **Steed (GS)** Chain-of-rotators (COR) Perturbed-anisotropic chain theory (PACT) Perturbed hard chain theory (PHCT)	For non-polar to mildly polar systems up to very high pressures. Requires little input so they are widely used in the hydrocarbons industry. Interaction parameters are required for low relative volatility and the mildly polar systems to enhance the accuracy. Some newer EOSs are being developed to handle highly polar systems but these are not in wide use. These include PACT, COR, and PHCT.
Liquid activity coefficient (LACT) **Margules** **Van Laar** **Wilson** **NRTL** **UNIQUAC**	For highly polar systems at low to moderate pressure. Requires experimental data on each binary pair.
Group contribution method **UNIFAC** ASOG	For highly polar systems at low to moderate pressure. Requires no experimental data just interaction parameters for the various groups that make up the molecules. These group interactions are provided with most programs.

5. Are there any pinch points, azeotropes, or partial miscibility which affect the key separation?
6. What is the required product purity?
7. What is the accuracy required? Is it for a preliminary study or the final design?
8. Can an experimental test be run?

From this information and reference material one can make the proper choice for a thermodynamic equilibrium model, but be aware of possible pitfalls. The LACT model will be the most accurate but only when used within the range of the experimental data. Extrapolations in pressure up to approximately 60% of the critical point on the binary are possible. Beyond that, the liquid activity is affected by pressure. As the system becomes compressible in the liquid phase the molar volumes become variable. The EOSs and group contribution models require little input data which makes them easy to use but these models are the most likely to give erroneous results when used on a system for which they are not intended. If there is any doubt a check can be made on a homologous system for which information is known. For example, if UNIFAC is the only method short of running experimental tests for triethylamine and methyl ethyl ketone but data exist for the binaries triethylamine and acetone and also methyl ethyl ketone and butylamine, then a check can be made of the validity of the UNIFAC method for the known binaries. If it reflects the data well, it should also simulate triethylamine and methyl ethyl ketone well.

16.2.2.2 Design parameters. When designing a distillation system the main parameters to consider are the operating pressure, product specifications, number of theoretical stages, minimum reflux, type of internals (i.e. trays or packing), and stage efficiency. Once these are determined the distillation system can be sized and a cost estimate performed.

The operating pressure of the distillation column determines the overhead condensing temperature and the boil-up temperature. If the pressure is set too low the condensing temperature may require mechanical compression or refrigeration both of which are very expensive. Likewise, if the pressure is too high the bottoms component temperature may exceed the heating medium temperature that is available, or cause product degradation. The optimum operating temperature should give an overhead temperature that allows either air or water cooling. This temperature should be approximately 20 to 40°F (10–25°C) above the cooling medium temperature. Thus if cooling water is available at 90°F (32°C) an operating pressure giving a condenser outlet of 120°F (49°C) would be preferred. The heating medium should be the lowest temperature available as this is usually the cheapest but at least 20 to 80°F (10–45°C) above the temperature of the bottoms component. Above 100°F (55°C) temperature difference between the heating medium and the bottoms component may cause film boiling in the reboiler which will seriously impede the heat transfer. The operating pressure also affects the column size. At higher pressures the column diameter required is reduced because the vapor phase is at a higher density and the diameter is directly dependent on the volumetric flow rate of the vapor. But as the pressure increases the relative volatility of the components decreases requiring more stages to make the separation. During the simulation it is a good idea to vary the pressure while staying within the applicable temperature constraints and determine if the diameter can be reduced with little effect of the number of stages. In pharmaceutical applications the product integrity may override the above constraints. The product may not withstand the temperature of the bottoms without degradation while using air or cooling water for condensing. The system pressure would then be set by the temperature constraints of the products. This may entail the use of refrigeration and vacuum equipment.

Determination of the number of theoretical stages, reflux ratio and feed point should first be done using a short cut method such as the Fenske equation[28], the Underwood equation[30,31], and the Erbar–Maddox equation[29]. The Underwood method will determine the minimum reflux given infinite stages. The optimum reflux is around 1.1 times the minimum. Since the number of stages required between minimum reflux and 1.1 times the minimum goes from a number less than 200 to infinity it is best, and less than a 2–6% cost increase, to be higher than 1.1 times the minimum reflux. The recommended reflux ratio is in the range of 1.15 to 1.25. The Fenske equation will determine the number of theoretical stages required and the Erbar–Maddox equation will determine the optimum feed location. These equations have some very limiting assumptions, such as equal molar overflow. The values determined can be used as a starting point for rigorous simulations or for preliminary studies to determine the relative cost difference between distillation schemes.

The rigorous simulation is performed on a computer. In a simulation every significant relationship involving material flow, energy transfer, and vapor–liquid equilibrium is converged under the assumed conditions. There are many PC-based simulator programs such as Simulation Sciences PRO II, Deerhaven Technical Software's PD-PLUS, ChemShareAspen and Chemstation to name a few. The simulation is run using the preliminary reflux ratio, number of stages, and feed location determined above. After this initial run a brief parametric study should be performed. The feed location should be raised and lowered to see the effect on separation. Likewise the reflux should be lowered until the separation degrades to determine the zone of confidence that the column would operate in. Once the optimum distillation column has been simulated the equipment can be sized.

In order to size the distillation system the column internal components must be determined. Two major classes of internal components exist, either trays or packing. Table 16.7 is intended to assist in this selection.

Table 16.7 Parameters to be considered in choice of internal components

Parameter	Type	Rating
Pressure drop per theoretical stage	Trays	Poor
	Random packing	Fair
	Structured packing	Good
Minimal liquid hold-up	Trays	Poor
	Random packing	Fair
	Structured packing	Good
Low liquid rate performance	Trays	Good
	Random packing	Poor
	Structured packing	Poor
High liquid rate performance	Trays	Poor
	Random packing	Good
	Structured packing	Good
Cost of internals	Trays	Good
	Random packing	Fair
	Structured packing	Poor
Cost of column plus internals	Trays	Good
	Random packing	Fair
	Structured packing	Fair
Liquid-to-vapor ratio range Distillation operations are normally low but absorber and scrubbers can be high	Trays	Poor
	Random packing	Good
	Structured packing	Good
Foaming systems	Trays	Poor
	Random packing	Fair
	structured packing	Good
Solids present	Trays	Good
	Random packing	Fair
	Structured packing	Poor
Corrosion resistant materials	Trays	Fair
	Random packing	Good
	Structured packing	Poor
Columns less than 46 cm diameter	Trays	Poor
	Random packing	Good
	Structured packing	Fair
Turndown	Trays	Good
	Random packing	Fair
	Structured packing	Fair

16.2.2.2.1 Tray Types. Trays are the most common distillation column contacting device. They have the highest overall operating range and their design methods are well known and proven. But as more work is done and knowledge is gathered the use of trays may decrease in certain applications as packing becomes more competitive.

The two main tray types used in pharmaceutical distillation today are the sieve and valve. Other types such as bubble cap, cross-flow, dual-flow, and baffle exist but are for special applications and will only be mentioned briefly.

Sieve trays are the simplest and, therefore, the least expensive tray. A sieve tray is made from a flat sheet of metal perforated across the entire surface with 0.32 cm (1/8″) to 2.54 cm (1″) diameter holes. The number and sizes of these holes are determined mainly by the vapor and liquid volumetric throughput. Although it is possible to design the tray this task is normally left to the tray supplier. This is also true for valve and bubble cap trays. Sieve trays are good general purpose devices but have minimal operating range. See the tray selection guide in Table 16.8, for characteristics.

Valve trays are similar to sieve trays in that the base is a perforated plate on which are placed the valves which are either movable disks or rectangular metal strips over the holes. The rising vapor lifts the coverings in proportion to its flow rate allowing the pressure drop across the tray to remain nearly constant until all the valves are fully opened. The valves are constructed with stops to prevent high vapor flow from lifting them out of the hole. These valves will also reseat themselves at low vapor rates to prevent liquid from leaking through the tray perforations as would happen with sieve trays. These movable valves allow a very wide operating range for this type of tray.

Bubble cap trays have been around for a long time. They are mainly found in older service or where very high operating range is required and will justify the extra cost or if a minimum liquid level is required on the tray regardless of the vapor rate.

In *dual-flow trays* both the rising vapors and falling liquids must share the same holes while flowing through the column. Dual-flow trays have no downcomers. These trays though less expensive than trays with downcomers have a very narrow operating range, and the efficiency is similar to sieve trays, but only if operated at the optimum design point after which efficiency falls rapidly. This limits their use to highly fouling services with minimal turndown[17].

In *baffle trays* the falling liquid forms a curtain of liquid as it rains down from the upper baffle to the one below. The rising vapor comes into contact with the liquid as it flows through this curtain. These trays have the same limitations as the dual-flow trays and are used for mainly fouling services. Sometimes, if only a few stages are required they may be used in vent gas scrubbing.

Table 16.8 Tray characteristics

	Sieve	Valve	Bubble	Dual-flow
Operating range (% of flood)	60–80	25–80	10–80	75–80
Efficiency	1	1	2	4
Fouling resistance	2	3	4	1
Cost factor	2	3	4	1
Foaming systems	2	2	3	2
Vent scrubbing	2	2	3	1
Viscous fluids	2	2	3	3
Low pressure drop	1	2	3	1

1–Excellent, 2–good, 3–fair, 4–poor.

16.2.2.2.2 Packing types. The biggest drawback to the use of packing is the lack of accurate generalized design methods. Until the late 1970s the design had been based primarily on past experience or vendor input. This led most engineers to specify trays because of the comfort factor. However, with effort by university laboratories, research institutes, and packing manufacturers there has been an increase in data available to the engineer to improve the comfort factor when specifying packing. Packing can be categorized into two groups. Random (or dumped) packing and structured packing.

The contacting mechanism for packing is such that the falling liquid forms a film on the packing surface that contacts the rising vapor. With this mechanism it is very important that the liquid wets the packing material to improve the formation of a thin film to obtain good mass transfer. In other words plastic packing would not be the best choice for an aqueous system. Also as the tortuosity of the vapor path increases the contacting surface increases thus decreasing the height of packing equivalent to a theoretical plate (HETP). This could increase the pressure drop and vapor velocity beyond design limits, so it is important to balance the packing type and size with the design limits of the system. Detail design and selection of packing are beyond the scope of this work. There are many excellent sources (see the Bibliography) to assist in determining the best packing type and size. Much work is in progress as new types of packing are developed.

Random packing as the name implies is placed in the distillation column in no set order. As a matter of fact the packing is actually dumped in place. For breakable packing more care is taken in the dump, such as dumping into the column filled with a liquid to reduce the settling velocity. The materials of construction for packing include most metals and plastics but also ceramics, thus greatly enhancing the range of corrosive material that can be handled.

The mass transfer for random packing can range from 50–183 cm (20–72″). As with all packing the larger the size the smaller the available surface area which reduces the transfer efficiency but increases the capacity. Care must be used in choosing a packing type with low HETP but still having the required capacity at an acceptable pressure drop. Many packing vendors have introduced new high efficiency random packing in the past decade. This has greatly increased the uses for random packing. Typically a packing can be found to supply an HETP of 76.2 cm (30″) and this should be used for preliminary design and quite possibly when a packing vendor is called and the design has been firmed up, this number can be lower and also guaranteed.

Structured packing comes in two types, corrugated textured sheets or gauze. The first type makes use of a constant configuration of geometrically arranged corrugated sheet throughout the column. The main feature that allows better mass transfer than random packing is the surface texture that is applied to the sheet[32]. The type of texture is unique to the manufacturer. This type of packing is available in metal, plastic, and ceramic, although the metal has the best texturing capabilities. Also by increasing the corrugation depth the capacity can be increased for a constant pressure drop but causes reduced mass transfer efficiency. Typical values for HETP that can be obtained range from 30.5–76.2 cm (12 to 30″). For preliminary designs 50.8 cm (20″) can be used.

Gauze-type structured packing is different in that the corrugated sheets are made from a woven fabric. This fabric can be either a woven metal or a plastic fabric. The gauze type packing is characterized by the highest mass transfer efficiency or as commonly noted, the lowest HETP available. The HETP value ranges from 45.7 cm (6 to 18″) depending on the system, with 30.5 cm (12″) being a good average number to be used in preliminary work.

On a cost per cubic meter basis, structured packing is more expensive than random packing. With higher capacity and better efficiency the cost per theoretical stage can be less. Structured packing may also be required to meet a lower pressure drop requirement or for low liquid loadings. Table 16.9 gives characteristics for these three types of packing.

16.2.2.3 Column sizing. Once the type of internal component is selected the column can be sized. The height of the column is determined by the tray efficiency or the height equivalent to a theoretical plate (HETP) for packing. The determination of this value for either packing or trays lags behind all the advances in computer simulation and thermodynamic equilibrium prediction. The main reason for the inability to predict this value is the difficulty in determining the diffusion coefficient of the components in the liquid and vapor phases. Therefore only simplified charts such as the O'Connell chart for trays or published data on a system such as those found in McFarland *et al.* [33] and Vital *et al.*[34] for trays or Bolles and Fair [35] and Vital *et al.*[36] for packing are available or lastly, the predictive method can be used. The procedure for calculating the efficiency of a tray is as follows:

Table 16.9 Packing characteristics[a]

Packing type	HETP (in)	*P*/theo stage (mmHg)	Liquid loading (GPM/ft²) Min	Liquid loading (GPM/ft²) Max
Random	30	1–2	0.4	70 (2″) 125 (3.5″)
Structured				
Textured sheets	20	1–1.5	0.2	12–18
Gauze	12	0.1–0.5	0.1	5–10

[a] The values for HETP assume median conditions for liquid loading and pressure drop along with moderate operating conditions.

1. Estimate using the following predictive methods:
   For trays— MacFarland[33], Sigmond, Van Winkle method and the AIChE method.
   For packing—Bolles and Fair method[35].
2. Search the literature for information on your system or a similar system.
3. Contact tray or packing vendors for their input.
4. Choose the value from actual data for the same system from a reliable source. If not, choose the lowest calculated value for tray efficiency or highest HETP closest to a similar system.
5. Add 5% to the number of trays or the height of packing as a safety margin.

The tray spacing required for good operation is between 45.7–76.2 cm (18 and 30″) with 60.9 cm (24″) being the norm. The tray spacing is set to minimize the entrainment of liquid with the vapor to the tray above. For a foaming system where entrainment is a greater possibility, the tray spacing or the column diameter is increased. From equation (16.4) or (16.5) below plus the miscellaneous spacings shown in Figure 16.17 for trayed columns and Figure 16.18 for packed columns the total height of the column can be determined.

$$H_t = 1.05(N_{ts}/\text{efficiency}) \times \text{TS} \tag{16.4}$$

$$H_p = 1.05(\text{HETP} \times N_{ts}) \tag{16.5}$$

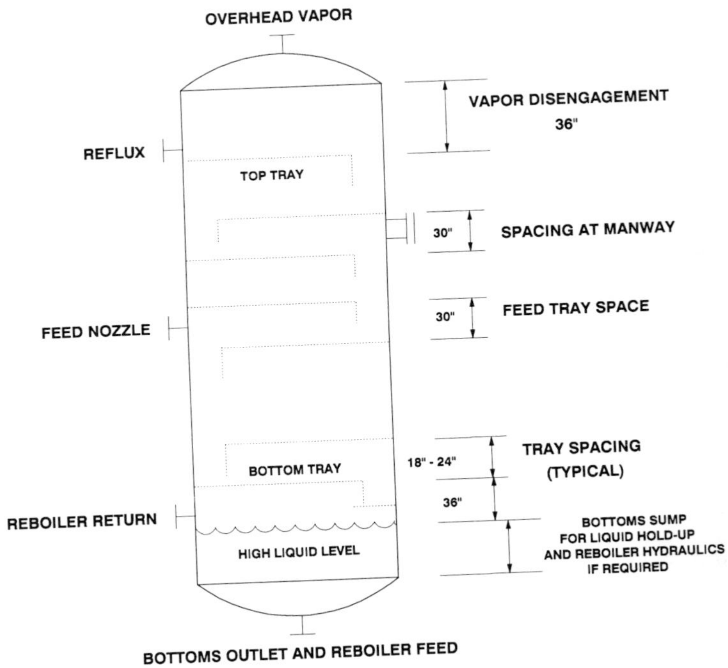

Figure 16.17 Typical spacing for a trayed distillation column.

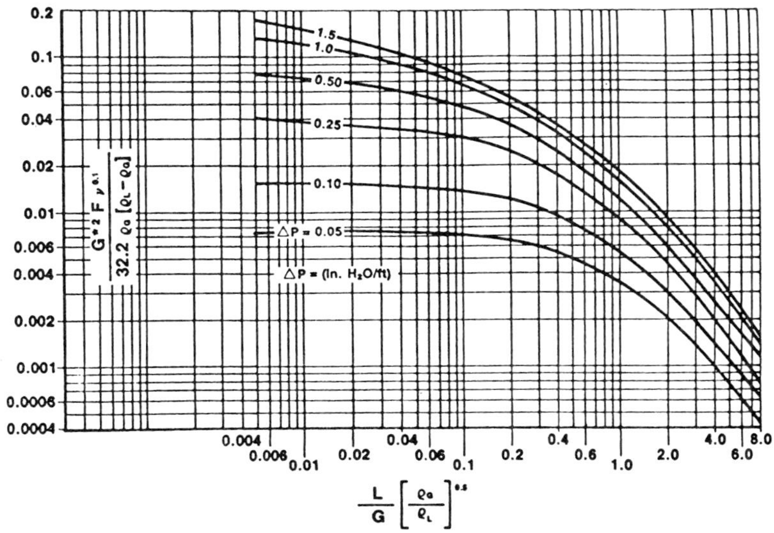

Figure 16.18 Generalized pressure drop correlation. G is the gas mass flow in lb/ft^2 – s (kg/m^2 – s), G^* is the gas mass flow in lb/ft^2 – h (kg/m^2 – h), L is the liquid mass flow in lb/ft^2 – h (kg/m^2 – h), F is the packing factor, ρ_v is the gas density in lb/ft^3 (kg/m^3), ρ_L is the liquid density in lb/ft^3 (kg/m^3), and v is the kinematic viscosity (cSt).

where HETP is the height equivalent to a theoretical plate, H_t is the height of trayed section, H_p is the height of packed section, N_{ts} is the number of theoretical stages, and *TS* is tray spacing, usually 60.9 cm (24″). The trayed or packed column diameter is determined based on vendor specific equations or the equation from Fractionation Research Inc. (FRI).

For preliminary sizes the trayed column can be estimated from the following equations:

Column diameter

$$D_t = 0.042(V)^{0.5}(1/\rho_v(\rho_l - \rho_v))^{0.25} \tag{16.6}$$

Tray efficiency

$$\log \text{eff} = 1.67 - 0.25(\log(\mu\alpha)) + 0.30(\log(L/V)) \tag{16.7}$$

Column height

$$H_c = H_t + \text{misc spacings} \tag{16.8}$$

where H_c is the total height of column (ft), D_t is the total height of column (ft), L is the liquid rate of tray with maximum vapor rate (lb/h), V is the

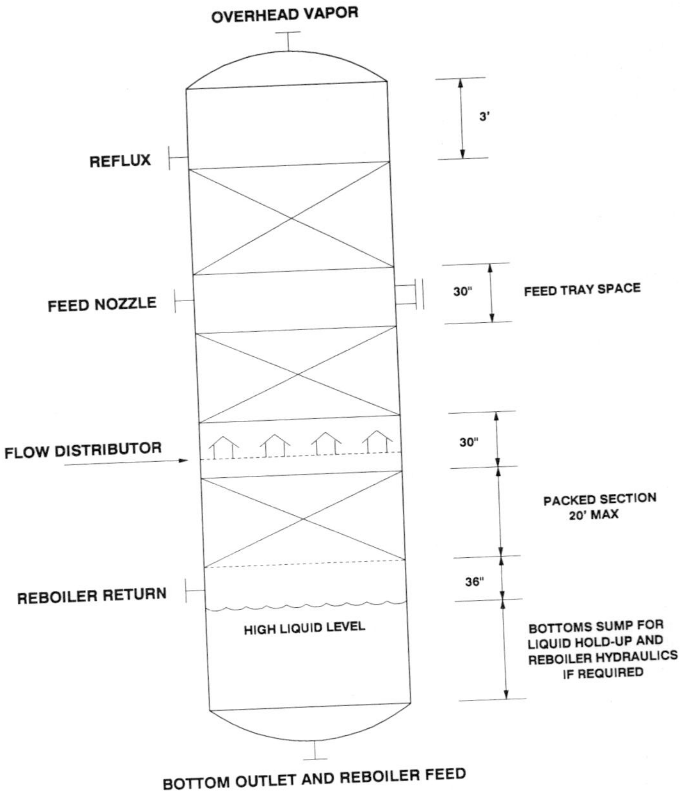

Figure 16.19 Typical spacing for a packed distillation column.

maximum vapor rate (lb/h), μ is the average liquid viscosity (cP), α is the average relative volatility of key components, ρ_v is the vapor density of V (lb/ft^3), and ρ_l is the liquid density of L (lb/ft^3).

Preliminary sizes for a packed column can be determined by the following method. From Figure 16.18 and the packing factor obtained from the vendor catalog choose a diameter that allows operation between the 2.54 cm (1.0″) and the 1.27 cm (0.5″) operating line. The column height can be determined using an HETP value from packing vendors or from Bolles and Fair[35] or Vital *et al.*[36]. Then by adding the miscellaneous spacings from Figure 16.18 the total height can be determined.

These calculations will give conservative sizes for the trayed and packed columns which will be appropriate for cost comparison and rough order of magnitude estimates.

References

1. Murrel, J.N. and Boucher, E.A. (1982). *The Properties of Liquids and Solutions*, John Wiley, New York.
2. Hildebrand, J.H. and Scott, R.L. (1962). *Regular Solutions*, Prentice Hall, New Jersey.
3. Sharp, K.A., Nichollis, A., Fine, R.F. and Honig, B. (1991). *Science*, **252**, 106.
4. Anderson, R., Cambio, R. and Prausnitz, J.M. (1962). *AIChE J.*, **8**, 67.
5. Anderson, R. and Prausnitz, J.M. (1961). *AIChE J.*, **7**, 97.
6. Barton, A.F. (1983). *Handbook of Solubility Parameters and other Cohesion Parameters*, CRC Press, Florida.
7. Garner, F.H. and Ellis, R.M. (1951). *Trans. Inst. Chem. Eng.*, **29**.
8. Van Winkle, M. (1967). *Distillation*, McGraw-Hill, New York.
9. Berg, L. (1969). *Chem. Eng. Prog.*, **65**, 52.
10. Hoffman, E.J. (1964). *Azeotropic and Extractive Distillation*, Interscience, New York.
11. Horsley, L.H. (1973). *Adv. Chem. Serv.*, **116**.
12. Scheibel, E.G. (1948). *Chem. Eng. Prog.*, **44**, 927.
13. Null, H.R. and Palmer, D.A., *Chem. Eng. Prog.*, **65**, 47 (1969).
14. Walas, S.M. (1988). *Chemical Process Equipment Selection and Design*, Butterworth, Boston, MA.
15. Bastos, J.C., Soares, M.E. and Medina, A.G. (1988). *Ind. Eng. Chem., Res*, **27**, 1269.
16. Thomas, E.R. and Eckert, C.A. (1984). *Ind. Eng. Chem., Process Res. Dev.*, **23**, 194.
17. Rousseau, R.W., *Handbook of Separation Process Technology*,
18. Diwekar, U.M., Malik, R.K. and Madhaven, K.P. (1987). *Comput. Chem. Eng.*, **11**, 629.
19. Quintero-Marmol, E. and Luyben, W.L. (1990). *Ind. Eng. Chem. Res.*, **29**, 1915.
20. Schweitzer, P.A. (1979). *Handbook of Separation Techniques for Chemical Engineers*, McGraw-Hill, New York.
21. Perry, R.H. and Chilton, C.H. (1973). *Chemical Engineers Handbook*, 5th edn, McGraw-Hill, New York.
22. UIC Inc. Bulletin (1990). *Short-path Distillation from Laboratory to Production.*
23. McHugh, M.A. and Krukonis, J.V. (1989). *Supercritical Fluid Extraction Principles and Practice*, AiChe Today Series, New York.
24. McHugh, M.A. and Krukanis, V.J. (1986). *Supercritical Fluid Extraction Principles and Practice*, Butterworth, Boston, MA.
25. Dobbs, J.M., Wong, J.M., Lahiere, R.J. and Johnston, K.P. (1987). *Ind. Eng. Chem., Res.*, **26**, 56.
26. *Chemical Week* (1987). June 24, p. 66.
27. Wilcox, R.F. and White, S.L. (1986). *Chem. Eng.*, Oct. 27, p. 141.
28. Fenske, M.R. (1932). *Ind. Eng. Chem.*, **24**, 482.
29. Ebar, J.H. and Maddox, R.R. (1961). *Petrol Refiner*, **40**(5), 183.
30. Underwood, A.J.V. (1945). *J. Inst. Petrol*, **31**, 111.
31. Underwood, A.J.V. (1946). *J. Inst. Petrol* **32**, 598, 614.
32. Brovo, J.L., Ashutosh, P.A. and Edgar, T.F. (1992). *Ind. Eng. Chem. Res.*, **31**, 604.
33. McFarland, S.A., Sigmund, P.N. and Van Winkle, M. (1972). *Hydro Proc.*, **51**, 111.
34. Vital, T.J., Grossel, S.S. and Olsen, P.I. (1984). *Hydro Proc.*, 147.
35. Bolles, W.L. and Fair, R.J. (1979). *Ind. Chem. Eng. Symp. Series*, **56**, 33.
36. Vital, T.J., Grossel, S.S. and Olsen, P.I. (1984). *Hydro Proc.*, 75.
37. Ewell, R.H., Harrison, J.M. and Berg, L. (1944). *Ind. Eng. Chem.*, **36**, 871.

Bibliography

Physical properties

Battino, R. and Clever, H.L. (1966). *Chem Rev.*, **66**, 395.
Thorough review and compilation of gas solubilities in liquids.

Boublik, T., Fried, V. and Hala, E. (1973). *The Vapor Pressure of Pure Substances*, Elsevier, Amsterdam.
Correlation and compilation of vapor pressure data for pure fluids. Normal and low pressure region.

Chu, J.C., Wang, S.L., Levy, S.L. and Paul, R. (1956). *Vapor–Liquid Equilibrium Data*, J.W. Edwards, Ann Arbor, MI.
Collection of vapor–liquid equilibrium data.

Claxton, G. (1958). *Physical and Azeotropic Data*, The National Benzene and Allied Products Association, London.
Compilation of azeotropic data as well as other physical properties including melting and boiling points.

Design Institute for Physical Property Data (DIPPR) (1990). *Data Compilation Tables of Properties of Pure Compounds*, AICHE, New York.
Source for most pure component properties for over 1000 compounds. Includes regressed equations for temperature dependent properties. Equations have upper and lower temperature limits and do not extrapolate well.

Faldix, P., *Fortsch. Verfahrenstech.*, **6,** 440 (1965); **7,** 650 (1967); **8,** 625 (1969); **9,** 561 (1971).
Bibliography for vapor–liquid equilibrium data.

Francis, A.W. (1961). *Critical Solution Temperature*, ACS, Washington, DC, Vol 31.
Collection of critical solution temperatures for various binary mixtures.

Fredenslund, A., Gmehling, J. and Rasmussen, P. (1977). *Vapor–Liquid Equlibria using UNIFAC*, Elsevier, New York.
Detailed and extensive information on the UNIFAC method for estimating activity coefficients with application to vapor–liquid equilibria at moderate pressures.

Gmehling, J. and Onken, U. (1977). *Vapor–Liquid Equilibrium Data Collection*, DECHEMA Chemistry Data Ser., Frankfurt, Vol. 1, pp. 1–10.
Comprehensive data collection for more than 6000 binary and multicomponent mixtures at moderate pressures. Data correlation and consistency tests are given for each data set.

Hala, E. *et al.* (1967). *Vapour–Liquid Equilibrium*, 2nd English ed., trans. George Standard, Pergamon, Oxford.
Excellent comprehensive survey, including a discussion of experimental methods.

Hala, E., Wichterle, I., Polak, J. and Boublik, E. (1968). *Vapor–Liquid Equilibrium Data at Normal Pressures*, Pergamon, Oxford.
Compilation of experimental data for binary mixtures.

Hicks, C.P. (1978). Bibliography of thermodynamic quantities for binary fluid mixtures, in *Chemical Thermodynamics*, M.L. McGlashan, Chemical Society, London, Vol. 2, chap. 9.
Literature references for vapor–liquid equilibria, enthalpies of mixing and volume change for binary systems.

Horsley, I.H. (1952–62). *Azeotropic Data*, Advances in Chemistry Series, No. 6 and No. 35, American Chem. Soc., Washington, DC.
Possibly the most extensive single source compilation of azeotrope data.

Jordan, T.E. (1954). *Vapor Pressure of Organic Compounds*. Interscience, New York.
Compilation of vapor pressure data for organic compounds; data are correlated with the Antoine equation and graphs are presented.

Landolt Bornstein (1975). *Zahlenwerte und Funktionen aus Naturwissenschaft und Technik*, New Ser. Group IV. Vol. III, Springer Verlag, Berlin. Also part 4c (Solubility of Gases in Liquids), 1976.
Vapor–liquid equilibrium, data and vapor pressure data. Updates earlier compilation of vapor–liquid equilibrium data.

Reid, R.C., Prausnitz, J.M. and Poling, B.E. (1986). *The Properties of Gas and Liquids*, 4th ed., McGraw-Hill, New York, NY.
Most recent edition of the famous series started by Reid, Prausnitz, and Sherwood. Source for estimation techniques for pure component physical properties along with a compilation of pure component properties for 618 compounds. Excellent starting point.

Riddick, J.A. and Bunger, W.B. (1970). *Organic Solvent*, 3rd ed., Wiley Interscience, New York, NY, Vol. II.
Various information on organic solvents.

Seidell, A. and Linke, W.F. (1958–65). *Solubilities of Inorganic and Metal-Organic Compounds*, Van Nostrand, Princeton, NJ, Vols. 1 and 2 and supplement.
Presents solubility data for gas–liquid, solid–solid, and liquid–liquid systems.

Stephen, H.T. (1963). *Solubilities of Inorganic and Organic Compounds*, Pergamon Press, Oxford.
Solubility tables at various temperatures for binary and ternary systems.

Wichterle, I., Lind, J. and Hala, E. (1973, 1976, 1982). *Vapor–Liquid Equilibrium Data Bibliography*, Elsevier, Amsterdam, Vols. 1, 2, and 3.
Thorough compilation of literature sources for binary and multicomponent data; includes

many references to the East European literature. Excellent starting point to find if published information exists for a system.

Equilibrium computation

Discusses the thermodynamic basis for computer calculations for vapor–liquid equlibria; computer programs are given.
Fredenslund, A. and Heideman, R.A., *Vapor–Liquid Equilibria in Complex Mixtures*, I. Chem. E. Symposium Ser. No. 104.
Review of models for the computation of vapor–liquid mixtures which include either electrolytes, polymers or polar components.
Prausnitz, J.M., Eckert, C.A., O'Connell, J.P. and Orye, R.V. (1967). *Computer Calculations for Multi-component Vapor–Liquid Equilibria*, Prentice-Hall, Englewood Cliffs, NJ.
Ray, P. and Hobson, G.U., *The Selection and Use of VLE Methods and Data*, I. Chem. Symposium Series No. 104.
Discussions and review of VLE model for various applications.

Distillation equipment design

Fair, J.R. and Bolles, W.L., *Performance and Design of Packed Distillation Columns*, I. Chem. E. Symposium Ser. No. 56.
Performance evaluation of packed column for the following parameter; flood velocity, pressure gradient and mass transfer for various commercial packings.
Glitch Inc. Bulletins
King, C.J. (1979). *Separation Processes*, McGraw-Hill, New York.
Discussion with equations for designing and sizing distillation equipment.
Koch Engineering Bulletins
Norton Chemical Process Products Co. Bulletins
Nutter Engineering Bulletins
Perry, R.H. and Chilton, C.H. (5th). *Chemical Engineers Handbook*, McGraw-Hill, New York.
Discussion with equations for designing and sizing distillation equipment.
Rousseau, R.W. (1987). *Handbook of Separation and Process Technology*, John Wiley, New York.
Discussion with equations for designing and sizing distillation equipment.
Schweitzer, P.A. (1979). *Handbook of Separation Techniques for Chemical Engineers*, McGraw-Hill, New York.
Discussion with equations for designing and sizing distillation equipment.
Shinskey, G.F. (1984). *Distillation Control*, 2nd ed. McGraw-Hill, New York.
Detailed discussion of control method for distillation systems.

17 High purity water

S. FISHKIN

17.1 Introduction

High purity water is a relatively recent term and may readily be defined if what is required or needed is known exactly. However, neither the science nor the art of so-called high purity water have progressed this far, and we have to deal in some combination of purity ranges and judgements of experience.

The most direct method of measuring the purity of water as it relates to high purity water is by resistivity (Ω-cm). Table 17.1 provides a conversion to conductivity ($\mu\Omega^{-1}$-cm), with dissolved solids (ppm), and grains per gallon (as $CaCO_3$). For the purposes of this chapter, so-called normal water (or normally treated water), including potable water, will be defined to have a resistivity of less than 0.01 MΩ (more than 50 ppm). Certain specific impure items like color, odor, or turbidity are more restrictive.

So-called high purity water (treated to be high purity water) will be defined to have a resistivity of greater than 0.1 MΩ (less than 5.0 ppm). It is well to note that many specific users have stated requirements up to 18 MΩ (28 ppb); a variation factor of 180.

Section 17.2 covers the general nature of feedwaters. Section 17.3 covers various types of purification systems used in most types of pretreatment systems for both 'pure' and high purity water. Sections 17.4 to 17.7 cover a typical specific treatment system type developed in response to the need for low 'bioburdens', high reliability of continuous operations, and detailed validation of the entire system. Most other system requirements will deviate in one or more areas, but the concepts, flow, and choices will be related.

In general, experience has indicated that, although this is a highly specialized field, it is important that the owner and end user of the system take an active role and have direct involvement in the design-making decision processes and in the specific high purity water/water for injection (WFI)/clean steam requirements analysis.

17.1.1 Water—the normal state of purity

Water is one of the most abundant naturally occurring substances, and excluding air, the one for which there is the most use and the most experience. But our knowledge and recognition of its complexities leave much to be desired.

Table 17.1 Fishkin conversion chart

Conductivity ($\mu\Omega^{-1}$-cm) at 25°C	Resistivity (Ω-cm) at 25°C	Dissolved solids (ppm)	Approximate grains/gallon (GPG as) $CaCo_3$
0.056	18 000 000	0.0277	0.00164
0.059	17 000 000	0.0294	0.00170
0.063	16 000 000	0.0313	0.00181
0.067	15 000 000	0.0333	0.00193
0.072	14 000 000	0.0357	0.00211
0.077	13 000 000	0.0384	0.00222
0.064	12 000 000	0.0417	0.00240
0.091	11 000 000	0.0455	0.00263
0.100	10 000 000	0.0500	0.00292
0.111	9000 000	0.0556	0.00322
0.125	8000 000	0.0625	0.00368
0.143	7000 000	0.0714	0.00415
0.167	6000 000	0.0833	0.00485
0.200	5000 000	0.100	0.00585
0.250	4000 000	0.125	0.00731
0.333	3000 000	0.167	0.00971
0.500	2000 000	0.250	0.0146
1.00	1000 000	0.500	0.0252
1.11	900 000	0.556	0.0322
1.25	800 000	0.625	0.0368
1.43	700 000	0.714	0.0415
1.67	600 000	0.833	0.0485
2.00	500 000	1.00	0.0585
2.50	400 000	1.25	0.0731
3.33	300 000	1.67	0.0971
5.00	200 000	2.50	0.146
10.0	100 000	5.00	0.292
11.1	90 000	5.55	0.322
12.5	80 000	6.25	0.368
14.3	70 000	7.14	0.415
16.7	60 000	8.33	0.485
20.0	50 000	10.0	0.585
25.0	40 000	12.5	0.731
33.3	30 000	16.7	0.971
50.0	20 000	25.0	1.46
100.0	10 000	50.0	2.92
111	9000	55.6	3.22
125	8000	62.5	3.63
143	7000	71.4	4.15
167	6000	83.3	4.85
200	5000	100	5.85
250	4000	125	7.31
333	3000	167	9.71
500	2000	250	14.6
1000	1000	500	29.2
1110	900	556	32.2
1250	800	625	36.8
1450	700	714	41.5
1670	600	833	48.5
2000	500	1000	58.5
2500	400	1250	73.1
3330	300	1670	97.1
5000	200	2500	146
10 000	100	5000	292

Its most natural and common purest state is when it evaporates into the atmosphere and rests in a cloud. Since the beginning of the science of water chemistry, this cloud state has been referred to as pure water with a simple chemical composition of H_2O. However, even in the comparatively pure state of a cloud, the water is not pure water, and its actual chemical composition is far more complex than just H_2O.

If we start with water evaporating into the atmosphere, impurities are occluded from the parent body. The water evaporating into the clouds and the rainwater falling, come into contact with the air in the atmosphere and absorb impurities from the air as they move through it, up to the clouds and down again. Contact with the ground, runoff ground liquids, ground solids and surface air each increase the impurity concentration of the evaporated water. This water from the normal purity cycle is the water that we are familiar with, that we use in the home and in commerce; that we taste, drink, cool, swim in, dilute, mix and react with. This water is treated in various ways to make it suitable for each specific purpose, and the simple treatment methods have been generally found acceptable.

Typical examples of the simple treatment methods are softening, chlorination, coagulation, filtration, precipitation, screening, settling, aeration, liming, carbon beds, demineralization, chemical treatment, disinfection, distillation, flocculation, fluoridation, degassing, sterilization, or combinations of these and others. Normal water contains suspended and dissolved impurities, turbidity and color, and bioburdens such as bacteria and pyrogens.

In the mid to late twentieth century, there has developed a need beyond normally treated water. In reasonably progressive succession came the special requirements for the high purity chemical industry, the semiconductor and computer chip industry, and the pharmaceutical, medical and biotechnical industry.

Small and/or minute quantities of impurities in raw materials and/or feedwater played a role in reducing the desired qualities of intermediates and products to unacceptable levels in these industries. A new art was developed bringing together varied combinations of scientific processes and equipment to produce waters which fall into our definition of high purity water.

17.2 Feedwaters and their characteristics

The equipment selection process starts with the raw water source and it is important that this source be profiled over a period of time. It is important when working with multiple water sources or surface waters to have a reliable water quality analysis curve since this will often dictate the kind of pretreatment required for a particular process purification technique. As an example, while most surface waters may be relatively low in dissolved ionizables, they usually contain organic and collodial matter which can be quite troublesome in the operation of ion exchange and distillation plants.

With normal treatment systems for process or drinking water, failure to check/test all aspects of the water and rely on prior data can present serious problems. Inflows to and outflows from treatment plants vary considerably and on occasion present a contamination problem in the product.

When high purity water is the product, failure to check/test all aspects of the influent water can have serious consequences. In general, those providing the data on treated or untreated water are not in a position to verify the data, source, reliability or date frame of the data. To give a simple example, in a particular system design the industry and the water company were asked to confirm the water pipeline pressure prior to design. The reply was 125 psig average and 90 psig minimum. After construction, in nine months of actual operation the pressure never exceeded 80 psig and was most often 60 to 70 psig. Fortunately the actual design of the system circumvented this problem, otherwise operation would have had serious problems and required modification. The chemical properties can vary widely by season, day and time of day. The problem can be more serious and circumventing it more difficult.

Data for a variety of waters across the United States have been included here for reference (see Tables 17.2 to 17.5). The references in the tables indicate where more specific and later data can be found. The data given are not intended to be used as a substitute for rigorous check/tests, but are intended as a rough guide for discussion. Appendix 2 lists useful associations.

Table 17.2 Comparison of mean trace mineral concentration in untreated and treated waters in the United States

	Untreated water[a] (μg/l)	Treated water[b] (μg/l)
Aluminum	74	179.1
Barium	43	28.6
Beryllium	0.19	0.1
Cadmium	9.5	12.0
Chromium	9.7	7.5
Cobalt	17	26
Copper	15	43
Iron	52	68.9
Lead	23	33.9
Manganese	58	25.5
Molybdenum	68	85.9
Nickel	19	34.2
Silver	2.6	2.2
Vanadium	40	46.1
Zinc	64	79.2

Source: Data taken from *Drinking Water and Health*, National Academy of Sciences, Washington, DC, 1977, pp. 210–211.

[a] 1577 Raw surface waters in United States. [b] 380 Finished waters. These are not necessarily the same waters.

Table 17.3 Community water supply study of 2595 distribution samples from 969 public water supply systems

	Standards		
	Limit (mg/l)	Maximum concentration (mg/l)	Percent exceeding
Cadmium	0.01	3.94	0.2
Chromium	0.05	0.08	0.2
Copper	1.0	8.35	1.6
Iron	0.3	26.0	8.6
Lead	0.05	0.64	1.4
Manganese	0.05	1.32	8.1
Silver	0.05	0.026	0.0
Zinc	5.0	13.0	0.3
Water samples below are not distribution samples, but finished water samples			
Arsenic	0.05	0.10	0.2
Barium	1.0	1.55	<0.1[a]
Selenium	0.01	0.07	0.4

Source: L.J. McCabe *et al.*, 'Survey of Community Water Supply Systems', in *J. Am. Water Works Assoc.*, November, 1970, pp. 678–687.
[a] This constituent was evaluated only on selected samples. The remainder were assumed not to exceed the limit.

Table 17.4 Mineral levels found in distribution samples from community water supply study

	pH 6.9 Percent exceeding average standard (mg/l)		pH 7.0–7.4 Percent exceeding average standard (mg/l)		pH 7.5–7.9 Percent exceeding average standard (mg/l)		pH 8.0 or more Percent exceeding average standard (mg/l)	
Number of samples	(425)		(556)		(550)		(595)	
Cadmium	0.0	0.001	0.05	0.008	0.2	0.001	0.0	0.000
Chromium	0.0	0.000	0.07	0.002	0.0	0.001	0.0	0.003
Copper	5.4	0.295	1.1	0.119	0.2	0.067	0.5	0.050
Iron	10.8	0.184	11.7	0.331	7.5	0.116	3.9	0.081
Lead	1.6	0.013	2.3	0.016	0.5	0.012	0.3	0.009
Manganese	9.9	0.026	11.7	0.003	6.8	0.019	4.9	0.012
Silver	0.0	0.000	0.0	0.000	0.0	0.000	0.0	0.001
Zinc	0.7	0.225	0.5	0.321	0.0	0.180	0.0	0.056

Source: L.J. McCabe, 'Problem of Trace Metals in Water Suplies: an Overview', in *Proceedings of the 16th Water Quality Conference*, University of Illinois, February 12–13, 1974.

There are some good reasons for the diversity of opinions concerning health effects of waterborne contaminants. The sheer number of potentially harmful materials found in today's water is a leading factor. In excess of 35 000 toxic compounds are now recognized by the Environmental Protection Agency (EPA). A substance is termed 'toxic' if it has a negative effect on the

Table 17.5 US public health drinking water standards

	1962		1975 Interim (effective Dec. 1976) (mg/l) (2)
	(mg/l)	(1)	
ABS	0.5	0.05	0.05
Arsenic	0.01	0.05	
Chloride	250		
Copper	1		
Carbon chloroform extract	0.2		0.7
Cyanide	0.01	0.2	0.2
Fluoride	0.7–1.2		2.4 (50°F (10°C)); 1–4 (90°F (32°C))
Iron	0.3		
Manganese	0.05		
Nitrate	45		10 (as N)
Phenols	0.001		
Sulfate	250		
TDS (total dissolved solids)	500		
Zinc	5		
Barium		1.0	1.0
Cadmium		0.01	0.01
Chromium		0.05	0.05
Lead		0.05	0.05
Selenium		0.01	0.01
Silver		0.05	0.05
Mercury			0.002
Chlordone			0.003
Endrin			0.0002
Heptochlor			0.0001
Heptochlor epoxide			0.0001
Lindane			0.004
Methoxychlor			0.1
Toxaphene			0.005
2,4-D			0.1
2,4,5-TP Silvex			0.001
Turbidity T.U.			1
Coliform count (membrane) No/100 mls			1
Residual chlorine			<0.2

(1) Grounds for rejection, radiological contaminant levels to be developed.

health of an organism. The following is a short list of toxic contaminants reported found in our water supplies:

Heavy metals	Chlorine	Trihalomethanes
Nitrates	Chlorine dioxide	Bromoform
Nitrites	Dimethylformamide	Coliform bacteria
Nitrosomines	Fluorine	*Giaria lamblia*
Pesticides	Chlordane	Viruses
Sulfates	Methylene chloride	Arsenic
Asbestos	Ozone	Acetaldehyde
Chloramine	Phthalates	Bromobenzene
Bromochloromethane	PCBs	

A pharmaceutical, medical, or related facility needs to be certain (absolutely) that these types of contaminants do not reach their products.

17.3 Pretreatment and operations

The purification methods here are used for both 'pure' water and high purity water.

17.3.1 Types of purification systems

As pointed out earlier, a combination of water purification techniques must be employed to remove all the impurities from water. Filters made of synthetic fibers (like multimedia filters) or activated carbon (or a blend of the two) are the predominant methods of point-of-entry water purification.

An important consideration in the system design is the need to define the quality acceptance and performance criteria for each component. The main goal of any water purification system design should be to properly apply and operate each component to achieve a system which produces the intended volume of water product, at the specific quality, and in a reliable and economic manner.

It should be recognized that a purification system is essentially a grouping of components arranged in such a fashion so as to process the quality of feedwater into a product of predetermined quality. Each component within the system has one of several specific functions:

- To protect other components of the system from damage or non-function
- To provide economical water pretreatment prior to processing to final quality
- To provide processing to the final quality
- To recycle water or restore quality to former levels of purity.

If equipment selection is approached with the above in mind, specifications, operational procedures, monitoring and preventive maintenance procedures will become a natural progression and allow for present or future validation.

Filtration is the most commonly used means of cleaning up water. Activated carbon filters and reverse osmosis filters are predominant. A variety of fiber filters are also used. Some filters employing powdered carbon are also still made, but these should be avoided since there have been problems with the powder being taken up by the water, complete with the contaminants it has filtered out, and passing into the water that is consumed.

17.3.2 Activated carbon

With an activated carbon filter, water passes through either granules or a solid block of carbon. Although the carbon does physically filter out some of the larger impurities that are suspended or dissolved in water, it is effective primarily in removing much smaller, and often more dangerous, impurities. No other filtration method is as effective as activated carbon in removing many of the extremely toxic organic chemicals often found in water. Activated carbon removes more of the toxics than other filters, and does it better.

Toxics are removed as the water passes through the carbon by a process called adsorption. Activated carbon is extremely porous, approximately one pound of the material has a surface area of one acre. This honeycomb of minute pores attacks and traps pollutants passing through it; the impurities are adsorbed. The effectiveness of an activated carbon filter is dependent on several factors:

- The amount of time water is exposed to the carbon (the longer the better);
- The manner in which the water flows through the medium (the more diffuse its path, the better);
- The number and range of size of the micropores in the carbon (the greater the range of micropores size, the greater the range of contaminants removed);
- The amount of water the carbon has already filtered.

When an activated carbon filter is installed, it may remove nearly 100% of the smaller molecule synthetic chemicals from processed water, but with time, as more of the micropores are filled, its efficiency will drop. When a filter has been used too long, some of the pollutants it has trapped will be bumped into the water that reaches the consumer displaced by chemicals with a greater affinity for their resting place on the filter.

In an overloaded carbon filter, for instance, molecules of the insecticide aldrin, which is very strongly attracted to carbon, could dislodge molecules of the carcinogen benzene, which has a very low affinity for carbon. The benzene would then be free to pass through into the finished drinking water supply. Lead and other materials that may become lodged in the filter are especially easily displaced. A greater portion of other pollutants will simply pass through a filter that is overdue for replacement or regeneration.

The risk that harmful bacteria will become established in the filter medium is also increased by waiting too long for replacement. Filters that use carbon that has been treated to resist bacterial growth are more resistant, but in areas with many bacterial contaminants in the water, even these can become contaminated. So it is a good idea to change filter media at recommended intervals, or, better yet, more often. Some experts recommend changing them twice as often. If a water filter has not been used for several hours, running water through it for ten seconds or more will reduce bacteria levels in finished water.

If chlorinated city water is used, a simple test can determine whether the filter is functioning correctly. Water that has been processed can be tested for the presence of chlorine with an inexpensive kit from: Hach Chemical

Company, PO Box 389, Loveland, Colorado 80539. If the chlorine is getting through, some of the toxic elements in the water are too.

The life of a carbon filter can be extended by using a (less expensive) pre-filter to remove some of the impurities from the water. Many carbon filters come with a pre-filter to remove sediment. Those with a pre-filter in a separate canister are superior to filters with carbon and pre-filter sharing a container, because they are easier and cheaper to maintain. These sediment filters will pick up many of the larger contaminants that would otherwise pass right through the carbon, making this combination a very good idea.

17.3.3 Reverse osmosis

Reverse osmosis filters force water through a plastic membrane, leaving the majority of pollutants behind. Most of the particulates, radioactive materials, and dissolved solids that might pass through a carbon filter are removed. In this respect, reverse osmosis (RO) is superior to an activated carbon filter, but misses many of the very small toxics that are so prevalent. Radon is among the toxics that can pass through a reverse osmosis filter. Membranes commonly used in RO are polyvinyl acetate, cellulose acetate, polysulfone, thin film composites, polyamide types, etc.

A good combination filtration system will consist of three filtration steps performed in three separate containers. A particulate filter strains out the sediment; next the reverse osmosis filter removes some of the smaller dissolved materials and the larger organic chemicals; the activated carbon then filters out the volatile organic chemicals and other very small contaminants. The inexpensive pre-filter in this sort of system will typically have to be replaced every few months, while the reverse osmosis membrane and the carbon filter will last a few years between replacements.

The filter membranes can be the weak link in reverse osmosis filtration. Some of the thinner membranes cannot be used with chlorinated water (without a carbon pre-filter to remove it), while other membranes will become choked with bacterial growth if they are used with unchlorinated water. The quality of the filter membrane material varies. Operating at peak efficiency, however, a reverse osmosis membrane will catch 90% of the toxics in the water that passes through it.

Small capacity is another disadvantage of reverse osmosis filters. Most historically have produced only enough water for drinking and cooking, a few liters per day. Industrial reverse osmosis systems today produce many liters per minute.

17.3.4 Distillation

Distillation has long been used to purify water. Distilled water is used in scientific experiments because of its exceptionally high degree of purity. Water passing through a typical distiller is heated to boiling, causing it to evaporate and then recondense as pure water, leaving pollutants behind, basically the same process that gives the earth fresh rainwater.

There have been some problems with volatile compounds evaporating when the water is boiled inside a distiller and passing with the steam into the area where the water vapor is recondensed. Some of the new distillers have overcome this problem. A distiller used in combination with a carbon filter can produce some of the purest water attainable.

The small capacity of most distillers, usually producing only a few liters per day, and the inconvenience of using many units are drawbacks. Some, although not all, distillers must be loaded by hand and turned on and off manually. All require periodic cleaning to remove buildup of scale. Distillers are also expensive to operate. Distillers with housings made of stainless steel have been reported to contribute aluminum to processed water. Rouging problems have also been reported.

Distillation is a process that removes water from its impurities by heat vaporization. The phase change process removes particulate matter, dissolved inorganics, organics including bacterial, virus, pyrogens as well as dissolved gases including carbon dioxide and oxygen.

Stills must be designed so that the gas (vapor) phase separates from the liquid without carrying over particulate as entrainment. The baffle system velocity must be sufficient to remove the entrainment. The purified steam must not be contaminated and the condenser design must allow removal of volatile impurities which might condense in the process. Single and multiple effect stills are used.

Vapor compression operates by compressing the steam produced in the evaporator to increase the temperature of the steam. This temperature increase is transferred into the evaporator coils and allows heat to flow from the steam to the boiling water. Once the boiler has begun to produce steam, based upon the operation of auxiliary heaters, the steam that is produced is re-used by pumping it around from the top of the evaporator to the steam coil of the evaporator. This heat which is required to boil the water (latent heat) is saved by the use of the compressor. After initial start-up, the system requires electricity for the compressor and thermal energy (sensible heat) to raise the feedwater temperature and maintain it during operation.

Thermocompression still is an outgrowth of the vapor compression still design. Instead of using a mechanical compressor to raise the steam temperature, a steam ejector, Venturi type, is used. A second ejector is added to balance the heat load by removing approximately one extra pound ($\frac{1}{2}$ kg) of steam from the evaporator.

17.3.5 Ion exchange

Demineralization, deionization (DI), or ion exchange is based on the removal of impurities from the water by means of synthetic resins which have an affinity for dissolved ionized salts. Most synthetic resins are made of styrenes cross-lined with divinylbenzene before being activated with either sulfuric acid to make them cation acceptors or amines to make them anoin acceptors. A DI system might be a two bed, multiple bed, or mixed bed type depending on raw water conditions and water quality required. Demineralized water systems are tailored to fit the end use, e.g. feed water to a still, USP (United States Pharmacopoeia) purified water, point of use polishing, deacidification, decolorizing, silica removal, high pressure boiler feed, carbon dioxide removal, and oxygen removal. RODI is a system using RO followed by DI to produce high purity water.

Ultrafiltration (UF). Membranes which do not have the capability selectively to remove salts but act as filters for large organic molecules can be considered ultrafilter membranes or molecular filters. While a microporous membrane filter removes particles according to pore size, an ultrafiltration (UF) membrane functions as a molecular sieve that separates dissolved molecules on the basis of size by passing a solution through an infinitesimally fine filter (1000–100 000 molecular weight (MW)). The UF is a selective permeable membrane that retains most macromolecules above a certain size, including colloids, microorganisms, and pyrogens. Smaller molecules, such as solvent and ions pass into the filtrate.

Electrodialysis involves cation and anion membranes arranged alternately in series between two electrodes. The cation membrane consists of synthetic plastic materials such as sulfonated polystyrene–divinylbenzene which have ion exchange properties allowing cations or positive ions to pass through the membranes. The anion membranes are also synthetic plastic barriers having ion exchange properties; however, these membranes are similar to anion exchange resins in structure (divinylbenzene–polystyrene resins converted to the amine form). Anion groups of ions with negative charges pass through anion membranes. Since ions are also attracted by direct current potentials, the basis of electrodialysis is the separation of the ions' mobility and using the selective and cation exchange membrane as a means of separating the ions from each other.

17.3.6 Other pretreatment methods

Several other methods are used in the pretreatment of water. Aeration can remove most of the volatile compounds from contaminated water. Simply exposing water to air in a thin film or a spray, will vaporize many of the harder-to-remove contaminants. Chloroform, radon, and many gasoline components are among the volatiles that will be mostly removed through aeration. Aeration is widely used in large water treatment plants to take solvents such as trichloroethylene (TCE) and other volatiles out of groundwater, and some units small enough for domestic use may soon become available.

Ultraviolet radiation is energy in the form of a short wavelength. It fits into the electromagnetic spectrum between the shortest visible wavelengths and the longest X-rays. It covers the region between 100 and 400 nm, though only wavelengths below 290 nm show a strong microbiocidal effect. There is a peak of bactericidal effect around 254 nm and mercury vapor lamps which emit in this range are used. This radiation is used to kill bacteria.

Mechanical filtration deals with water treatment means for reducing suspended matter in water by passing the solution through a suitable porous material. Media filters of the pressure type are widely used in industrial water conditioning. Water flows through a bed of a fine granular medium such as sand. The suspended matter gets caught in the discontinuities which occur within the bed. The bed is then backwashed to release the solids to waste after which the filter is returned to service. Media filters are capable of filtering out fine solids including colloids and give a finer filtration because greater area of medium is available. Particles tend to accumulate and media filters promote coagulation. Single medium filters usually operate less efficiently since clogging occurs in the top portion of the bed, rendering the bottom part useless.

Multimedia types combine more than one medium in a single vessel, with coarse media on top, intermediate porosity media below it, and fine media below that. As water flows downward through the vessel, it encounters media beds of decreasing porosity and thus higher filtering efficiency. Proper layering increases efficiency and reduces clogging.

Particulate membrane types, e.g. cellulose acetate, polyamide, polysulfone, thin film composite, etc., are made with controlled pore size in a variety of ranges. In the pharmaceutical/biotechnical fields the 0.22 and 0.45 micro-membranes are most often used because they will remove bacteria.

Organic filters (activated carbon and regenerative organic) are utilized to reduce organic coloring, odors, chlorine, and certain organic acids. They can cause operational problems in products and for reverse osmosis and ion exchange equipment.

Softening is essentially the process of removing calcium and magnesium hardness from a water supply. There are several methods, namely, sodium cycle cation exchange, hot process ion exchange softening, hydrogen cycle cation exchange, etc. The most common softening method in the industry is the sodium cycle cation exchange process which is highly effective, yet relatively simple. Water is passed through a column of the sodium form of a cation exchanger which replaces the objectionable calcium and magnesium ions with non-objectionable sodium ions.

Coagulation is a process that combines, by chemical means, finely divided particles of turbidity and color to effect rapid settling. The three essential steps that should be followed in the control of coagulation processes prior to filtration are mixing, flocculation, and sedimentation. The most widely used coagulant is aluminum sulfate (alum) which can be used over a pH range of 5.5–8.0. Others include coppers, sodium aluminate, and magnesium oxide.

Disinfection is used extensively to produce ranges of pure water with small ranges of live organisms. The most popular is chlorination, which is a process in which chlorine gas or chlorine compounds such as hypochlorites are added to water or sewage, usually for the purpose of disinfection. Chlorination kills disease-producing organisms, removes certain objectionable tastes and odors, oxidizes iron, manganese, hydrogen sulfide, and organic matter. Other disinfecting techniques used widely are shown in Table 17.6.

pH measures the acidity or alkalinity of a given water. Common water is most desirable at a pH of 8 to 8.5. High purity water and many waters where the minerals/concentration is low have an acidic low pH. A simplified pH range table is shown in Table 17.7.

Corrosion control. Water with a low pH is highly corrosive. In addition, high purity water has a very low concentration of ions. This leads to a very large driving force (differential) between the containment system and the high purity water increasing the tendency of the containment materials (corrosion) to go into solution.

The tables, figures and appendices included in this chapter show the effect of a RODI system on a particular feedwater, a conversion chart for pure water units, typical block flow diagrams, guideline considerations for conceptual design, the types of feedwater normally used and their typical impurities, and a glossary of water purification methods, terms and associations.

17.4 Major treatment for high purity water

Sections 17.4 to 17.7 will cover a typical specific treatment system type design developed in response to the need for low 'bioburdens', high reliability of continuous operations, and detailed validation of the entire system. Most other system requirements will deviate in one or more areas, but the concepts, flow, and choice will be related. Batch systems, however, will require a different, but similar, type of design analysis.

The major consideration for high purity water system design is to quantify the various items required, i.e. RODI, water for injection (WFI), and clean steam. It is imperative not only to define the process requirements, but the ancillary requirements as well. Each of the use and cleaning steps should be itemized in detail. In one application the preliminary sizing was based on the process requirements, but the total cleaning water required reached almost a factor of 5 versus the known process requirements. In another application, sink faucets accounted for 75% of the WFI design.

17.4.1 Introduction

High purity water systems are designed to serve the functions listed in Table 17.8. Supplementary design considerations are shown in Table 17.9. They should provide purity water at 10 to 18 MΩ quality; however, 0.5 to 10 MΩ

Table 17.6 Disinfection scoreboard

Technique	Effectiveness	Reliability	Residual	By products	Pretreatment	Process safety
Fine filters	Depends on size of filter openings: the finest filters remove most protozoa and bacteria but few viruses; suspended solids that may harbor microorganisms are also removed	Buildup on filter reduces flow rate; bacterial breakthrough possible with some types	None	Some filters can add bacteria; fiber filters may release fibers	Filtration with coarse filter to remove larger particles that would clog fine filter; addition of bactericide to control growth of bacteria on fiber itself	No hazards
Reverse osmosis	The best can remove nearly all microorganisms plus many other impurities	Leaks are possible; frequent back-washing required; mineral buildup can clog filters	None	None	Filtration with fine filter to remove larger particles; water should be low in minerals and have pH less than 7	No hazards
Chlorine (chlorine gas, calcium and sodium hypochlorite powders, hypochlorous acid solution)	Can kill almost all microorganisms if adequate dosage applied for sufficient time	Careful monitoring of dosing system required to ensure constant application rate	Leftover chlorine compounds	Chlorinated organics and many other potentially toxic chlorinated compounds are possible	Filtration with this fine filter and activated carbon to remove suspended solids that may harbor microorganisms as well as organic impurities that could react with chlorine to form undesirable by products: pH should be less than 7.5	Chlorine gas is exceedingly dangerous; other forms are corrosive and must be handled with care

Table 17.6 Continued

Technique	Effectiveness	Reliability	Residual	By products	Pretreatment	Process safety
Chloramines	Weak and slow-acting; generally not effective	Careful monitoring of dosing system required	Leftover chloramines	Chloramines may be considered undesirable; otherwise there are few by products of concern	Filtration with fine filter to remove suspended solids	Chloramines are generated on-site from chlorine and ammonia, both of which are hazardous
Chloride dioxide	Can kill almost all microorganisms if adequate dosage applied for sufficient time	Careful monitoring of dosing system required	Leftover chlorine dioxide (may not last long)	Chlorinated organics and other potentially toxic chlorinated compounds are possible but less abundant than with chlorine dioxide and the chloride ions formed when it breaks down may be harmful	Filtration with fine filter and activated carbon to remove suspended solids and organic impurities	Chlorine dioxide is generated on-site from chlorine which is hazardous
Bromine (liquid bromine, hypobromous acid, bromine resin)	Can kill almost all microorganisms if adequate dosage applied for sufficient time	Liquid forms require careful monitoring of dosing system; bromine resin requires little attention	Leftover bromine compounds	Brominated organics and other potentially toxic brominated compounds are possible	Filtration with fine filter and activated carbon to remove suspended solids and organic impurities	Liquid bromine is exceedingly dangerous; often liquid forms are corrosive and must be handled carefully; bromine resin poses no hazards if kept dry
Bromoamines	Can kill almost all microorganisms if adequate dosage applied for sufficient time	Careful monitoring of dosing system required	Leftover bromoamines	Bromoamines may be considered undesirable; otherwise there are few by products of concern	Filtration with fine filter to remove suspended solids	Bromoamines are generated on-site from bromine and ammonia, both of which can be hazardous

Table 17.6 Continued

Technique	Effectiveness	Reliability	Residual	By products	Pretreatment	Process safety
Iodine (iodine crystals, triiodide resin)	Can kill almost all microorganisms if adequate dosage applied for sufficient time	Little attention needed	Leftover iodine and iodine compounds	Water may become cloudy if iron bacteria present; iodinated organics and other potentially toxic iodine compounds may be formed but in much less abundance than with chlorine and bromine; iodine may be hazardous to some persons	Filtration with fine filter to remove suspended solids; pH should be less than 7.5	Iodine crystals must be handled with care; triiodide resin poses no hazards
Ozone	Can kill almost all microorganisms if adequate dosage applied for sufficient time	Electrical parts of ozone-generating device require periodic inspection; automatic valves should be provided to shut off water flow in the event of malfunction	None under normal conditions but ozone may persist for some time at very low water temperatures	Ozone will break down organics into simpler compounds, some of which are potentially toxic; use of ozone may increase bacterial growth in water lines	Filtration with fine filter to remove suspended solids	Ozone is toxic and potentially explosive, should be handled with care; high-voltage electricity or ultraviolet radiation used to generate ozone may also pose hazards

Table 17.6 Continued

Technique	Effectiveness	Reliability	Residual	By products	Pretreatment	Process safety
Silver	Weak and slow-acting; generally not effective	Little attention needed	Leftover silver	Silver may be a health hazard in high concentrations	Filtration with fine filters and activated carbon to remove minerals	No hazards
Heat	Several minutes exposure to temperatures greater than 70°C will kill almost all disease-causing bacteria and most viruses but boiling for 30 min is needed to kill almost all microorganisms	Heating equipment requires periodic inspection; automatic valves should be provided to shut off water flow in the event of malfunction	Heat until water cools	Heat may be undesirable	None required	Burn hazard
Ultraviolet radiation	Can kill almost all microorganisms if adequate dosage applied for sufficient time	Lamp covers must be cleaned periodically; lamps must be replaced annually; automatic valves should be provided to shut off water flow in the event of malfunction	None	None	Filtration with fine filter and activated carbon to remove suspended solids and organic impurities; iron removal if level is high	Direct exposure to UV light is hazardous (well designed systems prevent accidental exposure)

Source: Prepared by L. Jaslow, Environmental Consultant, Baltimore, MD.

Table 17.7 pH of water and some common household products

		pH Table	
Water will taste 'acid' because of the presence of bicarbonates, carbonates, and hydroxides		14	Lye
		13	
		12	
		11	Ammonia
		10	
Most desirable pH for water is 8 to 8.5	Alkaline	9	
		8	Baking soda
	Neutral	7	
		6	
Water with a low pH is apt to corrode plumbing	Acid	5	
		4	
		3	Vinegar
		2	
		1	
		0	Muriatic acid

Table 17.8 Functions of high purity water systems

DI supply for clean steam generators	
DI supply for water for injection still	
DI supply for cleaning	
DI supply for high purity water for users	
Sinks	Autoclaves
Labs	Reactors
Solutions make-up buffers	
WFI supply for cleaning	
WFI supply for injection/product users	
Autoclaves	
Reactors	
Solutions make-up buffers	
Clean steam supply for products/seals and high vacuum	
Clean steam supply for sterilization/sanitation	

quality is probably acceptable for most non-WFI services. Of prime importance are the special provisions required to ensure and maintain low 'bioburdens' and the high degree of reliability needed, i.e. 7 days per week, 24 hours per day continuous operation without shutdown. With proper attention to design, four shifts of operation required less than one shift of operator coverage in one particular plant.

Some high purity water systems in the lower ranges of purity, i.e. 0.1 to 5 MΩ quality, are tailored to fit the end use in the following typical functions:

- feed water to a still
- USP purified water
- point of use polishing
- deacidification of wines
- decolorizing sugar syrups
- silica removal

- high pressure boiler feed
- carbon dioxide removal
- oxygen removal

Table 17.9 Supplementary design considerations

a.	Inspection approvals
b.	Sampling
c.	Sanitization
d.	Sterilization
e.	Validation/CGMP (current good manufcturing practice)
f.	Bacterial growth areas
	Resin beds
	Eliminate potential stagnant areas
	Minimum flow
	Velocity
g.	Regeneration (manual/auto)
h.	System monitoring (some systems over 100 points)
i.	Computerized monitoring (including first out)
j.	Continuous flow
k.	Passivation
l.	Temperature requirements
	RODI-cold; WFI-hot
	RODI-hot; WFI-cold
m.	Shutdown — provide redundancy
n.	Sterility barriers

17.4.1.1 Sterility barriers. The entire system, including the supplementary sanitary utilities should be set up into a sterile/sanitary boundary classification system at the earliest opportunity. These sterility boundaries should be systematically outlined in the process flow diagram and on the P&IDs (to the fitting); they should be continually updated. This sterility boundary is a basic tool in defining for the designer, vendor, and contractor the degree of sanitary design/installation necessary to satisfy the sterility requirements of the entire system; classified as indicated in Table 17.10. The sterile barrier classification in Table 17.10 will also serve to assist in start-up and validation procedures and in the specification for related equipment.

With minor exceptions, all system areas should be designed for continuous flow at the recommended minimum velocities to prevent stagnation. Dead legs should be religiously avoided in all piping layouts. Shutdown periods

Table 17.10 Sterility barriers classifications

1. Sterile A	WFI for precleaning and cleaning
2. Sterile B	RODI for precleaning; WFI for cleaning
3. Sterile C	RODI and tap water for precleaning; WFI for cleaning
4. Sterile D	RODI for precleaning and RODI for cleaning
5. Non-sanitary	Tap water for precleaning; RODI for cleaning
6. Non-sanitary	Tap water for precleaning and tap water for cleaning
7. Sterilizing required	Regularly or occasionally
8. Sanitizing required	Regularly or occasionally

should be avoided except for sanitization, and necessary maintenance. The system should be designed to allow for reasonable shutdown without interference with the end users. The use of sanitary fittings throughout is strongly recommended, especially at the sterility barriers.

17.4.1.2 Additional design considerations. These are listed as follows:

- What products are presently in place and/or contemplated and what are their respective water requirements?
- What are the water quality requirements of the various products?
- Is there more than a single water quality specification in place based upon the particular end use, e.g.
  - (i) water quality from a bio-burden/endotoxin level
  - (ii) water quality from an ionizable level
  - (iii) are there any particular elements/heavy metal/organics that might be toxic to the process/product?
- Are there any government or industry regulations or criteria in place that have no direct/indirect accountability, and, if not, could such an accountability regulation be a reality in the future, e.g. FDA, Good Manufacturing Practice (GMP), USP?
- Will the water generated become a part of the process, and/or product, and/or waste?
- What are the minimum, average and peak loading requirements?
- Is there a significant requirement for high purity water for cleaning, washing, and rinsing?
- Are there batch loading versus continuous make-up requirements?
- Do we need single and/or multiple distribution loops?
- Are the loops hot or cold?
- Are there specific heat and cooling ramifications, e.g. point of use heaters/coolers?
- What are the clean air requirements?
- Is clean steam required for seals and/or ventilation?
- What are permit considerations for chemical and/or biological waste?
- What is the nature of the water supply; temperature, pressure, and quality?

In the general pretreatment operations in section 17.3, we have defined distillation in general. There are some proponents of the production of high purity of water who include distillation, and some proponents for whom distillation is the only purification step to high purity water. For this reason a more detailed discussion of the distillation process is included here.

17.4.1.3 Distillation. This is a proven and acceptable process for producing water that is suitable for injection meeting the requirements of USP XXI. It is the only single process capable of producing water free of organics, ionizables, bacteria, particulate matter, and dissolved gases. These are the major advantages of producing high purity water by distillation.

The disadvantages are listed here:

- Because of the latent heat of evaporation, 970 BTU/#, the cost per liter of water produced can in many instances be substantial.
- Multiple effect distillation can reduce operating costs but operating costs must be balanced with initial cost.
- Maintenance can be severe if proper pretreatment equipment is not utilized to protect the heat exchanger system from scaling.

Distillation is a process that removes water from its impurities by heat evaporation. In this phase change process particulate matter, dissolved organics, organics including bacteria, viruses, pyrogens as well as dissolved gases including carbon dioxide and oxygen are removed.

All stills must be designed so that the water in the gas (vapor) phase separates from the liquid phase without carrying over particulate contamination as entrainment. Likewise, the baffling system must be designed so that the velocity is sufficient to ensure removal of any of the entrained particles. In addition, the purified steam must be protected from all contamination potential and the condenser design must be such as to allow the removal of volatile impurities which might normally recondense in the distillation processes.

17.4.1.4 Types of water distillation processes.

Single effect: These stills are relatively simple, reliable, and can be fabricated of many different materials. The heating medium could be steam, electricity, direct combustion, superheated hot water, etc. The single effect atmospheric still consumes approximately 1000 BTU/# of distillate produced. In addition, the cooling water requirement would approach 7 lb of water/# of distillate.

Low velocity designed water stills operate at velocities of between 1–2 ft/s (0.3–0.6 m/s) so the evaporator must be built sufficiently wide and high to provide for ample disengaging space for the vapor and to minimize turbulence at the boiling interface.

High velocity designed stills (usually 15 gallon/h and above) operate on an entirely different principal. Water in the evaporator is boiled by the heat source at a much higher velocity (50–200 ft/s) at the surface of the water before being forced through a cyclone separator for removal of entrainment. Essentially, the high velocity designed still utilizes centrifugal force to accomplish separation before the vapor is delivered to the baffle section and the condenser.

Multiple effect: As the term implies, multiple effect distillation employs more than one evaporator. The number of evaporators is usually determined by an economic analysis based upon the heat available and the volume of water to be produced per hour. The principle of operation essentially consists of boiling water in a closed container which is called the first effect which must be designed to take high pressure. The steam produced in this first effect is then used to heat the second effect by passing through a baffle separator before being discharged under pressure to the heat exchanger coil of the second effect where the latent heat of vaporization is given up to boil the water in the second effect.

Since the steam from the first effect, which becomes distillate, is actually condensed by the boiling liquid in the second effect evaporator, there is obviously no requirement for a condenser and also cooling water. Thus, if one has a double effect still, one can save approximately one half the cooling water and steam that would be required for a single effect distillation plant. As the number of effects is increased, the cost of operation is proportionately decreased.

It is generally recommended that all high purity water systems be designed to operate with mainstream and backup equipment in the dynamic mode, i.e. no shutdown or stand-by positions. Each piece of equipment, including backups, operated in the forward, backwash or regeneration mode at all times. A continuous mode of operation is also generally recommended in both the pretreatment, treatment, and end-user systems whenever possible.

17.4.2 Pretreatment of RODI water

A simplified diagram is shown in Figure 17.1. Inlet water is adjusted to provide a temperature of 26°C (78°F) to the RO unit. Heating is most often needed in the winter. Cooling is not normally used. In a continuously recirculating mode of operation, there is sufficient heat from the RO pump to provide at least sufficient feedwater temperature rise. If the inlet water is chlorinated, the biogrowth in the pretreatment recirculating loop can be effectively controlled and reduced. The pure water bypass is continually mixed

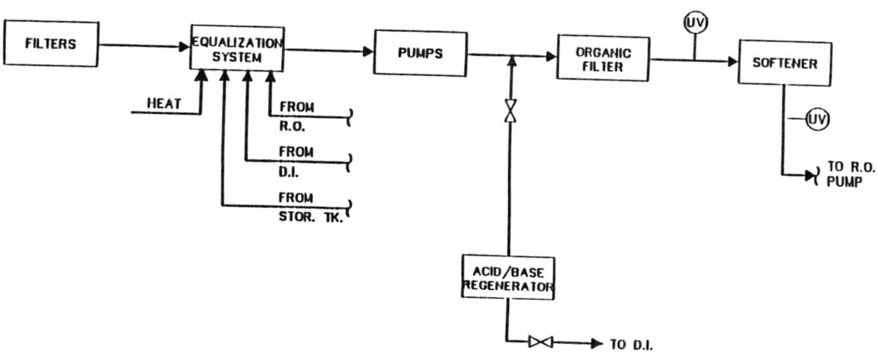

Figure 17.1 Simplified pretreatment of RODI water.

with incoming chlorinated water. In the absence of chlorine, other sanitization precautions will be required.

Incoming water is passed through redundant multimedia filters which are normally regenerable with feed water backwash on a time cycle basis. When not in the regeneration position, all filters should operate at partial capacity. The multimedia filters reduce particulates and the SDI. The operating filters should be capable of providing water for RODI manufacture and cycled regeneration water for either the organic filters or the softeners. We prefer manual regeneration for the organic filters, and can therefore avoid excessive peak loads on the filters by scheduling in-between backwashes.

The water is then fed to an equalization system. The variation in incoming water pressures and equipment pressures and equipment pressure drop are eliminated. In one system inlet water pressure had been projected at 90 to 125 psig. In actual operation this turned out to be 60 to 80 psig. A detailed step-by-step pressure analysis (clean versus dirty) is necessary to design the equalization system, with controlled fluctuation, so as not to affect adversely downstream operation. Variations to be designed for include clean versus dirty and forward flow versus backwash or regeneration. Pumps must be specifically selected to cover these variations and should be coordinated with pressure regulation to control pressure to the RO inlet, and with a recirculation loop for intermediate pressure control during backwash and regenerations.

The water is then passed through a set of resin bed softeners which exchange sodium ions for the calcium and magnesium ions in the water. Each bed should preferably be capable of handling full flow requirements. The softener is both effective and economical for the removal of calcium and magnesium, and for the exchange of dissolved iron.

The inclusion of the softener adds significantly to the reliability of the system. The removal of calcium and magnesium could be done effectively downstream. However, the reverse osmosis system has the potential to precipitate calcium, a potential scaling problem, preferably avoided for many raw waters.

The softener is regenerable on an automatic cycle basis with salt. The softener resin beds are a potential bacterial growth area and are similarly followed by filter and a UV sterilizer. At the outlet of the softener, pressure control is added to reduce the water pressure to a range ideal for feed to the reverse osmosis system.

17.4.3 RODI preparation

RODI preparation is shown diagrammatically in Figure 17.2. Reverse osmosis systems operate at a pressure in the range of 200 to 400 psig or higher, forcing water through membranes. The reverse osmosis process should reject about

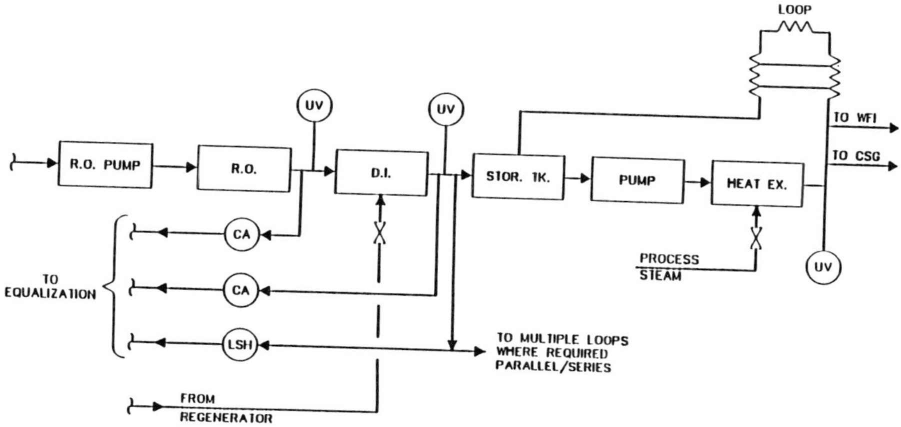

Figure 17.2 Simplified RODI preparation.

95 to 97.5% of the ionizable salts, 99% of organics with a molecular weight over 300, and is extremely effective in rejecting bacteria and pyrogens. However, the dirty side of the membranes rapidly builds up undesirable bacterial concentrations.

To alleviate this potential problem the membranes are normally automatically flushed on a continued cycle basis, say 3 to 8 min every 4 h; but full sanitization with a sanitization chemical like phosphoric acid is required on a reasonable cycle, based on a continual monitoring of pressure drop, conductivity, and bacterial count. The RO is designed to reject about 25% of RO feedwater at all times.

In addition to measuring the product quality, the monitoring meters should read directly in percent reject, so that the performance of the unit can be rapidly seen in the different modes, i.e. forward, flush, and recirculation. At forward flow, the rejection is 97% or above. During recycle, this rejection drops below 97%. In our quest for perpetual flow it is preferred to recirculate RODI at non-forward flow periods in the cycle and through any potential stagnant areas.

Recycle occurs when the storage tanks are not calling for water, mostly off the daytime shift. The quality of the feedwater after continued recycle has been known to improve 80% prior to entering the RO. During these periods of high quality RO feed, the percent reject has been known to drop below 94%. This combination, of course, provides a higher quality of water exiting the RO. The recycle aspect of the system adds heat to the recirculating water, mainly via the pumps. It is this operation that reduces the heat requirement to the system except for very cold water inlet periods. During the occasional times when the effect of the recycle mode overheats the water, this may be

Table 17.11 Water system analyses

Item	Plant feed	RO inlet	DI outlet
Turbidity	0	0	0
Color	0	0	0
pH	8.8	7–8.8	5–5.7
Alkalinity	16 mg/l	16	trace
Hardness	38 mg/l	1	trace
Calcium	10 mg/l	1	trace
Magnesium	3.2 mg/l	1	trace
Sodium	23 mg/l	41	trace
Potassium	3.1 mg/l	3.1	trace
Iron	0.04 mg/l	0.04	trace
Manganese	0.03 mg/l	0.03	trace
Sulfate	27 mg/l	27	trace
Chloride	49 mg/l	49	trace
Nitrogen (ammonia)	0.05 mg/l	0.05	trace
Nitrogen (nitrate)	0.30 mg/l	0.30	trace
Nitrogen (nitrite)	0.002 mg/l	0.002	trace
Copper	0.002 mg/l	0.002	trace
SDI (fouling index)	25	5	trace

offset by a small RO exit bleed, which is compensated for by additional fresh feedwater.

An additional advantage of the recycle system is that the recycled deionizer DI water lowers the pH of the water entering the RO inlet and this is very effective in lowering the scaling characteristics of the RO unit.

The RO system outlet water should be monitored for conductivity. Should the RO outlet water be unsatisfactory, it is diverted back to the equalization system and does not proceed to the deionizers until the problem is resolved.

Following the RO unit a system of two (or pairs of) mixed bed deionizers are used operating in the progressive mode and maintaining dynamic flow. The deionizer beds remove all but traces of dissolved organics. The DI water oput should be in the range of 14 to 18 MΩ and is fed to storage for use in the plant operations.

The two deionizer beds operate on the method of lead/lag in series. Bed A leads and takes most of the load and bed B lags and acts as a polisher. The lead bed A is exhausted first and is regenerated while bed B takes the full load. When bed A finishes regeneration, it is put back in service as the lagging or polishing bed (bed B now leads). When bed B needs regeneration the process is reversed, and so on. The DI outlet is monitored for conductivity and bioburdens as a guide to problems and regeneration requirements.

17.4.4 Regeneration

The deionizer beds are regenerated with acid and caustic on a cycle basis. The regeneration is designed for non-automatic operation with operator attendance. The organic resin beds are similarly acid/base regenerated non-automatically on a cycle basis. Acid/base equipment is thereby used only at controlled periods. This makes it safer to use and requires less than four shifts of operators.

A separate regeneration supply skid unit is used to regenerate either the organic filter or the deionizer unit, but not simultaneously, and only when there is an operator on duty. The regeneration equipment and the equipment being regenerated are generally housed together in a curbed area with separate drain so that the acid and base used are always properly contained.

17.4.5 Piping delivery systems

The overwhelming choice as the material of construction for high purity water delivery system is 316L stainless steel. However, careful attention needs to be directed to welding procedures, polish, cleaning, passivation inspection, and documentation.

The use of sanitary fittings is the preferred approach to joining components where access is required. The use of compression fittings should be avoided because they contain large crevice areas that make cleaning a problem and could advance the growth of bacteria. Flanges are not generally recommended because of alignment difficulties leading to an imperfect joint. Valves should be of the sanitary diaphragm design mechanically polished to at least a 180 grit followed by electropolishing. Electropolish provides several benefits; it removes peaks, rounds out regularities, improves cleaning by creating a smooth finish, passivates the weld surfaces, and improves the inspection process by highlighting imperfections in welds. The use of sanitary fittings is strongly recommended throughout.

One of the major considerations in piping design is the dead leg criteria. All dead legs tend to increase the potential for stagnation pockets, which, over a period of time tend to become bioburden breeding grounds. Dead legs should therefore be deleted, or their effect minimized. Some acceptable rules have been developed for this purpose and some vendors have developed special fitting and/or valves to accomplish this. One acceptable rule is the so-called '6D' rule where the strait run of the dead leg from the center line of the pipe, fitting or valve shall not exceed six times the smaller fitting diameter. On occasion, designers have used 5D, 4D, and 3D rules for some applications. It is imperative that the rule used is updated to keep in line with changing FDA guidelines.

17.5 RODI storage

The product from the deionizer system is fed to RODI storage tank(s) by level control. When the tanks are full of the measured deionizer product conductivity is high, the DI product is diverted back to the equalization system bypassing the RODI storage tank. In the event of trouble in RODI manufacturing, the RODI tank should be capable of supplying from 4 to 12 h of water before shutdown, or more, depending on operating requirements.

The storage tank system is designed with a recirculatory loop system for supplying water to end users. The loop has a set of pumps, operating and spare, an ultraviolet sterilizer, a heat exchanger, and a 0.2 μm filter. The spare pump is installed but the pipes are not connected to avoid stagnation problems. Loops in the extent of 50 end users and in excess of 2000 feet of sanitary pipe have been successfully used. If properly designed, installed, passivated, tested and sanitized, this quantity of loop can be readily handled. There are no data on an upper quantity of loop, but engineering judgment indicates that sanitization/sterilization may have to be increased as loop size is increased. All sanitary parameters must be maintained, regardless of size. The loop system is dynamic, i.e operates 24 hours/day, 7 days per week. Dead or stagnant spots are avoided and velocities are kept above minimum recommended levels. Each end user is fed via a sub-loop with the same criteria. The loop and end users should be carefully designed with a very detailed series/parallel configuration.

Sanitary piping and its ancillary considerations commence at the inlet to the RO unit and continue to completion. Standard vendor sanitary designs for RO inlet, RO outlet and general DI piping are not necessarily provided as completely sanitary, or completely sterilizable, even when so called for. These should be reviewed in detail to the fitting. Significant modifications may be necessary to bring these into proper specification. All RO outlet piping, and all DI piping (except the regeneration piping) are to be capable of both sanitization and sterilization. In some cases this requires modification of instrumentation and UV sterilizers. All sanitary piping and equipment starting with the RO are electropolished and/or passivated. Titanium is recommended in specific equipment areas to prolong equipment life. PVC piping or fittings are not recommended for any pharmaceutical or biotechnical application.

The storage tank and the loop systems should also be capable of both sanitization and sterilization in their entirety. A heat exchanger which is kept in full flow on the water side at all times may be used for this purpose. This heat exchanger water side is part of the RODI recirculating loop. When heated, the loop liquid recirculates hot RODI water for sanitizing. After the proper sanitization time–temperature operation, the RODI loop may be allowed to cool, or cooling water may be used to speed up the cooling process.

A comprehensive monitoring system of alarm and status lights are recommended to ensure the continuous operational mode. Full computer control is not necessary for this purpose, but computer monitoring of the main points is a very good way to ensure proper first out diagnosis. The important points can be selected for additional remote monitoring (not control) to a computer readout to an area with 24 h manual coverage. These monitoring readouts (alarms) are used to recognize and treat problems on a timely basis. They can also be designed as a diagnostic tool for the origin of the problems, i.e. a first out program.

The monitoring arrangements above should include the pretreatment, the RODI preparation, storage and loops as well as the additional end user systems, i.e. the water for injection system and the clean steam generator system. As an alternative, a properly designed DCS control system is even more desirable and reliable, usually at significantly greater expense.

17.6 Water for injection (WFI) system

RODI water is fed to the WFI still preheater by level control and on to an evaporator heated with the plant process steam. The evaporated water should be fed through filters/separators. The steam is condensed with cooling water, and then partially reboiled to remove dissolved gases. The distillate is fed to a WFI storage tank. A conductivity monitor diverts below specification distillate to drain or recycle. WFI production is controlled by an on/off level control in the WFI storage tank. Boiler controls are incorporated in the WFI still. The WFI system generally has less users than RODI; the WFI still may be designed to be down a good part of the time.

The still is vented automatically when on standby waiting for level control to request water (see Figure 17.3).

17.6.1 WFI storage loop

Fresh feed from the WFI still enters the WFI storage tank by level control to keep a constant level range in the tank. The level control on the tank is

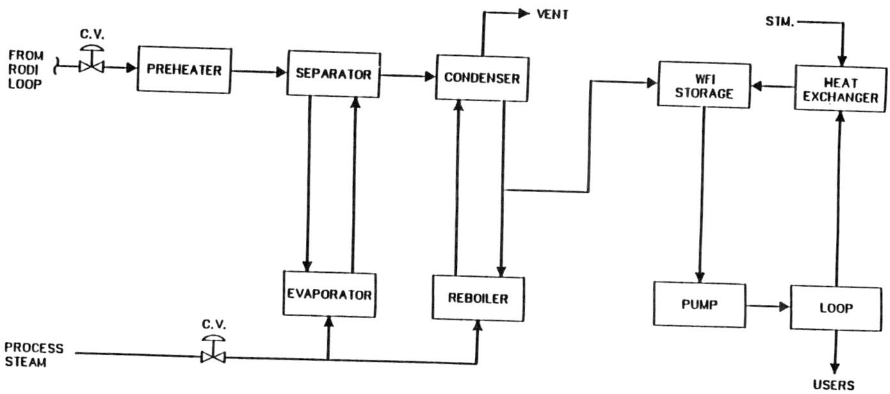

Figure 17.3 Simplified WFI system.

interlocked with an on/off switch on the WFI still so that WFI is produced only when the level switch calls for water.

WFI from the WFI storage tank is fed to WFI users via a WFI recirculating loop. The WFI recirculating loop velocity is designed to prevent stagnancy and is designed to run continuously.

Standard control methods are used for the WFI tank. The LSH point turns the WFI still off or on depending on whether the water level is at or below the LSH point. The LSL point is interlocked to the recirculation pump to shut it off and alarm should the level reach the LSL point.

The WFI storage tank water is maintained at about 82°C (180°F). This water may be temperature controlled (heated) at the return end of the WFI loop via a WFI heat exchanger (double tube sheet) or a hot jacket may be used. We prefer the heat exchanger. Where cool WFI is required, point-of-use coolers (double tube sheet) or a cool WFI loop is provided. Considerable control effort is needed for the point-of-use cooler design to meet the continuous flow non-stagnancy standards.

17.7 Clean steam generation

RODI water is fed to an evaporator heated by plant process steam in a coil. The evaporated water should be fed through a filter/separator. The separator product is the pure clean steam. The unit operates on an as-needed basis. When steam pressure drops, the unit starts up to prevent steam. There is a manual standby switch for periods when clean steam will not be needed, although it is expected that some clean steam will always be needed for seals, humidification, etc. This requirement places a high priority on the reliability of the operation. The unit incorporates boiler controls to aid in reliability.

The steam pipeline may be an open ended loop with no condensate return and may involve dozens of users and thousands of meters of pipe. All the piping and equipment for the clean steam and WFI system should be electropolished and/or passivated, amd titanium used in specific areas of the clean steam and WFI equipment to prolong life.

When water with very low ion concentration is boiled, a combination of high temperature and a high concentration differential potential occurs. This combination is highly corrosive. SS316L failures are known. Titanium, tinned copper and alloy 20 have been used. We know of one case where SS316L failed in 18 months and the titanium replacement showed no signs of corrosion after five years.

The monitoring systems mentioned previously for RODI also cover the important points of the WFI system and especially the clean steam system, which is often the most critical of the systems for continuous operation. A system (RODI, WFI, clean steam) similar to those described in this chapter was the first unit to go on stream at one plant. The RODI high purity water system operated manually for the first three months (deliberately) and for

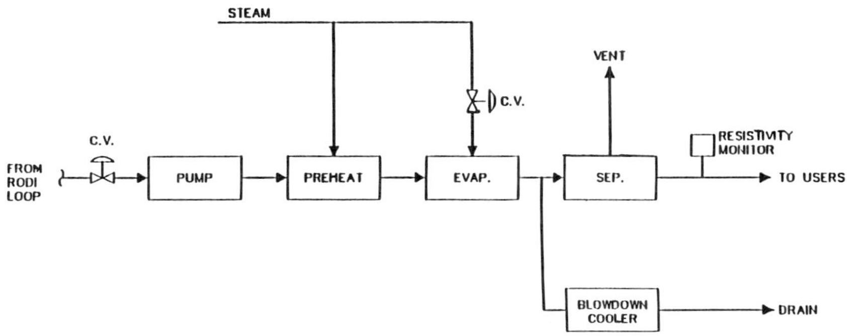

Figure 17.4 Simplified clean steam generation.

over four years has been in fully automatic operation except for the planned manual regeneration. Periods of three, four, and five days (long weekends) without operator attendance have been reported as a standard operating condition with no adverse effect on the overall operation (see Figure 17.4).

17.8 CGMP and validation

No review of high purity water can be complete without a minimum reference to CGMP (current good manufacturing practice) and validation. It is emphasized here because no work, conceptual or detailed, should be initiated without the 'early' incorporation of CGMP and validation compliance. The specific definition of 'early' in this paragraph refers to day one, hour one, etc.

17.8.1 Application of GMP (good manufacturing practice)

Good manufacturing practices of the GMP, is a body of criteria imposed by the FDA that has brought about some revolutionary approaches to the design of plants for 'clean' industries. The design, updating, and qualification of manufacturing facilities must not only consider current and proposed regulations, but also recognize the changing nature of these regulations and project future requirements.

Compliance with the FDA/GMP requires the ability to interpret the FDA regulations and their impact on each client's individual needs. Categories of design and operation that are heavily influenced by GMP are:

- Production and support equipment
- Piping design and layout
- Materials of construction
- Movement of materials and personnel
- Control of environmental conditions

- Clean-in-place and sanitation systems
- Product changeover

To implement succesfully the qualification of a biotechnology or pharmaceutical facility it is essential that the qualification team have a detailed understanding of client needs and the regulatory requirements of the FDA. Further, it is often necessary to review special features of a project with the FDA to ensure that they are carried out in a manner that can win approval. This involves:

- Regular contact with the FDA to stay up-to-date on the status of regulatory matters
- Coordination of contacts with the FDA for project reviews and discussions of special requirements
- Obtaining and distributing the latest FDA regulations and industry publications.

Current good manufacturing practices are:

- Concepts that embody: FDA guidelines/opinions, industry practices, and state of the art technology
- CGMP process considerations: equipment, specification, suitability for intended purposes, materials of construction, cleaning, sterilization and layouts
- Equipment arrangement
- System design: piping specifications, piping layout, CIP/SIP, biocontainment
- Instrumentation/control systems
- Quality of utilities: water, compressed air, other

17.8.2 Approach to validation

Validation capability should be designed into a pharmaceutical process and facility. Facilities for the manufacture of drugs, drug-related products, parenteral, oral, topical or ophthalmic dosage forms, blood-related products, and medical devices for human or animal use are regulated by the Food and Drug Administration (FDA), Department of Health and Human Services. A key element of FDA's regulatory review includes assurance that manufacturing facilities are in compliance with the following regulations and revisions:

- Current Good Manufacturing Practice (CGMP); Regulations for Finished Pharmaceuticals, 21 CFR, Parts 210 and 211.
- Current Good Manufacturing Practice (CGMP); Regulations for Blood and Blood Components, 21 CFR, Part 606.
- Good Manufacturing Practice; Regulations for Medical Devices, 21 CFR, Part 280.
- Good Laboratory Practice (GLP) for Nonclinical Laboratory Studies, 21 CFR, Part 58.

17.8.3 *Relationship between CGMPs and validation*

The purpose of a validation program is to establish documented evidence that provides a high degree of assurance that a specific process will consistently produce a product which meets predetermined specifications and quality characteristics. Another way of stating this is to say that the process in question is always under control and does what it purports to do. The concept of validation is not a requirement in addition to the CGMPs, but rather, embodied within the CGMPs are validation requirements stated in both specific and general terms. It is necessary, however, to interpret properly and apply the CGMPs to process systems and sub-systems to ensure that the total facility is in compliance with the CGMPs when it is operational. Services which are required include:

- Validation program scoping
- Documentation audit
- Validation master plan
- Computer control system validation
- Development of qualification documents/procedures
  - Installation qualification
  - Calibration specifications and procedures
  - Operational qualification
  - Performance/process qualification
- Development of standard operating and maintenance procedures (SOPs)
- Field calibration services
- Field execution of protocols
- Validation report
- Revalidation services

17.8.4 *Program implementation*

The validation team needs to be involved in the project from the design stage through construction and start-up. In order for a validation program to be most cost effective and to be implemented on a timely basis, inclusion during design, construction, and start-up is recommended. The objective is to ensure a timely implementation of the validation program as the systems within the facility become mechanically complete. To achieve this objective, execution of validation protocols should be scheduled as individual systems of a facility are made operational.

For the purpose of planning and implementation, a validation program can be broadly separated into three phases; the preliminary phase, the design development phase, and the field execution phase. These validation phases are readily coordinated with similar phased approaches utilized for the design and construction activities of a project. Below are enumerated activities included for each phase of the validation program.

17.8.4.1 Preliminary phase. The preliminary phase of validation may include the following:

- CGMP design audit
  - Process
  - Facility
- Validation scope definition
  - Process equipment
  - HVAC equipment
  - Sanitary piping
  - Computer control systems
  - Emergency power supplies
  - Utilities
- Master plan development
  - Validation approach
  - Identification of systems/instruments
  - Define documentation/protocol formats installation qualification (IQ), calibration, operational qualification (OQ), performance qualification (PQ), SOPs, PMPs final report)
  - Schedule
- Validation SOP/define responsibilites and procedures
- Development/audit of training program
- FDA review/liaison

17.8..4.2 Design development phase. The design development phase may include the following activities:

- Development of detailed protocols
- Coordination with design/construction effort
- Establishment of document control procedures and files
- Review of engineering design documents
- Documentation of control system software development/testing
- Master plan update
- Vendor audits
- Development of operating procedures

17.8.4.3 Field execution phase. The field execution phase may include the following activities:

- Execution of the validation protocols
- Calibration of instruments and controls
- Field testing of integrated control system
- Coordination of validation with the construction effort
- On-site construction review

- Turnover package coordination
- Final validation report
- FDA review/inspection

17.8.5 Validation scope

As noted under preliminary phase above (section 17.8.4.1), one of the first tasks to be accomplished during a validation effort is the definition of which elements of the process/facility in question will be validated. The list of elements which may require validation in a pharmaceutical facility includes the following:

Reactors
Mixers/blenders
Columns
Dryers
Tableting equipment
Dust collectors
Air conditioning systems
Clean rooms
Deionized water
Compressed air
Process vacuum
Processing areas
HEPA filters
WFI systems
Clean steam
CIP/SIP systems
Sterile gases
Biocontainment systems
Fermenters/bioreactors
Steam sterilization
Gas sterilization
Purification systems
Sterile filling OPS
Computer operations
Packaging areas
Sterile filters

A typical partial list of high purity water equipment to be validated follows:

RODI, clean steam and WFI installation qualification protocols

- Water tempering exchanger
- Multimedia filters
- Head tank
- Organics filter
- Water softeners
- Reverse osmosis unit(s)
- Ultraviolet sterilization unit(s)
- Deionizer unit(s)
- Regeneration feed package(s) (DI–UV package)
- RODI storage tank(s)
- Heater/cooler(s) for sanitization
- RODI circulation pump(s)
- Chemical sump and pump(s)
- WFI still
- Clean steam still

A typical index for the IQ validation of one of the items on the equipment list follows:

Installation qualification (IQ) for E-2006 water tempering exchanger (equipment installation qualification index) (p. 3 of 17)

1. Equipment documents
   Equipment information summary
   Equipment manual and drawing index
2. Support utilities and ancillary support systems
   Support utilities and ancillary support systems index
   Installation qualification data sheets
3. Equipment instrumentation
   Critical process instrumentation list
   Reference process instrumentation list
4. Change parts reference list
5. Spare parts reference list
6. Safety, caution, warning information
7. Lubricants and chemical additives

Please note that for this piece of equipment only, the IQ was 17 pages.

17.8.6 HVAC environmental certification

The Code of Federal Regulations, Title 21, Section 211.46 requires, in part, that equipment for adequate control over air pressure, microorganisms, dust, humidity, and temperature be provided where appropriate, and that air filters be used when appropriate on air supplies to production areas.

In general, installation protocols and operational protocols should be written for each air handling system used in the area of the facility to be qualified. The scope of documentation and services to be provided for high vacuum (HVAC) generally includes the heating and cooling equipment for the production area of the facility. The protocols here should be written strictly on a system basis including discreet elements such as air handler, condensing units, fans, filters, and ductwork in the same document.

All HVAC specifications and drawings should be continually updated to an as-built state during the construction of the facility, including HVAC control system documentation. HEPA filters should be tested on site prior to, and after, installation, and certified by the testing sub-contractor as meeting the required filtration limits.

Pressure differentials will be verified by using sensors calibrated to a National Bureau of Standards (NBS) traceable standard. Air flow will be measured to verify that each area receives the proper number of air changes per hour. Total generated flow from each will be verified to be within 10% of the specified level, and the distribution of this total flow to different areas will also be verified. These parameters will all be tested and recorded during actual facility operation. The operation of system controls will also be verified during this phase of testing.

The activities described above generally fall under the classification of operational qualification (OQ) of HVAC systems and sub-systems. Once this operation qualification is complete, a performance qualification (PQ) of the HVAC system should be carried out. This would involve verification of the design air classification of the room in question, both with the room at rest and during actual operating conditions.

HVAC SOPs and maintenance procedures should be written and refined during the field qualifications of these systems also. Sanitizing procedures may be appropriate in some cases.

Appendix 1 — Glossary

Acidic — having a high concentration of hydrogen ions; pH less than 7.0.
Activated alumina — porous particulate form of aluminum oxide with the capacity to adsorb fluoride and arsenic.
Activated carbon — a carbon substance used to purify water. The 'active' type of carbon is formulated mainly from coal and lignite. It is used in public water filtration systems and some household water filters. Running water through activated carbon eliminates odors, off tastes, and some impurities. Granular activated carbon, in a good filter, will also remove chlorine and organic chemicals.
ADI — acceptable daily intake. Estimate of a relatively safe daily exposure level to a substance (assumes that substance is not a carcinogen). Ingestion less than the ADI will probably cause no adverse health effect.
Adsorption — adhesion of molecules to another substance. For example, activated carbon adsorbs organic contaminants.
Aeration — a water treatment procedure in which water is combined with air by spraying, bubbling, or trickling. The method removes odors produced by algae, oxidizes iron and manganese, and improves the flat taste of cistern or distilled water. Most gases and dissolved compounds that easily evaporate or vaporize will be removed by aeration. These include hydrogen sulfide, many hydrocarbons, chloroform, and other chlorine compounds. The rate of removal will depend on temperature. At lower temperatures, many hydrocarbons, like chloroform, will be lost slowly.
Air stripping — a type of aeration used to remove volatile organic contaminants.
Alkaline — having a high concentration of hydroxyl ions (OH^-); pH greater than 7.0.
Alum — a chemical substance (aluminum ammonium sulfate) used by public water treatment plants to purify water. The use of alum can significantly increase the aluminum concentration of drinking water.
American Bottled Water Association — issues purity standards for commercially

sold bottled water. The agency helped bring about federal regulations for bottled water by coordinating rules that previously varied from state to state. For more information, write the American Bottled Water Association, 1411 West Olympic Boulevard, Los Angeles, CA 90015.

American Water Works Association—a non-profit, scientific, and educational organization of more than 31 000 people interested in drinking water supply. It issues publications, develops standards, and supports research programs in water works design, construction, operation, and management. American Water Works Association, 6666 West Quincy Avenue, Denver, CO 80235.

Aquifer—a porous rock formation that holds water. The formation can consist of sand and gravel or a layer of sandstone or cavernous limestone. At the bottom of an aquifer is a layer of bedrock which makes the formation fairly watertight. When a well is constructed, it taps the water contained in an aquifer.

Artesian wells—mineral waters naturally forced to the earth's surface by internal pressure. One of the oldest and best known artesian wells is Mountain Valley Water located in Hot Springs, Arkansas.

Asbestos—a mineral that separates easily into long, flexible fibers that resist fire and most solvents. Prolonged exposure to asbestos filters can result in pulmonary fibrosis (asbestosis), emphysema, and lung cancer. Although asbestos naturally occurs in certain water supplies in small concentrations, massive amounts of the mineral are being found in the drinking water of large American cities. It appears that the asbestos cement pipes used to carry the water are disintegrating, thereby releasing asbestos into the water.

Backwashing—a term used to describe the reversible flow of water (usually bottom to top) through the resin tank of water softeners, or in filters, to dislodge the accumulated sediment and mineral particles in the exchange bed or filtration medium.

Bacterial contamination—a category of microbial contaminants consisting of bacteria, viruses, and parasites that can cause waterborne diseases. Public drinking water supplies are tested daily for bacterial contamination. Chlorination is the usual bacterial treatment.

Biodegradation—degradation of organic matter by microbes.

Bottled water—potable water sold commercially. Some is mineral water, some spring water. Other bottled waters may be purified tap water. The four categories of purified water include: drinking water, distilled drinking water, fluoridated water, and purified water.

Brine—salt solution used in regenerating an ion-exchange water softener. Also refers to both surface and ground salty water. Also used as a coolant at temperatures below the freezing point of water.

Bromine—a chemical germ killer similar to chlorine used mainly to disinfect swimming pools.

Calcium—a mineral responsible, in part, for hard water. Studies have shown that the dissolved calcium in hard water limits the internal adsorption of the heavy metals lead, cadmium, and zinc.

Carcinogens — descriptive name given to substances that cause cancer.

Cascade aerators — water treatment equipment patterned after the babbling brook principle. Water is spread out so that it is shallow, then forced over many obstructions to produce turbulence. This allows the water to come into contact with air, which is the cleaning medium. Water is oxygenated and degassed of those gases with a partial pressure below those present in air.

Catchment basis — rain collecting surfaces on house or farm building roofs suitable for water drainage into a cistern.

Chemical solution feeders — automatic pumps used to add treatment chemicals to water at a constant ratio. For the homeowner who has several problems to treat, a chemical feeder is an essential component of his water system. Feeder pumps are usually wired to operate with well pumps and work by adding the chemical solution into the water line between the well pump and pressure tank.

Chlorine — or chlorinated lime is the chemical used to disinfect water. Since its use began in 1908, chlorine has been used to kill bacteria, parasites, and some viruses that cause waterborne diseases and to improve the taste and color of water. More recently, studies have shown that chlorine, when used to disinfect water, interacts with humic acid to form chloroform, a known cancer-causing agent. Chlorine is one of the most commonly used disinfectatnts. Table 17.6 indicates a detailed comparison of disinfectants.

Chloroform — a chemical once used as an anesthetic, but banned by the Food and Drug Administration in 1976 because of its carcinogenic properties. Chloroform can be found in drinking water as a result of the chlorination process.

Chloro-organics — a group of organic chemicals often found in the drinking water. The most common are compounds known as trihalomethanes (THMs), of which chloroform is the most prevalent. Others include vinyl chloride and carbon tetrachloride. All are suspected cancer causers. The Environmental Protection Agency has set 0.10 mg/l as the maximum allowable contaminant level for THMs.

Cistern — a tank, usually underground, for storing rainwater. This water can be used for drinking or it can be a practical source of additional water for laundering and hot water heating.

Clean steam — high purity steam usually made from high purity water. In addition to process uses, clean steam may be used for humidification and for clean seals.

Coagulation — clumping together of microscopic debris. A water treatment plant adds one or several chemical agents to water so that particles that are too small to settle out separately stick together in larger aggregates and then settle out.

Coliform bacteria — in drinking water, indicates that human or animal wastes are entering the water supply, and other harmful contaminants may also be getting in. Coliform bacteria, by themselves are not harmful since they are found in the intestines of humans. The coliform bacterial count in the drinking water acts as a gauge or measuring stick for other forms of bacterial contamination.

Corrosion—a natural process in which acid water gradually eats away the metals found in household plumbing. As a result, lead, copper, cadmium, and other metals that have an adverse effect on health enter the drinking water system. Other factors in water that can induce corrosion are oxygen concentrations, electrical conductivity, and temperature.

Corrosion control—water that is highly acidic is corrosive. Acid water can be identified by measuring pH. The lower the pH number under 7, the more acid the water. Corrosion can be estimated or reduced by raising the water pH with potassium carbonate or potassium hydroxide. Corrosion problems can also occur if water is highly alkaline. In this case, running the water through a water softener or an ion-exchange medium followed by reverse osmosis treatment will solve the problem.

Demineralization—see *Deionization units* (see also section 17.3).

Deionization units—water treatment equipment that works on the principle of substituting one ion for another in water. Water softeners are deionization units, with sodium ions substituting for calcium and magnesium ions in water. Other deionization units use hydrogen to replace unwanted element ions in water. These units have two ion exchange materials, one to remove positively charged ions and another to remove negatively charged ions. Water treated by deionization will contain few, if any, minerals. Deionization will remove copper, fluoride, hardness, soluble ion, nitrates, silica and silicates, sodium, sulfate, total solids, arsenic, and selenium. It will not remove bacterial or organic chemicals.

Discharge well—used to pump contaminated water out of an aquifer's supply area. Water is removed from the well and dumped elsewhere. Frequently dumped water is considered a hazardous waste and has to be treated as such.

Disinfection—the destruction or killing of disease-causing bacteria. Water can be disinfected by chlorine, ozone, iodine, bromine, silver, and ultraviolet light (see Table 17.6).

Distillation—a process whereby water is changed to vapor through heating and is then allowed to condense back into liquid form. Distilled water usually contains no solids, minerals, or trace elements (see section 17.3).

Downspout filters—a water filter placed in the downspouting to clean the rainwater before it enters the cistern. These filters need to be maintained regularly or they will create more pollution than they remove.

Dowsing—tthe ancient technique of locating underground water using a divining rod. Today the forked willow branch has been replaced by a fibreglass rod or a slim piece of copper tubing. As the dowser walks with the rod held in front of him, the presence of undergound water may be indicated when the rod dips downward. The dowsing phenomenon is believed to result from small magnetic field vibrations.

Effervescence—a term used to describe the natural bubbliness of certain bottled drinking water. The bubbliness is produced by the escaping carbon dioxide in the water. Only naturally occurring bubbling water can be sold as effervescent; bottled waters pepped up with carbon dioxide gas after leaving the spring do not fulfill the effervescent claim.

Effluent — the discharge resulting from waste water treatment plants (or any process). The effluent usually contains organic human matter, some trace elements, and organic and inorganic compounds from industrial sewers. Effluents are major point sources of pollution.

Electrodialysis — (see section 17.3).

Endotoxin — a substance given off by dying bacteria and known to cause fever when injected into a person's blood stream. Endotoxins can remain in drinking water after normal treatment. The health effects of endotoxin residue in water are not yet fully known, but are to be avoided.

Environmental Protection Agency (EPA) — the federal agency, created in 1970, responsible for monitoring pollutants in our environment and setting acceptable limits, for developing methods to clean the environment, and for enforcing the laws and standards it has developed. Environmental Protection Agency, 401 M Street, S.W., Washington, DC 20460.

Epidemiology — branch of public health or medicine that studies the occurrence of diseases in human populations. Differences in the incidence of adverse health effects among different populations may suggest the causes of disease.

Filter cake — for water purification–the scum that accumulates inside a filter during water treatment. When filter cake is thick, it has to be removed along with the top inch of the filter medium. (Filter cakes for processes may be the desirable layer).

Filtration — passing water through a porous medium that traps solid material in or on top of the medium.

Floc — material in coagulated water that has clumped together. When alum is used as the coagulant, the floc looks like small, stringy dirt strands.

Fluoridation — addition of fluoride to the water supply to reduce the incidence of dental caries.

Fossil water — underground water trapped millions of years ago and completely encased in a non-porous rock aquifer. Fossil water replenishes itself very slowly, so it could be completely used up before it is replaced. Tucson, Arizona, is an example of a city using fossil water as its drinking water source.

Fractional distillation units — water treatment equipment that will remove chloroform and organic chemicals as well as debris, bacteria, minerals, and other contaminants. The basic components of the distiller are the heater, boiling chamber, vapor rising column, and condensing chamber. Distillers operate by heating water until it turns to steam, then condensing the steam back into water. Distilled water is pure, but has no taste.

Free chlorine residual — uncombined chlorine remaining in water following the chlorination process. The free chlorine acts as a disinfectant during water distribution.

GMPs—Good Manufacturing Practices. Initially, loosely worded FDA standards that govern the equipment, facilities, and procedures used in producing pure grades of water. This has presently been extended to tighter definitions and made a requirement of most pharmaceutical, medical, and biotechnical operations. It is now an option for all types of industrial operations.

Grain—measurement of weight 0.0648 g or 0.000143 lb.

Grains per gallon—measurement of water hardness. 1 grain per gallon is approximately 17 ppm.

Granular activated carbon filters—water filters containing activated carbon (a form of charcoal treated by high temperatures and steam in the absence of oxygen as its medium). The filters are very effective in removing chloroform, chlorine, some pesticides, and other organic chemicals, as well as bad taste and odor. They can reduce the concentration of heavy metals, but will not remove fluoride, nitrates, or asbestos fibers. These filters work by adsorption, whereby the surface area of the activated carbon grabs and holds organics and other chemicals. Because bacteria and other organisms grow easily on the carbon surface, the filters must be monitored closely for signs of filter failure and the carbon should be replaced on a regular basis.

Groundwater—water that collects below the soil level and is contained by the rock formation or aquifer underlying the surface. Rain is the source of all groundwater. Most of this water is tapped by wells.

GWSS—Groundwater Supply Survey. The EPA's early 1980s study of groundwater contamination.

Half-life—time required for half of the original atoms of a radioactive element to decay. (Also, time required for half of the original molecules of a non-radioactive compound to decay or degrade.)

Halogenated organics—chemical compounds containing carbon and chlorine, or less frequently, bromine or iodine.

Hard/soft water—these terms refer to the amount of dissolved materials contained in water. Hard water has large amounts of minerals in it, while soft has very little. Calcium and magnesium are the elements largely responsible for making water hard. Water hardness is measured according to how many grains of dissolved materials are found in 1 gallon: water is slightly hard if it contains 1 to 3 grains per gallon; moderately hard if it contains 3 to 6 grains per gallon; hard when it contains 6 to 12 grains per gallon; very hard if it contains over 12 grains per gallon. Studies indicated that drinking hard water may be beneficial to your health, but additional research is needed in this area.

Health advisory level—estimate of the concentration of a hazardous substance below which there will probably be no observed acute health effect (assuming the substance is not a carcinogen). The health advisories were issued in 1985 by the EPA as guidelines for public health officials.

Heavy metals—chemically defined as those metals having a high specific gravity or density. Some heavy metals include calcium, lead, and mercury.

A special characteristic of heavy metal chemicals is their strong attraction to biological tissues and, in general, the slow elimination of these chemicals from biolological systems. Heavy metals are persistent; once in the system, they remain for relatively long periods of time, continuously inducing their toxic effect.

High purity water—a recent term defined herein as a pure water, treated to be high purity water, using state of the art equipment and having a resistivity of greater than 0.1 MΩ.

Hippocrates' sleeve—an ancient form of water treatment recommended by Hippocrates, the father of medicine, in which a cloth bag was used to strain impurities from water.

Hydrogen sulfide—a gas that can form in well water and produce a 'rotten egg' odor. Certain strains of bacteria feed on the sulfate (sulfur) found in the water and produce the gas as a metabolic byproduct. Hydrogen sulfide is very acidic and can cause water to corrode in pipes, filters, water tanks, and even concrete. In very high concentrations, the gas is flammable and poisonous; it can cause nausea, illness and, in extreme cases, death. The maximum contaminant level for this gas is 0.05 mg/l.

Hydrological cycle—or water cycle. A sequence of phenomena showing how water moves in a never-ending revolution. Water in the oceans continually evaporates, putting water vapor into the atmosphere. Precipitation (rain, snow) puts the water on the ground where it can become surface water (lakes, ponds, puddles), or surface runoff (streams and rivers), or seep into the ground and be stored as groundwater. Eventually the water will travel back into the ocean, to begin the cycle again., The exchange between the earth and atmosphere is accomplished by the heat of the sun, the winds, and the pull of gravity.

Hypochlorous acid—powerful disinfectant formed when chlorine is added to water.

Inorganic solutes—a category of water 'contaminants' comprised of the following materials; barium, cadmium, chromium, lead, mercury, silver, sodium, beryllium, cobalt, copper, magnesium, manganese, molybdenum, nickel, potassium, tin, vanadium, and zinc. While many of these minerals can be quite healthful in small amounts, others can be very toxic in small amounts. Furthermore, some substances that are beneficial in small amounts can be dangerous in large amounts. To complicate matters further, some adverse effects are caused by a combined total intake from water, food, and air.

Iodine—a chemical germicide effective in killing bacteria, viruses, cysts, and other contaminants in water. Iodine can be used as a short term means of water disinfection, but it is not recommended for use by pregnant women and not considered safe for long-term use.

Ion exchange—a common chemical reaction where one ion is substituted for another in solution. Water softeners work on the principle of ion exchange, as do deionization units. The salt nuggets or crystals the homeowner regularly adds to his water softener consist, in part, of sodium ions. When water is softened, its passes through an ion-exchange material, usually salt, where the calcium and magnesium ions are removed and replaced by sodium ions. The softener will also remove poisonous heavy metals like lead and cadmium and can remove most of any radioactivity present.

Leachate — the contaminant that forms beneath municipal solid waste landfills when there is enough rainwater left over after evaporation to carry dissolved wastes into the ground, eventually reaching the underground water beneath the landfill. Leachates usually contain measurable amounts of trace metals and organic acids. Water contaminated by landfill leachate usually has a foul odor.

Magnesium — the mineral, along with calcium, responsible for making water hard. Studies suggest that the large amounts of magnesium found in hard water can lower death rates from heart disease. However, water with magnesium concentrations of 700 mg/l or more can act as a laxative. Presently, there is no limit for magnesium concentration in drinking water.

Measuring elements in water — water hardness is measured in grains. A grain is defined as the average weight of a single, dry grain of wheat which is 1/7000 of a pound. Very hard water containing 12 grains of elements per gallon weighs 12/7000 of a pound. Other elements and impurities in drinking water are measured in terms derived from the metric system. Parts per million (ppm) and parts per billion (ppb) describe ratio measurements. For example, 100 parts per million (ppm) of salt in water means that there are 100 parts of salts for every million parts of water. One part per million (1 ppm) is equal to 1 milligram per liter (1 mg/l) and 1 part per billion is equal to 1 microgram per liter (μg/l). A liter is equal to 1.056 quarts.

MCLG — Maximum Contaminant Level Goal. Maximum level of a contaminant that would result in no known or anticipated adverse health effect over a lifetime of exposure. The MCLG is a suggested level set by the EPA as a guideline for water utilities. It used to be called recommended maximum contaminant level (RMCL).

Membrane filters — screen filters made of paper-like material that can remove bacteria and other particles from water.

Membrane filter test — one of two methods used to determine the number of bacteria (coliform) contaminating the water. In using the membrane filter technique, a positive coliform test is obtained if more than one coliform bacterium is filtered from 100 ml of water (see also *Multiple tube fermentation technique*).

Micron — one millionth of a meter (μm). Used as a measure of pore size in depth and screen filters.

Microwatt second per cm^2 — unit of UV radiation.

Millirem — one thousandth of a rem (roentgen equivalent man). A measure of the likelihood of damage that radiation may inflict on a body or organ.

Mineral water — a term to describe the natural occurrence of large amounts of elements in ground or well water. Mineralization of ground water can occur from the slow erosion of rock in the aquifer and can cause the water to taste bitter. Highly mineralized water is usually not suitable for drinking. Natural mineralization increases the depth of the well, and may eventually reach a point where the water is no longer drinkable.

Mixed-bed ion exchange — deionization method. The process removes much of the total dissolved solids and produces very pure water without the addition of sodium. Similar to, but different from softening.

Multimedia filters — water filters that combine several types of material to remove contaminants and pollutants. Conventional multimedia filters have the largest particle medium (like gravel) at the bottom, working upward to the top layer, which is made of the finest particle material (like silica sand). In this way, the large particles support the finer ones, keeping them from slipping down the drain. Inverted multimedia filters have the largest particle medium at the top, working downward to the smallest particles. The arrangement allows the larger coarse material to be filtered out of the water before reaching the smaller media particles. The use of inverted multimedia filters provides 'in-depth' filtration, where the filtration process takes place throughout the entire depth of the column. This is in contrast to conventional multimedia filters where the top few inches have to filter out the entire contents of the water.

Multiple tube fermentation technique — one of two methods used to measure bacterial contamination (cf. *Membrane filter test*) of water. This procedure samples a small volume of water (loosely 1.0 mm) and reports the coliform bacteria count as the estimated number ((most probable number, MPN) of coliforms in 100 ml of water. The test is based on gas production (fermentation) resulting from bacterial growth in several test tubes. If the estimated coliform count exceeds 1 bacterium per 100 ml, the water is considered unsuitable for drinking.

Mutagen — substance that causes changes in the genetic character of cells.

National Association of Water Companies — an organization representing 225 privately owned water companies. The association conducts research and keeps members informed in economic, legal, and regulatory developments. To contact, write National Association of Water Companies, 1019 19th Street, N.W., Suite 1110, Washington, DC 20036.

NBS — National Bureau of Standards

NIH — National Institute of Health

Natural water filtration systems — are being developed by scientists from around the world. Because each of the systems involves the use of a naturally occurring substance or product, they represent actual steps forward in water treatment. Some of these include:

Aquatic plants are being investigated because some have been discovered to help purify the waters in which they live. Water hyacinths, a much-despised water-weed that clogs canals and waterways, has the ability to absorb heavy metal and organic chemicals from the water. Canadian scientists are using cattails and bulrushes to treat municipal waste waters.

Chitin is a substance found in the hard covers of shellfish. Scientists have found a way to process chitin into chitosan, a white powder that effectively purifies water. It is even capable of removing radioactive wastes from water discharged by nuclear plants.

Elephant fish are being used in West Germany to monitor heavy metal concentrations in drinking water. The fish, which have the ability to emit small electrical impulses, give out fewer of these signals when the water contains heavy metals.

Filters made of fibers from shredded coconut husks are being used in Third World countries to remove suspended solids from the water. The water is then passed through a filter made of burnt rice husks in order to remove any residual turbidity.

Nitrates/nitrites/nitrosamines—a related trio of organic compounds known to cause infant illness (methemeglobinemia) and suspected of causing cancer. In water, the major sources of nitrates are septic systems, animal feed lots, agricultural fertilizers and manured fields, industrial waste waters, sanitary landfills, and garbage dumps. Once inside the body, the relatively harmless nitrates are changed into nitrites. In adults, the nitrites can combine with other substances in the stomach to become nitrosamines. By distilling their drinking water supply, homeowners can remove nitrates.

Non-point source pollution—a contaminant that comes from many different non-specific sources. For example, both fertilizer and pesticide runoffs from farmlands, contaminating streams or wells would be considered non-point source pollution.

Non-purgable total organic carbon (NPTOC)—a term devised by the Environmental Protection Agency when it tested the efficiency of granular activated carbon water filters. NPTOC is a substance that represents the bulk of organic compounds found in drinking water. These compounds are mostly natural materials that have a high molecular weight.

NOMS—National Organics Monitoring Survey, the EPA's 1976–77 study of organic contamination.

NPTOC—Non-purgeable total organic carbon. Measure of the total amount of non-volatile organics in a water sample (see above).

NORS—National Organics Reconnaissance Survey, the EPA's 1974–75 study of organic contamination.

Organic compounds—are natural or synthetic substances based on carbon. Organic compounds can become part of the water supply through water treatment methods, from industry, sewage treatment plants, runoff, and from spills and accidents. Some organic pollutants include: chloro-organics (chloroform, carbon tetrachloride), polychlorinated biphenyls (PCBs), polynuclear aromatic hydrocarbons (formed from combustion using fossil fuel and includes fuel mixtures, crankcase oil), herbicides, and pesticides.

Organization for Economic Cooperation and Development — an international organization comprised of 24 countries, 19 west European governments, Canada, the United States, Japan, Australia, and New Zealand. Its purpose is to help stimulate economic cooperation between member countries for economic growth, expansion of world trade, and coordination of aid to less developed areas in an effort to achieve world prosperity and to contribute to peaceful and harmonious relations among the peoples of the world. Contact by writing the Organization of Economic Cooperation and Development, Publication and Information Center, 1750 Pennsylvania Avenue, N.W., Washington, DC 20006.

Oxidizing filters — offer a convenient, but limited, one step solution to iron problems. Oxidizing filters work by converting dissolved iron into insoluble iron by passing the iron-heavy water through a manganese-treated material. As the iron forms rust particles, it is filtered out by a granular material in the filter. However, these filters do not work on iron bacteria that can clog the filter and mineral beds. Oxidizing filters must be backwashed and rinsed regularly to remove precipitated iron from the filter bed.

Ozone — a form of oxygen used widely throughout Europe to disinfect water. This unstable gas is an excellent germicide, but it is expensive, it must be produced electrically as needed, and it cannot be stored.

pH — this symbol represents the degree of acidity or alkalinity of compounds. pH measurement is expressed by a number from 1 to 14; 7 is the neutral point. Values from 7 to 1 are increasingly acid, while those from 7 to 14 are increasingly alkaline. The most desirable range for water pH in the home is from 8 to 8.5.

Pathogen — organism that causes disease in humans.

PCB — polychlorinated biphenyl. Toxic synthetic organic contaminant.

Point source pollution — a contaminant that can be traced to an individual source. For example, a factory dumping chemicals by pipe directly into a river.

Polyphosphates — a generic term designating food-grade phosphate chemicals used to treat iron problems. Polyphosphates are sold under the names Zeotone, Micromet, or Nalco M-1. When added to the water, they stabilize and disperse the iron so it cannot oxidize when exposed to air. This treatment works only for dissolved iron and is effective in a pH range of 5.0 to 8.0.

Precipitation and filtration — two water treatment procedures that combine to produce the most thorough answer to iron problems since both dissolved and oxidized iron are removed. Chlorine, potassium permanganate, or carbon dioxide is added to the water to oxidize the iron. The water is then stored in retention basins so the rust particles can drop to the bottom. Water is then drawn off the top and filtered through activated carbon filters to remove all insoluble iron. This combination is often used to remove other contaminants when a method can be found to precipitate them.

Pyrogens—fever producing agent of bacterial origin. A group of substances thought to be polysaccharides of microbial origin that produce an increase in body temperature when injected into humans and some animals. Have also been known to cause a severe chill. Pyrogens are often thermostable and are filterable with certain types of filtration.

Radioactive—emitting charged sub-atomic particles and/or radiation.

Reclaimed water—reused or recycled water.

Recycled water—water used for the same purpose over and over again.

Regeneration—the process of cleaning deionization and other equipment. Regeneration involves the removal of contaminant ions from the resin bed replacing them with the type of ions originally in the resin bed. The procedure involves flooding the resin with an extremely concentrated solution of the original ion.

Residual chlorine—a term applied to the additional amount of chlorine public water systems place in water to kill any bacteria that may be picked up as the water passes through the mains and service lines to homes.

Resins—the ion-exchange materials found in deionization units and water softening units. Examples of ion-exchange materials are zeolite and resin beds.

Reverse osmosis—a method of treating water that removes turbidity, particulate and colloidal matter, ionized and non-ionized dissolved solids, bacteria, viruses, pyrogens, aromatic hydrocarbons, most pesticides, and asbestos. It will not remove simple compounds like chloroform and phenol. Reverse osmosis is a type of pressurized filtration system. The water to be treated is made to flow over a semi-permeable membrane that may be similar in appearance to cellophane. Under pressure, often hundreds of kilograms, the water is forced through the membrane, leaving behind the impurities which are unable to seep through. Reverse osmosis units consist basically of the membrane and some filters to remove larger particles of incoming water. The energy needed to operate the unit is pressure, normally furnished via high pressure pumps.

RMCL—recommended maximum contaminant level set by the EPA. Now known as the maximum contaminant level goal (MCLG).

RO—see *Reverse osmosis*.

RODI—a water treated with reverse osmosis followed by deionization (RO and DI), to produce high purity water.

Roof washers—a tank, barrel, or garbage can that is connected to downspouting to catch the dirty water from the first part of the rain. Once the container has filled, water will overflow into a pipe that is connected to the cistern.

Sand filters—water filters that use sand as the medium for cleaning the water. Sand filters can remove clay, silt, colloids, and microorganisms. If the pores in the filter are small enough, bacteria and viruses can be strained out. Most sand filters are used as an initial treatment for pond or other natural waters.

Sanitization—generally used to kill most bacteria, often by chemicals and heat. In high purity water systems the method is to heat the water to 82°C (180°F) and maintain at 82°C (180°F) for 30 min or more.

SDWA — Safe Drinking Water Act. 1974 Congressional act governing the quality of drinking water, amended in 1986.
Sediment — substances like dirt in water that settle or filter to the bottom.
Soft water — water with low concentrations of calcium and magnesium ions. 'Naturally' soft water has relatively few minerals; 'softened' water has had the calcium and magnesium removed.
Softeners — a device used in the home and in commerce or industry to artificially soften water. Most home water softeners work on a principle called ion exchange, whereby calcium and magnesium ions in water are removed and replaced by sodium ions. Fully automatic softening equipment has a timing mechanism that starts and controls the various steps of the softening process. All the homeowner has to do is add salt regularly.
Soil and Health Foundation — a nonprofit organization which provides funds for research and disseminates information to help people grow more of their own food, better their diets, to upgrade their drinking water, and to improve their health. For a nominal fee, the foundation can test your water for trace element and heavy metal content. For a sample bottle and instructions, write to the Soil and Health Foundation, 33 East Minor Street, Emmaus, PA 18049.
Spray aerators — water treatment equipment that works by shooting the water high into the air, creating a fountain effect.
Springs — groundwater that flows to the earth's surface through a crack or fault in the earth.
Sterilization — elimination of all microbial life, usually by high temperature heating to 100–150°C (212–300°F) for a set period of time.
Tangential flow — technique of filtration used in RO water treatment in which the water flows across the upstream side of the filter, sweeping off accumulated debris, and only part of the water passes through.
TDS — total dissolved solids. Measurement of all substances in solution in a water sample.
Thermocompression — see section 17.3.
Titration process — a method used by laboratories to measure the amount of concentration of a substance in solution. Water hardness testing kits for homeowner use employ a simplified version of the titration process to count the grains of hardness per gallon (ppm).
Trace elements — naturally occurring metallic substances found in minute quantities. Our knowledge about the human need for trace elements is sparse because, until the beginning of the 1970s, scientists did little research in this field. Current information indicates that the elements chromium, cobalt, copper, magnesium, molybdenum, selenium, tin, and zinc are essential to human nutrition in small amounts. Barium, silver, beryllium, manganese, nickel, and vanadium are believed to be non-essential, but future research could change this outlook. Cadmium, lead, and mercury are harmful. Most of the trace elements we consume come from food, but the metals can be present in water in a form more readily absorbed by man. Metals in water

can come from many sources: the soil, industrial waste, municipal sewage and landfills, the water distribution system, and household plumbing.

Trihalomethanes (THMs)—a group of organic compounds produced in water as a result of the chlorination process. These substances form when chlorine interacts with natural humic acid or with algae. Chloroform is the most well known compound in this group. All THMs are known to cause cancer in laboratory animals, and are suspected of being cancer-causing agents in humans (see also *Chloro-organics*).

Turbidity—a term used to describe cloudy water. Undissolved materials, such as clay, silt, sand, organic and inorganic chemicals, plankton, and microorganisms combine to give water its cloudy appearance. Turbidity sometimes indicates the presence of sewage, industrial waste, or asbestos. Turbid water can provide an optimum medium for harmful bacteria; therefore, turbidity in water must be reduced to nearly zero for chlorine to be totally effective in destroying germs. Turbidity can be removed by filtration or coagulation.

USP—United States Pharmacopoeia. Independent organization that sets standards for the pharmaceutical industry.

Ultraviolet disinfection—bacteria and certain viruses can be killed by exposing the water to a quartz–mercury vapor lamp that emits ultraviolet rays of the proper wavelength. A home ultraviolet unit consists of one or more germicidal lamps sealed inside a steel cylinder. The cylinder is connected to the water supply line so the untreated water goes into one end of the cylinder, passes over the UV lamp, and then goes out the opposite end into the water line. Ultraviolet light does not work effectively with cloudy or dirty water or with water having a high iron content. Moreover, bacterial spores and some viruses are fairly resistant to ultraviolet treatment. Both the UV equipment and energy needed to run it are expensive in comparison with other treatment methods. It is especially effective where water passes through a bed like a carbon bed or a DI unit, and on pure water.

UV radiation—ultraviolet light. A form of energy that carries more energy than visible light and can destroy bacteria (see section 17.3).

Vapor compression—see section 17.3.

Virus—smallest and simplest type of microbial life. All viruses need hosts in order to reproduce.

Volatile—evaporating relatively quickly.

Water feed wells—a water source that can be built into ground.

Bored wells, constructed by a hand-driven or power auger, are deeper and smaller in width compared with dug wells. They are suited to sandy or soft-rock areas.

Drilled wells can be constructed through the hardest rock, to depths of up to 1000 ft. Drilled wells are constructed with cable tool or rotary equipment.

Dug wells usually go no deeper than 50 ft and range from 3 to 20 ft in diameter. They are more suited to geological areas that are sandy or have soft rock such as fractured limestone. Their major drawback is that they frequently bottom out during dry periods.

Driven wells are the cheapest and quickest way to get water in a sandy area. Well construction is accomplished with a pointed object being forced into the ground. These wells usually go deeper than 50 ft and range from 3 to 20 ft in diameter. They are more suited to geological areas that are sandy or have soft rock such as fractured limestone. Their major drawback is that they frequently bottom out during dry periods.

Jetted or hydraulic well drilling is most successful in sandy oils, but is no good in hard rock formations. The well hole is made by the force of a high velocity stream of water. The depth of these wells can reach 1000 ft with a diameter ranging from 4 to 12 in.

Water filters—water treatment equipment that, depending on the type of filter, can remove some turbidity, odor, taste, bacteria, corrosion, chloroform, chlorine, some pesticides, and other organic chemicals.
Water hardness—a term to describe water containing large amounts of calcium and magnesium (see also *Calcium; Hard/soft water; Magnesium*).
Water softener—device that removes inorganic ions such as calcium and magnesium and replaces them with inorganic sodium ions.
Water table—the upper limit of the portion of the ground wholly saturated with water. The water table separates two different layers of water in ground. The upper layer contains only a small amount of water for plant life; the bottom layer is the zone of saturation representing the water table. In between the two layers is a zone of aeration, where the spaces between the soil and rocks hold air. During periods of heavy rain the water moves into the zone of aeration and the water table rises. In drought, the level of water in the saturation layer drops, indicating a corresponding drop in the water table.
WFI—see *Water for injection*
Water for injection—presently, this is the highest grade of high purity water and is acceptable to the FDA for use in the preparation of human injectables. It is almost always made in a still by evaporation, and most often in a multistage unit with specially designed baffles to prevent carryover. Feed to the still is often high purity water. The product is subject to severe bioburden testing.

Appendix 2 — Associations

A variety of national associations provide information about drinking water and its quality. Manufacturers, government agencies, water quality professionals, and interested citizens have banded together in such groups to set industry standards, to coordinate research, and to publish relevant data for use by members and by the public. The following associations offer services to the public:

American Institute of Hydrology, PO Box 14251, St. Paul, MN 55114.
This association of professional hydrologists and hydrogeologists (who specialize in the flow of groundwater and the structure of aquifers) establishes certification standard and procedures. A registry of professional hydrologists and hydrogeologists is published annually.

American Water Resources Association, 5410 Grosvenor Lane, Suite 220, Bethesda, MD 20814.
Engineers, scientists, and businesses interested in water resources are members. A variety of publications are available.

American Water Works Association, 6666 W. Quincy Ave, Denver, CO 80235.
The AWWA's membership is composed primarily of professionals in the drinking water supply and waste water treatment fields. A wide variety of publications, films, slide shows, and 'Waternet', a computerized water-information service are primarily geared to full-time water industry workers; they are also a good source of new information about standard procedures, as well as new developments in the supply, treatment, and distribution of drinking water.

Association of State and Interstate Water Pollution Control Administrators, 444 N. Capital St., N.W., Suite 330, Washington, DC 20001.
The association is a national professional organization made up of state administrators of the Clean Water Act. General information about the quality of the nation's water and progress toward meeting the goals of the act are available.

Association of State Drinking Water Administrators, 1911 Fort Meyer Dr., Suite 803, Arlington, VA 22209.
Information and publications about the Safe Drinking Water Act and the states' administration of the Act are available from ASDWA.

Clean Water Action Project, 317 Pennsylvania Ave., S.E., Suite 200, Washington, DC 20003.
The project, which started as an outgrowth of Ralph Nader's water pollution taskforce, is a citizen's group working for strong water quality laws. Information on drinking water quality and the health effects of toxics is available.

International Association for Great Lakes Research, c/o Institute of Science and Technology, University of Michigan, Ann Arbor, MI 48109.
Scientists, engineers, and others interested in ongoing research of the quality of the Great Lakes are members. The association publishes a quarterly journal and sponsors an annual conference.

International Bottled Water Association, 113 N. Henry St., Alexandria, VA 22314.
The association's members are water-bottling companies, marketers of water products, and equipment manufacturers. Publications relating to bottled water, its marketing, and quality are available from IOWA.

International Water Research Association, University of Illinois, 208 N. Romine, Urban, IL 61801.
Individuals interested in the development of water resources are members. Publications include a magazine, books, and conference proceedings.

Lake Erie Cleanup Committee, c/o John Chascsa, 3568 Brewster Rd, Dearborn, MI 48120.
This association is composed of 59 citizens' groups from the USA and Canadian provinces surrounding the Great Lakes. Their primary goal is the restoration of the quality of Lake Erie's water. The group is affiliated to the National Wildlife Federation.

Lake Michigan Federation, 8 S. Michigan Ave., Suite 2010, Chicago, IL 60603. The federation conducts conferences, workshops, litigation, and public education programs relating to Lake Michigan. Organizations, individuals, and educational institutions are members. The group encourages citizen participation in public policy decisions that affect water quality.

National Demonstration Water Project, 1111 N. 19th St., Suite 400, Arlington, VA 22209.
NDWP is an alliance of local groups interested in the development of adequate, affordable drinking water supplies and waste water disposal systems in low-income rural areas. Publishes the *Rural Water News*, a bi-monthly publication.

National Water Supply Improvement Association, PO Box 1344, Springfield, VA 22151.
The desalination of seawater and the recycling of waste water are the primary focuses of this group of researchers, businesses, and individuals. Publications include a monthly newsletter and conference proceedings.

National Water Well Association, PO Box 16737, Columbus, OH 43085.
Aquifers and the use of groundwater in community drinking water systems are the predominant subjects of NWWA publications. The association is made up of water equipment manufacturers and water-supply professionals.

Passaic River Coalition, 246 Madisonville Rd, Basking Ridge, NJ 07920.
The Passaic river, which flows through southern New York and northern New Jersey, supplies drinking water to three-and-a-half million people. The coalition is made up of organizations, foundations, municipalities, and individuals concerned about the river's quality. A variety of publications relating to the river are available.

Water Pollution Control Federation, 601 Wythe Street, Alexandria, VA 22314.
The Federation is a non-profit educational organization concerned with the treatment of municipal and industrial waste water. Local associations that are WPCF members are located in most states, and in five Canadian provinces. The Federation publishes books detailing all phases of waste water treatment, a technical journal, a newsletter, and public information pamphlets summarizing the disposal of sewage and hazardous watses.

Water Quality Association, 4151 Naperville Rd, Lisle, IL 60532.

This association of water purification equipment dealers and manufacturers sets standards and handles consumer complaints regarding water-cleanup equipment. Pamphlets describing commonly encountered water quality problems and other water quality information are available.

Water Resources Congress, 3800 N. Fairfax Dr., Suite 7, Arlington, VA 22203. The group, which is composed of state and local government agencies as well as agricultural, industrial, labor, and civic organizations, promotes the improvement and development of the nation's rivers, harbors, and lakes. A variety of publications are available.

References

1. Brown, F.S. and Roettger, L.J. RO membrane offers myriad of applications, *Water Conditioning and Purification*, Publicom, 1987.
2. Dickson, J.M., Fundamental aspects of reverse osmosis, in *Reverse Osmosis Technology: Applications for High-Purity Water Production*, ed. Parekh, B.S., Marcel Dekker, NY, 1988, chap. 1.
3. Ganzi, G.C. and Parise, P.L., The production of pharmaceutical grades of water using continuous deionization post reverse osmosis, *J. Parent. Sci. Technol.*, **44**, 231 (1990).
4. Giorgio, R.J., Considerations in the design of hot circulating water for injection systems, *Pharm. Technol.*, Dec. 1978.
5. Meyrick, C.E., Practical design of a high purity water system, *Pharm. Eng.*, **9**, Sept./Oct. (1989).
6. Parekh, B.S., Get your process water to come clean, *Chem. Eng.*, January, 1991.
7. Parise, P.L., Parekh, B.S. and Waddington, G., *Ultrapure Water*, November, 1990.
8. *Technical Primer CG301*, Millipore Corporation, Bedford, MA, 1986.
9. *Technical Primer CR415*, Millipore Corporation, Bedford, MA, 1987.
10. United States Pharmacopoeia XXII, U.S. Pharmacopoeial Convention Inc., Mack Publishing, Easton, PA, 1990.
11. Water Refining Industrial, *Commercial and Industrial Products Catalog*, 1985.

Bibliography

CGMP, June 1976, Federal Register Part 212.

DiSessa, P.A., The generation of pharmaceutical grade water by distillation, *Pharm. Technol.*, November, 1980.

Fishkin, S., High purity water plant, *ISPE, Aseptic Facility Design Seminar*, February, 1990.

Kaloyanides, J., Anatomy of a pure water system for biotech facilities, *ISPE*, November, 1988.

18 The facility design process

S. LITMAN

18.1 Introduction

A 'clean room' can be defined as a space in which research, production, or other activities can occur. In order to afford safety to the personnel using the space and/or to provide a contaminant-limited or contaminant-free environment, certain specific space plans, finishes, environments, equipment, and services must be provided to support the desired activities.

There are many standards and guidelines that are important to the design of clean rooms. For each design situation, the needs and requirements of the situation must be assessed and the appropriate clean room design elements implemented to obtain the proper conditions required by regulations, good manufacturing practices, or good common sense.

Since FDA, NIH, and building code regulations are constantly evolving documents having varied interpretations and levels of interpretation by enforcing officials, a guide for the design of clean rooms can contain specifics, generalities, and other criteria that can conflict with each other. Additionally, facility constraints and client restrictions with respect to timeframe and budget can have an influence on the quality or level of detail of the clean room design features.

In this chapter, examples will be given of some of the generally accepted and commonly used design procedures and characteristics of clean room design, focusing on the nature of the physical environment in and around the clean room.

18.2 Conceptualization

It is not possible competently to design a space without the full understanding of all design parameters. The ideal time for this to occur is immediately after a 'client kick-off meeting'. During this first working meeting, the client begins to inform the architects/engineers of the design conditions, physical, environmental, personnel, and process requirements. The depth and degree of programming already performed by the client (or a design consultant) becomes the basis for further research or at the very least verification of the previously acquired information. Presumably, a process and instrumental diagram (P&ID), PFD, or process conceptualization diagrams have been developed, becoming the backbone of the physical design of the clean room.

Before the process flows and equipment relationships become too firmly established, the way in which the process fits into a space must be examined. The following is a sample of some of the physical questions that must be addressed before design of the clean room itselt can commence.

- Is the space allotted either a fixed entity within an existing facility, a new space (as large as the client's budget allows), or a combination of the two (addition and alteration)?
- Has sufficient information been developed in the following areas: time and material studies, methods of manufacturing the products, ways in which products or work-in-process products are transported from point A to point B, sizes of major tanks and equipment including utility requirements, quantities and methods of supplying utilities (bottles, tanks, central systems) the number of people, and shifts of work?
- Has the desired throughput been verified by marketing analysis or some other method of need determination; is there any known need for the product at this time or may the process and space be dismantled after a period of time implying a need for flexibility in the design of the facility? Will there be campaigning of different products or will the space be dedicated?
- Are there any special interior environmental conditions required to sustain the process?
- Are their multiple processes that can be grouped together environmentally?
- What are the containment requirements with respect to protection of product and/or protection of people?
- Are there any biological, chemical, radioactive, explosion issues that need to be addressed? Are there any biological classifications known or desired at this time; is the space 'classified' (with respect to the heating, ventilation, and air conditioning (HVAC) design)?
- Have any deadlines for process start-up been established and if so, how feasible do they seem and how do they affect design time and construction schedule? Do these deadlines impose any restrictions on selection of facility materials or process equipment?

Before the facility design can start, the answers to the above questions as well as many others relating not only to the process but to client goals and requirements need to be answered or at least brought into the design arena for discussion and answers. This information in conjunction with other elements of concern such as budget, space availability, aesthetic issues, and political importance of the project must all be available as early as possible to provide the architect with the best information at the program and conceptual development stage. If the client has already developed a conceptual floor plan or a preliminary program, they should at least be verified by the architect and process engineer to give everyone involved in the design of the space the same understanding of all information available.

18.3 Spatial relationship diagrams

The spatial relationship diagram is one of the most important conceptualization tools the architect uses to shape the development of the pharmaceutical, biotech, or any other type of facility. These elegantly simple 'bubble diagrams' can illicit a greater response than a long, detailed verbal description due to the clear depiction of the way in which spaces relate to each other.

Spatial relationship diagrams are used to group together activities and/or spaces having similar characteristics. In organizing the spaces this way, the obvious can become apparent, such as product quarantine needs to be adjacent to shipping/receiving, gowning needs to be adjacent to bathrooms/lockers, Class 100 work rooms should be placed adjacent to each other and connected to a Class 10 000 clean corridor etc. Room classifications (see section 18.7.2) are established based on separation, containment, or circulation requirements, implying definitive room locations. Approximate HVAC equipment sizes and mechanical room locations can be tested. Proximity to spaces served can be studied.

Spaces with special characteristics such as explosion hazard, corrosion hazard, or biohazard which usually imply the grouping or separating of these spaces apart from others can be addressed at this early stage of design. The ways in which one space serves as a node or a focus for other spaces can be studied. The current Good Manufacturing Process (cGMP) concept of providing ample space to operate the facility properly, and the creation of separate work areas, quarantine areas, staging areas, clean, and dirty storage areas can be addressed.

Bubble diagrams are used at many different levels of design from a macroscopic area relationship study of an entire facility to a microscopic activity study of the different bench top activities within a single laboratory. They become the skeletons of the design onto which the developing design fabric is layered. They also are used as a precursor to material/personnel/equipment flow studies, the road maps of the facility design.

18.4 Design development

After the conceptual spatial relationship diagrams are refined and design criteria are established, more definitive parameters can be assigned to the spaces. Dimensional requirements, for instance, are established reflecting any or all of the following:

- Earlier studies in which process equipment footprints are determined.
- Material/equipment/personnel flows, in which cross traffic and constrictions are minimized.

- Good manufacturing practices, allowing ample square footage to perform the activities desired within a specific space.
- State and local building code requirements (emergency egress, fire separation, etc.)
- HVAC systems and duct layouts, building utilities, process systems, and electrical systems requirements are solidified and planned out; square footage requirements are cross-checked with previously allotted areas in the floor plan making any necessary adjustments as early as possible in the design.
- Structural systems are also worked out to support the building shell, facility equipment and the process equipment.

Physical products of design development are building plans, elevations, sections, preliminary equipment arrangements, and revised updated material/personnel/equipment flow diagrams, as well as design criteria and preliminary specifications for all aspects of the facility. It should be noted that at this stage, it is usually a good idea to have established a dialogue with all building code enforcement and standards setting agencies in order to maintain tacit approval during this preliminary planning of the project and to avoid surprise requirements surfacing later during detail design.

18.5 Detail design — construction documents and specifications

When all parties involved in the project agree that the preliminary facility design best satisfies the client's parameters and requirements set forth earlier, detail design can proceed. Fine tuning of exact facility requirements are determined. Compliance with cGMPs is assured. All facility materials, finishes, and equipment are selected and specified. In short, the realization and physical culmination of the facility occurs.

The key to completing a project successfully with well coordinated detail design is not to lose sight of the goals established originally and design requirements and to minimize design changes during detail design. Coordination and rework of major design changes after detail work is well underway causes a chain reaction of stumbling blocks to expedient and error-free completion of the detail design documents.

18.6 Facility design elements

18.6.1 Components of biotechnical/pharmaceutical facilities

Many types of spaces within the biotechnical or pharmaceutical facility can be classified as 'clean rooms'. To define the different levels of cleanliness required, guidelines indicating the amount of contaminants or particles per area have been established by the Food and Drug Administration (FDA) for many activities and functions.

Biotechnical and pharmaceutical facilities are set up in such a way that as one proceeds from 'dirty' outside areas into 'clean' manufacturing, pilot plant, process, or research areas, the level of cleanliness increases and the amount of particulates in the air decreases. These increasing levels of cleanliness progression are established to provide either protection to the product being manufactured or protection to the environment. In fact, the key to a good pharmaceutical or biotechnical facility design is a logical flow of materials, personnel, and equipment coupled with a separation of adjacent spaces having different conditions, using architectural means and HVAC systems to create the necessary isolation. Typical spaces found within a biotechnical or pharmaceutical facility are listed below in progression from outside (uncontrolled) to inside (controlled) spaces:

General areas
- Shipping/receiving/quarantine
- Warehouse (often environmentally controlled)
- Boiler room, chiller room, utility areas, electrical areas, water treatment areas
- Component preparation areas
- Solution preparation areas
- Tank and/or equipment washing and staging
- Released raw material staging
- Quality assurance
- General office areas
- Amenities—cafeteria, break areas, small meeting areas, toilets/lockers
- Records storage
- General laboratories
- Autoclaving/depyrogenation
- Decontamination systems (kill tanks)

Dry product areas
- Weighing/formulating
- Mixing/blending/milling
- Granulation
- Drying
- Lyophilization
- Tableting/encapsulation/coating
- Labeling
- Inspection
- In-process staging
- Packaging

Wet products — parenterals, solutions

- Solution preparation, media preparation
- Distillation/filtration/chromatography/purification
- Fermentation/bioreaction
- Filling
- Capping
- Labeling
- Inspection
- Packaging

18.6.2 Separation/containment philosophies

Requirements for and methods of separation of spaces and containment within spaces of a biotechnical or pharmaceutical facility form the basis of the design of the facility. The cleanliness levels of a space could not be maintained without appropriate containment. Two approaches to separation/containment are 'protect the product' and 'protect the people, facility, or the world'. When working with live organisms, DNA, viruses, etc, NIH regulations require protection of the environment, meaning negatively pressurized spaces from which no organism can escape into the facility. This is accomplished by building a tight room and adjusting the HVAC system to exhaust more air than supplied. As an added precaution, live potentially dangerous materials should only be exposed under the protection of a biosafety cabinet. The important issue is the containment of the organisms within the area in which they are exposed.

Protecting the product is required with the research and manufacturing of pharmaceuticals and biologicals. Any product manufactured for human or animal use must be pure. By forcing more air into a space than out of the space, the natural movement of the air will be out of the space, thus keeping out any less clean air from outside the manufacturing room. It is very often the case that inside a positively pressurized room there is a biosafety cabinet, which is a negatively pressurized unit. This prevents the organisms from escaping into the room possibly harming the people and at the same time allows purified air within the room to be drawn past the organism in the exhausting biosafety cabinet. If the room in which the biosafety cabinet were located were negatively pressurized, then not-as-clean air from outside the room would be drawn in, possibly contaminating the organism.

18.6.3 Material/personnel/equipment flows

Maps of the general movement of raw materials, work-in-progress, products, and equipment are used to determine the proper location of clean room (and other types of) spaces in a facility on a macroscopic level of design; these diagrams can also be used to determine the optimum location of activities within a clean room on a microscopic level of design.

Material, personnel, and equipment flow diagrams are pictorial representations superimposed onto a floor plan of the path taken by the object or activity selected for study. The three categories, 'material, personnel and equipment', are overall groupings for movement studies. Usually, subdivisions within each of the three categories are mapped and studied to investigate completely all activities occurring in the facility. This map is made by tracing the path taken by the object or activity being studied onto a conceptual layout or floor plan. These paths are then analysed and tested against criteria set up by the owner, user, and architect at the outset of design. The logic and simplicity of these flow diagrams as well as the interrelationship of all the flows are studied in great detail; they are representations of the way in which the facility or suite will operate after it is built. The ideal facility exemplifies simple, direct flows with no crossovers or conflicts between any of the flow paths.

These flow diagrams can be developed at the earliest stage of conceptual design. As the design progresses, they should be revisited to verify that the design is proceeding towards the best solution. Ultimately, these flows are used as a check if final design is properly laid out. They are also used as presentation tools to FDA, the Center for Biological Evaluation and Research (CBER), the client, or other groups to illustrate the way in which the facility operates.

18.6.4 Conflicting requirements of FDA/CBER and building codes

One of the major problems encountered by the architect in designing clean room facilities is that of satisfying all applicable building codes as well as all applicable clean room 'codes' and guidelines. The issues of separation and containment usually imply the creating of many rooms and the placement of many doors in these rooms to separate properly and contain the materials or products within the room or to protect it from things outside the room. Additionally, pharmaceutical and biotechnical facilities typically require a great deal of utilities and services, many of which enter the rooms from overhead. The end result is usually many rooms with a large quantity of ducts and pipes running overhead and up and down in chases to mechanical room and/or basements.

Building codes usually stress simple corridors and exit access paths leading directly to the outside of the facility. One clause in the BOCA code, used in New Jersey and other states, says that when exiting a room, the path is allowed to lead through only one other room before entering the egress corridor. In biopharmaceutical facility layouts, a person leaving a sterile room must often travel through a sterile corridor, gown-out airlock, toilet/locker, vestibule, and corridor before reaching the exit door. Furthermore, many of the doors along this exit route do not swing in the direction of travel as is required in cases where there are 50 or more people using the means of egress. (Many of the doors along the means of egress are interlocked, thus requiring defeats or failsafes to ensure the free operation of the doors in the case of a fire.) These apparent violations of the building code are usually resolved after numerous discussions with building officials during which descriptions of clean room operations are explained and resolutions to the building code conflicts are developed.

Another difficult building code issue is that of fire separation partitions or firewalls. Any wall having a 1, 2, or 3 hour rating required to serve as fire separation, must be built from the floor slab up to the roof or floor deck above and sealed tightly around all penetrations passing through. Furthermore, ducts passing through a fire separation partition must have accessible firedampers. Pipes and conduits passing through a fire separation partition or firewall must be sealed around the opening with fire-retardant sealant. Firewalls and partitions usually add cost and complexity to the design.

Since they create additional barriers in the layout of spaces, fire partitions often have a detrimental affect on the movement of people, materials or equipment through the corridors of the facility, creating the need for additional corridors and doors. Although like clean room doors, fire rated doors are required to be closed at all times, vision panel size is limited and panic hardware is required. Firewalls and fire doors add cost and inconvenience to any facility, but are necessary to ensure good fire safety.

Building use group and construction classification are two other major elements of all building codes, having tremendous effect on the design of clean room facilities. The construction classification dictates the allowable building area for a given type of construction; if the area desired is greater than the allowed area for that type of construction, the use of spray-on fireproofing or fire separation partitions allows an increase in allowable building area. If fireproofing is required, cementitious types are more advisable for use since they are not friable (flaky), although they are more costly than the messy flaky types. Fire separation partitions can prove to be effective in increasing the allowable building area, as long as they can be positioned in areas of the layout requiring separation.

Building use group classifications determine the allowable area based on the type of activities occurring in the facility; the more hazardous the operation, the smaller the allowable building area. To increase the allowable area, construction types using fire protection can be used and fire separation partitions can be used.

The architect must strike an understanding between all regulating agencies to ensure that all codes and guidelines are carried out to the fullest extent possible. The requirements of building codes add a level of complexity and certainly add to the cost of design and construction of clean room facilities. Since all of these codes were established in the name of the health, welfare, and safety of the workers and users of the products of these facilities, it is the design consultants' responsibility to satisfy the codes to the greatest extent possible.

18.7 Biopharmaceutical design elements

18.7.1 Biological hazard classifications

The National Institute of Health has published guidelines indicating procedures and conditions required when recombinant DNA and RNA are present. In these codes, design guidelines and architectural finish materials for laboratories and manufacturing spaces as well as requirements for operational practices are discussed. Some ot the architectural requirements as they relate to clean room design have been excerpted and illustrated below.

Laboratory design standards are categorized by the classifications BL-1, BL-2, BL-3, and BL-4; manufacturing design standards are categorized by similar classifications GLSP, BL1-LS, BL2-LS, BL3-LS. The 'LS' suffix categories ('large scale') deal with research or production of quantities greater than 10 liters involving viable organisms containing DNA molecules (GLSP stands for good large scale production using more than 10 liters). In general, regulations for BL-1, for example, are similar to BL1-LS and so on, except for the 10 liters differentiation of the quantities of materials involved as mentioned above. Discussions herein will deal with 'BL' categories when discussing levels 1, 2, and 3. (As a point of clarity, BL-2, BL-3, and BL-4 are often called either 'P1, P2 and P3' or BSL-2, BSL-3 or BSL-4, respectively.)

The type of BL classification required is dependent on the specific classification of the biological materials being used within the spaces; the level of hazard or danger of the organism determines the proper type of facility design needed to provide adequate safety for the environment and laboratory or manufacturing personnel.

The following descriptions of the four different biosafety classifications are excerpted from the NIH Guidelines published in May 7, 1986 and the NIH proposed guidelines published in December 27, 1990.

18.7.1.1 BL-1 spaces. Biosafety Level 1 spaces are suitable for work involving agents of no known or minimal potential hazard to laboratory personnel or the environment. Laboratories are usually assigned to educational and teaching uses. Characteristics of BL-1 spaces are as follows:

- There is limited access to the laboratory when experiments are in progress, at the discretion of the laboratory director. The laboratory is not separated from the rest of the building.
- Work surfaces must be decontaminated once per day and after any spill of biologically viable material.

- Decontamination of liquids or solid wastes must occur before disposal.
- There is no eating in the area, however refrigerators and cabinets designated and used only for food are permitted in the area.
- People working in the laboratory must wash hands after working with recombinant DNA materials before leaving the laboratory. Therefore each laboratory should have a lavatory for handwashing.
- Lab. coats or gowns are recommended.
- No special equipment is required for BL-1 material manipulation.
- The laboratories should be designed to be easily cleaned. Laboratory furniture should be sturdy; spaces between benches and cabinets must be accessible for cleaning.
- Bench tops must be impervious to acids, bases, solvents, and moderate heat.
- If laboratory windows open, they should be fitted with screens.
- Contaminated materials that are to be decontaminated at a site away from the laboratory shall be placed in durable leakproof containers which are closed before leaving the laboratory.
- Containment equipment is recommended although not specifically required for the handling of biological agents in this level.

18.7.1.2 BL-2 spaces. Biosafety Level 2 spaces are suitable for work involving agents of moderate potential hazard to laboratory personnel or the environment. Laboratories are usually assigned to clinical, diagnostic, teaching, and research uses. Characteristics of BL-2 spaces are the same as for BL-1 except for the following:

- There is limited or restricted access to the laboratory when experiments using recombinant DNA are in progress.
- Lab. coats or gowns shall be used. When personnel leave the laboratory, coats shall either be covered with another coat or removed and discarded for cleaning. Gloves should be worn when dealing with animals.
- No special equipment is required for BL-2 material manipulation.
- The laboratories should be designed to be easily cleaned. Laboratory furniture should be sturdy and spaces between benches and cabinets should be accessible for cleaning.
- Experiments of lesser biological hazard can be carried out in the designated parts of the laboratory.
- A hazard warning sign showing the universal biohazard symbol shall be posted on the door of the laboratory when organisms containing recombinant DNA are being worked with.
- Biological safety cabinets shall be used; the class of the cabinet depends on the type of organism used inside.
- If centrifuging, grinding, mixing, or blending materials containing recombinant DNA, the work shall be done in a biological safety hood.
- An autoclave for decontaminating laboratory wastes must be available.

18.7.1.3 BL-3 spaces. Biosafety Level 3 spaces are suitable for work involving indigenous or exotic agents which may cause serious or potentially lethal disease to laboratory personnel or the environment, by result of exposure through inhalation. Laboratories are usually assigned to clinical, diagnostic, teaching, research, and production uses. Characteristics of BL-3 spaces are the same as for BL-2 except for the following:

- Laboratory doors shall be kept closed while experimentation is in progress.
- Strict policies and procedures for access of personnel, i.e. immunization shall be in effect.
- All DNA and recombinant material work shall be conducted in biosafety cabinets or other physical containment devices; no work shall occur in open vessels.
- Laboratory clothing shall not be worn outside the laboratory; decontamination shall occur before laundering. Gloves and masks shall be worn.
- The laboratory shall be completely separated from areas which are open to unrestricted traffic flow. Passage through two sets of doors (airlock) is a basic requirement. Physical separation of the containment laboratory from access corridors shall be provided by two sets of doors (airlock), which may include a clothes change room. A shower may be included as part of the airlock.
- Penetrations in walls, floors, and ceilings shall be sealed to allow for regular cleaning and decontamination of surfaces.
- Each laboratory shall have a sink located near the door and shall be controlled by foot, elbow, or automatic device.
- Any windows in the laboratory shall be closed and sealed.
- An autoclave for decontamination of laboratory wastes is preferably located within the laboratory.
- Ducted exhaust air ventilation system shall be provided creating a negative pressure in the laboratory with respect to the entry and exit airlock. Exhaust air cannot be recirculated but may be exhausted without filtration.
- Filtered exhaust air from Class I or Class II biosafety cabinets may be discharged directly to the outside or to the building exhaust system if the cabinets are tested and certified every 12 months.

18.7.1.4 BL-4 spaces. Biosafety Level 4 spaces are suitable for work involving dangerous or exotic agents which are known to cause serious or potentially lethal disease to laboratory personnel or the environment, by result of exposure through inhalation. Laboratories are usually assigned to clinical, diagnostic, teaching, research, and production uses. Characteristics of BL-4 spaces are the same as for BL-3 except for the following :

- Biological materials from Class III biosafety cabinets or the containment laboratory in a viable state shall be transferred to a non-breakable, sealed container then enclosed in a non-breakable, sealed secondary container and removed from the laboratory through a disinfectant 'dunk tank', fumigation chamber, or airlock designed for this purpose.

- Except for viable materials as described above, all other materials removed from the laboratory shall be autoclaved or decontaminated by gaseous methods.
- Personnel entering and leaving the facility shall do so only through airlock clothing change and shower areas; showers are required before entering and leaving the laboratory.
- A complete change of clothing is required upon entering and exiting.
- All supplies and materials shall be brought in and shall leave via the airlock.
- The laboratory shall be located either in a separate building or in a clearly demarked and isolated zone within a building.
- HEPA (high efficiency particulate air) filtration is required for all exhausts; heat treatment decontamination is required for all liquid wastes (showers may be decontaminated by chemical means.
- Drinking fountains cannot be located within the laboratory.
- Any windows (closed and sealed) within the laboratory shall be break resistant.
- Individual supply and exhaust air systems shall be provided complete with HEPA filtration and system failure alarms.
- Exhaust from Class III hoods must be through two HEPA filters in series.
- Emergency lighting and communication systems shall be provided.

As mentioned earlier, there are four additional biosafety levels describing conditions required for large scale production; these are the GLSP, BL1-LS, BL2-LS, and BL3-LS categories, dealing with operations of 10 liters or more of culture.

The physical characteristics of the three latter categories are the same as described above for BL-1, BL-2, and BL-3. The GLSP category is recommended for large scale research or production involving viable, non-pathogenic and non-toxigenic recombinant strains derived from organisms having an extended history of safe large scale use.

Characteristics for GLSP are similar to those of BL-1 except for the following:

- Viable organisms do not have to be handled in primary containment environments, but they should be handled in such a way as to safeguard the health of the individuals working with the materials.
- Sample collection, exhaust air, and discharge wastes shall be handled in such a way that the health and safety of the individuals working with the materials is maintained.

Knowing the biological classification of the material is important for the architect or space designer because many of the classification requirements have significant impact on the design of the facility. The two most critical design disciplines affected are architectural and HVAC. Air circulation volumes are directly affected by the biological classification.

These volumes in turn have a direct effect on the amount of floor, ceiling plenum, and wall space necessary to house the mechanical equipment and the ductwork necessary to satisfy the air requirements. Because of the evolving nature of the design of clean room facilities, as the design of the facility progresses, there often becomes a shortage of mechanical room space and ceiling plenum space, especially towards the end of the detail design phase. Therefore, it never hurts to plan for the largest amount of space practical from the earliest stages of the development of the design.

The number and types of spaces required to satisfy health and safety requirements (NIH and building codes) also have a large impact on the square footage and spatial arrangement of the floor plan. These factors also have a tremendous impact on the cost of the facility, therefore the design team must strive to meet all applicable codes, standards and guidelines without going overboard, unnecessarily 'gold-plating' the design.

18.7.2 Room classifications

Room classification is used as a measure of cleanliness for clean rooms. Cleanliness is achieved by controlling the size and the amount of particulates in the air. The types of classifications used are as follows:

- Class 100 000: has no more than 100 000 particles of 0.5 μm or larger per cubic foot of air.
- Class 10 000: has no more than 10 000 particles of 0.5 μm or larger per cubic foot of air.
- Class 100: has no more than 100 particles of 0.5 μm or larger per cubic foot of air.

In the biopharmaceutical realm in the United States, Class 1000 is rarely used and Class 10 (10 particles of 0.12 μm) and Class 1 are beyond the needs of biological work and therefore not used.

The desired cleanliness is achieved through HEPA (high efficiency particulate air) filters which can filter out particles to a 99.95% or 99.997% efficiency for Class 100 000, Class 10 000 and 100 spaces. HEPA filters are used on air supplies or returns, depending on 'BL' classifications and room requirements. They also are used in biosafety cabinets to 'scrub' the air inside to protect the product; these recirculating hoods are sometimes called 'run-around' hoods, since the air is recirculated within a room. Some typical activities requiring classification are as follows:

- Class 100 000: sterile corridors, wash rooms, clean sides of autoclave, or depyrogenation ovens, capping or fermentation rooms, weigh rooms, solution preparation and component preparation rooms.

- Class 10 000: rooms in which Class 100 biosafety hoods are located — sterile filling rooms, fermentation rooms, filtration rooms, purification rooms.
- Class 100: any space or area where a sterile product is exposed to the air. Since high velocities (90 ft/min), once through air circulation, and laminar air flow (air movement with no turbulence) are required, rooms classified as Class 100 should be kept to a minimum. The large size and numbers of HVAC units required to move the air through these spaces and the large sizes of ducts causes these spaces to cost a great deal of money to build and operate. This usually means that Class 100 spaces are restricted to being located either inside biosafety hoods, under ceiling mounted HEPA filter units surrounded by plastic curtains, or inside tunnels of process equipment where product is being transferred from filling to capping, for example.

18.7.3. Characteristic finishes of clean rooms

The finished surfaces in a clean room form the first barriers to the materials and operations within the space. Surfaces need to respond to the specific requirements of the space in order that the integrity of the environment is not compromised at any time or in any way during the operation of the clean room.

The general principles common throughout all types of clean rooms are the concepts of cleanability, durability, and material containment. To be cleanable and durable, a complete knowledge of all materials (including cleaners) to be used within the space must be understood. Proper surface protection can be specified to resist the chemicals used.

Flush vision panel frames and glazing, and flush door jambs are extremely important features in the minimizing of dust-holding surfaces. In setting these units into walls, all joints not razor tight and flush should be sealed with silicone caulk to form a tight seal. If vision panel units cannot be installed glazed flush, then sloping sills must be employed. Coupled with a good housekeeping program, dust and dirt will not collect to as great an extent as it would on a flat sill.

18.7.3.1 Floors. Standard epoxy terrazzo or trowelled epoxy floor 1/4 in (0.635 cm) thick complete with coved integral base 6 in (15.24 cm) high is commonly used. This material has high impact resistance, 8000–12 000 psi, compression strength and with the coved base makes an easily cleanable system. The strength of the floor surfacing is adequate for almost any tankage of equipment operation and can withstand the rolling of small wheels without damage.

A quartz grit can be broadcast into the surface of the curing epoxy floor to give it a slip-resistant finish. The small chips provide excellent traction for walking and pushing large tanks. Choices of many colors are available to make the space more appealing.

As with any floor topping installation, care must be taken to follow the manufacturer's guidelines and installation procedure. Surfaces must be clean, dry, and smooth before installation of the topping. Beneath a slab on grade on which the new epoxy is often installed, there should be a continuous polyethylene vapor barrier to prevent the migration of moisture through the concrete, lifting up the epoxy topping.

Trowelled epoxy should be applied in such a way to minimize patching and termination in an open area, since the material does not join well; epoxy terrazzo can be jointed provided a metal expansion strip has been installed. If the floor topping is being installed in a clean room through which an expansion or control joint is located in the concrete, specific manufacturer's instructions should be followed, however the joint is usually prepared with a fiberglass tape bridging material forming a strong bridge with the epoxy over the concrete joint.

Tank washing or other constantly wet areas are excellent applications for troweled epoxy floors. Seamless and impervious to almost all solvents and cleaners, they can be continued vertically onto a curb or onto a wall as a coved base to form a good waterproof basin. The coved base should be applied over a 6 in (15.24 cm) high plywood substrata secured to the base of the wall (with a heavy gage 10 in (25.4 cm) high steel bent plate continuously behind on metal stud walls) and finished flush with the wall surface or terminated with a radiused edge against the wall surface, the former being the most desirable detail.

Seamless vinyl floors are used where there is limited heavy equipment traffic or solvent spills. Sheets of 1/8 in (0.318 cm) solid vinyl are fully adhered to the concrete slab substrata then heat welded together making a jointless installation. Integral, welded coved wall bases are also used. This use of this type of clean floor is limited to laboratories, control rooms, corridors, and wash areas not having large, rolling tanks.

Seamless vinyl floors can also be used in wash areas on the floor and on the wall to provide a completely bump resistant and waterproof barrier. Since the solvent resistance of these floors is limited, careful attention needs to be given to the chemicals to be used in this area.

For corrosive, acidic, alkaline, or other hazardous chemicals, there are impervious, highly resistant coatings not unlike paints in their composition. Often no more than 1/8 in (0.318 cm) thick, these materials also have high impact resistance 8000–12 000 psi, and can be applied over new or existing concrete or epoxy floors.

Furan cements, vinyl ester resins, and other variations can also be used when extremely hazardous chemicals are being used. When used either as a floor coating or as the mortar with acid brick, these materials have their unique chemical resistance and are highly resistant floor toppings.

18.7.3.2 Walls. Clean room walls can be divided into two types: prefabricated and field assembled. Prefab. clean room walls can be used where the clean room is small, messy construction in an area cannot be tolerated, or the clean room space may be relocated in the near future. Finished materials of prefab. clean room walls are usually painted steel, having integrated cove base, doors, 'clean' visions panels, and ceiling system. Utility openings in these panels need to be coordinated and grouped carefully if the goal is to reuse the panels in the future. Standard details for wall penetrations using rubber-type gaskets are provided as part of the prefab. package. These wall systems are usually strong enough to support small HVAC units on top of the ceiling panels, but the noise vibration from the HVAC unit may be undesirable inside the space. It is usually better to locate HVAC units near these rooms on the floor.

Field assembled walls can be of masonry or gypsum board and metal stud construction. Both materials are capable of receiving any finish, but masonry must often receive a parge coat (of cement) to smooth out the joints effectively and other block irregularities, depending on the type of finish being applied to the block. Epoxy paint is a standard, durable material applied usually in 2 coats, each 2–4 ml thick over a prime coat. When applied over block, it forms a strong impact resistant coating, good for moderate traffic areas. When used over gypsum board, continuous horizontal wall bumpers would be needed if heavy tanks are being moved frequently. In fact, in any heavy traffic area where large tanks or equipment are being used, continuous horizontal wall bumpers should be used for any type of wall material to afford the maximum protection.

In areas where the chance of heavy wall damage due to traffic impact is limited, it is possible to use a Mipolam wainscot for protection of the wall. (Mipolam is a trademark name of sheet polyvinylchloride wall covering). Coming in a flexible textured sheet 1/16 in (0.519 cm) thick (and in a dozen or so colors), or in a 1/8 in (0.318 cm) thick semi-rigid high gloss sheet (available in limited colors) this material not only provides moderate protection against damage, but also gives a 'high-tech' look. Being a vinyl product, it is susceptible to some solvents and ultraviolet light. Since this product is approximately seven times more expensive than epoxy paint and in light of its limitations, its use should be carefully considered.

Stainless steel is often used as a wall finish either installed directly onto the substrata or fabricated into panels and hung onto the substrata. Joints between the panels must be caulked. Around autoclaves or other hot equipment, stainless steel can withstand the heat given off by the piece of equipment, and it fits in well with the appearance of the oven. Stainless steel is the most difficult and costly system to install, but can give the most impressive, rich look in the end.

18.7.3.3 Ceilings. The materials used for clean room ceilings can be either suspended gypsum board, suspended tile system, or laminar flow. Often, the ceiling in a clean room must be airtight to maintain differential air pressures and to maintain the cleanliness of the room. The choice of ceiling material and finish is determined by the need to wash down the room, the accessibility criteria, the pressurization requirements of the room, the number of HEPA filters, or the desired appearance of the room.

When a space must be hosed down or scubbed down, the ceiling must be as tight as possible. An epoxy painted suspended gypsum board ceiling will serve very nicely in this case. Provided the light fixtures and HVAC diffusers are well caulked, this will be the tightest ceiling possible. Problems will occur with the gypsum board ceiling when access doors are required which even though they can be obtained with four sided gaskets, are susceptible to leaks, especially if used frequently and the more that are required, the greater chance of leakage. Access doors also add to the cost of the job and detract from the clean appearance of the gypsum board ceiling.

When greater accessibility is required, a grid-type ceiling might be a more appropriate choice. Using an extruded aluminum grid with a foam gasket or gel gasket, painted steel or aluminum panels are clamped into the grid. Access doors are still required but they fit into the system and are far less visible than with gypsum board ceilings. These gasketed metal panel ceilings are just as tight as gypsum board ceilings when installed properly and have a very 'clean' look. Although accessibility is mainly through the access doors, panels may be removed if necessary. The main drawback of this type of ceiling is the price, which is approximately seven or eight times the price of an epoxy painted gypsum board ceiling.

Another type of suspended grid ceiling is the cold-rolled painted suspended aluminum grid with either vinyl coated gypsum board panels or mylar wrapped compressed fiberboard panels. Either panel may be set into the grid in a foam gasket and clipped into place. Neither ceiling is meant for a regular hosing down but can be used in spaces such as laboratories, return corridors, and other spaces designed as moderately clean but not sterile or aseptic. This type of ceiling is not as tight as the previous two and therefore its use in pressurized rooms is not advisable.

Laminar flow ceilings (always accompanied by low to the floor air returns in the walls) consist of complete HEPA filters as the ceiling material, punctuated by narrow 'teardrop' fluorescent lighting fixtures mounted on the grids between two adjacent HEPA filters. (The shape of these lighting fixtures allows a smooth aerodynamic flow downwards with the fewest possible vortexes; the notion of smooth airflow all the way down to the work plane is a basic design characteristic of laminar flow ceilings.) Access above this type of ceiling is very difficult since the above plenum is full of ductwork supplying the HEPAs. Often, walkways are built over these areas to maintain accessibility to equipment, valves, etc. above and to provide access for changing HEPA filters. The filters may be changed from inside the room, but this creates a dirty situation in the room and requires the space to be temporarily shut down.

18.7.4 Components used to achieve airtight separation

Separation/containment philosophies are basic to the functioning of clean room suites. The requirements of successive spaces are established, progressing from the uncontrolled environment to the controlled biopharmaceutical processing environment. Each door, wall, and ceiling become demising barriers we depend upon, allowing adjacent spaces to have different cleanliness and pressurization conditions.

As mentioned earlier, air pressurization and separation-containment philosophies are extremely important considerations in clean room design. In order to achieve the desired pressurization and separation, the architecture of the rooms must provide the support for the environmental conditions. Any penetration in the walls or ceiling is a possible source of air loss. Access doors in walls and ceilings must have secure latches and gaskets. Pipe, duct, or conduit penetrations in walls or ceilings must be caulked with silicone caulk or sealed with 'Cocoon'. Cocoon is a trademark name of a non-shedding, clean, sprayed vinyl material designed to seal openings too large to caulk. (Cocoon can also be used as a wall finish, although its textured look is not always desirable.) This material will seal tightly around almost any type of object and is especially good at closing penetrations.

In order to meed the demands of the design philosophy, the architect must carefully employ all necessary sealing devices to allow the demising barriers to function properly. Essentially, the extensive use of the appropriate caulking and gaskets ensures the separation of two adjacent environments. The following is a list of some of the ways in which clean, tight conditions, and environmental separations are achieved:

Ceilings

- Gel gasketed metal panels, lighting fixtures, and HVAC diffusers set into suspended extruded aluminum grid.
- Neoprene gasketed metal panels, lighting fixtures, and HVAC diffusers set into suspended extruded aluminum grid.
- Gypsum board cleaning with triple gasketed light fixtures and neoprene gasketed HVAC diffusers; gypsum board is finished with epoxy paint or Mipolam.
- Standard cold-formed aluminum grid with mylar or gypsum ceiling panels and gaskets and hold down clips. (This system is usually good for 'clean'

areas but not advisable for classified areas since it is the least capable of holding room pressures.)

- Silicone caulk around all penetrations through ceiling panels; caulk also along joint of contact of suspended grid edge molding and wall.
- Laminar flow HEPA filter ceiling with teardrop lights.

Walls

- Walls built of gypsum board, masonry, or prefabricated baked enamel or plastic panels; metal panels with kynar paint may be used for a 'clean' blast relief wall provided the smoothest possible liner panel profile is selected.
- Coved integral epoxy or seamless vinyl base for classified spaces. Standard vinyl cove base on gypsum board with wall runner stud set in caulk and caulk along floor, or vinyl cove base on masonry wall. Vinyl cove base is used only in unclassified spaces.

Doors and jambs

- Jamb details should be designed such that metal frames are finished flush with gypsum board in metal stud systems; a bead of silicone caulk is still required along the joint of the two materials. Jambs set in masonry openings should have the bead of silicone caulk along the metal frame–masonry joint and masonry jambs should be constructed using standard block shapes having rounded corners.
- Door frames should have flexible neoprene gaskets on two jambs and head, secured to the metal door frame.
- Bottoms of doors should have a drop seal (mechanically operated device that pops downward as the door reaches the closed position at the jamb), providing the most secure seal at the sill. Sweeps on the door bottom can leave marks in the floor and can wear out losing their ability to hold a pressure difference. A weatherseal-type of threshold has too great a lip and is either a tripping hazard or poses too great a barrier for equipment rolling through the door.
- Double doors must have an astragal with a neoprene gasket (not a brush-type gasket) secured to the door jamb or the astragal.
- Doors that swing in the direction of higher pressure to lower pressure usually close tighter and operate better in maintaining pressure differentials. Often, however, the building code will require a specific direction for the door swing contrary to the above. Space limitation may also preclude the desired direction of door swing.
- Doors should have heavy duty, adjustable closures. Exterior grade closers usually work well. They should have a smooth cover as opposed to a ridged cover to permit easy cleaning.
- The use of interlocking doors is important to prevent two or more doors from opening at the same time allowing air to mix on opposite sides of an airlock.

Door interlocks. Electronically controlled locking mechanisms are used to prevent the simultaneous opening of two sets of doors at the clean and dirty end of an airlock or between two spaces required to have separation. This system must fail in the open position to allow smooth egress in an emergency. Often, there are four or five doors being interlocked together; in this case, door vision panels or glass side lights as well as intercoms are important to prevent people from getting stuck inside a malfunctioning interlock group.

Wall and ceiling closures. All penetrations in walls and ceilings must be completely sealed off to allow the appropriate pressurization of the space. A continuous bead of silicone caulk in the joint around the wall penetration exposed to the room can be appropriate when the joint is no more than 1/4 in (0.635 cm) in thickness. When more than 1/4 in thick, Cocoon can be sprayed around the crack to bridge the joint after which the sealed joint can be painted; since Cocoon is messy, proper masking of the area is necessary. Stainless steel escutcheons can be used to cover a wall opening too large to Cocoon or caulk; they should be set in silicone caulk to seal the opening.

Vision panels. Whether located in doors or walls, all jambs are set such that there is no more than a 1/8 in (0.318 cm) lip where the frame overlaps onto the wall. Sills of frames should be sloping, or glazing should be flush with the face of the frame. A silicone caulk bead may be run completely around the joint of the metal frame and the wall (or door) to ensure a tight seal.

Service chases. These spaces are similar to narrow walk-in closets and should ideally be located on the floor, adjacent to the processing suites. They often serve as return air plenums and are negatively pressurized. More often than not, walk-in service chases cannot be provided due to floor area limitations. Instead, it is imperative to provide adequate numbers and sizes of service closets with wide opening doors for good access.

Pipe/duct chases are always major elements in clean room design and must be provided in ample size and location to accommodate all hidden services. Services requiring frequent connection should have valves and fittings located in clean chases (with sloping bottoms) recessed into the wall, or dropped through a soffit and run down the surface of the wall. Unistrut lightweight steel supports are almost always needed to carry services in these chases. In fact, the mechanical and piping engineers need to communicate closely with the architects about their space needs and chase locations.

18.8 Conclusion

Clean room design is a multifaceted, multidiscipline process requiring a great deal of coordination between the disciplines involved and the client. Because of the many code and regulatory requirements and because of the continuously evolving nature of clean room design processes, many details must be examined and studies made to ensure that the most effective, efficient design is implemented.

Bibliography

Federal Standard FED-STD-209D; *Clean Room and Work Station Requirements, Controlled Environment*, June 15, 1988

Gabler, E.M., Dustbusting: how they build the cleanest rooms on earth, *Architectural Record*, April 1987, pp. 128–133.

National Institute of Health, *Laboratory Safety Monograph*, U.S. Department of Health, Education and Welfare, June, 1978.

National Institute of Health, Federal Register Part III, *Guidelines for Research Involving Recombinant DNA Molecules; Notice*, May 7, 1986

National Institute of Health, Federal Register Part V, *Recombinant DNA Research; Proposed Actions Under Guidelines and Meeting of Advisory Committee; Notice*, December 27, 1990

Title 21 CFR; Federal Register Vol. 43, No. 190 Rules and Regulations; Fri. Sept. 29, 1978; Part 211 Subpart C, Section 211.42

19 Clean room testing and certification

S. NORWOOD

19.1 Introduction

This chapter aims to provide a basic knowledge of the testing and certification of clean room facilities, in an easy-to-comprehend progression to introduce the process to those not familiar with the subject. There are several in-depth papers and articles which can provide more information on the technical nature of the various tests. The standards referenced in this chapter also provide valuable insight to the actual testing equipment, procedures and reporting.

The testing and certification of clean rooms is nothing more than an evaluation of the rooms to ensure that they are providing the required environmental conditions. The certification is typically performed to demonstrate that the finished clean room is performing as required by the job specifications. The certification is also a means of demonstrating that the contractors have fulfilled their contractual obligations for the project.

However, an important distinction must be made between the certification and the final HVAC validation. Many different industries require clean rooms for their production including microelectronics, biotechnical, pharmaceutical, medical devices, aerospace, educational, and defense contractors. Validation is often required in addition to the certification for biotechnical, pharmaceutical, and medical devices. This includes what is known as the heating, ventilation, and air conditioning (HVAC) validation, which is a thorough evaluation of the entire mechanical system including the clean rooms. The validation is to ensure that every aspect of control and performance is tested, evaluated, and documented. This is not dealt with in this chapter, but further information about these requirements is available through the FDA (Food and Drug Administration). This chapter is designed for any user of a clean room who is not familiar with the entire certification process.

There are usually two to three phases of certification performed for a new clean room. The initial testing is called 'as-built', or 'static' certification. This is the testing of the clean rooms after completion, air and water balancing, and initial cleaning, but prior to installation of any process equipment or personnel occupation. The purpose of this testing is to ensure that the rooms have been constructed in accordance with specifications and to release the contractor(s) from the project. If any deficiencies are discovered, it also gives the contractor a chance to rectify the problems in fulfilment of his contract.

The second phase of certification is the 'at-rest' certification. This is performed after all process equipment is installed and running, but prior to personnel occupation. This phase is very important when the process equipment requires a large amount of exhaust or the equipment may be generating a large number of particles. Projects which do not have these constraints often delete this phase of testing.

The final phase of testing is called 'operational' or 'dynamic' testing. This certification is performed for the room in full operation with all equipment and personnel performing their normal duties. The number of tests performed in this phase is often reduced such that the testing process does not create contamination which would inhibit production.

19.2 Standards

The testing and certification is typically performed based on a specification for the job. Each particular job will have a section of the specification which details the testing process and acceptance criteria. These specifications are usually based upon current industry standards. The testing and certification of clean rooms has three standards which are most frequently used. Each of these, as well as several others, are discussed.

19.2.1 Institute of Environmental Sciences Standard IES-RP-CC-006-84

The Institute of Environmental Sciences (IES) publishes several standards which are frequently used in the clean room industry. The standard which relates to testing and certification of clean rooms is IES-RP-CC-006-84-T. This is a tentative issue of the standard, released for trial use in 1984. The use of the document is voluntary and it is currently under revision by a working group.

The standard is complete in the number of tests with which it deals and the detail of each test. It is also presented in a format which is easy to transform into a job specification. The purpose of each test, apparatus, procedure, and reporting method are all detailed. The user simply has to fill in the appropriate acceptance criteria to utilize the standard.

19.2.2 Federal Standard 209D

Federal Standard 209D (FS 209D) defines the airborne particulate count as a means of classifying different clean rooms by airborne cleanliness. The standard was originally published in 1963 as a means of qualifying manufacturing environments for contract work. In 1983, the standard was revised by the Institute of Environmental Sciences, the current revision was in October 1987. This standard is also in the process of revision for comparative metric classifications and therefore worldwide acceptance.

There is a large amount of confusion regarding this document and proper application for certification. The standard only deals with particle counting and not further certification testing. Since it is such a well known standard, many people assume that it defines a full certification, which it does not. It does recommend further testing, but does not give details required for these tests. Therefore, this standard is typically referenced for the particle count test, while other standards such as IES-RP-CC-006-84 are referenced for the remaining certification.

19.2.3 NEBB procedural standards for certified testing of clean rooms

The National Environmental Balancing Bureau published a standard in 1990 which deals with a wide range of certification tests. The standard is relatively new and has not received complete acceptance as yet, but is very good. It not only deals with the details of the testing, but also gives the basics of clean room design and operation, as well as several other types of equipment which frequently require certification such as biological safety cabinets (BSCs), exhaust hoods, and laminar flow hoods. This standard also provides several options for each test which helps in developing a certification specification.

19.2.4 Other standards

There are many additional standards which refer to various aspects of clean room testing. These would be advised for a complete understanding of the certification and for a reference library. Some of the most frequently used (see Bibliography) include:

- Food and Drug Administration 'Good Manufacturing Practices'. This deals with general guidelines for use of manufacturing space, and proper documentation of such use.
- American Society for Testing and Materials (ASTM) standards 24, 25, 50, and 51. These standards all deal with various methods of counting and sizing airborne particulate.
- American Society of Heating, Air Conditioning, and Refrigeration Engineers (ASHRAE) standard 52-76, deals with evaluating filters which have efficiencies below high efficiency particulate air (HEPA).
- Institute of Environmental Sciences, IES-RP-CC-001-86, describes the materials, construction, and testing of HEPA filters and provides for classification of the various filter grades.
- IES-RP-CC-002-86 provides testing required for laminar flow clean air devices.
- National Sanitation Foundation standard 49 provides a complete description of materials, construction, and certification of biological safety cabinets.
- SAMA (Scientific Apparatus Makers Association) standard 10 details the testing and certification of exhaust hoods.

19.3 Testing and certification companies

There are many different types of certification companies and services, each of which could have implications for the type of service required. The important fact to remember in choosing a company is to ensure that it will provide a complete unbiased service. However, it should not be assumed that a company will automatically be biased on the basis of the structure of the company. As with many services, some companies provide a better service than others.

19.3.1 Independents

The most common form of certification is by an 'independent' source. This means that the certification company is not involved in any way with the products, installation, or design of the facility being tested, and that it will not have any bias towards the evaluation of the facility.

19.3.2 Divisions of larger companies

Many larger clean room construction firms have divisions or personnel who provide testing and certification services. Some of the filter suppliers also provide certification services. These services are usually only for the work of the parent company, but some also perform certification for others. There are advantages and disadvantages to this structure. As a division of the parent company, there could be obvious biases towards the quality or integrity of the product or installation. This has been a well documented concern over the years, and can usually be quelled by close supervision by the owner. In other words, if the owner witnesses the actual testing and asks pertinent questions to ensure thorough and unbiased evaluation, the results should be considered valid.

This type of certifier will be familiar with the product or installation since they specialize in this company's work. This often means that recognition of potential problems is sometimes easier since there is a degree of trust between the divisions of a company.

19.3.3 Other companies

There are several other types of company which perform certification services. With the introduction of the National Environmental Balancing Bureau (NEBB) standard, mentioned previously, a large number of air balancing companies now perform certification. These are companies which historically dealt with the testing and adjusting of HVAC systems, but are usually not clean room related. The additional testing and precision measurement is a natural progression for a balancing company.

Companies which primarily certify BSCs or other hoods typically provide clean room certification services. The equipment and testing requirements are similar for both types of certification, but there are substantial differences. There are also validation companies, calibration laboratories, and equipment vendors who periodically perform clean room certification. A company which does not specialize in clean room certification should be thoroughly evaluated prior to utilizing their service.

19.4 Certification tests

The following tests are the more common certification tests which are performed. These are only brief descriptions of the tests, further information on each test can be found in the standards previously referenced.

19.4.1 HEPA leak testing

HEPA filters are designed and manufactured to provide a certain efficiency in filtration, usually 99.97%, 99.99%, etc. This should be verified at several steps through the progress of the project. If the filters are to be scan tested at some point in the project, then they should be a minimum of 99.99% efficient. They should also be scan tested at the factory prior to shipping.

19.4.1.1 Aerosol challenges. An important consideration in the scanning of filters is the suitable 'challenge'. The challenge is a high concentration of particles introduced upsteam of the filters to exaggerate any leaks in the filters. The most well known challenge is dioctyl phthalate (DOP). This is an oil which is aerosolized through the use of Laskin nozzles and compressed air. The nozzle is designed to shoot jets of compressed air through the oil which shears off and aerosolizes the particles. It has been losing popularity since studies found that DOP is a suspected carcinogen. Other oils which can be used in similar fashion include dioctyl sebecate (DOS), corn oil, and mineral oil. Caution should be used if corn or vegetable oil is the selected challenge because high concentrations of these substances in the filters can support potential growths of bacteria and mold. Several industries may not want an oil-based challenge used for this testing for fear that it could damage the product. These include thin film manufacturers, fiberoptics, and some specialized microelectronics processes. Alternative challenges are discussed below for these and other industries. It is very important to note that different challenge oils may produce slightly different particle size distributions and quantities when used in an aerosol generator. The aerosolization characteristics DOP, DOS, corn oil, polyplycol oil, and mineral oil are all about the same[1]. However, if the upsteam concentration cannot be measured, and the photometers' internal calibration light is to be used, a correction factor may be required for the mass concentration of the substitute aerosol[2].

Ambient challenge is sometimes used if it is a sufficient concentration. This is usually not the case if a photometer is to be the test instrument, but it may be acceptable if a particle counter is used for scanning. It is often a requirement that the prefiltration be removed to bring the concentration to acceptable levels. This is often a problem since the prefilters provide a degree of protection for the final HEPA filters. There have been concerns that unwanted substances would be allowed into the HVAC system when the prefilters are removed. However, the prefilters are usually only designed to filter a certain percentage of the larger particles. The particles which are usually a concern, such as mold and bacteria, would not be efficiently retained by these filters. However, if the HVAC system has HEPA filters at the intake, this is a valid consideration as the HEPAs would capture these particles.

Two newer challenges have been receiving a large degree of success in the microelectronics clean room industry. Silica dust has been used for scanning HEPAs for several years. The main drawbacks of the dust are the cost, ability to maintain a high concentration, and the very small particle size. If silica dust is used as a challenge, a particle counter with sensitivity to 0.1 μm is usually required as the mean particle size of the dust is 0.09 μm. A newer challenge is polystyrene latex spheres. This is a relatively inexpensive challenge capable of providing a high upsteam challenge.

The advantages and disadvantages of each challenge should be considered. Further information on which challenge would be best for the particular application under consideration could be supplied by the certification contractor. The various stages and types of tests are presented below.

19.4.1.2 Preinstallation. It is becoming increasingly popular to scan the HEPA filters prior to installing them in the ceiling or ductwork. This is a prudent step as only good filters are installed minimizing wasted time required to repair in place or replace filters found to be leaking in later testing. This also detects the leaks which would have occurred during the shipping process, which is usually the most common source of filter damage. Since the filters were tested prior to leaving the factory, the damage either occurred during shipping or in handling required for testing. It should be noted that the filters should be thoroughly inspected when they are delivered from the factory. Any visible dents in the boxes or corners of the boxes should be noted on the receiving board. This will allow a refund to be obtained from the trucking company if these filters have incurred damage.

The preinstallation test is performed using a portable test bench which supplies air through the filter. The amount of air should be similar to the design flow of the installed filters. The bench will have adaptors for the various filter sizes and shrouds which sit on top of the filter to isolate the clean air through the filter from the surrounding environment. The selected challenge is introduced upsteam of the filter and measured to ensure it is a sufficient concentration.

The required concentration will vary depending on the selected test instrument. The test instruments are discussed in the following section. Each individual filter is placed on the test bench, the shroud is clamped in place to eliminate any bypass leakage, and the challenge is introduced. The downstream side of the filter is scanned with the sample probe of the test instrument. The probe is a flat funnel with the size of the opening designed to be isokinetic, or, to pull the air into the sample probe at the same velocity as the surrounding air velocity. The probe is passed within an inch of the filter in overlapping strokes until the entire face of the filter and frame have been tested. The scanning rate for a one CFM instrument is 10 linear feet per minute (0.31 m/min). Any filters which are found to be leaking are set aside for corrective action. There should be a sufficient number of test stations to supply filters to the installation contractor. A typical certifier will scan roughly 50 2 ft × 4 ft (0.61 × 1.22 m) filters in a day. If the installer wanted to install 150 per day, three stations would be required.

The number and type of leak detected should be carefully monitored. If an excessive number of leaks are detected, the potential source of the leaks should be determined. If the shipping boxes were damaged, the handling of the filters should be investigated and the manufacturer's representative should be contacted for information.

The possibility of repair to the filters should be decided upon by the owner. Repair is usually accepted using small patches of pure silicone caulk. The patches should not exceed 5% of the active filter media, and one dimension should not exceed 1.5in (3.81 cm). However, if more than 10% of the filters require repair, the situation should be investigated. It has been found that over time the patches may not be effective. Since they do not support airflow, they have a resistance which puts pressure on the surrounding delicate media. Over time, this can cause additional leakage around the perimeter of the patch. A high leak rate means there will be a problem at some point with the filters. The responsible party should bear the cost of replacement to ensure proper performance over time. Once the filters have passed the preinstallation scan, they are supplied to the installation contractor for installation. It is often a good idea to track the filters by serial number for future reference. This is sometimes required by the FDA. It this is carried out, a method of identification of the installed filters by serial numbers should be implemented.

19.4.1.3 Scanning instruments. The most popular test instruments for scanning are photometers and particle counters. The instruments are similar in that they both work on the scattering of either light or a laser to count or size particles. The difference is that a photometer works on a percentage penetration basis, while a particle counter detects each individual particle. The main advantage of a photometer is that there is no doubt about the 0.01% leakage since the instrument displays percentage penetration. However, this instrument also requires a substantial upstream concentration to make resolution to 0.01% possible. Calculations made based on required upstream concentration indicate that as many as 1000 particles must be sensed in a burst to cause a leak to be detected.

Since the particle counter can count each individual particle, scanning can be accomplished with a lesser upstream concentration. The problem with this instrument is in defining a leak. Typically, the upsteam concentration is specified to be a minimum of 1 million particles per cubic foot at the selected particle size, and a repeatable reading of one or more particles per second is considered a leak. Assuming that the particle counter has a sampling rate of 1 ft^3/min, which is the minimum recommended, the calculated penetration for a leak of 0.01% is: 1000 000 per minute = 16 667 per second × 0.01% (0.0001) = 1.7 particles per second, or roughly 2 particles per second. The upstream concentration can be increased to reduce the impact of background or 'false' readings.

However, there is a substantial drawback to using a particle counter as the test instrument for scanning. A basic requirement of scanning is to measure and record the upstream concentration of particles. Since leaks are defined as a percentage penetration, the upstream must be known to determine what percentage penetration is allowed. Most particle counters can only measure 500 000 to 1 000 000 particles per cubic foot at the smallest particle size for the instrument. This is often not a sufficiently great concentration to perform a reliable scan. It is therefore suggested that an instrument known as a sample dilution device be used to obtain a reliable reading of upstream concentration. A sample dilution device is a means of filtering out a known portion of the sample to bring the total count within the resolution range of the instrument. This typically works by utilizing a pump working at 1 ft^3/min (0.0283 m^3/min), (assuming a 1CFM counter is being used), and HEPA filtering the air from this pump. The air is then fed through a valve which only allows 0.9 ft^3/min (0.0254 m^3/min) to flow on to the counter. The counter then pulls in the 0.9 ft^3 of filtered air along with 0.1 ft^3 (0.0028 m^3/min) of the upstream concentration sample. The resultant count is multiplied by 10 to yield the upstream concentration per cubic foot (or m^{-3}) of air. Additional readings should be taken to ensure a repeatable reading.

Either instrument is usually acceptable for the scanning. The application of the instrument usually determines which would be used. For instance, a particle counter is usually used when there are a large number of filters served by the same fan and the challenge cannot be isolated. The large amount of upstream challenge required for a photometer makes the use of a photometer impractical for this type of application. For clean rooms with a higher background count, Class 10 000 or 100 000, a photometer is often used. This is because the high number of ambient particles causes a large number of false readings on the particle counter. The photometer would be a better instrument for this application.

19.4.1.4 Final, in-place scan test. The completed ceiling filter assembly is usually required to pass a scan test. This test is especially important if the desired room class is Class 100 or cleaner. This test is performed by introducing a challenge upsteam of the filters and scanning the downstream side of HEPA. The same decisions must be made between the acceptable challenge and scanning instrument. The procedure is very similar to the preinstallation scan. The isokinetic probe is passed within 1 in (2.54 cm) of the filter face at a rate of 10 linear feet per minute (3.048 m/min) in overlapping strokes. A leak is usually defined as any reading which is 0.01% of the upstream concentration.

There are additional concerns encountered when scanning the installed filter. Because of the set up of the preinstallation test, the frame of the filter is often not tested. Since the filter sits on the test bench, the air is only flowing inside the frame. Depending on how the filter is installed, the frame itself may have to isolate the filter from the ceiling. A leak in the frame of the filter is often missed in the preinstallation scan and picked up in the final scan.

If the HEPA filters are to be installed into the housings, this presents similar problems. The seams or construction joints of the housing often run from the top of the housing to the bottom. The filter is installed halfway down this seam, usually with a clamping device. This means that the unfiltered air can enter this seam upstream of the filter, run along the channel of the seam, and exit through the bottom of the seam below the HEPA and into the room. This may seem improbable, but is a very common problem. When the resistance, or pressure drop, of the filter is considered, the air will find any possible way around the filter. For engineering purposes, air is considered a liquid. Like water, air will follow the path of least resistance, which is often not through the HEPA filter.

Care must be taken when using oil-based challenges for this test. There are often sensors and smoke detectors installed in the ductwork of the mechanical system. The oil-based challenges can leave a film on these sensors which could impact the accuracy of the sensor. This problem is often worse over time, as dust and particulate will stick to this film and cause a buildup on the sensor. If oil-based challenge is to be used, it should be introduced such that it will not come in contact with these sensors. This is especially important if a large amount of challenge is needed for the test. The sensors can often be either removed or covered for the test.

19.4.1.5 Ceiling assembly shroud test. A less standardized but still popular method of testing ceiling HEPA filters is by using a shroud test. This is a very difficult test to perform properly, but is specified for some jobs. The test is performed by capturing all of the air supplied through a HEPA filter in a shroud, or funnel, and sampling this air with a particle counter. The air is drawn into the funnel by using a fan at the bottom of the funnel which is adjusted to the flow of the filter. A challenge can be used to exaggerate any leakage present. There are two main problems with performing this test. The top of the funnel must be held tightly to the ceiling to ensure the filter is isolated from surrounding filters or ceiling tiles. This often requires a gasket to make the seal between the ceiling and the shroud. However, the vibration of the fan can cause particles to be shed from the gasket which provide false readings. The funnel must have a flow which is very close to the flow of the filter being tested. If the flow is different, the test can be upset. Surrounding air will be drawn into the funnel if the flow is set too high, and air which would be sampled can escape if the flow is insufficient. For areas where each filter does not have the same design flow rate, this test is very impractical.

19.4.1.6 Overall efficiency test. There are many instances where the filters are installed in ductwork or a housing which makes them inaccessible for scanning. For this type of application, an overall efficiency test is performed. This test requires a challenge to be introduced upsteam of the filter and samples measured downstream to calculate the efficiency of the filter. The same challenges and test instruments can be used for this test as with previously discussed scan tests. The main problem with this test is having enough length of ductwork to perform the measurements accurately. To ensure representative readings a long stretch of straight duct must be present on both sides of the filter. The typical requirement is to specify the point where the challenge is introduced and the point upstream where the challenge is measured. An additional 10 duct diameters are required from the filter to the downstream point where the penetration is measured. This is often not the case for in-line filters. The judgment and experience of the technician is important in determining if the test can be performed with any reliability.

Some options include taking readings at several locations across the ductwork to obtain a representative average, or having a small access door installed to allow the HEPA to be scanned. A fan, or other features in the ductwork, will reduce the length of duct required as they tend to mix the air to increase the reliability of the readings.

If the test is performed with a particle counter, you must first measure the background levels both upstream and downstream of the filter. For both test instruments, the test challenge should then be introduced upstream of the filter and measured. The challenge is left on and the penetration is measured downstream of the filter. If the test is performed with a photometer, this will either indicate pass or fail. If it is performed with a particle counter and sample dilution device, then these readings must be incorporated into calculations for overall efficiency. The background is subtracted from the readings with challenge to obtain the number of particles added both upstream and downstream of the HEPA. The downstream count is divided by the upstream count to provide a particle penetration reading and multiplied by 100 to yield percentage penetration. When this number is subtracted from 100 it is the percentage efficiency of the filter(s).

The acceptable penetration is usually based on the design of the filters being tested, either 99.97% or 99.99% efficient. This means that a penetration of less than either 0.03% or 0.01% would be acceptable.

19.4.2 *Velocity and uniformity*

Velocity and uniformity are measured to ensure that the airflow into the room is sufficient and as a secondary check of the air balance contractor. The velocity is typically measured at the centre of 2 × 2 ft (1.61 × 1.61 m) grids directly beneath each HEPA filter. The location of the velocity reading varies depending on the job. Typically it is measured either 6 or 12 in (15.24 or 30.48 cm) from the filter face. However, for Class 100 or cleaner areas, where the percentage of filter coverage is higher, additional readings are often required at the entrance and exit of the 'work zone'. The work zone is the volume of air defined by the horizontal planes (for vertical flow rooms), located 2–3 ft (0.61–0.91 m) below the ceiling and 2–3 ft (0.61–0.91 m) above the floor (depending on the particular room). This is the area where the highest degree of cleanliness is required for the process in the clean room. These additional measurements are made to ensure that the airflow is laminar, or unidirectional. In other words, the flow is smooth through this zone without areas of high or low flow.

19.4.2.1 Test instruments. There are several types of instrument that are typically used for measuring velocity. The most popular instrument is a hot wire anemometer. This instrument has a small wire in the sensing portion of the probe which is heated to a known temperature. The airflow across the probe cools this wire, and the instrument converts this cooling to velocity. This instrument is fairly reliable, and is comparatively inexpensive. However, there are limitations to its use. The typical range for a hot wire anemometer is 30 to 3000 FPM (feet per minute). This is the usual range of airflow velocity tested, but it is not uncommon to see readings outside this range. Also, the size of the sensing device has historically been a concern. The sensing device is often smaller than 1/2 in^2 (1.61 cm^2), which is a very small portion of airflow being tested. It is often a requirement to take several readings, or a profile, to have a reliable velocity reading with this instrument.

A second type of instrument is a rotating vane anemometer. This instrument has a fan-like propeller which is rotated by the airflow. The speed of rotation is converted to velocity. The range of velocity which this instrument can measure is similar to that of a hot-wire anemometer. The size of the vane will vary between manufacturers and models, but it is usually a more representative reading than can be obtained with a hot-wire anemometer. However, due to the size of the sensing vane, this instrument is not practical for measuring air in a duct. The best application for this instrument is for open spaces such as clean rooms. One of the concerns with this instrument is with the inertia of the vane as it rotates. This often causes the intrument to be very slow in responding to a change in velocity. If this instrument is to be used, the length of time for each reading should be increased to ensure a proper reading.

It is also possible to measure velocity with pressure sensing instruments. These instruments use different types of probes which measure the velocity pressure and convert this to velocity. The accuracy of these instruments is dependent on the accuracy of the pressure sensing device being used. An incline manometer, which can be used to measure pressure as well as velocity, is usually not very accurate. It is also very inconvenient as it must be set up and leveled for each reading. Newer electromanometers provide good adaptability for measurement of velocity. Some of the more sophisticated instruments will correct the reading for temperature and pressure which is important. There are several probes available for use with these instruments which allow most applications of velocity measurement to be tested accurately. This includes probes and pitot tubes for duct traverses, and probes and grids for measurement in clean rooms. The grid will provide a very good reading for a HEPA filter as it will measure velocity at several locations across the grid and display the average. Of course, it is very important to only use the attachment with the sensing instrument for which it is designed.

19.4.2.2 The K *factor.* One of the most common problems with velocity is that the acceptance criteria for the test are not determined properly. In the past, the typical recommended velocity was thought to be 90 FPM (27.4 m/min) for almost every application. While this is a good target velocity for many projects, the design of the project should not be set around this figure. Other factors which are also very important in determining the velocity include required room air exchange rate and required pressurization.

Once the required air flow is determined, and the number of filters is set, then the velocity can be determined. The velocity is directly related to the flow, which can be seen by the abbreviations for the units. Air velocity is typically represented in feet per minute, FPM, while total flow is referred to as cubic feet per minute, CFM. In other words, by cubing feet, velocity becomes flow. The amount of space through which the air passes (in square feet) is the area which converts the velocity to flow. This area is known as the K factor.

However, it is important to realize that a 2×4 ft (0.61×1.22 m) ceiling HEPA filter does not have a K factor of 8 ft^2 (0.74 m^2) which would naturally be assumed. The filters ar usually undersized to fit into a 2×4 ft (0.61×1.22 m) grid, which means the K factor is smaller. And since the filter medium is the only portion of the filter which actually supports airflow, this is the area which should be used for the K factor. The filter frame, the urethane sealant, caulking, and the center support bar (if applicable) must all be subtracted out to obtain an accurate K factor. This means that a 2×4 ft (0.61×1.22 m) filter would actually have a K factor of between 5.5 and 7 ft^2 (0.51 and 0.65 m^2) depending on the style of the filter.

The typical range of acceptable velocity readings is between 75 and 90 FPM (22.9 and 27.4 m/min). A velocity higher than this can cause turbulence which causes particles to remain airborne longer. If the lower range of velocity is desired, there must be sufficient filter coverage to provide the air exchange rate required for cleanliness.

19.4.2.3 Uniformity. A second aspect of velocity which is often required in a performance specification is the uniformity of the velocity readings. This is a set of calculations which analyze the individual velocity readings to determine if the range of readings is acceptable. This does not apply to all types of clean rooms. For instance, a room may be designed where each filter does not support the same amount of airflow. Naturally, the velocity would not be the same through each filter, either. When different size filters are used the uniformity is often skewed. However, if the area is designed to be Class 100 or cleaner, the uniformity is very important. This is a good means of assessing the amount of turbulence in the area. If the uniformity is good, there is generally a smooth, unidirectional flow of air through the room.

Therefore, the application of an acceptance for the uniformity of velocity should be based on the room design. If this analysis is desired, it is usually worded in the following manner:

> The average velocity for the room(s) shall be 90 FPM ± 5 FPM. A total of 90% of the individual readings shall be within 10% of the measured average velocity. The remaining 10% of the readings shall be within 30% of the measured average velocity.

19.4.3 Total air flow

The measurement of the total airflow into a room is a popular but unstandardized test. There are several air balancing standards which reference this test, but most clean room testing standards do not detail the test. This test provides a more representative measurement of the flow when compared to a calculation made from a velocity reading to determine the flow. This is important to evaluate fan performance and room exchange rate, as well as a check of the air balance contractor's work.

For most projects, this test would be considered optional. It is recommended for rooms where the HEPA filters are in a remote bank and the clean air is delivered through diffusers. A diffuser generally will not provide a uniform velocity to make the velocity reading valid. The total flow is the best method of providing representative depiction of airflow distribution. It is also a good test for rooms which are considered 'turbulent flow'. This means there is less than 80% filter coverage and the airflow patterns through the room are not unidirectional.

This test is performed with an instrument known as a balometer or a flow hood. This is a cone shaped hood which captures the air supplied through a filter or diffuser. There is a sensing grid at the base of the instrument which measures the flow based on pressure. Some of the more advanced flow instruments will also compensate the readings for temperature, density, and back pressure. The test is performed by holding the hood tightly to the ceiling so that it captures all of the air supplied. The flow hood kit should have several sized hoods to enable measurement of various sized filters and diffusers. The reading is taken for approximately 5 to 10 minutes depending on the instrument. The instrument reads directly to cubic feet per minute, which is the unit of measure. The individual flow readings can be added to provide the total air flow supplied by a fan or into a room.

The instrument can also be used to measure return or exhaust flows. However, a correction factor is usually required when using the instrument for reverse transition into the smaller sensing portion of the hood. When reverse flow is measured, there is no transition, the air simply flows directly into the base of the instrument. This causes a higher resistance than is typically seen for the diffuser, so the flow is somewhat restricted. The correction factor is determined by measuring velocity with the hood on and with the hood off, and calculating the difference. The correction factor is usually less than 10% depending on the size of the reverse flow being measured.

The total air flow is very important in determining the number of air changes for a room or area. The room air exchange rate is the number of times per hour the entire room volume is theoretically replaced with filtered air.

This is more a theoretical calculation than an exact figure as some areas of the room have higher flow rates than other sections. This is determined by dividing the room volume by the total delivered air flow (CFM), and multiplying the figure by 60 to convert minutes to hours.

19.4.4 Room differential pressure

The pressure relationships between the various rooms is an important consideration. This becomes more important for biotechnical and pharmaceutical facilities where segregation of areas is needed. The pressure relationships are measured using a manometer.

There are several types of manometer, the most basic are called incline manometers. The test is performed by simply connecting a tube firmly to one pressure port and measuring the differential across a closed door. This is repeated for each door, starting with the innermost, until readings are obtained for the entire area. Prior to starting, all doors should be closed, and traffic in the area should be restricted to ensure that an unseen door will not be opened to change the pressure relationships. The test should be continued until the pressure is measured to ambient, outside of the building. This is important as the clean room facility commonly does not have doors to ambient, just to the building itself. However, if the building itself is either positive or negative, it will have an impact on the differentials. Reading to ambient will allow for the building pressure to be considered.

19.4.5 Airborne particle counts

Particle counts are the sole test required to determine room cleanliness class. The test is usually performed in accordance with Federal Standard 209D. Entire papers can, and have, been written about this standard. I will be brief in simply discussing the application of the standard. The proposed revision to this standard is currently pending, and will have an impact on the test procedures.

Federal Standard 209D is designed to provide a standard class of air cleanliness for clean rooms and laminar flow equipment. In order to develop a specification, a design or target class must first be selected to which it is wished to certify the room. Tables 19.1 and 19.2 can then be used to determine the requirements for the testing based on the target class. The sample volume (size of each particle count) is determined from Table 19.1 for minimum values.

After determining the correct sample volume for the measured particle size, the number of samples for the room or area being classified must be determined. The minimum number of sample locations required for classification of a clean zone shall be the lesser of (i) the area of the entrance plane in square feet divided by 25, or, (ii) the area of the entrance plane divided by the square root of the anticipated or desired class designation. There shall be a minimum of 2 locations and 5 samples for each clean zone.

The acquired particle counts are then analyzed statistically to determine the random error by use of the one tailed Student's T distribution. This analysis determines the mean of the counts, standard deviation, standard error, and the upper confidence limit (UCL)

The clean room or clean zone shall meet the acceptance criteria if, (i) the average of the particle concentrations measured at each location falls at or below the class limit, and (ii) the mean of these averages falls at or below

Table 19.1 Minimum volume per sample (in cubic feet) for the air cleanliness and measured particle size shown*

	Measured particle size (μm)				
Class	0.1	0.2	0.3	0.5	5.0
1	0.6	3.0	7.0	20.0	NA
10	0.1	0.3	0.7	2.0	NA
100	NA	0.1	0.1	0.2	NA
1000	NA	NA	NA	0.1	3.0
10 000	NA	NA	NA	0.1	0.3
100 000	NA	NA	NA	0.1	0.3

* Source: see ref. 3, p. 4.

Table 19.2 Class limits (in particles per cubit foot) of size equal to or larger than particle sizes shown*

Class	Measured particle size (μm)				
	0.1	0.2	0.3	0.5	5.0
1	35	7.5	3	1	NA
10	350	75	30	10	NA
100	NA	750	300	100	NA
1000	NA	NA	NA	1000	7
10 000	NA	NA	NA	10 000	70
100 000	NA	NA	NA	100 000	700

* Source: see ref. 3, p. 9.

the class limit with a 95% UCL. The class limits for each particle size are presented in Table 19.2.

It is important to remember that it is possible to measure a particle size between the specified sizes by means of interpolation, but that the maximum or minimum size cannot be exceeded by means of extrapolation.

The particle counts, as well as the entire certification, are performed for as many as three phases. This includes the 'as built' or 'static', where no personnel or equipment are in the rooms, the 'at rest' where equipment is installed and running, but no personnel are present, and the 'operational' or 'dynamic' where the room is in full operation with all equipment and personnel carrying out typical duties. The phase of testing should be mentioned in the certification to Federal Standard 209D as it has an impact on the results. Rooms are often cleaner in the 'as-built' testing so the additional particulate load of the personnel and equipment does not cause the room to exceed Class.

19.4.6 Temperature and humidity control

The control of temperature is crucial for some clean rooms, but merely for comfort in others. The degree of control required for each area should be evaluated to ensure proper testing.

Several methods of testing temperature and humidity exist, the type of test should be selected based on the user's particular needs. It should be kept in mind that it is much harder to control temperature and humidity on hot humid summer days and cold dry winter days. If a room passes on a mild day, it may not pass on a more severe day. Most facilities where temperature and humidity are critical have intricate control systems which constantly monitor the conditions. The certification data should be cross checked to these sensors to ensure consistency. The monitoring sensors could then be a constant source of control capabilities.

The simplest form of testing is performed by taking readings at grid points across the room. This only requires a hand held sensor for temperature and humidity. The main drawback to this test is that normal fluctuations in temperature and humidity will not be reflected. A certain spot or grid point may appear to be cooler or hotter when compared to others. This could be true, but it could also be that the entire room was cooling or heating, and the reading at that time reflects the trend for only a portion of the readings. This may be a good test for a loosely controlled system, where the temperature and humidity control is for comfort only.

A second method which is very similar, is the one recommended by IES-RP-CC-006-84-T. This requires two hand held temperature and humidity sensors. The room is again divided into grid locations, usually the center of 10 × 10 ft (3.048 × 3.048 m) grids. The two probes are held together to ensure they read consistently. One probe becomes the reference probe, and is held at the center gridpoint throughout the testing. The reference temperature and humidity are recorded. With the second probe, each additional gridpoint is recorded as the difference between the reference probe and that gridpoint. This test provides a better study of the uniformity of temperature and humidity across a room.

A third method has become very popular, but is not directly specified in any industry standard. This involves using several temperature and humidity sensors and a data logger, which simply stores all of the readings in memory. The probes are again held together to ensure that they read consistently. They are then distributed across the room or at gridpoint locations. The data logger can usually be programmed to take a reading at various intervals for a certain length of time. For instance, readings can be continuously monitored at four locations with a single temperature and humidity reading taken every minute for each probe for a 2 h period. Or, readings can be recorded every 15 min for a 24 h period. The data logger is then hooked into a computer to dump the data into graph format displaying the temperature or humidity versus time for the specified period of time. This gives a very good indication of the swings in temperature and humidity, and the ability of the system to control over time.

The last method discussed is from an older standard, AACC-76. This is a complicated test and involves many sensors. Each temperature and humidity control zone is tested as a whole. This means that if one room is supplied by one air handler and has one sensor, it is a single control zone. If a large room has four air handlers and four sensors, then there would be four zones to test. The number of zones is equal to the number of control sensors in the space. The test requires that a probe be placed at all points of concern throughout this control zone. This includes at the intake to the outside (make-up) air duct, at the inlet and discharge of each coil (i.e. make-up unit coils, recirculation unit coils, etc.). Additional probes are placed in the area served by the air handler, on a grid pattern. These reading are all logged for a specified period of time, as in the previous test. The main drawback of this test is the extraordinary amount of time that it takes to set up the probes properly and the large number of probes required. A variation of this utilizes the building control sensors which are usually in many of these locations. In other words, the control system can collect many of the data points for the user.

19.4.7 HEPA filter pressure drop

A very important test which is often overlooked is the HEPA filter pressure drop. This is simply the resistance of the HEPA filter to the airflow. This test monitors the loading of the filters and determines how much life is left in the filters.

The test is performed using a manometer, or pressure gauge. A tube is connected to the high pressure tap, with a static tip to eliminate the effect of velocity on the reading. The tip is then placed on the upstream side of the filter with the low pressure port exposed to room pressure. This is usually accomplished by removing a screw from the centerboard of the filter which reveals a port for testing such as this (or for checking the challenge when scanning the filter). However, some HEPA filters do not have these ports, which makes this test difficult. For large open plenum systems, a single port can be made for an entire plenum. This should be done very carefully to prevent the possibility of damaging the filters or leaving a port which allows unfiltered air to enter the clean room through it.

It is important to remember that the resistance is directly related to the amount of airflow. As the flow increases, so does the resistance. This is important because as a filter loads, not only is there an increase in static pressure, but also a reduction in air flow. It is therefore good practice to calculate the resistance per cubic foot of air. This will correct the resistance at the various flow rates and allow for comparison.

A filter is considered loaded when either the initial resistance has doubled, or the air flow required cannot be maintained. Some filters are changed more frequently than this to conserve electricity. However, a large saving is required to off-set the expense of the filters and testing required.

It is important to perform this test at least initially for a new facility to draw a baseline for future results. Then after 7 or 10 years, when loading is a concern, there will be data to compare with the current results. It should be noted that HEPA filters load at an increasing rate over time. For instance, the first 5 years of life may see no increase in static pressure. By 7 years a noticeable increase is evident, and each additional year shows a larger increase.

19.4.8 Airflow parallelism and dispersion

This test is intended for unidirectional flow clean rooms, usually Class 100 or cleaner. The purpose of the test is to measure the parallelism of the airflow within the clean room, and the ability of the room to limit the dispersion of particulate. There are several common methods for performing this test and the most popular will be discussed here.

The method recommended by IES-RP-CC-006-84-T uses a smoke source and a particle counter. The test is performed at the center of a grid pattern, usually 10 × 10 ft (3.048 × 3.048 m). The smoke is introduced at the entrance to the work zone, usually 1 to 3 ft (0.31–0.91 m) below the ceiling depending on ceiling height. The probe of the particle counter is advanced towards the source from all directions at the exit to the work zone, usually 2 to 3 ft (0.61–0.91 m) above the floor. A plumb bob (which is nothing more than a weight on a string) is suspended from the point of introduction to determine perfect vertical position. The point where the counts increase as the probe is advanced is measured as the offset from vertical. This is measured using either a pointer and stand to measure in degrees, or with a tape measure to measure in inches (cm). After the offset from vertical has been measured in all directions then the test is repeated for the next grid point. The main problem with this test is that it requires the introduction of smoke in the clean room, and it is time consuming.

One of the most popular alternatives to this test involves hanging streamers to the ceiling grid to reflect airflow patterns. The plumb bob and either pointer or tape is again used to determine offset from vertical. The main problem with this test is that different streamers will reflect different patterns. Some of the more common streamers are polyethylene streamers (with a thickness of under 0.5 mil), certain types of thread which are very loosely wound, tyvek paper (which does not shed particles), cassette tape, and even dental floss. Any of these can be used, but there are concerns with some. The tyvek is very thick and tends to exaggerate the veritcal by its weight. Audio cassette tapes are usually coated with magnetic particles which are detrimental to many clean rooms. Dental floss is very difficult to unwind, and often will not hang vertically but will roll up on itself. The main consideration in choosing a material is that it should have good surface area to weight ratio to ensure that any patterns will affect the streamer.

A third method involves generating helium bubbles, using a special machine, which float in the air streams. The main problems with this test are that the bubbles are not introduced isokinetically (at the same velocity as the air being tested). The bubbles are also very small and difficult to see, which makes detecting stray bubbles difficult.

The last method involves using the fog produced by dry ice to visualize airflow patterns. The fog itself is very cool and would sink quickly to exaggerate vertical flow. The fog should therefore be warmed to roughly room temperature prior to introduction to ensure proper depiction. There should also be a control on the velocity of the fog output to ensure that it is isokinetic. These same concerns are valid in the use of a theatrical fog generator, which is fairly popular. The cleanliness (or lack of it) of the theatrical fog has also been a problem since rooms require complete cleaning after its use.

As can be seen, there are many methods for performing this test. Each has its specific benefits and drawbacks. As with any test, it should be determined exactly what it is wished to demonstrate with the test, and which of the methods will best provide the demonstration.

19.4.9 Room recovery

The purpose of this test is to document the ability of the clean room to cleanse itself from any internal contamination source. The test is performed by introducing a contaminant and recording the time which the room requires to return to background levels. For rooms with unidirectional flow, the smoke is introduced at the entrance to the work zone, and the particle counter is located directly downsteam of the source at the exit to the work zone. For turbulent flow rooms, the smoke is introduced at the center of grid points, and the counter is positioned at the nearest return grill.

There are several problems with this test and it has become less popular in recent years. Of course, the introduction of smoke is the obvious problem. There are also variables in the test which can dramatically affect the recovery time. The distance from the smoke source to the counter is one variable. The amount of smoke introduced is a second. Typical standards recommend generating smoke for one minute. Different generators will produce different quantities of smoke in this period. As long as the same generator is used, and the initial concentration is recorded, this variable can be reduced. However, most particle counters will not count extremely high concentrations of particles, which again limits the accuracy of the test.

The usefulness of the test is also a point of contention for many people. The results are meant to determine how long the room will take to recover from internally generated particles. In other words, if a problem occurs, how long a wait is necessary before reusing the room. However, the nature of the problem will probably vary, sometimes significantly, from the parameters of this test. And, if a problem does occur, should a cleaning be required prior to reusing the room to remove large particles?

19.4.10 Induction leak test

The induction leak scan is designed to detect any source of unfiltered air penetrating the clean room through the walls and features of the walls. These features include wall joints or cracks, outlets, pass troughs, sprinklers, speakers, or any other interruption or penetration of the clean room envelope. The recommended method of performing this test is by introducing smoke on the opposite side of the wall under test and scanning all features of the wall with a particle counter.

This test is often very difficult to perform because the background particle counts of the room interfere with possible leak readings. In other words, if particles are detected while scanning, it must be determined if they are the typical particles allowed in a room of this class, or if they are the smoke challenge penetrating the room. This test requires a good deal of experience to be performed properly. For rooms with a positive pressure plenum, it is a good idea to perform the scan with the smoke in the plenum. The smoke should not penetrate from any area which is negatively pressurized to the room under test, and if a surrounding room has a higher pressure, the chances are that it will be of a cleaner class. The only area which is typically positive is the plenum, which is often the problem. For example, electrical conduits can run through the plenum and become pressurized, allowing particulate to travel through them unfiltered, into the room.

This is another test where judgment must be used as to the value of this test in a particular project. For instance, it would be very difficult to detect a leak in a Class 100 000 room, yet a positive plenum Class 10 room could be severely compromised by a problem that this test would detect.

19.4.11 Noise level testing

Noise levels are often a concern in clean rooms due to the large amount of air being moved and the equipment required to perform this. The specialized clean room walls and floor do not absorb sound, which complicates this problem. There are two basic methods of measuring the airborne sound pressure levels, either ‘A’ scale readings or octave band analysis.

The ‘A’ scale is a method of measuring sound where the actual pressures are weighted to reflect what the human ear actually hears. This test is typically performed at the center of 10 × 10 ft (3.048 × 3.048 m) grids with the probe held 4–5 ft (1.22–1.52 m) above the floor. The details of the test will vary depending on the type of instrument being used, but it should be set to accept random noise from all directions. According to standards, the test should be performed with all clean rooms fans off to obtain background levels, then repeated with the equipment on to determine total added sound pressure. However, once a clean room is complete, it is often very difficult to arrange to have the equipment turned off and risk contamination. A typical compromise is to measure the total sound, and if it exceeds a preset design figure, then the equipment is shut off to determine the background. Certain pieces of equipment can then be turned on to try to pinpoint the problem. If the total readings are at or below the design figure, then the levels are accepted.

An important fact about sound pressure is that there is a logarithmic relationship between the level and the distance from the source. As the measurement approaches the source of the sound, the intensity increases logarithmically. This means that air handling equipment located in close proximity to the room will increase noise in the clean room. The integral fan/HEPA filter modules which have become popular over recent years have a high noise because the fan is directly above the filter. Therefore, these fan/filter units are typically not used for any room cleaner than Class 1000.

The second form of sound testing is called octave band analysis. In this test, the sound pressure levels are measured for eight bands, or frequencies, measured in hertz. The procedure is basically the same, but the eight individual band readings are taken at each location. These eight points are graphed, and the acceptance is based on standard curves known as NC curves. This test is good in that it can identify exactly which frequencies of sound are present. This is important since damage to hearing is possible with lower pressures in the 4000 Hz range.

19.4.12 Light level tests

Light level readings are performed to ensure that proper levels of lighting are maintained either for worker comfort or for some particular product exposure. The test is typically performed at the center of 10 × 10 ft (3.048 × 3.048 m) grids with the probe held at a 48 in (122 cm) height. The test instrument is called a photometer (however, it is a different photometer from that used for the scan tests). This instrument gives a direct readout in footcandles, which is the unit of measure for this test. A range of 100 footcandles (*x* cd) is usually sufficient for employee comfort. Some microelectronics or photoresist clean rooms require special lighting which may have a different acceptance. There are also some special pharmaceutical products where exposure to light must be controlled.

19.4.13 Vibration testing

Vibration testing is sometimes required along with the certification. Many microelectronics manufacturers must control the vibration to ensure that the small diameters of the chips can be attained. This testing is a science in itself, and could fill entire an chapter. In an effort to be brief, I will simply explain the testing process.

The selection of the test equipment is important in order to have the resolution and accuracy for the minute readings often required. Baseline data are often collected prior to design of the facility to enable the vibration control to be designed into the project. Final readings are then measured at the locations of critical pieces of equipment whose performance could be compromised by vibration. The readings are recorded in frequency of the vibration, amplitude of the various peaks, and direction on the *X*, *Y*, or *Z* axis.

19.5 Selecting a certification contractor

Selecting a proper certification contractor is a very important decision in the clean room project. Unfortunately, since it is one of the last services required, it is often decided hastily in the final stages of the project.

The certification is often the only means of verifying that the project is complete and the required performance levels are met. This is an important point, especially when the cost of the clean room is considered. The selection of a certification contractor is usually done using a bid document and job specifications.

19.5.1 Responsibilities

The final certification can be performed in many different ways for a clean room project. If the certification is to be performed by a division of one of the contractors or suppliers, the following bid information would not apply. However, it is prudent at least to check on the pricing of this type of certification to ensure it is cost effective. If the certification is to be performed by an independent contractor, the company or person responsible for the certification must be determined.

The certification is usually performed either for the general contractor or for the owner of the project, but this is not always the case; the certification could be the responsiblity of the installation contractor, the ceiling contractor, the mechancial contractor, or even the architect. There is a conflict of interest when a contractor employs a company to certify his work. The certification is performed to ensure that the requirements of the owner are met, so naturally it follows that the owner should employ the certifier.

19.5.2 Specifications

The job specifications should be custom tailored to the job to ensure proper evaluation of the project. Some of the items which should be considered include which tests are important for performance evaluation, the exact testing instrumentation and procedure, and the desired acceptance criteria. This acceptance criteria must also be referenced in the construction specifications to ensure that the contractors are guaranteeing certain performance levels. It should also be noted that the performance will be tested, and if deficiencies are found, it is the responsibility of the contractor to rectify the problems. When the contractors know that their work will be tested, the quality is often better.

At the very least, the acceptance criteria should be listed in the job specification, along with delegation of responsibility. For instance, the balancing or mechanical contractor would be responsible for providing an average velocity of 90 FPM (27.4 m/min) and the required levels of pressure. The filter supplier or installation contractor would guarantee that the installed filters are free from any type of leakage of 0.01% or greater. This allows the job to progress while proper specifications are developed.

The final certification specification should include all of the following items.

- All areas to be tested, including a matrix if some areas require more tests than others. It is often a good idea to include drawings of the mechanical layout, room layout, and reflected ceiling plan with some specifications.
- The tests to be performed, purpose of each test, equipment required for these tests, exact test procedure, acceptance criteria, and reporting requirements.

Once the certification specification is developed, it can be incorporated into a bid document, which can be used to solicit proposals.

19.5.3 Bid documents

There are usually several additional documents which combine with the job specification to form the bid package. Items such as scheduling requirements, any site safety information, and minimum insurance coverages are included along with the formal request for proposal.

The qualification of the certifier is an important consideration in the bid package, since it should be ensured that the selected contractor is qualified and capable of providing the required service. Some of the more common requirements are that the company has been in business for a number of years, and that it has completed several projects of similar size and scope in the near past, and all equipment is properly calibrated. Additional conditions include experience of each of the technicians as well as the project manager, and references to contact regarding previous work. The qualifications of the technicians are extremely important as they will probably be collecting most of the test results. The user should ensure that the people working on the project have sufficient experience for the user's particular needs.

The structure of the request should allow for the certifier to provide a lump sum bid based solely on the job requirements. This will allow for easy comparison between the various bids. However, there should also be a section for 'suggested alternatives' to the specification and a breakout of the bid by test to allow for thorough evaluation of the bids. The alternatives allow for various companies to suggest methods of testing or test equipment which could provide an equal or better evaluation at a lower cost. The alternatives must be explained in detail to ensure that the result will be valid and the certification will not be compromised. The alternatives can then be compared to the individual test prices of the other bidders which are broken out per test. The document should request hourly rates, both straight time and off-hours, for each class of labor in the event that additional repair, adjustments or retesting are required.

19.5.4 *Interpretation of the bids*

The ease of interpretation of the various proposals will depend on the complexity of the certification required, and the complexity of the request for proposal. Any alternatives proposed should be thoroughly evaluated. Since the individual tests are broken out in the bid, the effect of the alternatives can be compared between the various bidders. The references should be contacted to determine if the bidder satisfactorily performed previous certification. It should also be verified that the 'qualification' projects were either similar in size and scope or larger. All insurance coverages, qualifications, and equipment calibration should be considered. A personal interview with the certifier prior to award of the project is also a prudent step.

19.6 The certification

19.6.1 *Scheduling the certification*

Once the certifier has been chosen, the actual certification can be scheduled. The scheduling of the project should be arranged far enough in advance to ensure a proper time frame to perform the work. For larger projects, two weeks to a month notice may be required, smaller projects may require less notice. This will also depend on the size and work load of the certification contractor.

Some of the other work which should be complete prior to the final certification include all construction, air and water balancing (it may be helpful to obtain copies of the field results from the balance contractor for reference during the certification rather than waiting for the final report), verification of the controls system, and at least a rough cleaning of the space. The final cleaning is often delayed since a fine tune of the system may be required which could involve removing HEPA filters or lifting ceiling tiles.

If some of theses trades are still in progress, it could have an impact on the certification results. For instance, the balance contractor requires the fan to be shut off for some of the readings, which would cause many tests to fail. If the controls system is not fully operational, the certification may depict results that will change as the system is implemented. If the construction is not complete, this could also affect the readings. Some common problems encountered in the past include the absence of windows or door sweeps which affect pressure, incomplete ceilings, and other small items which can have an impact on the readings.

Once the certification has begun, access to the clean rooms should be kept to a minimum to prevent interference with the test results. Any work required which could have an impact on the readings should be reviewed with the certifier to minimize the impact of the work. The certification could be performed in another area, or a test which would not be affected could be performed.

19.6.2 Working with the certifier

It is a good practice to work closely with the certification contractor to ensure a thorough understanding of the performance of the clean rooms. This could be as simple as meeting the certifier at the start or end of each day, or actually witnessing the certification and asking pertinent questions. This becomes even more important if there is any question of bias in the certification contractor. It is also good practice to check the calibration stickers or documents for the equipment being used for the testing.

19.6.3 Dealing with test failures

The design and construction of clean rooms is a difficult process. There are often many problems encountered in the construction process which were not anticipated by the design. The smallest oversight could have a dramatic impact on the certification results. As projects become cleaner or more complex, these issues become more crucial. There are also many design assumptions which need to be verified by the final certification. For instance, for design purposes each door is assumed to leak 25 CFM (2.3 m^3/min) which is therefore added to the total make up for each room. What happens if the doors actually leak 50 CFM (4.65 m^3/min) each, or do not leak at all?

For these reasons and many more, it should not be expected that every test will pass with flying colors. The user should be prepared for the possibility that one or more of the tests will fail, and to take corrective action to resolve the problem. This is not to say that every project has problems, many do not. However, being prepared for possible problems should make dealing with them easier.

As mentioned previously the responsibility for resolving potential problems should be addressed in the specifications for the project. The resolution of problems often requries the certifier to work with other trades to test and retest until the problem is resolved . It may be prudent to consult with the design team if the scope of the problem is large to ensure proper resolution.

19.6.4 The final report

Once the testing and certification are complete, the final certification report is compiled. The report should include all test results in an easy to comprehend format. The test results are often reported on an area layout or a reflected ceiling plan. This allows for the results to appear in the exact location of the test and is important to allow for correlation of the results from different tests. For instance, if the particle counts are higher for one area of a room, the velocity section can be consulted to ensure that there is proper flow in the area.

The test procedures and reporting format should be explained for each test. This allows for a better understanding of the results, and the ability to repeat the exact test in the future to compare results over time. Any required analysis such as Federal Standard 209D calculations or uniformity analysis should be displayed in full. There should be an executive summary or comments section where any corrective actions required could be explained in detail. This section could also include any relevant events which impacted the test results. The final report should also include the calibration forms for all of the test equipment used on the project.

References

1. Cadwell, G.H. Jr., *Filtration Update*, Institute of Environmental Sciences Technical Presentation.
2. First, M.W., Rudnick, S.N. and Yan, X. (1992). *Use of Alternative Liquids in Laskin Nozzle Generated Aerosols for Filter Testing*, Harvard Air Cleaning Laboratory, Harvard School of Public Health, Boston, MA.
3. *Federal Standard 209D*, Institute of Environmental Sciences, Mount Prospect, Illinois, June 1988.

Bibliography

American National Standards Institute, *Procedure for Measuring and Rating Steady-State Room Noise*, New York.

American Society for Testing and Materials, (ASTM), *Standards 24, 25, 50, and 51*, Philadelphia, PA. These standards all deal with various methods of counting and sizing airborne particulate.

American Society of Heating, Air Conditioning, and Refrigeration Engineers, (ASHRAE), *Standard 52-76*, Atlanta, GA. This deals with evaluating filters which have efficiencies below HEPA filters.

Associated Air Balance Council, *Field Service Manual*, Washington, DC.

Food and Drug Administration, *Good Manufacturing Practices*, Washington, DC. This deals with general guidelines for use of manufacturing space, and proper documentation of such use.

IES-RP-CC-002-86 provides testing required for laminar flow clean air devices.

Illumination Engineering Society of North America, *Lighting Handbook*, New York.

Institute of Environmental Sciences, *Federal Standard 209D*, Mount Prospect, IL, 1984.

Institute of Environmental Sciences, IES-RP-CC-001-86-T, Mount Prospect, IL, 1986. This describes the materials, construction and testing of HEPA filters, and provides for classification of the various filter grades.

Institute of Environmental Sciences, IES-RP-CC-006-84-T, Mount Prospect, IL, 1984.

National Environmental Balancing Bureau, *Procedural Standards for the Certified Testing of Cleanrooms*, Vienna, VA.

National Sanitation Foundation standard 49 provides a complete description of materials, construction, and certification of biological safety cabinets.

National Toxicology Program, *Technical Report on the Carcinogenesis Bioassay of DOP*, Research Triangle, NC.

SAMA Standard 10 details the testing and certification of exhaust hoods.

20 Regulatory considerations

W.S. KLETCH

20.1 Historical considerations

People and organizations often need a crisis to stimulate them to action. The US Congress action on food and drug safety requirements is a prime example. In the early 1900s the public outcry from the sale of putrid, decomposed, meat caused the US Congress to enact The Meat Inspection Act and the original Food and Drugs Act of 1906. Industry lobbying caused these laws to have little or no enforcement power, and accordingly they were ineffective.

In 1937 the 'Elixir of Sulfanilamide' incident, in which 107 people died as a result of taking a product which was formulated with a toxic solvent, caused the US Congress to enact the Federal Food, Drug and Cosmetic (FDC) Act of 1938. This act was the basis of the current FDC statute and provided, among other things, the requirement that drugs be safe, the setting of tolerances for unavoidable poisonous substances, the authority for factory inspections, and the provision for court injunctions for recalcitrant offenders.

As a result of the thalidomide investigation in the early 1960s, Congress enacted the Kefauver–Harris Drug Amendments in 1962, which required manufacturers to demonstrate the effectiveness of their products prior to marketing. In 1972, the FDA was charged with the regulation of biologics, including serums, vaccines, and blood products. Recognizing the increasing role of medical devices, similar legislation was approved in 1976, requiring that the safety and effectiveness of devices be demonstrated as well.

While at times the enforcement philosophy has waned, the Fines Enhancement Laws of 1984 and 1987, and the current congressional oversight of the FDA, has caused a strict enforcement philosophy to evolve. The industry should be forewarned by the tone of FDA podium pronouncements and this flurry of congressional activity. The history of the FDA is replete with notable milestones and clearly documents the need for regulatory requirements[1].

The FDA act makes reference to the authority of the Secretary of the Department of Health and Human Services to promulgate regulations for the efficient enforcement of the Act, and Government regulations do abound (see Table 20.1). Regulations have been declared for biological products, including human blood and blood products[2], for cosmetics and for medical devices[3], all of which have evolved from the drug regulations[4]. Requirements from the drugs regulations can be found in all of these 'second generation'

Table 20.1 FDA areas of responsibility

Federal Food, Drug, and Cosmetic Act, As Amended
Public Health Service Act, Biological Products
Public Health Service Act, Radiation Control for Health and Safety
Public Health Service Act, Quarantine and Inspection
Public Health Service Act, Pacemaker Reimbursement Review and Reform
Lead Based Paint Poisoning Prevention Act
Fair Packaging and Labeling Act
Federal Anti-Tampering Act
Tea Importation Act
Filled Milk Act
Federal Import Milk Act
Saccharin Study and Labeling Act
Orphan Drug Act
Drug Price Competition and Patent Term Restoration Act
Pesticide Monitoring Improvements Act
AIDS Amendments
Food and Drug Administration Act

regulations. A point that is often forgotten in the review of these requirements is that they are not intended to be 'how-to' documents; rather, they identify concepts which must be considered in manufacturing drugs, biologics, and devices. Accordingly, each product class has its own specific requirements. The drug good manufacturing (GMP) regulations, being the original, will be the basis for this discussion. However, a selected reading list is included for more specific guidance on the other regulations.

The drug GMP regulations neatly break out many of the critical elements related to manufacturing. These elements will be discussed in turn. It is important to note that regulations are written in such a way as to give essential guidance, not stifle innovativeness and yet be dynamic in approach.

20.2 Front-end considerations

One would think the need for trained and qualified personnel would be self-explanatory. The regulations require personnel to have qualifications commensurate with their assigned tasks. If this were readily accepted, the requirement for documenting qualifications and training would not be necessary, and reports of incidences of product failure due to 'operator error' would not be prevalent[5].

For example, for the fiscal year 1990, 405 product recalls were logged by the FDA. One way or another, these product failures were all traced back to 'operator error'. Whether the failures resulted from acts of omission or commission, the bottom line remains the same: a tremendous expense that needs to be overcome. If one considers time and materials lost, future product acceptance, corporate image and civil and/or criminal penalties, the expense may be incalculable. In this real life scenario where all that can go wrong quite often does, a positive reaction plan must be in place. This should include definition of levels of responsibility for those 'what if' situations.

The building and facility need to incorporate design and construction features that facilitate the intended operations, including process integrity and accountability, to prevent or minimize contamination, and to provide for housekeeping and maintenance[6]. Facility design factors should also consider people flow, product flow, material (including components and waste), and equipment flow. Ideally, people, material, and equipment movement should be unidirectional, either from clean to less clean areas or dirty to clean areas (depending upon process requirements) with the highest degree of cleanliness required where the product or product contact surfaces are exposed. Applying the concept of unidirectional movement precludes the contamination potential created by crossovers of 'clean and dirty' people, equipment, or material.

Product movement is generally from a less clean area to a more clean area as the product quality and purity attributes become better defined. Officially, the FDA recognizes two area classifications. Critical areas where product or product contact surfaces are exposed are typically designated as Class 100 as defined in Federal Standard 209. The other recognized area is the controlled area, typically designated as Class 100 000. The FDA further suggests that Class 100 areas be located within the next higher class area, Class 10 000. The requirements of critical areas are further defined in terms of air pressure differentials; a pressure differential of at least 0.05 in (0.127 cm) of water between adjacent unequally clean areas is suggested. Sufficient airflow to sweep potential contaminants from critical areas is also a prime consideration. Skilled and experienced design of the heating, ventilating and air conditioning system (HVAC) is paramount in meeting these criteria.

Another integral part of facility design requires that structural surfaces be of such materials as to be durable, non-shedding, and cleanable. The FDA requirement is less specific and the concept of demonstrated suitability is the key. Providing an aesthetically pleasing environment to minimize operator fatigue and to maximize operator efficiency should also be considered[7].

The phrase, "control by design or, when all else fails, by procedure" has been recently embraced by industry. This attempt to compensate with personnel for a poorly designed facility is fraught with pitfalls, yet continues in practice. A well thought out design with qualified and trained personnel is the only reasonable procedure.

20.3 Process requirements

A manufacturing process is an integrated process incorporating many critical elements including equipment, components, and procedures.

Equipment should be designed to its intended operation (mix or blend adequately, dry thoroughly, sterilize, filter etc.). It should be of durable construction and should not be additive, reactive, or absorptive with the formulation ingredients. Further, the equipment must be of suitable design and construction to facilitate and demonstrate adequate cleaning[8]. Again the question should arise — is this not self-explanatory? However, there are incidents reported for the FDA as FD 483 observations that suggest not. For example, dry powder formulations are mixed in polyethylene bags, much like kneading bread, or formulation is put into a fiber drum and simply rolled down a hallway.

On the list of critical elements, we should next consider the components. Components include all materials used in the preparation of the product, including the immediate container and closure and even those materials which may not appear in the final dose form (solvents, drying agents, buffer solutions, filters, etc.). Component specifications must be established and confirmed in order to establish suitability for their intended use and promote batch to batch uniformity with regard to quality, strength, and purity.

Storage of components should be in such a way as to prevent mix ups in use, adulteration, contamination, degradation, etc., and provide for visual and physical examination as well as a system of controlled release for manufacturing. Controlled release is appropriate for inventory control and to provide a paper trail in the event of a product failure, to facilitate the required investigation and possibly the resolution or prevention of additional failures[9].

20.4 The recipe

The regulations go to great lengths in the area of written procedures. The master production record and production batch records are described in detail[10]. The required information is essential for batch to batch uniformity and maintenance of product quality. This information is outlined here.

20.4.1 Master production record

- Name, strength, description of dosage form,
- Name and weight of each active component,
- Complete list of components,
- Accurate statement of weight or measure,
- Statement of any excess of components,
- Statement of theoretical weight or measure including acceptable minimums/maximums,
- Description of drug container, closure, packaging materials, and documented labeling,
- Complete manufacturing and control instructions, sampling, testing procedures, specifications, and special precautions to be followed.

20.4.2 Batch production record

A batch production record should include an accurate reproduction of the master production record and documentation that each significant step of the process has been accomplished. The following summarizes these requirements.

- Dates of production,
- Identification of major equipment and processing lines,
- Specific batch or component identification,
- Statements of weight and measures,
- Special instructions/safety precautions,
- In-process sample test results,
- Pre- and post-inspection of the packaging area,
- Yield reconciliation — process and labeling,
- Labeling control records,
- Description of the container/closure,
- Documentation of sampling performed,
- Identification of persons performing/supervising significant process steps,
- Documentation of investigations made as a result of unexpected occurrences,
- Documentation and results of finished product inspections.

20.5 Validation requirements

The most popular FDA requirement is probably validation. There are many variations on the definition of validation. A simple one is that validation consists of establishing documented evidence that a system does what it is intended to do time after time and this evidence has as its basis a preplanned protocol.

The concept of validation is not really new and was practiced by many quality-minded firms prior to the formalization of the GMP regulations. The added emphasis of the GMP regulations has resulted in the identification of key terms which signal the need to develop documented evidence that a system or process operates as designed. Some of these key words are:

adequate	appropriate	assure	assess
meaningful	design	qualify	verify
reliability	system	suitable	reproducible

Validation as practiced today is a complex concept which includes, in addition to the process, all of the attendant quality assurance programs and utilities which are alluded to in the GMP regulations. As stated earlier, the GMP regulations are evolving in nature, therefore, the validation requirements must also be evolving, which in turn further complicates the concept[11]. Initially, validation was specified for sterilization processes. As technology advanced and the benefits of validation and a more quality conscious society came of age, it was natural for a philosophy which calls for building quality into a process and product from the start to evolve. This is the bottom line of validation. The FDA looks upon validation as a life cycle phenomenon, while many firms who have looked upon it as a single incident have incurred the ire of the FDA.

20.6 Finishing requirements

Requirements for packaging and labelling operations include written procedures describing the receipt, identification, storage, review, and release of labeling. Procedures to reconcile labels issued and used, as well as for destruction of obsolete and outdated labels, need to be in place[12].

The regulations address warehousing procedures which include quarantine of products until quality attributes have been confirmed and the product is released for distribution, storage under appropriate conditions of temperature, humidity, light, and a distribution procedure that creates a paper trail suitable for recalling product from the marketplace in case of a product failure.

Laboratory testing is another area addressed by the regulations. The establishment of specifications and test methods is required in order to demonstrate that components and products meet pre-established quality attributes. Test methods need to be specific and provide a degree of robustness to make them suitable to support or be the basis of regulatory action. Products which are identified in the United States Pharmacopoeia (USP) are recognized as drugs and must meet minimum specifications outlined in the USP and be analyzable by those USP methods. Comparable methods demonstrated to be as good as the official method can be employed[13].

So far we have discussed personnel, general facility design, equipment, components, the production process, packaging and labeling, and holding and distribution. The requirements pertaining to the supporting utilities will be discussed now, namely USP purified water, USP water for injection (WFI), and compressed gases (air, nitrogen, etc.).

The requirements for these systems are not well defined by regulation. These requirements are better defined by 'podium pronouncements' and precedent findings. For example, the June 1, 1976, proposed guideline for large volume parenterals has not and probably never will be adopted as a final order. But as a result of presentations made by FDA staff the edicts of this document are commonly practiced and generally quoted as a primary reference for water and water systems. Requirements for batch versus circulating loop high purity water systems as well as maintenance of the system with regard to quality and purity were also addressed in the 1976 proposal.

Welded stainless steel piping is accepted universally as a water system component. Use of other materials requires the user to demonstrate their suitability for the intended use, i.e. that they are not reactive, additive, absorptive, and do not otherwise affect the component (water) in a deleterious way. It is not recommended that innovative materials be used if the goal is to put a product on the market quickly. Using the tried and approved materials of construction and procedures will be more fruitful. Other requirements suggested in this proposed document, i.e. HVAC and sterilization criteria, are widely employed but beyond the scope of this discussion.

20.7 Records and report requirements

The GMP requirements also specify the types of records and reports to be generated and maintained[14]. The types of records include the following.

Equipment cleaning and use logs document equipment cleaning and use, i.e. which products were processed in specific equipment, when the equipment was cleaned, and by whom. Considering the basic FDA requirement, "products shall not be adulterated," the importance of this operation becomes very clear. This information could be extremely useful in evaluating the source of product contamination and gives insight into rectifying an 'out of control' condition.

Inventory records for active as well as inactive components, product containers, closures, and labeling are required, all with the intent of controlling the use of such items and providing a complete manufacturing history which could be used to investigate, resolve the seriousness of, or rectify the cause of a product failure.

Production review records are generated from an annual review of all the batches of the product manufactured in the past year. An often echoed phrase, "we have been doing this process this way for years," is the target of this requirement. A close investigation of this process will often, in fact, show that subtle deviations or variations have crept into the process and may have either gone unnoticed or been accepted as an inconsequential change. This type of review guards against this and also can elucidate gradual shifts of process parameters and of the quality attributes. It would seem reasonable, in this age of keen competition, that a company would want to evaluate past performance with an eye to improving efficiency. Yet, this review had to be mandated as a requirement for the sake of maintaining quality.

Laboratory records documenting the testing, methods employed, and results obtained as well as 'method validation studies' demonstrating method comparability are required. Testing product and not documenting the 'how and why' does little to establish and maintain product quality. This, as with the other required records and reports, can be invaluable in investigating, evaluating, and resolving unexpected occurrences.

A complaint file could be used as an overall barometer of a firm's quality commitment and in many instances is the starting point of regulatory scrutiny.

A record of returned and salvaged product is required and much like the complaint file, this could be used as an overall barometer of a firm's quality commitment.

The need for flexibility of these requirements was suggested earlier. An area currently under discussion for possible modification is the area of product and component sterilization. Some products are manufactured under 'clean' conditions, filled into their final container, sealed, and sterilized[15]. The question of maintaining product integrity for products manufactured under these conditions is moot. The validated terminal sterilization remains the concern for sterility failure at earlier points. The other side of the coin has each product component, including the product, sterilized individually and then brought together under aseptic conditions. This process is termed aseptic processing as opposed to terminal sterilization. As can readily be seen, there are several chances for sterility process failures and compromised component or product integrity to occur through improper handling of the previously sterilized materials which could in turn jeopardize the sterility of the ultimate product[16].

In a case brought against the FDA (US versus Norwich), the opinion rendered by the court stated that sterile meant the absence of living organisms. This creates a problem in that there is no non-destructive test for determining sterility; in other words each unit manufactured would have to be opened and tested with no product remaining for sale. As a resolution to this problem it was and is generally accepted that sterility is defined in terms of a probability of sterility. A process that can demonstrate a probability of producing a non-sterile unit of 1×10^{-6} is considered to produce a sterile product.

This requirement can easily be demonstrated for terminal sterilization processes. However, published data suggest that with aseptic processing, a probability of only 1×10^{-3} or 1×10^{-4}, is readily attainable. Yet products manufactured by either process are labeled as sterile, with no distinction made between the methods of manufacture.

The FDA is in the process of establishing a policy that states that all products that can be terminally sterilized must be terminally sterilized. It should be noted that other countries have already taken a similar stand.

20.8 Philosophical considerations

For the most part, regulations promulgated by the FDA are the FDA's perception of the industry standards currently practiced. The basis of this perception is found in the numerous establishment inspections, recognized problems in the industry, specific product failures, and general common sense as applied to manufacturing processes.

It should be pointed out that this promulgation process is involved and time consuming. The first step is to identify the need, gather the basic information and then make public notice of intent in the Federal Register, where comments are invited with regard to the notice. The comments received are reviewed and published in the Federal Register with the FDA discussion of the comments. Once again comments are requested before the final order is published as a regulation. This process can easily cover years[17].

In recent times efforts to streamline the promulgation process have resulted in the issuance of 'guidelines', which are not legal requirements as such. Similar to the use of stainless steel pipe, the FDA will easily accept the results of studies performed in accord with these 'guidelines', however, industry is not obliged to follow these 'guidelines'. The industry is invited to discuss their innovativeness with the FDA prior to the expenditure of money and effort on an activity which may later be unacceptable to the FDA. Caution must be advised with regard to asking broad and general questions of the FDA, because this type of query will usually result in a very conservative answer which will cover many not here-to-fore considered, and perhaps not important, considerations. It is more beneficial to present specific situations along with prospective resolutions of the situation and solicit FDA concurrence[18].

20.8.1 FDA inspections

A discussion of this sort would not be complete without addressing the topic of FDA inspections by the field investigation staff. First it should be noted that the bulk of FDA field work is done by investigators and not inspectors. Section 704 of the Federal Food, Drug and Cosmetic Act provides for the right of entry at reasonable times of operation for the purpose of evaluating the degree of compliance with the requirements of the Act, meaning that the FDA has the right to come inside a company's manufacturing facility and look around. Subsequent legislation provided for court relief in the form of inspection warrants for the facilities that refused inspection, if necessary.

It is safe to say that an FDA inspection disrupts the normal course of business. Given the complexity of the regulation, service companies have been established to help prepare the 'investigatee' on how to deal with this stressful event. Many large corporations have enough facilities to find it worthwhile to have individuals specially trained and armed with corporate SOPs to deal with these situations. Some strictly follow these SOPs to the letter of the law with respect to what the FDA is entitled to see and do in pursuit of their mission. It is not recommended here that the FDA should be prohibited from their mission on the basis of legal technicalities: for although in these circumstances it might appear that one has won the battle, in reality one may have lost the war.

Courteous attention should always be afforded to the FDA investigator. The long term effects of providing the required information should be considered. For example, the issue of permitting the investigator to take photographs or to have unescorted free movement about the facility is one matter, but permitting entry into critical areas or controlled areas without proper training and permitting spontaneous conversation with operating personnel could be dangerous to the product, operation, and the investigator[19].

It is generally acknowledged that, in an adversarial situation, the more difficult one makes the job the more determined the opposition will become. The FDA does have the power and might prevail if not afforded cooperation. This is not and probably will not be its choice of action but, if need be, it can go to the courts, demonstrate a possible danger to the public's health and gain the authority of the US court system in championing its cause. No organization should want this kind of scenario.

20.9 Prospective strategy

The FDA mission is to ensure that only safe and efficacious drugs are marketed in the United States. The FDA is a very large organization in regard to the absolute number of individuals involved but a rather small organization considering the task to be accomplished. In the process of bringing a drug to market in the United States, a company will become intimitately involved with the FDA. Successful interaction with this regulatory agency can be enhanced by recognizing the following points.

The FDA 'bottom line' is that a safe and efficacious product will be available for marketing. Its database comes from the efforts of many firms, and is very broad but not necessarily deep. This places those performing research and day-to-day problem solving in the industry in a unique position as 'primary experts' and offers them the opportunity to 'teach' the agency scientists.

A fact that is often lost on firms dealing with the FDA is that this huge bureaucracy is made up of people who share the same traits as other people — anxieties, prejudices, egos. Establishing a rapport of confidence and character is just as important as the quality of the data to be evaluated by the agency.

Familiarization with the project, the regulations, and other available guidance documents should be the basis for formulating a positive strategy. Explaining the requirements of the firm's project and how the regulations have been applied and then soliciting FDA's concurrence is a positive strategy. Continuously requesting direction from the FDA can be taken as a sign of frustration and confusion which has resulted from a poorly planned project. The FDA has said they will assist a firm but will not fulfill an industrial function, i.e. QA/QC (quality assurance/quality control), design, or engineering.

Another point of positive strategy is that the firm's involvement with a project gives it a keen insight into the project. The FDA review of the project will be short by comparison, and accordingly the data presented for review should be concise and straightforward, utilizing tables, charts, and summaries where appropriate. The data presented should be limited to those that will support the current discussion.

References

1. *Federal Food, Drug, and Cosmetic Act*, as Amended, and Related Laws, U.S. Government Printing Office, Washington, D.C. 20402, 1989.
2. Title 21, Code of Federal Regulations, Part 600, *Biological Products:* General, April 1, 1992.
3. Title 21, Code of Federal Regulations, Part 800, *Medical Devices:* General, April 1, 1992.
4. Title 21, Code of Federal Regulations, Part 211, current *Good Manufacturing Practice for Finished Pharmaceuticals*, April 1, 1992.
5. Title 21, Code of Federal Regulations, Part 211, current *Good Manufacturing Practice for Finished Pharmaceuticals*, April 1, 1990, Sec.211.25.
6. Title 21, Code of Federal Regulations, Part 211, current *Good Manufacturing Practice for Finished Pharmaceuticals*, April 1, 1990, Sec.211.56.
7. Title 21, Code of Federal Regulations, Part 211, current *Good Manufacturing Practice for Finished Pharmaceuticals*, April 1, 1990, Sec.211.65.
8. Title 21, Code of Federal Regulations, Part 211, current *Good Manufacturing Practice for Finished Pharmaceuticals*, April 1, 1990, Sec.211.67.
9. Title 21, Code of Federal Regulations, Part 211, current *Good Manufacturing Practice for Finished Pharmaceuticals*, April 1, 1990, Sec.211.80.
10. Title 21, Code of Federal Regulations, Part 211, current *Good Manufacturing Practice for Finished Pharmaceuticals*, April 1, 1990, Sec.211.100.
11. *Guideline on General Principles of Process Validation*, DHHS, PHS, FDA, 5600 Fishers Lane, Rockville, MD 20857, HFD-320, May 1987.
12. Title 21, Code of Federal Regulations, Part 211, current *Good Manufacturing Practice for Finished Pharmaceuticals*, April 1, 1990, Sec.211.122.
13. Title 21, Code of Federal Regulations, Part 211, current *Good Manufacturing Practice for Finished Pharmaceuticals*, April 1, 1990, Sec.211.160.
14. Title 21, Code of Federal Regulations, Part 211, current *Good Manufacturing Practice for Finished Pharmaceuticals*, April 1, 1990, Sec.211.180.
15. *Validation of Steam Sterilization Cycles*, Technical Monograph No. 1, 1978, Philadelphia, Parenteral Drug Association.
16. *Guideline on Sterile Products Produced by Aseptic Processing*, DHHS, PHS, FDA, 5600 Fishers Lane, Rockville, MD 20857, HFD-320, June 1987.
17. L.R. Holton, Regulations, proposals, guidelines, advisory opinions and letters of advice: what legal significance?, *J. Parenteral Sci. Technol.*, May/June, 138–141 (1991).
18. FDA Guidelines. A listing is available by writing to FDA, Center for Drug Evaluation & Research HFN-365, 5600 Fishers Lane, Rockville, MD 20857, USA.
19. A.P. Shroff, GMP inspections and compliance audits, *Pharm. Eng.*, September/October, 31–36 (1990).

Bibliography

H.L. Avallone, GMP inspections of biopharmaceutical manufacturing facilities, *Pharm. Eng.*, (1989).

H.L. Avallone, development and scale-up of pharmaceuticals, *Pharm. Eng.*, 38–41 (1990).

M.G. Beatrice, Regulation, licensing and inspection of biological products, *Pharm. Eng.*, 29–35 (1991).

F.J. Carleton and J.P. Agalloco, *Validation of Aseptic Pharmaceutical Processes*, Dekker, 1986.
Current Good Manufacturing Practice in the Manufacture, Processing, Packing, or Holding of Large Volume Parenterals, Federal Register, Vol. 41, No. 106, June 1, 1976, 22202-22219.
Federal Food, Drug, and Cosmetic Act, as Amended, and Related Laws, U.S. Government Printing Office, Washington, D.C. 20402, 1989.
S.W. Fitzpatrick *et al.*, Implement good manufacturing practices for the production of biopharmaceuticals, *Chem. Eng. Prog.*, 26–31 (1990).
M.J. Groves, W.P. Olsen, and M.H. Anisfeld, *Sterile Pharmaceutical Manufacturing*, Interpharm, 1991.
Guideline on General Principles of Process Validation, DHHS, PHS, FDA, 5600 Fishers Lane, Rockville, MD 20857, HFD-320, May 1987.
Guideline on Sterile Products Produced by Aseptic Processing, DHHS, PHS, FDA, 5600 Fishers Lane, Rockville, MD 20857, HFD-320, June 1987.
D. Hill and M. Beatrice, Facility requirements for biotech plants, *Pharm. Eng.*, 35–41, (1989).
D. Hill and M. Beatrice, Biotechnology facility requirements: Part I Facility and systems design, *BioPharm*, 20–26 (1989).
D. Hill and M. Beatrice, Biotechnology facility requirements: Part II Operating procedures and validation, *BioPharm*, 28–32 (1989).
W.S. Kelley, Issues in the development of biotechnology products, *Pharm. Technol.*, 48–54 (1986).
D. Larkin, GMP history and implementation in the design of bulk pharmaceutical facilities, *Pharm. Eng.*, (1989).
J.W. Levchuk, Good manufacturing practices and clinical supplies, *J. Parenteral Sci. Technol.*, 152–155 (1991).
W.J. Lhoest, Design and selection of pharmaceutical production equipment in the scope of modern automated plants: Part I, *Pharm. Eng.*, 46–58 (1991).
W.J. Lhoest, Design and selection of pharmaceutical production equipment in the scope of modern automated plants: Part II, *Pharm. Eng.*, 47–50 (1991).
P.D. Parkman, CBER's Responses to the Vice-Presidential Initiatives to Expedite Drug Development.
C.A. Perkowski, Designing biopharmaceutical facilities, *Mech. Eng.*, 54–55 (1989).
J.X. Phillips, FDA inspections—what does the investigator look for during inspections of parenteral manufacturers, *Pharm. Eng.*, 35–38 (1989).
T.J. Schoemaker, Organization of validation as part of the total quality care in biotechnology facilities, *Pharm. Eng.*, 23–27, (1991).
W.L. Schwemer, Validation: foundation of GMP, *Pharm. Eng.*, 44–46 (1990).
A.C. Stocker and J.A. Embree, Creating cost effective biotech facilities, *Pharm. Eng.*, 37–45 (1991).
D.R. Stoudt, Compliance and preparation for FDA inspections, *BioPharm*, 39–41 (1988).
Title 21, Code of Federal Regulations, Part 210, current *Good Manufacturing Practice in Manufacturing, Processing, Packing, or Holding of Drugs:* General, April 1, 1990.
Title 21, Code of Federal Regulations, Part 600, *Biological Products:* General, April 1, 1990.
M. Vasilevsky and C.W. Cherry, Documentation requirements in the biotechnology industry, *BioPharm*, 32–35, (1990).

21 Validation

B. ECKMAN

21.1 Federal regulation of food and drugs — the past is prologue

Until the end of the 19th century, US Federal Government involvement in the regulation of the food and drug production industries was minimal. Science had reached the point of minimizing bacterial contamination in food (through pasteurization for instance), by about 1860. Injectable drugs were not commonly given at this point.

In 1906, Congress passed the Food and Drug Act, largely in reaction to unclean conditions in meat packing plants which had been exposed by the muckraking journalists of the era. This Act was largely geared to prevent adulteration of the final product for human consumption.

In 1938 Congress passed the Food, Drug and Cosmetic Act in reaction to the 1937 recall of tainted batches of Elixir of Sulfanilamide. The elixir was formulated using a toxic substance, diethylene glycol, and one hundred users of this elixir died before the remainder of the batch could be recalled and destroyed. This 1938 Act was designed to ensure product safety for human consumption.

In 1962, thalidomide, a fertility drug, turned out to cause birth defects in children born to women who had taken the drug. In reaction to this tragedy, the Kefauver–Harris Amendments to the 1938 Food, Drug, and Cosmetic Act were passed. These amendments were responsible for the creation of the first Good Manufacturing Practice (GMP) regulations and introduced the concept of product efficacy to the regulatory mind set.

The above history is given by Chapman[1]. The modern era of pharmaceutical regulation began in the early to mid 1970s when several members of the public died following injection of a drug formulation which turned out to be non-sterile. While the material in question was sterilized in an autoclave for the correct time at the correct chamber temperature, no attempt was made to prove that the entire volume was subjected to proper sterilization temperatures for a sufficient duration. The incident led to the modern idea of process validation as defined in the February 1976 proposed cGMP (current Good Manufacturing Practice) regulations.

The 1976 proposals were adopted in 1979. These regulations, however, were general in nature and had only fleeting references to 'validation'[2]. The Food and Drug Administration (FDA) in 1983 therefore issued the draft version of their 'Guideline on General Principles of Process Validation'. The final version of this document was promulgated in 1987[3].

Over the past few years the validation concept has been extended to computer systems. The methods have been refined through development by commercial pharmaceutical manufacturers with the FDA providing a 'cross pollination' function. This work attempts to define in one chapter the basics of validation as now practiced and to introduce a coherent overall concept of how to validate a process or facility.

21.2 Basic concepts and definitions

What is validation? The formal regulatory definition is given in the section on Definitions (1) at the end of the chapter. An important point to note is that the product characteristics must be specified before the procedures are set up and run. These characteristics have traditionally included, for example, potency, purity, and dissolution rate (for tablets). Also, absolute certainty is not required, only a high degree of assurance.

Alternate clarifications and definitions of the validation concept are given in the section on Definitions (2). One common thread which runs throughout all these concepts is the requirement for documentation. The types of document required will be defined later in this chapter. It is also clear that the validation activity should include the organized presentation of test data to prove that the process in question produces the product sought in a reproducible manner.

As the importance of validation began to be stressed by the FDA in the early 1980s, it became evident that an agreed terminology for the field was needed. The seminal work serving to define validation terms was that of Chapman[4]. This work defined the terms to be explained in this chapter in a way all could understand and use them. Validation is a unique field in that it can involve everyone from research microbiologists to mechanical engineers. Therefore, an agreed terminology is critical. (As a matter of practical course, when two parties set out on a validation effort, agreement on the exact meaning of the various validation terms is essential to minimize confusion and control costs.) A lexicon of validation terms is almost always included in any validation master plan (explained later in this chapter) and one is included at the end of this chapter.

The regulatory authority of the FDA, a division of the Department of Health and Human Services, is given in the Code of Federal Regulations (CFR)[5]. Part 210 of the code defines, in general terms, the cGMPs in manufacturing, processing, packing, or holding of drugs. Part 211 covers the cGMP for finished pharmaceuticals. The section on Definitions (3) shows a typical cGMP reference to validation. This section codifies the necessity of process validation. The section on Definitions (4) gives the cGMP requirement for validation of the sterilization process often used in pharmaceutical manufacturing. Another part of the CFR includes a reference to the need of suppliers of equipment/raw materials to validate their function (CFR 211.84) and finally the need to validate computers used in the pharmaceutical industry is defined in CFR 211.68.

Regulations covering biologicals are given in Part 606 of CFR title 21[6]. Regulations covering medical devices are given in Part 820 of CFR title 21[7]. Many other guideline articles are available in the literature[8–31] regarding validation of specific processes or pieces of equipment.

Unlike many other federal regulations, the validation regulations are stated in general terms, with a goal clearly defined, but with limited information on the methodology to be used to achieve that goal. The FDA has begun to expand on the purely regulatory language by issuing non-mandatory 'guidelines'[3]. The basic 'Guideline on General Principles of Process Validation' clarifies definitions and General Terms of Process Validation and gives examples of successes and failures in the area of process and equipment validation including medical devices.

Another aspect where food and drug regulations differ from other federal regulations is the format of documents to be submitted to the regulatory authority (FDA) for approval. There are no standard preprinted formats for the various protocols and procedures to be defined shortly. The user is free to define any suitable format which can be defended to regulatory personnel.

The basic tool of validation documentation is the validation protocol (shown in section on Definitions (5)). This is the vehicle by which the specific aspects of validation for a piece of equipment, group of related equipment, or an entire process is documented and certified. Together IQs, OQs and PQs, once filled out and certified, constitute validation.

The three basic types of protocols are the installation qualification protocol (IQ) (Definition (6)), the operational qualification protocol (OQ) (Definition (7)), and the process (or performance) qualification protocol (PQ) (Definition (8)). These three documents are defined on page 595.

Examples of several sheets from an IQ protocol are given in Figures 21.1–21.4. As can be seen, this IQ is a sophisticated inspection document wherein all critical aspects of the installation of a piece of equipment are certified as being correct. It is quite common also to include instrumentation and auxiliary support equipment in this document. Further, the existence and retrievability of manufacturers' drawings, manuals, etc. should also be noted in this document. As can be seen in Figure 21.1, the cover sheet, these documents are strictly controlled with respect to revision and must be certified by all personnel with responsible charge for the equipment in question. The sample 'Table of Contents' (Figure 21.2) shows what type of information is often put into these documents. Remember, there are no hard requirements in this area and other formats are equally acceptable.

The format shown includes room for signature when this document is

PROTOCOL FOR THE INSTALLATION QUALIFICATION OF

SYSTEM NO. U4

CLEAN STEAM SYSTEM

PREPARATION APPROVAL

	NAME/TITLE	SIGNATURE	DATE
Protocol Prepared By	____________	____________	________
Protocol Reviewed By	____________	____________	________

PROTOCOL APPROVED FOR EXECUTION

DEPARTMENT	NAME/TITLE	SIGNATURE	DATE
Validation	____________	____________	________
Engineering	____________	____________	________
Manufacturing	____________	____________	________
Quality Assurance	____________	____________	________

APPROVAL FOR EXECUTED PROTOCOL

DEPARTMENT	NAME/TITLE	SIGNATURE	DATE
Validation	____________	____________	________
Engineering	____________	____________	________
Manufacturing	____________	____________	________

Figure 21.1 Installation qualification for clean steam system U4—approvals.

issued as a protocol (format for data collection) and when it is issued as a final completed report (executed protocol). Some validation personnel prefer the protocol document to include only initial format approval signatures. In this case, a second report approval cover sheet would be added to the document when it is published as a report.

Figure 21.3 shows a typical installation qualification sheet for a heat exchanger in a pure steam generator. Note typical identification data such as serial number, manufacturer, and purchase order number are recorded and the design of the unit is confirmed for quantity and size of nozzles.

Examples of an OQ protocol are given in Figures 21.5–21.7. This document also has a controlled, certified cover sheet. A typical 'Table of Contents' is presented in Figure 21.6. Figure 21.7 shows typical OQ test information for a heat exchanger (inlet and outlet temperatures and pressures) in a pure

TABLE OF CONTENTS

SECTION	DESCRIPTION
1.0	Purpose of Protocol
2.0	Key Reference Engineering Drawings and Specifications
3.0	Description of System
4.0	Procedures
5.0	Acceptance Criteria
6.0	List of Attachments
	6.1 Installation Qualification Forms
	6.2 Boundary Drawing(s)
	6.3 Deviation Reports
7.0	Summary and Conclusion Report
8.0	Manufacturers' Documentation
	8.1 Drawings
	8.2 Spare Parts List
	8.3 Preventative Maintenance

Figure 21.2 Installation qualification for clean steam system U4—table of contents.

steam generator which is functionally a small sub-system made up of heat exchangers, pressure vessels, and analytical and control instrumentation.

The goal of an OQ test is to prove that a piece of equipment or sub-system operates as it is designed, given the proper inlet and outlet streams it would encounter when the process is run. (While IQs are usually written on individual pieces of equipment, OQs often include 2–3 related pieces.) An example of this is an air compressor, its receiver tank, and the desiccant dryer system which treats the air before it is distributed throughout the pharmaceutical facility. The air flow through each piece of equipment could be measured separately or the entire train could be simply qualified at once. Also, while the moisture content at the compressor outlet may fluctuate and lead to many different readings, the level at the dryer outlet should remain at a constant low dew point which can be read over a preselected duration of time and effect of ambient moisture level noted if ambient humidity variations are anticipated to be a problem.

Process (or performance) qualification is defined in the section on Definitions (8). These documents typically have cover sheets, similar to those illustrated under IQ and OQ. PQs are concerned with proving that the process being

HEAT EXCHANGER

SYSTEM NAME: Clean Steam System SYSTEM DESIGNATION: U4
ITEM NUMBER:________ ITEM NAME: Double Tube Sheet Evaporator
P&ID:________

The purpose of this protocol step is to document that the heat exchanger is installed, meets equipment specifications and is mechanically complete.

The responsibility of the individual executing this protocol will be to verify that this heat exchanger is in accordance with the equipment arrangement drawing, the P&ID, and the manufacturer's specifications by initialling each item below.

To verify equipment:	CHECKED
1. Ensure manufacturer's specifications, drawings and manuals exist and where they are located (List on FORM IQ-8-2)	___
2. Ensure purchase order is available (List on FORM IQ-8-2)	___
3. Locate equipment in field.	___
4. Ensure type of exchanger and nameplate agrees with the manufacturer's specifications and/or drawings.	___
5. Ensure material of construction per manufacturer's drawing.	___
6. Ensure that the following, if applicable, are installed:	
A. drains	___
B. safety devices	___
C. instrumentation	___
D. vents	___
7. Ensure field piping connections completed.	___
8. Ensure nozzle quantity and size per manufacturer's drawing and/or nozzle schedule attached. Complete and initial the nozzle schedule attached. (FORM IQ-8-2)	___

COMMENTS:________

CONDUCTED BY:________ DATE:________
REVIEWED BY:________ DATE:________

Figure 21.3 IQ for clean steam system U4—heat exchanger protocol

investigated actually works as it is supposed to. Once all IQ and OQ runs are complete and certified, the entire flow of material through the pharmaceutical plant is started, raw material, intermediates, and final products are sampled for chemical purity, bacterial levels, and activity (for final product and elsewhere as appropriate).

It is important here that product meets predetermined acceptance criteria. As a product is produced for sale to the public or for human clinical trials, a dosage and purity have been set based on laboratory tests and/or animal trials. Dosage and purity may have to be adjusted during clinical trials but should be fixed by the time bulk production has begun.

HEAT EXCHANGER

SYSTEM NAME: Clean Steam System SYSTEM DESIGNATION: U4
ITEM NUMBER: ____________ ITEM NAME: Double Tube Sheet Evaporator
P&ID: ______________________________

MANUFACTURER INFORMATION

MANUFACTURER: ______________________________
MODEL #: ______________ SERIAL #: ______________
HEAT TRANSFER AREA: ______________ RATING: ______________ BTU/HR
DESIGN PRESSURE: ______________ DESIGN TEMPERATURE: ______________
PURCHASE ORDER NO: ______________ LOCATION: ______________

HEAT EXCHANGER NOZZLE SCHEDULE

IDENTIFICATION	SERVICE	LOCATION	SIZE	COMMENTS	CHECKED

VENDOR SPECIFICATION DOCUMENTS, MANUALS AND DRAWINGS

DESCRIPTION	LOCATION

SPARE PARTS LIST ______________ ______________

CONDUCTED BY: ______________________________ DATE: ______________
REVIEWED BY: ______________________________ DATE: ______________

Figure 21.4 IQ for clean steam system U4—heat exchanger manufacturer information.

It is generally felt advisable by the FDA to complete at least three repetitions of each PQ run in order to verify the reproducibility of the results and to begin to develop a feel for the 'acceptability limits'. For example, if a secondary metabolite is produced in a 37°C fermentation, how far can the fermenter temperature vary from this ideal before the yield of the desired product drops to unacceptable levels. The fermentation should be run at, say 33–40°C and the temperature range of suitable yield be recorded. The range will be important to understand later when the commercial process set point fluctuates and a protein product worth thousands of dollars per gram is being produced.

The PQ concept was originally applied to the sterile filling and packaging

PROTOCOL FOR THE OPERATION QUALIFICATION OF

SYSTEM NO. U4

CLEAN STEAM SYSTEM

PREPARATION APPROVAL

	NAME/TITLE	SIGNATURE	DATE
Protocol Prepared By	______	______	______
Protocol Reviewed By	______	______	______

PROTOCOL APPROVED FOR EXECUTION

DEPARTMENT	NAME/TITLE	SIGNATURE	DATE
Validation	______	______	______
Engineering	______	______	______
Manufacturing	______	______	______
Quality Assurance	______	______	______

APPROVAL FOR EXECUTED PROTOCOL

DEPARTMENT	NAME/TITLE	SIGNATURE	DATE
Validation	______	______	______
Engineering	______	______	______
Manufacturing	______	______	______

Figure 21.5 Operation qualification for clean steam system U4—approvals.

of injectable formulations or capsules/tablets where purity of material to be filled, water (or solvent) and the cleanliness/sterility of the apparatus used were easily measured and defined. The concept has been modified slightly to apply to fermentation where raw materials can be as ill-defined as 'bovine serum' and products are protein mixtures whose 'purity' may depend on the state of the art of chromatography at the time it reaches the market. (A mixture of proteins from fermentation may be used as an injectable drug without detailed knowledge of the protein molecular structure.)

One important aspect of qualification is the verification that process lines and equipment are cleanable (in place or remotely) or sterilizable (again, in place or remotely). Cleanability has classically been measured by treating a surface with water of increasing purity such as tap water, deionized water, reverse osmosis polished deionized water, and/or water for injection (pyrogen-

TABLE OF CONTENTS

SECTION	DESCRIPTION
1.0	Purpose of Protocol
2.0	Key Reference Documents
3.0	Description of System
4.0	Procedures
5.0	Acceptance Criteria
6.0	List of Attachments
	6.1 Operation Qualification Forms
	6.2 Boundary Drawing(s)
	6.3 Deviation Reports
7.0	Summary and Conclusion Report

Figure 21.6 Operation qualification for clean steam system U4 — table of contents.

free water). One or more steps may involve use of detergents, acid, caustic, etc. but all should be followed by a rinse of suitable purity water. The degree of cleanliness traditionally has been measured by measuring the conductivity of the rinse water following the cleaning step. The cleanliness is validated when the conductivity approaches sufficiently closely the conductivity of the incoming rinse water. For guaranteed reproducibility, a clean-in-place (CIP) system with spray ball should be installed.

Verification of cleanliness where the primary contaminant is a non-water soluble fatty or greasy substance is more problematic. One suggested method is to collect all waste water and analyze how much a 'doped' load of contaminant is covered by direct chemical analysis of the rinse solutions.

A variation of this doping method is to wipe down the surface in question with rags and soak the rags in appropriate solvent before analyzing for the contaminant. This is not a long term operational method of verification of cleaning but can be used to validate a CIP system for instance and may be preferable to simple conductivity measurement.

Once a process line or piece of equipment is sufficiently clean, it may be sterilized if process considerations dictate. Sterilization in-place applies to sanitary stainless lines and large non-portable process vessels. Sterilization here usually involves the injection of pure steam made from water for injection into lines/vessels until selected thermocouples/RTDs read sterilization temperature (121°C) plus a proper safety margin (3–4°C at least). Temperatures should be read at the coldest conceivable points. (For instance, temperature

HEAT EXCHANGER

SYSTEM NAME: Clean Steam System SYSTEM DESIGNATION: U4
ITEM NUMBER: ________ ITEM NAME: Double Tube Sheet Evaporator
P&ID: ________

The purpose of this protocol step is to document that the heat exchanger is suitable for operation and meets operating specifications.

The responsibility of the individual executing this protocol will be to verify that this heat exchanger performs within its assigned limits by initialling each item below.

To verify heat exchanger:

1. Ensure heat exchanger provides correct duty. See EQUIPMENT TEST PROCEDURES, FORM OQ-16-1 for this heat exchanger.

	INLET	
	Specified	Observed
PROCESS TEMPERATURE:		
PRESSURE:		
HEAT TRANSFER TEMPERATURE:		
PRESSURE:		

COMMENTS: ________

CONDUCTED BY: ________ DATE: ________
REVIEWED BY: ________ DATE: ________

Figure 21.7 OQ for clean steam system U4—heat exchanger protocol.

in a line should be measured as far from a steam injection point as possible to allow for temperature loss and steam flow through the line.) In a vessel, temperature should be measured at the vessel drain and vented in sample/instrument nozzles as appropriate.

Remote sterilization usually involves use of an autoclave (wet sterilization) or a depyrogenation oven (dry sterilization). Here the important temperatures to record are those inside the mass of material being sterilized (the load). All load configurations to be sterilized must be run multiple times and the temperatures verified to be sufficiently high for a sufficient period of time to prove sterilization. The time/temperature curves must be verified using spore strips (biological indicators). These are pieces of paper impregnated with a highly heat resistant spore-forming microorganism which turn color upon sufficient heat exposure. To satisfy oneself of the accuracy of these indicators at least some strips should be 'plated out' and suitable media incubated to look for growth.

Special aspects of installation, operation, and process qualification are involved when validating an HVAC (heating, ventilating and air conditioning) system or a computer control system. The special requirement of these systems are discussed in sections 21.9 and 21.10.

While 'process qualification' may apply to any process, 'performance qualification' often is used to describe the long term measured operation of a system such as a water-for-injection (WFI) system. Here, the system must run for long periods of time without having any significant bacterial growth or contamination by pyrogens.

To validate a WFI system, the storage tank, pump, and recirculating lines may be initially sterilized as described previously. The still is activated until the tank and lines are full. (Air is vented from the system as appropriate.) The lines are then recirculated over a period of time, usually about 60 days, with samples taken approximately once per day from selected spots in the system. These samples are analyzed for pyrogens and bacterial contamination and compared to the firm's pure water standards. (The US Pharmacopoeia Standard modified as appropriate is often used for this purpose.) If all samples are satisfactory with no unexplained failures the test may be considered successful.

The difference between a performance qualification and a process qualification lies in the 'load' given to the system. An autoclave is loaded with bacteria and is expected to kill these microorganisms as part of its normal function. A WFI system is not loaded with bacteria before it is tested, however, because it is not designed to remove or kill bacteria. It is simply designed to perform properly starting with an acceptable sterilization or sanitization procedure being performed on the empty tanks/lines. The system should be challenged by turning on various WFI drops simultaneously and verifying that the main loop does not run dry and suck back contaminants from the outside world.

Definition (9) gives the definition of calibration and Definition (10) defines preventive maintenance, two concepts important to the overall validation effort. Calibration holds particular significance in the pharmaceutical industry. Calibration here refers to standardization with a NIST (formerly National Bureau of Standards (NBS)) traceable standard. Of course the calibrator should have all proper certification documents readily available at all times.

Instruments should never be out of calibration. If an instrument is out of calibration, many questions are raised about whether the process is under control. If not, how long has it been out of control? Critical instruments should be calibrated at least once per year and preferably once every six months. Electronic instruments/controls, including distributed control systems, programmable logic controllers, personal computers, and one loop electronic analog controllers should all be calibrated to show proper response from the controller upon an electronic variance of the measured variable compared to its setpoint. A 4–20 mA signal generator could be used for this and the

4–20 mA generated by the control device noted (for instance, 4 mA when a valve should be closed, 20 mA when it should be open, or 12 mA when the valve should be half open). The response of the valve to these signals should also be verified.

There are many references in the cGMP regulations to preventive maintenance (CFR 211.42, 211.67, etc.)[5]. The preventive maintenance procedure (see Definition (10)) illustrates some of the basic points to consider when establishing a preventive maintenance program. All equipment requires maintenance to some extent. Each piece of equipment should have its own preventive maintenance procedure. This document is usually written based on the vendor's operating/instruction manual for the unit in question. It should define maintenance or inspection activities to be carried out weekly, monthly, quarterly, twice per year, once per year, etc. An excellent discussion of preventive maintenance is given by Tomson[10].

The maintenance procedure should include a regular cleaning procedure, or this can be a separate document. It certainly should include instructions on replacement of consumable parts such as O-rings, seals, or bearings and should include instructions on lubrication if these apply. This document could also be used to define the stock of spare parts to be kept on site (or refer to a local distributor who stocks appropriate parts). Another aspect of maintenance to be defined is the extent to which the operating pharmaceutical company will rely on the equipment vendor or other outside contractor for maintenance or will prefer to perform all maintenance functions internally. Of course the ultimate responsibility for maintenance always falls on the owner/operator of the plant in question.

21.3 Validation approach

There are several different approaches to validating a pharmaceutical facility. These (shown in Definitions (11–13)) are prospective validation, concurrent process validation, and retrospective validation. By far, the most desirable of these is prospective validation. This approach, planning for validation and running qualification tests before the process/facility produces a drug for public sale, is often used in new construction situations where there is no systematic archive of product/process data to fall back on. The prospective approach is also mandatory when producing sterile product unless extremely detailed time/temperature data are retrievable. Prospective validation will be discussed in more detail below.

Concurrent process validation is similar to prospective validation except the operating firm will sell the product made during the qualification runs to the public at its market price. This approach involves a higher degree of risk than prospective validation and probably should be limited to drugs which do not differ significantly from well-established products such as 'orphan drugs' which may be produced in quantities as small as one batch per year.

Retrospective validation relies on data collected by the drug manufacturing firm over a long period of time to prove that the process has always remained in control. This approach found some acceptance in the mid 1980s when the FDA began to enforce more strictly the ideas of validation on the pharmaceutical industry in general. However, in the 1990s this retrospective approach is fast becoming obsolete, with non-sterile bulk production the one facet of the industry where it still has some proponents.

The overall approach known as prospective process validation is illustrated in Figure 21.8. The generation of IQ and OQ documents has been discussed previously as has calibration. The remainder of the blocks on the diagram

PROSPECTIVE PROCESS VALIDATION

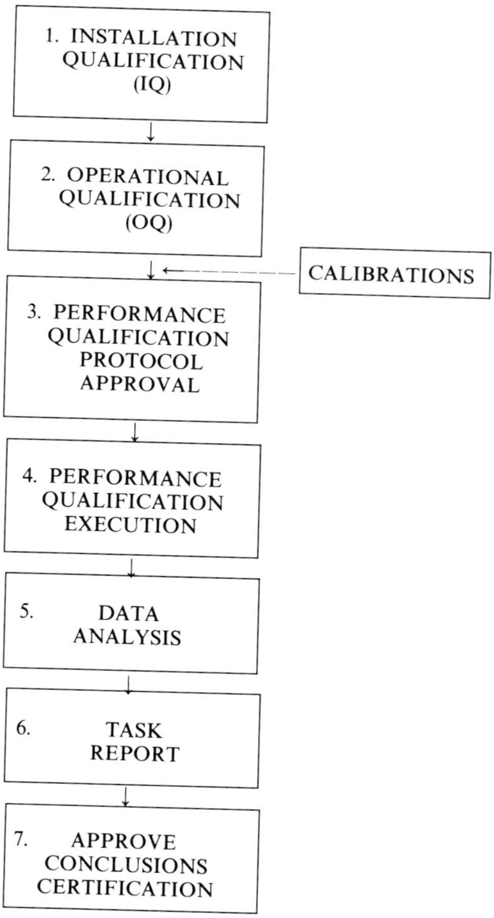

Figure 21.8 Schematic representation of prospective process validation.

define the topics to be discussed in the remainder of the chapter after a brief description of the basic validation planning tool, the Master Plan. (This chart is an update of Chapman's seminal work[4]. This particular version of this chart has never been published. The original chart from Chapman[4] included the terms 'VALIDATION PROTOCOL APPROVAL' and 'VALIDATION EXECUTION' in boxes 3 and 4. These terms were replaced with 'PERFORMANCE QUALIFICATION' in order to eliminate confusion with the overall validation process which covers all boxes on Figure 21.8.

21.4 Validation master plan

The main purpose of a validation master plan is to define clearly in one treatise what is to be validated in the pharmaceutical facility in question and how it will be validated. As previously described, this document also defines which systems are to be challenged and which must simply be run over time without failure.

The matter of which systems/equipment are to be validated in a facility is not one to be taken lightly. In a fermentation-based parenteral protein process, one could make the argument that all steps going back as far as media preparation should be validated because each can adversely affect the quality of the final product if not properly carried out. Since microbial contamination is of the utmost concern, it is quite logical that any step involving a sterilization should be validated. Minimally, all equipment should be qualified — an IQ and an OQ should be carried out as explained elsewhere.

On the other hand, in a non-sterile powder producing facility, the raw material is often nothing more than an industrial chemical with or without some assurance of purity. While normal industrial QA/QC practices are applicable to this product, it is not controlled to the point of being validatable. While one could validate the cleaning of various unit operations in the bulk plant, it is much more common simply to validate the last purification step where the bulk chemical becomes a 'pharmaceutical'. Again, all equipment should be qualified. IQ and OQ forms may be shorter here than for a sterile facility.

One aspect of bulk pharmaceutical production which many producers choose to validate is the computer control system for the process. Control computers tend to be seen as a 'black box' into which regulatory authorities cannot see and therefore it makes all concerned more at ease if the computer to be operated is certified and validated. Validation of computer systems is covered in detail elsewhere in this chapter (see section 21.10).

Once again, there is no one standard accepted for a master plan. Any format which is logical and self-consistent can be defended. The author heartily believes in authoring one of these plans whenever a validation effort of any consequence is begun. The small cost savings realized by not completing a master plan is almost always eaten up by inefficiency and rework later at the execution stage.

One important function of a master plan is that it lets everyone involved in a validation effort agree on the basic scope of process/utilities/facility to be validated including definitions of terms to be used on the project. A well written master plan should include at least some sample forms to be used on the job and preferably all such forms.

A schedule of validation activities should be included, overlaid on the facility construction schedule if the validation is for a newly constructed or renovated facility. In a large construction effort, the validation responsibilities of various disciplines/departments may be delineated. These departments may include, but are not limited to:

- Process engineering
- Instrument engineering
- Production/operations
- Mechanical/HVAC
- Architectural
- Quality assurance/quality control
- Electrical
- Piping
- Construction management

If a design consulting engineering/construction firm(s) is employed, the master plan can also serve to divide the validation responsibilities between the owner and his hired professional workers. These are day-to-day tasks only. The ultimate responsibility of validating and operating a pharmaceutical production facility is always the owner's.

One function the master plan may serve is to introduce appropriate inspection/regulatory personnel at the US Food and Drug Administration to the facility or process renovation to be validated. The schedule provided allows FDA personnel to schedule their future workload or to comment on the proposed process or facility design. A preliminary meeting with the FDA may also allow the owner to determine whether a facility handling recombinant microorganisms should be classified as Biological Level 1, 2, 3 or 4. (The level of containment of microorganisms/cell containing waste is defined by these various levels[32].)

21.5 Prequalification phase activities

As a convenient reference, the various activities involved in validating a facility may be divided into several phases as defined in the PMA treatise[33]. These are the prequalification phase, the qualification phase, the validation phase, and on-going activities. The prequalification phase of validation more or less parallels the 'front end' or 'conceptual design' phase of an engineering effort.

By the end of this phase, a facility layout including all discrete rooms, corridors, and areas should be complete. Cipriano[21] gives an FDA perspective on this subject for biotechnology-based production facilities. Overlaid onto this layout, should be the following information, preferably in individually removable discrete layers or multiple colors:

- Material flow
- Raw material flow
- Product flow
- Personnel flow
- Restrictions on personnel movement, gowning/isolation considerations
- Equipment flow
- Clean equipment storage
- Dirty equipment flow to wash
- Clean equipment flow to sterilization

In conjunction with the above facility layout, an overlay or separate drawing featuring HVAC design considerations such as air particulate level classification (typically Class 100 000, Class 10 000, or Class 100) and air pressure differentials should be completed. In general, air pressures should be designed such that clean air leaks into dirty air and not the other way around. In a typical fermentation/product recovery facility, the recovery area and inoculum areas should be pressurized relative to the fermenter vessel area. This is because open sterile product or aseptic live culture must be handled in the recovery or inoculum rooms, respectively.

The architectural aspects of the facility should be more or less defined at the end of the prequalification phase. Provisions should be made to seal airtight all walls which are to isolate areas of differing pressure and/or air classification. Doors, which serve a pressure retention function should be fitted with closures and even alarms if experience shows that personnel are likely to prop these open with 'broom handles' or other similar devices. (Bear in mind, many doors in a pharmaceutical facility have handles on only one side to prevent contaminated personnel from back entering a clean production area.)

In more sophisticated facilities, a security computer is programmed only to allow proper personnel to enter each secure area. In a modification of this security approach, the exit door of an airlock can be interlocked with the entrance door to the airlock. In this way both cannot be open at once and let relatively 'dirty' air back into a 'clean' production area.

Wall and floor finishes should be specified during the prequalification phase also. These finishes represent the trade-off between ease of cleanability with the cost of such relatively expensive materials as terrazzo flooring. Of course the degree of cleanliness desired is dependent on the type of process to be carried out in a given room. Even the type of flush mounted ceiling lights or wall mounted electrical switches will be dictated by the degree of cleanliness desired in the room.

At the start of the prequalification phase, the process to produce a new (or improved) drug formulation has probably been proven at the bench scale but not scaled up to an industrial or production scale. This scale-up has to be addressed along with its many ramifications.

What purity of water is necessary to formulate, media, buffers, CIP solution, etc.?[13] Will pyrogen free WFI be necessary or will microorganism controlled reverse osmosis effluent polished by a deionizer suffice? In some preliminary steps, tap water or distilled water my even serve the purpose.

To what extent will redundant utility generators such as air compressors, vacuum pumps, steam generators, or WFI stills be necessary? Can the downtime of a unit be handled through use of a large storage tank? Can defective parts in a 'down' machine be replaced quickly, or will days be spent waiting for vendors' maintenance/repair personnel to arrive? (See Moes[34] for discussion of these and other start-up/validation considerations.)

Another area of process design to be decided upon is the degree of computerization of controls for the process/facility. Should a central distributed control system (DCS) be installed or should all operations be controlled locally by personal computers or even pneumatic control loops? The decision has great ramifications regarding the degree of computer validation work which will be necessary.

The validation of a DCS which controls process throughout a facility can be long and complicated. In fact, many firms avoid this problem by using the DCS only in a data collection mode. This leaves the validation of control loops at the local level, a much more straightforward task.

The master validation plan as previously described will be authored at the late stage of the prequalification phase. At this point in the project, it is often advisable for the owner and any appropriate representatives/consultants to arrange a face-to-face meeting with FDA personnel. With this timing, any necessary changes can be incorporated without having to tear out walls, floors, etc.

21.6 Qualification phase activities

The qualification phase of validation more or less parallels the 'detailed design/engineering' phase of an engineering effort. The first validation activity required in the qualification phase is the development of system descriptions. These are extremely detailed documents which can include lists of interlocks and preliminary operating procedures/instructions. These documents should be 'blessed' by operating/production personnel and quality assurance personnel as well as safety personnel after a proper hazard/operation (HAZOP) review.

Batch processes must be especially well defined at this point since they require a step-by-step definition of what valve should be opened in reaction to what measured variable or passage of time.

These system descriptions are normally incorporated into IQ and OQ protocols, and eventually standard operating procedures (SOP). They may be used as a basis for creating software logic if the process is to be computer controlled.

In a proper engineering effort, process and instrumentation diagrams (P&IDs) should be completed to their 'issued for design' revision during this phase of the project. Equipment should have been ordered by this point and operating manuals should have been requested at the earliest possible date for use in writing IQs and OQs. At this point, typical manuals/equipment drawings will suffice in order to begin authoring IQ protocols and OQ protocols and process and performance qualification protocols. (As certified manuals/drawings become available, the information they contain should be incorporated as soon as possible.)

Possibly even before the authoring of system descriptions, validation/regulatory personnel should have input to the project equipment specifications and P&IDs in order to ensure that the systems will be validatable. For instance, sanitary process lines should be sloped to a proper degree and a sterile product storage tank (and transfer lines) should be of polished type 316 stainless steel with pure steam hard piped in. There should be a minimum of high points and low points in these lines and proper sanitary valves should be specified. System cleanability should also be verified at this stage.

One important note about SOPs, the first SOP which should be written at this qualification stage is the SOP which defines how SOPs and other documents should be formatted and written. This planning will avoid inefficiency and rework later.

During the qualification stage, acceptance criteria for various process measurements will have to be agreed upon by all parties. What resistivity of pure water will suffice? Ten million ohm-cm?, or will 15 million ohm-cm water be necessary? What bacterial count will suffice for various types of water in the plant? What bubble test procedure will be used for 0.2 μm sterilizing filters?

These are only typical acceptance criteria and any facility will have many others which will have to be considered. Many of these criteria will depend on how the product to be manufactured was developed in the research laboratory. Was water conductivity or bacterial contamination carefully controlled at the bench scale level? Was the exact composition/purity of sugar or other ingredients used to make fermentation media critical or was a 'serum' used?

In summary, it is essentially impossible to list even all the types of criteria to be considered. The user is the best judge of what measurements are best taken to measure and control the process and the user should react accordingly.

21.7 Validation phase (performance qualification) activities

The (validation) performance qualification phase of validation normally takes place after the facility (or renovation) in question has been completely mechanically constructed and all equipment has been run for at least a few hours to demonstrate mechanical operability. This is the time when personnel are stationed in the newly constructed facility and begin to verify in the field the information contained in the signed-off protocols.

As noted previously, instruments should be calibrated, and this calibration documented properly by the time the validation field work is begun. Ideally, all qualification report data should be collected by the regular operating personnel for the facility who will be responsible for the steady state operation of the facility. If these personnel are not yet available, engineering and/or scientific personnel, either internal or external to the operating company, may be used, or professional validation personnel may be contacted.

The collection of validation data is an excellent way for operating personnel/technicians to become familiar with the facility in which they will be working. If external short term personnel are used, there should be provisions made to transfer the knowledge/experience they obtain to permanent staff before they leave the job site.

Installation qualification data collection can begin in phases as various sections of a facility are deemed mechanically complete. In this way, the validation effort may be merged with the engineering construction effort to compress the facility construction/start-up/validation schedule to the maximum possible extent.

The IQ data collection effort, if done properly, can serve as a double check on the engineering/design effort of a facility, and certainly can be combined with the generation of as-built documents such as P&IDs, piping drawings, equipment layout drawings, equipment specs, etc. to ascertain that these drawings reflect the as-built condition of the facility. The FDA prefers to see a good set of as-built drawings to document a facility on day one of its validated operation. These drawings serve as a ‘baseline’ from which changes will be measured. If these changes are significant enough a revalidation effort will be necessary. This process is described further under section 21.8 (On-going activities).

Once the IQ is complete for a piece of equipment (or train of related equipment), the operational facility may begin. This exercise involves actually running equipment and taking quantitative measurements of its performance. These measurements may include such considerations as flow of cooling water to an air compressor, pressure drop through the unit, and temperature rise through the unit. These measurements may also be used to balance water flows in a multiple branch recirculating system.

It is good engineering practice to verify the compliance of the equipment being tested to all other safety/health codes during this OQ phase also. Noise levels could be documented here as could the temperature of a waste water going to a building floor drain (for compliance with plumbing codes). If toxic chemicals are used, the ambient concentration of these elements could also be verified at this OQ stage.

OQ activities for certain types of equipment used in the pharmaceutical industry have become highly specific exercises which are often done by professionals specializing in this area. Temperature distributions in autoclaves or depyrogenation ovens[35] are examples of such special tasks. Here as many as 20 or 30 thermocouples must be placed in various predetermined load configurations (specific ways of arranging glassware/plasticware etc. in the chamber) to verify that the coldest spot in the load reaches the proper (dry or wet) sterilization temperature. These studies also help to determine the proper amount of time that the chamber must be held at a nominal set point temperature (as measured by the instrument which came with the unit) in order to reach proper sterilization temperature throughout the load.

The FDA has developed a *de facto* requirement that all experimental runs be carried out through at least three identical repetitions. This three repetition rule should be followed when conducting field measurements of this nature and all runs properly documented. This information should be archived in order to establish a baseline in a manner similar to that described under IQ. These multiple runs also begin to provide a quantitative measurement of the variability of the operation in question (both accuracy and precision of measurements). Armed with this knowledge of baseline data uncertainty, trends in steady state data collected later can be discerned more easily.

Other specialized operations might include a lyophilizer (freeze dryer)[36], a filling machine[37], or a tableting machine. If in doubt about how to validate, outside expertise should be consulted. There is never a need to 'reinvent the wheel' when the start-up schedule is at risk.

The OQ data collection effort will serve as a double check on the process calculations carried out during the design of the facility. In an extreme example, it may be discovered that it really was not possible to send 600 gallons per minute through that 2 inch line. More realistically, one may discover a need to throttle back certain flows or add more sophisticated flow control equipment, for instance.

Once the OQ activities are complete, the overall process qualification effort may begin. This typically involves verification of actual kill of microorganisms by sterilization through use of biological indicators. These are often spore strips which turn color when all spores are dead. Sterilization of tanks, lines, glassware, etc. all need to be verified during PQ with multiple runs again being required.

The cleaning-in-place procedures to be used during plant operation also need to be validated during the PQ stage. As explained previously, the degree of cleanliness achieved is normally measured by measuring the conductivity of the final rinse water passed through the equipment to be cleaned. If intermediate rinses are used, a conductivity feedback may also be helpful. The exact solvent/detergent/acid/caustic combination necessary to clean a given soil should also be verified at this stage and adjusted if necessary.

Other PQ activities could conceivably include verification of the activity of a vaccine formulation or verification of the amount of moisture in a freeze-dried preparation. All processes have intermediate points where chemical purity analyses may be meaningfully carried out.

All reports generated in the field should be circulated and signed as approved by all responsible disciplines as listed on the cover sheet. IQs and PQs where data have been collected should include graphical representation of these data in readily understandable form. The results should always be analyzed and a summary page should be circulated with each report. One should not have to read 50 pages to decide that an autoclave has produced the proper temperature distribution.

All reports finally must be certified by the previously defined select validation committee. This process ensures that management 'buys in' to the methodology used to validate the process/facility, and in the eyes of the FDA, eliminates the 'executive deniability' defense.

The validation committee should include as high a level of management as practical. It would not be unusual for the chief executive officer (CEO) of a 40 man firm to sign such a document. It also is more common for the head of operations at a job site to sign in lieu of the CEO of a multinational corporation. Common sense has to rule in this case.

21.8 On-going activities

Even though all validation paperwork as previously described has been signed, sealed and delivered, the validation effort for a facility has just begun. The validated systems must undergo an on-going evaluation as defined in the lexicon. Applications for producing a new drug (NDA) have to be filed with the FDA for each new product and in the case of recombinant fermentation processes, an establishment license must also be obtained.

Once all the paperwork is complete, the FDA will actually arrive at the facility and inspect selected random aspects of its design and operation. While the FDA probably will never cover everything, it may cover anything and so it is best to have everything in order.

The topic of an FDA inspection is a treatise unto itself and will not be covered here. However, once the FDA give the go-ahead to produce product for public consumption, validation responsibilities do not cease. As mentioned before, proper change control documents for all certified qualifications and procedures must be in place. Part of this procedure should include the retraining of operators when the process changes significantly. Also all controlled documents should be recirculated for approval when a change becomes necessary.

Preventive maintenance as defined in section 21.2 must be carried out at regular intervals and certified, as should a periodic audit of computer software for unauthorized changes.

Probably most importantly, data should be analysed regularly for trends, particularly negative trends. See Figure 21.9 for an example of a trend analysis for a sterilization temperature measured at the same point in each load over six months of operation. The levels marked are defined in the lexicon given at the end of this chapter.

As can be seen from this fictional case, the sterilization temperature was set on day one at 260°F (127°C), well above the 250°F (121°C) necessary to sterilize the load in the time called for in the SOPs. For whatever reason the same temperature has fallen slowly but steadily over time to the point that it has dropped below the preselected alert level of 255°F (124°C). If these data are logged or plotted on a computer, this 255°F level should alarm the operator but not shut down the process.

If this was a tank being steam sterilized automatically by injection of steam into its jacket, the temperature being measured could be falling due to a slow build-up of steam or cooling water contaminants on the inside jacket surface (exterior of the tank). The alert level alarm could motivate operators to shut down the production process and clean out the vessel jacket (or more likely,

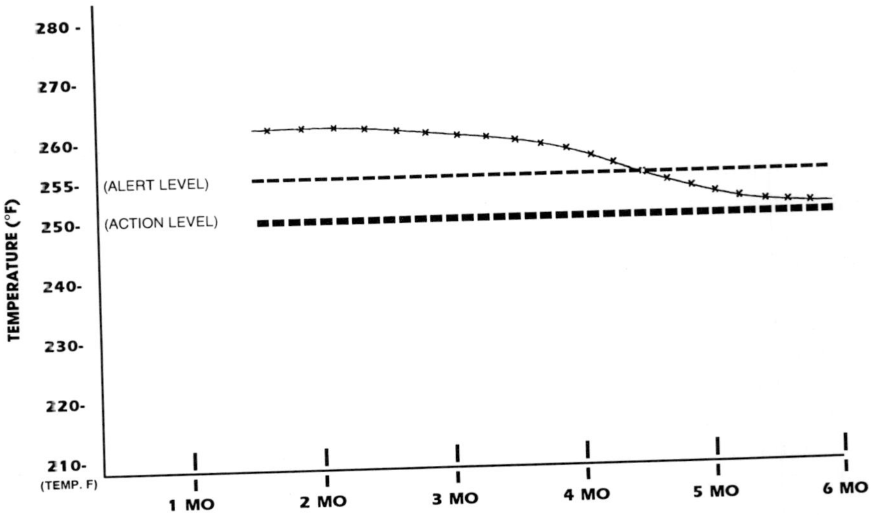

Figure 21.9 Sterilization temperature at the center of a given load configuration—a long term data analysis.

in the real world, wait for the next scheduled shutdown, in this way the temperature keeps falling but is more closely monitored).

Of course the temperature cannot be allowed to fall forever. Once it reaches 250°F (121°C), the process will automatically shut down and alarms will sound. If this occurs during a period of critical production, many negative ramifications could result.

Perhaps the most crucial on-going validation activity is the interaction with the FDA particularly regarding inspection and possible product recall. To use the vernacular "your ducks should always be in order". Validation never ends. Paperwork should be updated today, not at the end of the week when a free hour may be available.

It is not a bad idea to run 'mock' FDA inspections through use of in-house QA personnel or audit consultants. An experienced competent consultant can spot many of the same violations or deficiencies that FDA personnel would spot. The cost of a consultant for a day or two is certainly a lot less than the cost of being shut down by FDA (or even answering extensive 483 Form write-ups).

21.9 Validation of HVAC systems

One aspect of pharmaceutical facilties which must be validated is the heating, ventilating, and air conditioning (HVAC) system. The exact degree of validation depends on the degree of sterility required in the process. A non-sterile bulk facility would require much less HVAC validation than a sterile fermentation/protein recovery/filling operation. HVAC validation will be covered only briefly in this section. The subject is covered in other works[21,22,38–42].

In general, one of the most important parameters in controlling ambient air in a pharmaceutical facility is 'class' (number of particles greater than 0.5 μm contained in a cubic foot of air in the area in question). This general concept of controlled environment work space is presented in Federal Standard 209E[43]. (The letter suffix of this document changes when it is revised — the next revision will be 209F). This standard was in reality authored for the space industry by NASA, but due to lack of anything more appropriate it has become a *de facto* pharmaceutical industry standard. The cGMPs call for a Class 100 environment where sterile filling of pharmaceuticals is done into open containers. This design has become standard and must be designed into any new sterile filling area which is being constructed. Note that 209E does not address viable particulates, just total particulates. In fact, it is defined in terms of 0.5 μm particles when bacteria can be as small as 0.2 μm.

When validating particulate levels, an initial background count of viable and non-viable particulates should be done in the production area in question before any live cells or media, etc. are introduced into the area. The particle count during steady state operation, both 'at rest' and with full complement of production workers and equipment in motion should be taken and verified to be within the design level of the area (Class 100 000 for instance). Once the design level has been validated spot checks for particulate level should be run at regular intervals to prove that the operation remains validated.

The temperature of a pharmaceutical production area should be precisely controlled to preset limits and validated under the same at-rest and operating conditions as described above. Temperature should in general be controlled to approximately ±/± 2°F (1.1°C) at all times. (ASHRAE standards for instance only define temperature control at 90 or 99% levels and may not suffice for validation purposes.) The temperatures of various rooms in a facility may be different depending on such factors as the degree of gowning of the occupants of the room.

Relative humidity is important for the correct operation of such pharmaceutical operations as tableting or packaging of bulk dry powder. Moisture level should be measured and controlled as needed in rooms housing these operations. Moisture level is usually adjusted by direct injection of steam into the recirculating room air. Therefore the chemical additives in this steam must be carefully watched for compatibility with drug products.

Clean air should leak into dirty air and not the other way around. While 209E calls for minimum pressure differentials of 0.05 in (0.127 cm) of water column, this level is not always achievable. Levels of 0.02–0.03 in (0.051–0.076 cm) of water will probably suffice to ensure proper airflow.

The absolute airflow through a room is related to differential pressures. A Class 100 000 room may require 20–30 air changes per hour while a Class 100 room may require 50–60 air changes per hour. The recirculating air rate may be dictated by either particulate removal or heat removal considerations. Once-through airflow (as opposed to recirculated flow) has gained popularity recently and may be the only alternative if a build-up of explosive vapors is to be prevented.

Particles are generally removed from the inlet duct into a room through use of a high efficiency particulate air (HEPA) filter. Particles as small as 0.1 μm may be removed to an efficiency of 99.9995% using HEPA filters.

The validation of a HEPA filter in place begins with the so-called DOP (dioctyl phthalate) test. This test was developed for gas masks during World War I and involves use of DOP which forms extremely uniform round particles when injected into air. The particles are created to a known concentration on the upstream perimeter of the filter. The results of this DOP test should be part of the permanent validation record of an HVAC system and should be done by professionals experienced in the technique. Alternatives to this test have recently been suggested[52] since DOP is a known human carcinogen.

As with process validation, HVAC validation should begin with a proper installation qualification document. It makes sense with these systems to IQ an airhandler, ductwork, blowers, and possibly even an outside air markup unit all in one document due to the tight nature of their interaction. The OQ of an HVAC system should include a thorough set of leak tests. Also the cleanability of the ductwork (without disassembly) should be verified. The cleanability would involve extra chemical addition or handhole ports or even manways into the duct if it is big enough.

The PQ for HVAC equipment has been described previously in this section. Preventive maintenance for HVAC equipment should be formalized under controlled procedure as described under process equipment. Arambulo[42] suggests the various types of validation documents necessary for a HVAC system in various parts of a fermentation-based pharmaceutical production facility. While these suggestions are not cast in stone, they are at least a good starting point for a discussion on the subject.

Biosafety cabinets and possibly fume hoods or laminar flow hoods will require validation if the final product formulation (or a closely related intermediate) is handled in them[41]. Validation of these units involves filter testing and flow verification as described under room validation. If one of these controlled units has a large intake, or there are many in a room, their interaction with the airflow in the room may be critical and worthy of special validation consideration.

21.10 Computer system validation

Much has been written over the past few years concerning the subject of computer system validation. One of the seminal works in the field was 'Validation Concepts for Computer Systems Used in the Manufacture of Drug Products'[33]. The Pharmaceutical Manufacturers' Association committee has continued its work by authoring works which help define further the methodology to be used in validating computer systems[2,44–47]. These works cover such topics as software testing, change control, and installation qualification.

Several other works[48–52] refer to various strategies and aspects of software/computer validation. Due to space limitations, this topic will not be addressed in detail in this chapter. However, the reader is referred for further information to the excellent papers mentioned above, or to one of the many symposiums or seminars given on this subject every year. Many computer validation terms are also included in the lexicon[33], so that the reader may be conversant with at least the basics of the subject.

Definitions

1. PROCESS VALIDATION is establishing documented evidence which provides a high degree of assurance that a specific process will consistently

produce a product meeting its pre-determined specifications and quality characteristics[3].

2. *Alternate validation definitions.* "Validation is attaining and documentation of sufficient evidence to give reasonable assurance, given the current state of science, that the process under consideration does, and/or will do, what it purports to do."

(BYERS, 1980)

Thus, validation is defined as the process of testing (fact gathering) to prove that a manufacturing system meets its pre-established requirements, giving it strength to stand up to a legal challenge (FDA inspection).

(LOVEJOY, 1986)

Validation—"In today's pharmaceutical industry, whether you are thinking about a computer system, a water treatment system or a manufacturing process, validation means nothing more than well-organized, well documented, common sense".

(CHAPMAN, 1986)

3. *FDA regulatory basis for validation.* 21CFR211.110 'Sampling and testing of in-process materials and drug products' states[5]:

"Control procedures shall be established to monitor the output and *validate* the performance of those manufacturing processes that may be responsible for causing variability in the characteristics of in-process material and drug product."

4. *Validation requirement for sterilization process.* 21CFR211.113 'Control of microbiological contamination' states[5]:

"Appropriate written procedures, designed to prevent microbiological contamination of drug products purporting to be sterile, shall be established and followed. Such procedures shall include *validation* of any sterilization process."

5. *Validation protocol.* A prospective experimental plan that, when executed, is intended to produce documented evidence that the system has been validated[5].

6. *Installation qualification (IQ).* Documented verification that all key aspects of the installation adhere to appropriate codes and approved design intentions and that manufacturer's recommendations are suitably considered[5].

7. *Operational qualification (OQ).* Documented verification that the system or subsystem performs as intended throughout all anticipated operating ranges of pressures, temperatures, times, and other operational parameters[5].

8. *Process (performance) qualifications.* Documented verification that the process performs as intended throughout all anticipated operating ranges of pressures, temperatures, times, and other operational parameters[5].

9. *Other validation related documentation calibration reports.* Calibration: demonstrating that a measuring device produces results within specified limits of those produced by a reference standard device over an appropriate range of measurements. This process results in corrections that may be applied if maximum accuracy is required.

10. *Preventive maintenance procedures.* To satisfy cGMPS, the following items must be part of a preventive maintenance program:

Numbering system to identify all equipment requiring servicing
Equipment history file
Preventive maintenance schedule reminder file
Equipment failure emergency action records
Work order forms
Preventive maintenance standard operating procedure file
Engineering/procedure change control request, work order and completion
Operating logs and cross-reference file
Audit procedure system

11. *Prospective validation.* Establishing documented evidence that a system does what it purports to do based on a preplanned protocol[3].

12. *Concurrent process validation.* Establishing documented evidence that a process does what it purports to do based on information generated during actual implementation of the process[3].

13. *Retrospective validation.* Establishing documented evidence that a system does what it purports to do based on review and analysis of historic information[3].

Lexicon[33]

Acceptability limits:	A specific set of parameter values, deviation from which renders the product quality questionable.
Action levels:	A specified parameter value deviation from which requires documented action to return the system or process to a state of control.

Alert levels:	A specified parameter value which signals a deviation from the norm and a potential action condition. This level is still within the proven acceptable range but requires close monitoring to ensure the action level is not reached.
Application software:	A program adapted or tailored to the specific requirements of the user for the purpose of data manipulation, data archiving, or process control.
Calibration:	Demonstrating that a measuring device produces results within specified limits to those produced by a reference standard device over an appropriate range of measurements. This process results in corrections that may be applied if maximum accuracy is required.
Certification:	Documented testimony by qualified authorities that a system qualification, calibration, validation, or revalidation has been performed appropriately and that the results are acceptable.
Change control:	A formal monitoring system by which qualified representatives of appropriate disciplines review proposed or actual changes that might affect a validated status to determine the need for corrective action that would ensure that the system retains its validated state.
Computer program:	A collection of logically interrelated statements or instructions that when executed by a computer makes possible the performance of a predefined task.
Computer system:	A group of hardware components assembled to perform in conjunction with a set of software programs, which are collectively designed to perform a specific function or group of functions.
Control parameters:	Those operating variables that can be assigned values that are used as control levels.
Control parameter range:	Range of values for a given control parameter that lies between its two outer limits or control levels.
Critical parameter:	An operative variable whose performance has a direct and measurable impact on product quality.
Documentation:	Manuals, written procedures or policies, records, or reports that provide information concerning uses, maintenance, or validation of a process or system involving either hardware or software. This material may be presented for electronic media.
Environmental factors:	Those outside influences on a computer system that

may affect its operation, including temperature, moisture, electrical interference, electromagnetic flux, power supply, and so on.

Flow chart: A graphical representation of the logical steps involved in a procedure or program.

Functional testing: A process for verifying that the software performs the intended functions.

Hardware: Physical apparatus that makes up a computer. This term is also used to describe the various pieces of equipment in the computer system, including the central processing unit, the printer, the modem, the cathode ray tube (CRT), and so on.

HIPO chart: A software charting technique used to define and document programming systems in a manner that facilitates communication between the programmer and the user. HIPO is an acronym for hierarchy plus input-process-output.

Installation qualification (IQ): Documented verification that all key aspects of hardware installation adhere to approriate codes and approved design intentions and that the recommendations of the manufacturer have been suitably considered.

Life cycle: An approach to computer system development that begins with identification of the user's requirements, continues through design, integration, qualification, validation, control and maintenance, and ends only when commercial use of the system is discontinued.

Module: A unit task or subdivision.

On-going evaluation: A term used to describe the dynamic process employed after a system's initial validation to maintain the validated state of a computer system.

Operating system: A set of programs provided with a computer that function as the interface between the hardware and the applications program.

Operating variables: All factors, including control parameters, that may potentially affect process state of control and/or fitness for use of the end product.

Operational qualification (OQ): Documented verification that the system or subsystem performs as intended throughout representative or anticipated operating ranges.

Peripheral: A device attached to a computer, such as a terminal, a printer, a disk drive, a magnetic tape unit, a plotter, and so on.

Procedure: An approved document which gives a specific set

of instructions which when followed will consistently produce a result defined by a specification.

Process:	The controlled system.
Process validation:	Establishing documented evidence that a process does what it purports to do.
Protocol:	A prospective experimental plan that, when executed, is intended to produce documented evidence that the system has been validated.
Pseudocode:	Common language (such as English) in an arbitrary code-like format used to describe meanings of routines and subroutines written in a programming language. In other words, an English translation of a computer program.
Quality assurance:	The activity of providing, to all concerned, the evidence needed to establish confidence that the quality function is being performed adequately.
Revalidation:	Repetition of the validation process or a specific portion of it.
Software:	A collection of programs, routines, and subroutines that control the operation of a computer or a computerized system.
Software development standards:	Written policies or procedures that describe practices a programmer or software developer should follow in creating, debugging, and verifying software.
Software diagnostic procedures:	Tests for predicting errors or malfunctions in programs.
Source code:	An original computer program expressed in human-readable form (programming language), which must be translated into machine-readable form before it can be executed by the computer.
Utility programs:	Special program frequently supplied by the producer of the operating system. They perform general functions such as making backup copies of programs and copying files from tape to disk.
Validation protocol:	A prospective experimental plan that, when executed, is intended to produce documented evidence that the system has been validated.

References

1. K.G. Chapman, *Pharm. Tech.*, **8**, 22 (1984).
2. K.G. Chapman and J.R. Harris, *Pharm. Tech.*, **13**, 60 (1989).

3. Center for Drugs & Biologics and Center for Devices and Radiological Health, *Guidelines on General Principles of Process Validation*, Food and Drug Administration, May, 1987.
4. K.G. Chapman, *Pharm. Tech.*, **7**, 51 (1983).
5. Department of Health & Human Services, *Human and Veterinary Drugs*, FDA, CFR Title 21, Parts 210 to 211 (revised 4/1/90).
6. Department of Health and Human Services, *CFR Title 21 Parts: 606*, Food and Drug Administration-Biologicals.
7. Department of Health and Human Services, *CFR Title 21 Parts: 803, 820*, Food and Drug Administration — Medical Devices (rev. 4/1/90)
8. *Guide to Inspection of Bulk Pharmaceutical Chemical Manufacturing*, April, 1984. (Division of Field Investigation (HFO-500, Office of Regional Operations of Regulatory Affairs) and (Division of Drug Quality Compliance (HFN-320) Associate Director for Compliance, Office of Drugs, Center for Drugs and Biologics)
9. *Guide to Inspection of Computerized Systems in Drug Processing*, February, 1983. (Division of Drug Quality Compliance (HFN-320) Associate Director for Compliance, Office of Drugs, National Center for Drugs and Biologics)
10. J. Tomson, *Pharm. Eng.*, **8**, 35 (1988).
11. F.G. Harrison, *BioPharm*, **2**, 24 (1989).
12. A.R. Casola, *Pharm. Eng.*, **9**, 32 (1989).
13. *PMA Guidelines for the Production, Packing, Repacking or Holding of Bulk Pharmaceutical Chemicals*, 2nd ed., (revised June 1978).
14. A. Tankha and H. Wilford, *Pharm. Eng.*, **7**, 17 (1987).
15. E.L. Inman, *et al.*, *J. Chromatogr. Sci.*, **25**, 252 (1987).
16. R.G. Werner and H. Langlouis, *Gau Arzneim, Forsch* **39**, 108 (1988).
17. F.G. Harrison, *BioTechnol.*, **3**, 43 (1985).
18. PMA's Validation Advisory Committee, *Pharm. Technol.*, **9**, 78 (1985).
19. B. Loftus and R.A. Nash, *Drugs Pharm. Sci.*, **23**, XVII, 286 and 286P (19??).
20. T.H. Meltzer, *Ultrapure Water*, **7**, 40 (19??).
21. P.A. Cipriano, *Pharm. Eng.*, **2**, 36 (1982).
22. D. Hill and M. Beatrice, *BioPharm*, **2**, 20 (1989).
23. D. Hill and M. Beatrice, *BioPharm*, **2**, 28 (1989).
24. P. Wolber and M. Dosmar, *Pharm. Technol.*, **11**, 38 (1987).
25. M.J. Carter and R.A. Scotland, *BioPharm*, **2**, 24 (1989).
26. H. Avallone, *Pharm. Eng.*, **8**, 50 (1988).
27. K.J. Halloran, L.M. Lagade and D.E. Martynuk, *Pharm. Technol.*, **13**, 33 (1989).
28. I.R. Berry, *Drug Development and Industrial Pharmacy*, **14**, 377 (1988).
29. S.S. Chrai, *Pharm. Technol.*, **13**, 84 (1989).
30. A.G. Lord, *Pharm. Eng.*, **8**, 30 (1988).
31. D. Larkin, *Pharm. Eng.*, **4**, 27 (1989).
32. *NIH Guidelines for Research involving Recombinant DNA Molecules*, May 7, 1986 and August 24, 1987.
33. Pharmaceutical Manufacturers Association (PMA) Computer System Validation Committee, *Pharm. Tech.*, **10**, 24 (1986).
34. H. Moes, *Proceedings Pharm. Tech. Conference* '88, September 26–28.
35. R. Baird, *Pharm. Eng.*, **8**, 31 (1989).
36. E.H. Trappler, *Pharm. Eng.*, **13**, 56 (1989).
37. H. Seyfarth and H. Barnickel, *Drugs Made in Germany*, **33**, 58 (1990).
38. International Pharmaceutical Federation, *J. Parenteral Sci. Technol.*, **44**, 272 (1990).
39. A.H. Rolston and J.V. Ricigliano, *Pharm. Eng.*, **8**, 25 (1988).
40. J.A. Kuhn, *Pharm. Eng.*, **7**, 31 (1987).
41. A.S. Arambulo, *Pharm. Eng.*, **7**, (1987).
42. *Federal Standard Clean Room and Work Station Requirements*, Controlled Environment Fed. Std. — 209E.
43. PMA's Computer System Validation Committee, K.G. Chapman and J.R. Harris (principal authors) *Pharm. Technol.*, **13**, (1989).
44. PMA's Computer System Validation Committee, J.S. Alford and F. Cline, Jr. (principal authors), *Pharm. Technol.*, **14**, (1990).

45. PMA's Computer System Validation Committee, E.J. Subak, Jr. (principal author), *Pharm. Technol.*, **13**, (1989).
46. PMA's Computer System Validation Committee, J. Agalloco (principal author), *Pharm. Technol.*, **14**, (1990).
47. L.J. Goren, *BioPharm*, **1**, 29 (1988).
48. L.J. Goren, *BioPharm*, **2**, 38 (1989).
49. H. Bozenhardt and J. Agalloco, *Pharm. Eng.*, **7**, 14 (1987).
50. A.R. Bluhm, *Pharm. Technol.*, **13**, 00, (1989).
51. F.A. Rowley, J.A. Hitchner, A.J. Pagliero and R.S. Bertolani, *Pharm. Technol.*, **14**, 52 (1990).
52. *Validation of Aseptic Pharmaceutical Processes*, eds F.J. Carleton and J.P. Agalloco, Marcel Dekker, New York, 1986.

22 Project execution for the design and construction of a biotechnology facility

S.K. YU

22.1 Introduction

Since the mid-1960s progress in the science of molecular biology and genetic engineering has fueled an explosive growth in the biotechnology industry. Biotechnology is the manipulation of microorganisms, including recombinant microbes or cells to produce desirable materials, such as drugs, or desirable actions, such as cleaning up a polluted water stream.

The pharmaceutical industry has benefited greatly from the advance of biotechnology. Biotechnology enables scientists to synthesize large quantities of complex chemicals, such as human insulin, interferon, and growth hormones, which would otherwise be uneconomical to produce by conventional chemical processes.

As technology advances, new pharmaceutical products are created from all over the world at a rapid pace. In order to compete profitably, pharmaceutical companies are looking for ways of expediting the transition of new products from the development laboratory to the market place. One promising approach is the use of a multipurpose biotechnology facility[1] which will allow a corporation to handle all the products in their various stages of development.

The guidelines and standards for building a multipurpose biotechnology facility are currently being fomulated. In the pharmaceutical industry most of the existing facilities are built according to the traditional chemical pharmaceutical facility standards for clean facility requirements, process tubing, and equipment material of construction, etc.

Since the field of biotechnology is progressing rapidly, facility and process requirements change constantly. In this chapter, the important elements and issues that are involved in the design and construction of a biotechnology facility will be discussed.

22.2 Project management

One of the major issues for a biotechnology facility project is project management. It must be emphasized that a project should be well planned at all phases. Because of the evolutionary state of biotechnology, all the parties involved in the project, i.e. the owner, the architect and engineering firm, the construction management firm, and even the FDA (Food and Drug Administration), are somewhere on the learning curve. Therefore changes to different parts of the design should be anticipated and properly managed. Management should have a realistic view of the complex nature of the project and set the project deadlines accordingly. Although the 'fast track' approach, starting construction with an incomplete design package, is quite common in the biotechnology industry, management should weigh the benefits against the penalties of this approach. If the scope of the project is not well defined, 'fast track' may lead to cost overrun and even delay in the project schedule. The course of a biotechnology facility construction project can be grouped into six different stages as follows:

1. *Project initiation*—At this stage management identifies a business opportunity and a need for a biotechnology facility. The decision to initiate the project is based on the company strategic plan. It is important to emphasize that management should understand the complexity of the project and allocate the resources and set the deadlines for the completion of the project appropriately.
2. *Preliminary engineering design and cost estimate*—At this stage an architect and engineering (A&E) firm is selected for the conceptual design, and the cost estimate of the facility. The total investment includes the cost of engineering design and construction capital. A detailed project schedule is developed at this point.
3. *Detail design*—At this stage all the construction documentation, i.e. drawings and specifications are completed. The documents are used for construction permit applications and for construction contract bidding purposes. A construction management firm or general contractor is selected, and a detail construction schedule is finalized. The generation of the validation protocols also starts at this stage.
4. *Construction*—At this stage all the trades are mobilized for the construction activity. As the start of construction, the ground rules are set for all field activities, such as area for equipment and material, housekeeping, safety, and coordination among trades. The degree of owner participation depends on the terms of the construction contract.
5. *Start-up*—At this stage the facility with all the equipment and utilities will be started up and transferred to the owner. It is necessary to make sure that all vendors provide start-up service for their respective equipment, and that the start-up activities are documented. Also during start-up the high purity piping systems are cleaned and passivated. At the end of this stage all punch list items are resolved.
6. *Validation*—After the facility is formally accepted, the validation activities occur as follows: check the completeness of equipment documentation,

calibrate all critical instruments, perform installation and operation qualification on all equipment and systems that are essential to the bioprocess operation and perform process validation.

The above list indicates that a biotechnology facility project is extremely complex, and to be successful a team approach from the owner is necessary. A team is composed of personnel from different disciplines, such as mechanical, process, and validation. The responsibility of the team is to define the project scope, supervise the design and construction activities. The team approach ensures a broad range of expertise to deal with the complexity of the project.

One important group that needs to participate in the project is the validation team. During the design stage it is important to have the input of the validation group, and to incorporate all the necessary elements into the design for conformance to cGMP[2]. The key is to provide adequate instrumentation, sampling points, and other access points for the validation team to perform its work.

As with any other project, the most important project management function is to keep the project on track with respect to the scope of the work, project schedule, and cash flow. Other items that need proper management are permit applications, contractor's insurance requirements, work quality, housekeeping, and safety on the job site.

22.3 Defining project scope

The first step in any design is defining the scope of the project. This should be a team effort and involve all relevant parties. All scope items have to be justified and explicitly documented in written form, such as the function and size of the facility, the type of processes, the size of the facility, the number of personnel, the capacity of various utilities, and the future expansion plan.

The scope should be based on facts, because erroneous rule of thumb estimates may lead to scope changes at a later stage of the project. The worst case scenario would be when the errors are not caught during design and have to be corrected during construction. The penalty for a sloppy project scope is delay in the project schedule and cost overruns. It should be emphasized that an accurate and realistic project scope paves the way for a smoother project in the long run.

Due to the evolving nature of biotechnology, however, changes are inevitable during the design and construction of the project. It is important to set up a system to evaluate and handle changes as they occur. One major factor affecting the scope is the introduction of new personnel into the project. For example, when new personnel join the project and bring with them new design specifications, it is important to evalute any suggested changes before incorporating them into the scope.

22.4 Process design

A biotechnology facility for cell culture can generally be classified into the following areas:

- Process area.
  - Media and buffer preparation
  - Inoculum
  - Cell culture and harvesting
  - Purification
- Building and process utilities
- Laboratory area for quality assurance/control
- Shipping and receiving, and warehouse
- General office area for administration

The design of the biotechnology facility is process driven; this means that the facility is designed to service the bioprocess. The centerpiece of the facility is the cell culture process, with its bioreactors[3]. Figure 22.1 is a schematic representation of a biotechnology facility. The bioreactors are located at the

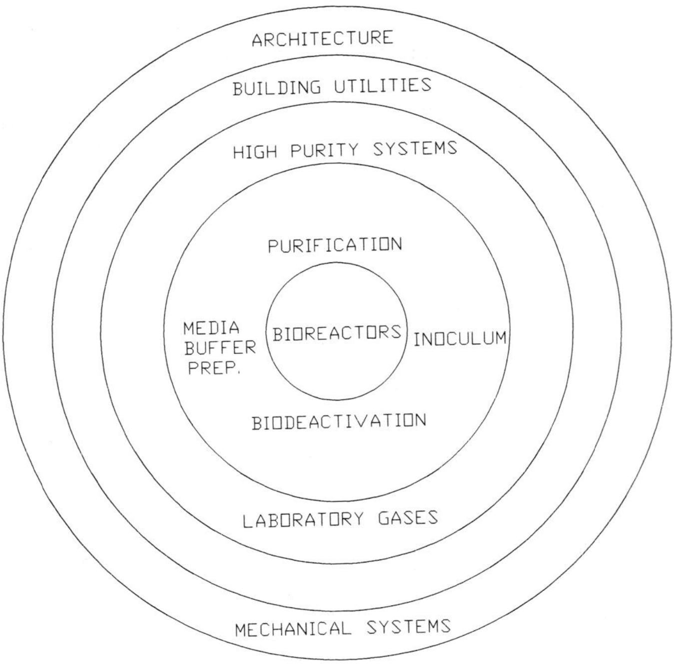

Figure 22.1 Elements of a biotechnology facility.

center of a circle. Surrounding the bioreactors are the support facilities, i.e. inoculum area, media and buffer preparation area, purification area, and biodeactivation area.

Extending the circle outward, there are the high purity water and steam systems, and the laboratory gases that service all the process areas. Outside this circle there are the plant utilities, i.e. process water, cooling waters, plant steam, compressed air, and vacuum.

At the outer circle, there are the architectural and mechanical elements of the building, comprising the isolated HVAC (heating, ventilation, and air conditioning) and the floor layout. These two elements control the flow of raw materials, products, personnel, and air movement within the facility. Proper design isolates different functional areas to prevent cross contamination.

Figure 22.2 is a flow sheet of a typical cell culture or bacteria fermentation process. The process starts with the inoculum operation, which initiates the propagation of a specific cell line. The inoculation starts with a small volume, i.e. a 250 ml flask, which then progressively scales up to several liters in volume. The large volume cell mass is then transferred aseptically into a large bioreactor for further scale-up. The size of the large scale bioreactor depends on the individual process. Successive scale-up might reach a volume of several thousand liters.

After the fermentation is completed, the bioreactor is harvested for its product. Most microbial fermentation products are intracellular, which means that the products remain inside the cell. The harvesting process requires distruption of the cell to release the product. Equipment such as a homogenizer, or a high speed ball mill, is frequently used for cell disruption. The resulting cell mixture is then centrifuged to recover the broth which is then usually

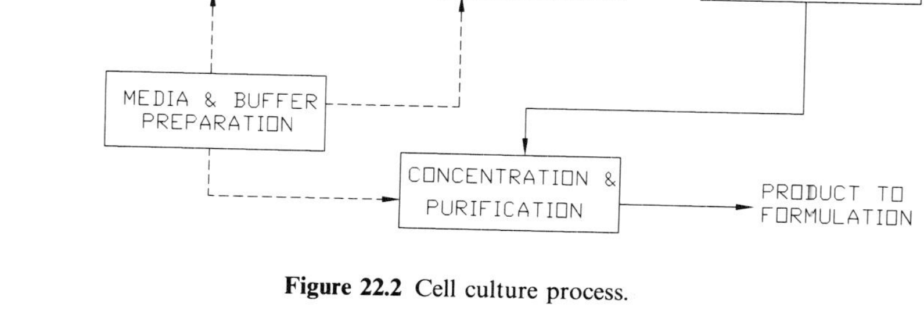

Figure 22.2 Cell culture process.

clarified by tangential membrane filtration equipment. All remaining biomass material from the fermentation process is deactivated either by heat or chemicals before it is discharged to the sewer.

The broth from fermentation is concentrated by an ultra filtration system. The concentrated material is sent to the purification area for final product recovery. Due to the biological nature of the product, the purification operation is often carried out at below room temperature, usually in a cold room. The operation involves equipment such as affinity chromatography, high performance liquid chromatography (HPLC), ion exchange, and gel filtration columns. The final product is filtered through a 0.2 μm depth filter, and aseptically filled under a Class 100 biosafety cabinet into a bulk container to be used for formulation.

Most bacteria fermentation products are intracellar and require extensive purification. Because of the expensive purification equipment and the strict aseptic requirements of the operation, the purification process is a major factor in the total product cost. In some cases purification accounts for about 60% of the total product cost. For cell culture operation, the major cost factor swings toward the fermentation process, because of the expensive serums.

The media and buffer solutions for the inoculum, fermentation, and purification operations are usually provided from one central area. This area is generally equipped with batching equipment, i.e. weight scale and powder dissolution tanks. The glassware and tanks are cleaned and sterilized in a separate area provided with glassware washers, tank washing stations, hot air sterilization oven, and autoclave.

22.4.1 Bioreactors

Different types of bioreactor design are available in the industry. The type of reactor being used will depend on the type of cell and nature of the bioprocess.

Most commercial cell culture operations are batch process, in which the bioreactors are continuous stirred tanks. Continuous stirred tank bioreactors incorporate most of the traditional chemical reactor design elements such as agitator, baffles, gas sparger, and heat transfer jacket. The stirred tank design is versatile and adaptable for most biological systems.

A newer design, called air lift, incorporates a draft tube in the reactor vessel. The mixing action is provided by injecting gas into the bottom of the reactor creating a top to bottom convective circulation inside the vessel. The air lift reactor is suitable for shear sensitive mammalian cells and is often chosen for large scale operations, because of its low shear, efficient internal flow, and efficient oxygen transfer characteristics.

Cells that require a solid surface on which to grow, as do the majority of mammalian cells, may require a fluidized bed reactor in which the cells are immobilized onto a pelletized substrate, generally referred to as 'microcarrier beads'. The pellets are suspended in the reactor by the upward flowing nutrient solution.

There are experimental continuous cell culture processes that immobilize whole cells in an inert matrix such as a crosslinked ion exchange resin or a hollow fiber polymeric membrane. Cells that excrete extracellular product are required for this type of operation. The potential of this new technology is the lower cost of producing biological products due to the high cell density nature of the reactor.

Continuous stirred tank reactor design requires information on the effects of various scale-up parameters, such as geometric similarity, shear rate, and power input per volume on the growth rate of the microorganism. The mass transfer and heat transfer rate characteristics of the reactor are also critical to the design. Proper design enables the circulation of nutrient and oxygen to every part of the bioreactor that sustains the growth of the cells and ultimately the yield of the bioprocess. Precise control of the system temperature and pH is important for the survival of the cell. Most microorganisms and cells thrive within a narrow span of temperature and pH.

Bioreactor design should provide an element of foam control in the vessel. Due to the protein nature of the process material, foaming in the bioreactor is a common problem. Foam decreases the efficiency of the fermentation process by carrying nutrient away from the bulk of the broth and denaturing the protein products at the gas–liquid interface. In the worst case scenario, foam can contaminate the reactor by migrating out through the sterile vent filter and flowing back into the vessel.

The bioreactor should be constructed of 316L stainless steel with an internal surface finish of better than 20 microinch RA (arithmetic mean roughness value). It is important that the internal surface of the vessel be free of crevices and dead spaces where microbial films can grow. All sample ports, material inlet and outlet ports, and agitator seals, should be provided with steam purge assemblies to maintain aseptic conditions in the bioreactor.

22.4.2 *High purity tubing systems*

22.4.2.1 Material specifications. Stainless steel grade 316L is used commonly in the pharmaceutical and biotechnology facilities for process and high purity utility systems such as deionized water, water for injection, and pure steam. For some monoclonal antibody cell culture systems that are sensitive to metallic contamination, polymeric material such as Kynar may be used.

The surface finish of the stainless steel tubing, fitting, and vessels has to be specified precisely. Most facilities require the surface to be better than 20 RA, as measured by the profilometer. It is important to ask the stainless steel tubing or vessel fabricator to provide the procedure used to achieve the final surface finish. Such a procedure should include the type and number of mechanical polishes required. It is very important that the vendor has a quality control program to meet the user's specifications.

There is no single opinion on the subject of material quality and surfaces finish requirements for stainless steel tubing. The steel mill offers two types of tubing, welded and seamless. Welded tubing is less expensive, and to achieve the final finish several progressive mechanical polishes and a final electropolish operation will be required. The main problem with welded tubing is in the seam area. Defects in the seam are unavoidable, no matter how extensively the mechanical polish is performed. Generally after the final electropolish all the inherent imperfections such as silicon, sulfur, and other non-metallic inclusions are magnified along the seam. The common form of defect is pitting along the seam.

Seamless tubing costs about three times that of the welded tubing, but it is not a perfect solution to the problem of welded tubing. A small percentage of base metal defects are still present, but normally the defects are about 10% those of the welded tubing. In some cases, the seamless tubing lacks dimensional integrity which presents fit-up problems during fabrication. Additionally seamless tubing is not readily available for sizes above 2 in (5.08 cm).

Recently several speciality tubing/fitting vendors that have been servicing the semi-conductor industry are now promoting a higher standard of tubing for the pharmaceutical and biotechnlogy industry. These vendors buy the welded tubing from the mill and perform mechanical polish and electropolish on the tubing to meet customer specifications. The higher quality is achieved by a program of quality inspection, and higher quality comes at a higher cost. The owner should balance the additional cost against specific process requirement to determine which type of tubing is appropriate for the project.

22.4.2.2 Piping system design. To minimize microorganism contamination, all tubing and equipment design should conform to the 'six diameter' rule of thumb used by the FDA to minimize dead leg or area in the system where water, product, or cleaning solution will be stagnant. The rule states "do not have an unused portion greater in length than six diameters of the unused pipe measured from the axis of the pipe in use". All side ports, use points, and sampling ports should be designed in conformance to the 'six diameter' rule. In most cases special fittings, such as short outlet tees and short outlet valves are specified.

The most important aspect of the piping system design is to provide adequate slope to avoid having areas of stagnant liquid in the piping system. The minimum slope of 1/8 in/ft should be used. During fabrication, the slope should be inspected and documented. To ensure that the slope requirement is satisfied, none of the other piping and duct work systems must interfere with the path of the high purity piping system. The piping runs should slope to a low point, which is usually a use point. Most facilities provide transfer stations for product line connections. It is important to provide low point drain valves at the transfer panel for complete drainage after each use.

The sampling point requirements have been discussed. This is critical for the validation process. Sufficient sampling points in all high purity systems are needed for the validation process to take place. To ensure that these requirements are satisfied it is important to enlist the input of the validation group at the early stage of the design. Sampling points are generally required near the use points to test the integrity of the piping system.

To provide a safe and effective pressure for the deionized water (DIW) and water for injection (WFI) loops, the maximum pressure at the first use point should be 70 psig or less, and a minimum of 30 psig pressure at the tank return. The above guidelines provide a safe pressure range at all use points. Also excessive high pressure in the high purity water loop increases the chance of valve leakage. High loop pressure and high flow velocity may cause water to hammer into the loops whenever the recirculating pump turns on and off. Therefore, it is important to calculate precisely the pressure drop for each loop. The loop pressure can be reduced by dividing the loop into several smaller loops.

22.4.2.3 Fabrication. For a high purity tubing system, automatic orbital machine welding is required. Qualification of the welding process is necessary to ensure that it can consistently produce acceptable welds. The qualification process starts with the owner and contractor agreeing on a set of criteria for the acceptable weld. The owner and contractor should produce test coupons for use as an acceptable standard in the field to resolve quality disputes during the fabrication process.

Two levels of qualification are needed, i.e. machine qualification and welder qualification. Each machine must be qualified for different size tubing and different locations of operation, i.e. field versus shop fabrication. Each welder will be qualified on different machines and different sizes of tubing. The acceptance criteria for both qualifications must be agreed upon between the owner and the contractor prior to the start of fabrication.

The most common weld problems are lack of fusion and discoloration at the weld. Proper alignment of the electrode and adequate purging of the weld zone are important to minimize these occurrences. The contractor needs to provide a clean area either in the shop or in the field for the stainless steel fabrication work. Work should not begin without owner's inspection and approval of the fabrication area. The area should include sufficient racks made of non-carbon steel material for all piping and fittings. To ensure the quality of the stainless steel system, the weld fabrication should be subject to 100% inspection. To support validation, every phase of the welding operation must be documented properly.

It is important for the contractor to supply a complete fabrication schedule, which should include coordination of the material received, manpower requirements, welding machine requirements, and machine qualifications prior to fabrication.

22.5 Passivation

Passivation denotes a process of immersing stainless steel in an acid such as nitric acid to remove any embedded iron from the base metal and the weld areas. It also promotes the formation of a corrosion resistant oxide layer on the weld surface.

After the high purity stainless steel tubing systems are fabricated, they must be cleaned and passivated before use. Passivation is often performed by a specialized contractor. Prior to passivation, the tubing systems are cleaned by recirculating a solution of 1% sodium hydroxide by weight through the systems at about 160°F (71°C). After cleaning, the passivation is done by circulating a solution of 10% nitric acid by weight through the piping systems at about 130°F (54°C).

To avoid the problem associated with the disposal of nitric acid, some companies have experimented with a passivation solution of organic acid and chelating agent such as citric acid and ethylenediamine tetraacetic acid (EDTA). While some data seem promising, further studies and data are needed before the effectiveness of this approach is fully evaluated and accepted by the industry.

Passivation is a crucial element that is often overlooked in the tubing system design. Passivation of a high purity system is a complicated task. Passivation contractors often use flexible hoses to complete the chemical recirculation path from the chemical solution tank to the high purity system and back to the tank. The chance of acid leakage increases with the number of flexible tubing interconnections. To promote safety and expedite the passivation process, it is wise to consider the passivation circuit in the design stage, and to provide ports and rigid jumpers for feeding and circulating the cleaning and passivation solutions. If it is possible, select a floor system that is compatible with the caustic and acid solutions. Passivation cannot be an afterthought, it must be well thought out at the design stage.

22.6 Cleaning and sterilization requirements

In a biotechnology facility, sterility of the equipment is most important. After each use the equipment is cleaned and then steam sterilized. To provide a consistent cleaning procedure, an automatic clean-in-place (CIP) system is required. All process equipment and vessels, including bioreactors can be designed for CIP.

A CIP system usually consists of two to three tanks, for caustic, weak acid, and purified water. The tanks are connected to a main manifold that feeds the equipment or vessel to be cleaned. Spray balls are placed in the equipment to provide complete solution coverage of all surfaces. The cleaning solution pumps through the spray ball into the vessel, and returns to the chemical tank by another pump located downstream of the vessel. The CIP system has a heat exchanger to maintain the solution temperature. Generally a CIP process consists of the following cycles:

- Initial rinse — all the residual contents from the vessel are rinsed off and pumped to drain.
- Detergent wash — caustic cleaning solution is recirculated through the tank for a period of time and then pumped to drain.
- Acid rinse — a weak acid solution is recirculated through the vessel to neutralize the residual caustic solution.
- Recirculated rinse — high purity water, such as deionized water, is recirculated through the vessel for a period of time.
- Final rinse — high purity water, such as deionized water or water for injection if needed, is pumped through the vessel, and then pumped to drain, or held for reuse as the initial rinse in the next wash cycle.

After the vessel is cleaned, it is steam sterilized by injecting pure steam into the vessel or equipment. The temperature is maintained at about 121°C for a period of time as determined by the validation process. As with CIP, the steam sterilization procedure should be automated to ensure that it is consistent and is documented.

22.7 Biocontainment

The Institute of Health[4] has provided a set of guidelines for laboratory scale research involving the uses of organisms containing recombinant DNA molecules. The guidelines designate four biosafety levels from BL1 to BL4. BL1 pertains to microorganisms not known to cause disease in healthy adult humans, while BL4 pertains to dangerous and exotic microorganisms that pose a high risk of infection to humans.

For research experiments or production scale involving more than 10 l of culture, the NIH has provided four biosafety levels: GLSP, BL1-LS, BL2-LS, BL3-LS. The LS suffix refers to large scale laboratory research. The BL designation corresponds to the equivalent laboratory scale containment requirements. Currently there is no guideline for large scale research which requires BL4 containment.

The level GLSP (good large scale practice) has been recently added to the NIH guidelines. This level pertains to operations involving non-pathogenic recombinant strains derived from host organisms that have a history of safe large-scale use. This designation requires only appropriate laboratory procedures for biocontainment. Most biotechnology processes fall into the BL1-LS and BL2-LS categories.

There are two aspects of containment. Primary containment protects the personnel and the immediate laboratory environment from exposure to infectious agents. Secondary containment protects the environment external to the laboratory from exposure to infectious materials.

The following summarize the key elements for each level of large scale physical biocontainment:

GLSP

- Provide written instruction and training for personnel handling viable organism cultures.
- Provide emergency response plan for handling culture spills.
- Provide handwashing sink and protective clothing.
- Validate equipment containment capability.

BL1-LS:

- Include all the elements of GLSP.
- All vessels containing microorganisms should be a closed system. Vessel vents should be equipped with 0.2 micro filters.
- Provide the capability for *in situ* deactivation of the culture fluids in the closed vessels. This can be accomplished by injection of steam or by addition of chemicals.
- Provide a closed system for collecting culture samples to prevent release of aerosols.

BL2-LS:

- Include all the elements of BL1-LS.
- Provide double mechanical seals with steam condensate flush for agitators, and steam seals for all vessel entry port valves.
- Provide instrumentation for monitoring the integrity of each closed system.
- Provide permanent identification on each closed system.
- Provide a universal biohazard sign on each closed system.

BL3-LS:

- Include all the elements of BL2-LS.
- Provide a liquid waste decontamination system in the containment area for the discharges from the floor drains, pure steam condensate, and process sinks.
- The containment area should be designed to prevent release of culture fluids outside the area after a spill.
- The containment area HVAC system should be isolated. The supply air should be HEPA (high efficiency particulate air) filtered. The exhausted air should not be recirculated. The air pressure in the area should be negative in relation to the adjacent facility.
- The walls, ceiling, and floors should be sealed and constructed of material that can be readily cleaned and decontaminated.

22.8 Coordination

Coordination is a process of bringing together all the elements of the facility so that they are properly connected in correct locations and do not interfere with each other. Coordination is required in the design stage and in the construction stage.

During the design process, coordination is important for the completeness of the design. Coordination addresses the utility and space requirements of the equipment. The designer should only use the latest dimensional and utility requirement data from the vendors. For utility requirements the following check list should be considered: electric power, control power, cable length from the equipment to the control panel, steam pressure and flow rate, cooling water pressure and flow rate, compressed air, laboratory gases.

For space requirements the following check list should be considered: equipment dimensions, space for maintenance around the equipment, drain locations, equipment locations in relation to laboratory casework, building columns, and walls.

Cleaning and sterilization equipment such as glassware, washers, and autoclaves often must be installed between two rooms for pass-through operation. The designer should provide detail drawings for the installation of the sealing flanges around the equipment.

One major concern is the piping arrangement above the ceiling. In the clean areas the ceiling and walls are sealed, which creates a problem for maintenance access. Some designs provide a utility corridor in the middle of the facility where the piping systems are located. These design features are more expensive initially but may be justified.

For most designs, the piping and duct work systems are not exposed, i.e. they are located behind the wall and above the ceiling. The designer should coordinate and group all the valves at specific access panels along the wall or above the ceiling. Ideally all maintenance access should be located outside the clean area. Isolation valves should be provided on utility piping branches that allow shutting off an area for maintenance purposes. Separate elevation planes should be provided for various piping systems, duct work, and conduits above the ceiling. This will minimize interference during construction and facilitate trouble shooting and maintenance. It is critical to check the proper sloping of all high purity piping such as DIW, pure steam, and WFI.

22.9 Facility design

Facility design involves the building layout, HVAC systems, and utility services. The unique aspect of a biotechnology facility is that it must conform to good manufacturing practice (GMP) requirements and National Institute of Health (NIH) containment guidelines. During the design stage all the necessary elements must be incorporated to conform with the above guidelines and regulations. All critical construction activities must be documented to prove that the facility is constructed according to the design. The accuracy and necessity of any deviations must be documented and verified.

After construction the facility and its process equipment will be validated to ensure that the design is in conformance with GMP and that the construction is in conformance with design specifications. The importance of validation is to prove that the process equipment and its various support services in the facility are performing consistently and are under control at all times.

22.9.1 *Architectural layout*

The facility layout is especially important for multipurpose biotechnology facilities. Physical separation of each functioning area is most vital. It is not sufficient to rely on standard operating procedures (SOPs) to prevent cross contamination of different production areas. If the facility is producing material at various stages of development, i.e. clinical and licensed products, then physical separation is an absolute necessity. This requirement starts with the architectural layout which provides walls to separate each area.

The facility should provide absolute control of all flows, air, material, equipment, and personnel within the process areas. Material flow starts and ends at the warehouse. The warehouse should provide different areas for material storage, such as a quarantine area for raw materials, separate product storage areas, rejected product area, etc. Strict control of the warehouse is very important for conformance to GMP.

For the process area, space must be provided for in-process material storage, dirty equipment, and clean equipment storage areas. The layout should permit equipment flow from the clean area to the process area. After the equipment has been used, it goes to the cleaning and sterilization area, and finally back to the clean area for re-use.

Personnel flow should be controlled by a security system that allows only authorized personnel access to specific areas. The design should provide door interlocks to prevent cross contamination between different areas, and back flow or personnel from the dirty areas into the clean areas.

The stringent requirements of isolation and flow control also complicate the facility layout. A simple item such as 'mean of egress' could become a problem. Egress problems in the design can hold up construction permits. All factors have to be balanced correctly, personnel, products, and equipment flow versus isolation and egress requirements for the different areas.

22.9.2 *HVAC system*

A unique feature of a high purity facility is the isolated HVAC system that provides clean air for each classified clean room area. To achieve isolation of the air flow, separate air handlers may be required. Terminal HEPA filters must be provided for the supply air into each process area. For critical areas such as purification, a velocity of 100 ft/min (30.48 m/min) and 50 air changes per hour may be required. The temperature is usually maintained at about 68°F (20°C), with relative humidity of 45%. For the final bulk sterile filtration process where the product may be exposed to the environment, it is often performed in a Class 100 biosafety cabinet.

To prevent cross contamination, the direction of air flow in each area is controlled by room pressurization. Air pressure differential between adjacent areas is generally set at an increment of 0.03 in (0.076 cm) of water column (WC). The conventional designation for successive increments are: 0, +1, +2, +3. The pressure in an uncontrolled area is used as reference and designated as '0'. The cleanest area, such as purification, is designated as +3, which means that the pressure in that area is controlled at 0.09 in (0.229 cm) WC.

Generally for biocontainment the air flows from the cleanest area to the dirtiest area. The pressure in the containment area for cell culture should be negative in relation to other adjacent process areas. To maintain the sterility in the cell culture area the pressure in the access airlocks should be more negative. Depending on the biocontainment level required, the exhausted air may have to be HEPA filtered, to prevent microorganisms from escaping from the facility. To control room pressurization, air locks are provided for personnel and equipment passing through. The high air change rate and flow requirements will incur high operating costs. Therefore, energy efficiency equipment should be considered in every case.

Other design items which require attention are location of the air intake, and sound level in the work areas. To prevent contamination, the intake of the air handlers should be located in an area away from the direct exhaust of any laboratory roof fans. To conform to OSHA standards, the noise level of the high velocity air flow duct systems should not exceed 85 db. Sound attenuation devices may be required to control the sound level.

During construction it is important to treat the HVAC system with the same attention as is given to the high purity stainless steel tubing system. The design specification should require the vendor and contractor of the HVAC system to provide clean ductwork for installation. The contractor should provide a procedure for cleaning and protecting the ductwork and submit it for the owner's approval prior to the start of field fabrication.

22.9.3 Utility systems

Utilities for a biotechnology facility can be divided into two groups, the high purity utilities and the plant system utilities as follows:

- High purity utilities
  - Deionized water (DIW)
  - Water for injection grade water (WFI)
  - Pure steam
  - Laboratory gases — nitrogen, carbon dioxide, oxygen
  - Pharmaceutical grade clean air
- Plant system utilities
  - Plant steam
  - Compressed air
  - Vacuum system
  - Tower cooling water
  - Chilled water
  - Electrical supply

In a biotechnology facility the high purity utilities are the most critical. The deionized water usually conforms to the United States Pharmacopoeia (USP) purified water requirements. The USP requirements address the maximum acceptable level of several types of contaminants allowed in the water, i.e. particulates, inorganics, organics, and bacteria. The USP purified water can be used for most operations except parenteral preparations.

A typical process for deionized water consists of several pretreatment steps, such as multimedia filtration, activated carbon filtration, and water softening. Following the pretreatments, the water is purified by a reverse osmosis system that removes most of the dissolved solids and bacteria. A deionization system removes the remaining ions in the water. To control microbial growth, an ultraviolet sterilizer is provided at the discharge of the deionization system.

The deionized water is also used to generate WFI water and pure steam. The WFI water is generated by further distilling the deionized water. The FDA bacteria action limit for the WFI water is 10 CFU/100 ml. The WFI water is used for most media and buffer preparations. Pure steam is generated by evaporating the deionized water and it is used to sterilize process equipment and piping.

Both the DIW and the WFI systems are designed as continuously recirculating loops. While the DIW loop is normally not heated, the WFI loop is maintained at a minimum of 80°C.

The type and capacity of the utility systems should be well thought out. Correct diversity factors should be used for capacity determination. It is important to obtain as much input as possible from all parties involved in the project. It is a challenge to provide a utility system that is reliable in terms of its capacity, ease of maintenance, and minimum downtime. A cell culture process often lasts for weeks, therefore it is important that all the building equipment and utilities continue to function with minimum downtime. To provide this assurance, auxiliary systems such as emergency power, redundant equipment, etc. are required.

22.9.4 *Example of a typical facility*

Figure 22.3 presents a layout of a hypothetical fermentation facility. The drawing is intended for illustration only; it is not drawn to scale. The layout incorporates most of the features for BL-2 and BL-3 biocontainment and GMP requirements for material, equipment, and personnel flow control. The support facilities such as adminstration offices, laboratories, and utility areas are not shown on the drawing. The facility is grouped into five major areas:

1. Glassware and tank wash areas
2. Media and buffer preparation areas
3. Inoculum area
4. Fermentation and harvesting areas
5. Purification area

The whole facility is enveloped and isolated by a return corridor by which the material and equipment enter and leave the facility via airlocks. The return corridor is also used for moving dirty glassware and equipment back to the wash areas and for personnel exiting the facility.

Within the facility, the five major areas are accessed through a clean corridor. The process materials, clean equipment, and personnel flow from the media and buffer preparation areas across the clean corridor into the other three areas. The inoculum, fermentation, and purification areas are isolated with additional airlocks. Material and equipment enter each area through one airlock, while personnel enter through a separated airlock. This separation provides a strict control on the flow paths of material and equipment. All traffic flows are controlled by one way doors and an automatic door interlocking system.

To enter the facility the worker puts on one level of gowning and goes through the airlock by the locker rooms. From the clean corridor, the worker can access directly the media/buffer preparation areas. To enter the critical areas inoculum, fermentation, and purification, additional gowning may be required.

The inoculum area is provided with a warm room for incubating cultures, and other equipment such as biosafety cabinet, freezer, and CO_2 incubator. The propagated cell line may be transferred to the fermentation area by a sealed container or a portable tank, for aseptically inoculating the bioreactors.

The fermentation area is adjacent to the inoculum area. It has an office for production control, several bioreactors, a biodeactivation system, and harvesting equipment such as an ultrafilter and a centrifuge. The harvested product is transferred to the purification area through piping or portable tanks.

The purification area is provided with a cold room for low temperature chromatographic separation and recovery of the product. It also includes a biosafety cabinet for sterile filtration and filling of bulk products.

The dirty equipment recycles back to the glassware wash or tank wash

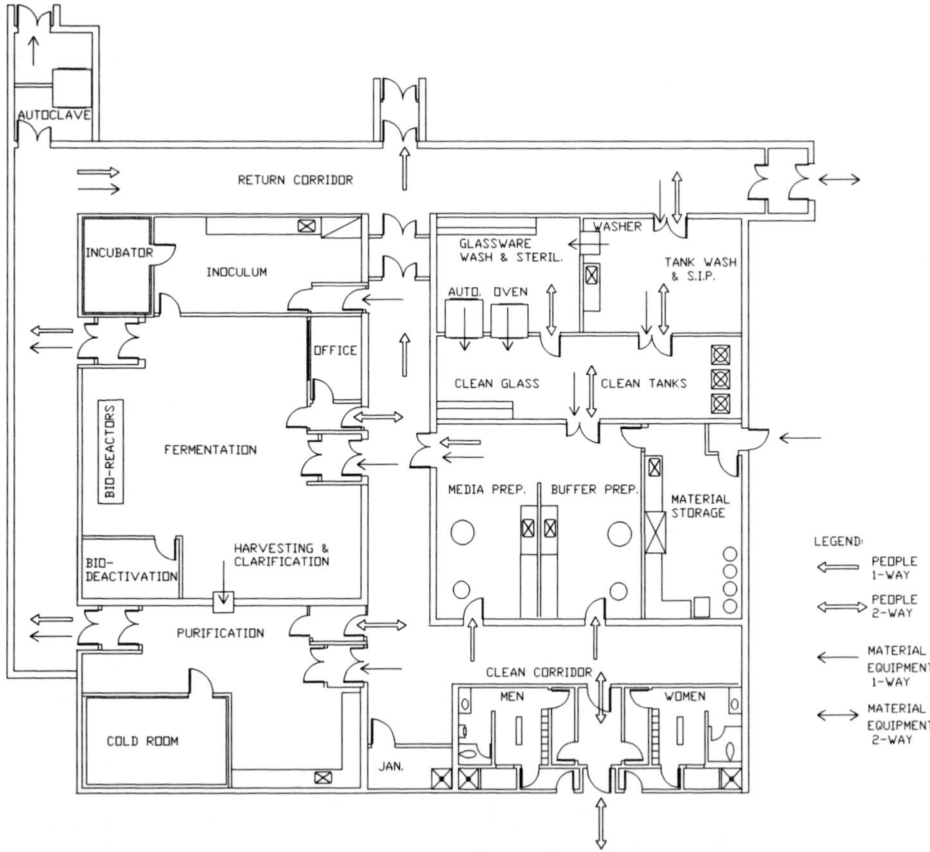

Figure 22.3 Fermentation facility (not drawn to scale).

area for cleaning and sterilization. Other contaminated materials are sterilized in an autoclave located in the return corridor, before leaving the facility for disposal. Separate areas are provided for cleaning and sterilizing the glassware and tanks. The glassware is cleaned in a washer, then sterilized in either an autoclave or a hot air oven. The dirty tanks are washed and steam sterilized in the tank wash room. Areas are provided for storing the cleaned glassware and tanks.

The media and buffer preparation are in separate areas. Common to both is the material storage room, in which the powder ingredients are preweighed for batching. The media and buffer solutions generated in this area may be transferred to the inoculum, fermentation, or purification area, by portable tanks or piping.

22.9.5 Facility cost

The cost of a biotechnology facility is high compared to other types of structure such an an office building, which cost about US$150 per square foot. This is an order of magnitude estimate of the building not including furniture and movable equipment.

For a biotechnology facility with clean rooms and containment features, the cost is in the range US$200/ft^2 to US$300/ft^2. The isolated HVAC and its control systems account for 20% to 30% of this total cost. This cost does not include the cost of process equipment, which is a significant variable. Depending on the process requirement, the unit cost of the facility including the process equipment may be more than US$400/ft^2.

22.10 Instrumentation and control

To ensure that the facility will run consistently, automatic process control should be an integral part of all operations, especially for critical processes such as bioreactor operation, clean-in-place operation, steam sterilization, and biodeactivation.

A typical control system is a distributed system using PLC for field closed loop and sequential control. The critical process parameters such as temperature, pressure, resistivity, pH, gas concentrations, and agitation rate are being monitored and controlled. The local PLCs are linked on a network to a central supervisory computer system that provides data acquisition and trend analysis capabilities.

The central computer also provides the ability to monitor the operation of the facility remotely by connecting the computer to a modem. The process from first inoculation of the cell line to the last purification operation occurs over several weeks. Interrupting the process at any critical junction such as cell culture or purification will cause delays in the production schedule and the loss of valuable product. It is necessary for the operating personnel to monitor the process continuously and take corrective action quickly when problems occur. While the computer provides an instant access to the bioprocess operation, it also opens access for unwanted tempering. Therefore it is important to provide security in the computer system to prevent tempering of any process parameters from a remote location.

22.11 Construction

During construction design mistakes and oversights become apparent. This can be the most difficult part of the project. The owner should supervise every aspect of the construction. Construction submittals should be reviewed carefully taking into consideration the impact of decisions on other elements of the facility. Equipment layout, equipment installation, and maintenance access must be considered. Other items which need to be addressed are temporary equipment storage, protection of the equipment in the field, start-up, and testing procedures. All critical construction activities should be inspected and documented. The construction schedule should be updated continuously. The progress of the construction activities should be tracked with respect to the original schedule. Any deviation from the original should be investigated for its cause and its impact on the final project completion date. Construction problems and deviation from design specifications should be documented on a punch list and communicated to the construction contractor on a continual basis. The owner should follow up all punch list items until they are resolved.

Effective communication between the construction management firm, the architect and engineering firm and the owner is very important. A weekly meeting is required to resolve construction schedules, construction quality, safety, housekeeping, and punch list items.

22.12 FDA regulations and licensing requirements

The biotechnology industry is regulated by a division of the FDA, the Center for Biologics Evaluation and Research (CBER). CBER strictly controls the manufacturing of the biological product by two licenses, the product license, and the establishment license[5].

This dual licensing requirement is rooted in the historical development of biological products. Traditional biological products are seen as complex mixtures that are difficult to characterize. The dual license ensures the identity of the biological product by requiring the manufacturer to control and document the facility, equipment, and manufacturing processes.

The licensing process starts with the investigational new drug (IND) application. IND may be submitted for one or more phases of an investigation. A pre-IND meeting with CBER is often necessary to discuss ways of avoiding locking into an inefficient process during the clinical trial period. After CBER agree on the clinical protocals, the manufacturer then submits IND for one or more of the three phases of clinical studies described below:

Phase 1 — initial introduction into human to determine metabolism and pharmacologic actions.
Phase 2 — evaluation of the effectiveness of the drug for a particular indication in patients with the disease and determination of side effects.
Phase 3 — additional information to evaluate the overall benefit–risk relationship of the drug.

After phase 2 has been completed and the facility has been validated, the manufacturer may concurrently submit to CBER the Product License Application (PLA) and the Establishment License Application (ELA)

22.13 Validation

The validation process is extensive, it would require a book to cover the subject thoroughly. The following are the elements required from the design and construction activities to support validation. To validate a new facility, two sets of qualification procedures are required, the Installation and Operation Qualification (IOQ) and the Performance Qualification (PQ). All critical equipment and systems relating to the cell culture process have to be validated by the IOQ and PQ procedures.

To perform the IOQ and PQ procedures, numerous documents are required from the design and the construction stages. The documents prove that the facility design is in conformance to cGMP and that all critical equipment and systems are installed according to specifications.

A general list of documentation required to support validation is as follows:

- Design and construction specifications and drawings
- Equipment purchase orders
- Operation and maintenance manuals and drawings
- Inspection and test reports
- Material of construction certifications
- Instrumentation calibration reports
- Equipment start-up reports
- Equipment and system acceptance documentation
- IOQ and PQ final reports

References

1. F. Bader, A. Blum, B. Garfinkle, D. MacFarlane, T. Massa and T. Copmann, *BioPharm*, 32–40, September (1992).
2. 21 CFR Part 211, *Current Good Manufacturing Practice For Finished Pharmaceuticals.*
3. C. Beck, H. Stiefel and T. Stinnett, *Chem. Eng.*, 121–129, February (1987).
4. *National Institute of Health, Federal Register*, **51**, 16952–16985 (1988).
5. M. Beatric, *Pharm. Eng.*, **10**, 29–35 (1991).

23 Bulk pharmaceutical and biopharmaceutical plant design considerations

R. LESNIK

When designing and building a bulk pharamaceutical plant, there are many things to be considered that are different from biopharmaceutical or finished pharmaceutical facilities. In this chapter, we shall review some of the design parameters and considerations and shall contrast biopharmaceutical with bulk pharmaceutical plants.

23.1 Basis of comparison

There are basic differences, as well as many similarities, between bulk pharmaceutical and biopharmaceutical facilities. Biopharmaceuticals are microbial or mammalian cell generated products, usually manufactured in batch fermenters. Bulk pharmaceuticals may be derived from a wider variety of sources, including fermentation, organic synthesis, or extraction of naturally occurring materials. Both types of operation require multipurpose flexible processing operations, that can handle a diversity of operating steps and equipment, although biopharmaceutical processing generally is on a smaller scale. It is unusual for bioreactor systems to be larger than 2 000 l, although production systems are as large as 10 000 l. Bulk manufacturing reactor systems can be as large as 35 000 l, with some bulk fermentation equipment reaching 100 000 gallons (380 000 l) in size. Both operations require supporting recovery and purification systems to take the crude reactor or fermenter product to a purified bulk stage. In some cases the bulk product is sterile, but in the large majority of cases it is not. Sterilization of the product is needed prior to packaging in the final dosage form. Because of the need for purification steps, direct support space for purification equipment and process utilities is required close to the reactor operations.

Bulk pharmaceutical and biopharmaceutical facilities are normally designed to handle multiproduct campaigns, where several batches of one product or intermediate are manufactured before cleaning the system and beginning another product. Production volumes, which are relatively small, dictate this approach to operations. Even in large bulk systems, the newer active pharmaceutical ingredients are more potent than older drugs, and thus require lower therapeutic dosages resulting in lower overall production quantities.

A bulk facility may be designed for a single product, but that product may require several dissimilar manufacturing steps before reaching the final bulk material. This also leads to campaigning of operations. A few biopharmaceutical and many bulk pharmaceutical products are manufactured in continuous systems. Continuous fermentation equipment is available for bioreactions and many types of continuous reactors are used in bulk organic synthesis and purification. The trend is away from continuous systems in bulk operations due to the reduced product quantities needed and the specialization that those systems imply. It is uneconomical to install expensive equipment that will be idle for a large percentage of the operating year. On the other hand, more continuous bioreactors are being manufactured, because the number and quantities of biopharmaceuticals are increasing.

The major points of discussion in the remainder of the chapter will be the following:

- Facility layout
- Processes and unit operations
- Equipment and piping systems
- Types of material and materials handling
- Heating, ventilation and air conditioning system (HVAC) and utilities systems
- Waste disposal and handling
- Safety considerations
- Regulatory considerations

23.2 Facility layout

23.2.1 Bulk pharmaceuticals

Bulk pharmaceutical manufacturing facilites may be as varied as the kinds of chemical reactions that they utilize. It is important to remember that a bulk plant is basically a chemical production facility, where the products happen to have an end use as a pharmaceutical. There are no 'universal' types of equipment arrangements that work for all processes. Some typical bulk pharmaceutical processes include fermentation and isolation of fermentation products, synthetic chemical reactions, extraction of naturally occurring materials, and microencapsulation or other physical processes. In many locations a large portion of the plant may be outdoors, with only the final product isolation, drying, and packaging in an enclosed area.

Bulk plants may utilize batch, continuous, or semi-continuous processes or possibly a combination of all three. The type of processing is generally a function of the required output of the facility, with high volume products frequently being made in a continuous system, although batch processing is also used for large product volumes. Batch plants provide versatility, allowing many different products and intermediates to be manufactured with a relatively few different types of equipment. It is not easy to design a multipurpose continuous process or reactor.

In preparing an equipment arrangement for a bulk plant, there are several important items to be kept in mind. Process equipment should be located in such a way so as to minimize piping runs. The longer the runs, the more costly they are to install and the more energy is required to drive any transfer pumps. Product losses in transfer lines obviously increase with the length of the line, so this is especially important in low volume processes. Gravity is an excellent method of product transfer. It is often advantageous to arrange a series of reaction and isolation vessels vertically, dropping the batch from one tank to another, or from a crystallizer to a centrifuge. At times, a three-story or even a four-story building can be the most efficient envelope for the production equipment, and possibly the most economical.

In multistory arrangements it is important to provide the operators with easy access from floor to floor. At times it may be desirable to provide mezzanines rather than complete floors so that the operators can see the tops and bottoms of the equipment. Proper arrangement and proper access can reduce the number of operators that are needed to run the plant. Even though many of the production operations are now being run by control computers, operators are still needed to monitor field equipment.

Equipment should be arranged so that the process flows smoothly from one step to the next. Criss crossing the building to move from step to step is inefficient and can create serious operating problems. It is also advisable for the operator to be able to see both ends of a transfer process if possible. In an ideal case, the process would start with raw materials at one end of the building and end with finished product at the other, passing through sequential steps along the way. This usually is not possible, but is should be one of the initial goals.

Another important aspect of equipment arrangement is minimization of the possibility of cross-contamination. This is also important in equipment design, particularly when a piece of equipment is being campaigned, or used for several different products at different times. Equipment must be capable of being easily and completely cleaned, but the piping and other process auxiliaries must also be cleanable. Clean-in-place (CIP) systems for vessels and piping are being installed more frequently than in the past. These systems provide a degree of automation and consistency to the cleaning of process equipment that reduces the dependence on manual cleaning methods. Charging areas or other isolation areas where raw materials, intermediates, or other materials are exposed to the surroundings must be sufficiently isolated from each other to prevent their mixing.

Flammables should be processed in outdoor equipment where possible. If production equipment must be inside, an effort should be made to reduce the area where flammables will be processed. Explosion-proof electrical equipment is more expensive than non-explosion-proof, and the flammables processing area will need explosion-proof lighting, outlets, heaters, etc. in addition to the process motors. There are also constraints on the size of the processing areas, with regard to area-to-volume ratios, created by the explosion relief area requirements of the various building codes.

The storage of flammables can also affect the layout. The amount of flammable material that can be stored within an area is regulated by code. If more than specified amounts are to be stored, the area may require an upgraded classification, necessitating the use of explosion-proof electricals and the addition of a sprinkler system.

Building designs for bulk manufacturing plants can vary greatly from plant to plant, but most are multistory buildings with large processing areas separated into 'wet' (reaction, purification, isolation) and 'dry' (drying, size reduction, packaging) processing sections. Many manufacturers include the final product isolation step, usually centrifugation as part of the dry area. Bulk plants differ from biopharmaceutical or finished dosage form facilities in that a large part of the production area may be located outdoors, or in a structure with screens in place of exterior walls. Rooms tend to be larger and fewer than in biopharmaceutical plants.

Building interior finishes are not the same as those needed for biopharmaceutical or pharmaceutical production plants. Only the dry area, where the active ingredient may be exposed to the surroundings, requires special attention. Here, as in biopharmaceutical or finished dosage form plants, we will utilize floor, wall, and ceiling finishes that are easily cleaned, non-porous and non-flaking and will eliminate places where dust or product may collect.

This is not to say that the production areas should be dirty; they should be designed so that good housekeeping can be easily maintained. They just do not need the same degree of interior finishes that the isolation areas require. Floors should be sloped toward drains so that water does not collect in pools, and should be finished with materials that will resist any chemical spills that might occur. Many plants require the isolation of some or all of the floor drains to prevent the loss of valuable products or the potential contamination of the waste treatment facilities. Provisions must also be made for recovering or treating the isolated waste, a seemingly obvious but not always recognized factor. All areas must be protected from birds, insects, and rodents.

In some cases full floors may not be desirable. Mezzanine areas for support of equipment, with platforms for operator access may be all that are required. There are many plants that have upper floors only around the perimeter of the building, with the entire center going from floor to roof. This allows the roof, as well as the walls, to be utilized for explosion relief. At times floors will be constructed from steel or corrosion-resistant grating materials, such as FRP. If raw materials or products are exposed to the surroundings on these levels, gratings are not a good approach, since material can fall or drip to the levels below, creating potential safety and housekeeping problems.

Bulk manufacturing plants, like biopharmaceutical and finished pharmaceutical plants, should be designed with expansion in mind. The potential routes of expansion, the additional types and number of pieces of equipment that might be needed, the quantities of additional utilities and the number of additional personnel should be considered during the design stages. Too often, the expansion of a facility is restricted or made more expensive by the failure of the owner and/or the designer to consider expansion possibilities. Oversizing utility distribution lines and providing some extra space for new utilities equipment is often a small price to pay in the initial plant for the ability to expand. Digging up undersized underground utility lines to replace or add to them is an expensive task that could frequently have been avoided by proper planning. Production areas can often be expanded by moving into warehouse space and then building new warehouse if the building has been planned for expansion.

We should briefly touch upon the aesthetics of the facility. It is understood that most companies do not want to make an 'architectural statement' with a bulk manufacturing plant, but attractive buildings can be designed and built with little or no cost premium. Building facades can hide rooftop equipment, siding treatments using various color schemes and the positioning of large or bulky equipment away from the central line of view all can contribute to a pleasing appearance. A building that blends in with its surroundings is a good advertisement for the company.

23.2.2 Modular construction

Since many bulk manufacturing facilities frequently use similar equipment to run the various process steps, a new concept, modular design and construction, can often be the optimum design/construct approach. A production 'module' could consist of a jacketed reactor, condenser, receiver, transfer pump or pumps, polishing filter, centrifuge or other solid product isolation equipment, and modular heating/cooling units. Although a given module might not contain all of these items, it could contain other equipment common to several processing steps. The major difference between many of the modules would only be the size, with a reaction module ranging from 200 gallons to 10 000 gallons (7500 – 38 000 l) and containing similar equipment, piping, and controls. The modular approach has several advantages:

- Facility design is less costly, since a module must only be designed once, and then repeated for as many times as necessary.
- Plant operation is simplified, since the operating and control systems for different modules will be the same.
- Changeover from one product to another is done more quickly, since the piping, electrical, and instrument connections will be similar within each module.

- Equipment of different sizes, e.g. a 1000 gallon (3800 l) train, a 2000 gallon (7500 l) train, etc. can be designed as similar modules, again providing savings in design time and cost while allowing production batch sizes to be varied as needed to fulfill marketing requirements.
- Installation time can be greatly reduced by constructing the modules off-site while the building is being constructed, or in the case of a renovation, while production operations are continuing. The preconstructed modules can be installed more quickly than 'stick-built' systems and can minimize production downtime for installation. This is particularly important when a production 'window' exists during a planned shutdown or vacation period.

Obviously, the modular design concept cannot be universally applied to all bulk plants, but it should be considered when planning a multiuse batch operation.

23.2.3 Biopharmaceuticals

The design of biopharmaceutical facilities is driven by several factors. The possiblity of cross-contamination either from other products or from cruder steps of the same process must be reduced to an absolute minimum. There must be segregation of contained processing areas operating dissimilar steps within the same process, for example fermentation and purification, as well as segregation of similar steps running different processes. In most cases the separation will be physical, compared with bulk plants where containment within a vessel is frequently the norm. There must be increasing zones of cleanliness of the surroundings as the process goes through increasing degrees of purification, culminating in classified areas as high in quality as Class 100. There must be a secure means of isolating and decontaminating hazardous wastes. The facility must provide for operator protection from hazardous materials and for product protection from contamination by operators.

Raw material, personnel, equipment, and product flow patterns must be carefully defined within the facility to accommodate the concerns listed above. The resulting manufacturing plant generally consists of several small, self-contained manufacturing suites, each housing separate fermentation, isolation, and purification equipment and provided with separate access air locks. The personnel flow is ideally all one-way, with 'clean' and 'dirty' corridors for passage between the various areas. Operators working in a given suite should not be able to pass into other suites without degowning and regowning. The transfer of small equipment and other items is done via pass-throughs in the walls of the suites.

Process utilities and HVAC equipment for all of the suites can be located in a central mechanical room, and a quality control laboratory for either in-process samples or finished goods is also centrally located. As in bulk facilities, there must be warehousing areas for raw materials, in-process materials and finished goods, with identified or isolated quarantine areas set aside for raw materials and finished products. There will also be a frequent need for low temperature storage areas, ranging from 4°C to −70°C, each also including isolated quarantine areas.

Building finishes for biopharmaceutical plants resemble those of finished pharmaceuticals. The walls, floors, and ceilings must be easily cleaned and must be made of non-flaking, non-porous materials. Gypsum board walls supported by metal studs are common, with epoxy paint or solid vinyl coatings on the gypsum board. Other systems, such as smooth FRP, are also in use. Vision panels in the walls should be flush with the wall or should have sloping sills to prevent particle accumulation. Any electrical outlets or switches should also be flush mounted.

Floors may be epoxy-terrazzo, troweled epoxy, or solid vinyl-coated concrete, with coved moldings to avoid sharp corners where dirt or particles can accumulate. The floor coating is dependent on the type of traffic that an area is expected to encounter. Epoxy-terrazzo and troweled epoxy are rugged and can withstand equipment being rolled through the area. Solid vinyl is susceptible to penetrations by hard objects, such as vessel wheels, but is easy to patch. Ceilings may be expoxy-painted gypsum board or lay-in tile. The tile should be faced with Mylar or other non-flaking material to prevent potential contamination. Coved ceilings and walls are also being used. Lights should be flush mounted in the ceilings and access should be provided from above, where possible.

Demountable packaged clean room systems, some complete with their own HVAC equipment, are available from several vendors. These systems usually are provided with walkable ceilings to provide access to the high efficiency particulate air (HEPA) filters and other equipment mounted above them. They tend to be more expensive than built-in-place systems, but they do have the advantage of being movable. It is rare however, that anyone actually moves a portable clean room.

Floor drains are a source of potential contamination in biopharmaceutical facilities. It is possible to provide open floor drains in the fermentation area where large quantities of aqueous wastes may be generated. As the process proceeds through the various purification steps, open floor drains are eliminated. The wastes are piped through the closed drains to holding tanks for decontamination prior to being discharged to the treatment facilities. Sealing of the floor penetrations where the drain connections are made is a major problem, since the lines are frequently subjected to large changes in temperature when the systems are steam sterilized. The condensate from steam sterilization must be removed from the operating suites, but the expansion and contraction of the drain pipe can cause cracking in the seal between the pipe and the floor.

23.3 Processes and unit operations

23.3.1 Bulk pharmaceuticals

The typical bulk pharmaceutical process can be diagrammed as a series of operating steps as shown below:

- Solution/reagent preparation
- Reaction
- Crude filtration
- Crystallization
- Crude isolation
- Recrystallization/filtration
- Pure bulk isolation
- Drying
- Size reduction
- Bulk packaging

An organic synthesis process for a bulk pharmaceutical usually contains some or all of the above steps. Each of the steps shown may also refer to an intermediate product en route to the final pharmaceutical. The intermediates may or may not be isolated, and in some reaction sequences as many as fifty intermediates can be prepared. There may also be sterile bulk production that will not undergo further purification, but these operations are rare. The general manufacturing steps will be discussed in this section and equipment will be covered later.

23.3.1.1 Solution or reagent preparation. The first step involves the manufacture of various reagent solutions that will be used to run the chemical reaction. Often the reactants can be mixed in the reaction vessel, but there are many occasions when one or another of the materials must be added at a controlled rate or during some specific portion of the reaction. The solid reagents are charged to a solution vessel in measured amounts, either from preweighed containers, or directly from the storage drums, bags, or bins. Liquid materials or solvents are either charged into the vessel from preweighed containers, weighed in directly from drums, or are metered through transfer piping from bulk storage tanks. The solution vessels are usually on the top floor of the building and the liquid solutions are often charged by gravity into the reactor.

23.3.1.2 Reaction. The synthesis reaction takes place either in a batch or continuous reaction vessel, generally in a solvent other than water. A batch reactor may be of a variety of materials of construction and may be of almost any size. It is agitated, generally utilizing an electrically-driven agitator, with the agitator blades being of many possible configurations. The type of agitator depends on the service for which it is designed. Often, reaction vessels will have agitators with adjustable or removable blades to provide a wider degree of flexibility in multiproduct or multistep trains. Continuous reactors are usually tubular in shape, with a heated or cooled jacket. They may be a series of heat exchangers of various types or possibly smaller agitated vessels operated in an overflow mode. Tubular reactors can have static mixers in line to provide continuous mixing of reactants. Continuous reaction systems are not common in bulk pharmaceutical plants because of their lack of flexibility. Unless a plant is designed to run one product for the majority of the year, a continuous reactor will be idle most of the time. They are not easily adapted to changing products.

23.3.1.3 Crude filtration. Crude filtration is generally the next operation. The crude product, either a solid or a solution at this stage, is isolated from unreacted reagents and other impurities and transferred to a crystallization vessel. Filtration may be replaced by extraction in some cases. Certain systems lend themselves to easy isolation of the desired product by liquid–liquid extraction, rather than liquid–solid separation. The filtration can be accomplished with a line filter operated batchwise, or via a drum, belt or other continuous filter. The choice depends upon the relative quantities of the materials present in the system. If there are large amounts of impurities, a continuous belt or drum filter may be the method of choice. With small amounts of materials to be removed, a plate filter could be a proper selection.

23.3.1.4 Crystallization. The solid product is then crystallized from the mother liquor, although an intermediate may not be dried or even isolated. There are occasions where the entire batch is transferred to another reactor for subsequent reactions and product isolation. This also can be done either in a batch or a continuous crystallizer. Batch systems are far more common, but many continuous crystallizers are in use for certain products. The crystallization process is often simply one of cooling the batch to below the crystallization temperature, seeding the batch with a small amount of the desired product, and agitating for the proper period of time to obtain the maximum amount of product. Alternatives to crystallization are precipitation, where the product is made insoluble by changing the pH, or by addition of another solvent to precipitate the desired material.

23.3.1.5 Crude isolation and purification. The crude product is isolated, solids by filtration and liquids by extraction or distillation, and purified. The most common means of solids purification is by recrystallization, usually from a different solvent from the one used in the reaction step. Changing the solvent will remove additional impurities since they will normally have different solubilities as the solvents change. The crude product is heated to solution and a decolorizing agent is added along with a filter aid. The solution is filtered through a polishing filter to remove the solids and is cooled to crystallize.

23.3.1.6 Pure product isolation, drying, and packaging. The purified product is isolated, usually in an automatic batch centrifuge, and transferred to a dryer. Dryers may be batch or continuous, vacuum or atmospheric, static or rotating, depending upon the product characteristics and the amount and type of solvent present. At the end of the drying step the product is milled or screened to obtain the desired particle size, then packaged in drums for further pharmaceutical processing. The material at this stage is a pure active pharmaceutical ingredient.

Liquid products are usually purified by distillation. This step is usually continuous, with the pure liquid being condensed either in the overheads stream or at a takeoff point a small distance above the bottom of the distillation column. The condensed liquid is subcooled and collected in a receiving vessel, from where it is packaged into drums.

23.3.1.7 General bulk manufacturing parameters. Some major generalizations can be made for bulk pharmaceutical processing operations that show clear differences from biopharmaceuticals. Except possibly for the final purification, the product isolation steps show minimal concern for microbial and pyrogen content. There is minimal steam-in-place of vessels and piping, with sterilization of equipment not being required. Clean-in-place is used for the majority of the equipment and piping systems.

Most bulk pharmaceutical processing is solvent based, rather than water based. This leads to a lesser concern for microbes and pyrogens, but to a greater emphasis on safe handling of flammable materials and potection from potential fires.

Much of the processing equipment is glass lined for corrosion protection. With the large variety of solvents and raw materials used in bulk pharmaceuticals, the degree of corrosion problems is geometrically increased over biopharmaceutical processing. Add to this the increased pH range (1–14), the wider range of operating temperatures (−100 to 600°F (−87–365 °C)) and pressures (30 in Hg vacuum–500 psig+ (−1.44 to 35.2 kg cm^{-2})) and it is evident that bulk pharmaceutical processes are far more severe than biopharmaceuticals. As with solvents, the severe conditions act to kill microbial organisms, also reducing the concern for microbial contamination of the final product.

Analytical testing is 'classical'. Most products and intermediates are reasonably well defined, and the degree and type of impurities present can be more readily identified. The chemistry of the systems is better understood and analytical testing can begin from a broader knowldege base.

23.3.2 *Biopharmaceuticals*

The typical biopharmaceutical process can also be shown diagrammatically as a series of operating steps as shown below:

- Seed generation
- Fermentation
- Cell recovery/lysis
- Crude purification
- Intermediate purification
- Final purification
- Sterile bulk packaging

23.3.2.1 Seed generation. The first step in biopharmaceutical manufacturing is seed generation. Small quantities of active materials are cultured or grown to provide the feedstock for the larger fermentation reaction. Often, a series of small fermentations is run, with the most active cultures transferred to larger vessels prior to the full scale fermentation step.

23.3.2.2 Fermentation. Fermentation on a production scale can range from a few liters to thousands of liters, although most of the larger fermenters are about 2000 l in capacity. Critical parameters include agitation, pH control, oxygen transfer, and heat transfer. Many of the products are shear sensitive, so the agitation must provide heat transfer and oxygen uptake, while avoiding shear damage to the cells. Fermentation reactions are usually run at about room temperature, although some may run at 35–40°C.

Smaller production operations may take place in stirred or shaken flasks. Scale-up of these reactions may take the form of increasing the size of the flasks, increasing the number of flasks used, or running the reaction in a larger fermentation vessel, eliminating the flask or carboy operation entirely. In many cases, however, the regulatory submissions have been made for stirred or shaken flask processing and an amendment is necessary to change to a fermenter.

Some fermentation processes lend themselves to continuous operations. A fermentation is started, and nutrients are continually added to the system, while product and byproducts are continuously removed at a controlled rate. The system must be kept in balance. A balanced system can run for several months without other changes being made to it.

23.3.2.3 Cell recovery/lysis. The active biopharmaceutical may be either within or external to the cells. In most cases it is internal, so the cell must be ruptured to release the active ingredient. The cell recovery or cell lysis can be done in a mill or homogenizer, where the batch is recirculated through a high-shear centrifugal pump from the reaction vessel. The cells are ruptured, releasing the product. Cell lysis may also be accomplished in a high-speed centrifuge, where the liquid is separated from the lysed cells. The solids are discharged and discarded, while the product solution is collected for purification. In certain operations, cell lysis can be completed chemically. Addition to the proper solution can cause cell lysis, and the solids can be separated from the liquids by centrifugation or filtration.

23.3.2.4 Purification. Purification of biopharmaceuticals is often a lengthy and complex process. By their nature, naturally occurring materials contain a more complex mixture of materials than synthetic chemical products. It is important to characterize accurately the quantities and identities of all the components in the crude product or intermediate, in order to determine the correct method of purification. Chromatography is the most commonly used purification technique. Passing the solution through a chromatographic column system packed with the proper gel absorbent, will allow the various components to separate in layers on the gel. A buffer solution, usually saline or phosphate-buffered saline, is then put through the column to elute the desired product from the gel. Ultraviolet detectors are used to determine when the desired product is being removed, so that it can be collected and isolated from the impurities. A series of purifications is necessary to manufacture a product pure enough for pharmaceutical use. The chapter in this volume on biological separations provides an in-depth look at this phase of biopharmaceutical processing.

23.4 Equipment and piping systems

23.4.1 Bulk pharmaceuticals

23.4.1.1 Reactors. There are a variety of things to be considered when designing the reactors for a bulk manufacturing operation, whether it is a batch or a continuous process. Aside from the obvious, such as proper materials of construction to resist attack from the chemical reactants and identification of the need for heating and/or cooling, there are many other parameters to review. Some parameters that apply to a batch production operation are discussed here, although some are also applicable to continuous systems. The following questions should be asked.

How big should the reaction vessel be? Large enough to satisfy the annual production requirements, naturally, but there are other factors as well.

Should all of the production be run in one vessel, allowing for downtime in determining its size, or should the production be divided into two or more vessels, allowing continuity of operation in case of the failure of one vessel or system?

How many shifts per day or how many days per week should the plant operate? How should the plant be staffed? The operating philosophy will be a major factor in reactor sizing. Many bulk operations run a 7-day, 4-shift basis, enabling the plant to run more but possibly smaller batches to meet the production requirements.

How much does the product cost? If the product is very expensive, it might be prudent to limit the financial loss if a batch was ruined by some unforeseen failure during production. This would reduce the reaction batch to an 'economical' size.

Are there any unusual heating or cooling requirements in the process? Vessel size can be limited by the heat transfer area available in a jacketed vessel. If there are difficult heating or cooling problems, such as a large exotherm during the reaction, the jacket may be unable to provide enough heat transfer area by itself and additional surface area, either internally or externally, must be provided. In some cases this is either impractical or undesirable, so the vessel size would be controlled by its surface to volume ratio. The normal reaction vessel has a height-to-diameter ratio of from 1:1 to 1.5:1, but this might be varied to alter the heating surface.

Are there any unusual processing problems that may be encountered, such as excessive foaming, that would require extra freeboard? If there are, the reactor must be sized to account for them or the height-to-diameter ratio must be adjusted as mentioned before.

What kind of agitation is needed? If the reaction mixture is very thick or very viscous the size of the agitator may grow to unrealistic proportions if the reactor is very large, necessitating an extremely high power motor. If the agitator is long, it might need a secondary support from the vessel bottom. Unusual types of agitators may be required to provide homogenization, dispersion, emulsification, or other physical process functions. The size of the reactor may be limited by any of these factors.

Are there parameters other than temperature, level, or pressure that must be controlled during the operation? If pH control, for example, is needed, how can the reagents be added and what amount of lag time is acceptable before causing problems? If the control must be quick, the size of the vessel, the degree and efficiency of agitation and the location of the measuring instrument are all important considerations.

How frequently will the reactor be cleaned and how difficult is the cleaning procedure? If a product cakes on the walls and must be removed after each batch with either a high pressure or other CIP cleaning system or with an operator's labor, the ability to reach all parts of the vessel interior is essential.

Can the internal pressure on the vessel be relieved through the normal nozzles available on the vessel, or will special connections be required? If a vessel is processing a flammable liquid with a low latent heat, a larger relieving area will be needed. That area is dependent upon the exposed surface area of the vessel and the potential exotherm.

These are some of the factors involved in sizing a reactor. Undoubtedly, there are many others, that can be considered.

23.4.1.2 Product isolation equipment. Now that the product has been synthesized, it must be isolated. We will now discuss the isolation of solid and liquid bulk pharmaceutical products from reaction mixtures. There are many different methods that are used for both intermediates and finished products. First we will review liquid separations, which are less common than solids, and include the following isolation unit operations:

- Phase separation/settling
- Extraction
- Distillation
- Absorption

Each may be run as either batch or continuous processes.

Phase separation. Phase separation is the simplest method of liquid product separation, and can be used when the liquid is immiscible with the reaction mixture. In a continuous separation only a small decanter is needed, even for large quantities of product. The light and heavy liquids are continuously decanted and collected in surge tanks either for reuse or additional processing. We can only isolate a crude product via phase separation, since there is always some mutual solubility between the immiscible layers.

Extraction. A product may be run in one reaction medium, then extracted with another solvent to isolate the product stream from unreacted raw materials, byproducts or catalysts. Most extractions are now run in continuous equipment when production levels are high enough, although batch extractions (usually countercurrent) are still being run. Batch extractions are performed in a series of agitated vessels, where, after settling, the light and heavy layers move countercurrently to the next vessel in the train. Fresh solvent is introduced at the end of the system, where the concentration of the desired product is lowest, to ensure removal of the maximum amount of product and to minimize losses.

Continuous extractions are generally run in agitated columns of various designs. Each design tries to maximize the contact between the two liquid phases to extract most efficiently the product from the reaction stream. An interface must be maintained at a predetermined level in the column and the heavy and light streams must be removed in the correct ratio to the feed and solvent addition streams. It is usually necessary to provide a draw-off point near the interface to remove any buildup of rag layer or emulsion that may form.

Some common extraction columns are the Karr column, constructed from perforated plates on a central mixing shaft; the rotating disc contactor with a rotating disc on a central shaft centered between discs on the wall of the column; the Schieble column consisting of paddle-type agitators on the central shaft, rotating between discs on the column wall; and the Oldshue–Rushton column also with paddle agitators, but with a different disc design than in the Schieble column.

At times a product may be reacted in one solvent, extracted into a second and possibly into a third in order to purify it. Impurities that are carried over into the first extraction may be left behind when the second and third extractions take place. Extraction is an energy-efficient method of separation, provided that the extraction solvent does not require distillation later on in order to remove it from the product.

Distillation. Distillation systems are as varied as the products that they isolate. They can be batch or continuous and can be run under vacuum, elevated pressure, or atmospheric pressure. Depending on the requirements of a specific system, the feed to a continuous column can be liquid or vapor and can be introduced at any point in the column.

Column internals can consist of trays of various designs, such as sieve trays, valve trays, or bubble cap trays, or one or more of a number of types of packing, from Pall rings to many of the high-efficiency packings. The type of distillation selected is dependent upon several factors, some of which are listed below.

- Ease of separation of components
- Temperature sensitivity of the product
- Cleanliness of the feed stream
- Number of products to be separated
- Size of tower required
- Cost of the distillation system

If production volumes warrant continuous operation, it is the most efficient distillation method. Cleaner products can be isolated and 'mixed cuts' are not taken as they are in batch operations. This generally leads to a more energy-efficient operation, since redistillation of the mixed stream is not necessary. Distillation columns can be controlled using a number of operating parameters such as feed rate, bottoms temperature, or overheads temperature. The reflux rate may be fixed or variable, depending upon the consistency of the feed composition.

Absorption. Liquid products or solutions of liquid products may be isolated or purified by absorption in one of two ways. The stream to be purified can be passed through a bed of absorbent where impurities in the feed are removed, leaving a clean product. The absorbent may be activated carbon, or an ion-exchange resin, depending on the impurities. The absorbent may be regenerable or may be discarded, again depending on the system. In the other case, the product is absorbed onto the absorbent and the remainder of the feed stream is collected for disposal or further treatment. The product is then eluted off the absorbent with another solvent.

Centrifugation. Centrifugation to separate liquids is a special case of phase separation. The centrifuge, by use of centrifugal force, speeds up the settling operation. Tubular centrifuges or disc-bowl centrifuges are frequently used for liquid–liquid separations. The disc-bowl machine provides a series of small settling chambers to effect the separation, similar to a cream separator.

23.4.1.3 Solid products.

Separations. Product isolation refers to the separation of a solid product from a liquid, rather than the physical operation of more than one solid. There are three main methods of performing this isolation:

1. Evaporation
2. Settling and decantation
3. Filtration

A concentrate or a dry product may be isolated directly by *evaporation* to remove some or all of the liquid. This can only be done if the stream contains very few impurities or if the product purity is not required to be high. There are several types of evaporators available, with the most energy efficient being multiple-effect types or those with vapor recompression. Thin-film or wiped-film units are often used in this service, but evaporation is not a common separation technique because of high equipment and operating costs and purity problems.

As with liquids, *settling and decantation* is the simplest and least expensive means of solids isolation. Unfortunately, most solid materials do not settle quickly or completely enough to prevent product loss in the decantation step or to produce a dry enough product to move on to the next operation.

There are many *filters* available for product isolation, but they all fall into three basic categories:

1. Vacuum filters
2. Pressure filters
3. Centrifuges

Vacuum filters may be of several different configurations, but the basic principle involves the drawing off of the liquid through a cloth or screen by means of an external vacuum source. A filtrate receiver is provided to collect the filtrate, which could be either the product or the reaction mixture. The cake is usually washed to remove remaining mother liquor or product and the wash can be either combined with the filtrate or treated separately. Enough wash to displace twice the mother liquor should be the minimum volume used. We will briefly discuss the available vacuum filters here, the Nutsche, belt, and rotary filters.

The Nutsche filter is a larger version of the laboratory Buchner funnel. It is used for small, batch-type operations and can be either a manual or an automatic system, complete with the ability to agitate, reslurry, refilter, dry, and discharge. There is no exposure of the operator to any toxic fumes from the solvent.

The automatic Nutsche filter is convenient, but it is expensive and the operating time cycle can be quite long if the unit is used as a dryer. More than one unit may be needed to avoid a bottleneck. In some installations the discharge system has been an operating problem, mainly due to the seals.

Belt filters are frequently the choice for large volume production. They provide a large surface area for filtration and permit adequate washing of the filter cake. Belt filters are best used in aqueous systems, because enclosing a belt filter to keep solvent vapors out of the operating area can be very expensive. The product slurry is dropped onto the filter at one end, and the liquid is drawn through the belt by both gravity and vacuum. Wash is applied via sprays further along the belt and can be diverted to a different receiver from the filtrate if desired. The final part of the belt is a drying area, where the filter cake is sucked as dry as possible. Wet cakes containing as little as 20% moisture can be produced on a belt filter. The cake discharge can be diverted to any type of receiver or dropped directly to a dryer.

The rotary vacuum filter is a common machine for isolating large volume products or for filtering large quantities of liquids. The filter can be precoated and fitted with an advancing scraper blade to provide continually a fresh precoated surface for filtration. Most designs have series of internal pipes through which the vacuum is drawn, causing the filter to pass through the cloth. The vacuum is not continuous on the vacuum side of the filter, so materials that filter very quickly have a tendency to fall back into the feed trough. There are other designs where the entire drum is subjected to the vacuum, which helps eliminate the problem. Filter cake is discharged by the use of strings or belts being pulled away from the drum and/or by a pulse of air to blow the cake off the cloth. Again, as with the belt filter, the rotary filter is best used in aqueous systems. A rotary filter can be built to be vapor tight, but the cost increases very quickly and sealing problems are common.

As with vacuum filters, there are several types of pressure filter including the plate and frame filter press, plate filters (horizontal and vertical), and rotary filters. In these machines, pump pressure or gas pressure is used to force the liquid through the cloth or screen, providing a greater driving force than a vacuum filter.

Perhaps the most common pressure filter is the plate and frame filter press. The filter is made up of alternating solid plates and hollow frames, with the plates being covered by filter cloth. Slurry is fed into the press and the liquid passes through the cloth and out of the drain ports in the plates. The solids collect in the hollow frames. A wash can be run through the filter at the end of the feed cycle and the cake can be blown dry with either air or inert gas.

Filter presses have, in the past, been mostly manual operations, but modern machines are automated. Wicking of the filter medium and the subsequent leakage of liquids tends to be a problem with plate and frame filter presses. A recessed-plate filter press is also available, in which the hollow frames are replaced by recessed plates. The recesses in adjacent plates form the space in which the solids are collected.

There are many different plate filters on the market, but they all fall into two basic categories. There are horizontal plate filters and vertical plate filters. The names describe them. In a vertical plate filter, the plates extend into the slurry and the liquid is forced through the cloth, into the plate and out of the filtrate line. Both sides of the plate provide filter surface. The solid product may be removed in a number of ways, from manual removal to blowback and discharge from the bottom of the filter tank. One of the major problems with vertical plate filters is that they leave a fairly large unfiltered portion or heel in the filter. If a large quantity of product has been passed through the filter, the heel is insignificant, but if the batch is small, the unfiltered slurry may represent a large percentage of the total product.

Several companies manufacture a horizontal plate filter. In this type, the slurry is pumped through holes into a bundle of plates that are covered with filter cloth or filter paper. The liquid passes into the plate and then into a perforated center shaft and out of the filter. The solids remain on the plates. Plate filters are most often used to remove solid impurities or spent catalysts rather than for solid product isolation. They can be fitted with a scavenger plate which fits the tank bottom to minimize or eliminate the unfiltered heel. Horizontal plate filters with automatic discharge mechanisms are also available. They usually have a motor-driven center shaft, which when spun will throw the cake off the plates by centrifugal force. The cake will pass through a discharge chute to a receiver. The filter should be backwashed periodically to remove material that can remain behind after spinning and it must also be opened for cleaning at regular intervals.

A rotary filter can be run under pressure as well as vacuum. The housing must be designed for this service, frequently leading to large cost increases over a vacuum filter. Another major consideration is product discharge. In a vacuum filter the product is discharged from the roll at atmospheric pressure. In a pressure filter the product must be dropped through a pressure let-down valve before being discharged to atmosphere.

As with other filters, there are several kinds of centrifuges. We mentioned liquid–liquid centrifuges earlier, but there are also liquid–liquid–solid and liquid–solid separators. Centrifuges can be continuous or batch, manual or automatic, vertical or horizontal, suspended or underdriven, solid bowl, or perforate basket. The type of centrifuge selected is based upon several factors. These include concentration of solids in the slurry, particle size distribution, and degree of cake dryness required.

Solid bowl centrifuges are centrifugal separators that utilize centrifugal force to speed up the settling and decantation of solids and liquids. There are various considerations and methods of solids discharge. Some use a screw or augur to remove the solids while others rely only on particle movement due to the forces generated. Most solid bowl centrifuges are spun on a horizontal axis with solids discharging through a nozzle in the bottom of the machine. The filter cake can not be washed in a solid bowl centrifuge, but can be washed in a screen bowl unit. This is a combination of a solid bowl and perforate-basket centrifuge. G forces can reach 5000–10 000 in solid bowl or disc-bowl centrifuges.

Perforate-basket centrifuges may be built in any of the styles mentioned earlier, but most batch machines are on a vertical axis with a bottom product discharge. The cycle is generally feed, spin, wash, and discharge. These functions can be performed either manually or on a PLC or a series of timers. A cloth or screen is used as the filter medium, usually supported by a backing screen to prevent plugging of the basket holes. Perforate-basket centrifuges can produce filter cakes with as little as 5% moisture, depending upon the product. They can be used for isolating products from flammable solvents since they can be blanketed with inert gas. Continuous perforate-basket centrifuges are really semi-continuous. They are batch units of various sizes that run on very short cycles, sometimes as short as one minute. Feeding, washing, dewatering, and discharge are automatically controlled.

There are two basic theories about the washing of centrifuge cakes that are popularly held. One is that the cake should be spun free of mother liquor prior to starting the wash cycle. The other is that the wash should be started as soon as the feed is complete, but before the liquor is spun out. Since centrifuge washing is a displacement wash, with the wash being in a plane, it will not mix with the mother liquor, so theory two is probably correct. Spinning the cake out first, can cause the dry cake to crack in places and the wash will channel, leaving more mother liquor in the product.

23.4.1.4 Piping systems. Bulk pharmaceuticals are manufactured using a variety of piping materials and systems. Process piping may range from carbon steel to various stainless steels to glass or glass lined to a host of plastic materials. The methods of connecting the piping together and the types of valves used are also numerous. Piping can be welded in a number of ways, flanged, screwed or coupled with various types of connectors. Process and utility valves can also be of several types, with the most common being ball valves, gate valves, globe valves, and diaphragm valves. Other valves, such as pinch valves, needle valves, butterfly valves, etc. are used as required by the particular process system.

The major concerns in selecting the correct piping system involve the compatibility of the materials and the valves with the process fluids. There is usually little emphasis on maintaining sterility within the system, or in providing specific materials and valves solely for their ability to be cleaned or sterilized in place. Bulk pharmaceutical piping must often be able to withstand extremes of temperature and pressure or a range of pH or other corrosive conditions. Piping systems must not contaminate the product by dissolving or leaching into the liquid streams, nor develop into safety hazards due to internal or external corrosion. For these reasons, bulk process piping generally has thicker walls than the sanitary tubing common in either finished pharmaceutical or biopharmaceutical processing. Schedule 40 pipe is common in the process industries, with Schedule 80 sometimes required. Sanitary tubing, on the other hand, is roughly equivalent to Schedule 5 pipe.

23.4.2 Biopharmaceutical equipment and piping

23.4.2.1 Fermentation systems. There are many types of fermentation systems available in the market today. They range in size from a few hundred milliliters to several thousand liters. Small-scale fermentations are often carried out in stirred flasks or carboys. The agitator is either a magnetic stir bar or a lab model propeller. Magnetic stirring is gentler and is more applicable in low-shear fermentations. Propellers tend to be high shear and can damage sensitive products.

As the scale of the equipment increases, the systems become more complex. Dozens of vendors can supply a fermentation skid, complete with reaction vessel, addition tanks, heating-cooling systems, clean-in-place and steam-in-place connections and process controls. For many purposes a packaged system will work well, but most manufacturing firms will purchase a package and make some modifications to suit their product's exact requirements. The ability easily to sterilize the fermenter and its auxiliary equipment is critical. Product contamination must be eliminated.

Control systems are vitally important in a fermenter. The pH, temperature, dissolved oxygen concentration, airflow, agitator speed, and other parameters must be accurately controlled and monitored to ensure reproducible fermentation results. Most systems, even smaller ones, will be computer controlled. The parameters will be adjustable by the operator, so that fermentation temperature, pH, or other factors may be varied according to the requirements of the particular product. A multipurpose fermenter, capable of running several different products is desirable but not always possible.

23.4.2.2 Centrifuges. The majority of the centrifuges used to isolate the product from the cell debris are vertical solid bowl designs. They may also be a series of plates, such as in a disc-bowl design. Centrifuges run at high G forces, up to 25 000, in order to rupture the cells and to separate the liquids from the solids. Liquids overflow from the top of the machine while solids are either continuously removed or collected in the bottom of the machine and removed on a batch basis. Flow rate through the units can range from a few to a few hundred liters per minute. The equipment is constructed of 316L stainless steel of sanitary design. There are no rough surfaces to which material can adhere. The major manufacturers of solid bowl centrifuges are Sharples and Bird.

Another type of centrifuge, used for smaller batches, is similar to a laboratory unit. The rotor contains openings for tubes or jars, up to about 1 l in size. The jars are placed in the machine and spun rapidly for a specified period of time to separate the solids from the liquid portion of the batch. These types of centrifuges can be used in low-volume blood processing, to separate the plasma from the red cells. The centrifuges can be refrigerated to maintain temperatures down to 5°C or less. Major manufacturers are Sorvall and Beckmann.

23.4.2.3 Diafilters/ultrafilters. Intermediate and final product purification is frequently done in ultrafiltration and diafiltration units. Diafilters are tangential flow filters, where the product flows across the filter surface, keeping it clean, as well as flowing through the filter membranes. Diafilters are classified by pore size, so that the desired molecular weight material can either be retained in the system (retentate) or permeated through the membranes (permeate). Diafiltration is used to remove impurities or unreacted raw materials or to replace one buffer solution with another.

The product solution is diafiltered against a given quantity of buffer in order to perform the purification step. The solution is pumped through a holding tank, using a peristaltic pump to reduce shear, and recirculated through the tangential flow filter. The filter itself may be a series of horizontal plates or a tubular design. In either case the filtration principle is the same. The majority of the fluid is recycled back to the holding tank, with a small percentage, perhaps 10–15%, is discarded to waste. The level in the feed tank is kept constant by continuous addition of buffer. Alternatively, the level in the feed tank can be allowed to drop to a given point, at which time a new charge of buffer is added and the cycle repeated until the required amount of fresh buffer has been added to the system. The final product concentration is adjusted with fresh buffer if necessary, and the final product is transferred to a holding vessel prior to final filling into vials.

Ultrafiltration is similar to diafiltration in that it utilizes a membrane for filtration. In this case, the material is passed through the filter only once, with the permeate being collected in one vessel while the retentate is collected in another. Ultrafiltration is sometimes used in water purification, although it is not an accepted method of making water for injection (WFI) as shown in the United States Pharmacopoeia (USP).

23.4.2.4 Autoclaves/depyrogenation ovens. Autoclaves and depyrogenation ovens are used to ensure sterility of processing equipment and packaging materials. Glassware is washed in various kinds of vial or ampoule washing equipment and depyrogenated in batch ovens or continuous tunnels. These ovens run at temperatures above 250°C to ensure sterility of the components. A continuous tunnel must be properly sized to allow the correct time to heat, sterilize, and cool the glass prior to filling. The air supply to the oven is HEPA-filtered to ensure that a minimum of particulate reaches the clean and sterilized vials.

Steam sterilizers or autoclaves are used to sterilize other materials that cannot withstand the temperatures reached in a depyrogenation oven. An autoclave is usually operated at 121°C and 15 psig ($1\,kg\,cm^{-2}$) saturated steam pressure. It is important to validate the temperature uniformity and steam penetration to all parts of the sterilizer. Rubber stoppers, parts for aseptic filling equipment, plastic parts, etc. are autoclaved.

The FDA is now requiring terminal sterilization of all products, whether aseptically filled or not, except in cases where the manufacturer can prove that high temperatures will damage the product. Terminal sterilization means that all filled products must be resterilized after filling, even though the product and all packaging components were previously sterilized and then aseptically filled. Terminal sterilization provides much greater assurance of true sterility in the products that reach the market. The degree of assurance is about one million times greater than with aseptic filling.

23.4.2.5 Piping systems. The approach to piping systems is different in biopharmaceutical operations from that in bulks. There is a far greater emphasis on cleanliness and sterility and the avoidance of microbial or pyrogen contamination of products. Systems are designed for clean-in-place (CIP) and steam-in-place (SIP) operations, with less concern for the ability to withstand corrosion or extreme operating conditions.

The most common type of piping found in biopharmaceutical systems is sanitary 316L stainless steel tubing, usually electropolished after being mechanically finished to a 180- or 240-grit polish. When dealing with sterile systems, it is of utmost importance to ensure that the transfer piping and valves do not provide places for bacterial growth. Pipe sections are connected either by welding on automatic machines or by the use of special sanitary clamps (Tri-Clamps). Most biopharmaceutical processes are aqueous and are run at temperatures below 40°C. Except for some filtration applications, such as reverse osmosis (RO) systems, they are commonly run at low pressures and can be constructed of thin-walled tubing, compared with Schedule 40 process piping.

The non-sterile side of the biopharmaceutical process may contain plastic piping of various types, or in small systems, be composed of flexible hoses. Carbon steel or galvanized steel is only used for utility piping, and most of the gas systems are copper on the non-sterile side of the sterilizing filter and stainless steel on the sterile side.

Current industry practice for 'clean' utilities (clean steam, water for

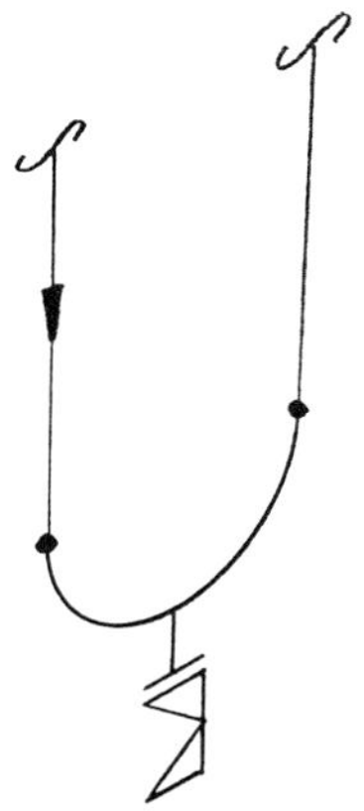

Figure 23.1 Diagrammatic representation of water for injection piping.

injection) utilizes sanitary 316L stainless steel tubing joined via automatic orbital welding. All welds are inspected with a boroscope, and many installers videotape the inspection for future reference. Sanitary diaphragm valves are used in sterile service, since they have no crevices where bacterial growth can occur. Some ball valves are used for vessel discharges, but except for clean steam uses, are not used in other piping systems. Dead legs, or places within piping systems where stagnant fluids can accumulate, must be kept to a minimum of six pipe diameters in length. The WFI piping is constructed in a continuous loop, with the use points teed off a secondary drop, as shown in Figure 23.1. This type of piping is not used in bulks.

Transfers between vessels or production suites, CIP and SIP lines or utilities supplies to the various production equipment are commonly made through transfer panels. These panels are of sanitary construction and are located on the walls of the production or filing suites. Sanitary connections (Tri-Clamp type) are provided at each of the outlets. To transfer process materials from one area to another, or to provide utilities to a vessel, a hose connection will be made. For providing CIP or SIP or for other process transfers, a sanitary stainless steel U-tube will be provided. Proximity switches are often installed at the panel to ensure that the connection has been properly made and that the correct fluid is going to the correct destination.

23.5 Materials and materials handling

23.5.1 Bulk pharmaceuticals

Bulk pharmaceuticals utilize a wide variety of reagents and raw materials, usually in much larger quantities than in biopharmaceuticals. Toxic, corrosive, and highly potent materials are commonly used, along with all types of solvents. Starting materials are received in various containers, from small vials or cans to bins and trucks or rail cars. Major concerns include the storage of raw materials, transport of solid materials from the shipping containers to the storage equipment, transport from storage to the reactors, and handling of the final intermediates or products.

Solids transfer systems using dense-phase or dilute-phase airveying, vacuum, screw conveyors, and belt conveyors are common. The method of transfer is dependent upon the physical properties of the material and the potential hazards to both the operator and the environment that it represents. Airveying and vacuum transfer consist of a fan or blower, sometimes more than one, as a motive force, transfer ductwork, and a product collection and discharge item at the storage silo or the reactor. Bag collectors, filled with cloth filter bags, or cyclones can be used at the end point of the system. Again, this depends upon the type of application. Bin transfer, using computerized guidance systems, is also becoming more common in the industry.

Bulk production often involves the use of large quantities of activated carbons or filter aids to decolorize or to filter solutions. The handling of these materials creates concerns that are not present in biopharmaceutical operations, mainly due to the amount of material used. They also affect the plant layout to some degree, because of the desire for isolation of those materials from the remainder of the process.

A carbon handling area is an example of a place that should be isolated and curbed when possible. Charcoal treatment of solutions is a common process step, often done batchwise, but the disposal of the wet cake from a manually cleaned pressure filter is a messy operation. The carbon cake should be confined to a washable area to prevent the inadvertent transfer of carbon particles throughout the manufacturing area.

23.5.2 *Biopharmaceuticals*

Aside from bulk fermentation, which is a specialized case of biopharmaceutical production, the majority of the materials used for biopharmaceuticals are handled in small quantities. Vials, flasks, and small cans of raw materials are common, and drums of materials usually are considered to be a large amount. Many of the materials, usually all of the actives, are of natural origin and can be hazardous or toxic. The remainder of the reactants are usually innocuous salts and high-purity water, used for buffer solutions. Although some solvents, such as methanol or ethanol are used, the majority of the processing is done in water.

As a result of the scale of operations, many material transfers are made by hand, simply by pouring materials from one container to another or from a container into a reactor. With many of the materials being hazardous, these transfers are usually done inside a protective cabinet or under a hood.

In handling larger quantities, it is important to maintain a closed system to protect both the operator and the surroundings, and to ensure that all of the product has been transferred. Peristaltic pumps or nitrogen pressure are common methods of liquid transfer. The use of centrifugal pumps is often avoided because of the high shear and the potential pump seal problems inherent in their design.

Water is one of the most important materials used in biopharmaceutical production. The quality of water may vary, but it is uniformly higher than that used for bulks. While some solutions may be prepared with only deionized water, commonly used in bulk production, most of the water has also been treated via reverse osmosis or ultrafiltration to remove biological materials. The high quality feed water can then be fed to a multieffect distillation unit to produce pyrogen-free water for injection, USP (WFI) or purified water (USP). Most WFI systems are maintained above 80°C to reduce potential biological contamination, although some systems are successfully operated at room temperature. The types of storage vessels, transfer pumps and piping filters, and instrumentation required in WFI systems are discussed elswhere in this book.

23.6 HVAC and utilities systems

23.6.1 HVAC

Bulk plant heating, ventilating, and air-conditioning (HVAC) requirements are significantly different from those in finished pharmaceutical plants. Depending upon the nature of a particular biopharmaceutical system, the HVAC requirements could be similar either to a bulk or a finishing facility. The air flow requirements in bulk processing areas range from two to six air changes per hour, depending upon whether or not flammable materials are processed in the area and upon the heat load being generated by the equipment and the personnel. In finishing facilities the number of air changes is much higher, from 20 to 700 air changes, because of the need to maintain low particulate counts in areas where product might be exposed. Room classification is defined as the number of particles greater than 5 μm in size per cubic foot of air. Areas are classified from Class 100 to Class 100 000, with areas higher than Class 100 000 being called Unclassified. Biopharmaceutical plants can be at either end of the spectrum, although the trend is towards the finished pharmaceutical level.

Both bulk and biopharmaceutical plants have a major concern with protection of the operators and the environment from the materials being handled within the facility. Increasing the number of air changes per hour will not provide this protection, so the problem in biopharmaceutical facilities is addressed by providing positive or negative pressure differentials between rooms or suites where hazardous materials are being processed. Other methods, such as scrubbers, adsorbers, incinerators, etc. are used in bulk plants.

Biopharmaceutical facilities do have some of the same area classification concerns as finished pharmaceuticals, because in many cases the final product is being filled within the plant. This sometimes creates a paradox when dealing with room pressurization. The filling area is the most positive with respect to other areas in order to protect product integrity. A hazardous materials processing area is the most negative in order to protect the surroundings. Filling a sterile hazardous material apparently requires both. The solution is to provide a positive airflow at the filler, but a negative airflow from outside of the filler to the rest of the surroundings. The operators are protected by masks and other protective clothing.

A key element in the HVAC design for biopharmaceutical plants is the isolation of the various area HVAC systems. Each type of process area will be served by its own air handler. The fermentation areas and possibly cell recovery can be grouped within one system, but the purification suites will be serviced with another. Each filling room will have its own HVAC system, and any other non-production areas will be on other separate systems. In cases where more than one fermentation or purification suite is constructed, each of the suites will have its own HVAC unit. The number of air handling systems is obviously much higher in a biopharmaceutical facility than in a bulk plant, where isolation by function and byproduct is not of such great concern.

Bulk pharmaceutical plants do not normally require classified areas. Manufacturing areas for early steps in a multistep process are usually not air conditioned, only ventilated for operator comfort. At a minimum, the product isolation areas are normally air conditioned and isolated from the adjacent production areas, although recent trends are toward increasing concern for room classification in areas where the final step or final product isolation are being run. Humidity control in the isolation room tends to be a function of the characteristics of the product. If the product that may be exposed to the surroundings is hygroscopic or otherwise moisture sensitive, the area will be humidity controlled. If it is not, no humidity control is necessary. Because of these substantial differences in philosophy and regulatory requirements, the cost of HVAC equipment is substantially higher for finished pharmaceutical and biopharmaceutical facilities than for bulk plants.

23.6.2 Steam

There are major differences between bulk and biopharmaceutical plants in such a common utility as steam. Bulk plants use steam of various distribution pressures for process heating. Operating pressures of about 110 psig ($7.75\,kg\,cm^{-2}$) are common in most plant distribution systems, although pressures as high as several hundred psig are needed in special high-temperature process applications and as low as 15 psig ($1.06\,kg\,cm^{-2}$) are used in building steam heating. Boilers may be of many types, and the feed water is softened or sometimes even deionized to prevent scaling in the boiler tubes. Various treatment chemicals are added to the feed water to prevent scaling and corrosion in the boiler and in the distribution piping system. Process steam condensate is frequently collected and returned via a deaeration system for reuse.

Process steam piping systems are normally constructed of Schedule 40 carbon steel. They are connected with screwed fittings, although they are sometimes flanged. Globe valves, gate valves, or ball valves are commonly used as shut-offs, and diaphragm valves are the most common type of control valve. Brass is the valve material of choice. Bourdon-type pressure gauges are used throughout the system for steam pressure indications.

Biopharmaceutical facilities have only limited use for process steam. Its main purpose is to drive the water for injection still and the clean steam generator and to heat the autoclave, WFI storage tank and vent filter jackets. Process steam at about 115 psig (8 bar) is required for maximum efficiency of the still and steam generator.

Clean steam is often used in biopharmaceutical facilities. It is usually produced from a water for injection feed to a clean steam generator, although it can be taken from the first stage of a multieffect evaporator producing WFI by distillation. Clean steam is used for sterilization of components, some process materials, piping, and equipment. It is distributed at about 25 psig (1.75 kg cm^{-2}) so that surfaces or items to be sterilized may reach a temperature of 121°C. No additives are used in the feed water or in the clean steam generator. Condensate from the system may be collected, but it normally is used for other purposes and not returned to the clean steam generator as feed water.

The material of construction of the clean steam generator and its distribution piping system is Type 316L stainless steel. The piping is welded using automatic welding equipment, with all welds inspected by boroscope. Stainless steel ball valves may be used on the headers or as primary shut-offs, but the secondary valves at the users are diaphragm type. Pressure gauges are also diaphragm type so that only a sanitary material will be in contact with the fluid. Clean steam systems must be validated to verify that they are operating as designed.

23.6.3 Water

Water is the most common solvent in both bulk and biopharmaceutical manufacturing. Its quality can vary from raw process water to ultra pure water for injection (USP). How pure to make the water depends upon its end use.

Bulk plants require lower water purity than biopharmaceutical plants. The bulk plant water will be used as a reaction or purification solvent, and will often have other non-sterile materials dissolved in it. The closer we come to the pharmaceutical material, the purer the water will be, but deionized water with a resistivity of about 1–2 mΩ is usually sufficient for most applications.

Pretreatment of water may take many forms, depending upon the particular impurities dissolved in the raw water supply. There will generally be a sand or multi-media filter to remove particulates, a carbon filter to remove organics, and either a softener to remove hardness (for boiler feeds) or a deionizer system to remove dissolved ionic impurities. The deionizer may consist of anion and cation bed demineralizers followed by a mixed bed polisher, or may be only one or more mixed bed units in series. Further purification, such as through a reverse osmosis (RO) unit or an ultrafilter is not usually necessary. There is normally no requiremnt for using purified water (USP) or sterile water in a bulk manufacturing operation.

The biopharmaceutical case is very different. Although early process steps may use only deionized or RO water, as we reach the final production steps WFI is generally required. The production of WFI can be via a multistage distillation or a two-stage reverse osmosis system. The feed water to either of these systems must be pretreated in the same manner as the deionized water in bulks, although other pretreatment (RO or ultrafiltration) may be desirable for distillation feeds. Sterilization methods such as exposure to ultraviolet light may also be employed.

WFI is maintained in a 316L stainless steel storage tank at a minimum temperature of 82°C. Some users prefer multiple storage tanks to allow each 'batch' of WFI to be tested, or to provide back-up capability in case of an operating problem. Operating problems however, rarely occur with the storage tanks, and microbiological testing takes longer than one day, so all WFI usage is actually run 'at risk'. Use of more than one storage tank complicates the supply piping to the circulation pumps as well as the supply lines from the still, the return lines from the systems, and the level controls on the tanks that start and stop the still. A validated system, running at the conditions for which it has been validated, is the best approach to avoiding operating problems.

The distribution system is constructed of a continuous 316L stainless steel loop, welded using automatic equipment, inspected with a boroscope and with no 'dead legs' greater than six pipe diameters in length. A 'dead leg' is a section of pipe where water is not normally flowing. Circulating pumps must be of sanitary design, the piping must slope to be completely drainable, and the system must have the capability of being steamed out for sanitization. Design water flow rates for the WFI loop should be about 7 ft/s (2.13 m/s). Clean steam injection piping must be supplied at appropriate intervals in the system, usually 100 ft (30.48 m), and steam traps must be provided at low points.

23.6.4 Air and other gases

The major difference between bulk and biopharmaceutical air, nitrogen, and other gas systems is the sterility requirement. Bulk systems use oil-free compressed air, often dried to a dew point of −40°F (−40°C) (instrument air), and filtered at the points of use to remove possible scale or other impurities in the supply lines. They may also use air produced in a lubricated compressor, fitted with coalescing filters to ensure oil removal. The distribution piping is copper or carbon steel and the distribution pressure is normally about 100 psig (7.04 kg cm^{-2}). The pressure is reduced to 15 psig (1.05 kg cm^{-2}) for instrument control. Small silica gel drying cartridges may also be installed at the use points to ensure that wet air will not corrode instrument internals.

Biopharmaceutical systems will always use air from non-lubricated compressors. The distribution piping may be either copper or carbon steel and the distribution pressure will be the same as in bulk systems, but there will be one or more sterilizing filters in the system either prior to or at the points of use. Piping downstream of the final sterilizing filter will be sanitary 316L stainless steel tubing, welded and finished as per other sanitary systems. Nitrogen will be distributed in the same manner, again with final sterile filtration.

Flow instrumentation and controllers on biopharmaceutical gas systems must be located upstream of the sterilizing filters. There are no sanitary design instruments for flow measurements, nor are many sanitary pressure reducing, control, or relief valves available. Pressure gauges must also be sanitary, with a diaphragm made of an approved material acting as a barrier between the sanitary process fluid and the instrument. In bulk pharmaceutical systems, simpler gauges may be used provided that they suit the process conditions.

The sterilizing filters and housings used in biopharmaceutical gas systems must themselves be sterilizable, preferably in place. Clean steam must be supplied upstream of the filters to allow sterilization of the system. The piping must be properly valved to prevent backflow of steam into the gas system and the filters must have vents and drains.

23.7 Waste disposal and handling

Bulk pharmaceutical waste disposal is often a matter of disposal of large quantities of various kinds of mixtures of materials, whether they be liquids, solids, or gases. Biopharmaceutical waste disposal often involves decontamination of toxic materials, both liquid and solid, and the elimination of hazardous materials in waste air streams.

23.7.1 Bulk pharmaceuticals

One of the common problems encountered in dealing with bulk wastes is characterization of the streams. In multiproduct or multiprocess facilities, there are many combinations of solid and liquid wastes to be disposed of. The streams vary from day to day or even from hour to hour, so that there may never be exactly the same flow and composition of wastes leaving the plant. It may be theoretically possible to calculate the rates and compositions of the waste streams, but it is more practical to sample at frequent intervals and to develop a daily composite composition for the plant effluents. In this way, a 'worst case' scenario can be determined and the waste treatment facility can be designed to handle that worst case.

23.7.1.1 Liquid wastes. The treatment of liquid wastes is a subject that is in itself the topic of many books. There are some common methods of liquid waste treatment that are applied in most bulk plants. First, it is important to isolate process from sanitary and storm drains. Process drains within the building may be further isolated so that floor drains, which collect normal washdowns, are separated from equipment drains, which may contain hazardous or flammable materials. The process drains are directed through separation equipment to separate organic and aqueous phases. This may include skimmers, centrifugal separators, or gravity separators. The immiscible organics are collected in holding tanks and then sent to solvent recovery or to disposal.

Aqueous wastes are passed through settling basins or settling ponds for solids removal. The size of the settling area is based on the quantity of liquids and the percentage of solids that it contains. Overflow from the settling basin is then adjusted for pH. Most municipal treatment facilities will allow pH ranges between 5–9 to enter the plant. pH control is automatic. The pH of the stream is continuously monitored as it flows out of the settling basin and is adjusted with either acid or base as necessary to maintain the proper pH. Dilute sodium hydroxide is generally the base that is used and either dilute hydrochloric or sulfuric acid is the acid adjustment material.

In some waste systems solids removal and pH control is not enough. There may be BOD (biological oxygen demand) or COD (chemical oxygen demand) levels in the treated effluent to make it unacceptable for discharge either to the municipal treatment plant or to the surrounding waters. Streams containing methanol or ethanol commonly have high BOD levels. In these cases, secondary treatment will be necessary. BOD removal can often be accomplished by aeration either using enclosed equipment or using fans to blow large quantities of air through the effluent to oxidize the BOD. Tertiary treatment using aerobic or anaerobic bacteria may also be required before the effluent is sufficiently free of contaminants for discharge.

23.7.1.2 Solid wastes. Common solids that must be safely disposed of include catalysts, activated carbon, filter aids, fermentation solids, water treatment sludges, and containers. Spent precious metals catalysts can be recovered by separation of the metal from the carrier. The carbon is incinerated in a reducing atmosphere and the recovered metal is replated onto fresh substrate. Other solid wastes also can be incinerated, with only the ash remaining to be sent to a secure landfill. Filter aids and fermentation filter cakes are often incinerated. Sludges isolated from secondary and tertiary water treatment are usually landfilled. Drums, liners, bags, and other packaging materials can be recycled for reuse. Fiber drums can be reused for intermediate or crude product storage, by placing plastic drum liners inside the drum. The liners can also be cleaned and recycled in some of the new plastics recycling facilities currently being built.

Waste minimization is an important element in the design of any new bulk manufacturing facilities. All processes should be carefully studied to avoid the generation of large quantities of waste materials. Raw materials and intermediates should be purchased in reusable storage containers, solvents should be recycled for reuse in the process and potential waste streams should be investigated for recoverable materials. Minimization is the best approach to waste treatment.

23.7.1.3 Air pollution control. Bulk pharmaceutical plants have major air pollution control problems, usually resulting in major expenditures for control equipment. Because of the diverse nature of the processes being run within the plants, there are serious concerns with odors, organic vapors, sulfur compounds, and nitrogen oxides. Odor control is a serious problem when plants are located in populated areas, especially in fermentation operations or processing steps where mercaptans or other sulfur-containing compounds are used. The odors can be controlled by adsorption on activated carbon, by incineration of the air streams or by chemical oxidation, such as with potassium permanganate.

Volatile organics (VOCs) control and control of fugitive emissions are currently major initiatives with the EPA (Environmental Protection Agency). Fugitive emissions are those from many minor sources, such as pump seals, flanges, piping connections, hoses, etc. The aggregate of these fugitive emissions comprises a large quantity of air pollutants, so their containment and control is important. VOC control can be accomplished in many ways. Scrubbers, absorbers, and adsorbers use many different circulating liquids for vapor removal. Contact and non-contact cyrogenic systems and mechanical refrigeration systems condense the organic vapors and remove them from the effluent air streams.

23.7.2 Biopharmaceuticals

Waste disposal problems in biopharmaceutical production are different from those in bulks, but also require serious consideration.

23.7.2.1 Liquid wastes. Liquid process wastes from biopharmaceutical operations must be separated from sanitary and storm drains. Because of the potential hazards to the operators and to the surroundings, the process vessels are usually piped directly to contained drain systems. Most processing areas will not have open floor drains, and floor wash liquids are frequently picked up with a wet–dry vacuum cleaner.

Although there are some exceptions, liquid wastes are usually decontaminated prior to disposal to the municipal treatment plant. Decontamination can be either thermal or chemical, batch or continuous. The liquids to be treated are collected in a holding tank and transferred to the decontamination system. Two or more holding tanks should be provided, so that one is always available for use. Thermal treatment involves heating the liquid to about 125°C for sufficient time to kill the hazardous organisms. This can be done either in a tank large enough to process the effluent for a given time period of production or in a continuous heater. If a continuous decontamination process is selected, the holding time within the heated section of the system must be sufficient to kill the organisms. Continuous systems are preferable when large quantities of material must be processed.

Chemical decontamination may also be used. The chemical is usually a chlorinated or oxygenated material, such as dilute sodium hypochlorite or other such contaminant. A calculated quantity of the treatment chemical is added to the batch and held for the requisite time to kill the organism. The chemical can also be added to a continuous system. In either chemical or thermal treatment systems, whether they are batch or continuous, it is important to provide a recorder for temperature measurement or for concentration of decontaminant chemical in the system. There are instances where the use of chemical decontamination is more damaging to the municipal treatment plant than no prior treatment at all. If the effluent contains blood, for example, a chemical disinfectant may destroy the bacteria in a municipal plant, while the addition of a relatively small quantity of blood will not have any effect.

23.7.2.2 Solid wastes. There are usually smaller quantities of solid wastes in biopharmaceutical plants than in bulk plants, but solids also must be decontaminated at times. An example of solids to be decontaminated are vials of rejected biologically active materials. As with liquids, the decontamination process may be either thermal or chemical. Thermal processing is done in a steam sterilizer or autoclave. The solids are placed in containers and heated to a minimum of 121°C with 15 psig (1.05 kg cm^{-2}) steam to kill the organisms. Materials may also be finely ground in a mill and chemically decontaminated with sodium hyplochlorite. The treated solids can then be sent to a secure landfill or to the municipal system.

23.7.2.3 Air pollution control. Many biopharmaceutical processes are run under strict conditions of environmental containment. Air supplied to the production area is HEPA-filtered and, in cases where BL-3 or BL-4 materials (see section 23.8.2) are being used, the exhaust air is also HEPA-filtered.

The concern in these cases is the possibility of discharge of viable orgnisms to the surroundings. Use of HEPA filters on the exhaust system will prevent accidental discharges. The filters are changed regularly, using a bag-in/bag-out system to eliminate exposure of the used filters to the environment.

23.8 Safety considerations

23.8.1 Bulk pharmaceuticals

Safety considerations in bulk pharmaceutical processing include dust and vapor hazards, fire and explosion hazards, toxic and highly potent chemical handling, and isolation and protection of personnel. Fire and explosion hazards have been major design parameters for many years. Electrical codes have been developed to assign operations to various categories of hazard, in order to prevent fires and explosions due to electrical arcs, sparking, or hazardous concentrations of dust or flammable vapors. Motors have been classified according to their use and explosion-proof equipment, including room lighting, must be installed where required for safety.

The isolation and protection of personnel from toxic compounds has also been a design factor for many years. Various test procedures and equipment to determine the concentration of hazardous materials (phosgene, dimethyl sulfate, arsine, cyanide, etc.) in the atmosphere are available and are commonly used. The recognition and classification of highly potent compounds however, is relatively new. As pharmaceutical research moves forward, the potency of new drugs is increasing at a dramatic rate. Micrograms, or in extreme cases nanograms, of material are now sufficient for therapeutic doses, where milligrams were once required. The requirement for isolation of the operators from those materials has also increased dramatically. In the bulk plants of the future, a realistic goal will be to eliminate totally operator contact with processing operations by use of computer controls and robotics. This will protect both the operator from the product and the product from the operator.

23.8.2 Biopharmaceuticals

Safety considerations in biopharmaceutical facilities are different from those in bulk plants with regard to dust, vapor, fire, and explosion hazards. Dust or solvent concentrations high enough to reach explosive limits are not encountered in biopharmaceutical plants. Fire hazards are more from normal office hazards than from chemicals, since most processing is done in aqueous systems. The major hazards involve the protection of operating personnel and the surrounding environment from toxic, carcinogenic, mutagenic, and other biologically active materials.

The National Institute of Health (NIH) has classified biological hazards from BL-1 to BL-4, with BL-1 being the least restrictive. As the biosafety levels increase, the degree of protection must also increase. Each level requires all of the safeguards of the prior ones, plus additional protection. BL-1 hazards are relatively innocuous and are controlled by procedural methods. The materials involved are those that might be found in university laboratories. BL-2 requires protective clothing along with the procedures. BL-3 is not commonly encountered, but requires isolation of the area, biosafety cabinets for operations, isolation of drains or other waste streams, and HEPA-filtered air. HEPA filtration of exhaust air from the area is commonly installed, although not specifically required. BL-4 is rare and requires shower areas for personnel. It can also require incineration of the exhaust air, depending upon the degree of virility of the organisms used.

Personnel protection for most biopharmaceutical operations involves the use of protective clothing for operators as well as its disinfection at the end of the operation. Disinfectant sprays for uniforms and separate shower areas for personnel exiting the biosafety area may also be provided. It is important to isolate those areas where biologically active materials will be used for adjacent operations within the building. The handling of particularly toxic or carcinogenic materials should be done under conditions of extreme isolation and should be done by remote or robotic operations if possible.

23.9 Regulatory considerations

Contrary to popular belief, both bulk pharmaceuticals and biopharmaceuticals must be produced under the current Good Manufacturing Practices (cGMPs). Biopharmaceutical operations are also regulated by NIH as are some bulk processes. All pharmaceutical manufacturers must adhere to cGMPs, as they are interpreted by the FDA (Food and Drug Administration). The key word is current. What major manufacturers are currently doing in their operations is frequently interpreted by FDA as what all manufacturers should be doing, so the industry tends to gravitate to the highest level of compliance.

In many cases regarding bulk operations, the Pharmaceutical Manufacturers Association (PMA) guidelines are considered to be the cGMPs. It is not necessary physically to isolate all of the steps in bulk production for protection from the environment, only the steps where the final product may be exposed. It is necessary to follow the cGMPs with regard to batch identity and record keeping, qualification of raw materials, quality specifications for raw materials and intermediates, and other quality and record keeping aspects. The cGMPs are the law in the USA, not guidelines to be interpreted by a manufacturer as he sees fit. Violation is a criminal offense. Guidelines issued by FDA are frequently interpreted by FDA inspectors to be regulations, although they may not yet have been issued as part of the Federal Register.

Validation of bulk facilities is not specifically required in the Code of

Federal Register (CFR) 210 and 211, but many major pharmaceutical companies are beginning to validate the final steps of bulk drug manufacturing processes. A substantial number are also preparing installation qualification (IQ) and operational qualification (OQ) protocols for all bulk processing steps. Validation is required for drug manufacturing operations, where the final dosage form is being prepared. It is certain that equipment qualification and process validation will become part of the regulations in the future. The only question is when in the future will it happen?

Biopharmaceutical companies are required to validate most of their processing steps. There is some debate as to the point where a biopharmaceutical product ceases to be a bulk material and when it becomes the final dosage form. FDA is becoming more stringent on their definitions and more process validation is being required.

Both bulk and biopharmaceutical manufacturers must be familiar with new drug applications (NDAs), establishment license applications (ELAs), and product license applications (PLAs). The requirements are the same for both types of manufacturers. Biologicals are regulated and inspected by the Center for Biological Evaluation and Research (CBER) of the FDA, while drugs are under the control of the Center for Drug Evaluation and Research (CDER). Drugs are usually chemical entities, although some natural products are classified as drugs, while others are biologicals. The requirements of each branch of the agency are slightly different, with CDER requiring the filing of an NDA and CBER requiring ELAs and PLAs. The contents of each document are essentially the same, although in different formats. If a manufacturer wants to relocate a biological process he must file a new ELA to cover the new facility.

CBER is also more stringent than CDER on control of the methods of manufacture for intermediates used in biological processes. The PLA must include a description of the manufacturing process for the various intermediates, if they are to be used to produce a biological product regulated by CBER. This can cause some problems if the material is manufactured overseas, where the supplier may be unwilling to divulge his manufacturing secrets, or if one of the intermediates is not necessarily a cGMP material and is not made in an approved facility. CDER only requires the identification of the manufacturer, although they do require an amendment to the NDA if a new supplier is used for raw materials or intermediates.

It is important to recognize the fact that both bulk and biopharmaceutical production is closely regulated, both in the United States and abroad. In fact, some of the facilities regulations with regard to air classifications in finishing areas are more stringent in the United Kingdom and in Europe than in the United States. If a manufactuer expects to market his products worldwide, he must be acutely aware of the regulatory requirements in all of his marketplaces. Compliance with one set of regulations does not automatically provide compliance with all of them.

24 Optimization of protein recovery using computer-aided process design tools

D.P. PETRIDES and E.S. SAPIDOU

24.1 Introduction

The successful design, scale-up, and optimization of protein recovery and purification processes is a difficult task that requires collaboration of many disciplines. Computer-aided process design tools can facilitate this task by introducing a common language of communication among the various research and development groups and by providing a total systems approach. Such tools enable chemical/biochemical engineers rapidly to develop, evaluate, and optimize flowsheets for the production and recovery of biological products. This chapter, through an example, demonstrates how such a software design tool[3] can be used in an industrial environment to facilitate process development and flowsheet optimization.

24.2 Computer-aided bioprocess design

Process simulation and computer-aided design tools have been successfully used for almost four decades in the oil and chemical industries to scale-up and optimize integrated processes[1]. Process simulation has found limited applications in the bioprocess industry until now because first, the biotechnology industry is relatively new and primarily dominated by scientists that are not process oriented; second, most biological products, especially the high value ones, are complex and labile materials (e.g. proteins and polysaccharides), poorly characterized with respect to physical properties (e.g. diffusivity, solubility, viscosity, etc.), and these values are difficult to predict from thermodynamic and microtransport models; third, several important bioprocess unit operations are poorly understood and predictive models which could be used for design do not exist; fourth, most bioprocesses operate in batch or semi-continuous mode introducing time as an independent variable.

Despite these difficulties, the scientific challenge and the potential reward from the commercialization of biotechnology has stimulated interest in recent years in developing process simulation tools[2–4]. The benefits of the use of such tools vary depending on the type of product, the size of the investment, the location of the facility, and other parameters.

24.2.1 Product and project selection

During project selection, process simulation tools can be used to evaluate rapidly a large number of project ideas and discard the unpromising ones. This can drastically reduce engineering time because experience has shown that less than 5% of new project ideas are ever commercialized[3].

24.2.2 Research and development planning

Process simulation tools can be used to plan and prioritize pilot plant work. More specifically, before serious process development is initiated, the engineer develops a conceptual model of the entire process that requires scale-up on the computer. Through the initialization steps, the user identifies all the missing information (e.g. partition coefficients for extractors, rejection coefficients for membrane filters, etc.) that is needed to develop an accurate model of the entire process. Then, based on the missing information, the engineer designs the appropriate experiments to gather that data and complete the modeling work. Having a good model of the entire process, the engineer starts asking 'what-if' questions and carrying out sensitivity analysis and optimization with respect to key design variables. If simulation yields any unexpected results, experiments are planned to test them. This systematic procedure of designing and running pilot plant experiments saves development time and yields more efficient and robust processes.

24.2.3 Environmental and safety issues

Environmental and process safety problems are best solved when such issues are considered during the early stages of process development and built into the initial process design. Computer-aided process design tools, when equippped with the appropriate features, enable engineers to consider the environmental impact of new products and processes at the early stages of process development when it is still easy to make process modifications[5]. This is very important especially for therapeutic biological products because after a new product is approved as safe and effective by the Food and Drug Administration (FDA) it is very costly to make process modifications. Using such tools, for instance, the engineer can systematically select extraction solvents or chaotropic agents for solubilization of inclusion bodies that meet all process requirements and at the same time are environmentally benign. The engineer also can readily evaluate process modifications (i.e. addition of recycle streams, replacement of chemicals, etc.) that minimize waste generation.

24.2.4 Improved communication

When computer-aided process design tools are used by all groups involved in process development and design, they introduce a common language of communication that drastically reduces duplicate work. In a sense, the use of such tools constitutes a first step towards concurrent engineering and integrated product development with reduced cost.

24.2.5 Retrofit of existing facilities

Process design tools are not only useful for designing new processes but also for retrofitting existing ones. More specifically, such tools can be used to evaluate process modifications of existing manufacturing facilities in order to increase capacity, reduce production cost, reduce waste generation (and consequently cost of waste treatment and disposal), improve process safety, etc.

The new generation of computer-aided process design tools has incorporated results from the recent advances in computer science. They feature interfaces that are fully graphical and eliminate the need for development of FORTRAN-like input files required by most mainframe process simulators. The initialization of a flowsheet (see Figure 24.1) and the display of results is done through appropriate dialog windows. Figure 24.2, for instance, shows the first input–output dialog window of a high-pressure homogenizer. The user of such tools is expected to be a chemical/biochemical engineer or a chemical engineering student with some experience in process design, but

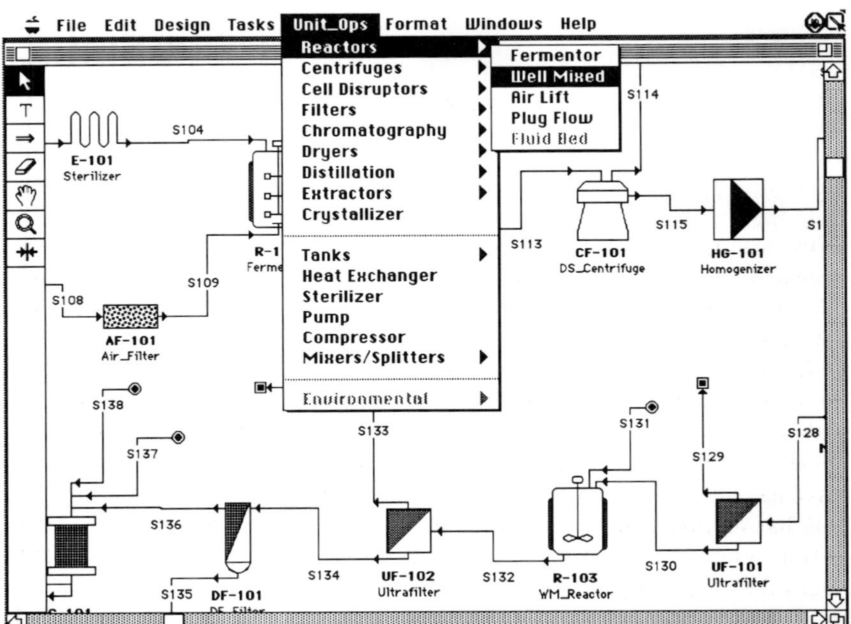

Figure 24.1 Main window of a bioprocess software design tool.

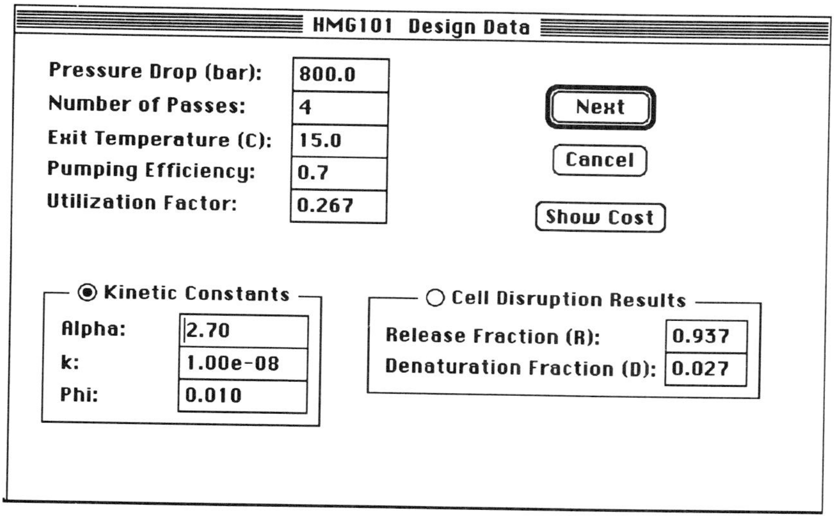

Figure 24.2 First input–output dialog window of the high-pressure homogenizer.

not necessarily with programming experience. An illustrative example demonstrating computer-aided process design follows.

24.3 Analysis of β-galactosidase production using process simulation

24.3.1 Background

This example analyzes the production of an intracellular protein, β-galactosidase, by *Escherichia coli*. However, the analysis of the production of any other intracellular protein that does not form inclusion bodies would be very similar.

E. coli is the best studied bacterium by microbiologists and molecular biologists and has been used extensively for cloning and expression of eukaryotic recombinant proteins. Human insulin, human and various animal growth hormones are among the recombinant proteins produced by *E. coli* that have already been found or are on the way to commercial success. *E. coli* has minimal nutrient requirements, grows fast (doubling time as little as 20 min) and is not pathogenic. *E. coli*, however, as a prokaryotic microorganism lacks the machinery of post-translational modifications that several complex mammalian proteins require. Consequently, genes of those proteins cloned in *E. coli* are expressed, but the product is inactive and forms inclusion bodies. If these post-translational modifications can be done *in vitro*, this drawback of *E. coli* is not serious, but for products where *in vitro* modifications cannot be done, the use of eukaryotic microorganisms, such as yeast or mammalian cells, is required.

In this example, the production of β-galactosidase, a natural product of *E. coli*, is analysed. This enzyme is normally produced by *E. coli* up to 1–2% of total cell protein under conditions of induction of the lac operon. Using genetic engineering, however, the level can go up to 20–25% of total protein. In the base case of this analysis, an easily attainable level of 10% was assumed. β-galactosidase is formed intracellularly and allows growth of *E. coli* on lactose (a disaccharide present in milk). There are several types of β-galactosidase formed by different microorganisms. The one formed by *E. coli* has a molecular weight of 540 000 and an isoelectric point of 4.8.

24.3.2 *Market analysis*

β-galactosidase has found limited application until now. It is mainly used in the utilization of cheese whey. More specifically, immobilized reactors with β-galactosidase have been developed to convert lactose found in cheese whey to glucose and galactose, yielding a sweetened product which can be used as an additive to ice cream, egg-nog, yogurt, and other dairy products. A possible future application of β-galactosidase is in the treatment of milk products. An increasing number of people have been found to be lactose intolerant and cannot digest milk or milk products. Production of lactose-free milk products will allow those people to digest them. The lactose from cheese whey can also be utilized by *E. coli* as a carbon source for the production of other biological products.

24.3.3 *Design basis*

Even though the actual industrial production of β-galactosidase is low, an annual production of about 5 metric tons of pure enzyme is the design basis for this case study. This allows the size of the plant and the corresponding costs to be high enough and representative of the costs of a relatively large plant producing an intracellular recombinant protein. The plant batch time is 72 hours and 96 batches are run per year.

24.3.4 *Process description*

24.3.4.1 Block diagram. A block diagram of the production of β-galactosidase by *E. coli* is shown in Figure 24.3. The upstream section includes the process steps of media preparation, media sterilization, and fermentation. The downstream section is divided into the primary recovery stages and the high resolution stages. The first process step of the primary recovery stages is cell harvesting. As mentioned previously, this enzyme is intracellular and disruption

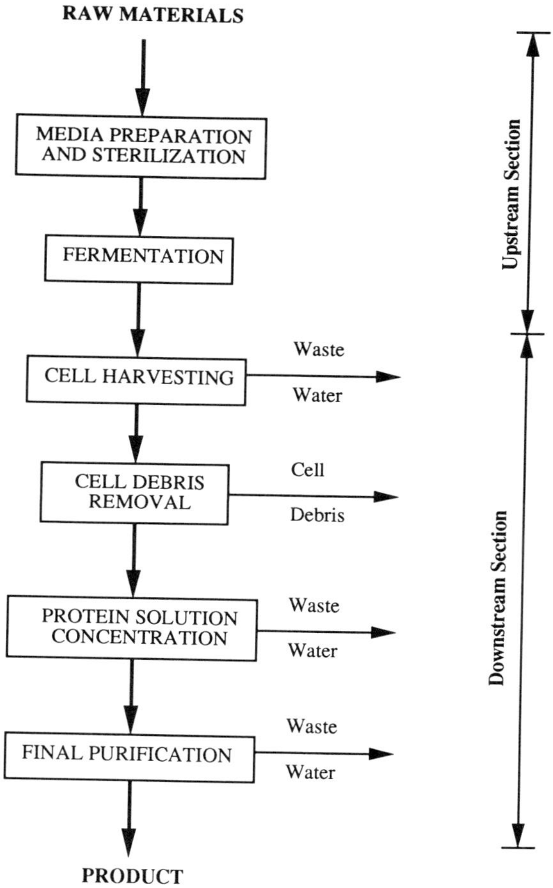

Figure 24.3 Production of β-galactosidase: process block diagram.

of cells by a high pressure homogenizer is required to release the product. After cell disruption, the cell debris particles are removed by a combination of a disk-stack centrifuge and a dead-end polishing filter. The high resolution stages include two ultrafiltration units for protein solution concentration and two chromatography units for product purification. More details for the various unit operations are presented in the description of the process flowsheet (Figure 24.4).

24.3.4.2 Process flowsheet. Fermentation media are prepared in a stainless steel tank, V-101, and are continuously sterilized in STR101. The filling of the production fermenter, FRM101, with sterilized media takes 4 h. Two

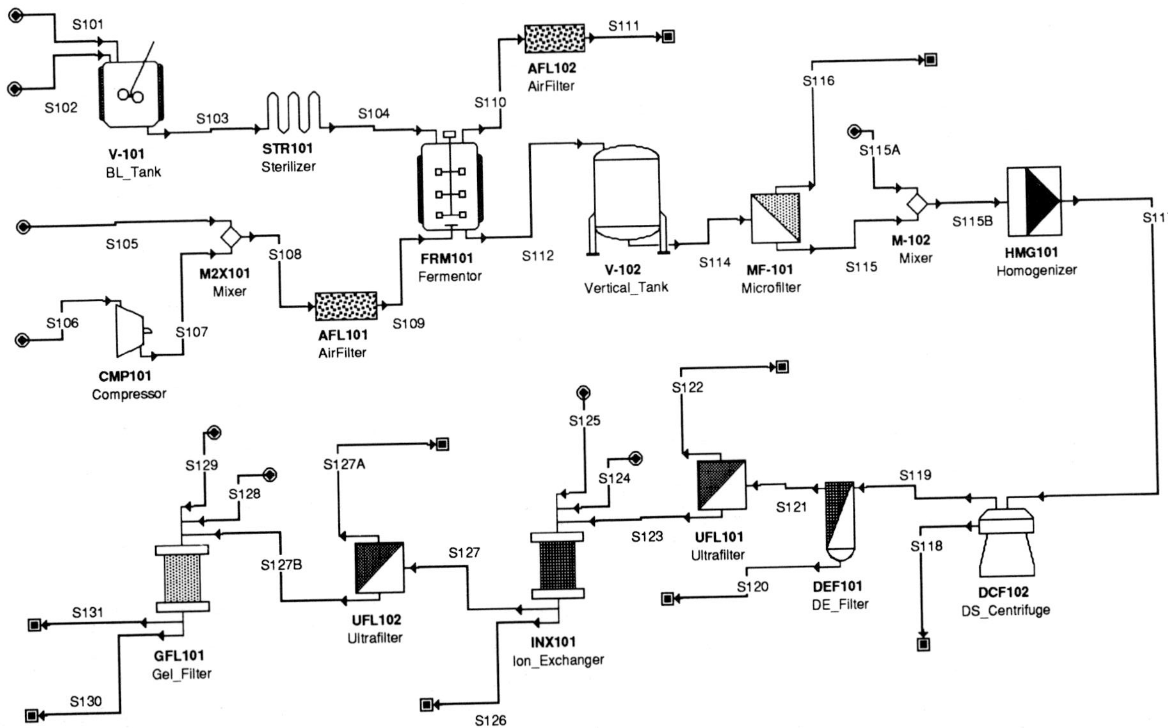

Figure 24.4 Production of β-galactosidase: process flowsheet.

seed fermenters (not shown on the flowsheet), one of 0.57 m^3 and the other of 5.7 m^3 total volume respectively, are used to provide inoculum for the production fermenter of 57 m^3. An axial compressor, CMP101, and an absolute filter, AFL101, provide sterile air to the fermenters at an average rate of 0.4 VVM. The fermentation time in the production fermenter is about 18 h and the turn around time is 12 h. The final concentration of *E. coli* in the FRM101 is about 35 g/l (dcw). The fermentation temperature is 37°C. After the end of fermentation, the broth is cooled to 5°C by a heat exchanger while being transferred to a holding tank, V-102. This tank isolates the upstream from the downstream section of the plant and serves as a buffer to allow semi-continuous operation in the downstream section.

The first step of the downstream section is cell harvesting. It is carried out in a hollow-fiber membrane microfilter, MF-101, of 0.45 μm pore diameter. A two-fold volumetric reduction is achieved in two cycles of 4 h each. The concentrated broth is mixed in line mixer, M-102, with a buffer that stabilizes the product after its release. A high pressure homogenizer, HMG101, breaks the cells and releases the intracellular product. The broth undergoes two passes through HMG101 at a pressure drop of 800 bars. Cell debris particles, created by homogenization, are removed by a disk-stack centrifuge, DCF102, over a period of 8 h. A dead-end polishing filter, DEF101, removes any remaining particulate material, which can block the chromatography units and impair the final purification.

The dilute protein solution is concentrated by a hollow-fiber membrane ultrafilter, UFL101, of 100 000 molecular weight cut-off to a total protein concentration of 5–6% w/w. The concentration is accomplished in two 8 h cycles (6 h process time + 2 h turn around time). The ultrafiltration unit significantly reduces the amount of liquid that needs to be processed by the expensive ion exchange column, INX101. It also facilitates product purification because low molecular weight proteins and other solutes pass through the membrane, while the product remains in the concentration stream (S123). INX101 is used for high resolution purification. Anion exchange (fast flow mono Q from Pharmacia) is needed because β-galactosidase is negatively charged in the range of pH where it is stable. Four 6-h cycles are required. The ion exchange step dilutes the protein solution, so another ultrafilter, UFL102, is used to reconcentrate it to approximately 5% w/w. The second concentration also requires two 8-h cycles (6 h process time + 2 h turn around time). The final purification to 99.5% purity is done in a gel filter, GFL101 using Sephadex G25 or Sepharose CL resin (from Pharmacia). Gel filtration is accomplished in six 4-h cycles. The final product is a solution. If a solid form is desired, a lyophilization (freeze drying) step must be added.

In this example, it was assumed that β-galactosidase (at the level of 10% of total protein) is produced by *E. coli* in its native form. If β-galactosidase formed inclusion bodies, flowsheet modifications would be necessary to handle the new process requirements. Specifically, after the separation of inclusion bodies from cell debris, solubilization and refolding of the product would be required prior to final purification.

24.3.5 Process scheduling

Figure 24.5 shows a process schedule specified by the user and displayed by the program. The Gantt chart displays a bar for each process step. The dark part of a bar represents process time while the gray part represents turn around time. Multiple rectangles (e.g. INX101 and GFL101) represent multiple cycles of that process step within the same plant batch each time. The cycle times of both adjacent process step and overall batches may overlap. The scheduled plant batch time of this process is 72 h (three days). The base case assumes two batches per week or a total of 96 batches per year. The process scheduling information specified by the user is used by the program in calculating the material and energy balances as well as in the sizing and costing of equipment.

24.3.6 Material balances

Using computer-aided design tools, the engineer specifies only the flow rates of the feed streams to the flowsheet. The program estimates the flow rates and compositions of all other streams. The results of the material balances

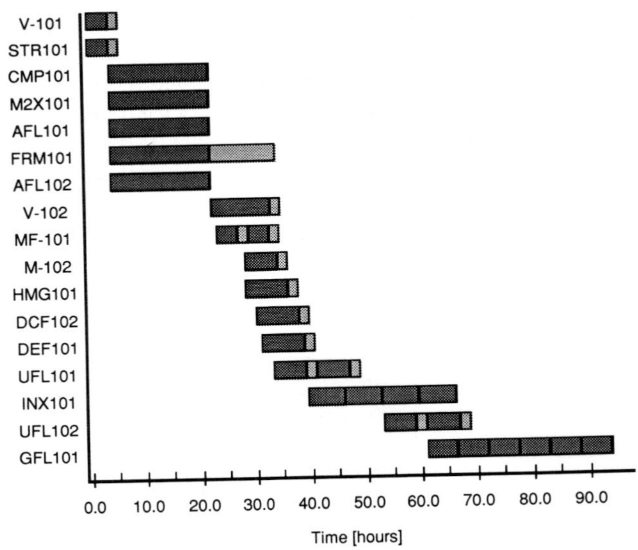

Figure 24.5 Process scheduling Gantt chart, plant batch time is 96 h and there are 80 batches per year.

are shown in Table 24.1. The flow rate of 'biomass' is given on a dry basis. The components 'salts' represents all other nutrients besides 'glucose' and 'NH_3'. Figure 24.6 shows the overall material balance per batch. Note the high amount of waste water (including all the chromatography buffers) that is generated. The overall product recovery yield is about 72%.

24.3.7 *Process economics*

24.3.7.1 Capital cost estimation. Table 24.2 lists the purchase cost of all the equipment that is included in the flowsheet. To estimate the purchase cost such programs make use of curve-fitted data from a number of equipment vendors. Note that fermenters and compressors are the most expensive pieces of equipment in the upstream section, while centrifuges and chromatography units dominate the downstream section. Table 24.3 provides a summary of the fixed capital investment. The total fixed capital investment for a plant of this capacity (5000 kg/yr) is about US$108.7 million. The user has the option to modify the multipliers that are used to estimate the various items of the fixed capital investment.

24.3.7.2 Operating cost estimation. Table 24.4 provides a summary of the total annual cost which, including depreciation, is about US$30.3 million. Figure 24.7 shows a breakdown of the annual operating cost. Note that the direct fixed capital-dependent cost items dominate the annual operating cost (73.5%). Therefore, efforts to minimize the production should focus on better utilization of equipment. The second most important item is the labor cost (10.4% or 14.8% including administration and overhead). Tables 24.5 to 24.9 present in more detail the various items of operating cost.

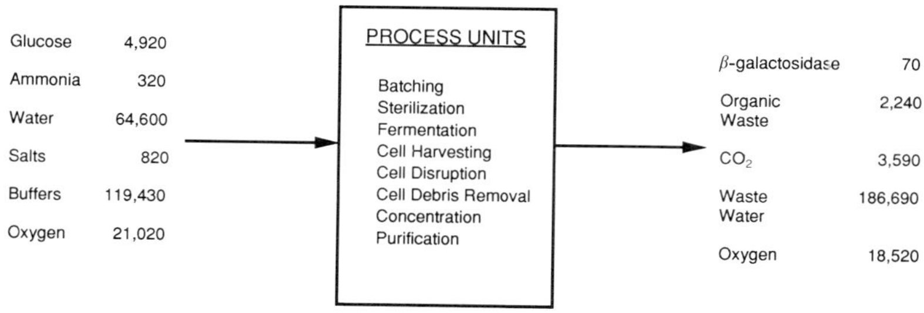

Σ**in = Σout = 211,110 kg/batch**

Figure 24.6 Overall material balances (kg/batch).

Table 24.1 Stream report (batch time = 96 h)

Stream name	S101	S102	S103	S104	S105
Source	Feed	Feed	V-101	STR101	Feed
Destination	V-101	V-101	STR101	FRM101	M2X101
Stream properties					
Activity (U/ml)	0.0	0.0	0.0	0.0	0.0
Temperature (°C)	25.0	25.0	25.0	37.0	25.0
Pressure (bar)	1.0	1.0	1.0	1.0	1.0
Component flowrates (kg/batch)					
β-Gal	0.0000	0.0000	0.0000	0.0000	0.0000
Debris	0.0000	0.0000	0.0000	0.0000	0.0000
H_2O	51 266.0312	0.0000	51 266.0312	51 266.0312	0.0000
CO_2	0.0000	0.0000	0.0000	0.0000	0.0000
Glucose	0.0000	4915.9209	4915.9209	4915.9209	318.8333
NH_3	0.0000	0.0000	0.0000	0.0000	0.0000
Salts	0.0000	821.6611	821.6611	821.6611	0.0000
Protein	0.0000	0.0000	0.0000	0.0000	0.0000
Biomass	0.0000	0.0000	0.0000	0.0000	0.0000
Total	51 266.0314	5737.5821	57 003.6134	57 003.6134	318.8333

Stream name	S106	S107	S108	S109	S110
Source	Feed	CMP101	M2X101	AFL101	FRM101
Destination	CMP101	M2X101	AFL101	FRM101	AFL102
Stream properties					
Activity (U/ml)	0.0	0.0	0.0	0.0	0.0
Temperature (°C)	25.0	40.0	39.9	39.9	37.0
Pressure (bar)	1.0	6.0	3.5	3.5	2.2
Component flowrates (kg/batch)					
β-Gal	0.0000	0.0000	0.0000	0.0000	0.0000
Debris	0.0000	0.0000	0.0000	0.0000	0.0000
H_2O	0.0000	0.0000	0.0000	0.0000	0.0000
CO_2	0.0000	0.0000	0.0000	0.0000	3593.9316
Glucose	0.0000	0.0000	0.0000	0.0000	0.0000
NH_3	0.0000	0.0000	318.8333	318.8333	34.5947
O_2	21 026.0957	21 026.0957	21 026.0957	21 026.0957	18 525.7598
N_2	69 216.1641	69 216.1641	69 216.1641	69 216.1641	69 216.1641
Salts	0.0000	0.0000	0.0000	0.0000	0.0000
Protein	0.0000	0.0000	0.0000	0.0000	0.0000
Biomass	0.0000	0.0000	0.0000	0.0000	0.0000
Total	90 242.2598	90 242.2598	90 561.0931	90 561.0931	91 370.4512

Stream name	S111	S112	S114	S115	S115A
Source	AFL102	FRM101	V-102	MF-101	Feed
Destination	Product	V-102	MF-101	M-102	M-102
Stream properties					
Activity (U/ml)	0.0	0.0	0.0	0.0	0.0
Temperature (°C)	37.0	37.0	37.0	25.0	25.0
Pressure (bar)	2.2	2.2	2.2	2.2	1.0
Component flowrates (kg/batch)					
β-Gal	0.0000	0.0000	0.0000	0.0000	0.0000
Debris	0.0000	0.0000	0.0000	0.0000	0.0000
H_2O	0.0000	53 894.8359	53 894.8539	23 017.5000	13 333.3330
CO_2	3593.9316	0.0000	0.0000	0.0000	0.0000
Glucose	0.0000	98.3183	98.3183	27.6664	0.0000
NH_3	34.5947	0.0000	0.0000	0.0000	0.0000
O_2	18 525.7598	0.0000	0.0000	0.0000	0.0000
N_2	69 216.1641	0.0000	0.0000	0.0000	0.0000
Salts	0.0000	147.1968	147.1968	41.4206	0.0000
Protein	0.0000	0.0000	0.0000	0.0000	0.0000
Biomass	0.0000	2053.9048	2053.9048	2053.9048	0.0000
Total	91 370.4512	56 194.2564	56 194.2564	25 140.4920	13 333.3334

Table 24.1 Continued

Stream name	S115B	S116	S117	S118	S119
Source	M-102	MF-101	HMG101	DCF102	DCF102
Destination	HMG101	Product	DCF102	Product	DEF101
Stream properties					
Activity (U/ml)	0.0	0.0	2270.8	0.0	2364.7
Temperature (°C)	25.0	25.0	15.0	25.0	25.0
Pressure (bar)	1.6	2.2	1.6	1.0	1.3
Component flowrates (kg/batch)					
β-Gal	0.0000	0.0000	96.9324	14.2892	82.6432
Debris	0.0000	0.0000	612.9314	600.6729	12.2586
H_2O	36 350.8320	30 877.3398	36 350.8320	6068.3208	30 282.5156
CO_2	0.0000	0.0000	0.0000	0.0000	0.0000
Glucose	27.6664	70.6519	27.6664	4.0784	23.5880
Salts	41.4206	105.7762	41.4206	6.1060	35.3146
Protein	0.0000	0.0000	924.6200	136.3022	788.3179
Biomass	2053.9048	0.0000	10.7998	134.9977	−124.1979
Nucleic acids	0.0000	0.0000	408.6210	60.2366	348.3844
Total	38 473.8254	31 053.7644	38 473.8254	7025.0036	31 448.8219

Stream name	S121	S122	S123	S124	S125
Source	DEF101	UFL101	UFL101	Feed	Feed
Destination	UFL101	Product	INX101	INX101	INX101
Stream properties					
Activity (U/ml)	2233.7	0.0	7922.3	0.0	0.0
Temperature (°C)	25.0	25.0	25.0	25.0	25.0
Pressure (bar)	1.3	1.3	1.3	1.0	1.0
Component flowrates (kg/batch)					
β-Gal	81.8168	0.0000	79.3623	0.0000	0.0000
Debris	0.0000	0.0000	0.0000	0.0000	0.0000
H_2O	31 697.4336	23 375.7285	8321.7070	0.0000	0.0000
CO_2	0.0000	0.0000	0.0000	0.0000	0.0000
Glucose	24.1629	17.8193	6.3436	0.0000	0.0000
Salts	36.1754	26.6781	9.4973	0.0000	0.0000
Protein	780.4346	274.5195	508.3696	0.0000	0.0000
Biomass	0.0000	0.0000	0.0000	0.0000	0.0000
Inx-buffer	0.0000	0.0000	0.0000	23 640.0566	35 460.0859
Nucleic acids	344.9005	254.3519	90.5487	0.0000	0.0000
Total	32 964.9250	23 949.0963	9015.8287	23 640.0564	35 460.0846

Stream name	S126	S127	S127A	S127B	S128
Source	INX101	INX101	UFL102	UFL102	Feed
Destination	Product	UFL102	Product	GFL101	GFL101
Stream properties					
Activity (U/ml)	56.4	14 058.8	0.0	41 330.9	0.0
Temperature (°C)	25.0	25.0	25.0	25.0	25.0
Pressure (bar)	1.1	1.2	1.2	1.2	1.0
Component flowrates (kg/batch)					
β-Gal	3.9681	75.3941	0.0000	73.8863	0.0000
Debris	0.0000	0.0000	0.0000	0.0000	0.0000
H_2O	8321.7070	0.0000	0.0000	0.0000	0.0000
Glucose	6.3436	0.0000	0.0000	0.0000	0.0000
Salts	9.4973	0.0000	0.0000	0.0000	0.0000
Protein	485.4929	22.8766	0.0000	24.3845	0.0000
Biomass	0.0000	0.0000	0.0000	0.0000	0.0000
Inx-buffer	54 372.1289	4728.0112	3217.4458	1510.5654	0.0000
Gfl-buffer	0.0000	0.0000	0.0000	0.0000	20 111.3555
Nucleic acids	90.3223	0.2264	0.1540	0.0723	0.0000
Total	63 289.4613	4826.5085	3217.6000	1608.9085	20 111.3561

Table 24.1 Continued

Stream name Source Destination	S129 Feed GFL101	S130 GFL101 Product	S131 GFL101 Product
Stream properties			
Activity (U/ml)	0.0	57.5	15 435.4
Temperature (°C)	25.0	25.0	25.0
Pressure (bar)	1.0	1.0	1.1
Component flowrates (kg/batch)			
β-Gal	0.0000	3.6943	70.1919
Debris	0.0000	0.0000	0.0000
H_2O	0.0000	0.0000	0.0000
CO_2	0.0000	0.0000	0.0000
Glucose	0.0000	0.0000	0.0000
Salts	0.0000	0.0000	0.0000
Protein	0.0000	24.1407	0.2438
Biomass	0.0000	0.0000	0.0000
Inx-buffer	0.0000	1510.5654	0.0000
Gfl-buffer	40 222.7109	56 311.7969	4022.2712
Nucleic acids	0.0000	0.0723	0.0000
Total	40 222.7123	57 850.2698	4092.7071

Overall material balance (kg/batch)
(hours per batch = 96)

Component	In	Out	Δ (out–in)
β-Gal	0.000000	92.970023	92.970023
Debris	0.000000	612.931430	612.931430
H_2O	64 599.364800	67 228.172579	2628.807779
CO_2	0.000000	3593.931431	3593.931431
Glucose	4915.920960	98.318325	−4817.602635
NH_3	318.833280	34.594728	−284.238552
O_2	21 026.095680	18 525.759962	−2500.335718
N_2	69 216.165120	69 216.165120	0.000000
Salts	821.661120	147.196782	−674.464338
Protein	0.000000	928.582360	928.582360
Biomass	0.000000	10.799818	10.799818
Inx-buffer	59 100.141082	59 100.141082	0.000000
Gfl-buffer	60 334.068418	60 334.068418	0.000000
Nucleic acids	0.000000	408.620953	408.620953
Total	280 332.250460	280 332.253010	0.002550

24.3.7.3 Profitability analysis. Table 24.10 provides a summary of the profitability analysis. The total investment, including working capital, is approximately US\$109.5 million. The production cost of β-galactosidase for a plant of this capacity is about US\$5500/kg. For a selling price of US\$12 000/kg, the project is marginally profitable with a return on investment of 29%, a payback time of 3.45 years and an internal rate of return (IRR) after taxes of 13.5%. The net present value (NPV) of the project is 45.8, 26.2, and −10.3 million US dollars for discount interest of 8, 10 and 12%, respectively. Table 24.11 shows the results of the cash flow analysis for the

Table 24.2 Major equipment specification and fob cost (1992 prices)

Quantity/stand-by	Description	Unit cost (US$)	Cost (US$)
1/0 V-101	Blending tank Vol = 83.83 m³, SS304, 17, 81 kW	63 000	63 000
1/0 STR101	Continuous sterilizer Capacity = 17.81 m³/h	352 000	352 000
1/1 CMP101	Centrifugal compressor DPress = 5.0 bar, SS316, 510.24 kW	934 000	1868 000
1/0 AFL101	Cartridge air filter air inlet, SS316 housing	4000	4000
1/0 FRM101	Agitated fermenter Vol = 95.01 m³, SS316, 213.76 kW	510 000	510 000
1/0 FRM101	Agitated seed fermenter Vol = 9.50 m³, SS316, 21.38 kW	168 000	168 000
1/|0 FRM101	Agitated seed fermenter Vol = 0.95 m³, SS316, 2.14 kW	42 000	42 000
1/0 AFL102	Cartridge air filter exhaust gas, SS316 housing	11 000	11 000
3/0 V-102	Vertical storage tank Vol = 27.55 m³, SS304, 1.01 bar	14 000	42 000
3/0 MF-101	Membrane microfilter Area = 73.0 m², 14.61 kW	76 000	228 000
1/0 HMG101	High pressure homogenizer Cap = 11 992.5 l/h, 266.50 kW	50 000	50 000
14/0 DFC102	Disc-stack centrifuge Σ = 2 889 532 m², 84.04 kW	321 000	4 494 000
1/0 DEF101	Dead-end filter Area = 34.3 m², Pore size = 0.45 μm	26 000	26 000
2/0 UFL101	Membrane ultrafilter Area = 62.6 m², 12.53 kW	73 000	146 000
19/0 INX101	Ion exchange unit $L = 0.25$ m, $D = 1.00$ m	178 000	3 382 000
1/0 UFL102	Membrane ultrafilter Area = 16.8 m², 3.35 kW	35 000	35 000
11/0 GFL101	Gel filtration unit $L = 0.50$ m, $D = 0.98$ m	181 000	1 991 000
	Cost of unlisted equipment 20.0% of total		3 353 000
Total equipment purchase cost			16 765 000

entire life of the project. A number of assumptions were made for the above analysis: first, a 15 year project horizon, second, a 10 year straight-line depreciation with a 5% salvage value, third, a 30 month plant construction, fourth, 40% income taxes, fifth, 40% of direct fixed capital is financed at 12%, sixth, 100% of working capital is financed at 15%.

24.3.8 Optimization and sensitivity analysis

After a model for the entire process has been developed on the computer, the user can use such tools to ask and readily answer 'what if' questions and

Table 24.3 Fixed capital estimate summary (1992 prices)

Total plant direct cost (TPDC) (physical cost)		
1. Equipment purchase cost	(PC)	US$16 765 000
2. Installation	(0.49 × PC)	8277 000
3. Process piping	(0.45 × PC)	7544 000
4. Instrumentation	(0.45 × PC)	7544 000
5. Insulation	(0.03 × PC)	503 000
6. Electrical	(0.10 × PC)	1676 000
7. Buildings	(0.45 × PC)	7544 000
8. Yard improvement	(0.15 × PC)	2515 000
9. Auxiliary facilities	(0.40 × PC)	6706 000
		TPDC = 59 074 000
Total plant indirect cost (TPIC)		
10. Engineering	(0.25 × TPDC)	14 768 000
11. Construction	(0.35 × TPDC)	20 676 000
		TPIC = 35 444 000
Total plant cost (TPDC + TPIC)		TPC = 94 518 000
12. Contractor's fee	(0.05 × TPC)	4 726 000
13. Contingency	(0.10 × TPC)	9 452 000
		Σ(12 + 13) = 14 178 000
Direct fixed capital (DFC)	TPC + 12 + 13	108 696 000

Table 24.4 Annual operating cost (1992 prices)

DFC-dependent items (DFC = US$108 696 000)		
Depreciation		US$10 326 000
Maintenance material	(0.03 × DFC)	3261 000
Insurance	(0.01 × DFC)	1087 000
Local taxes	(0.02 × DFC)	2174 000
Factory expense	(0.05 × DFC)	5435 000
		22 283 000
Labor-dependent items		
a. Operating labor	(31 802 h × 24.0 $/h)	763 000
b. Maintenance labor	(0.01 × DFC)	1087 000
c. Fringe benefits	(0.40 × (a + b))	740 000
d. Supervision	(0.20 × (a + b))	370 000
e. Operating supplies	(0.10 × a)	76 000
f. Laboratory	(0.15 × a)	114 000
		3150 000
Administration and overhead expense	(0.6 × (a + b + c))	1332 000
Raw materials		1487 000
Other consumables		1705 000
Utilities		238 000
Waste treatment		117 000
Running R&D		0
Running royalties		0
Sales cost		0
Total annual operating cost		
including depreciation		30 312 000
excluding depriciation		19 986 000

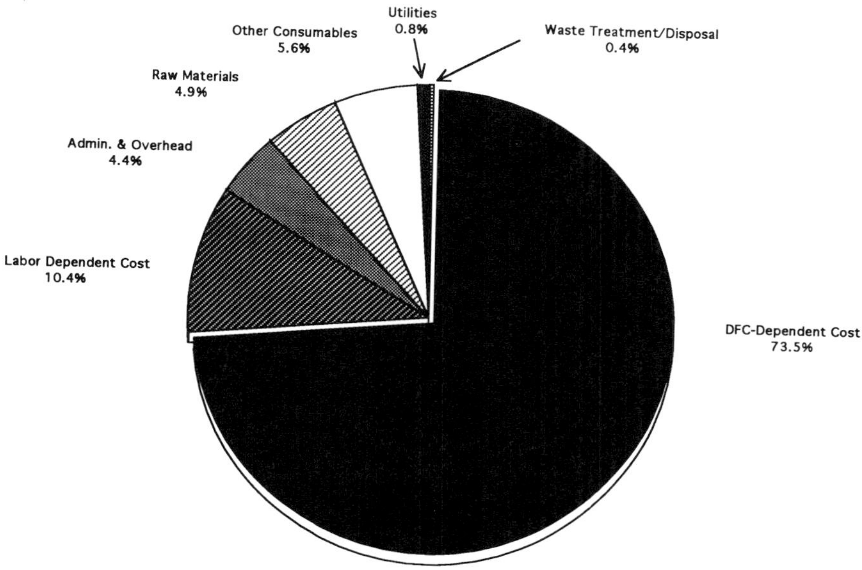

Figure 24.7 Breakdown of the annual operating cost.

Table 24.5 Labor requirement estimate summary

Equipment name	Number of units	Operator hours (per year)
Warehouse		2000.0
V-101	1	96.0
STR101	1	48.0
CMP101	1	14.4
M2X101	1	14.4
AFL101	1	14.4
FRM101	1	240.0
AFL102	1	14.4
V-102	3	156.0
MF-101	3	288.0
M-102	1	6.4
HMG101	1	80.0
DCF102	14	1120.0
DEF101	1	64.0
UFL101	2	256.0
INX101	19	12 312.0
UFL102	1	128.0
GFL101	11	8949.6
Packaging		2000.0
Other		4000.0
Total		31 801.6

Table 24.6 Raw materials cost (1992 prices)

Component	Unit cost (US$/kg)	Annual amount (kg)	Cost (US$/yr)
H_2O	1.000e − 03	5167 949.50	5000
Glucose	4.500e − 01	393 273.69	177 000
NH_3	3.000e − 01	25 506.66	8000
Salts	1.600e + 00	65 732.89	5000
Inx-buffer	1.500e − 01	4728 011.00	709 000
Gfl-buffer	1.000e − 01	4826 725.50	483 000
Total			1487 000

Table 24.7 Other consumables cost (1992 prices)

Equipment	Unit cost (US$/m²)	Annual amount (m²)	Cost (US$/yr)
Membrane or filter cloth			
MF-101	200.00	105.18	21 000
DEF101	180.00	381.54	69 000
UFL101	200.00	80.18	16 000
UFL102	200.00	10.73	2000
Subtotal			108 000
Chromatography resins			
INX101	200.00	2955.01	591 000
GFL101	200.00	5027.84	1006 000
Subtotal			1597 000
Total			1705 000

Table 24.8 Waste treatment/disposal cost (1992 prices)

Waste component	Unit cost (US$/kg)	Annual amount (kg)	Cost (US$/yr)
Debris	1.000e + 00	49 035	49 000
H_2O	2.000e − 04	5378 254	1000
Glucose	8.000e − 01	7865	6000
Protein	8.000e − 01	74 287	59 000
Biomass	1.000e + 00	864	1000
Total			117 000

carry out sensitivity analysis with respect to key design variables. In this case study, the effect of a number of parameters on the production cost was examined.

24.3.8.1 Effect of cell harvesting. The main objective of cell harvesting by microfilter MF-101 is to reduce the volume of broth and consequently reduce the size and cost of equipment further downstream. However, there is also a negative effect associated with cell harvesting. In the removal of cell debris

Table 24.9 Utility requirements (1992 prices)

Equipment	Power (kW/unit)	Annual amount (kWh)	Cost (US$/yr)
Electricity (US$ 0.07/kWh)			
V-101	17.8	8551	1000
CMP101	510.2	734 750	51 000
FRM101	235.1	564 336	40 000
MF-101	14.6	42 072	3000
HMG101	266.5	213 200	15 000
DCF102	84.0	941 242	66 000
UFL101	12.5	32 074	2000
UFL102	3.4	4290	0
Unlisted equipment (5.0% of total)			10 000
General load (8.0% of total)			16 000
Subtotal			204 000
Equipment	**Duty (kcal/h/unit)**	**Annual amount (10^6 kcal)**	**Cost (US$/yr)**
Chilling (US$25.0/$10^6$ kcal)			
V-101	−15 341.9	−7	0
MF-101	−14 892.8	−43	1000
HMG101	−232 981.5	−186	5000
DCF102	−71 919.5	−805	20 000
UFL101	−10 764.7	−28	1000
UFL102	−2879.8	−4	0
Subtotal			27 000
Steam (US$4.40/1000 kg)			
STR101	562 535.6	270	2000
Subtotal			2000
Equipment	**Duty (kcal/h/unit)**	**Annual amount (1000 kg CW)**	**Cost (US$/yr)**
Cooling water (25°C → 35°C) (US$0.04/1000 kg)			
STR101	−320 645.3	15 391	1000
CMP101	−207 107.2	29 823	1000
FRM101	−311 568.2	74 776	3000
Subtotal			5000
Total			238 000

particles by the centrifuge DCF102, the solids in the heavy phase comprise only about 20–30% of total volume. The remainder is liquid, with dissolved product protein. The higher the extent of cell harvesting, the higher the product protein concentration in stream S117 (feed to centrifuge) and the higher the product loss in the heavy phase of the centrifuge. The cost effect of cell harvesting dominates in this case. The result may change if the flowsheet is modified by dilution and recentrifugation of the heavy phase of the centrifuge (stream S118) to recover some of the lost product.

Table 24.10 Profitability analysis (1992 prices)

A. Direct fixed capital	US$108 696 000
B. Working capital	836 000
C. Total investment (A + B)	109 532 000
D. Production rate (kg/year)	
β-Gal	5503.048
E. Production cost (US$/kg)	
β-Gal	5508.220
F. Selling price (US$/kg)	
β-Gal	12 000.000
G. Revenue (US$/year)	
β-Gal	66 037 000
H. Annual operating cost	30 312 000
I. Gross profit (G − H)	35 725 000
J. Taxes (40%)	14 290 000
K. Net profit (I − J + depreciation)	31 761 000
Gross margin	0.54
Return on investment	29%
Payback time (years)	3.45

24.3.8.2 Effect of dilution after cell harvesting. An analysis related to the previous one is the effect of dilution after cell harvesting (flow rate of stream S115A) on process economics. Dilution of the concentrated broth with a buffer solution after cell harvesting but prior to cell disruption ensures the stability of the product after its release by cell disruption. Dilution can have both a negative and a positive effect on process economics. The higher the dilution, the larger the downstream equipment size and cost, and consequently the higher the production cost. However, the higher the dilution, the lower the product loss in centrifuge DCF102 and consequently the lower the

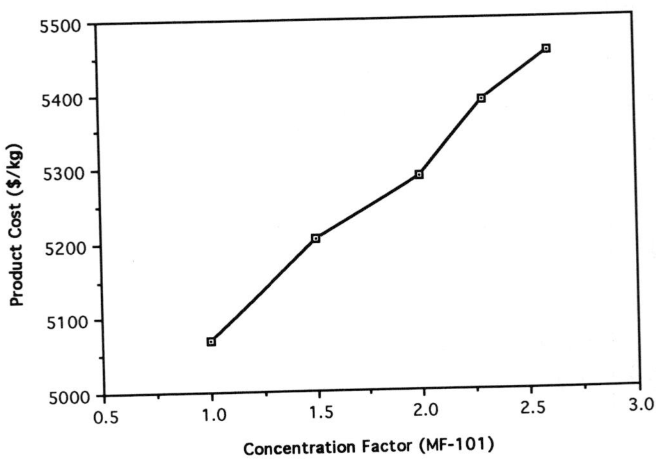

Figure 24.8 Effect of cell harvesting on process economics.

Table 24.11 Cash flow analysis (1992 prices)*

Year	Capital outlay	Sales revenues	Operating cost	Gross profit	Loan payment	Depreciation	Taxable income	Taxes 40.0%	Net profit	Net cash-flow
1	−32 609	0	0	0	0	0	0	0	0	−32 609
2	−43 478	0	0	0	0	0	0	0	0	−43 478
3	−33 445	16 509	14 963	1546	7916	10 326	−16 696	0	−6370	−39 815
4	0	33 018	16 638	16 381	7916	10 326	−1861	0	8465	8465
5	0	52 830	18 647	34 183	7916	10 326	15 941	6376	19 891	19 891
6	0	52 830	18 647	34 183	7916	10 326	15 941	6376	19 891	19 891
7	0	66 037	19 986	46 051	7916	10 326	27 809	11 124	27 012	27 012
8	0	66 037	19 986	46 051	7916	10 326	27 809	11 124	27 012	27 012
9	0	66 037	19 986	46 051	7695	10 326	28 030	11 212	27 144	27 144
10	0	66 037	19 986	46 051	7695	10 326	28 030	11 212	27 144	27 144
11	0	66 037	19 986	46 051	7695	10 326	28 030	11 212	27 144	27 144
12	0	66 037	19 986	46 051	7695	10 326	28 030	11 212	27 144	27 144
13	0	66 037	19 986	46 051	0	0	46 051	18 420	27 631	27 631
14	0	66 037	19 986	46 051	0	0	46 051	18 420	27 631	27 631
15	6271	66 037	19 986	46 051	0	0	46 051	18 420	27 631	33 901

IRR before taxes = 0.193
IRR after taxes = 0.135

Interest	0.080	0.100	0.120
Net present value	45 792	26 254	10 343

	Direct fixed capital	Working capital	Up front R&D	Up front royalties
Amount	108 696	836	0	0
Equity (%)	60.0	0.0	80.0	50.0
Debt (%)	40.0	100.0	20.0	50.0
Interest (%)	12.0	15.0	15.0	15.0
Loan time (yrs)	10.0	6.0	6.0	6.0

Depreciation method: straight-line

* The above amounts are given in thousands of US dollars.

production cost. The results of the effect of dilution on process economics are shown in Figure 24.9. As can be seen, the combined effect of dilution initially is positive but after a certain level (20 000 l/batch) it becomes negative.

24.3.8.3 Effect of protein solution concentration. After cell debris removal, the dilute protein solution is concentrated by ultrafiltration in UFL101 and UFL102 to reduce the volume of material to be processed by the expensive chromatography units. In this case study, the effect of protein solution concentration by UFL102 on process economics was analyzed. The higher the concentration factor, the lower the capital and operating cost of the gel filtration unit, GFL101. The results of the analysis are shown in Figure 24.10. The *y*-axis on the left shows that the product cost is reduced drastically by increasing the concentration factor. The *y*-axis on the right shows the total protein solution concentration (as % w/w) in the feed stream to gel filtration. Chromatography manufacturers recommend that the total protein concentration in the feed stream to a chromatography unit should not exceed 5% w/w. This information can be used to set an upper limit to the extent of concentration.

24.3.8.4 Effect of process scheduling. In the base case, the plant batch time was set at 72 h (see Figure 24.5) and 96 batches were run per year. This represents less than two batches per week and is a rather conservative assumption. Since the direct fixed capital-dependent cost is the dominant component of the production cost (Figure 24.7), increasing the number of batches per year through better equipment utilization will provide significant savings. The results of the analysis are shown in Figure 24.11. Savings of up

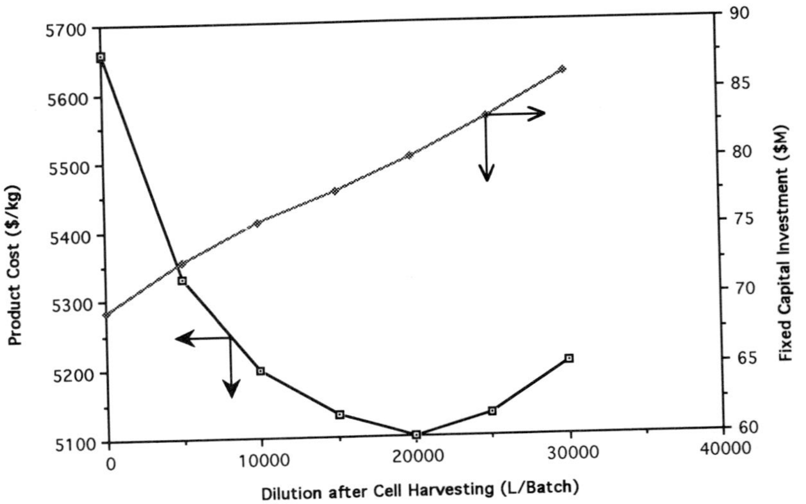

Figure 24.9 Effect of dilution after cell harvesting on process economics.

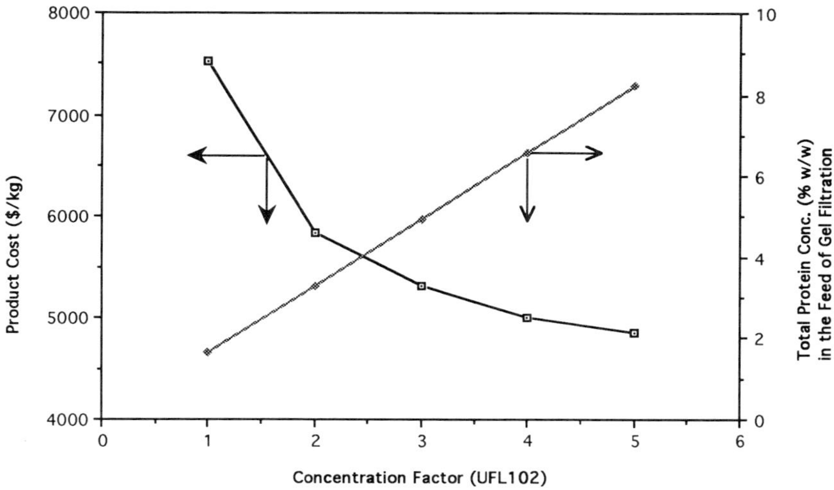

Figure 24.10 Effect of protein solution concentration on process economics.

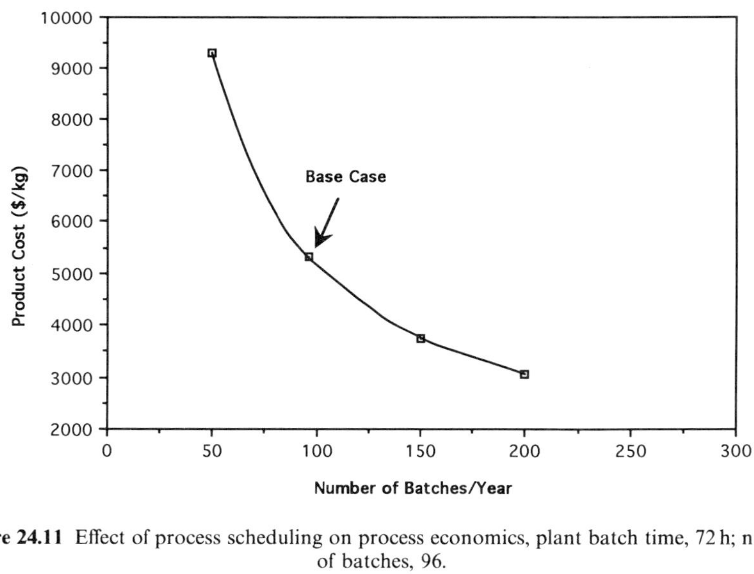

Figure 24.11 Effect of process scheduling on process economics, plant batch time, 72 h; number of batches, 96.

to 40% of the production cost can be realized by doubling the number of batches (overlapping batches are feasible in this case) per year.

24.3.8.5 Effect of the economy of scale. To examine the economy of scale, the production cost of β-galactosidase was estimated as a function of plant

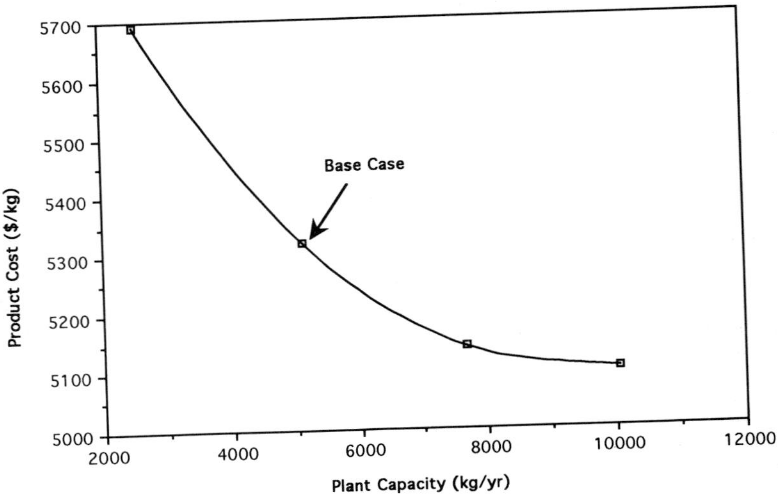

Figure 24.12 Effect of the economy of scale on process economics.

capacity (Figure 24.12). By doubling the plant capacity but keeping the process schedule the same, the production cost is reduced only by 5%, showing that economy of scale does not play a significant role in this case because higher production requires multiple production lines in parallel.

24.4 Conclusions

Computer-aided bioprocess design tools can play a major role in the development of biochemical processes and commercialization of biological products. At the early stages of project selection, such tools can be used to screen the large number of project ideas from a profitability viewpoint and help focus development efforts on the most promising projects. During process development, design tools can be used to analyze and evaluate alternative processing schemes, interpret experimental results, and help design experimental protocols. During final design and plant construction, they can be used to optimize the entire process from a total-systems point of view. All these benefits substantially reduce cost and development time associated with a biological product.

References

1. D. Petrides, C.L. Cooney, L.B. Evans, R.P. Field and M. Snoswell (1989). Bioprocess simulation: an integrated approach to process development, *Computers and Chem. Eng.*, **13**, 553–561.

2. L.B. Evans and R.P. Field (1988). Bioprocess simulation: a new tool for process development, *BioTechnol.*, **6**, 200–224.
3. Intelligen, Inc. (1993). *Manual of BioPro Designer — An Advanced Computing Environment for Designing Integrated Biochemical Processes.*
4. C.A. Siletti (1988). *Computer-Aided Design of Protein Recovery Processes*, Ph.D. Thesis, Massachusetts Institute of Technology.
5. D. Petrides, K.G. Abeliotis and S.K. Mallick (1994). EnviroCAD: a design tool for efficient synthesis and evaluation of integrated waste recovery, treatment and disposal processes, *Computers & Chem. Eng.*, **18**, S603–S607.
6. C.L. Cooney, D. Petrides, M. Barrera and L. Evans (1988). Computer-aided design of a biochemical process, in *The Impact of Chemistry on Biotechnology*, eds M. Phillips, S.P. Shoemaker, R.D. Middlekauff and R.M. Ottenbrite, ACS Symposium Series 362.
7. C.L. Cooney, J.E. Strong, D. Petrides and L. Evans (1987). Computer-aided design of biochemical recovery processes, in *Biochemical Engineering — A Challenge for Interdisciplinary Cooperation*, eds H. Chmiel, W.P. Hammes and J.E. Bailey, Gustav Fischer Verlag, Stuttgart – New York.
8. J.M. Douglas (1988). *Conceptual Design of Chemical Processes*, McGraw-Hill.
9. D.P. Petrides (1990). *Computer-aided Design of Integrated Biochemical Processes; Development of BioDesigner*, Ph.D. Thesis, Department of Chemical Engineering, MIT.
10. D. Petrides (1994). BioPro designer: An advanced computing environment for modeling and design of integrated biochemical processes, *Computer & Chem. Eng.*, **18**, S621–S625.
11. E.M. Rosen and A.C. Pauls (1977). Computer aided chemical process design: the FLOWTRAN system, *Computers & Chem Eng.*, **1**, 11–21.

25 Off-site construction

A. TERGEVORKIAN

In the construction of pharmaceutical and biotechnology facilities, 'off-site construction' refers to that portion of the job which is first assembled and constructed at a shop and then transported to the actual job site. Historically, in process industries, this type of construction is referred to as modular construction. The term 'module' usually refers to skid mounted equipment or packaged equipment. The classical definition of a skid mounted unit is a completely fabricated, assembled, and shop tested unit with all equipment, instruments and controls, piping and tubing, structural supports, and reinforcing steel in place and protected for shipping. There are many examples of off-site construction/modular units. They range from single purpose equipment packages (e.g. packaged boilers) to fully integrated barge transported plants.

In the pharmaceutical and biotechnology industries, off-site construction is by no means limited to skid mounted or packaged-type equipment. The application of off-site construction techniques can range from entire modular clean rooms to reactor vessels and hi-tech laboratories. Some examples of off-site construction or modular units are listed below.

- Clinical trial facilities for introducing new pharmaceutical product.
- Containment facilities which are designed to protect the people, process, and the environment.
- Multiproduct and multiuser facilities which are flexible systems without cross contamination.
- Sterility testing labs which need very high tolerances and environmental controls.
- R&D facilities which convert into modularized systems allow changing equipment configurations, for on-going process development.
- Vivariums, animal rooms requiring even airflow dispersal and minimization of drafts.

The above examples clearly illustrate that the term 'off-site construction' is more suitable for shop fabricated units in pharmaceutical industries than the term 'modular units'. An off-site construction project is an engineering/construction project that is an alternative to traditional design and field construction. The cost savings and schedule benefits of the modular or off-site constructed units are proven. Off-site construction offers high corporate profits and improved ROIs. With proper implementation, off-site construction can provide a competitive advantage.

The pharmaceutical and biotechnology industry, in comparison to the process industry, is far more technology, research, and process driven than it is construction or market driven. The pharmaceutical companies are facing capital limitations and schedule pressures that were almost non-existent a few years ago. Also, they are facing constructability issues, pharmaceutical potency issues, and progressively tightening regulations. With respect to cost and schedule effectiveness of the modular approach and the above mentioned reasons, there is an emerging movement in the industry to build as much of a plant away from the ultimate plant site as is possible. The ultimate configuration of the off-site constructed unit will be defined by the driving force of the particular project and client. It should be noted that the off-site constructed element is preassembled and validated only to the extent required to meet the defined project criteria and driving forces. The construction of a modular unit where everything within the module boundary is 100% completed and tested is no longer critical to the success of the concept.

Some of the driving forces that cause pharmaceutical companies to do more work away from the final plant site are:

- Ever more stringent cleanliness and containment requirements are driving up workmanship, quality, validation, and labor content in pharmaceutical facilities. Emerging automation and computerization in plants are causing increased system sophistication and much more field labor content.
- There is a declining craft skill base, both in terms of number of craftsmen and the skill level of craftsmen, who know how to work in these construction environments.
- Local construction labor costs may be high relative to the assembly labor cost in strategically located assembly facilities.
- As pharmaceutical facilities become more complicated in terms of containment, cleanliness ratings etc., there are restrictive field construction sequences which generally extend construction and commissioning cycles. Simply put, the craftsmen must construct certain parts of the plant in a specific sequence in order that structures, containments, process equipment, systems, instrumentation, and related items be in the right places and at the right time.
- There are strong schedule pressures to enter the marketplace quickly after clinical trials are completed in order to recoup high development costs and limited patent protection windows.
- Because of these factors there is an increasing pressure to reconfigure existing facilities without incurring inordinate cost or extended construction schedule.
- Possible government constraints on pharmaceutical pricing will instill added competitive pressures in this marketplace. Already companies that

were historically cavalier with regard to capital budgets are becoming sensitive to construction costs.

- More stringent government regulatory requirements (the creeping current good manufacturing practices (cGMP) definition) will drive higher quality expectations into every element of the plant, which will drive up sophistication, labor content, and schedule span.
- Reduced start-up cost.
- Facilities where expansion and modifications are achieved by addition of new modular units may experience fewer interruptions of operation.
- Reduced congestion, construction infrastructure, and outside labor on the plant site, all of which can contribute to higher productivity and improved site safety.

25.1 Engineering of off-site construction

The first step taken by the process and project design team is to undertake an initial screening evaluation which is performed during the early stages of process flow diagrams development. The evaluation concentrates on types of equipment used, their sizes, and arrangement constraints. Its purpose is to determine whether most of the individual components can fit within nominal module transportation envelopes. A qualitative assessment of equipment and associated piping may also provide an early estimate of off-site construction incentives, such as cost, schedule, downstream flexibility, etc. This study will establish an 'evaluate more/no go' decision. The next step is for the owner and a qualified module designer/constructor to perform a feasibility study. The following process data should be available:

- Preliminary process flow diagrams
- Preliminary utility flow diagrams
- Conceptual process and instrument diagrams
- Equipment types
- Approximate equipment sizes
- Process arrangement issues
- Plant access issues

The feasibility study produces a module equipment general arrangement drawing. From this information, the design team can study whether a module arrangement can satisfy the process requirements and at the same time be time and cost effective. This will establish a 'go/no go' decision or possibly an 'evaluation more' decision.

Next is the conceptual arrangement study. The information generated during this phase is directly usable in final off-site design provided the engineering data are accurate. The following information is required for this phase:

- Design basis process flow diagrams
- Design basis utility flow diagram
- Preliminary P&IDs (process and instrumental diagrams) (75% complete)
- Peliminary equipment specifications
- Preliminary equipment 'cut sheet'
- Preliminary sizing on 'in-line' devices

With this information, a more complete module layout showing control station locations, pipe routings, operating, and maintenance access is established. A complete review of this layout will establish the ground work for a final detail design of the off-site constructed system where the above mentioned design information is no longer preliminary, but is issued for construction.

Off-site constructed units in the pharmaceutical and biotechnology industries are designed to meet GMP and performance specifications. Engineering for an off-site constructed facility focuses on customizing and optimizing components to meet the needs of the process and the constraints imposed by the facility. Throughout the design stages, there are many design considerations involved including:

- Room size and layout
- People, process, material, and equipment flow
- Access, gown-up, and entry areas
- Utility penetrations and exact locations of headers
- Procedures for operators
- Spill containment
- Doors, windows, pass-throughs
- Architectural requirements
- Temperature, humidity, pressure requirements
- Airflow and direction
- Cleanliness level
- Future expansion or relocation
- Ensuring all equipment and components can be operated safely
- Proper venting system for equipment and pressure safety devices
- Equipment design and layouts meeting the GMP requirements
- Routine operating and maintenance access
- Infrequent maintenance access
- Major turn around access.

All these design considerations should be based on good engineering and operating judgements, and a thoughtful reasoning and logic in applying historic practices. In addition, the following items are to be considered in the same way as any other on-site construction:

- Corporate standards
- Plant standards
- Standard design practices
- Standard specifications
- Standard access criteria
- Plant layout standards

With respect to all these engineering stages and design considerations, it is evident that off-site construction or modular projects demand a higher level of planning, project administration, and control in order to coordinate design efforts. Engineering and design cost is usually 5–10% more in an off-site construction project due to the level of coordination required.

25.2 Off-site construction project cost elements

As mentioned previously, off-site construction historically has proven to be cost effective in comparison to on-site field construction. The following often costs more in off-site modular construction:

- Project management, control and administration,
- Engineering and design in project, structural and modeling,
- Structural steel,
- Freight and shipping preparation.

The cost saving in off-site modular construction is usually in the following categories:

- Quality assurance
- Civil design
- Overhead and benefit burdens
- Building, structural, and architectural
- Construction costs (piping and HVAC (heating, ventilation, and air conditioning))
- Electrical and instrumentation
- Painting and coating
- Insulation
- Indirect and distributed costs
- Owner distributed project costs

Other categories not listed above have approximately the same cost. It is evident that the overall cost savings favor modular construction. However, the off-site/modular engineering costs can be driven up by inability to freeze the process at an approriate point in the design sequence and inadequate planning or project definition resulting in design recycles. With proper management, the cost impact for such items can be prevented resulting in efficient project execution.

25.3 Shipping and installation

Off-site construction units are shipped by truck, rail, barge, or ship depending on the plant site destination and the most easy and practical way of transporting the unit to the plant site. There are some shipping limitations which should be considered in the engineering design stages. For example, for transportation via trucks, they should not be larger than 12–13 ft (3.66–3.96 m) wide, 10–11 ft (3.05–3.35 m) high, and 60 ft (18.29 m) long. Weight limits are typically 80 000 pounds (36 287 kg). Rail transportation extends this envelope to 14–15 ft (4.27–4.57 m) wide, 14–15 ft (4.27–4.57 m) high and over 100 ft (30.48 m) long with no limitation on weight. There are no limitations on water transportation, but generally the cost is much higher. Shipping costs normally run from $\frac{1}{2}$ to 2% of the value of the modular system. Shipping preparation of modular units is an ability based on experience. It is not a science. There are no set specifications or procedures involved. Modules are designed to withstand normal operating loads, as well as rigging stresses and transportation forces.

The installation of off-site constructed units is complex and requires comprehensive erection planning, design reviews, and communication between the designer/contractor and the erector. Many pharmaceutical modular installations involve mating modules together resulting in a multimodular plant. A well written erection and assembly specification is mandatory for all multimodule projects.

25.4 Summary

Off-site construction or modular construction where a large amount of work from a relatively uncontrolled field environment is relocated in a carton assembly facility improves labor efficiency and reduces the cost. Off-site construction is the true answer to today's high cost of pharmaceutical facility construction. Off-site construction favorably addresses a number of business and project driving forces which are emerging today. With an off-site construction mentality, the design and construction details and approach are tailored to meet the driving force of a project. Pharmaceutical companies must carefully assess the benefits of off-site construction due to its cost effectiveness and time savings. Well designed off-site constructed facilities meet all criteria for flexible manufacturing, expandability, relocatability, adaptability, and versatility. Off-site construction or the modular approach is the answer to reducing the cost of pharmaceutical construction.

26 Waste water treatment in the pharmaceutical industry

R. CHEN and K. CROWELL

There is a wide range of waste water streams produced as byproducts of chemical synthesis, fermentation, and formulation operations in the pharmaceutical and biotechnology industries. Treatment of such streams in strict compliance with all applicable regulatory requirements is an increasingly vital aspect of ethical and OTC pharmaceutical manufacturing. Environmental authorities in most countries are raising the penalties significantly for non-compliance by means of heavy fines and criminal sentencing. Additionally, in the USA, the Food and Drug Administration (FDA) is placing increasing emphasis on environmental considerations in the new drug approval process. In short, waste water treatment can no longer be considered as peripheral and secondary to pharmaceutical manufacturing but must be considered as an integral part of pharmaceutical process design.

Waste water treatment processes may be broadly divided into physical and biological treatment operations.

Biological treatment is typically applied where plant discharge is *not* to a publicly owned treatment works (POTW) or where pretreatment requirements are imposed by the receiving POTW. Biological treatment utilizes the physiological processes of microbial organisms to reduce the pollutant loading of waste water streams to an acceptable level.

A variety of physical treatment technologies has been applied to waste water streams that are not cost effectively treated by biological means. Physical processes that will be addressed in this article include activated carbon adsorption, air and steam stripping, chemical precipitation, ion exchange, and membrane separation processes.

26.1 Biological treatment

Biological treatment processes that have found wide application in pharmaceutical manufacturing operations and that will be discussed in this article include the several variations of suspended activated sludge processes, trickling filters, and sequencing batch reactors. The inactivation of potentially infectious or recombinantly derived biological wastes will be briefly addressed. Activated sludge and trickling filter processes generally require waste water equalization and neutralization as pretreatment operations.

26.1.1 Equalization

Equalization is a means of controlling or minimizing fluctuations in the characteristics of the waste water. It also provides optimum conditions for downstream treatment processes such as neutralization and biological treatment system. In industrial waste water treatment, the equalization basin is sized based on the fluctuation of waste water flow, plant production schedule, periodic dumps or spills, as well as the collection of contaminated rainwater after a heavy storm.

Adequate mixing is necessary in the equalization basin to ensure uniformity of waste characteristics through mixing as well as preventing the settleable solids from depositing in the basin. Mixing in the equalization basin is provided by mechanical agitation or diffused air aeration. For mechanical surface aerators, 15–20 hp per million gallons (0.0029–0.0039 W/l) can be employed as a rule of thumb. Diffused air requirements are approximately 1.5 (cm) cfm per 1000 gallons ($0.4\,ft^3/min^{-1}\,Kl^{-1}$) of equalization volume.

Another objective of equalization is to allow the alkaline and acid wastes to react with and neutralize each other thus producing a more neutral waste stream. The neutralization reaction in the equalization basin reduces the amount of base or acid required to control the pH of the waste water in the neutralization step.

The equalization basin serves to provide continuous feed to the treatment system downstream, for example, for neutralization and biological treatment during periods when the manufacturing plant is not operating. It should also provide sufficient damping of organic load fluctuation in the waste water to prevent shock loading of the biological system. In addition, toxic materials present in dumps or spills can be effectively diluted and metered into the biological treatment system without any adverse effect on the operation of the system.

26.1.2 Neutralization

Upon equalization, an industrial waste water containing acidic or alkaline material is required to be neutralized before discharging to a POTW, or to chemical or biological treatment. In order to ensure optimum biological activity, the pH of the waste water should be adjusted to 6.5–8.5. Neutralization is the reaction of compounds with active hydrogen ions and hydroxyl ions forming water and salts. For example, the neutralization of caustic soda with sulfuric acid produces sodium sulfate, a salt, and water. When the waste water is saturated with calcium ions, addition of sulfuric acid may make the waste water oversaturated with calcium sulfate, resulting in the formation of scaling in piping, valves, equipment, and instruments.

Neutralization can be accomplished by batch or continuous mode. Batch treatment is usually only limited to facilities which generate 60 000 gallons

(16 000 l) per day or less. Continuous mode is typically carried out in two stages. The first stage is a roughing neutralization stage and the second stage is for final pH adjustment. When the waste water handled requires neutralization from either acidic or alkaline state, it is necessary to provide facilities for both caustic and acid additions.

26.1.3 Types of activated sludge systems

26.1.3.1 Conventional. This is a plug–flow-type of activated sludge system which includes an aeration basin, a clarifier, and a sludge recycle system. The recycle sludge (RAS) and influent enter the aeration basin at the inlet and leave the basin at the opposite or outlet end. The solids in the mixed liquor are separated out in the clarifier and recycled back to the aeration basin. The principal application of conventional treatment systems has been for low strength sanitary waste waters, as such systems are relatively sensitive to shock loadings and to the presence of refractory organics. Aeration in the basin can be achieved by diffused air or mechanical aerators. The food to microorganism (F/M) ratio and mean cell residence time of the system are typically 0.2–0.4 and 5–15 days, respectively. Typical hydraulic retention time is 4–8 h. Air feed is typically 500–900 ft^3/lb (500–900 $ft^3\,lb^{-1}$) BOD_5, if coarse bubble diffusers are used and 400–600 ft^3/lb (6–12 $m^3\,kg^{-1}$) BOD_5 if fine bubble diffusers are used. If mechanical aerators are used, agitation should be adequate to supply 1.0–1.2 lb O_2/lb BOD_5 (1.0–1.2 kg O_2/kg BOD_5). A properly operated conventional activated sludge system can achieve a BOD removal efficiency of 85–90%. The RAS of a conventional system is generally 25–50% of the influent flow. Conventional activated sludge systems typically produce well flocculated and easily dewatered sludge. Sludge production is typically 0.35–0.55 lb sludge/lb BOD_5 (0.35–0.55 kg sludge/kg BOD_5).

26.1.3.2 Complete mix. Here the influent and the RAS are introduced at several points near the center of the aeration basin from a header or central channel. Effluent from the aeration basin is collected from both sides of the basin by means of two effluent channels. A clarifier is used to separate the solids in the basin effluent before the sludge is recycled back to the head of the system. Excess sludge is sent to the sludge handling facility of the treatment plant. Aeration can be accomplished by diffused air or mechanical aeration. In this system, both the hydraulic and organic loadings as well as the oxygen demand on the aeration basin are uniform throughout the basin, and resistance to shock loading is much better than that of conventional constructed sludge processes. Conversely complete mix systems have a greater susceptibility to the growth of filamentous, bulking organisms and hence poorer sludge setting and dewatering properties. Complete mix systems typically have an F/M ration of 0.2–0.6, a mean cell residence time of 5–15 h, and a hydraulic residence time of 3–5 h. Loadings can be higher than for a conventional system with 50–120 lb $BOD_5/1000\,ft^3$/day (0.8–2.0 kg BOD_5/m^3/day) being typical. Oxygen consumption and sludge production rates, on a per lb BOD_5 removed basis, are the same as those of conventional systems. BOD_5 reductions of 85% to 95% are achievable.

26.1.3.3 Contact stabilization. In this process, the influent is first mixed with the return activated sludge for approximately 30 min or for long enough for the organics to be absorbed by the flocs of the activated sludge in a contact basin. Effluent from the contact basin is fed to a clarifier where the activated sludge is settled out and then transferred to another aeration/stabilization basin. Inside the stabilization basin, the activated sludge is aerated for 3–6 h before it is mixed with the influent in the contact basin. The absorbed organics are converted to carbon dioxide, water, and new cells by means of biochemical oxidation. Excess sludge generated from the process can be wasted from the bottom of the clarifier or from the stabilization basin outlet. Typically the F/M ratio is 0.2–0.6, mean cell residence 4 time 5–15 days, and BOD_5 loading 60–75 lb $BOD_5/1000\,ft^3$/day. 80% to 70% BOD_5 reduction can be accomplished.

Total aeration requirements in this process are typically 50% of those of a corresponding conventional plug–flow or tapered aeration system. When used for the treatment of industrial waste water, this process may be similarly useful for waste water in which the organics are mostly present in insoluble form. Hence, laboratory treatability studies should always be performed when this process is used for the treatment of a new waste water stream.

26.1.3.4 Extended aeration. The flow scheme of this process is similar to that of a conventional plug–flow system except that hydraulic residence times are typically 18 to 30 h rather than 6–8 h. This high residence time, the activated sludge system operates in an endogenous respiration phase in which there is inadequate organic material in the system to support all the microorganisms present due to a low BOD loading. Hence a low F/M ratio of 0.05 to 0.15 and high mean cell residence times of 20 to 30 days are used. BOD_5 loading is typically 10–12 lb $BOD_5/1000\,ft^3\ day^{-1}$ (0.15–0.4 kg $BOD_5/m^3\ day^{-1}$) and air supply is typically $2000\,ft^3$/lb BOD_5 ($125\,m^3$/kg BOD_5). Extended aeration is resistant to shock loading and long residence times allow degradation of many organic contaminants to occur. Under this type of operating condition, an extended aeration process can produce low sludge and a highly treated effluent. Sludge production rates of 0.1–0.2 lb/lb BOD_5 (0.1–0.2 kg/kg BOD_5) and BOD_5 reductions of 75% to 95% are typical of extended aeration. This is the 'workhorse' process of industrial waste water treatment.

26.1.3.5 Tapered aeration. Tapered aeration is a modified version of the conventional plug–flow process in which the air supplied to the aeration basin is designed to meet the demand. At the inlet of the basin where the return activated sludge first comes in contact with the influent, the oxygen demand is the greatest, and thus the amount of air injected is the highest. Generally, this is accomplished by spacing the diffusers closer together at the inlet resulting in a higher oxygenation rate there. As the mixed liquor moves toward the outlet of the aeration basin, the oxygen demand decreases and the spacing of the diffusers increases to decrease the oxygenation rate accordingly. This arrangement allows for a more efficient utilization and hence overall air supply requirements. Other parameters are similar to those of the conventional process.

26.1.3.6 Step aeration. This is another modification of the conventional activated sludge process. Influent to the aeration basin is split up into four or more equal streams and are then fed into four parallel channels separated by baffles. Each channel is equivalent to a separate step and these steps are connected together in series, similar to four small plug-flow systems arranged in series. The first step is commonly used to reaerate the return activated sludge preventing the sludge from being oxygen starved when it comes into contact with the influent. Parameters of the process are similar to those of the conventional process although slightly higher loadings, typically 40–60 lb $BOD_5/1000\,ft^3\,day^{-1}$ (0.65–0.95 kg $BOD_5/m^3\,day^{-1}$) can be treated as consequence of the greater process efficiency.

26.1.4 Biotower/trickling filter

Biotowers or trickling filters have been used for the treatment of both domestic and industrial waste water for many years. Conventional trickling filters consist of a bed of rock or gravel over an underdrainage system. Waste water is continuously applied to the media which results in the development of a gelatinous film of microorganisms or biological growth attached to the media surface. Both the soluble and colloidal organics in the waste water are adsorbed by the film as the waste water trickles over the media. With the oxygen diffusing into the biological mass, a biochemical reaction occurs which oxidizes the organic material to carbon dioxide, water, and additional new cells. When enough biological mass is built up on the media, the gelatinous film sloughs off and readily settles out in a clarifier which results in the removal of the majority of BOD in the waste water.

With the advent of the plastic media, trickling filters have been constructed to heights of as much as 30 ft (9.14 m). This is because plastic packing can be stacked many times higher than can conventional rock filters. Such plastic-filter medias, operating with recirculation in single-stage or two-stages in series, can have hydraulic and organic loadings of 200–1000 gpd/ft^2 (8150–40 750 l/day m^{-2}) and 25–300 lb BOD_5/1000 ft^3/day (0.4–5.0 kg BOD_5/m^3/day) of packing per day, respectively. Conventional rock filter beds have been designed with standard hydraulic and organic loading rates of 25–100 gpd/ft^2 (1000–4000 l/day/m^2) and 5–25 lb BOD_5/1000 ft^3/day (0.08–0.4 kg BOD_5/m^3/day) of BOD/1000 ft^3 day.

Another advantage of the plastic-media filters is that under higher BOD loading conditions, they do not get clogged up like the conventional rock filters. In addition to a lower land area requirement, they also withstand sudden increases in BOD and hydraulic loadings without significant reduction in the treatment efficiency. However, with higher organic loadings, odor problems may develop which may make it necessary to install odor control equipment.

Plastic-media filters have been used as rough filters in both pilot plant and full scale units in the pharmaceutical industry. When used as a pretreatment step for an activated sludge process, loadings of up to 1000 lb COD/1000 ft^3/day (16 kg COD/m^3/day) are commonly used. With a properly designed biotower, high strength pharmaceutical waste water with COD of 4000–6000 mg/l can be reduced to 2000–3000 mg/l as a pretreatment step.

26.1.5 Sequencing batch reactor (SBR)

This treatment technology is very similar to the conventional activated sludge process except that it is operated in a batch mode. This process is typically applied to small volume waste water streams (hundreds to tens of thousands of gallons/day). It is particularly useful where waste water characteristics may be expected to vary substantially on a daily basis to ensure proper treatment. It consists of a single reactor which is used for equalization, aeration, and clarification on a sequential basis. Essentially, the batch reactor is a single tank equipped with an inlet for influent, aeration equipment, a sludge draw-off mechanism at the bottom to waste sludge, a decant mechanism for the removal of supernatant upon settling, and a control system to time and sequence the process.

The five basic steps of operation cycle in a typical SBR system are: fill, react, settle, draw, and idle.

1. **Fill.**—This step allows influent to enter the reactor tank via the influent inlet. Ordinarily, the fill step is controlled by either volume or time. If controlled by volume, the liquid level in the reactor is allowed to rise from 25% of capacity (end of idle) to 100%. If controlled by time, this step generally lasts approximately 25% of the full-cycle time.
2. **React.**—In this step of the process, the aeration equipment is used to supply air to the batch treatment process. Oxidation of organic material occurs in the same manner as in the conventional activated sludge process.

Depending on the organic loadings, generally this step takes up approximately 35% of the total cycle time.

3. **Settle.**—This step allows the solids to be separated from the liquid, providing a clarified effluent to be discharged from the system. Again this step is controlled by a timer and is usually preset at 0.5–1.0 h which permits the sludge blanket to remain below the withdrawal mechnism during the draw step.
4. **Draw.**—The treated effluent is discharged from the top of the reactor in this step and the reactor is made ready for the next batch.
5. **Idle.**—This step is only used where there is more than one reactor in the SBR system. It allows the reactor to remain idle until the fill step is complete in the other reactor.

Periodically, excess sludge is wasted during the settle or idle step. As in the conventional activated sludge process, the frequency and the amount of sludge to be wasted is a function of the performance requirements. Consequently, the sludge age and sludge volume are controlled by sludge wasting. If extended aeration is practiced in the SBR system, endogenous respiration would occur in the react phase. If the BOD loading is such that the number of pounds (kilos) of cells produced per day is equal to the number of pounds (kilos) of cells destroyed per day by endogenous respiration, the need of sludge wasting would be eliminated.

The operation of SBR more or less follows that of a conventional activated sludge system. In terms of F/M ratio, it can be operated in the range of 0.15–0.40. For pretreatment of industrial waste water, the F/M ratio can be 1.0 or higher.

When compared to a conventional continuous flow system, SBR has the unique feature of being very flexible. The advantages of SBR over other conventional processes are outlined as follows:

1. The SBR tank can serve as an equalization tank, a reactor tank, and a clarifier. Thus the amount of plant surface area required for the SBR installation is drastically reduced when compared with a conventional flow system.
2. The time sequence for each step can easily be changed depending on the organic loading as well as the quality of the effluent desired.
3. The SBR system can tolerate high peak hourly flows and/or substantial organic shock loads without compromising the effluent quality because the SBR reactor tank also serves as an equalization tank.
4. The solids in the mixed liquor do not get washed out by hydraulic surges as they can be held in the tank as long as necessary.
5. Pumping of return activated sludge is not required because the mixed liquor is always in the reactor tank. As a result, both capital and operating costs are reduced.

6. The operating level in the SBR tank can be adjusted to treat various sizes of batches. During low flow condition, a smaller batch can be collected and treated as desired.
7. The settle step occurs under ideal quiescent conditions during which short circuiting cannot take place. In addition, the larger reactor size provides small surface settling rates, which allow the settling of even small floc particles that may be washed out in continuous flow systems.

26.1.6 Biological waste inactivation

Contaminated biowaste from pharmaceutical and biotechnology manufacturing, research laboratories and pilot plants must be inactivated prior to disposal as required by federal, state, local, and corporate regulations and guidelines. Typical discharge sources consist of fermentation-contained drains, process centrifuges, and any other streams that are suspected to contain infectious or recombinant microorganisms.

In general, decontamination is achieved by steam sterilization and sometimes supported by chemical addition, or pit adjustment. The equipment used for liquid waste decontamination is generally a skid-mounted unit consisting of waste collection and decontamination tanks, pumps and piping, agitator, vent filters, and drain system. The decontamination tank also should have a view port, a light port, a steam spraying port, steam jacket, safety pressure relief valve, appropriate instrumentation, and a removable spray ball. The decontaminated waste stream is then discharged to the plant process sewer to be mixed in with other waste water from the plant.

26.2 Physical treatment

A variety of physical treatment technologies have been applied to pharmaceutical waste water streams. In general physical treatment processes are cost effective for the treatment of very small volume streams, for the treatment of streams containing biologically resistant (refractory) or toxic substances, or for streams where economical recovery of a raw material is feasible. Physical treatment processes that will be addressed in this chapter include activated carbon adsorption, air and steam stripping, chemical precipitation, ion exchange, and reverse osmosis processes.

26.2.1 Activated carbon

Adsorption is a common and widely used pollution control unit operation in the pharmaceutical industry. The operation is based on the affinity of organic contaminants for the surface of a solid adsorbent medium. Activated carbon is by far the most commonly used adsorbent material in waste water treatment applications. The utility of the technique is due to the broad range of organic compounds that may be effectively removed through its use. Cost effective applications of activated carbon to the treatment of pharmaceutical waste waters include treatment of very dilute streams such as contaminated stormwater, groundwater or vessel washouts; as a tertiary treatment step for plant waste water treatment facilities; as an efficiency improving addition to aerobic or anaerobic waste water treatment processes; or for small volume applications requiring broad range effectiveness such as laboratory bench drains.

While adsorption is effective at removing and concentrating a wide range of pollutants, some environmentally acceptable ultimate fate for the pollutant must then be found. It is also usually technically feasible and economically desirable to regenerate the carbon for subsequent re-use. A number of methods of carbon regeneration exist with the optimum choice for a given application being determined by a number of system and site specific factors such as the volume of carbon used, specific contaminants removed, availability of low cost steam and contaminant solubility, among others. For smaller applications many carbon vendors will accept used carbon back and will regenerate it for the user for a fee. Hence an activated carbon system will have two principal sections, adsorption and regeneration, both of which must be carefully considered during design.

The mechanism of adsorption may be based on either physical or chemical interaction (or a combination of the two) between the organic adsorbate and the solid carbon surface. Chemisorbed species may be very difficult or impossible to remove from the carbon because of the presence of stronger intermolecular forces between the organic substance and the carbon than between the organic substance and waste water. Typically, the major driving force is the 'desire' of the hydrophobic organic to agglomerate on any available surface and hence find itself in a more favorable, lower energy thermodynamic state.

Activated carbon in this view is effective primarily due to the vast surface area per unit volume it provides (typically a football fields' equivalent in area per 4 to 5 g). The distribution of pore diameters in the carbon is also important since this factor controls access to the interior surface area at the molecular level. A number of different raw materials including coal, wood, coconut shells, peat, and coke are used to produce activated carbon and the final products differ widely in specific surface areas, pore size distributions, and specific molecular affinities. All of these properties may have an impact on carbon effectiveness for a given application. Carbons suitable for waste water adsorption typically have specific surface areas of 500–2000 m^2/g and a pore size distribution having a majority of pores in the 5–100 nm range (Hutchins, 1980). Carbons suitable for gas phase adsorption applications by contrast typically have a majority of pores in the 0.3–10 nm range. Activated carbon is available in granular and in powdered form. Granular carbon is used in beds and in canisters, principally for the treatment of continuous or semi-continuous waste water and air streams. Powdered carbon is used for batch adsorption applications and especially in conjunction with biological treatment, as in the PACT process developed and marketed by DuPont Co. and Zimpro-Passavant Inc.

While published data on adsorption of specific compounds exist (see for example Dobbs and Cohen, 1980 or Shields, 1985), design of an adsorption system should always be based on adequate experimental studies of the specific waste water it is intended to treat (Calgon Carbon, 1988). A wide range of factors including the presence of surfactants or complexing agents, presence of suspended solids, pH, temperature, presence of other organics with a higher affinity for carbon, water hardness, and the specific type of carbon can all have a major impact on the adsorption process and can lead to results quite different from those of published studies on single component systems.

In general the higher the molecular weight and the lower the water solubility of a contaminant the more readily it is adsorbed. Organic compounds having three carbons or more are usually adsorbed unless they are very water soluble (Hutchins, 1980). Branched chain hydrocarbons are more adsorbable than straight chain species. Polar compounds are often poorly adsorbed. Compounds with hydroxyl, amino, or sulfonic acid functional groups are also often poorly adsorbed. Methyl isobutyl ketone (MIBK), toluene, and hexanol are common solvents in the pharmaceutical industry that are readily removed from waste water streams by carbon adsorption while methylene chloride, methanol, and isopropanol (IPA) are poorly removed (Dobbs and Cohen, 1980).

Equilibrium data in the form of isotherms is developed from bench scale column experiments and will determine expected contaminant percent removal, carbon usage rates, and the mode of contaminant 'breakthrough' (Calgon Carbon, 1988). Breakthrough occurs when the contaminant effluent concentration exceeds its design value. Breakthrough is related to the kinetics of adsorption of a given contaminant and may be described through the concept of an adsorption zone. A species that is rapidly adsorbed will deplete the carbon in a bed as a moving front; this is described as a shallow adsorption zone. The effluent concentration will go from its design level to the feed concentration soon after breakthrough. A compound that is less rapidly adsorbed will form a concentration gradient that migrates down through the carbon bed. This is described as a deep adsorption zone. In this case breakthrough will occur more gradually, allowing more response time in the event of a breakthrough. Breakthrough behavior is an important factor that must be considered in overdesign factors, monitoring instrumentation, and mode of operation (Voice, 1989).

In general it is wise to provide considerable excess capacity in designing an adsorption process. This will allow not only for uncertainties in the experimental data but also for the possibility of a contaminant 'slug' resulting from a spill or process upset in the plant and for the reduced adsorptive capacity of regenerated versus virgin carbon.

Carbon adsorption processes typically are supplied with multiple beds to allow for operating efficiency, flexibility, and reliability. These beds may be piped in parallel or series and operated as either upflow or downflow units.

Parallel units are manifolded together and are placed on stream at predetermined intervals or if a breakthrough event occurs. Once a new column goes on line the depleted bed may be replaced or regenerated. Parallel systems are optimal for relatively low required percentage removal processes and where high pressure drop is expected. A shallow adsorption zone is also favorable for the use of a parallel system because effluent quality will remain high until the carbon is nearly depleted.

A deep adsorption zone on the other hand favors a series arrangement, since in such a system effluent quality will become unacceptable long before the adsorptive capacity of the carbon is reached. By providing a second bed in series, effluent quality from the system can be maintained long after breakthrough of the first bed has occurred. Typically such a system is provided with three beds; a primary, a secondary, and a back-up. When the primary bed carbon is exhausted the secondary bed becomes the new primary, the back-up becomes the new secondary and the old primary is taken off line while the carbon is regenerated or replaced.

Downflow operation permits simultaneous filtration of suspended solids but sufficient pumping head must be provided to overcome the backpressure as the bed becomes plugged with suspended solids. Frequent backwashing of downflow beds is also typically required for the same reason. Upflow beds can often be implemented as gravity flow systems because the bed expands with the flow and little build-up of suspended solids is experienced. Backwashing frequency is also typically much lower. Post adsorption solids filtration may however be necessary.

Adsorption system design must also consider how and where (on or off-site) carbon regeneration or disposal will be accomplished. Typical regeneration techniques include thermal reactivation in a multiple hearth furnace or rotary kiln at temperatures of 1200–1900 °F (650–1040 °C); steam stripping of the organics with subsequent condensation and phase separation; ozone mediated *in-situ* oxidation or chemical regeneration. A recent advance is the use of a hot inert permanent gas (such as N_2) for regeneration, which allows for easy recovery of the organic substance by condensation with no phase separation step required. Choice of method will depend on the properties of a given system and the characteristics of the plant size.

26.2.2 *Air and steam stripping*

Air and steam stripping are applicable to pharmaceutical waste water streams contaminated with volatile organic compounds (VOCs). The VOC is transferred from the water to air or steam under the influence of the concentration gradient existing between the two phases.

In general air stripping is the simpler and less costly of the two techniques. Steam stripping however is effective for the removal of compounds that cannot be treated by air stripping. Air stripping is useful for treatment of very dilute (< 100 ppm) streams containing volatile and insoluble organics such as trichloroethylene, toluene, or methylene chloride. Steam stripping is capable of treating much more concentrated waste streams (up to several weight percentage) and of removing less volatile or more soluble compounds.

26.2.2.1 Air stripping. The efficiency of an air stripping treatment process will be governed by the Henry's law constant and overall mass transfer coefficient of the system.

Henry's law:

$$p_A = H_A x_A \tag{26.1}$$

where p_A is the vapor pressure of compound A (in atm), H_A is Henry's law constant (in atm), and x_A is the liquid phase mole fraction of compound, describes the phase equilibrium relationship existing in dilute systems such as those encountered in air stripping applications. Henry's constant is actually a function of temperature and a weak function of composition and pressure as well. Fortunately however, air stripping processes are normally run under ambient conditions and hence the published data, generally given at 20 or 25°C is useful. Henry's law may also be expressed on a concentration basis:

$$p_A = H_A C_A \tag{26.2}$$

where the nomenclature is as above and C_A is the concentration of A in the waste water.

The maximum possible percentage removal for a given contaminant by air stripping can be estimated as:

$$\% \text{ removal} = \frac{C_{Ai} - C_{Ao}}{C_{Ai}} \times 100 = 1 - \frac{1}{\dfrac{1 + H_A R_v}{RT}} \times 100 \tag{26.3}$$

where C_{Ai} is initial concentration (mol/m_3), C_{Ao} is final concentration (mol/m^3), H_A is Henry's constant ($atm\,m^3/mol$), R_v is volumetric air/water ratio $= V/L$, V is air fed (cfm), L is water fed (cfm) $=$ GPM/7.4805, $\boldsymbol{R}$ is gas constant $= 8.206 \times 10^{-5}$ ($atm\,m^3/mol\,K$), and T is temperature (K).

Air stripping is typically an effective treatment method for streams where the contaminant has a Henry's constant of 10 atm or more and is present in concentrations of 100 ppm or less in the waste water (Li and Hsiao, 1991).

Historically air stripping was nearly always carried out in packed column equipment, which was normally fabricated from FRP. In recent years, however, low profile air strippers have been gaining significant acceptance. Low profile strippers require substantially greater air flows than packed columns and hence have higher capital (blower) and operating (power) costs. Their application is thus most practical for low concentration or easily strippable streams. Low profile units do have a significant advantage when the waste stream contains high levels of inorganic ions such as iron, calcium, or magnesium or high levels of suspended solids. Under such conditions packed beds tend to foul rapidly and to require frequent and expensive cleaning or replacement. Low profile units are also desirable when limited site plot plan area height or accessibility conditions prevail (Reese, 1992).

Packed column sizing is largely controlled by the mass transfer coefficient of the system which in turn is dependent on composition, on the specific packing used by the pollutant concentration in the feed, by the required stripping efficiency, and by the required air and water volumetric flow rates.

A wide variety of packings have been installed in air strippers including conventional packings such as Raschig and Pall rings and Berl saddles as well as packings designed specifically for air stripping. Plastic packing is acceptable since air stripping is commonly carried out at ambient temperature. Packing selection is usually made on the basis of cost and surface area versus pressure drop trade-offs.

The overall mass transfer coefficient, K_{La}, is a function of the specific packing selected and the diffusivity of the contaminant through air and water phases:

$$\frac{1}{K_{La}} = \frac{1}{k_{La}} + \frac{p_{mw}}{k_{Ga}H_A} \tag{26.4}$$

where K_{La} is the overall mass transfer coefficient (mol/m^3 s), k_{La} is liquid phase mass transfer coefficient (mol/m^3 s), k_{Ga} is gas phase mass transfer coefficient (mol/m^3 s), p_{mw} is molar density of water = 55 600 mol/m^3, and H_A is Henry's constant (m^3/mol).

In general the overall mass transfer coefficient is determined experimentally from pilot scale runs. It is important that such pilot testing be performed on the actual waste water to be treated if at all possible.

Given a mass transfer coefficient, air and water flow rates, and influent and effluent contaminant concentrations, the required column height and diameter can be estimated. Air stripping towers are typically 15–60 ft (4.57–18.29 m) in height with diameters ranging from 1–10 ft (30.48–304.8 cm). Water loading rates are normally in the range of 40–250 lb/ft^3 min (640–4000 kg/m^3) and volumetric air/water ratios are typically in the range of 10 to 300.

Design of a stripping system must also consider further treatment of the air stream, air blowers, tower internals, and controls. Typically contaminated air from the stripper is routed to a fume incinerator but vapor phase carbon adsorption or low temperature condensation may also be useful in some circumstances. Current trends in air stripping include the development and application of improved packings and the application of waste water preheating to achieve more difficult separations.

26.2.2.2 Steam stripping. Steam stripping is useful for removal of organics such as acetone or methanol that have too low a volatility or too high a solubility to be effectively air stripped. Steam stripping is also useful for the treatment of much more concentrated streams, successfully treating concentrations of up to several weight percent, while air stripping is limited to streams of about 10 ppm. Capital and operating costs of steam stripping are substantially greater than are those of air stripping however. Steam stripping can typically achieve contaminant removal of 99% or better and is effective for the removal of most organics with a normal boiling point of 150°C or lower.

Steam stripping is in effect merely a variant of conventional distillation in which energy is supplied by the injection of live steam rather than by means of a reboiler and in which no rectification section is present. The bottoms produced is clean water while the overheads is a mixture of organic vapors and steam. These overheads are condensed and if not completely miscible, decanted, with aqueous phase being recycled to the column feed and the organic being recycled to the process or collected for proper treatment and disposal.

For treatment of a stream containing a miscible solvent, such as methanol, steam stripping is essentially only a waste concentration step since the concentrated aqueous overheads stream must still be disposed of. If large volumes of solvent are involved the addition of an appropriate entrainer and subsequent recovery of the solvent via azeotropic distillation should be considered. Such a system can be quite expensive however since at least one, and probably several, additional columns with their associated heat exchange equipment and controls will be required to separate the entrainer from the solvent.

While packed towers are usually employed in air stripping, steam strippers are normally designed as trayed towers. This makes the removal of suspended solids or fouling or polymerizing materials from the waste water stream before it enters the column imperative. Failure to consider the potential for tray plugging is one of the most frequently encountered problems in the application of steam stripping. The presence of highly volatile materials in the waste water can lead to unintended air emissions unless the decanter is positively vented to a fume incinerator or other secondary control device.

Steam strippers are generally designed by the use of computer simulation programs, although preliminary estimates can be prepared by modifications of the McCabe–Thiele or Fenske–Underwood–Gilliland methods. In any case, the development of reliable equilibrium data will be critical to the completion of a successful design. In general such data must be obtained through pilot scale testing on the actual waste water stream to be treated. This is typically done by the potential supplier of the full scale system, for a fee.

26.2.3 Heavy metals removal

While heavy metals, such as chromium, copper, nickel, or cadmium are not typical constituents of pharmaceutical waste water, there are processes, particularly in the chemical intermediates segment of the industry, which, for example, use heavy metals as catalysts or which use chromic acid as an oxidizing agent.

Such streams should be treated at the process source. This will minimize the waste water volume to be treated and hence the equipment size of the required treatment, but will more importantly minimize complications in the removal process. A number of organic compounds, and also ammonium, form complexes with heavy metals and this makes removal much more difficult. Heavy metal-bearing streams must be treated prior to any biological treatment that the stream will also require. Heavy metals are toxic to microorganisms and will, even in very low concentration, reduce biological treatment plant efficiency. Metals will also concentrate in the sludge forcing it to be classified and managed as a RCRA hazardous waste material.

By far the most commonly applied treatment for heavy metal waste water streams is hydroxide precipitation. Carbonate precipitation is capable of producing a very high quality effluent, as is iron coprecipitation cementation. In specific cases ion exchange and reverse osmosis technology may be useful. Hydroxide precipitation uses lime or liquid sodium hydroxide as reactants to form relatively insoluble metal hydroxides.

$$M + 2\,OH- = M(OH)_2\,(ppt) \qquad (26.5)$$

The hydroxide solids are then settled, filtered, and removed as a sludge. Achievable effluent concentration levels depend on the metal being treated, precipitant used, degree of pH (or ORP) control maintained, and on the presence or absence of substances inhibitory to precipitation, but concentrations of less than 1 ppm are typical. Control of pH is significant since many metal hydroxides have solubility minima at a specific pH (Lanouete, 1977).

Quicklime (CaO), hydrated lime ($Ca(OH)_2$), and liquid sodium hydroxide (50%) are commonly used as precipitating agents. Capital and operating costs and precipitate filterability are important considerations in the selection of a reactant. In general, lime is much cheaper material and forms a more easily settled, filtered, and handled sludge. Lime however is itself more difficult to handle and the capital and maintenance costs of a lime addition system will be much higher, particularly for a small scale system like those typically found in the pharmaceutical industry.

In the case of cadmium- or lead-bearing waste waters, sodium carbonate (soda ash, Na_2,CO_3) precipitation, as the metal carbonate, should be considered since sludge handling is usually easier and effluent quality is often better than would be the case if hydroxide were used.

Chromium-bearing waste waters require pretreatment since hexavalent chromium will not react with hydroxide. Hexavalent chromium must be reduced to the trivalent form by reaction with ferrous sulfate, sodium metabisulfate, or sulfur dioxide (Eckenfelder, 1989; Chung, 1989). The reduction must be carried out at a pH below 3.0. Achieving complete reduction is important since any remaining hexavalent chromium will remain in solution in the effluent. The pH is then raised to pH 4.5 to 8.0 and hydroxide precipitation is carried out.

Once formed the hydroxide particles must first be settled and then filtered to a handleable sludge. Minimization of sludge volume is significant since this material will need to be managed as a hazardous waste. Settling can be carried out in conventional clarifiers or more typically in inclined parallel plate units. Clarifier overflow is then passed through a sand bed or mixed media filter to further purify the effluent. Underflow, typically at 1 to 2% solids is fed to a dewatering device such as a vacuum or candle filter, a centrifuge, a sludge (sand) drying bed, or most commonly to a plate and frame filter press. Plate and frame presses generally require more operation attention but achieve higher solids content, typically 50–60%.

The solubilities of metal sulfides are much lower than those of the corresponding hydroxide, and hence sulfide precipitation is capable of achieving a much cleaner effluent, typically 50 ppb or better (0.05 ppm). A further potential advantage of sulfide precipitation is that hexavalent chromium can be directly treated without first being reduced. The disadvantages of sulfide precipitation are greater complexity of operation, more expensive reagents, and the hazard potential of generating toxic hydrogen sulfide gas (H_2S). Reagents that have been used include sodium sulfide (Na_2S), sodium hydrosulfide (NaHS), and ferrous sulfide (FeS). Ferrous sulfide has the advantage of being less likely to form H_2S, but is an unstable compound and must be generated at the point of use. The Permutit proprietary Sulfex process incorporates ferrous sulfide and a unique reactor configuration into a sulfide precipitation system.

Iron coprecipitation and cementation, as incorporated into Unocal's Unipure process for example, is also effective at attaining low effluent heavy metals concentrations. In this process soluble ferrous iron is precipitated by oxidation. The iron precipitate binds and removes heavy metals including hexavalent chromium.

Ion exchange can be useful for heavy metal removal particularly for nickel, zinc, copper, or chromium removal where the metals can be recovered from the regenerating solution and recycled back into the process or sold. Ion exchange has also been applied to treatment of streams containing complexing agents or other compounds that would interfere with a precipitation process.

Reverse osmosis uses semi-permeable membranes and high pressure to produce a clear permeate and a retentate solution containing salts and ions including heavy metals. The technique is effective if the retentate solution can be re-used in the process. The equipment tends to be expensive and fouling of the membranes has been a common problem. Considerable research is being carried out on the membrane process, however and it is likely to be more commonly applied in the future.

Bibliography

General

Arbuckle, J.G., Vanderver, T.A. and Randle, R.V. (1991). Water pollution control, in *Environmental Law Handbook*, 11th edn, Government Institute, Rockville, MD.
Blumenshein, C. (1992). *Pollution Eng.* **24**, 68.
Cheremisinoff, P.N. (1987). *Wastewater Treatment Handbook*, Pudvan Publishing, Northbrook, IL.
Eckenfelder, W.W. (1989). *Industrial Wastewater Pollution Control*, McGraw-Hill, N.Y.
Goldman, J.C. and Bowen, P.T. (1992). *Pollution Eng.*, **24**, 57.
Metcalf and Eddy Inc. (1991). *Wastewater Engineering*, 3rd edn G. Tchobanogloss and F.L. Burton (revisors), N.Y.
Shields, E.J. (1985). *Pollution Control Engineers Handbook*, Pudvan Publishing, Northbrook, IL.
WPCF (1977). *Wastewater Treatment Plant Design*, Lancaster Press, Lancaster, PA

Carbon adsorption

Calgon Carbon Corp. (1988). *The Laboratory Evaluation Of Granulated Activated Carbon.*
Chereminisoff, P.N. and Ellerbusch, F. (eds) (1978). *Carbon Adsorption Handbook*, Ann Arbor Science, Ann Arbor, MI.
Delthorn, R.T. and Mazzoni, A.F. (1986). *Water Technol.*, **9**, 8.
Dobbs, R.A. and Cohen, J.M. (1980). *Carbon Adsorption Isotherms for Toxic Organics*, USEPA 600/8-80-023.
Hutchins, R.A. (1980). *Chem. Eng.*, **87**, 2.
Irving-Monshaw, S. (1990). *Chem. Eng.*, **97**, 2.
Process Design Manual for Carbon Adsorption (1973). USEPA 625/1-71-002a.
Voice, T.C. (1989). Activated carbon adsorption, in *Standard Handbook of Hazardous Waste Treatment and Disposal*, H.M. Freeman, ed. McGraw-Hill, N.Y.

Steam and air stripping

Ellerbe, R.W. (1979). Steam distillation/stripping, in *Handbook of Separation Techniques for Chemical Engineers*, P.A. Schweitzer, ed., McGraw-Hill, N.Y.
Gosset, J. (1985). US Air Force Report ESL-TR-85-18.
Li, K. and Hsiao, K. (1991). *Chem. Eng.*, **98**, 7.
Reese, J. (1992). *Environ. Protection.*
Rousseau, R. (ed.) (1987). *Handbook of Separation Process Technology*, John Wiley, N.Y.
Speece, R., Lee, Y. and Nirmalakhandan, N. (1987). *AWWA J.*
Treybal, R. (1980). *Mass Transfer Operations*, 3rd edn, McGraw-Hill, N.Y.

Heavy metals removal

Bazza, R.V., Kelleher, C.M. and Yeliger, M.B. (1987). *Eighth Conference on Pollution Control for the Metal Finishing Industry.*
Chung, N.K. (1989). Chemical precipitation, in *Standard Handbook of Hazardous Waste Treatment and Disposal*, H.M. Freeman, ed. McGraw-Hill, N.Y.
Lanouete, K.H. (1977). Heavy metals removal, in *Chemical Engineering Deskbook*, McGraw-Hill, N.Y.
McAnally, S. (1984). *Separation Sci. Tech.*, **19**, 191.
Patterson, J.W. (1975). *Wastewater Treatment Technology*, Ann Arbor Science, Ann Arbor, MI.

Index

Page numbers appearing in *italics* refer to tables and page numbers appearing in **bold** refer to figures.